64 Springer Series in Solid-State Sciences
Edited by Peter Fulde

W0042379

Springer
Berlin
Heidelberg
New York
Barcelona
Budapest
Hong Kong
London
Milan
Paris
Santa Clara
Singapore
Tokyo

Springer Series in Solid-State Sciences

Editors: M. Cardona P. Fulde K. von Klitzing H.-J. Queisser

Managing Editor: H. K.V. Lotsch

Volumes 1–89 are listed at the end of the book

90 **Earlier and Recent Aspects of Superconductivity**
Editors: J. G. Bednorz and K. A. Müller

91 **Electronic Properties of Conjugated Polymers III** Basic Models and Applications
Editors: H. Kuzmany, M. Mehring, and S. Roth

92 **Physics and Engineering Applications of Magnetism** Editors: Y. Ishikawa and N. Miura

93 **Quasicrystals** Editors: T. Fujiwara and T. Ogawa

94 **Electronic Conduction in Oxides**
By N. Tsuda, K. Nasu, A. Yanase, and K. Siratori

95 **Electronic Materials**
A New Era in Materials Science
Editors: J. R. Chelikowsky and A. Franciosi

96 **Electron Liquids** By A. Isihara

97 **Localization and Confinement of Electrons in Semiconductors**
Editors: F. Kuchar, H. Heinrich, and G. Bauer

98 **Magnetism and the Electronic Structure of Crystals** By V.A. Gubanov, A.I. Liechtenstein, and A.V. Postnikov

99 **Electronic Properties of High-T_c Superconductors and Related Compounds**
Editors: H. Kuzmany, M. Mehring, and J. Fink

100 **Electron Correlations in Molecules and Solids** 3rd Edition By P. Fulde

101 **High Magnetic Fields in Semiconductor Physics III** Quantum Hall Effect, Transport and Optics By G. Landwehr

102 **Conjugated Conducting Polymers**
Editor: H. Kiess

103 **Molecular Dynamics Simulations**
Editor: F. Yonezawa

104 **Products of Random Matrices** in Statistical Physics By A. Crisanti, G. Paladin, and A. Vulpiani

105 **Self-Trapped Excitons**
2nd Edition By K. S. Song and R. T. Williams

106 **Physics of High-Temperature Superconductors**
Editors: S. Maekawa and M. Sato

107 **Electronic Properties of Polymers**
Orientation and Dimensionality of Conjugated Systems Editors: H. Kuzmany, M. Mehring, and S. Roth

108 **Site Symmetry in Crystals** Theory and Applications By R. A. Evarestov and V. P. Smirnov

109 **Transport Phenomena in Mesoscopic Systems** Editors: H. Fukuyama and T. Ando

110 **Superlattices and Other Heterostructures**
Symmetry and Optical Phenomena
By E. L. Ivchenko and G. E. Pikus

111 **Low-Dimensional Electronic Systems**
New Concepts
Editors: G. Bauer, F. Kuchar, and H. Heinrich

112 **Phonon Scattering in Condensed Matter VII**
Editors: M. Meissner and R. O. Pohl

113 **Electronic Properties of High-T_c Superconductors**
Editors: H. Kuzmany, M. Mehring, and J. Fink

114 **Interatomic Potential and Structural Stability**
Editors: K. Terakura and H. Akai

115 **Ultrafast Spectroscopy of Semiconductors and Semiconductor Nanostructures** By J. Shah

116 **Electron Spectrum of Gapless Semiconductors**
By J. M. Tsidilovski

117 **Electronic Properties of Fullerenes**
Editors: H. Kuzmany, J. Fink, M. Mehring, and S. Roth

118 **Correlation Effects in Low-Dimensional Electron Systems**
Editors: A. Okiji and N. Kawakami

119 **Spectroscopy of Mott Insulators and Correlated Metals**
Editors: A. Fujimori and Y. Tokura

120 **Optical Properties of III–V Semiconductors**
The Influence of Multi-Valley Band Structures
By H. Kalt

W. Ludwig C. Falter

Symmetries in Physics

Group Theory Applied to Physical Problems

Second Extended Edition
With 91 Figures

Springer

Professor Dr. Wolfgang Ludwig
Professor Dr. Claus Falter

Westfälische Wilhelms-Universität, Institut für Theoretische Physik
Wilhelm-Klemm-Straße 10, D-48149 Münster, Germany

Series Editors:

Professor Dr., Dres. h. c. Manuel Cardona
Professor Dr., Dres. h. c. Peter Fulde*
Professor Dr., Dres. h. c. Klaus von Klitzing
Professor Dr., Dres. h. c. Hans-Joachim Queisser

Max-Planck-Institut für Festkörperforschung, Heisenbergstrasse 1, D-70569 Stuttgart, Germany
* Max-Planck-Institut für Physik komplexer Systeme, Bayreuther Straße 40, Haus 16,
D-01187 Dresden, Germany

Managing Editor:

Dr.-Ing. Helmut K.V. Lotsch

Springer-Verlag, Tiergartenstrasse 17, D-69121 Heidelberg, Germany

Library of Congress Cataloging-in-Publication Data

Ludwig, W. (Wolfgang), 1929-
Symmetries in physics : group theory applied to physical problems / W. Ludwig, C. Falter. -- 2nd ed.
p. cm. -- (Springer series in solid-state sciences ; 64)
Includes bibliographical references and index

1. Symmetry (Physics) 2. Symmetry groups. 3. Chemistry, Physical and theoretical. I. Falter, C. (Claus), 1944- . II. Title.
III. Series.
QC174.L83 1995
530.1'522--dc20 95-39620
CIP

ISBN-13: 978-3-540-60284-2 e-ISBN-13: 978-3-642-79977-8
DOI: 10.1007/978-3-642-79977-8

This work is subject to copyright. All rights are reserved, whether the whole or part of the material is concerned, specifically the rights of translation, reprinting, reuse of illustrations, recitation, broadcasting, reproduction on microfilm or in any other way, and storage in data banks. Duplication of this publication or parts thereof is permitted only under the provisions of the German Copyright Law of September 9, 1965, in its current version, and permission for use must always be obtained from Springer-Verlag. Violations are liable for prosecution under the German Copyright Law.

© Springer-Verlag Berlin Heidelberg 1988, 1996

The use of general descriptive names, registered names, trademarks, etc. in this publication does not imply, even in the absence of a specific statement, that such names are exempt from the relevant protective laws and regulations and therefore free for general use.

Typesetting: ASCO Trade Typesetting Limited, Hong Kong
SPIN: 10487490 54/3144 – 5 4 3 2 1 0 – Printed on acid-free paper

Preface

One of the most efficient methods in physics is based on the discussion of the symmetry of a physical system. Group theory and, in particular, representation theory are the mathematical tools for handling the symmetries of such a system. Of course, these fundamental ideas have not changed since the first edition of this book. Only the applications have been extended to new systems and are permanently expanding. Quasicrystals are very new systems in solid-state physics and are discussed in Appendix I. On the other hand, in particle physics, supersymmetry theories have been developed, which combine bosons and fermions. This means that in a study of corresponding models, there is a need for anticommutators (Grassmann algebra) as well as for commutators (Lie algebra). Since these theories are still being developed and since a discussion of the details would go far beyond the scope of this book, we only give some indications in Appendix J.

Other changes are only related to the correction of minor errors or misprints. Often-used symbols and abbreviations are listed in Appendix K.

As already mentioned in the Preface to the First Edition, a short version of the solutions to the exercises may be obtained from the authors.

We thank Prof. Dr. M. Stingl for some advice, Mrs. Schockmann for preparing the manuscript, and especially Dr. H. Lotsch of Springer-Verlag for his good cooperation.

Münster, October 1995 *W. Ludwig · C. Falter*

Preface to the First Edition

The majority of physical systems exhibit symmetries of one kind or another. These symmetries can be used to simplify physical problems (indeed, sometimes a result cannot be achieved in any other way) and also to understand and classify the solutions. The mathematical tools required for this, i.e. group theory, and in particular representation theory, together with their applications to physical problems, were treated by us in a series of seminars and lectures, which now form the basis of this book.

Our main objective is to prepare the necessary mathematical foundations so that they can be used in physics. Most statements are illustrated by examples, which are in many cases simple but occasionally more complicated (especially in connection with space groups). The method of symmetry projections is applied more widely than in most texts of a similar standard, but because this method is a suitable and powerful tool for the systematic reduction of representation spaces to irreducible spaces, and thus for the determination of the eigenstates of the system, it deserves to be better known. This theory finds applications in many areas of physics in which symmetry plays a role. We consider finite, discrete symmetries as well as continuous symmetries and also symmetry breaking, with examples taken from atomic, molecular, solid-state and high-energy physics.

This text is intended mainly for students who have attended basic courses in physics and for researchers working in physics. However, the occurrence of symmetry properties is by no means restricted to physics, so this book should also be useful for people primarily interested in other subjects such as chemistry and physical chemistry. Many problems are included in the text as exercises; a booklet of solutions may be obtained from the authors.

We are very grateful to Dr. W. Zierau, who gave us much good advice, and to K. Stroetmann, H. Rakel and J. Backhaus for help in preparing the manuscript, the subject index and in proofreading. We are especially indebted to Dr. H. K. V. Lotsch of Springer-Verlag for encouragement and cooperation and to Miss D. Hollis, who improved the style of our sometimes rather "German" English.

Münster, October 1987 *W. Ludwig · C. Falter*

Contents

... διὸ δὴ καὶ χώραν ταῦτα ἄλλα ἄλλην
ἴσχειν, πρὶν καὶ τὸ πᾶν ἐξ αὐτῶν
διακοσμηθὲν γενέσθαι. Καὶ τὸ μὲν δὴ
πρὸ τούτου πάντα ταῦτ'εἶχεν
ἀλόγως καὶ ἀμέτρως....
οὕτω δὴ τότε πεφυκότα ταῦτα πρῶτον
διεσχηματίσατο εἴδεσί τε καὶ ἀριθμοῖς.

Platon, *Timaios*, 53a, b

1. Introduction

Physical systems in general possess symmetry properties. An essential point in the discussion of such systems is to find the relevant symmetries and to classify the properties or the states of the systems with respect to these symmetries. Group theory provides the mathematical tools for the description of symmetries. Within representation theory, methods are developed that allow classification of the physical states of a system with respect to the irreducible representations of the symmetry group.

The symmetries may be of very different natures for different types of objects such as particles (elementary particles, atoms, molecules), many-particle systems (crystals, liquids, fluids), all kinds of fields and macroscopic bodies.

We may distinguish between universal and special symmetries. Examples of *universal symmetries* are the space-time symmetries of systems, that is, the invariance of equations with respect to Poincaré or Lorentz transformations. In many-particle systems, the symmetry with respect to an interchange (permutation) of identical particles is universal. The charge and gauge symmetries of fields also belong to this group of symmetries. In quantum field theory the symmetries may be discrete as well as continuous. Well-known examples of discrete symmetries are the invariances under CPT transformations. The continuous symmetries may be divided into those that do not depend on space-time coordinates (first kind) and those that do (second kind). Invariance of a field theory under gauge transformations of the first kind leads to conservation laws. The number of these laws is equal to the number of parameters involved in the transformation. In the second kind of transformations (local gauge transformations) the parameters depend on the coordinates. Invariance of the theory under such transformations gives rise, in addition to the conservation laws, to interacting fields in the Lagrangian density of the particle fields. Examples are the electromagnetic field, the Yang-Mills fields, and also the gravitational field. Symmetries of this type are also called *dynamical* symmetries. In these cases the interaction is determined by symmetry. On the other hand, the symmetry of an interaction is not always obvious and can only be seen from the phenomena caused by it.

Special symmetries are often of a geometrical nature. Then there are a number of symmetry operations that transform the physical system into itself (spatially). Crystalline symmetries, for example, belong to this category. The number of such operations is finite (or at least enumerable) in general.

The invariance properties of physical systems in space and time, as well as gauge invariances, define the physically conserved quantities, that means observables like momentum, energy, angular momentum, and charges. These quantities then obey conservation laws. This is one of the reasons why symmetry is so important in physics.

In this book, using group theoretical methods we discuss the connection between symmetry and the physical state and show how to simplify a physical problem by using a "given" symmetry. The most important tool in this respect is the representation theory of groups; with its help we can define projectors allowing determination of the symmetry-adapted states. Another essential theorem is that of Wigner and Eckart. It allows statements on matrix elements and transitions, especially in connection with the representation of tensor operators.

In Chaps. 2–10 we consider groups with a countable (discrete, mainly finite) number of elements. This comprises the geometric symmetry groups whose operations leave the distances between two points and the angles between two directions invariant. Apart from this, permutation groups belong to this category, and also further symmetries that sometimes occur in physics.

In the second part (Chaps. 11–14), we discuss continuous symmetry groups and the Lie algebras corresponding to them. Most universal symmetries are included in these groups. Because of the limited size of this book, the essential statements have been explained with the $\mathscr{SU}(n)$ groups, however, a transfer of methods and procedures to other groups in general is possible without difficulty. In the appendix we discuss the Lorentz group, which has infinite-dimensional unitary representations, as an example of a noncompact group.

As an application of the $\mathscr{SU}(n)$ groups we consider some aspects of modern gauge theories. Whereas previously one used to start with phenomenological equations, to investigate the interactions and then found the symmetries of the system, we will follow the recent development where one starts with a possible symmetry group for the gauge transformations and the gauge invariance then determines the form of the field equations and the interactions. Thus, in the development of physics it was of no special importance that the system "charged particle–electromagnetic field" is gauge invariant (under an Abelian gauge transformation). But having realized this fact, one can look to see what other gauge-invariant theories are possible if the conserved quantities ("charges") have been previously specified. This leads to new, non-Abelian gauge groups and new interacting fields that couple to the "charges" of the particles. The procedure therefore is the following. One has to look for the "charges" as the conserved quantities, from these one can derive the corresponding gauge invariance of the first kind.

The corresponding gauge invariance then specifies field Lagrangians and interacting fields. The principles of these theories are discussed with some examples in Chap. 14, without claiming to be complete.

For an understanding of the theory of continuous groups, especially Chaps. 12 and 13, the results of Sect. 5.5 are necessary. However, this comparatively difficult section is dispensable for many problems in connection with point and space groups; for these the considerations in Sect. 6.1 are sufficient.

To keep the size manageable we had to restrict ourselves in other respects. Mathematical proofs are given explicitly only as far as they are necessary for an immediate understanding. In many cases they are "simple" enough to be done in the form of an exercise. Thus the reader is strongly advised to solve the exercises; sometimes they are indeed necessary for a handling of the mathematics. In our opinion it was essential to develop the mathematical theory in such a way as to allow direct application to physical problems. Thus statements and theorems are always illustrated with definite examples; then the methods can be immediately transferred to other problems.

One main aim is to show that group theory makes it possible to treat problems from all parts of physics (and molecular chemistry), from classical mechanics to quantum field theory, due to the symmetry inherent in physical systems. Indeed, for many physical theories developed during recent decades, group theory is the central key. In order to demonstrate this we have chosen examples from solid-state as well as molecular physics, including electronic as well as vibrational spectra, and also examples from atomic, nuclear and elementary particle physics. The physical background and the basic relations of the different topics are assumed to be known.

In the applications we often have to use the irreducible representations of the group elements. It was not possible here to give all the irreducible representations of space groups explicitly. For this we have to refer to the existing books of tables, but at the same time we have to state that many things have been tabulated only incompletely. Then the reader has to calculate the irreducible representations, the reduction coefficients, the Clebsch-Gordan coefficients, etc., by himself. The methods are given.

The notation has been standardized in many respects. Where this is the case, we have adapted the generally accepted notation. But there are some fields (e.g. space groups) where several different notations are used. In such cases we had to choose. But the correspondence between different notations can always be established by comparing the definitions. The tables in the Appendix (especially in Appendix A) always allow a comparison.

2. Elements of the Theory of Finite Groups

Most groups which are essential in solid-state physics are finite groups, or at least can be looked upon as being finite; this is the case for the translation group of lattices. Therefore we first have to explain the concepts of the abstract theory of finite groups. This is done in this first section, where we give the basic notations and their relations. All this is illustrated by a simple example.

2.1 Symmetry and Group Concepts: A Basic Example

As an introductory example, we consider an equilateral triangle to which we additionally assign a set of points 1 to 6 (Fig. 2.1). The basic concepts will be illustrated by means of this example, which represents the symmetry of an NH_3 molecule.

The triangle and the set of points are transformed into themselves if the system is rotated about the centre of the triangle by multiples of the angle $2\pi/3$. The axis of rotation, perpendicular to the plane of the triangle, is called a threefold axis, since after three rotations (always through the basic angle $2\pi/3$) the initial situation is restored. These symmetry operations about a threefold axis are denoted by c_3, c_3^2, $c_3^3 = e, \dots$, where the *rotation* has always to be taken counterclockwise (*positive sense*) and e is the *identity operation* (unit operation), which does not move the triangle. Apart from these rotations, there are reflections σ_v, σ_v', σ_v'' transforming the triangle into itself. These mirror planes contain the threefold axis. We can illustrate the operations best with the mappings produced by them.

e: points and triangle are invariant (do not move)

$$\begin{aligned}
&c_3: 1 \to 3 \to 5 \to 1\ ; \quad 2 \to 4 \to 6 \to 2\ ; \quad A \to B \to C \to A\ . \\
&c_3^2: 1 \to 5 \to 3 \to 1\ ; \quad 2 \to 6 \to 4 \to 2\ ; \quad A \to C \to B \to A\ . \\
&\sigma_v: 1 \leftrightarrow 2\ ; \quad 3 \leftrightarrow 6\ : \quad 4 \leftrightarrow 5\ ; \quad B \leftrightarrow C\ . \\
&\sigma_v': 1 \leftrightarrow 6\ ; \quad 2 \leftrightarrow 5\ ; \quad 3 \leftrightarrow 4\ ; \quad A \leftrightarrow C\ . \\
&\sigma_v'': 1 \leftrightarrow 4\ ; \quad 2 \leftrightarrow 3\ ; \quad 5 \leftrightarrow 6\ ; \quad A \leftrightarrow B\ .
\end{aligned} \tag{2.1.1}$$

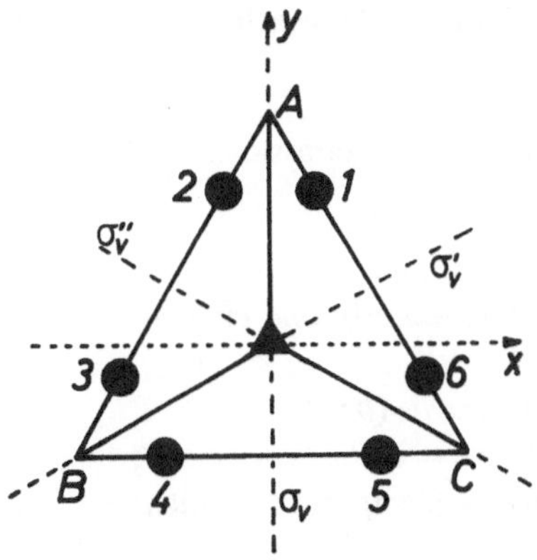

Fig. 2.1. Arrangement of points, or triangle, having $\mathscr{C}_{3v}$ symmetry. ▲: threefold axis; σ_v, σ_v', σ_v'': mirror planes containing the rotation axis

Table 2.1. The composition (group) table of the symmetry operations of the triangle in Fig. 2.1

a \ b	e	c_3	c_3^2	σ_v	σ_v'	σ_v''
e	e	c_3	c_3^2	σ_v	σ_v'	σ_v''
c_3	c_3	c_3^2	e	σ_v''	σ_v	σ_v'
c_3^2	c_3^2	e	c_3	σ_v'	σ_v''	σ_v
σ_v	σ_v	σ_v'	σ_v''	e	c_3	c_3^2
σ_v'	σ_v'	σ_v''	σ_v	c_3^2	e	c_3
σ_v''	σ_v''	σ_v	σ_v'	c_3	c_3^2	e

With this scheme the effect of successive symmetry operations is also easily depicted; we define the operation in the rightmost position to be performed always *first*, the operation second from the right, second, and so on. The execution of two successive operations is called the *product* (operation), e.g. $\sigma_v c_3$. Generating the corresponding image of $\sigma_v c_3$, we realize that this is identical with that of σ_v', i.e. $\sigma_v c_3 = \sigma_v'$. Accordingly, we find every product of the elements in (2.1.1) to be contained in the set $\{e, c_3, c_3^2, \sigma_v, \sigma_v', \sigma_v''\}$. We further realize that the *products* are *not* always *commutative*; for example, $c_3\sigma_v = \sigma_v''$. For every operation there obviously exists a *reciprocal* or *inverse* operation denoted by c_3^{-1}, σ_v^{-1} and so on. Clearly,

$$e^{-1} = e \ , \qquad c_3^{-1} = c_3^2 \ , \qquad (c_3^2)^{-1} = c_3^{-2} = c_3 \ , \qquad \sigma_v^{-1} = \sigma_v \ , \quad \text{etc.} \tag{2.1.2}$$

The general behavior in constructing products is represented by the so-called *composition* (*multiplication, group*) *table* of $(a \cdot b)$ (Table 2.1). We find that in each row and each column of the table every element of the set occurs exactly once. In addition, the inverse elements are readily specified: b and a are inverse to each other if $a \cdot b = e$.

Obviously this example is the geometric realization of a mathematical structure, which is called a group:

Formally, a pair $(\mathcal{G}, \cdot)$ with a set $\mathcal{G}$ of elements and a composition $\cdot$ defines a group if

1) there exists an internal *composition* law $\cdot$ on $\mathcal{G}$;
2) for every pair of elements (a, b) there exists *exactly one element* $c \in \mathcal{G}$ with $c = a \cdot b$;
3) the composition law is *associative*: $(a \cdot b) \cdot c = a \cdot (b \cdot c)$;
4) there exists an *identity (unit) element* with $a \cdot e = a$ for every $a \in \mathcal{G}$;
5) for every element $a \in \mathcal{G}$ there exists an *inverse (reciprocal) element* in $\mathcal{G}$ with $a \cdot a^{-1} = e$. (2.1.3)

A group is completely defined by its composition table. In our case, composition means the execution of successive operations also denoted as multiplication and written $a \cdot b$ or simply ab. If *all* the multiplications in a group commute, it is an *Abelian group* or *commutative group*. The group table is then symmetric with respect to the principal diagonal. Conditions (2.1.3) require only the existence of a right-identity and right-inverse element. However it follows immediately that these are also left-identity or left-inverse elements as

$$a^{-1} \cdot a = (a^{-1} \cdot a) \cdot (a^{-1} \cdot (a^{-1})^{-1}) = a^{-1} \cdot (a \cdot a^{-1}) \cdot (a^{-1})^{-1}$$
$$= a^{-1} \cdot (a^{-1})^{-1} = e \qquad (2.1.4)$$

and

$$e \cdot a = (a \cdot a^{-1}) \cdot a = a \cdot (a^{-1} \cdot a) = a \cdot e = a \ . \qquad (2.1.5)$$

Similarly it follows that identity and inverse elements are unique. For, if e and f are both identities, then

$$e = e \cdot f = f$$

because of (2.1.5) and if a^{-1} and $\bar{a}$ are both inverse elements, then

$$\bar{a} = \bar{a} \cdot (a \cdot a^{-1}) = (\bar{a} \cdot a) \cdot a^{-1} = e \cdot a^{-1} = a^{-1} \ .$$

The inverse of a product is given by

$$(a \cdot b)^{-1} \cdot (a \cdot b) = e \rightarrow (a \cdot b)^{-1} \cdot a = b^{-1} \rightarrow (a \cdot b)^{-1} = b^{-1} \cdot a^{-1} \ . \qquad (2.1.6)$$

Any nonempty subset of $\mathcal{G}$ satisfying (2.1.3) with the same composition law is called a *subgroup* $\mathcal{U} \subseteq \mathcal{G}$. (2.1.7)

Every group $\mathcal{G}$ possesses $\mathcal{U} = \mathcal{G}$ and $\mathcal{U} = \{e\}$ as trivial subgroups. If more exist, we speak of nontrivial or proper subgroups.

The group described by (2.1.1, 2) and Table 2.1 is denoted by

$$\mathscr{C}_{3v} = \{e, c_3, c_3^2, \sigma_v, \sigma_v', \sigma_v''\} \ . \tag{2.1.8}$$

Subgroups of $\mathscr{C}_{3v}$ are, for example,

$$\mathscr{C}_3 = \{e, c_3, c_3^2\} \qquad \text{and} \qquad \mathscr{C}_s = \{e, \sigma_v\} \ . \tag{2.1.9}$$

The group $\mathscr{C}_{3v}$ is non-Abelian, but $\mathscr{C}_3$ and $\mathscr{C}_s$ are Abelian. The subgroups can be seen directly from the multiplication table; they form a closed set with respect to the composition law.

The triangle, or the set of points, in Fig. 2.1 can also be mapped onto itself by other operations. For example the reflections σ_v can be replaced by twofold rotations c_2, c_2', c_2'', which are rotations by $2\pi/2 = \pi$ about axes lying in the plane of the triangle. The group table does not change formally. Such groups, in which elements and multiplications can be mapped uniquely one to one, i.e. the group table remains unchanged, are called *isomorphic groups*. There are further groups isomorphic ($\cong$) to $\mathscr{C}_{3v}$ which will be described later. Isomorphic groups can express different physical systems (Sect. 3.1). The group containing one threefold main axis and three twofold axes perpendicular to the threefold axis (angle $\pi/3$ between the twofold axes) is called the dihedral group

$$\mathscr{D}_3 = \{e, c_3, c_3^2, c_2, c_2', c_2''\} \ . \tag{2.1.10}$$

It is isomorphic to $\mathscr{C}_{3v}$:

$$\mathscr{D}_3 \cong \mathscr{C}_{3v} \ . \tag{2.1.11a}$$

Starting with an arbitrary given point (e.g. point 1 in Fig. 2.1), we can produce the set of points in Fig. 2.1 by applying special symmetry operations one or more times. In the examples these operations are $p = c_3$ and $q = \sigma_v$ or c_2. Such elements are called *generating elements* or simply *generators of the group*. Elements of a group $\mathscr{G}$ are called generators if any element of $\mathscr{G}$ can be represented by finite products of these generators. The choice of generators is not a unique one. We could also choose alternatively $q = \sigma_v'$ or c_2' or σ_v'' or c_2''. Sometimes it is useful to take more generators than necessary. Any group with a finite number of elements possesses a minimal system of generators, which is called the *basis of the group*. The number of elements in the basis is the rank of *the (finite) group*.

As an example we consider a group defined by two generators p and q with the *generating relations*

$$p^3 = e \ ; \qquad q^2 = e \ ; \qquad (q \cdot p)^2 = e \ . \tag{2.1.12}$$

The group then contains the elements

$$\mathscr{G}_6 = \{e, p, p^2, q, qp, qp^2\} \ . \tag{2.1.13}$$

Table 2.2. The group table of the group with generating relations (2.1.12)

	e	p	p^2	q	qp	qp^2
e	e	p	p^2	q	qp	qp^2
p	p	p^2	e	qp^2	q	qp
p^2	p^2	e	p	qp	qp^2	q
q	q	qp	qp^2	e	p	p^2
qp	qp	qp^2	q	p^2	e	p
qp^2	qp^2	q	qp	p	p^2	e

Because of (2.1.12), all the other powers and products are identical with one of these six elements. The group table is consequently Table 2.2, which is isomorphic Table 2.1, i.e.

$$\mathcal{G}_6 \cong \mathcal{C}_{3v} \cong \mathcal{D}_3 \ . \tag{2.1.11b}$$

The powers 3 und 2 in (2.1.12) are called the *orders of the elements* p and q, respectively. Generally the order $n = \operatorname{ord} p$ of an element p is defined by

$$p^n = e \tag{2.1.14}$$

with the *smallest* natural number $n \geqslant 1$. This has to be distinguished from the *order* $g = \operatorname{ord} \mathcal{G}$ *of a finite group* $\mathcal{G}$, which is the number of elements in $\mathcal{G}$.

A group is *cyclic*, if there is an element $p \in \mathcal{G}$ such that

$$\mathcal{G} := \{p^n | n \in \mathbb{Z}\} \ . \tag{2.1.15}$$

In cyclic groups, $\operatorname{ord} p = \operatorname{ord} \mathcal{G}$. Such a group is finite if and only if there exists a smallest natural number $n \neq 0$ with $p^n = e$. For example, $\mathcal{C}_3$ is a cyclic group. Obviously any element with its powers forms a cyclic subgroup $\mathcal{U}_p \subseteq \mathcal{G}$:

$$\mathcal{U}_p = \{p^n | p \in \mathcal{G}; n \in \mathbb{Z}\} \ . \tag{2.1.16}$$

Consequently any element $a \in \mathcal{G}$ is contained in (at least) one subgroup $\mathcal{U}_p \subseteq \mathcal{G}$. A cyclic group is always Abelian.

Finally we point out another possible way to represent the elements of $\mathcal{C}_{3v}$, i.e. the transformation of vectors, which means a one-to-one mapping of a system of vectors. Let us consider the vertices of the triangle in Fig. 2.1 in a Cartesian coordinate system with the origin at the centre of the triangle. The vertices then have the coordinates

$$\boldsymbol{x}_A = (0, 1) \ ; \qquad \boldsymbol{x}_B = (-\sqrt{3}/2, -1/2) \ ; \qquad \boldsymbol{x}_C = (\sqrt{3}/2, -1/2) \ . \tag{2.1.17}$$

Consequently the symmetry operations are represented by the matrices

$$e = \begin{pmatrix} 1 & 0 \\ 0 & 1 \end{pmatrix}, \quad c_3 = \begin{pmatrix} -1/2 & -\sqrt{3}/2 \\ \sqrt{3}/2 & -1/2 \end{pmatrix}, \quad c_3^2 = \begin{pmatrix} -1/2 & \sqrt{3}/2 \\ -\sqrt{3}/2 & -1/2 \end{pmatrix},$$

$$\sigma_v = \begin{pmatrix} -1 & 0 \\ 0 & 1 \end{pmatrix}, \quad \sigma_v' = \begin{pmatrix} 1/2 & \sqrt{3}/2 \\ \sqrt{3}/2 & -1/2 \end{pmatrix}, \quad \sigma_v'' = \begin{pmatrix} 1/2 & -\sqrt{3}/2 \\ -\sqrt{3}/2 & -1/2 \end{pmatrix}, \tag{2.1.18}$$

where the composition law is matrix multiplication. Applying c_3 to $\boldsymbol{x}_A$ we obtain

$$\boldsymbol{x}_B = c_3 \cdot \boldsymbol{x}_A \;, \tag{2.1.19}$$

with corresponding relations for the other operations.

Exercise 2.1. The group $\mathscr{C}_{nv}$ describes rotations about an n-fold main axis and reflections in n mirror planes containing this axis with an angle of π/n between the planes. Verify all the statements of this section using this group.

Exercise 2.2. The group $\mathscr{D}_n$ contains instead of n mirror planes n twofold axes perpendicular to the main axis with an angle π/n between the axes. Discuss this group as in Exercise 2.1.

Exercise 2.3. Show that $\mathscr{C}_{nv} \cong \mathscr{D}_n$. Discuss the physical difference of these two groups, possibly with the help of Sect. 3.1.

2.2 General Theorems on Group Theory

In many applications it is useful to separate the elements of a group into classes; this is achieved by *equivalence relations* or *equivalence classes*. As an example we consider a set of elements which are to be arranged according to their "length." Elements a, b of equal length are said to be equivalent ($\sim$) or a "is as long as" b. In general, any relation between two elements a, b of a set $\mathscr{M}$ is said to be an equivalence relation, if it has the following properties:

1) It is reflexive:

$$a \sim a \tag{2.2.1}$$

(any element is equivalent to itself, a "is as long as" a),

2) It is symmetric: if $a \sim b$, then $b \sim a$,
3) It is transitive: if $a \sim b$ *and* $b \sim c$, then $a \sim c$.

With the help of these relations, we can divide all the elements $a \in \mathscr{M}$ into classes (with elements of equal "length") by taking from $\mathscr{M}$ all those elements a_i which are equivalent to one $a \in \mathscr{M}$: $a_i \sim a$, which may be stated as: An equivalence class $[\![a]\!]$ of $a \in \mathscr{M}$ contains all the elements $a_i \in \mathscr{M}$, which are equivalent to a.

$$[\![a]\!] := \{a_i | a_i \in \mathscr{M} \wedge a \in \mathscr{M} \wedge a_i \sim a\} \;. \tag{2.2.2}$$

Any element of $[\![a]\!]$ can be taken for the definition of an equivalence class, for if $a \sim b$, then

$$[\![a]\!] = [\![b]\!] \; .$$

From this it follows immediately, that if a and b are not equivalent to each other ("they differ in length"), $[\![a]\!]$ and $[\![b]\!]$ have *no* common element:

The separation of a set into equivalence classes is always a division into disjoint (sub-) sets. (2.2.3)

For a proof we easily verify that obviously $[\![a]\!] \neq \emptyset$, since a is always equivalent at least to itself. If $a \sim b$, then $[\![a]\!] = [\![b]\!]$ (see above). If now a is not equivalent to b ($a \nsim b$) and if $[\![a]\!] \cap [\![b]\!] = x \in \mathcal{M}$, then because of the definition of $[\![a]\!]$ and $[\![b]\!]$, $x \sim a$ as well as $x \sim b$ and therefore $a \sim b$ because of the transitivity; but this is in contradiction to $a \nsim b$. Hence, if $a \nsim b$ then $[\![a]\!] \cap [\![b]\!] = \emptyset$. Consequently[1]

$$\mathcal{M} = \sum_i [\![a_i]\!] \qquad \text{with } [\![a_i]\!] \cap [\![a_j]\!] = \emptyset \qquad \text{for } a_i \nsim a_j \; . \tag{2.2.4}$$

The isomorphism of groups (Sect. 2.1) is also an equivalence relation on a set of groups since

$$\begin{aligned} &1)\ \mathcal{G} \cong \mathcal{G} && \text{(reflexive)} \\ &2)\ \mathcal{G} \cong \mathcal{G}' \Leftrightarrow \mathcal{G}' \cong \mathcal{G} && \text{(symmetric)} \\ &3)\ \mathcal{G} \cong \mathcal{G}' \text{ and } \mathcal{G}' \cong \mathcal{G}'' \Rightarrow \mathcal{G} \cong \mathcal{G}'' && \text{(transitive).} \end{aligned} \tag{2.2.5}$$

Such equivalence classes are important for a classification of groups. One possible classification starts with a well-defined subgroup $\mathcal{U} \subseteq \mathcal{G}$. What about the other classes, called cosets in this case? With $\mathcal{U} \subseteq \mathcal{G}$ and $a \in \mathcal{G}$, the subset

$$a\mathcal{U} := \{a \cdot p \,|\, p \in \mathcal{U} \subseteq \mathcal{G}; a \in \mathcal{G}\} \tag{2.2.6a}$$

is called a *left coset* and

$$\mathcal{U}a := \{p \cdot a \,|\, p \in \mathcal{U} \subseteq \mathcal{G}; a \in \mathcal{G}\} \tag{2.2.6b}$$

a *right coset*. If $a = e$, the coset is equal to the subgroup $\mathcal{U}$ itself. Every coset contains the same number of elements, because of (2.2.6) and (2.1.3). The division (2.2.6) is a partition into equivalence classes; elements of the subgroup are equivalent to each other. The reader should check the equivalence relations.

Because of (2.2.3), we find that none of the left (right) cosets of a group have a common element and every element is a member of exactly one left (right) coset.

---

[1] We also use $\sum$ in the case of a set "summation": $\bigcup_{i=1}^{r} [\![a_i]\!]$ if the sets are disjoint.

Every coset has the same number of elements, which is equal to the order of the generating subgroup $\mathcal{U}$. From this we derive immediately the *Euler-Lagrange* theorem[2]: The order of a subgroup $\mathcal{U}$ is always a divisor of the order of the group $\mathcal{G}$.

Further, with (2.1.14) and (2.1.16) there follows *Fermat's theorem*: Any element a of a finite group of order g satisfies $a^g = e$.

The number of cosets of a finite group $\mathcal{G}$, as generated by $\mathcal{U}$, is called the index of $\mathcal{U}$ in $\mathcal{G}$ (ind $\mathcal{U}$); therefore

$$\operatorname{ord}\mathcal{G} = g = \operatorname{ind}\mathcal{U}\cdot\operatorname{ord}\mathcal{U} = n_i\cdot n_u \ , \tag{2.2.7}$$

and ind $\mathcal{U}$ is also a divisor of g. The partition of $\mathcal{G}$ with respect to cosets can be written as

$$\mathcal{G} = \sum_{i=1}^{n_i} a_i\cdot\mathcal{U} = \sum_{j=1}^{n_i} \mathcal{U}\cdot a_j' \ , \qquad a_1 = e \ . \tag{2.2.8}$$

The set of cosets $\{a_1\mathcal{U}, a_2\mathcal{U}, \dots\}$ is denoted as the *quotient set* $\mathcal{G}/\mathcal{U}$ ($\mathcal{G}$ over $\mathcal{U}$). Let us consider again $\mathcal{G} = \mathcal{C}_{3v}$ and choose $\mathcal{U} = \mathcal{C}_3$. Then

$$\mathcal{C}_{3v} = e\cdot\mathcal{C}_3 + \sigma_v\cdot\mathcal{C}_3 = \{e, c_3, c_3^2\} + \{\sigma_v, \sigma_v', \sigma_v''\}$$

and

$$\mathcal{C}_{3v}/\mathcal{C}_3 = \{\mathcal{C}_3; \sigma_v\cdot\mathcal{C}_3\} \cong \mathcal{C}_s \ , \tag{2.2.9}$$

$$g = n_i\cdot n_u = 6 = 2\cdot 3 \ .$$

We could also choose a further subgroup of order 2 ($\mathcal{C}_s$) and decompose $\mathcal{C}_{3v}$ with respect to $\mathcal{C}_s$. The index of $\mathcal{C}_s$ in $\mathcal{C}_{3v}$ is 3.

$$\begin{aligned}\mathcal{C}_{3v} &= e\cdot\mathcal{C}_s + c_3\cdot\mathcal{C}_s + c_3^2\cdot\mathcal{C}_s \\ &= \{e, \sigma_v\} + \{c_3, \sigma_v''\} + \{c_3^2, \sigma_v'\} \ ,\end{aligned}$$

but

$$\begin{aligned}\mathcal{C}_{3v} &= \mathcal{C}_s e + \mathcal{C}_s c_3 + \mathcal{C}_s c_3^2 \\ &= \{e, \sigma_v\} + \{c_3, \sigma_v'\} + \{c_3^2, \sigma_v''\}\end{aligned}$$

and

$$\mathcal{C}_{3v}/\mathcal{C}_s = \{\mathcal{C}_s, c_3\cdot\mathcal{C}_s, c_3^2\cdot\mathcal{C}_s\} \ . \tag{2.2.10}$$

---

[2] Using the Euler-Lagrange theorem possible structures of finite groups can be investigated, however not completely.

This example shows that in the first case (index 2) the decomposition into left cosets is identical to that into right cosets; this obviously holds for all subgroups of index 2. In the second case (index 3) the decomposition into left cosets is different from that into right cosets.

A subgroup $\mathscr{U} \subseteq \mathscr{G}$ is an *invariant* (*normal*) *subgroup* (normal divisor) $\mathscr{N}$, if the decomposition of $\mathscr{G}$ with respect to $\mathscr{U}$ into left cosets agrees with that into right cosets:

$$\left.\begin{array}{l} a\mathscr{U} = \mathscr{U}a \\ \text{or} \quad a\mathscr{U}a^{-1} = \mathscr{U} \end{array}\right\} \quad \text{for all } a \in \mathscr{G} . \tag{2.2.11a}$$

Since this relation has to be valid for all the elements $b \in \mathscr{U}$, (2.2.11a) is equivalent to

$$aba^{-1} = b' \quad \text{for all } b, b' \in \mathscr{U}, a \in \mathscr{G} . \tag{2.2.11b}$$

From this discussion we derive:

All subgroups $\mathscr{U} \subseteq \mathscr{G}$ with index 2 and all $\mathscr{U} \subseteq \mathscr{G}_a$ in Abelian groups $\mathscr{G}_a$ are invariant subgroups. (2.2.12)

The latter follows from the commutativity of all the elements of $\mathscr{G}_a$.

The quotient set of $\mathscr{G}$ with respect to an invariant subgroup $\mathscr{N}$,

$$\mathscr{G}/\mathscr{N} = \{a_i\mathscr{N}\} = \{\mathscr{N}a_i\} , \qquad a_i \in \mathscr{G} , \tag{2.2.13}$$

forms a group in which the left (or right) cosets are the group elements: the *factor group* or, better, *quotient group* $\mathscr{G}/\mathscr{N}$. In this quotient group the multiplication of two elements $a_i\mathscr{N}$ and $a_j\mathscr{N}$ is welldefined by

$$a_i\mathscr{N} \cdot a_j\mathscr{N} = (a_ia_j)\mathscr{N} . \tag{2.2.14}$$

The identity element is $e\mathscr{N} = \mathscr{N}$, the inverse element of $a_i\mathscr{N}$ is equal to $a_i^{-1}\mathscr{N}$ because

$$a_i\mathscr{N} \cdot a_i^{-1}\mathscr{N} = (a_ia_i^{-1})\mathscr{N} = \mathscr{N} .$$

The order of $\mathscr{G}/\mathscr{N}$ is $g/g_{\mathscr{N}} = n_i$, i.e. equal to the index of $\mathscr{N}$ in $\mathscr{G}$. Of course, $\mathscr{G}/\mathscr{N}$ is not a subgroup of $\mathscr{G}$; the elements are sets with a composition law according to (2.2.14). Obviously a group $\mathscr{G}$ can be mapped completely onto a quotient group in the following sense: all the elements $x_i \in a_i\mathscr{N}$, $x_i \in \mathscr{G}$ are mapped onto the element $a_i\mathscr{N}$ of the quotient group $\mathscr{G}/\mathscr{N}$: $a \to a\mathscr{N}$. Since all the cosets contain the same number of elements $x_i \in \mathscr{G}$, all the elements of $\mathscr{G}$ are mapped "homogeneously" into $\mathscr{G}/\mathscr{N}$, which therefore is a *homomorphic image* of $\mathscr{G}$ and a special case of what is called *homomorphism*.

In general, a homomorphism between two groups means:

A map f of a group $(\mathcal{G}, \cdot)$ onto a group $(\mathcal{G}', \circ)$ is a homomorphism (epimorphism) $f: \mathcal{G} \to \mathcal{G}'$, if for all $a_1, a_2 \in \mathcal{G}$,

$$f(a_1) \circ f(a_2) = f(a_1 \cdot a_2) \ . \tag{2.2.15}$$

The image of a subgroup $\mathcal{U} \subset \mathcal{G}$ in a group homomorphism $f: \mathcal{G} \to \mathcal{G}'$ is a subgroup $\mathcal{U}' \subset \mathcal{G}'$, and vice versa. A mapping of $\mathcal{G}$ onto a subgroup $\mathcal{U} \subset \mathcal{G}$ is called an *endomorphism* $f: \mathcal{G} \to \mathcal{U} \subset \mathcal{G}$. A one-to-one (unique) group homomorphism is said to be a group *isomorphism*; $\mathcal{G}$ and $\mathcal{G}'$ are then isomorphic to each other: $\mathcal{G} \cong \mathcal{G}'$, see (2.1.11a, b), and (2.2.5). In the special case $\mathcal{G} \equiv \mathcal{G}'$, the isomorphism is called *automorphism.*

Examples:
homomorphism *of* $\mathcal{C}_{3v}$ onto $\mathcal{C}_2$,

$$f: \{e, c_3, c_3^2\} \to e \ , \qquad \{\sigma_v, \sigma_v', \sigma_v''\} \to c_2 \ ; \tag{2.2.16a}$$

endomorphism of $\mathcal{C}_{3v}$ onto $\mathcal{C}_s$,

$$f: \{e, c_3, c_3^2\} \to e \ , \qquad \{\sigma_v, \sigma_v', \sigma_v''\} \to \sigma_v \ ; \tag{2.2.16b}$$

isomorphism of $\mathcal{C}_{3v}$ and $\mathcal{D}_3$,

$$\begin{aligned} f: e \to e \ , \qquad & c_3 \to c_3 \ , \qquad c_3^2 \to c_3^2 \ , \\ \sigma_v \to c_2 \ , \qquad & \sigma_v' \to c_2' \ , \qquad \sigma_v'' \to c_2'' \ ; \end{aligned} \tag{2.2.16c}$$

automorphism of $\mathcal{C}_{3v}$,

$$\begin{aligned} f: e \to e \ , \qquad & c_3 \to c_3^2 \ , \qquad c_3^2 \to c_3 \ , \\ \sigma_v \to \sigma_v \ , \qquad & \sigma_v' \to \sigma_v'' \ , \qquad \sigma_v'' \to \sigma_v' \ . \end{aligned} \tag{2.2.16d}$$

The set of elements being mapped onto the identity element $e' \in \mathcal{G}'$ in a homomorphism $f: \mathcal{G} \to \mathcal{G}'$ is the *kernel of the homomorphism.* In the examples (2.2.16a, b), $\mathcal{C}_3$ is the kernel of $f: \mathcal{C}_{3v} \to \mathcal{C}_2$ and $\mathcal{C}_s$, respectively.

From this discussion, a theorem on homomorphism[3] can be deduced:

If $f: \mathcal{G} \to \mathcal{G}'$ is a group homomorphism and $\mathcal{N}_K$ is the kernel of f, then

1) $\mathcal{N}_K$ is an invariant subgroup of $\mathcal{G}$, and
2) $f(\mathcal{G}) \to \mathcal{G}'$ is isomorphic to $\mathcal{G}/\mathcal{N}_K$.
3) Inversely, if $\mathcal{N}_K$ is an invariant subgroup of $\mathcal{G}$ and $f: \mathcal{G} \to \mathcal{G}/\mathcal{N}_K$ with $f(a) = a \cdot \mathcal{N}_K$ for all $a \in \mathcal{G}$, then f is a homomorphism with $\mathcal{N}_K$ as a kernel. (2.2.17)

[3] Besides the theorem on homomorphism, there are two further theorems on isomorphism, which we will not need in the following. The reader is referred to specialist books on this topic.

For the group $\mathscr{C}_{3v}$, these statements are already contained in (2.2.16a, b) and (2.2.9).

A group is called a *simple* group, if it has *no* proper invariant subgroup, and a *semisimple* group if it has *no* proper Abelian invariant subgroup. (2.2.18)

The group $\mathscr{C}_{3v}$ is neither simple nor semisimple since $\mathscr{C}_3$ is a proper Abelian invariant subgroup.

For the construction of groups the combination of two groups to form another larger group is important. One possibility is given by the *outer direct product*:

If $\mathscr{G}_1$ and $\mathscr{G}_2$ are two subgroups of a group $\mathscr{G}$ with a well-defined multiplication; then $\mathscr{G} = \mathscr{G}_1 \times \mathscr{G}_2$ is the outer direct product of $\mathscr{G}_1$ and $\mathscr{G}_2$, if

1) for all the $a_i \in \mathscr{G}_1$ and $b_j \in \mathscr{G}_2$, $a_i b_j = b_j a_i$,
2) the intersection is $\mathscr{G}_1 \cap \mathscr{G}_2 = \{e\}$,
3) $\mathscr{G}_1 \times \mathscr{G}_2 = \mathscr{G}_2 \times \mathscr{G}_1 = \mathscr{G}$. (2.2.19)

Every element $p \in \mathscr{G}$ can then be written in a unique way as a product $p = a_i b_j = b_j a_i$. The order g of $\mathscr{G}$ is then the product of the orders of $\mathscr{G}_1$ and $\mathscr{G}_2$: $g = g_1 \cdot g_2$. As $\mathscr{G}_1$ *and* $\mathscr{G}_2$ are invariant subgroups of $\mathscr{G}$, $\mathscr{G} = \mathscr{G}_1 \times \mathscr{G}_2$ cannot be simple.

In many cases it is appropriate to use a more general definition of an outer direct product. The product can be formed by two completely independent groups, e.g. a rotation group and a permutation group (Sects. 3.1, 5).

The outer direct product of two groups $\mathscr{G}_1 = \{a_i\}$ and $\mathscr{G}_2 = \{b_k\}$ is then the group $\mathscr{G}_1 \times \mathscr{G}_2$ with the elements (a_i, b_k) and the multiplication

$$(a_i, b_k) \cdot (a_j, b_l) = (a_i a_j, b_k b_l) \ . \tag{2.2.20}$$

The elements (a_i, e) form a group which is isomorphic to $\mathscr{G}_1$, and the elements (e, b_k) form a group which is isomorphic to $\mathscr{G}_2$; the assumptions of the outer direct product are then satisfied. In the special case $\mathscr{G}_1 = \mathscr{G}_2 = \mathscr{G}$, the elements

$$(a_i, a_i) \in \mathscr{G} \otimes \mathscr{G} = \hat{\mathscr{G}} \cong \mathscr{G} \tag{2.2.21}$$

define a group which is isomorphic to $\mathscr{G}$. It is called the *inner product* of $\mathscr{G}$.

Besides the direct product there exists a *semidirect product* of two groups:

If $\mathscr{G}$ is a group with subgroups $\mathscr{G}_1$ and $\mathscr{G}_2$, $\mathscr{G}_1 \cap \mathscr{G}_2 = \{e\}$ and

1) $b_j \mathscr{G}_1 = \mathscr{G}_1 b_j$ for all $b_j \in \mathscr{G}_2$, $a_i \in \mathscr{G}_1$,
2) all $p \in \mathscr{G}$ are represented uniquely by $p = a_i b_j$, then $\mathscr{G} = \mathscr{G}_1 \boxed{S} \mathscr{G}_2$ is the semidirect product of $\mathscr{G}_1$ and $\mathscr{G}_2$. (2.2.22)

Because of 1), $\mathscr{G}_1$ is an invariant subgroup of $\mathscr{G}$, but $\mathscr{G}_2$ generally is not. An example is given by

$$\mathscr{C}_{3v} = \mathscr{C}_3 \boxed{S} \{e, \sigma_v\} \ .$$

As $\{e, \sigma_v\}$ is not an invariant subgroup, cf. (2.2.10). $\mathscr{C}_{3v}$ cannot be a direct product.

Exercise 2.4. Discuss all the statements of this section in connection with the groups $\mathscr{D}_{4h}$ and $\mathscr{T}$ (defined in Sect. 3.1.).

Exercise 2.5. Show that $\mathscr{T}$ does not possess an invariant subgroup of index 2.

2.3 Conjugacy Classes

For any two elements $a, b \in \mathscr{G}$ we say that $a' = bab^{-1}$ is a transformation of a with b into a', where a' is the *conjugate element*. If b runs through all the elements of $\mathscr{G}$, we obtain a set

$$K_a = \{a_i | a, b \in \mathscr{G} \text{ and } a_i = bab^{-1}, a \text{ fixed}\} \tag{2.3.1}$$

of elements, which are all conjugate to a. The conjugation is an equivalence relation (2.2.1) since

1) $e, a \in \mathscr{G}$, any element $a \in \mathscr{G}$ is conjugate with itself (reflexivity)
2) $b \in \mathscr{G}$ and $b^{-1} \in \mathscr{G}$, therefore $a' = bab^{-1}$ *and* $a = b^{-1}a'(b^{-1})^{-1}$ (symmetry)
3) with $a' = bab^{-1}$ and $a'' = ca'c^{-1}$ it follows that $a'' = cbab^{-1}c^{-1} = cba(cb)^{-1}$ and $cb \in \mathscr{G}$ (transitivity).

Equation (2.3.1) defines a division of $\mathscr{G}$ into disjoint, equivalence classes, which are called *conjugacy classes* or just *classes*. These have to be strictly distinguished from cosets. Any element is thus a member of exactly one class, and $\mathscr{G}$ can be divided into disjoint classes:

$$\mathscr{G} = \sum_{i=1}^{r} K_i \ , \qquad r\text{: number of classes in } \mathscr{G} \ . \tag{2.3.2}$$

From the definition it follows that:

i) All the elements of one class have the same order. Thus, in one class we have only rotations of the same manifold. The reverse does not hold! (2.3.3)
ii) An element that commutes with all the elements of $\mathscr{G}$, is said to be *self-conjugate* (invariant, isolated) and forms a separate class. The identity element always forms a separate class. In Abelian groups every element forms a separate class ($r = g$). (2.3.4)
iii) The classes of a direct product $\mathscr{G} = \mathscr{G}_1 \times \mathscr{G}_2$ are $K_{ij} = K_{1i} \times K_{2j}$, where K_{1i} and K_{2j} are all the classes of $\mathscr{G}_1$ and $\mathscr{G}_2$, respectively. The number of classes of $\mathscr{G}$ is then $r = r_1 \cdot r_2$. (2.3.5)

Statements about the number r of classes of a group $\mathscr{G}$ can be made only indirectly. Looking for the group properties, we can show that:

All the elements $y_i \in \mathscr{G}$ commuting with a fixed element $a \in \mathscr{G}$ form a group $\mathscr{V}_a$, called the *normalisator*. (2.3.6)

With a, all the powers of a are members of $\mathscr{V}_a$. If a is a self-conjugate element then $\mathscr{V}_a = \mathscr{G}$, otherwise, $\mathscr{G}$ can be decomposed into (left) cosets of $\mathscr{V}_a$:

$$\mathscr{G} = \mathscr{V}_a + x_1 \mathscr{V}_a + \cdots + x_{r-1} \mathscr{V}_a \ . \tag{2.3.7}$$

Forming all the conjugate elements of a with x_i,

$$eae^{-1} = a \ ; \qquad x_1 a x_1^{-1} = a_1 \ ; \qquad \ldots ; \qquad x_{r-1} a x_{r-1}^{-1} = a_{r-1} \ , \tag{2.3.8}$$

we can see that these are different from each other and thus they are members of a class K_a with r_a elements. Let v_a be the number of elements in $\mathscr{V}_a$, then we obviously have

$$r_a = g/v_a = \operatorname{ind} \mathscr{V}_a \ , \tag{2.3.9}$$

i.e. the *number of elements* r_a in a class K_a is a divisor of the group order g. However, the *number of classes* r in general is not a divisor of g. Since any element is a member of only one class, we have

$$\sum_1^r r_a = g \tag{2.3.10}$$

and with (2.3.9)

$$\sum_1^r 1/v_a = 1 \ . \tag{2.3.11}$$

Since all $v_a \leqslant g$, the number of classes of finite groups is limited.

For the group $\mathscr{C}_{3v}$, conjugate elements and classes can be derived easily from the group table: $K_e = \{e\}$, $K_3 = \{c_3, c_3^2\}$, $K_\sigma = \{\sigma_v, \sigma_v', \sigma_v''\}$; $r_e = 1$, $r_3 = 2$, $r_\sigma = 3$ are divisors of $g = 6$, and further $r_e + r_3 + r_\sigma = 6 = g$ and $\sum(1/v_a) = 1/6 + 1/2 + 1/3 = 1$. The number of classes, $r = 3$, in this case is (accidentally!) also a divisor of $g = 6$.

Multiplication of two classes is defined by the multiplication of every element of one class with every element of the other class, taking the order into account, of course. We have

$$K_i \cdot K_j = \sum_{l=1}^r h_{ij,l} K_l \ , \tag{2.3.12}$$

that is, the set which originates from class multiplication consists of complete

Table 2.3. Example of a class table

	K_e	K_3	K_σ
K_e	K_e	K_3	K_σ
K_3	K_3	$2K_e + K_3$	$2K_\sigma$
K_σ	K_σ	$2K_\sigma$	$3K_e + 3K_3$

classes, the manifold of which is given by the class multiplication coefficients $h_{ij,l}$. This can easily be verified with the group table. In this way we obtain a class table (e.g. for $\mathscr{C}_{3v}$; Table 2.3). From this table the $h_{ij,l}$ can be found, e.g. $h_{211} = 0$, $h_{212} = 1$, $h_{213} = 0$, $h_{221} = 2$, $h_{222} = 1$, $h_{223} = 0$.

We will mention a theorem which is useful for some applications.

A subgroup $\mathscr{U} \subset \mathscr{G}$ is an invariant subgroup if and only if it consists of complete classes of conjugate elements of $\mathscr{G}$. (2.3.13)

This follows from the above discussion and (2.2.11a). In the group $\mathscr{C}_{3v}$, only $\mathscr{C}_3$ consists of complete classes, $\{e, \sigma_v\}$, etc. do not.

All the elements $z_i \in \mathscr{G}$, which commute with all $a \in \mathscr{G}$, that is $z_i = az_ia^{-1}$, form an invariant subgroup of $\mathscr{G}$, which is called the *centre* of $\mathscr{G}$. The centre of $\mathscr{C}_{3v}$ consists of e only, that of $\mathscr{C}_3$ is the complete $\mathscr{C}_3$ itself.

Exercise 2.6. Show that if σ is a reflection at a plane $\boldsymbol{n}$ and if c is a rotation about a fixed axis, then $c\sigma c^{-1} = \sigma'$ is a reflection at a plane that results from the rotation c of the vector $\boldsymbol{n}$.

Exercise 2.7. Show that if c and c' are rotations about fixed axes, then $cc'c^{-1} = c''$ is a rotation about an axis that results from a rotation of c' about c.

3. Discrete Symmetry Groups

Those discrete groups which play the central role in solid-state physics are the point groups and their extensions (double, colour groups), the translation groups, and the combination of both (the space groups). These groups and the meaning of their elements are discussed in the following sections. Apart from these, the symmetric (permutation) groups are important in physical systems. They are described in Sect. 3.5.

3.1 Point Groups

3.1.1 Symmetry Elements

We first consider those symmetry operations that keep one point fixed and leave the distance between two points of the space (in general the space $\mathbb{R}_3$) unchanged. The points of the space are described in a Cartesian coordinate system, the origin of which is chosen to be the fixed point.

Some of the symmetry operations of the type mentioned above are already known: rotations about axes through the fixed point, reflections in planes containing the fixed point and products of these operations. The symmetry element corresponding to a *rotation* through an angle $2\pi/n$ is denoted by c_n where the direction of the rotational axis has to be specified. The element has the order n, the axis is an n-fold one. Note that

$$c_n^n = e \qquad \text{or} \qquad c_n^m = c_n^{n+m} \ . \tag{3.1.1}$$

Reflections in a plane are denoted by σ; they have the order 2, that is

$$\sigma^2 = e \qquad \text{or} \qquad \sigma^{-1} = \sigma \ . \tag{3.1.2}$$

The orientation of a rotation axis or a mirror plane is specified by the direction of the axis or the normal of the plane, relative to a coordinate system; for example, $c_n(z) = c_{nz}$ means an n-fold axis in the z-direction, $\sigma(z) = \sigma_z$ means a mirror plane parallel to the xy-plane. In general, the axis of rotation is taken to be the z-axis; in the case of several axes, the main axis is in the z-direction.

The symmetry elements are operators in $\mathbb{R}_3$ acting on points (vectors) of space. We choose as a convention, that the coordinates of a point (position

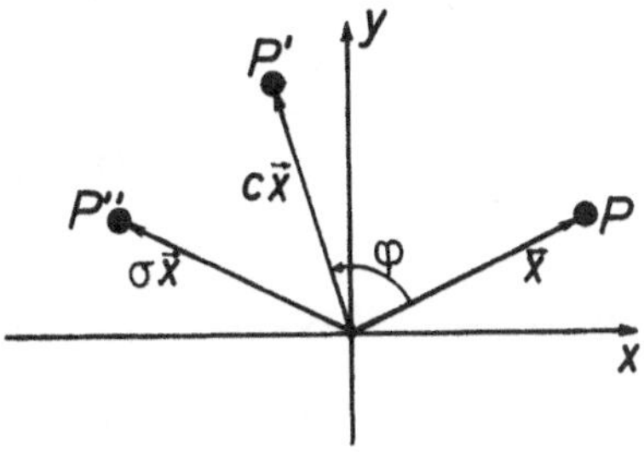

Fig. 3.1. Rotation $c\boldsymbol{x}$ of a vector or point (P onto P') and reflection $\sigma\boldsymbol{x}$ in the yz-plane (P onto P'')

vector) are always taken with respect to a *fixed coordinate system*[1]. Thus, through a rotation d a point P with vector $\boldsymbol{x} = (x, y, z)$ is transformed into another point P' with vector $\boldsymbol{x}' = (x', y', z')$; the coordinate system remains fixed. This procedure is described by

$$\boldsymbol{x}' = d\boldsymbol{x} \; ; \qquad d = c(\varphi) \; . \tag{3.1.3a}$$

If the rotation axis points in the z-direction, and if the vector is rotated through an angle φ in the positive sense (counterclockwise), then (Fig. 3.1) we have

$$\begin{pmatrix} x' \\ y' \\ z' \end{pmatrix} = \begin{pmatrix} \cos\varphi & -\sin\varphi & 0 \\ \sin\varphi & \cos\varphi & 0 \\ 0 & 0 & 1 \end{pmatrix} \cdot \begin{pmatrix} x \\ y \\ z \end{pmatrix} \; . \tag{3.1.3b}$$

A reflection in the yz-plane has the matrix

$$\sigma(x) = \sigma_x = \sigma_v = \begin{pmatrix} -1 & 0 & 0 \\ 0 & 1 & 0 \\ 0 & 0 & 1 \end{pmatrix} \; . \tag{3.1.4}$$

Further symmetry elements can be derived from the basic elements, rotation and reflection. The product of three reflections in three orthogonal planes (e.g. $x = 0$, $y = 0$, $z = 0$) yields (space) *inversion*:

$$i = \sigma(x) \cdot \sigma(y) \cdot \sigma(z) \; ; \qquad i = -\delta_{ik} \; . \tag{3.1.5}$$

A product of rotation and reflection is named *mirror rotation* s_n: it describes a rotation through $2\pi/n$ about an n-fold axis combined (simultaneously) with a reflection σ_h at a plane perpendicular to the rotation axis:

$$s_n = c_n \cdot \sigma_h = \sigma_h \cdot c_n \; , \qquad s_n^m = \begin{cases} c_n^m & m \text{ even} \\ c_n^m \cdot \sigma_h & m \text{ odd} \end{cases} \; . \tag{3.1.6a}$$

[1] Sometimes another convention is used in which vectors remain fixed and the coordinate system is rotated. In that case, φ has to be replaced by $-\varphi$. Once chosen, the convention has to be observed strictly.

Obviously

$$s_n^n = e \qquad \text{for } n \text{ even}$$

$$s_n^n = \sigma_h \,, \qquad s_n^{2n} = e \qquad \text{for } n \text{ odd} \tag{3.1.6b}$$

The order of s_n, therefore, is $2n$ for odd n. Instead of a mirror rotation often the

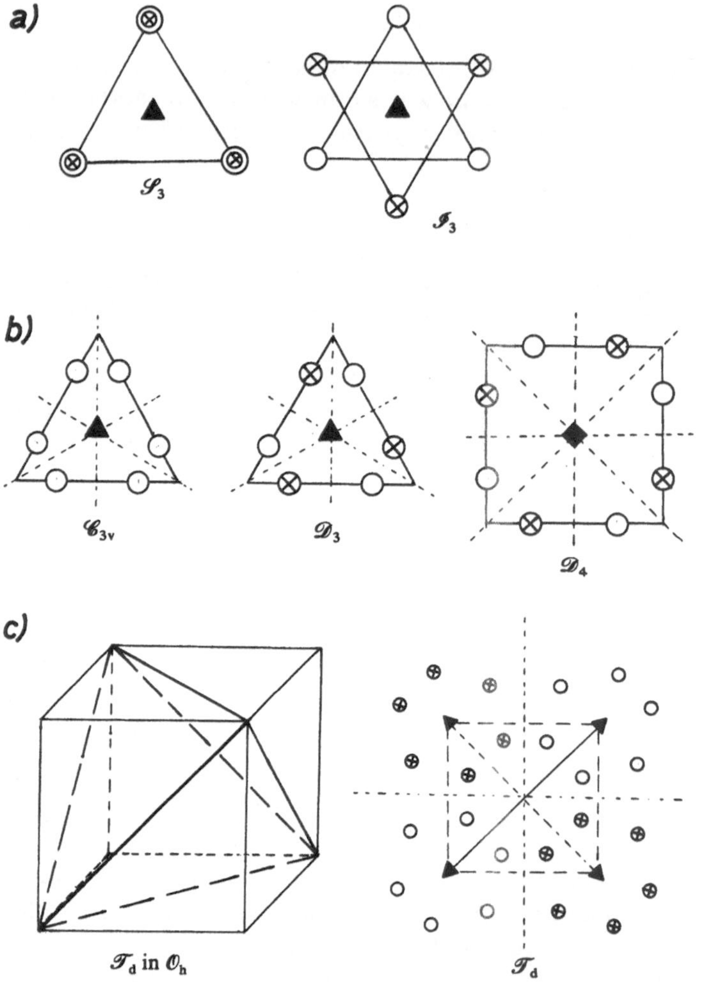

Fig. 3.2a–c. Stereograms and illustrations of different point groups. ○: Points a distance z above the plane of the paper; ⊗: Points at $-z$, i.e. a distance z below the plane of the paper. (**a**) Operations and groups $\mathscr{S}_3$ and $\mathscr{I}_3$. As the stereograms indicate, $\mathscr{S}_3$ and $\mathscr{I}_3$ are identical to $\mathscr{I}_6$ and $\mathscr{S}_6$, respectively. (**b**) Stereograms for $\mathscr{C}_{3v}$, $\mathscr{D}_3$ and $\mathscr{D}_4$ symmetry. For $\mathscr{C}_{3v}$ all the points lie in one plane. The stereograms show that there are two classes of twofold secondary rotation axes for $\mathscr{D}_4$, but only one class for $\mathscr{D}_3$. The axes of one of these classes can be transformed into each other by respectively c_4 and c_3 rotations. (**c**) Relation between hexahedron and tetrahedron and stereogram for $\mathscr{T}_d$ symmetry

roto-inversion i_n is used:

$$i_n = c_n \cdot i = i \cdot c_n \ , \qquad i_n^m = \begin{cases} c_n^m & m \text{ even} \\ c_n^m \cdot i & m \text{ odd} \end{cases} \tag{3.1.7}$$

with a relation corresponding to (3.1.6b). Operations (3.1.6a) and (3.1.7) are not independent, as can be easily seen (Fig. 3.2a and Exercise 3.2). In particular, we have

$$\begin{aligned} &i_1 = i = s_2 \ , \qquad i_2 = s_1 = \sigma_h \ , \\ &i_4^m = s_4^{4-m} \ , \qquad i_3^m = s_6^{6-m} \ , \qquad i_6^m = s_3^{6-m} \ . \end{aligned} \tag{3.1.8}$$

Up to now we have distinguished between the reflections σ_v and σ_h. In principle, there are three kinds of reflections:

σ_h: reflections in a plane perpendicular to the n-fold main rotation axis (h: horizontal),
σ_v: reflections in planes containing the main axis (v: vertical),
σ_d: reflections in planes containing the main axis and bisecting the angle between neighbouring additional axes (d: diagonal).

The notations σ_h, σ_v and σ_d are due to Schönflies; Hermann and Mauguin use $1/m$ and m. These are all the symmetry elements that occur in point groups.

3.1.2 Proper Point Groups

A (three-dimensional) rotation group is a group of symmetry operations that leave one given point and all the angles and distances in Euclidean space ($\mathbb{R}_3$) unchanged. If there are only rotations in this group (no reflection), it is a *proper rotation group*, being *isomorphic* to a subgroup of $\mathscr{SO}(3) = \mathscr{R}(3)$ of all orthogonal (3×3) matrices[2] with determinant 1. The direct product of a proper group with the inversion group $\mathscr{C}_i = \{e, i\}$ is an *improper rotation group*, which is isomorphic to a subgroup of $\mathcal{O}(3) = \mathscr{C}_i \times \mathscr{SO}(3)$ of all orthogonal matrices. There exists a homomorphism $f\colon \mathcal{O}(3) \to \mathscr{SO}(3)$ with the kernel $\mathscr{N}_K = \mathscr{C}_i = \{e, i\}$; the quotient group of $\mathcal{O}(3)$ with respect to $\mathscr{C}_i$ is $\mathcal{O}(3)/\mathscr{C}_i \cong \mathscr{SO}(3)$, cf. (2.2.17).

Finite *subgroups* of $\mathscr{SO}(3)$ and $\mathcal{O}(3)$ are called *proper* and *improper point groups*, respectively. To begin with, we consider proper point groups of finite order. They describe the symmetries of molecules and of (unit cells in) crystals. There are two notations used for point groups: that of Schönflies has special symbols, that of Hermann-Mauguin (international notation) uses the main symmetry elements (or generators) of the group. We shall mainly use the notation of Schönflies (Table 3.1).

[2] Matrices are said to be *orthogonal* if $d\tilde{d} = 1$ ($\tilde{d}$ is the transpose of d) or $\sum_k d_{ik}\tilde{d}_{kj} = \sum_k d_{ik}d_{jk} = \delta_{ij}$. If $\det d = +1$, d is a pure rotation, if $\det d = -1$, d is a roto-inversion. Matrix groups or groups of linear transformations in general will be discussed in Chaps. 11–14.

i) The *cyclic* (rotation) *groups* $\mathscr{C}_n$ (= n) contain n elements, namely the powers of the generating element c_n.

$$\mathscr{C}_n = \{c_n^m | m = 0, \ldots, n-1\} \ . \tag{3.1.9}$$

They describe arrangements with an n-fold axis (n-edged prism, regular polygon of n vertices); the groups are Abelian, every element forms a class by itself.

ii) The *dihedral groups* $\mathscr{D}_n$ (= $n22$ for even n, = $n2$ for odd n) describe symmetry arrangements with one n-fold main axis and in addition n 2-fold axes perpendicular to the main axis at angles of π/n to each other.

$$\mathscr{D}_n = \{c_n^m; c_2, c_2', \ldots | c_n^n = c_2^2 = (c_n c_2)^2 = e\} \ . \tag{3.1.10}$$

A regular prism with n surfaces possesses this symmetry. The order of $\mathscr{D}_n$ is $2n$, the generating system is $\{c_n, c_2 \perp c_n\}$. The group is not Abelian for $n \geqslant 3$; the number of classes is equal to $(n + 6)/2$ for even n, $(n + 3)/2$ for odd n (Fig. 3.2b).

Apart from these groups, there are only three additional ones with more than one main axis. These are as follows:

iii) The *tetrahedral group* $\mathscr{T}$ (= 23) possesses four 3-fold axes through a vertex and the centre of the opposing face of a tetrahedron, and three 2-fold axes which are perpendicular to each other and pass through the centres of opposing edges.

$$\mathscr{T} = \{c_3, c_3', \ldots; c_2, c_2', c_2'' | c_3^3 = c_2^2 = (c_3 c_2)^3 = e\} \ . \tag{3.1.11}$$

The order is 12, the number of classes 4 and the number of generators 2 (c_3 and c_2).

(iv) The *octahedral group* $\mathscr{O}$ (= 432) is described by the axes of a regular cube or an octrahedron. There are three 4-fold axes, perpendicular to each other, six 2-fold axes and four 3-fold axes. The order is 24, the number of classes 5. As a system of generators we can choose

$$\mathscr{O}: c_4^4 = c_{2d}^2 = (c_4 c_{2d})^3 = e \ ; \qquad c_4 c_{2d} = c_3 \qquad \text{or} \tag{3.1.12a}$$

$$\mathscr{O}: c_4^4 = c_3'^3 = (c_4 c_3')^2 = e \ ; \qquad c_4 c_3' = c_{2d} \ . \tag{3.1.12b}$$

The elements have the following meanings: c_4, rotation about z-axis; c_3, rotation about the cube diagonal $x = y = z$; c_3', rotation about the cube diagonal $-x = y = -z$; c_{2d}, rotation about the face diagonal $x = z$, $y = 0$. The tetrahedal group is a subgroup $\mathscr{T} \subset \mathscr{O}$ with index 2. This can be seen by embedding a tetrahedron in a cube so that the vertices of the tetrahedron coincide with the vertices of the cube (Fig. 3.2c).

(v) The *icosahedral group* $\mathscr{Y}$ (= 532) is the group of the regular icosahedron and dodecahedron. In physical problems it is discussed only recently (Bor-icosahedron, Fullerenes and the theory of quasicrystals, see Appendix I). The group consists of 6 fivefold, 10 threefold and 15 twofold axes, with altogether 60 elements in 5 classes. There are again two generators,

$$\mathscr{Y}: c_5^5 = c_2^2 = (c_5 c_2)^3 = e \; , \tag{3.1.13}$$

for which the directions of the axes have to be chosen in an appropriate way.

3.1.3 Improper Point Groups

The improper point groups can easily be derived from the proper ones. There are two methods: (1) the formation of outer direct products with the inversion group $\mathscr{C}_i = \{e, i\}$ and (2) the decomposition of the point group $\mathscr{G}$ into cosets with respect to an invariant subgroup of index 2, $\mathscr{G} = \mathscr{N} + a\mathscr{N}$, and formation of $\hat{\mathscr{G}} = \mathscr{N} + ia\mathscr{N}$ with the inversion i. Both methods together provide all possible improper point groups.

1a) From $\mathscr{C}_n$ we obtain $\mathscr{C}_{n\mathrm{h}}$ (even n) and $\mathscr{S}_{2n}$ (odd n).
 b) From $\mathscr{D}_n$ we obtain correspondingly $\mathscr{D}_{n\mathrm{h}}$ (even n) and $\mathscr{D}_{n\mathrm{d}}$ (odd n).
 c) From $\mathscr{T}$, $\mathscr{O}$, $\mathscr{Y}$ we obtain $\mathscr{T}_\mathrm{h}$, $\mathscr{O}_\mathrm{h}$, $\mathscr{Y}_\mathrm{h}$.
2a) From $\mathscr{C}_{2n}$ with invariant subgroup $\mathscr{C}_n$ we obtain $\mathscr{S}_{2n}$ (even n) and $\mathscr{C}_{n\mathrm{h}}$ (odd n).
 b) From $\mathscr{D}_n$ with invariant subgroup $\mathscr{C}_n$ we obtain correspondingly $\mathscr{C}_{n\mathrm{v}}$.
 c) From $\mathscr{D}_{2n}$ with invariant subgroup $\mathscr{D}_n$ we obtain correspondingly $\mathscr{D}_{n\mathrm{d}}$ (even n) and $\mathscr{D}_{n\mathrm{h}}$ (odd n).
 d) From $\mathscr{O}$ with invariant subgroup $\mathscr{T}$ we obtain $\mathscr{T}_\mathrm{d}$.
 e) $\mathscr{T}$ and $\mathscr{Y}$ do not possess an invariant subgroup of index 2.

Some of these groups are physically identical (not only isomorphic, see 1a and 2a or 1b and 2c). According to (3.1.8), this is also true for mirror rotation and roto-inversion groups. Thus

$$\mathscr{C}_i = \mathscr{I}_1 := \mathscr{I} = \mathscr{S}_2 \; , \quad \mathscr{I}_2 = \mathscr{S}_1 \; , \quad \mathscr{I}_4 = \mathscr{S}_4 \; , \quad \mathscr{I}_3 = \mathscr{S}_6 \; , \quad \mathscr{I}_6 = \mathscr{S}_3 \tag{3.1.14a}$$

and also

$$\mathscr{C}_s = \mathscr{C}_{1\mathrm{h}} \; , \quad \mathscr{C}_{3i} = \mathscr{S}_6 \; , \quad \mathscr{C}_{3\mathrm{h}} = \mathscr{S}_3 \; . \tag{3.1.14b}$$

Isomorphic groups are

$$\begin{aligned} &\mathscr{D}_n \cong \mathscr{C}_{n\mathrm{v}} \; , \quad \mathscr{C}_i \cong \mathscr{C}_s \cong \mathscr{S}_2 \; , \quad \mathscr{S}_4 \cong \mathscr{C}_4 \; , \\ &\mathscr{D}_2 \cong \mathscr{C}_{2\mathrm{h}} \; , \quad \mathscr{C}_6 \cong \mathscr{C}_{3\mathrm{h}} \cong \mathscr{C}_{3i} \; , \quad \mathscr{D}_4 \cong \mathscr{D}_{2\mathrm{d}} \; , \\ &\mathscr{D}_6 \cong \mathscr{D}_{3\mathrm{h}} \cong \mathscr{D}_{3\mathrm{d}} \; , \quad \mathscr{T}_\mathrm{d} \cong \mathscr{O} \; . \end{aligned} \tag{3.1.15}$$

The physically important groups are shown in Table 3.1. In this table we also mention those groups which we obtain as continuous groups in the limit $n \to \infty$.

Table 3.1. Point groups

System	Notation			Number	Symmetry elements		Number of	
	Schönflies	Hermann-Mauguin full	Hermann-Mauguin short		Generators	Others	classes	space groups
Triclinic	$\mathscr{C}_1$	1	1	1	Only the identity		1	1
	$\mathscr{C}_i$	$\bar{1}$	$\bar{1}$	2	i	–	2	1
Monoclinic	$\mathscr{C}_2$	2	2	2	c_2	–	2	3
	$\mathscr{C}_s$	m	m	2	σ_h	–	2	4
	$\mathscr{C}_{2h}$	$\frac{2}{m}$	$2/m$	4	c_2, i	σ_h	4	6
(Ortho-)rhombic	$\mathscr{D}_2$	$2\,2\,2$	222	4	c_{2z}, c_{2x}	c_{2y}	4	9
	$\mathscr{C}_{2v}$	$m\,m\,2$	$mm2$	4	c_{2z}, σ_x	σ_y	4	22
	$\mathscr{D}_{2h}$	$\frac{2}{m}\frac{2}{m}\frac{2}{m}$	mmm	8	c_{2z}, c_{2x}, i	$c_{2y}, \sigma_h, \sigma_x, \sigma_y$	8	28
Tetragonal (quadratic)	$\mathscr{C}_4$	4	4	4	c_4	–	4	6
	$\mathscr{S}_4$	$\bar{4}$	$\bar{4}$	4	s_4	–	4	2
	$\mathscr{C}_{4h}$	$\frac{4}{m}$	$4/m$	8	c_4, i	s_4, σ_h	8	6
	$\mathscr{D}_4$	$4\,2\,2$	422	8	c_{4z}, c_{2x}	$3c_2'$	5	10
	$\mathscr{C}_{4v}$	$4\,m\,m$	$4mm$	8	c_{4z}, σ_x	$3\sigma_v'$	5	12
	$\mathscr{D}_{2d}$	$\bar{4}\,2\,m$	$42m$	8	s_{4z}, c_{2x}	$c_{2y}, 2\sigma_d$	5	12
	$\mathscr{D}_{4h}$	$\frac{4}{m}\frac{2}{m}\frac{2}{m}$	$4/mmm$	16	c_{4z}, c_{2x}, i	$3c_2', s_4, \sigma_h, 2\sigma_v, 2\sigma_d$	10	20
Trigonal (rhombo-hedral)	$\mathscr{C}_3$	3	3	3	c_3	–	3	4
	$\mathscr{C}_{3i}$	$\bar{3}$	$\bar{3}$	6	s_6	–	6	2
	$\mathscr{D}_3$	$3\,2$	32	6	c_{3z}, c_{2y}	$2c_2'$	3	7
	$\mathscr{C}_{3v}$	$3\,m$	$3m$	6	c_{3z}, σ_x	$2\sigma_v'$	3	6
	$\mathscr{D}_{3d}$	$\bar{3}\frac{2}{m}$	$\bar{3}m$	12	c_{3z}, c_{2y}, i	$s_6, 2c_2', 3\sigma_d$	6	6

Hexagonal	$\mathscr{C}_6$	6	6	6	c_6	–	6	6
	$\mathscr{C}_{3h}$	$\bar{6}$	$\bar{6}$	6	s_3	c_3, σ_h	6	1
	$\mathscr{C}_{6h}$	$\frac{6}{m}$	$6/m$	12	c_6, i	s_3, s_6	12	2
	$\mathscr{D}_6$	$6\,2\,2$	622	12	c_{6z}, c_{2x}	$5c_2'$	6	6
	$\mathscr{C}_{6v}$	$6\,m\,m$	$6mm$	12	c_{6z}, σ_x	$5\sigma_v'$	6	4
	$\mathscr{D}_{3h}$	$\bar{6}\,m\,2$	$\bar{6}m2$	12	c_{3z}, c_{2x}, σ_h	$s_3, 2c_2', 3\sigma_v$	6	4
	$\mathscr{D}_{6h}$	$\frac{6}{m}\frac{2}{m}\frac{2}{m}$	$6/mmm$	24	c_{6z}, c_{2x}, i	$5c_2', 6\sigma_v', \sigma_h, s_3, s_6$	12	4
Cubic (regular)	$\mathscr{T}$	$2\,3$	23	12	c_{2z}, c_3'	$2c_2', 3c_3''$	4	5
	$\mathscr{T}_h$	$\frac{2}{m}\bar{3}$	$m3$	24	c_{2z}, c_3', i	as $\mathscr{T}$ and $3\sigma_h, 4s_6$	8	7
	$\mathscr{T}_d$	$\bar{4}\,3\,m$	$\bar{4}3m$	24	s_{4z}, c_3'	$2s_4', 3c_3'', 6\sigma_d$	5	6
	$\mathscr{O}$	$4\,3\,2$	43	24	c_{4z}, c_3'	$2c_4', 3c_3'', 6c_2'$	5	8
	$\mathscr{O}_h$	$\frac{4}{m}\bar{3}\frac{2}{m}$	$m3m$	48	c_{4z}, c_3', i	as $\mathscr{O}$ and all elements times i	10	10
Icosahedral groups	$\mathscr{Y}$	$5\,3\,2$	53	60	c_5, c_2	$5c_5', 14c_2', 10c_3$	5	–
	$\mathscr{Y}_h$	$\bar{5}\,\bar{3}\frac{2}{m}$	$53m$	120	c_5, c_2, i	as $\mathscr{Y}$ and all elements times i	10	–
Cylindrical (uniaxial-isotropic)	$\mathscr{C}_{\infty v}$	$\infty\,m$	∞m	∞	$c(\varphi)$			
	$\mathscr{D}_{\infty h}$	$\frac{\infty}{m}\frac{2}{m}$	∞/mm	∞	$c(\varphi), c_{2x}$			
Isotropic	$\mathscr{K}$	$\infty\infty$	$\infty\infty$	∞	$c(\vartheta, \varphi)$			
	$\mathscr{K}_h$	$\frac{\infty}{m}\frac{\infty}{m}$	$m\infty m$	∞	$c(\vartheta, \varphi), i$			

These are the orthogonal rotation groups $\mathcal{O}(n)$ and $\mathcal{SO}(n)$ with $n = 2, 3$. Sometimes it is useful to consider these groups as limits of the discrete groups for the symmetry of a cylinder and sphere, respectively. The point groups and their symmetries can be illustrated by stereograms. This is demonstrated in Fig. 3.2 using $\mathcal{C}_{3v}$ and $\mathcal{D}_3$ as examples; they are isomorphic, but they describe different symmetries.

Exercise 3.1. Determine the matrices for different symmetry transformations in analogy to (3.1.3a and 4). The operations are: the rotation $c(\varphi)$, the mirror-rotation $s(\varphi)$, the roto-inversion $i(\varphi)$, all through an angle φ about the z-axis, the inversion i, the reflections σ_h and $\sigma(\phi)$ (normal of mirror plane in direction ϕ perpendicular to the z-axis) and rotation $c_2(\phi)$ about an axis perpendicular to the z-axis (see Exercises 2.6 and 7).

Exercise 3.2. By comparison of the corresponding matrices, show that $s_2 = c_2 \cdot \sigma_h = i$.

Exercise 3.3. Draw stereograms for the different point groups.

3.2 Colour Groups and Magnetic Groups

Physical objects which are described by the symmetry elements of point groups (Fig. 2.1 or corresponding vectors in $\mathbb{R}_3$) do not have an internal (hidden) physical structure. However, these objects may have internal degrees of freedom besides the external geometrical degrees of freedom. In the following we restrict the internal degree of freedom to two values, which can be looked upon as two "colours" (black-white) or two "signs" (plus-minus) or two possible directions of a magnetic moment (spin parallel or antiparallel to a given direction). A change of such a value can be described by an operator r which changes colour or sign, etc., that is, black into white or vice versa. By enlarging the symmetry groups with r we obtain new groups besides the ordinary geometrical point groups. They can be characterized as follows (Fig. 3.3), where $r^2 = e$:

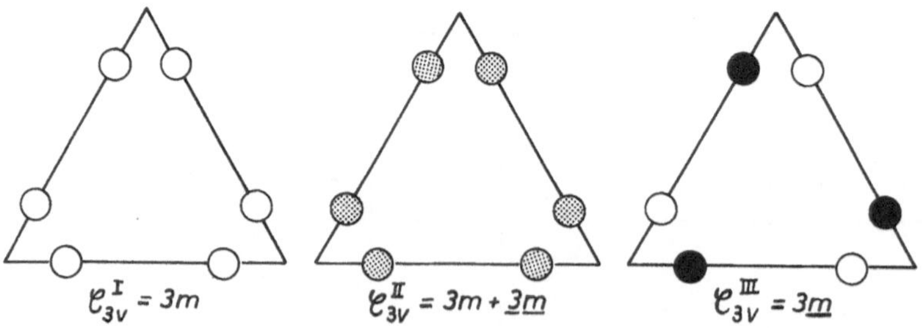

Fig. 3.3. Colour point groups of the Ist, IInd and IIIrd kinds explained for $\mathcal{C}_{3v}$. In $\mathcal{C}_{3v}^{I}$ all the objects have the same colour, in $\mathcal{C}_{3v}^{II}$ they are grey and in $\mathcal{C}_{3v}^{III}$ half are white and half black

I. *Colour or Shubnikov groups of the* Ist *kind* describe objects with a definite colour, e.g. white. The operator r is not a symmetry element of the system because it would change the colour into black. The group is equal to the geometric point group

$$\mathcal{M}_{\mathrm{I}} = \mathcal{G} \ ; \qquad r \notin \mathcal{M}_{\mathrm{I}} \ . \tag{3.2.1}$$

The number of possible $\mathcal{M}_{\mathrm{I}}$ is equal to the number of point groups $\mathcal{G}$. This also holds for the

II. *Colour or Shubnikov groups of the* IInd *kind*, also denoted as grey (or major) groups. They can be defined by the decomposition into cosets

$$\mathcal{M}_{\mathrm{II}} = \mathcal{G} + r \cdot \mathcal{G} \ ; \qquad r \in \mathcal{M}_{\mathrm{II}} \ . \tag{3.2.2a}$$

Here $\mathcal{G}$ is an invariant subgroup of $\mathcal{M}_{\mathrm{II}}$ with index 2, r is an element of $\mathcal{M}_{\mathrm{II}}$ and the order of $\mathcal{M}_{\mathrm{II}}$ is $2g$. The objects of the system described by $\mathcal{M}_{\mathrm{II}}$ possess simultaneously both colours (i.e. a mixed colour which is grey), so that a change black $\leftrightarrow$ white does not have any consequences. The operation r leaves the internal degree of freedom unchanged. As r commutes with all the elements of $\mathcal{G}$, $\mathcal{M}_{\mathrm{II}}$ can be written as a direct product

$$\mathcal{M}_{\mathrm{II}} = \mathcal{G} \times \{e, r\} \ . \tag{3.2.2b}$$

III. *Colour or Shubnikov groups of the* IIIrd *kind* (colour groups in the strict sense, minor groups) do not possess r as a separate element, but they contain elements that are coupled with r.

$$\mathcal{M}_{\mathrm{III}} = \mathcal{N} + r(\mathcal{G} - \mathcal{N}) = \mathcal{N} + ra'\mathcal{N} \ , \qquad a' \in \mathcal{G} - \mathcal{N} \ , \tag{3.2.3}$$

where $\mathcal{N}$ is an invariant subgroup of $\mathcal{G}$ with index 2. The objects of the system are black or white, and equal in number, so that certain symmetry operations transform the system into itself only with a simultaneous change of colour. The order of $\mathcal{M}_{\mathrm{III}}$ is that of $\mathcal{G}$. Since r commutes with all the elements of $\mathcal{G}$ we have

$$\begin{aligned} &(ar)b_i(ar)^{-1} = ab_ia^{-1} = b_j \ , \\ &(ar)(b_ir)(ar)^{-1} = ab_ira^{-1} = b_jr \ , \qquad a,\, b_i,\, b_j \in \mathcal{G} \ , \end{aligned} \tag{3.2.4}$$

which means the classes of $\mathcal{M}_{\mathrm{III}}$ correspond to those of $\mathcal{G}$; they have either only elements of $\mathcal{N}$ or of $r \cdot (\mathcal{G} - \mathcal{N})$. A point group $\mathcal{G}$ may have different invariant subgroups of index 2. Therefore the number of $\mathcal{M}_{\mathrm{III}}$ groups is larger than that of the common point groups (Table 3.2).

Figure 3.3. shows the colour groups derived from $\mathcal{C}_{3\mathrm{v}}$. The group $\mathcal{C}_{3\mathrm{v}}^{\mathrm{I}}$ has the elements given in (2.1.8), whereas

$$\mathcal{C}_{3\mathrm{v}}^{\mathrm{II}} = \{e, c_3, c_3^2\sigma_{\mathrm{v}}, \sigma_{\mathrm{v}}', \sigma_{\mathrm{v}}'', r, rc_3, rc_3^2, r\sigma_{\mathrm{v}}, r\sigma_{\mathrm{v}}', r\sigma_{\mathrm{v}}''\} \tag{3.2.5a}$$

Table 3.2. Colour groups of the IIIrd kind

System	$\mathscr{G}$	$\mathscr{N}$	$\mathscr{M}_{\mathrm{III}}$	Colour changing elements: Generator	Colour changing elements: Others
Triclinic	$\mathscr{C}_1$	–	–	–	–
	$\mathscr{C}_i$	$\mathscr{C}_1$	$\underline{\bar{1}}$	i	–
Monoclinic	$\mathscr{C}_2$	$\mathscr{C}_1$	$\underline{2}$	c_2	–
	$\mathscr{C}_s$	$\mathscr{C}_1$	$\underline{m}$	σ_h	–
	$\mathscr{C}_{2h}$	$\mathscr{C}_i$	$\underline{2}/\underline{m}$	c_2	σ_h
		$\mathscr{C}_2$	$2/\underline{m}$	i	σ_h
		$\mathscr{C}_s$	$\underline{2}/m$	i	c_2
(Ortho-) rhombic	$\mathscr{D}_2$	$\mathscr{C}_2$	$2\underline{2}\underline{2}$	c_{2x}	c_{2y}
	$\mathscr{C}_{2v}$	$\mathscr{C}_2$	$\underline{m}\underline{m}2$	σ_x	σ_y
		$\mathscr{C}_s$	$mm\underline{2}$	σ_x	c_{2z}
	$\mathscr{D}_{2h}$	$\mathscr{C}_{2h}$	$m\underline{m}\underline{m}$	c_{2x}	$c_{2y}, \sigma_x, \sigma_y$
		$\mathscr{D}_2$	$\underline{m}\underline{m}\underline{m}$	i	$\sigma_h, \sigma_x, \sigma_y$
		$\mathscr{C}_{2v}$	$\underline{m}mm$	i	c_{2x}, c_{2y}, σ_h
Tetragonal (quadratic)	$\mathscr{C}_4$	$\mathscr{C}_2$	$\underline{4}$	c_4	–
	$\mathscr{S}_4$	$\mathscr{C}_2$	$\underline{\bar{4}}$	s_4	–
	$\mathscr{C}_{4h}$	$\mathscr{C}_{2h}$	$\underline{4}/m$	c_4	s_4
		$\mathscr{C}_4$	$4/\underline{m}$	i	σ_h, s_4
		$\mathscr{S}_4$	$\underline{4}/\underline{m}$	i	σ_h, c_4
	$\mathscr{D}_4$	$\mathscr{D}_2$	$\underline{4}2\underline{2}$	c_{4z}	$2c_2'$
		$\mathscr{C}_4$	$4\underline{2}\underline{2}$	c_{2x}	$3c_2'$
	$\mathscr{C}_{4v}$	$\mathscr{C}_{2v}$	$\underline{4}m\underline{m}$	c_{4z}	$2\sigma_v'$
		$\mathscr{C}_4$	$4\underline{m}\underline{m}$	σ_x	$3\sigma_v'$
	$\mathscr{D}_{2d}$	$\mathscr{D}_2$	$\underline{\bar{4}}2\underline{m}$	s_{4z}	$2\sigma_v'$
		$\mathscr{C}_{2v}$	$\underline{\bar{4}}\underline{2}m$	s_{4z}	$2c_2'$
		$\mathscr{S}_4$	$\bar{4}\underline{2}\underline{m}$	c_{2x}	$c_{2y}, 2\sigma_v'$
	$\mathscr{D}_{4h}$	$\mathscr{D}_{2h}$	$\underline{4}/mm\underline{m}$	c_{4z}	$s_{4z}, 2c_2', 2\sigma_v'$
		$\mathscr{C}_{4h}$	$4/m\underline{m}\underline{m}$	c_{2x}	$3c_2', 4\sigma_v'$
		$\mathscr{D}_4$	$4/\underline{m}\underline{m}\underline{m}$	i	$\sigma_h, 4\sigma_v', s_{4z}$
		$\mathscr{C}_{4v}$	$4/\underline{m}mm$	i	$4c_2', \sigma_h, c_{4z}$
		$\mathscr{D}_{2d}$	$\underline{4}/\underline{m}m\underline{m}$	i	$2c_2', \sigma_h, 2\sigma_v', c_{4z}$
Trigonal (rhombohedral)	$\mathscr{C}_3$	–	–	–	–
	$\mathscr{C}_{3i}$	$\mathscr{C}_3$	$\underline{\bar{3}}$	i	s_6
	$\mathscr{D}_3$	$\mathscr{C}_3$	$3\underline{2}$	c_{2y}	$2c_2'$
	$\mathscr{C}_{3v}$	$\mathscr{C}_3$	$3\underline{m}$	σ_x	$2\sigma_v'$
	$\mathscr{D}_{3d}$	$\mathscr{C}_{3i}$	$\bar{3}\underline{m}$	c_{2y}	$2c_2', 3\sigma_v'$
		$\mathscr{D}_3$	$\underline{\bar{3}}\underline{m}$	i	$s_6, 3\sigma_v'$
		$\mathscr{C}_{3v}$	$\underline{\bar{3}}m$	i	$s_6, 3c_2'$
Hexagonal	$\mathscr{C}_6$	$\mathscr{C}_3$	$\underline{6}$	c_2	c_6
	$\mathscr{C}_{3h}$	$\mathscr{C}_3$	$\underline{\bar{6}}$	σ_h	s_3
	$\mathscr{C}_{6h}$	$\mathscr{C}_{3i}$	$\underline{6}/\underline{m}$	c_2	σ_h, c_6, s_3
		$\mathscr{C}_6$	$6/\underline{m}$	i	σ_h, s_3, s_6
		$\mathscr{C}_{3h}$	$\underline{6}/m$	i	σ_h, c_6, s_6
	$\mathscr{D}_6$	$\mathscr{D}_3$	$\underline{6}2\underline{2}$	c_{2z}	$3c_2', c_6$
		$\mathscr{C}_6$	$6\underline{2}\underline{2}$	c_{2x}	$5c_2'$

Table 3.2. (continued)

System	$\mathscr{G}$	$\mathscr{N}$	$\mathscr{M}_{\text{III}}$	Colour changing elements Generator	Others
Hexagonal (cont.)	$\mathscr{C}_{6v}$	$\mathscr{C}_{3v}$	$\underline{6}m\underline{m}$	c_{2z}	$3\sigma_v', c_6$
		$\mathscr{C}_6$	$6\underline{m}\underline{m}$	σ_x	$5\sigma_v'$
	$\mathscr{D}_{3h}$	$\mathscr{D}_3$	$\underline{\bar{6}}\underline{m}2$	σ_h	$3\sigma_v', s_3$
		$\mathscr{C}_{3v}$	$\underline{\bar{6}}m\underline{2}$	c_{2x}	$2c_2', \sigma_h, s_3$
		$\mathscr{C}_{3h}$	$\bar{6}\underline{m}\underline{2}$	c_{2x}	$2c_2', 3\sigma_v'$
	$\mathscr{D}_{6h}$	$\mathscr{D}_{3d}$	$\underline{6}/\underline{m}m\underline{m}$	c_{2x}	$2c_2', c_{2z}, 3\sigma_v', \sigma_h, c_6, s_3$
		$\mathscr{C}_{6h}$	$6/m\underline{m}\underline{m}$	c_{2x}	$5c_2', 6\sigma_v'$
		$\mathscr{D}_6$	$6/\underline{m}\underline{m}\underline{m}$	i	$\sigma_h, 6\sigma_v', s_6, s_3$
		$\mathscr{C}_{6v}$	$6/\underline{m}mm$	i	$6c_2', \sigma_h, s_6, s_3$
		$\mathscr{D}_{3h}$	$\underline{6}/mm\underline{m}$	i	$3c_2', c_{2z}, 3\sigma_v', c_6, s_6$
Cubic (regular)	$\mathscr{T}$	–	–	–	–
	$\mathscr{T}_h$	$\mathscr{T}$	$\underline{m}3$	i	$3\sigma_h, 4s_6$
	$\mathscr{T}_d$	$\mathscr{T}$	$\underline{\bar{4}}3\underline{m}$	s_{4z}	$2s_4, 6\sigma_d$
	$\mathscr{O}$	$\mathscr{T}$	$\underline{4}3$	c_{4z}	$2c_4, 6c_2'$
	$\mathscr{O}_h$	$\mathscr{T}_h$	$m3\underline{m}$	c_{4z}	$2c_4, 3s_4, 6c_2', 6\sigma_d$
		$\mathscr{O}$	$\underline{m}3\underline{m}$	i	$3s_4, 4s_6, 3\sigma_h, 6\sigma_d$
		$\mathscr{T}_d$	$\underline{m}3m$	i	$3c_4, 4s_6, 3\sigma_h, 6c_2'$
Icosahedral	$\mathscr{Y}$	–	–	–	–
	$\mathscr{Y}_h$	$\mathscr{Y}$	$53\underline{m}$	i	All the improper elements

and, according to (2.2.9),

$$\mathscr{C}_{3v}^{\text{III}} = \{e, c_3, c_3^2, r\sigma_v, r\sigma_v', r\sigma_v''\} \ . \tag{3.2.5b}$$

Colour groups are important if used as magnetic groups. Then r is the time reversal operator ϑ (reserval of motion). It reverses the direction of a physical event, in particular, the direction of a current (density times velocity). A magnetic moment corresponds to an electric current and behaves the same way.

The concept of colour groups can be expanded to multicolour groups. The internal degree of freedom then may have several possible values. The details will not be used in this book.

Exercise 3.4. Determine the colour groups of the IIIrd kind belonging to $\mathscr{D}_4$ and $\mathscr{D}_6$ (two in both cases) and the corresponding classes.

3.3 Double Groups

In many physical systems the objects (atoms, electrons, nuclei) possess a spin, which has to be taken into account. The case where the spin is 1/2 (electrons) is of special interest because this leads us beyond physics with integer quantum numbers for the angular momentum. In nonrelativistic quantum mechanics the spin is described by enlarging the (scalar) wave function to a spinor wave function φ_ν with two components ($\nu = 1, 2$). A spinor is characterized by its behavior under rotations. A general rotation of a vector $\boldsymbol{x}$ in space $\mathbb{R}_3$ is given by

$$\boldsymbol{x}' = c(\alpha, \beta, \gamma)\cdot \boldsymbol{x} \qquad \text{with} \tag{3.3.1}$$

$$c(\alpha, \beta, \gamma) = \begin{pmatrix} \cos\alpha\cos\beta\cos\gamma - \sin\alpha\sin\gamma & -\sin\alpha\cos\beta\cos\gamma - \cos\alpha\sin\gamma & \sin\beta\cos\gamma \\ \cos\alpha\cos\beta\sin\gamma + \sin\alpha\cos\gamma & -\sin\alpha\cos\beta\sin\gamma + \cos\alpha\cos\gamma & \sin\beta\sin\gamma \\ -\cos\alpha\sin\beta & \sin\alpha\sin\beta & \cos\beta \end{pmatrix} .$$

The rotation[3] is specified by its Euler angles α, β, γ: first rotation about the z-axis with $0 \leqslant \alpha < 2\pi$, then rotation about the y-axis with $0 \leqslant \beta < \pi$ and finally rotation about the z-axis with $0 \leqslant \gamma < 2\pi$. In such a rotation the transformation of a given (two component) spinor is

$$\varphi' = D^{1/2}(c(\alpha, \beta, \gamma))\cdot \varphi \qquad \text{with} \tag{3.3.2}$$

$$D^{1/2}(c(\alpha, \beta, \gamma)) = \begin{pmatrix} e^{i(\alpha+\gamma)/2}\cos(\beta/2) & e^{-i(\alpha-\gamma)/2}\sin(\beta/2) \\ -e^{i(\alpha-\gamma)/2}\sin(\beta/2) & e^{-i(\alpha+\gamma)/2}\cos(\beta/2) \end{pmatrix} .$$

This (2×2) matrix reproduces itself only after rotation by 4π, not by 2π, and therefore $\varphi' = -\varphi$ after rotation by 2π, but $\varphi' = \varphi$ after rotation by 4π. To every physical rotation $c(\alpha, \beta, \gamma)$ belong two matrices $D^{1/2}$; they differ in the rotation angle by 2π: α and $2\pi + \alpha$, etc. The number of parameters (α, β, γ) is equal in both cases, i.e. the matrices differ only in sign (Exercise 3.8). The matrices (3.3.1) form the *(real) special orthogonal group* $\mathscr{SO}(3)$ already mentioned in Sect. 3.1.1. The matrices (3.3.2) form the *special (unimodular) unitary group* $\mathscr{SU}(2)$; the matrices obey $DD^+ = 1$ and $\det D = 1$ (Sect. 4.1.). Two elements of $\mathscr{SU}(2)$ correspond to one element of $\mathscr{SO}(3)$, as discussed above. We have therefore a two-to-one homomorphism $\mathscr{SU}(2) \to \mathscr{SO}(3)$ in which the matrices (see Sect. 11.1)

$$\begin{pmatrix} 1 & 0 \\ 0 & 1 \end{pmatrix} \quad \text{and} \quad \begin{pmatrix} -1 & 0 \\ 0 & -1 \end{pmatrix} \in \mathscr{SU}(2) \tag{3.3.3a}$$

correspond to the matrix

[3] According to our convention (Sect. 3.1.1) the coordinate system remains fixed while the objects are transformed.

$$\begin{pmatrix} 1 & 0 & 0 \\ 0 & 1 & 0 \\ 0 & 0 & 1 \end{pmatrix} \in \mathcal{SO}(3) \; , \tag{3.3.3b}$$

i.e. (3.3.3a) is the kernel of the homomorphism.

A *double group* is related to a point group as $\mathcal{SU}(2)$ is to $\mathcal{SO}(3)$. This means that if $\mathcal{G}$ is a subgroup of $\mathcal{SO}(3)$ of order g, then the double group $\mathcal{G}^{\mathrm{D}}$ (order $2g$) belonging to $\mathcal{G}$ has a group table that can be derived from the corresponding matrices of $\mathcal{SU}(2)$. The double group $\mathcal{G}^{\mathrm{D}}$ is a subgroup of $\mathcal{SU}(2)$. Quite formally the double group can be described by adding a new element c_0. If $c(\alpha)$ is the rotation through an angle α, then the elements of the double group have to obey

$$\begin{aligned} c(\alpha + 2\pi) &= c(\alpha)\cdot c(2\pi) \neq c(\alpha) \; , \\ c(\alpha + 4\pi) &= c(\alpha)\cdot c^2(2\pi) = c(\alpha) \; , \end{aligned} \tag{3.3.4a}$$

with

$$c(2\pi) = c_0 \; , \qquad c_0^2 = e \; . \tag{3.3.4b}$$

The rotation through 2π is not identical with e. If $\mathcal{G} = \{e, a_1, a_2, \ldots\} \subset \mathcal{SO}(3)$, the corresponding double group is given by

$$\mathcal{G}^{\mathrm{D}} = \{e, a_1, a_2, \ldots ; c_0, c_0 a_1, c_0 a_2, \ldots\} \; , \tag{3.3.5}$$

which is isomorphic to a subgroup of $\mathcal{SU}(2)$. Note that $\mathcal{G}$ itself is not a subgroup of $\mathcal{G}^{\mathrm{D}}$ because the elements of $\mathcal{G}^{\mathrm{D}}$ that correspond to the elements of $\mathcal{G}$ do not form a closed set in $\mathcal{G}^{\mathrm{D}}$ (e.g. $c_2^2 = c_0 \notin \mathcal{G}$). For example, if

$$\mathcal{G} = \mathcal{C}_n = \{e, c_n, \ldots, c_n^{n-1}\} \qquad \text{and} \qquad c_0 = c(2\pi) = c_n^n \; ,$$

then

$$\mathcal{G}^{\mathrm{D}} = \mathcal{C}_n^{\mathrm{D}} = \{e, c_n, \ldots, c_n^{n-1}, c_n^n, \ldots, c_n^{2n-1}\} \; ,$$

which is a cyclic group of order $2n$. Clearly, $\mathcal{C}_n$ is not a subgroup of $\mathcal{C}_n^{\mathrm{D}}$ since $c_n^n = c_0 \notin \mathcal{C}_n$ in $\mathcal{C}_n^{\mathrm{D}}$.

Having determined with this method the double groups of the proper point groups $\mathcal{C}_n$, $\mathcal{D}_n$, $\mathcal{T}$, $\mathcal{O}$, $\mathcal{Y}$, the improper groups follow as in Sect. 3.1.3.

The elements a_i and $c_0 a_i \in \mathcal{G}^{\mathrm{D}}$ in general belong to different classes; i.e. if K_i is a class of $\mathcal{G}$, then the double group has the corresponding classes K_i and K_i^{D}. However there are *exceptions*:

If c_2 (which includes c_2, c_4^2, c_6^3, etc.) is a rotation through an angle π, then c_2 and $c_0 c_2$ belong to *one* class if and only if there exists in $\mathcal{G}$ *either* another twofold rotation perpendicular to the axis of c_2 *or* a mirror plane that contains the axis of c_2. (3.3.6a)

Table 3.3. Double groups. Here we have listed only those double groups whose number of classes is *not* twice that of $\mathcal{G}$

	Classes in	
	$\mathcal{G}$	$\mathcal{G}^{\mathrm{D}}$
$\mathcal{D}_2, \mathcal{C}_{2v}$	4	5
$\mathcal{D}_{2h}$	8	10
$\mathcal{D}_4, \mathcal{C}_{4v}, \mathcal{D}_{2d}$	5	7
$\mathcal{D}_{4h}$	10	14
$\mathcal{D}_6, \mathcal{C}_{6v}, \mathcal{D}_{3h}$	6	9
$\mathcal{D}_{6h}$	12	18
$\mathcal{T}$	4	7
$\mathcal{T}_h$	8	14
$\mathcal{T}_d, \mathcal{O}$	5	8
$\mathcal{O}_h$	10	16
$\mathcal{Y}$	5	9
$\mathcal{Y}_h$	10	18

If σ is a reflection, then σ and $c_0\sigma$ belong to *one* class if and only if there exists in $\mathcal{G}$ *either* another mirror plane perpendicular to the first one *or* a twofold rotation with its axis in the plane of σ. (3.3.6b)

Both statements follow from the noncommutativity of the corresponding elements (see below). The double groups of Table 3.3 belong to this category. For example,

$$\mathcal{D}_2 = \{e, c_{2x}, c_{2y}, c_{2z}\} \quad \text{and} \quad \mathcal{C}_{2v} = \{e, c_{2z}, \sigma_x, \sigma_y\}$$

are Abelian with 4 classes each. The double groups are

$$\mathcal{D}_2^{\mathrm{D}} = \{e, c_{2x}, c_{2y}, c_{2z}, c_0, c_0c_{2x}, c_0c_{2y}, c_0c_{2z}\} ,$$

$$\mathcal{C}_{2v}^{\mathrm{D}} = \{e, c_{2z}, \sigma_x, \sigma_y, c_0, c_0c_{2z}, c_0\sigma_x, c_0\sigma_y\} ,$$

which are not Abelian, with the five classes

$$\{e\} , \quad \{c_0\} , \quad \{c_{2x}, c_0c_{2x}\} , \quad \{c_{2y}, c_0c_{2y}\} , \quad \{c_{2z}, c_0c_{2z}\} \quad \text{and}$$

$$\{e\} , \quad \{c_0\} , \quad \{c_{2z}, c_0c_{2z}\} , \quad \{\sigma_x, c_0\sigma_x\} , \quad \{\sigma_y, c_0\sigma_y\} .$$

We obtain the decomposition into classes by conjugation, e.g.

$$c_{2x}^{-1}c_{2z}c_{2x} = c_0c_{2x}c_{2z}c_{2x} = c_0c_{2z} , \quad \text{i.e.} \quad c_{2z} \quad \text{and} \quad c_0c_{2z}$$

belong to one class. For other groups we have, e.g.

$$c_{2x}^{-1}c_{3z}c_{2x} = c_0c_{2x}c_{3z}c_{2x} = c_0c_{3z}^2 \ ,$$

$$c_{2x}^{-1}c_0c_{3z}c_{2x} = c_0c_{2x}c_0c_{3z}c_{2x} = c_{3z}^2 \ , \qquad \text{etc.}$$

In this way we find that in these double groups $\{c_{3z}, c_0c_{3z}^2\}$ and $\{c_{3z}^2, c_0c_{3z}\}$ each form a class but *not*, as in simple point groups, $\{c_{3z}, c_{3z}^2\}$. Similar relations hold for other elements and double groups. In applications this has to be observed very strictly (see character tables).

Exercise 3.5. Determine the classes of the double groups $\mathscr{T}^{\mathrm{D}}$ and $\mathscr{O}^{\mathrm{D}}$.
Exercise 3.6. Prove the statements (3.3.6).
Exercise 3.7. Review the theory of spinors and spinor transformations.
Exercise 3.8. Representing the rotation $D^{1/2}(c)$ by the rotation axis $\boldsymbol{n}$ and the angle φ, and using the Pauli spin matrices, calculate the product of two arbitrary rotations $\boldsymbol{n}_1, \varphi_1$ and $\boldsymbol{n}_2, \varphi_2$.

3.4 Lattices, the Translation Group and Space Group

3.4.1 Normal Space Groups

In solids the physical objects (atoms, ions, molecules) are very often arranged in regular periodic structures, which are called *lattices* (or *Bravais lattices*). The ideal lattice structure is looked upon as being *infinite*. Such a lattice can be described as a periodic repetition of identical *unit* or elementary *cells* (Fig. 3.4). Every unit cell is determined by three non-coplanar basis vectors $\boldsymbol{a}^{(i)}$. Different lattices may be distinguished according to the relative lengths of the $\boldsymbol{a}^{(i)}$ and the angles between them. Obviously, different symmetry operations belong to different structures. But, not all point group symmetries are compatible with a lattice structure. In order to get an agreement between point group symmetry and lattice structure, the whole space has to be completely filled by polyhedra or polygons (Fig. 3.5).

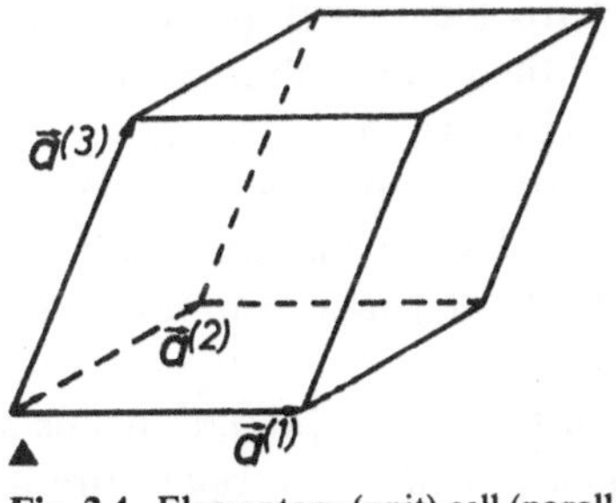

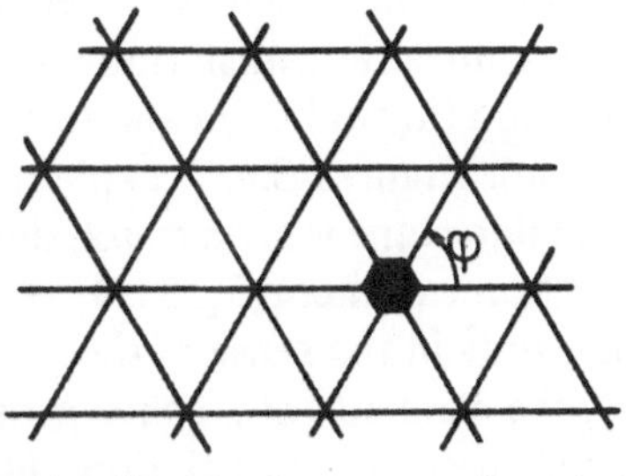

Fig. 3.4. Elementary (unit) cell (parallelepiped) defined by three non-coplanar basis vectors $\boldsymbol{a}^{(i)}$

Fig. 3.5. The determination of possible symmetry axes in a lattice. Triangles yield a sixfold axis: $m = 3 \Rightarrow \varphi = \pi/3$, $n\varphi = n\pi/3 = 2\pi \curvearrowright n = 6$

A plane has to be covered by equilateral polygons with m vertices. A point group symmetry demands an n-fold rotation axis through one vertex, which transforms the set of polygons into itself. Then, an integral multiple of the angle between two edges in the equilateral m-polygon, $\varphi = (m-2)\pi/m$, has to be equal to 2π, therefore

$$n\varphi = n(m-2)\pi/m = 2\pi \qquad \text{or} \qquad n = 2m/(m-2) \ . \tag{3.4.1}$$

Since m and n have to be integers, we have

$$\begin{aligned} m &= \infty \ \ 6 \ \ 4 \ \ 3 \ \ 2 \\ n &= \ 2 \ \ 3 \ \ 4 \ \ 6 \ \ \infty \quad \text{(-fold axes).} \end{aligned} \tag{3.4.2}$$

In an infinite lattice we have only 2-, 3-, 4- and 6-fold axes ($n = \infty$ means isotropy about an axis). In $\mathbb{R}_3$ a corresponding consideration provides the same result. In any case, mirror planes are also possible symmetry elements (see also Appendix I).

Table 3.4 shows the possible lattice structures and the maximal point group symmetries compatible with them. The lattice structure implies that the inversion is a necessary element. Starting with the most primitive type of lattice (P and R in Table 3.4) further types can be derived by implanting "centred" lattice points (C, F, I in Table 3.4). The centred lattices have the same point group symmetry as the primitive ones. There are 7 of these *holosymmetric* point groups, which define 7 *crystal systems*. Notation can be taken from Table 3.4.

All the vertices of a lattice or the position of all unit cells are given by

$$\boldsymbol{R}^{\boldsymbol{h}} = \sum_{i=1}^{3} h_i \boldsymbol{a}^{(i)} = h_1 \boldsymbol{a}^{(1)} + h_2 \boldsymbol{a}^{(2)} + h_3 \boldsymbol{a}^{(3)} \ , \qquad h_i \in \mathbb{Z} \ . \tag{3.4.3}$$

When we assign an identical set of points (atoms), a *basis*, to every unit cell, we obtain an ideal crystal structure. The symmetry operations which transform the crystal structure into itself form the *space group* $\mathscr{R}$. The space group comprises *translations* (displacements of the atoms relative to the lattice structure) and *point symmetry operations*. But there might also be combinations of point symmetry operations and translations. Such "new" elements are glide planes (a simultaneous reflection at a plane and translation parallel to the plane) and screw axes (a simultaneous rotation through $2\pi/n$ and translation by $m\boldsymbol{a}/n$, $m = 1, 2, \ldots, n-1$; $\boldsymbol{a}$: basis translation). Figures 3.6, 7 explain these operations. The translation connected with a glide plane is always $\boldsymbol{a}/2$, since $\sigma^2 = 1$! Screw operations always have the form n_m. Some of them ($3_1, 3_2; 4_1, 4_3; 6_1, 6_5; 6_2, 6_4$) are enantiomorphic, that is, they differ only in the sense of the screw (right-left).

Accordingly, the elements of a space group can be written $\{d|\boldsymbol{t}\}$, d means a point symmetry element, $\boldsymbol{t}$ a translation. The elements $\{d|\boldsymbol{t}\}$ are operators in $\mathbb{R}_3$-space (cf. Sect. 3.1.1) which act on points (position vectors) of the space:

$$\boldsymbol{x}' = \{d|\boldsymbol{t}\} \cdot \boldsymbol{x} := d\boldsymbol{x} + \boldsymbol{t} \ . \tag{3.4.4}$$

Table 3.4. Crystal systems and Bravais lattices

System	Relations	Number of lengths	Number of angles	Symbol	Symmetry, holohedry: Schönflies	Hermann-Mauguin full	Hermann-Mauguin short
Triclinic	$a \neq b \neq c \neq a$ $\alpha \neq \beta \neq \gamma \neq \alpha$	3	3	Γ_t P	$\mathscr{C}_i$	$\bar{1}\bar{1}\bar{1}$	$\bar{1}$
Monoclinic	$a \neq b \neq c \neq a$ $\alpha = \gamma = \frac{\pi}{2} \neq \beta$ or $\alpha = \beta = \frac{\pi}{2} \neq \gamma$	3	1	Γ_m P C Γ_m' A	$\mathscr{C}_{2h}$	$\bar{1}\frac{2}{m}\bar{1}$ or $\bar{1}\bar{1}\frac{2}{m}$	$2/m$
(Ortho-) rhombic	$a \neq b \neq c \neq a$ $\alpha = \beta = \gamma = \frac{\pi}{2}$	3	–	Γ_o P Γ_o' C Γ_o'' F Γ_o''' I	$\mathscr{D}_{2h}$	$\frac{2}{m}\frac{2}{m}\frac{2}{m}$	mmm
Tetragonal (quadratic)	$a = b \neq c$ $\alpha = \beta = \gamma = \frac{\pi}{2}$	2	–	Γ_q P Γ_q' I	$\mathscr{D}_{4h}$	$\frac{4}{m}\frac{2}{m}\frac{2}{m}$	$4/mmn$
Trigonal (rhombo-hedral)	$a = b = c$ $\frac{\pi}{2} \neq \alpha = \beta = \gamma < \frac{2\pi}{3}$	1	1	Γ_{rh} R	$\mathscr{D}_{3d}$	$\bar{3}\frac{2}{m}$	$\bar{3}m$
Hexagonal	$a = b \neq c$ $\alpha = \beta = \frac{\pi}{2}; \gamma = \frac{2\pi}{3}$	2	–	Γ_h P or H	$\mathscr{D}_{6h}$	$\frac{6}{m}\frac{2}{m}\frac{2}{m}$	$6/mmn$
Cubic (regular)	$a = b = c$ $\alpha = \beta = \gamma = \frac{\pi}{2}$	1	–	Γ_c P Γ_c' F Γ_c'' I	$\mathscr{O}_h$	$\frac{4}{m}\bar{3}\frac{2}{m}$	$m3m$

P: primitive; C, A: centred on faces C and A, respectively; F: face-centred (on all faces); I: body-centred R rombohedral; H: hexagonal.

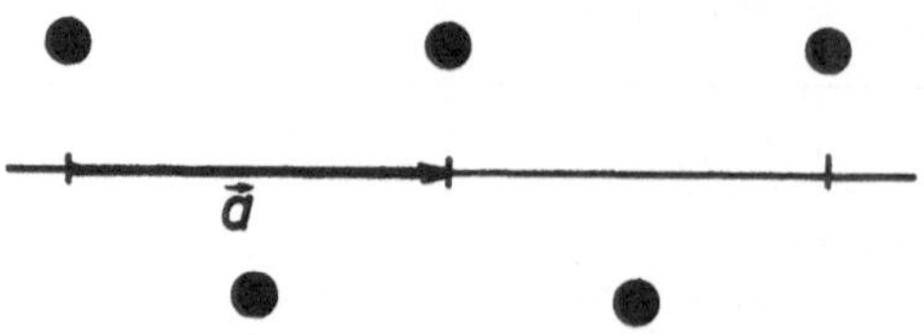

Fig. 3.6. Glide plane. The translation by $a/2$ is combined with a reflection at a plane containing a

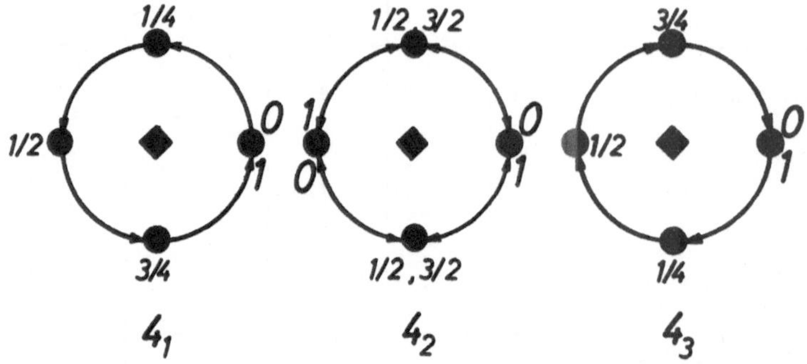

Fig. 3.7. Screw axis 4_m. The fractions give the height above the plane of the paper in units of the screw period. The elements 4_1 and 4_3 are enantiomorphic (left- and right-handed screw axes, respectively)

With two successive operations (3.4.4) we obtain the rule of multiplication

$$\{d_2|\boldsymbol{t}_2\}\cdot\{d_1|\boldsymbol{t}_1\} = \{d_2 d_1|\boldsymbol{t}_2 + d_2\boldsymbol{t}_1\} \ . \tag{3.4.5}$$

$\{d|\boldsymbol{t}\}^{-1} = \{d^{-1}|-d^{-1}\boldsymbol{t}\}$ defines the inverse element , (3.4.6)

$\{e|\boldsymbol{t}\}$ is a pure translation by $\boldsymbol{t}$, (3.4.7a)

$\{d|0\}$ is a pure rotation or reflection , (3.4.7b)

$\{e|0\}$ is the identity element . (3.4.7c)

Translations that transform the empty lattice (3.4.3) (without particles, atoms) into itself, are denoted by $\{e|\boldsymbol{R}^h\}$. They form the discrete, countably infinite *translation group* $\mathbb{T}$ of the lattice or the crystal. If there is only one atom in the unit cell (not necessarily at the vertices) we speak of a *Bravais crystal.* In the centred lattices (A, F, I) the basis vectors are then different from those of Table 3.5; they are chosen in such a way that every lattice point can be reached by a multiple of the basis vectors (Fig. 3.8). The cell defined in this way is the true *unit cell* (primitive, elementary cell). Another choice leads to the *Wigner-Seitz cell,* which reflects the symmetry of the crystal best (Fig. 3.8). A definite lattice point $\boldsymbol{R}^h = 0$ is connected with all “neighbouring points”. Then the planes $(\boldsymbol{r}\cdot\boldsymbol{R}^h) = \boldsymbol{R}^{h2}/2$, which are perpendicular to the connecting lines and bisect them, have to be determined. The smallest polyhedron defined by these planes is the Wigner-Seitz cell. It is invariant with respect to the holosymmetry of the lattice. The

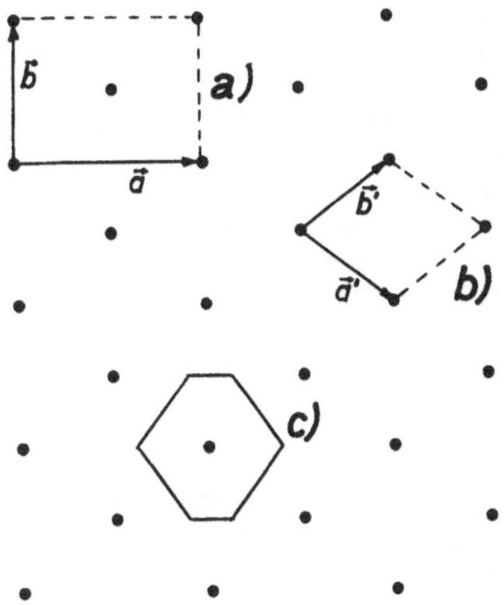

Fig. 3.8a–c. Possible unit cells of a centred lattice. (**a**) Primitive basis vectors $\boldsymbol{a}$, $\boldsymbol{b}$ defining a cell with two particles. (**b**) "Centred" basis vectors defining the proper unit cell with 1 particle. (**c**) The Wigner-Seitz cell, reflecting best the symmetry of the lattice ($\mathscr{C}_{2v}$ or holohedral $\mathscr{D}_{2h}$). Here $\boldsymbol{a}' = (\boldsymbol{a} - \boldsymbol{b})/2$; $\boldsymbol{b}' = (\boldsymbol{a} + \boldsymbol{b})/2$, $\boldsymbol{a} = \boldsymbol{a}' + \boldsymbol{b}'$; $\boldsymbol{b} = \boldsymbol{b}' - \boldsymbol{a}'$

volume of the unit cell is given by the scalar triple product $V_z = \boldsymbol{a}^{(1)} \cdot (\boldsymbol{a}^{(2)} \times \boldsymbol{a}^{(3)})$ of the basis vectors.

The translation group $\mathbb{T}$ is an Abelian invariant subgroup of the space group $\mathscr{R}$

$$\{d|\boldsymbol{t}\} \cdot \{e|\boldsymbol{R}^h\} \cdot \{d|\boldsymbol{t}\}^{-1} = \{d|\boldsymbol{t} + d\boldsymbol{R}^h\} \cdot \{d|\boldsymbol{t}\}^{-1} = \{e|d\boldsymbol{R}^h\} \ . \tag{3.4.8}$$

Again, $\{e|d\boldsymbol{R}^h\}$ is a translation covering all elements $\{e|\boldsymbol{R}^h\}$ of $\mathbb{T}$ when $\boldsymbol{R}^h$ assumes all possible values.

The pure point group elements $\{d|0\}$ are said to be the point group $\mathscr{G}$ of the lattice. The lattice, $\boldsymbol{R}^h$ according to (3.4.3), is not changed under any operation $d \in \mathscr{G}$ [see (3.4.8) and Tables 3.1, 4]. A complete description of the space group $\mathscr{R}$ needs also a specification of possible nonprimitive translations $\boldsymbol{s}(d)$ connected with glide planes and screw axes. Since $\mathbb{T}$ is an invariant subgroup of $\mathscr{R}$, the space group can be decomposed into cosets with respect to $\mathbb{T}$. The quotient group $\mathscr{R}/\mathbb{T}$ is isomorphic to the point group $\mathscr{G}$, cf. (2.2.13, 17),

$$\mathscr{R} = \sum_{d \in \mathscr{G}} \{d|\boldsymbol{s}(d)\} \cdot \mathbb{T} \ . \tag{3.4.9}$$

If all the $\boldsymbol{s}(d)$ can be put equal to zero by the choice of an appropriate coordinate system (Exercise 3.11), then $\{d|0\} \cong \mathscr{G}$ is a subgroup of $\mathscr{R}$. Since $\mathbb{T}$ is an invariant subgroup of $\mathscr{R}$ and since

$$\mathscr{G} \cap \mathbb{T} = \{e|0\} \ , \qquad \{d|0\} \cdot \mathbb{T} = \mathbb{T} \cdot \{d|0\} \ , \qquad \{d|\boldsymbol{R}^h\} = \{e|\boldsymbol{R}^h\} \cdot \{d|0\} \ ,$$

all the assumptions of a semidirect product are satisfied. These space groups are called *symmorphic* groups and can be written as

$$\mathscr{R} = \mathbb{T} \ \boxed{\text{s}} \ \mathscr{G} \qquad \text{(symmorphic)} \ . \tag{3.4.10}$$

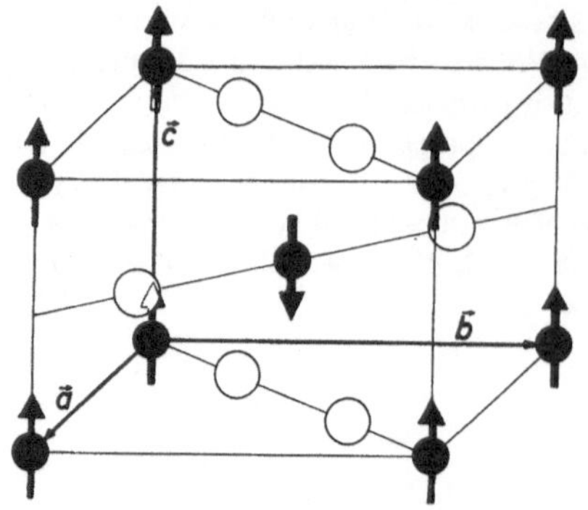

Fig. 3.9. Rutile structure ($a = b$) with two molecules MnO_2 in the cell. ●: magnetic ions (Mn, Fe, Ni, Co); ○: nonmagnetic ions (O, F). In the antiferromagnetic phase with its regular arrangement of the magnetic moments (shown here) the space group is $P\underline{4}_2/mn\underline{m}$. In the paramagnetic phase the magnetic moments are randomly oriented, the magnetic ions being "grey" just like the nonmagnetic ones. The symmetry is then $P4_2/mnm$ and in addition time reversal is valid (Sect. 3.4.2)

If, however, the $s(d)$ cannot be taken equal to zero, the space group is a *non-symmorphic* one.

As an example we discuss the group $\mathscr{D}_{4h}^{14} = P4_2/mnm$. It is the symmetry group of the paramagnetic phase of some magnetic materials, e.g. MnF_2, FeF_2, CoF_2, NiF_2 and MnO_2 (Fig. 3.9). The permanent magnetic moments of the ions are distributed randomly. The structure is also that of TiO_2, and is therefore called rutile structure. It belongs to the tetragonal system, the holosymmetry is $\mathscr{D}_{4h}$ with a primitive Bravais lattice. The basis vectors are orthogonal with lengths a, a, c; the volume of the unit cell is a^2c. It has a nonprimitive translation $s(d) = (\boldsymbol{a}^{(1)} + \boldsymbol{a}^{(2)} + \boldsymbol{a}^{(3)})/2$ (see Exercises 3.9, 11).

3.4.2 Colour and Magnetic Space Groups

Colour space groups are defined analogously to colour point groups. This holds especially for groups of types I and II.

I. These colour space groups describe systems with well-defined colours, that is,

$$\mathscr{M}_{\mathrm{I}} = \mathscr{R} \quad \text{and} \quad r \notin \mathscr{M}_{\mathrm{I}} \ . \tag{3.4.11}$$

II. Correspondingly, we have for the *grey* colour space groups

$$\mathscr{M}_{\mathrm{II}} = \mathscr{R} + r \cdot \mathscr{R} \quad \text{and} \quad r \in \mathscr{M}_{\mathrm{II}} \ . \tag{3.4.12}$$

III. Colour space groups of type III, *black-white* groups in the strict sense, have to be divided into two kinds. In IIIa, the element which changes colour is connected with an element of the point group, while in IIIb it is connected with an element of the lattice translation.

IIIa. In this case

$$\mathscr{M}_{\mathrm{IIIa}} = \mathscr{N} + r(\mathscr{R} - \mathscr{N}) \ , \qquad r \notin \mathscr{M}_{\mathrm{IIIa}} \ , \tag{3.4.13}$$

where $\mathscr{N}$ is an invariant subgroup of $\mathscr{R}$ with index 2 *and* $\mathscr{R} - \mathscr{N}$ does not

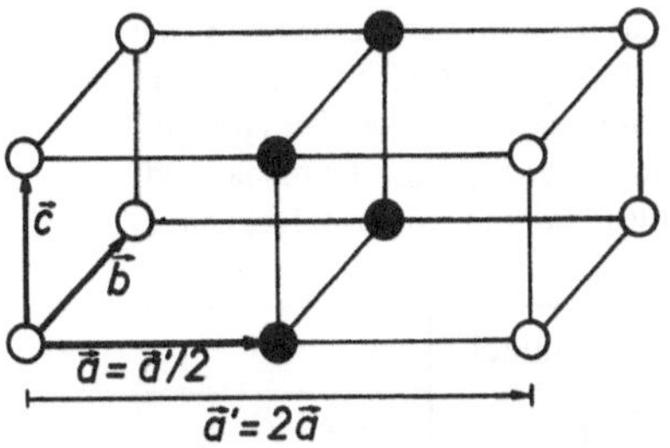

Fig. 3.10. Magnetic (colour) lattice corresponding to a primitive lattice. It is denoted by P_{2a} or $P_{a'}$, depending on the basis of the normal lattice from which it is derived

contain pure lattice translations. The corresponding lattices are the usual Bravais lattices (Table 3.4).

IIIb. The element r which changes colour is associated with the lattice; therefore we have first to define a black-white lattice. This can be achieved in analogy to point groups by combining an extra translation $\{e|\boldsymbol{t}\}$ (or a certain superposition of translations) with an element r that changes colour. After applying such a translation twice, the original colour has to be restored again (Fig. 3.10)! In this way we get the lattices of Table 3.5 of the type

$$\mathbb{T}_{\text{IIIb}} = \mathbb{T} + r\{e|\boldsymbol{t}\}\mathbb{T} \ .$$

The space group is then

$$\mathcal{M}_{\text{IIIb}} = \mathcal{R}' + r\{e|\boldsymbol{t}\}\mathcal{R}' \ , \qquad (3.4.14)$$

where $\mathcal{R}'$ is a space group[4] that does not contain the element $\{e|\boldsymbol{t}\}$.

These space groups occur in connection with magnetic structures. The element r that changes colour is then the time reversal operator ϑ (Sect. 3.2). Crystals with magnetic properties belong to these space groups as follows:

dia- and paramagnetic phases: all $\mathcal{M}_{\text{II}}$
antiferromagnetic phases: all $\mathcal{M}_{\text{I}}$, $\mathcal{M}_{\text{IIIa}}$, $\mathcal{M}_{\text{IIIb}}$
ferro-(ferri-)magnetic phases: some $\mathcal{M}_{\text{I}}$, $\mathcal{M}_{\text{IIIa}}$, altogether 31 point groups and 275 space groups, but no cubic group.

The antiferromagnetic phases of MnF_2, FeF_2, CoF_2 (3.4.13) belong to type IIIa, space group $P\underline{4}_2/mn\underline{m}$; the magnetic and nonmagnetic unit cells are identical (Fig. 3.9, Exercise 3.10). The number of possible groups is given in Table 3.6.

3.4.3 Double Space Groups

In Sect. 3.3 we extended point groups to double groups describing the transformations of spinors by including an element c_0, see (3.3.4). The elements of

[4] The group $\mathcal{R}'$ is an invariant subgroup of $\mathcal{R}$ with index 2 with respect to translations: $\mathcal{R} = \mathcal{R}' + \{e|\boldsymbol{t}\}\mathcal{R}'$.

Table 3.5. Colour or magnetic lattices. The basic translations $\boldsymbol{a}$, $\boldsymbol{b}$, $\boldsymbol{c}$ correspond to the definitions given in Table 3.4. Columns 2 and 3 describe the colour lattices starting from Bravais lattices in which half the lattice points have been replaced by white ones and half by black ones. Columns 4 and 5 describe the colour lattices starting from a unicoloured (e.g. white) Bravais lattice in which additional lattice points of another colour (black) have been added at the "positions" $\boldsymbol{t}$

System	Symbol	Basis vectors	Symbol	t
Triclinic	P_{2a}	$2a, b, c$	P_a	$a'/2$
Monoclinic	P_{2a}	$2a, b, c$	P_a	$a'/2$
	P_{2c}	$a, b, 2c$	P_c	$c'/2$
	P_A	$a, b+c, b-c$	A_b	$b'/2$
	A_{2a}	$2a, b, c$	A_a	$a'/2$
	A_P	$a, b+c, b-c$	P_A	$(b'+c')/2$
(Ortho-) rhombic	P_{2a}	$2a, b, c$	P_a	$a'/2$
	P_C	$a+b, a-b, c$	C_a	$a'/2$
	P_F	$a+b, b+c, c+a$	F_s	$(a'+b'+c')/2$
	C_{2c}	$a, b, 2c$	C_c	$c'/2$
	CP	$a+b, a-b, c$	P_C	$(a'+b')/2$
	C_I	$a+b, b+c, c+a$	I_C	$(a'+b')/2$
	F_C	$a, b+c, b-c$	C_A	$(b'+c')/2$
	IP	$a+b, b+c, c+a$	P_I	$(a'+b'+c')/2$
Tetragonal (quadratic)	P_{2c}	$a, b, 2c$	P_c	$c'/2$
	P_C	$a+b, a-b, c$	P_C	$(a'+b')/2$
	P_I	$a+b, b+c, c+a$	I_c	$c'/2$
	IP	$a+b, b+c, c+a$	P_I	$(a'+b'+c')/2$
Trigonal (rhombohedral)	R_R	$a+b, b+c, c+a$	R_I	$(a'+b'+c')/2$
Hexagonal	P_{2c}	$a, b, 2c$	P_c	$c'/2$
Cubic (regular)	P_F	$a+b, b+c, c+a$	F_S	$(a'+b'+c')/2$
	IP	$a+b, b+c, c+a$	P_I	$(a'+b'+c')/2$

double space groups $\mathscr{R}^D$ can be realized analogously: the rotation parts in the elements $\{d|\boldsymbol{t}\} \in \mathscr{R}$ are replaced by the matrices $\pm D^{1/2}(d)$, that is, $\mathscr{R}^D$ has twice as many elements as $\mathscr{R}$. We write $\{d^{\pm}|\boldsymbol{t}\} \in \mathscr{R}^D$ for short for the two elements related to $\{d|\boldsymbol{t}\} \in \mathscr{R}$. Then $\mathscr{R}^D$ possesses the double translation group $\mathbb{T}^D$ with the elements $\{e^{\pm}|\boldsymbol{R}^h\}$ as an invariant subgroup whereas $\mathbb{T}^D$ itself possesses the usual translation group $\mathbb{T}^+$ $(\cong \mathbb{T})$ with $\{e^+|\boldsymbol{R}^h\}$ as an invariant subgroup. Here we define $e^+ = e$; $e^- = c_0$. As $\mathbb{T}^D$ and $\mathbb{T}^+$ are Abelian groups,

$$\mathbb{T}^D = \mathbb{T}^+ \times \varepsilon \,, \qquad \text{with} \qquad \varepsilon = \{\{e^+|0\}, \{e^-|0\}\} \,, \tag{3.4.15}$$

is a direct product. Compared to $\mathscr{R}$, $\mathscr{R}^D$ has the additional elements ($d^+ = d$; $d^- = \bar{d}$)

Table 3.6. Point and space groups in $\mathbb{R}_n$

	$\mathbb{R}_1$	$\mathbb{R}_2$	$\mathbb{R}_3$
Crystal systems	1	4	7
Bravais lattices	1	5	14
Crystal classes (point groups)	2	10	32
Space groups	2	17	230
of which, non-isomorphic	2	16	219
Symmorphic space groups	2	13	73
of which, non-isomorphic	2	12	73
Magnetic lattices	2	10	36
of which, truly magnetic	1	5	22
Magnetic crystal classes	3	21	90
of which, of IInd kind	2	10	32
of which, of IIIrd kind	1	11	58
Magnetic space groups	5	63	1421
of which, of type II	2	17	230
of which, of type IIIa	1	26	674
of which, of type IIIb	2	20	517

$$\{d^-|t\} = \{e^-|\boldsymbol{R}^h\}\cdot\{d|s(d)\} \ ; \qquad t = \boldsymbol{R}^h + s(d) \ . \tag{3.4.16}$$

The group $\mathscr{R}$ is not a subgroup of $\mathscr{R}^{\mathrm{D}}$, since the elements of $\mathscr{R}$ are not closed with respect to $\mathscr{R}^{\mathrm{D}}$.

Exercise 3.9. Determine the symmetry elements of $\mathscr{D}_{4h}^{14} = P4_2/mnm$ (Fig. 3.9). Since $\{d|t\} = \{d|\boldsymbol{R}^h + s(d)\} = \{e|\boldsymbol{R}^h\}\cdot\{d|s(d)\}$ the specification of $\{d|s(d)\}$ is sufficient. Determine first the generators $\{c_{4z}|s\}$, $\{c_{2x}|s\}$, $\{i|0\}$.

Exercise 3.10. Following Exercise 3.9, determine the symmetry elements of the magnetic group $P\underline{4}_2/mn\underline{m}$ (Fig. 3.9). The magnetic generator is $r\{c_{4z}|s\}$.

Exercise 3.11. The description of the symmetry elements of a space group depends on the choice of the origin (of the basis). Show that the space group elements $\{d|t\}$ change into $\{e|\boldsymbol{a}\}^{-1}\cdot\{d|t\}\cdot\{e|\boldsymbol{a}\} = \{d|t + d\boldsymbol{a} - \boldsymbol{a}\}$ under a translation of the origin by a fixed vector $\boldsymbol{a}$. Correspondingly, nonprimitive translations (in connection with glide planes and screw axes) change from $s(d)$ into $s(d) + d\boldsymbol{a} - \boldsymbol{a}$.

3.5 Permutation Groups

Permutation or symmetric groups $\mathscr{P}_n$ are very important for the abstract theory of finite groups and also for physical applications. The term symmetric group is the more common one, nevertheless we will use the term permutation group because the term symmetric is less specific, as it is used in various contexts. These groups describe the behaviour of physical quantities under permutation of identical objects, e.g. under permutations of the n variables of identical particles in the argument of a wave function $\psi(x_1, x_2 \dots x_n)$. Numbering the objects, in this case the n position coordinates, by $1, 2, \dots, n$, a specific permutation is represented by

$$p = \begin{pmatrix} 1 & 2 & \dots & n \\ p_1 & p_2 & \dots & p_n \end{pmatrix} . \tag{3.5.1a}$$

Object 1 has to be replaced by p_1, etc., where $p_1 \dots p_n$ are the numbers 1 to n in an arbitrary sequence; the order of the columns in (3.5.1a) is irrelevant. With $n = 4$, the permutation

$$p = \begin{pmatrix} 1 & 2 & 3 & 4 \\ 3 & 1 & 4 & 2 \end{pmatrix} \tag{3.5.1b}$$

means that 1 has to be replaced by 3, 2 by 1, etc.

If q is a second permutation, the "product"

$$qp = \begin{pmatrix} 1 & 2 \dots n \\ q_1 & q_2 \dots q_n \end{pmatrix} \cdot \begin{pmatrix} 1 & 2 \dots n \\ p_1 & p_2 \dots p_n \end{pmatrix} = \begin{pmatrix} p_1 & p_2 \dots p_n \\ s_1 & s_2 \dots s_n \end{pmatrix} \cdot \begin{pmatrix} 1 & 2 \dots n \\ p_1 & p_2 \dots p_n \end{pmatrix}$$
$$= \begin{pmatrix} 1 & 2 \dots n \\ s_1 & s_2 \dots s_n \end{pmatrix} \tag{3.5.2a}$$

again is a permutation. Equation (3.5.2a) defines the rule of composition (multiplication), which is associative, the existence of identity and inverse elements is obvious, so the permutations of n objects form a group of order $n!$. In the example (3.5.1b) we have

$$qp = \begin{pmatrix} 1 & 2 & 3 & 4 \\ 2 & 3 & 4 & 1 \end{pmatrix} \cdot \begin{pmatrix} 1 & 2 & 3 & 4 \\ 3 & 1 & 4 & 2 \end{pmatrix} = \begin{pmatrix} 1 & 2 & 3 & 4 \\ 4 & 2 & 1 & 3 \end{pmatrix} ; \tag{3.5.2b}$$

$$p^{-1} = \begin{pmatrix} 1 & 2 & 3 & 4 \\ 3 & 1 & 4 & 2 \end{pmatrix}^{-1} = \begin{pmatrix} 1 & 2 & 3 & 4 \\ 2 & 4 & 1 & 3 \end{pmatrix} . \tag{3.5.2c}$$

The permutation can be written more easily by using "*cycles*". After the number of the object, the number by which it is replaced is written, and so on:[5]

$$p = (1 \quad 3 \quad 4 \quad 2) , \tag{3.5.3a}$$

$$qp = (1 \quad 2 \quad 3 \quad 4)(1 \quad 3 \quad 4 \quad 2) = (1 \quad 4 \quad 3)(2) , \tag{3.5.3b}$$

$$p^{-1} = (1 \quad 3 \quad 4 \quad 2)^{-1} = (1 \quad 2 \quad 4 \quad 3) . \tag{3.5.3c}$$

This has another advantage: the cycles contain only those objects that are interchanged with each other. A permutation can be written as a product of

---

[5] The permutations p and qp should be applied to a function $\psi(x_1, \dots, x_4)$, in order to show the result.

independent cycles; the number of any object occurs only in exactly one cycle. *Cycles containing just one number* (3.5.3b) imply no permutation and can be dropped.

Any permutation is characterized by its structure of cycles, the ordering of the cycles being irrelevant. A permutation p may be built up of ν_1 1-cycles, ν_2 2-cycles, ..., ν_n n-cycles (the subscript of ν is the number of objects in the cycle). Then we say that p has the *cycle structure*

$$(\nu) = (1^{\nu_1}, 2^{\nu_2}, \ldots, n^{\nu_n}) := (\nu_1, \nu_2, \ldots, \nu_n) \ , \tag{3.5.4}$$

with

$$\sum_{i=1}^{n} i \cdot \nu_i = n \ . \tag{3.5.5}$$

For example,

$$(143)(2)\colon (\nu) = (1^1, 2^0, 3^1) \ ; \qquad 1 \cdot 1 + 2 \cdot 0 + 3 \cdot 1 = 4 \ .$$

If $s \in \mathscr{P}_n$ is any element and $p = z_1 \cdot z_2 \cdot \cdots \cdot z_m$ (z_i is a cycle of p), then a conjugate element of p is

$$q = sps^{-1} = (sz_1s^{-1})(sz_2s^{-1})\ldots(sz_ms^{-1}) \ . \tag{3.5.6}$$

This shows us that q has the same cycle structure as p:

> All elements with the same cycle structure and only these belong to the same class. (3.5.7)

For $p = (14)(23)$ and $s = (12)(3)(4) = s^{-1}$, we have

$$q = [(12)(14)(12)] \cdot [(12)(23)(12)] = (24) \cdot (13) \ .$$

Consequently, the type and number of classes are determined by the solutions of (3.5.5) with integer $\nu_i \geqslant 0$. These solutions can be classified with the help of *partitions*

$$[\lambda] = [\lambda_1, \lambda_2, \ldots, \lambda_n] \ , \qquad \lambda_1 \geqslant \lambda_2 \geqslant \lambda_3 \geqslant \cdots \geqslant 0 \ , \tag{3.5.8}$$

where

$$\begin{aligned} \lambda_1 &= \nu_1 + \nu_2 + \cdots + \nu_n \ , \\ \lambda_2 &= \qquad\ \nu_2 + \cdots + \nu_n \ , \\ &\vdots \\ \lambda_n &= \qquad\qquad\qquad \nu_n \ , \end{aligned}$$

and because of (3.5.5)

$$\sum_{i=1}^{n} \lambda_i = \lambda_1 + \lambda_2 + \cdots + \lambda_n = n \; . \tag{3.5.9}$$

Thus, $[\lambda]$ is a decomposition (partition) of n. The inverse relation of (3.5.9) is given by

$$\nu_i = \lambda_i - \lambda_{i+1} \qquad \text{for} \qquad i \leqslant n-1 \; ; \qquad \nu_n = \lambda_n \; . \tag{3.5.10}$$

The number of classes in $\mathscr{P}_n$ is determined by the number of partitions $[\lambda]$ of n. (3.5.11)

The *number of elements* r_ν in a class with cycle structure (ν) is given by

$$r(\nu) = r_\nu = \frac{n!}{1^{\nu_1}\nu_1!2^{\nu_2}\nu_2!\ldots n^{\nu_n}\nu_n!} \; . \tag{3.5.12}$$

As an example we consider the group $\mathscr{P}_3$ of order $g = n! = 6$. The partitions of $n = 3$ are

$$3 = 1 + 1 + 1 \; , \qquad [\lambda] = [1,1,1] \; , \qquad (\nu) = (1^0, 2^0, 3^1) \; ,$$

$$3 = 2 + 1 + 0 \; , \qquad [\lambda] = [2,1] \; , \qquad (\nu) = (1^1, 2^1, 3^0) \; ,$$

$$3 = 3 + 0 + 0 \; , \qquad [\lambda] = [3] \; , \qquad (\nu) = (1^3, 2^0, 3^0) \; .$$

Thus, there is one class K_3 with 3-cycles (123) and (321), one class K_2 with 2,1-cycles (12)(3), (13)(2) and (23)(1), and one class K_1 with 1-cycles (1)(2)(3). According to (3.5.12), we have $r_3 = 2, r_2 = 3, r_1 = 1$. This group is isomorphic to the point groups $\mathscr{C}_{3v}$ and $\mathscr{D}_3$ (Exercise 3.12).

The *theorem of Cayley* establishes the relationship of groups of finite order to the permutation groups:

Every group $\mathscr{G}$ of order $g = n$ is isomorphic to a subgroup of $\mathscr{P}_n$. (3.5.13)

This can be seen from the one-to-one correspondence of an element $a_i \in \mathscr{G}$ with a permutation

$$a_i \rightarrow \begin{pmatrix} a_1 & a_2 & a_n \\ a_i a_1 & a_i a_2 & a_i a_n \end{pmatrix} \; .$$

This correspondence represents a group homomorphism. As a model of finite groups we can use the subgroups of $\mathscr{P}_n$. But the structure of the $\mathscr{P}_n$ is (at least for larger n) very complicated, so that no general statements exist. The inverse element p^{-1} of an element p belongs to the same class as p, since the inverse of a cycle is the cycle in the reverse sequence (3.5.3c).

Cycles of two numbers are called *transpositions*. Every cycle, and consequently every permutation, can be written as products of transpositions. This is because every permutation can be constructed by successive interchanges of pairs of objects. However, here the number (of one object) may occur in several transpositions of such a product. This decomposition is not unique and the commutation of two transpositions in general is another permutation, if one number occurs in both transpositions. For example,

$$(132) = (12)(13) \ , \qquad \text{but} \qquad (13)(12) = (123) \ .$$

It can be shown that the transpositions (12), (13), ..., (1n) or else (12), (23), (34), ..., ($n-1, n$) can be chosen as generators of the permutation group.

A permutation is *even* if it is represented by an even number of transpositions, and *odd* if the number of transpositions is odd. The even permutations form an invariant subgroup (index 2) of $\mathscr{P}_n$, the *alternating group* $\mathscr{A}_n$. The quotient group $\mathscr{P}_n/\mathscr{A}_n$ is isomorphic to $\mathscr{P}_2 = \{(1)(2); (12)\}$. The odd permutations do not form a group, since, for example, the identity element is missing.

Exercise 3.12. Show that $\mathscr{P}_3$, $\mathscr{C}_{3v}$ and $\mathscr{D}_3$ are isomorphic to each other by establishing the one-to-one correspondence between the different elements (Fig. 2.1.) and comparing the group tables. Determine even and odd permutations and show that the even ones are an invariant subgroup isomorphic to $\mathscr{C}_3$.

3.6 Other Finite Groups

Apart from the point, space and permutation groups discussed above, other finite (discrete) groups play hardly any role in physics. Thus we will just mention some groups that are occasionally useful.

The group of smallest order that is not cyclic is *Klein's four group* $\mathscr{V}$ of order $g = 4$. It is Abelian with two generators

$$\mathscr{V} := \{e, a, b, ab \,|\, a^2 = b^2 = (ab)^2 = e\} \ . \tag{3.6.1}$$

It is isomorphic to $\mathscr{D}_2$.

The Pauli spin matrices do not themselves constitute a group. However, they can be completed to build a group which is isomorphic to the *quaternion group* $\mathscr{Q}$

$$\mathscr{Q} \cong \{\pm e, \pm i\sigma_1, \pm i\sigma_2, \pm i\sigma_3 \,|\, \sigma_i\sigma_j = i\varepsilon_{ijk}\sigma_k; \, \sigma_i^2 = e\} \ . \tag{3.6.2}$$

Quaternions are an extension of complex numbers with properties similar to those of spin matrices

$$\mathscr{Q} := \{\pm e, \pm i, \pm j, \pm k \,|\, i^2 = j^2 = k^2 = -e, i \cdot j = k, \text{etc.}\} \ . \tag{3.6.3}$$

The quanternion group $\mathscr{Q}$ consists of 5 classes; it is not isomorphic to one of the simple point groups, but it is isomorphic to $\mathscr{D}_2^{\mathrm{D}}$ or $\mathscr{C}_{2v}^{\mathrm{D}}$.

By a definition different from (3.6.2), it is possible to form a group from Pauli spin matrices that is isomorphic to $\mathscr{D}_4$,

$$\mathscr{D}_4 \cong \mathscr{F}(2) = \{\pm e, \pm\sigma_1, \pm\sigma_2, \pm i\sigma_3\} \; . \tag{3.6.4}$$

Similarly, groups can be formed using Dirac's γ_i-matrices, namely

$$\mathscr{F}(4) = \{\pm\gamma_1^{\alpha_1}\gamma_2^{\alpha_2}\gamma_3^{\alpha_3}\gamma_4^{\alpha_4} | \alpha_j = 1, 2; \gamma_1^2 = \mathrm{e}\} \tag{3.6.5}$$

and

$$\hat{\mathscr{F}}(4) = \{\pm\gamma_1^{\alpha_1}\gamma_2^{\alpha_2}\gamma_3^{\alpha_3}\gamma_0^{\alpha_0} | \alpha_j = 1, 2; \gamma_i^2 = \mathrm{e} \text{ for } i = 1, 2, 3; \gamma_0^2 = -\mathrm{e}\} \; , \tag{3.6.6}$$

where $\gamma_0 = i\gamma_4$. Both $\mathscr{F}(4)$ and $\hat{\mathscr{F}}(4)$ contain 32 elements, each group is divided into 17 classes. But they are not isomorphic to each other. The groups $\mathscr{F}(2)$ and $\mathscr{F}(4)$ can be extended by adding further elements which have similar properties to the γ_i. These groups are said to be *Fermi groups* $\mathscr{F}(2n)$ of order $g = 2^{2n+1}$; they will not be discussed in detail.

A general problem of group theory is the determination of all the nonisomorphic groups of a given order g [see the theorem of Cayley, (3.5.13)]. This problem has been solved only for "small" g; for example, for $g = 128$ the number of nonisomorphic groups is unknown. For physical applications, however this question is not of interest, because in physics the groups and their elements are assumed to be given.

4. Representations of Finite Groups

The most effective tools in applying group theoretical methods to physical problems are the representations of the groups in "physical" spaces and the characters of these representations. Of all the possible representations, those by linear operators, or more specifically, by matrices, are the most essential ones. The basic spaces are the linear (vector) spaces, which are discussed first in this chapter. Then the properties of different representations are outlined with special emphasis on the irreducible ones. Finally we investigate product representations of groups.

4.1 Linear Spaces and Operators

Within the theory of group representations we need many fundamental ideas of algebra. The basic concepts will be briefly discussed here, using simple examples, but extensive discussion is left to special mathematics books on linear algebra.

4.1.1 Linear and Unitary Spaces

The concept of a *linear space* $\mathscr{L}$ is explained best with simple examples. Vectors $\boldsymbol{x}$ in $\mathbb{R}_3$ can be added together and multiplied by (scalar) numbers c_i. In both cases, we again obtain a vector in $\mathbb{R}_3$, or a linear combination,

$$\boldsymbol{x} = c_1\boldsymbol{x}_1 + c_2\boldsymbol{x}_2 + c_3\boldsymbol{x}_3 = \sum_{i=1}^{3} c_i\boldsymbol{x}_i \in \mathbb{R}_3 \ , \qquad (4.1.1)$$

which again is a vector in $\mathscr{L} = \mathbb{R}_3$. The set of functions that are solutions of a linear differential equation of 2nd order

$$\sum_{i=1}^{n} \frac{\partial^2 y}{\partial x_i^2} + \frac{1}{c^2}\frac{\partial^2 y}{\partial t^2} = 0$$

with appropriate boundary conditions can be added together and multiplied by numbers. The result is again a solution (*principle of superposition*)

$$y(\boldsymbol{x}, t) = \sum_{i=1}^{m} c_i y_i(\boldsymbol{x}, t) \in \mathscr{L} \qquad \text{with} \qquad \boldsymbol{x} \in \mathbb{R}_n \ . \qquad (4.1.2)$$

Furthermore, all the eigenfunctions of a quantum-mechanical system belonging to an eigenvalue E and being degenerate form a linear space.

The general definition of a linear space $\mathscr{L}$ is as follows:

A set of objects $\{y\}$ or elements, e.g. vectors, points, functions, matrices, with two rules of composition (addition $+$ and multiplication $\cdot$) is said to be a *linear* or *vector space* over the field of scalars $\mathbb{K}$ (of real $\mathbb{R}$ or complex numbers $\mathbb{C}$) if for all $y \in \mathscr{L}$ and all $c_i \in \mathbb{K}$

(i) $\mathscr{L}$ is an Abelian group $(\mathscr{L}, +)$ with respect to the addition see Sect. 2.1;

(ii) there is a multiplication with numbers $c \in \mathbb{K}$ defined in $\mathscr{L}$ such that

$$cy \in \mathscr{L} \text{ if } y \in \mathscr{L} \ ,$$

$$c_1(c_2 y) = (c_1 c_2)y \qquad \text{(associativity)},$$

$$c(y_1 + y_2) = cy_1 + cy_2 \qquad \text{and} \qquad (c_1 + c_2)y = c_1 y + c_2 y \quad \text{(distributivity)},$$

$$1 \cdot y = y \qquad \text{(existence of an identity element)}. \tag{4.1.3}$$

Any y that can be represented [see (4.1.2)] as

$$y = c_1 y_1 + \dots + c_n y_n \quad \text{with} \quad c_i \in \mathbb{K} \ , y_i \in \mathscr{L} \tag{4.1.4}$$

is a *linear combination* or a linear form of the y_i.

A set of n elements

$y_i \in \mathscr{L}$ is *linearly independent*[1] if

$$\sum_{i=1}^{n} c_i y_i = 0 \qquad \text{only if } c_i = 0 \text{ for all } i = 1, \dots, n. \tag{4.1.5}$$

Conversely, if (4.1.5) is not satisfied, the y_i are linearly dependent. Examples of linearly independent elements are

functions: $\{x, x^2\}$ or $\{\sin x, \cos x\}$;

vectors: $\{\boldsymbol{a}^{(i)} | i = 1, 2, 3 \ ; \boldsymbol{a}^{(i)} \in \mathbb{R}_3 \quad \text{not coplanar}\}$;

$$\text{matrices:} \quad \left\{ \begin{pmatrix} 1 & 0 \\ 0 & 0 \end{pmatrix}, \begin{pmatrix} 0 & 1 \\ 0 & 0 \end{pmatrix}, \begin{pmatrix} 0 & 0 \\ 1 & 0 \end{pmatrix}, \begin{pmatrix} 0 & 0 \\ 0 & 1 \end{pmatrix} \right\} \tag{4.1.6}$$

$$\text{or} \quad \left\{ \begin{pmatrix} 1 & 0 \\ 0 & 1 \end{pmatrix}, \begin{pmatrix} 0 & 1 \\ 1 & 0 \end{pmatrix}, \begin{pmatrix} 0 & -\mathrm{i} \\ \mathrm{i} & 0 \end{pmatrix}, \begin{pmatrix} 1 & 0 \\ 0 & -1 \end{pmatrix} \right\} .$$

---

[1] This definition holds provided that $y_i \neq 0$ for all i; in principle we have to distinguish between "element" zero and "number" zero. However, in the following this seems to be unnecessary.

A subset $\mathscr{L}' \subseteq \mathscr{L}$ is a subspace of $\mathscr{L}$, if for all $y_1, y_2 \in \mathscr{L}'$, and $c \in \mathbb{K}$ there are also $y_1 + y_2 \in \mathscr{L}'$ and $cy_1 \in \mathscr{L}'$. All vectors in a plane (on a straight line) form a subspace $\mathbb{R}_2$ ($\mathbb{R}_1$) of $\mathbb{R}_3$. A subset $\mathscr{B} \subset \mathscr{L}$ is a *basis* of $\mathscr{L}$ if all the elements of $\mathscr{B}$ are linearly independent and every element $y \in \mathscr{L}$ can be represented as a linear combination:

$$y = \sum_i^n c_i y_i \in \mathscr{L} \; ; \qquad y_i \in \mathscr{B} \; . \tag{4.1.7}$$

A basis is not a subspace! The cardinality n of $\mathscr{B}$ is the dimension of $\mathscr{L}$: $\dim \mathscr{L} = n$.

Examples:
The vector space

$$\mathbb{R}_n = \{\boldsymbol{x} | \boldsymbol{x} = \sum_{i=1}^n c_i \boldsymbol{e}^{(i)}; \; c_i \in \mathbb{R} \text{ or } \mathbb{C}\} \tag{4.1.8a}$$

has n linearly independent basis vectors $\boldsymbol{e}^{(i)}$ and $\dim \mathbb{R}_n = n$. With $\mathbb{K} = \mathbb{R}$ and $n = 3$ we have the Euclidean space $\mathbb{R}_3$.

The vector space of all linearly independent polynomials f_ν, $\nu = 1, 2, \ldots, \infty$ in an interval $[a, b]$ is infinite dimensional. All $(n \times n)$ matrices form a vector space of dimension n^2. A possible basis is formed by all matrices with one "1" and zeros at all other places

$$\begin{pmatrix} 1 & 0 & 0 & \cdot & \cdot & \cdot \\ 0 & 0 & 0 & \cdot & \cdot & \cdot \\ \cdot & \cdot & \cdot & \cdot & \cdot & \cdot \end{pmatrix}, \quad \begin{pmatrix} 0 & 1 & 0 & \cdot & \cdot & \cdot \\ 0 & 0 & 0 & \cdot & \cdot & \cdot \\ \cdot & \cdot & \cdot & \cdot & \cdot & \cdot \end{pmatrix}, \quad \begin{pmatrix} 0 & 0 & 1 & \cdot & \cdot & \cdot \\ 0 & 0 & 0 & \cdot & \cdot & \cdot \\ \cdot & \cdot & \cdot & \cdot & \cdot & \cdot \end{pmatrix}$$

etc.

Also, the matrices in (4.1.6) form a basis in a $2 \times 2 = 4$ dimensional space.

In an n-dimensional vector space there are n linearly independent vectors. The choice of the basis is not unique. By linear combination of basis elements we can get a new basis (transformation of the basis). Having chosen a basis $\mathscr{B} = \{\boldsymbol{e}^{(i)}\}$, the x_i of the decomposition

$$\boldsymbol{x} = \sum_{i=1}^n x_i \boldsymbol{e}^{(i)} \tag{4.1.8b}$$

are the *components* of $\boldsymbol{x}$ with respect to $\mathscr{B}$. We write

$$\boldsymbol{x} = (x_1, x_2, \ldots, x_n) \; . \tag{4.1.8c}$$

If appropriate, we can look upon the $\boldsymbol{e}^{(i)}$ as being axes (or directions of axes) of a *coordinate system* which will then be chosen mutually orthogonal (see below).

Every space $\mathscr{L}_n$ can be represented as a *direct sum* of subspaces $\mathscr{L}^{(i)}$. If every $\boldsymbol{x} \in \mathscr{L}_n$ can be decomposed uniquely into a sum of vectors $\boldsymbol{x}^{(i)}$ with $i = 1, \ldots,$

$m \leqslant n$; $x^{(i)} \in \mathscr{L}^{(i)} \subseteq \mathscr{L}_n$, where the intersections of the $\mathscr{L}^{(i)}$ have only the zero vector $\mathbf{0} = 0$ in common, then $\mathscr{L}_n$ is the direct sum of the linearly independent $\mathscr{L}^{(i)}$,

$$\mathscr{L}_n = \mathscr{L}^{(1)} \oplus \mathscr{L}^{(2)} \oplus \cdots \oplus \mathscr{L}^{(m)} \; ; \qquad x = x^{(1)} + \cdots + x^{(m)} \; . \tag{4.1.9a}$$

Obviously

$$\mathbb{R}_3 = \mathbb{R}_2 \oplus \mathbb{R}_1 = \mathbb{R}_1 \oplus \mathbb{R}_1 \oplus \mathbb{R}_1 \; ; \qquad \mathbb{R}_2 = \mathbb{R}_1 \oplus \mathbb{R}_1 . \tag{4.1.9b}$$

In addition to the direct sum there is a direct product of vector spaces (tensor product). Let $\mathscr{L}_m^{(1)}$ and $\mathscr{L}_n^{(2)}$ be two vector spaces of dimension m and n, respectively, with bases $\{e^{(1,i)}|i = 1, \ldots, m\}$ and $\{e^{(2,j)}|j = 1, \ldots, n\}$. Then the space of the direct product (tensor space) $\mathscr{L}_m^{(1)} \otimes \mathscr{L}_n^{(2)}$ has dimension $m \cdot n$ and basis $e^{(ij)} := e^{(1,i)} \cdot e^{(2,j)}$. If the $\{e^{(i,\rho_i)}|\rho_i = 1, \ldots, n;\ i = 1, \ldots, m\}$ are the bases of m n-dimensional spaces[2], the basis of the m-fold direct product $\mathbb{R}_n^{(1)} \otimes \cdots \otimes \mathbb{R}_n^{(m)}$ is

$$e^{\rho_1 \cdots \rho_m} = \{e^{(1,\rho_1)} \cdot e^{(2,\rho_2)} \cdot \ldots \cdot e^{(m,\rho_m)} | \rho_i = 1, \ldots, n \; ; \; i = 1, \ldots, m\} \; . \tag{4.1.10}$$

This product space just defines the space of *tensors of m-th* rank (Sect. 5.5, 6.1). If $m = 2$ and $n = 2$ we can use the matrices of (4.1.6) as a basis $e^{\rho_1 \rho_2}$.

A linear space $\mathscr{L}$ over a field $\mathbb{C}$ is a unitary[3] (vector) space $\mathscr{H}$ if we add the definition of an inner (scalar) product. The inner product is defined as a mapping of a pair of two elements $x, y \in \mathscr{L}$ onto a complex number $c \in \mathbb{C}$,

$$(x, y) = \langle x|y \rangle = c \in \mathbb{C} \; . \tag{4.1.11}$$

With $\lambda \in \mathbb{C}$ we have

$$\begin{aligned} &\langle x|y\rangle = \langle y|x\rangle^* \; , && \langle \lambda x|y\rangle = \lambda^* \langle x|y\rangle \; , \\ &\langle x|y + z\rangle = \langle x|y\rangle + \langle x|z\rangle \; , && \langle x|x\rangle \geqslant 0 \; , \end{aligned} \tag{4.1.12}$$

where the equality holds if and only if x is the zero-element $x = 0$. From (4.1.12) it also follows that

$$\begin{aligned} &\langle x + y|z\rangle = \langle x|z\rangle + \langle y|z\rangle \; , \\ &\langle x|\lambda y\rangle = \lambda \langle x|y\rangle \; , && \langle 0|x\rangle = \langle x|0\rangle = 0 \; . \end{aligned} \tag{4.1.13}$$

Furthermore, if $\langle x|z\rangle = \langle y|z\rangle$ for all $z \in \mathscr{H}$, then $x = y$.

By means of the scalar product (4.1.11) we can define the norm of a vector in $\mathscr{H}$,

$$\|x\| = (\langle x|x\rangle)^{1/2} \geqslant 0 \; . \tag{4.1.14}$$

---

[2] The dimensions of the spaces may be different too: $n \to n_1, \ldots, n_m$.

[3] Unitary space is also called pre-Hilbert space. Every unitary space is a normalized and metric space as well, but not vice versa.

Thus the unitary space is normalized. The norm can be interpreted as the *distance of a point* x from the origin or as the length of a vector x. Two vectors $x, y \in \mathscr{H}$ are said to be *orthogonal* ($x \perp y$) if their scalar product vanishes

$$\langle x|y\rangle = 0 \curvearrowright x \perp y \ . \tag{4.1.15}$$

A basis $\{e^{(i)}\}$ of the unitary space is *orthonormalized* (is an orthonormal basis) if for all the basis vectors

$$\langle e^{(i)}|e^{(j)}\rangle = \delta_{ij} \ , \qquad i, j = 1, \ldots, n \ . \tag{4.1.16}$$

Such a basis can always be constructed from a linearly independent one (Schmidt). In $\mathbb{R}_n$, the vectors $\{e^{(i)}\} = \{(1, 0, \ldots, 0); (0, 1, \ldots, 0); \ldots ; (0, 0, \ldots, 1)\}$ form an orthonormal basis. The scalar product of two vectors $|y\rangle = \sum y_i|e^{(i)}\rangle$ and $\langle x| = \sum x_j^* \langle e^{(j)}|$ is then

$$\langle x|y\rangle = \sum_{i,j} x_j^* y_i \langle e^{(j)}|e^{(i)}\rangle = \sum_i x_i^* y_i \ . \tag{4.1.17a}$$

If $x_i, y_i \in \mathbb{R}$, then (4.1.17a) corresponds to the usual definition of a scalar product in Euclidean space ($\mathbb{R}_3$); (4.1.14) is the usual distance or length)

$$\|x\| = (\langle x|x\rangle)^{1/2} = \left(\sum_i x_i^2\right)^{1/2} \ . \tag{4.1.17b}$$

If we have to deal with components having two subscripts, see (4.1.19, 20), the scalar product is [see (4.1.20)]

$$\sum_{ij} a_{ij}^* b_{ij} = \sum_{ji} (a^+)_{ji} b_{ij} = \mathrm{Tr}\{a^+ b\} \tag{4.1.17c}$$

and the norm is

$$\|a\| = (\mathrm{Tr}\{a^+ a\})^{1/2} \ . \tag{4.1.17d}$$

4.1.2 Linear Operators

Now, we assume $\mathscr{L}$ and $\mathscr{L}'$ to be two vector spaces over the same field $\mathbb{K}$. Then a *mapping* $A\colon \mathscr{L} \to \mathscr{L}'$ is said to be *linear* or a *homomorphism* if for all $x, y \in \mathscr{L}$ and all $c \in \mathbb{K}$ [cf. (2.2.15)]

$$\begin{aligned} A(x + y) &= Ax + Ay \qquad &&\text{(additivity of } A) \ , \\ A(cx) &= cAx &&\text{(homogeneity of } A) \ . \end{aligned} \tag{4.1.18}$$

Here $\mathscr{L}'$ may be a subspace of $\mathscr{L}$ or the complete $\mathscr{L}$. The latter means that the mapping is an *endomorphism.* The set of *all* linear mappings $A\colon \mathscr{L} \to \mathscr{L}'$ is denoted by $\mathscr{M}(\mathscr{L}, \mathscr{L}')$. A mapping A is also called a *linear transformation* or a *linear operator.*

If $\{e^{(i)}\}$ is an orthonormal basis of $\mathscr{H}$ and $A: \mathscr{H} \to \mathscr{H}$ is a linear mapping $A \in \mathscr{M}(\mathscr{H}, \mathscr{H})$, then the numbers

$$a_{ik} := \langle e^{(i)} | Ae^{(k)} \rangle \in \mathbb{R} \text{ or } \mathbb{C} \tag{4.1.19}$$

are said to be a matrix representation of A in the basis $\{e^{(i)}\}$. The number

$$\mathrm{Tr}\{A\} = \sum_i a_{ii} = \sum_i \langle e^{(i)} | Ae^{(i)} \rangle \tag{4.1.20}$$

is the *trace of A*.

Taking $\mathscr{H} = \mathbb{R}_3$, let the mapping $d: \mathbb{R}_3 \to \mathbb{R}_3$ be a rotation about the z-axis through an angle φ of a Cartesian coordinate system with basis $\{e^{(i)}\} = \{(1, 0, 0); (0, 1, 0); (0, 0, 1)\}$. The representation of the operator d is thus given by the well-known rotation matrices

$$d_{ik} = \begin{pmatrix} \cos\varphi & -\sin\varphi & 0 \\ \sin\varphi & \cos\varphi & 0 \\ 0 & 0 & 1 \end{pmatrix}, \tag{4.1.21}$$

which transform the vector[4] $(x, 0, 0)$ into $(x\cos\varphi, x\sin\varphi, 0)$. The trace of this transformation (of matrix d_{ik}) is $\mathrm{Tr}\{d\} = 1 + 2\cos\varphi$.

Once we have chosen a fixed orthonormal (ON) basis in $\mathscr{H}$, the action of an operator $A \in \mathscr{M}(\mathscr{H}, \mathscr{H})$ on a vector $x \in \mathscr{H}$

$$y = Ax \in \mathscr{H} \tag{4.1.22}$$

(y is the image of x under A) can be described in matrix representation. Using the decomposition (4.1.8b) into components[5]

$$\langle e^{(j)} | y \rangle = \sum_i y_i \langle e^{(j)} | e^{(i)} \rangle = y_j \tag{4.1.23}$$

we have the y_j. With (4.1.19 and 23) we have from (4.1.22)

$$\begin{aligned} \langle e^{(i)} | y \rangle &= \sum_k \langle e^{(i)} | Ae^{(k)} \rangle \langle e^{(k)} | x \rangle \text{ or} \\ y_i &= \sum_k a_{ik} \cdot x_k \, . \end{aligned} \tag{4.1.24}$$

Here use has been made of the completeness of the basis: $1 = \sum |e^{(k)}\rangle\langle e^{(k)}|$.

A *unitary linear (basis) transformation* S requires the existence of an inverse S^{-1}. With its help the ON basis $\{e^{(i)}\}$ can be transformed into another one

[4] As usual, vectors are considered to be columns if placed to the right of a matrix, rows if placed to the left. In the text they are written as rows.

[5] For vectors we use x as well as $|x\rangle$, and sometimes (in $\mathbb{R}_3$) also $\boldsymbol{x}$. A misunderstanding should not occur. Often the Dirac notation $|x\rangle$ is convenient.

$\{e'^{(i)}\}$. In the new basis, the components $x, y \in \mathscr{H}$ as well as the operators A have a different representation.

With S: $\{|e^{(i)}\rangle\} \rightarrow \{|e'^{(i)}\rangle\}$ we have

$$|e'^{(i)}\rangle = \sum_j S_{ji}|e^{(j)}\rangle \qquad \text{with} \qquad S_{ji} = \langle e^{(j)}|e'^{(i)}\rangle \tag{4.1.25a}$$

and

$$|x\rangle = \sum_i x'_i|e'^{(i)}\rangle = \sum_{i,j} x'_i S_{ji}|e^{(j)}\rangle = \sum_j x_j|e^{(j)}\rangle$$

or

$$x_j = \sum_i S_{ji}x'_i \ . \tag{4.1.26a}$$

Since the $\{e^{(i)}\}$ as well as the $\{e'^{(i)}\}$ are assumed to be an ON basis,

$$\langle e'^{(i)}|e'^{(k)}\rangle = \left\langle \sum_j S_{ji}e^{(j)} \middle| \sum_l S_{lk}e^{(l)} \right\rangle \tag{4.1.27a}$$

$$= \sum_{jl} S^*_{ji} S_{lk}\langle e^{(j)}|e^{(l)}\rangle = \sum_j S^*_{ji} S_{jk} = \delta_{ik} \ ,$$

$$S^*_{ji} = (S^+)_{ij} = (S^{-1})_{ij} \ . \tag{4.1.28a}$$

The representation of an operator is

$$\begin{aligned} a'_{ik} = \langle e'^{(i)}|Ae'^{(k)}\rangle &= \left\langle \sum_j S_{ji}e^{(j)} \middle| A \sum_l S_{lk}e^{(l)} \right\rangle \\ &= \sum_{jl} (S^{-1})_{ij} a_{jl} S_{lk} \ . \end{aligned} \tag{4.1.29a}$$

Instead of (4.1.25a–29a), we can write for short

$$e' = eS \ , \tag{4.1.25b}$$

$$x = Sx' \qquad \text{or} \qquad x' = S^{-1}x = S^+x \ , \tag{4.1.26b}$$

$$S^+S = SS^+ = 1 \ , \tag{4.1.27b}$$

$$S^+ = S^{-1} \ , \tag{4.1.28b}$$

$$A' = S^{-1}AS = S^+AS \ . \tag{4.1.29b}$$

With the above assumptions the *transformation* S is a *unitary* one: $S^+ = S^{-1}$. The operators A and A' are *equivalent* (*similar*) to each other, see (4.1.41b). Under a transformation of the basis the trace of a linear operator remains unchanged:

$$\mathrm{Tr}\{A'\} = \mathrm{Tr}\{S^{-1}AS\} = \mathrm{Tr}\{SS^{-1}A\} = \mathrm{Tr}\{A\} , \tag{4.1.30}$$

since $\mathrm{Tr}\{AB\} = \mathrm{Tr}\{BA\}$.

4.1.3 Special Operators and Eigenvalues

If two operators $A, A^+ \in \mathscr{M}(\mathscr{H}, \mathscr{H})$ satisfy for all $x, y \in \mathscr{H}$

$$\langle x|Ay\rangle = \langle A^+x|y\rangle , \tag{4.1.31a}$$

then A^+ is the *adjoint* operator to A (and vice versa). After commutation of x with y we obtain with (4.1.12)

$$\langle y|Ax\rangle^* = \langle A^+y|x\rangle^* \qquad \text{or} \qquad \langle Ax|y\rangle = \langle x|A^+y\rangle . \tag{4.1.31b}$$

The representation of A^+ by matrices is obtained from (4.1.19)

$$\langle e^{(i)}|A^+e^{(k)}\rangle = \langle Ae^{(i)}|e^{(k)}\rangle = \langle e^{(k)}|Ae^{(i)}\rangle^* ,$$

consequently,

$$(a^+)_{ik} = a^*_{ki} = \tilde{a}^*_{ik} , \tag{4.1.32}$$

where $\tilde{a}$ is the *transpose matrix* of a.

In the special case of $A^+ = A$, *A is self-adjoint* or *Hermitian*. That is, using (4.1.32) we have

$$a_{ik} = \tilde{a}^*_{ik} = a^*_{ki} . \tag{4.1.33}$$

For real a_{ik}, the matrix is *symmetric*. If $A^+ = A^{-1}$, A is an *unitary* matrix: the adjoint operator is equal to the inverse one. Scalar products do not change under unitary transformations

$$\langle Ax|Ay\rangle = \langle x|A^+Ay\rangle = \langle x|A^{-1}Ay\rangle = \langle x|y\rangle . \tag{4.1.34}$$

Hermitian and unitary operators play a most important role in quantum mechanics, and also in other physical problems.

Vectors $|x\rangle \in \mathscr{L}$, $|x\rangle \neq 0$, which do not change their "direction" under a mapping $A|x\rangle$ with $A \in \mathscr{M}(\mathscr{L}, \mathscr{L})$, that is

$$A|x\rangle = \lambda|x\rangle ; \qquad \lambda \in \mathbb{K} , \tag{4.1.35}$$

are denoted as *eigenvectors* $|x\rangle = |x^{(\lambda)}\rangle$ of A with *eigenvalue* λ. Equation (4.1.35) is called the eigenvalue equation for the operator A.

The general solution of the eigenvalue problem (4.1.35) leads to a homogeneous linear system of equations

$$(A - \lambda e)|x\rangle = 0 , \tag{4.1.36a}$$

or written in components

$$\sum_k (a_{ik} - \lambda\delta_{ik})x_k = 0 \ . \tag{4.1.36b}$$

Such a system has nontrivial solutions $|x\rangle \neq 0$ only if the determinant vanishes

$$\det(A - \lambda e) := \det(a_{ik} - \lambda\delta_{ik}) = 0 \tag{4.1.36c}$$

(secular equation). Eigenvalues of Hermitian operators are real:

$$\langle x^{(\lambda')}|Ax^{(\lambda)}\rangle = \lambda\langle x^{(\lambda')}|x^{(\lambda)}\rangle = \langle Ax^{(\lambda')}|x^{(\lambda)}\rangle = \lambda'^*\langle x^{(\lambda')}|x^{(\lambda)}\rangle$$

or

$$(\lambda - \lambda'^*)\langle x^{(\lambda')}|x^{(\lambda)}\rangle = 0 \ . \tag{4.1.37}$$

Since $\|x^{(\lambda)}\| \neq 0$, this yields for $\lambda = \lambda'$ that $\lambda = \lambda^*$, i.e. λ is real, and for $\lambda \neq \lambda'$

$$\langle x^{(\lambda')}|x^{(\lambda)}\rangle = 0 \ .$$

This means that eigenvectors of Hermitian operators with different eigenvalues are mutually orthogonal. If some of them belong to the same eigenvalues, the corresponding eigenvectors can be orthogonalized (Schmidt). Eigenvalues of unitary operators are complex numbers of absolute value 1: Since $A|x^{(\lambda)}\rangle = \lambda|x^{(\lambda)}\rangle$, also $A^{-1}|x^{(\lambda)}\rangle = \lambda^{-1}|x^{(\lambda)}\rangle$, and since $A^+|x^{(\lambda)}\rangle = \lambda^*|x^{(\lambda)}\rangle$ and $A^{-1} = A^+$, we have

$$\lambda^{-1} = \lambda^* \text{ or } |\lambda| \equiv 1 \text{ or } \lambda = e^{i\varphi},\ \varphi \in \mathbb{R} \ . \tag{4.1.38a}$$

Eigenvectors of unitary operators belonging to different eigenvalues are also orthogonal, because with (4.1.34) and (4.1.12) it follows that

$$\begin{gathered} A|x^{(\lambda)}\rangle = \lambda|x^{(\lambda)}\rangle \quad \text{and} \quad A|x^{(\lambda')}\rangle = \lambda'|x^{(\lambda')}\rangle \ , \\ \langle Ax^{(\lambda')}|Ax^{(\lambda)}\rangle = \langle x^{(\lambda')}|x^{(\lambda)}\rangle = \lambda'^*\lambda\langle x^{(\lambda')}|x^{(\lambda)}\rangle \ . \end{gathered} \tag{4.1.38b}$$

Therefore, $\langle x^{(\lambda')}|x^{(\lambda)}\rangle = 0$ for $\lambda' \neq \lambda$.

As an example we consider $A = d$, i.e. a rotation in $\mathbb{R}_3$, see (4.1.21). The length of $\boldsymbol{x}$ is unchanged, $\langle dx|dx\rangle = \langle x|x\rangle$, therefore $d^+ = d^{-1}$ and d is unitary. Since d is also real, $(d_{ik})^+ = d^*_{ki} = d_{ki} = \tilde{d}_{ik}$, that is, $d^{-1} = \tilde{d}$ or $d\tilde{d} = e$. Real unitary matrices are called othogonal. From $\det(d\tilde{d}) = \det d \cdot \det \tilde{d} = (\det d)^2 = 1$, we have

$$\det d = \pm 1 \tag{4.1.39}$$

(Sect. 3.1.). All the eigenvalues have absolute value 1, according to (4.1.38). The secular equation $\det(d - \lambda e)$ is a cubic equation in λ with real coefficients, so all the roots λ are either real or form pairs of complex conjugates. Consequently,

similar to d is

$$\Lambda = \begin{pmatrix} 1 & 0 & 0 \\ 0 & e^{i\varphi} & 0 \\ 0 & 0 & e^{-i\varphi} \end{pmatrix} , \qquad 0 \leqslant \varphi \leqslant \pi . \tag{4.1.40}$$

The eigenvectors follow from (4.1.38) considering (4.1.40), e.g. for $\lambda_1 = 1$,

$$d|x^{(1)}\rangle = 1|x^{(1)}\rangle .$$

This means $|x^{(1)}\rangle$ is not changed by the rotation d. It thus has the direction of the rotation axis (Exercise 4.1).

By means of the eigenvectors, d can be brought into the diagonal form (4.1.40). Equation (4.1.38b) gives

$$\sum_j d_{ij} x_j^{(k)} = \lambda^{(k)} x_i^{(k)} = \sum_l \lambda^{(k)} x_i^{(l)} \delta_{lk} , \tag{4.1.41a}$$

where k, l number the eigenvalues and eigenvectors, and i, j number the components of the vectors in an arbitrary basis. With $x_i^{(l)} = S_{il}$; $\Lambda_{lk} = \lambda^{(k)} \delta_{lk}$, it follows from (4.1.41a) that

$$dS = S\Lambda \qquad \text{or} \qquad \Lambda = S^{-1} dS , \tag{4.1.41b}$$

i.e. d is mapped onto the diagonal form Λ by a similarity transformation S, see (4.1.29b). In the above case [d according to (4.1.21)],

$$S = \begin{pmatrix} 0 & 1/\sqrt{2} & 1/\sqrt{2} \\ 0 & -i/\sqrt{2} & i/\sqrt{2} \\ 1 & 0 & 0 \end{pmatrix} . \tag{4.1.41c}$$

The eigenvectors are complex, as the elements of d and Λ can be. Applying a further transformation S', (4.1.40) leads to an other real representation of d, e.g.

$$d' = S'^{-1} \Lambda S' , \qquad \Lambda = S^{-1} dS , \tag{4.1.42}$$

$$S' = \begin{pmatrix} 1 & 0 & 0 \\ 0 & 1/\sqrt{2} & -i/\sqrt{2} \\ 0 & 1/\sqrt{2} & i/\sqrt{2} \end{pmatrix} , \qquad d' = \begin{pmatrix} 1 & 0 & 0 \\ 0 & \cos\varphi & \sin\varphi \\ 0 & -\sin\varphi & \cos\varphi \end{pmatrix} .$$

This again describes a rotation through an angle φ about an axis belonging to $\lambda^{(1)}$. The rotation axis is now the x-axis, whereas in (4.1.21) it was the z-axis. From (4.1.42) we have

$$d' = S'^{-1} S^{-1} dSS' = (SS')^{-1} dSS' , \tag{4.1.43}$$

that is, a rotation about the z-axis can be transformed into a rotation about the

x-axis (through the same angle φ) by an orthogonal similarity transformation. The rotations d and d' are connected by conjugation. This means for the elements of $\mathscr{SO}(3)$ that all rotations through an angle φ with arbitrary directions of the rotation axes are members of the same class. Since $d(\boldsymbol{n}, -\varphi) = d(-\boldsymbol{n}, \varphi)$, also all rotations with the same $|\varphi|$ are in one class. However, for double groups, $d(\boldsymbol{n}, \varphi)$ and $c_0 d(\boldsymbol{n}, \varphi) = d(\boldsymbol{n}, \varphi \pm 2\pi)$ are members of the same class only if $|\varphi| = |\varphi \pm 2\pi|$, which means that $\varphi = \mp\pi$. Thus the only possible class is the one with the elements $c(\pi) = c_2$ (Sect. 3.3).

Exercise 4.1. Calculate the eigenvectors of (4.1.21) with the eigenvalues as in (4.1.40) and express the rotation angle φ in terms of the matrix elements of d. Show that $|x^{(2)}\rangle = |x^{(3)}\rangle^*$ and $\langle x^{(i)}|x^{(j)}\rangle = \delta_{ij}$.

Exercise 4.2. Show that each improper orthogonal matrix d possesses one eigenvalue $\lambda = -1$.

4.2 Introduction to the Theory of Representations

4.2.1 Operator Representations by Matrices

Applying the methods developed in Sect. 4.1, the linear representations $D(\mathscr{G})$ of a group $\mathscr{G}$ can now be defined: A *representation* $D(\mathscr{G})$ is a homorphism of $\mathscr{G}$ onto a group of nonsingular linear operators P that map a linear space $\mathscr{L}$ onto itself.

$$D(\mathscr{G}): a \in \mathscr{G} \rightarrow P(a) \equiv P_a \ . \tag{4.2.1}$$

The space $\mathscr{L}$ is called the *space of the representation* of $\mathscr{G}$. A representation $D(\mathscr{G})$ which maps $\mathscr{G}$ isomorphically (one to one) onto operators P is said to be faithful. The trivial representation (REP) in which every element of $\mathscr{G}$, i.e. the complete $\mathscr{G}$, is mapped onto the unit operator, is called the *identity representation* (*unit* REP).

The dimension of the representation space $\mathscr{L}$ is the *dimension of the representation.* If we introduce a basis in $\mathscr{L}$, then the operators P_a are defined by their *matrix representation* (4.1.19); therefore, if $\dim \mathscr{L} = n$, we have a homomorphism of $\mathscr{G}$ onto the $(n \times n)$ matrices that belong to P_a. This is the most appropriate case with respect to applications and we will mainly discuss this kind of REP. However, for general considerations the definition in terms of operators in a linear space is often more advantageous, since it is independent of a basis. In the following applications, $\mathscr{L}$ is always a space of functions or vectors (in the sense of displacements, indicated by arrows). We have already treated an example of a vector space in (2.1.18), where $\mathscr{G} = \mathscr{C}_{3v}$, $\mathscr{L} = \mathbb{R}_2$ and $D(\mathscr{G})$ was a faithful REP by (2×2) matrices. The basis was Cartesian. A generalization to a space $\mathscr{L} = \mathbb{R}_n$ is obvious. Instead of the 2-tuple (x_1, x_2) we have to take the n-tuple $(x_1, x_2, \ldots, x_n)$, which means vectors in $\mathbb{R}_n$. The operators of the REP are then $(n \times n)$ matrices with a Cartesian basis. Consequently a vector $\boldsymbol{x} \in \mathbb{R}_n$ is mapped onto a

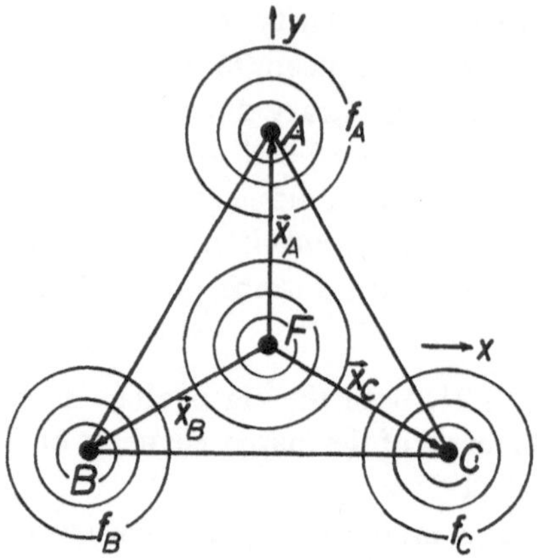

Fig. 4.1. Functions that transform according to $\mathscr{C}_{3v}$. The concentric circles indicate contour lines of equal values of the functions. As appropriate functions we can choose for example $f_i = \exp[-(\boldsymbol{x} - \boldsymbol{x}_i)^2]$, $i = A, B, C$ and $F = \exp(-x^2 - y^2) = \exp(-|\boldsymbol{x}|^2)$

vector $\boldsymbol{x}' \in \mathbb{R}_n$ by the operator P_a with $a \in \mathscr{G}$, which is written as

$$x_i' = \sum_{k=1}^{n} (P_a)_{ik} x_k \equiv \sum_{k=1}^{n} a_{ik} x_k \;, \qquad i = 1, 2, \ldots, n \;. \tag{4.2.2}$$

If $\mathscr{L}$ is a linear space of functions (e.g. the Hilbert space of quantum-mechanical states) the action of P_a with $a \in \mathscr{G}$ on functions has to be defined: a function $f'(\boldsymbol{x})$ is assigned to every function $f(\boldsymbol{x})$ with $\boldsymbol{x} \in \mathbb{R}_n$, via the operator P_a according to

$$P_a f(\boldsymbol{x}) := f'(\boldsymbol{x}) := f(a^{-1}\boldsymbol{x}) \;, \qquad a \in \mathscr{G} \;, \qquad \boldsymbol{x} \in \mathbb{R}_n \;. \tag{4.2.3}$$

The new function $P_a f(\boldsymbol{x}) = f'(\boldsymbol{x})$ assumes for $\boldsymbol{x}$ the value that is assigned to f for $a^{-1}\boldsymbol{x}$. If the function f is centered at the particle A, the f' "belongs" to a particle B, obtained by transformation of A by a. For example, in $\mathscr{C}_{3v}$ symmetry we have for a function $f_A(\boldsymbol{x})$

$$P_{c_3} f_A(\boldsymbol{x}) = f_B(\boldsymbol{x}) = f_A(c_3^{-1}\boldsymbol{x}) \;, \qquad \text{i.e. } P_{c_3} f_A = f_B \;.$$

Contour lines transform point-wise like vectors (Fig. 4.1). Correspondingly, we can verify

$$\begin{gathered}
P_e \begin{cases} f_A & f_A \\ f_B = f_B \\ f_C & f_C \end{cases} \quad
P_{c_3} \begin{cases} f_A & f_B \\ f_B = f_C \\ f_C & f_A \end{cases} \quad
P_{c_3^2} \begin{cases} f_A & f_C \\ f_B = f_A \\ f_C & f_B \end{cases} \\
P_{\sigma_v} \begin{cases} f_A & f_A \\ f_B = f_C \\ f_C & f_B \end{cases} \quad
P_{\sigma_v'} \begin{cases} f_A & f_C \\ f_B = f_B \\ f_C & f_A \end{cases} \quad
P_{\sigma_v''} \begin{cases} f_A & f_B \\ f_B = f_A \\ f_C & f_C \end{cases}
\end{gathered} \tag{4.2.4}$$

This example makes it clear that the functions $\{f_A, f_B, f_C\}$ are transformed into each other by the operations of $\mathscr{C}_{3v}$. The function space $\mathscr{L}$ is *invariant* under the operations of the group $\mathscr{G}$. The operators P_a, $a \in \mathscr{G}$ constitute a faithful representation of $\mathscr{G}$. Application of two operators yields

$$P_b f(\boldsymbol{x}) = f'(\boldsymbol{x}) \qquad = f(b^{-1}\boldsymbol{x})$$

$$f'(a^{-1}\boldsymbol{x}) = f(b^{-1}a^{-1}\boldsymbol{x})$$

$$P_a f'(\boldsymbol{x}) = f'(a^{-1}\boldsymbol{x}) = f(b^{-1}a^{-1}\boldsymbol{x}) = f((ab)^{-1}\boldsymbol{x}) = P_{ab} f(\boldsymbol{x}) \ ,$$

therefore

$$P_a P_b f(\boldsymbol{x}) = f((ab)^{-1}\boldsymbol{x}) = P_{ab} f(\boldsymbol{x})$$

and finally

$$P_a P_b = P_{ab} \ . \tag{4.2.5}$$

Equation (4.2.5) necessarily requires the definition (4.2.3). Only in this way can the composition law (multiplication) of the group elements be transferred to the operators.

The matrices of the representation with respect to the functions of $\mathscr{L}$ can now be constructed. Let $\{f_1, \ldots, f_n\}$ be a basis in the function space $\mathscr{L}$, i.e. n linearly independent functions, then $P_a f_i$ with $a \in \mathscr{G}$ is necessarily a linear combination of the functions f_i, since $\mathscr{L}$ is invariant. Therefore

$$P_a f_i = \sum_{j=1}^{n} D_{ji}(a) f_j \qquad \text{with } i = 1, \ldots, n = \dim \mathscr{L} \tag{4.2.6}$$

and all the g $(= \operatorname{ord} \mathscr{G})$ elements $a \in \mathscr{G}$. The order ji in (4.2.6) is essential! The matrix $D(a)$ describes the action of P_a within the basis $\{f_i\}$. Thus the matrices $D(a)$ for every $a \in \mathscr{G}$, or P_a, constitute a (not necessarily faithful) matrix REP of $\mathscr{G}$:

$$D(\mathscr{G}) = \{D(a) | a \in \mathscr{G}\} \ . \tag{4.2.7}$$

The composition of two matrices is their multiplication

$$D(a) \cdot D(b) = D(ab) \ , \tag{4.2.8a}$$

because, using (4.2.6) (P_a linear),

$$P_a P_b f_i = P_a \sum_{j=1}^{n} D_{ji}(b) f_j = \sum_{j=1}^{n} D_{ji}(b) P_a f_j$$

$$= \sum_{j=1}^{n} \sum_{l=1}^{n} D_{ji}(b) D_{lj}(a) f_l = \sum_{l=1}^{n} D_{li}(ab) f_l = P_{ab} f_i \ .$$

In particular, we have

Table 4.1. Representations of the group $\mathscr{C}_{3v}$ with two different basis systems

a	e	c_3	c_3^2	σ_v	σ_v'	σ_v''
$D(a)$	$\begin{pmatrix}1&0&0\\0&1&0\\0&0&1\end{pmatrix}$	$\begin{pmatrix}0&0&1\\1&0&0\\0&1&0\end{pmatrix}$	$\begin{pmatrix}0&1&0\\0&0&1\\1&0&0\end{pmatrix}$	$\begin{pmatrix}1&0&0\\0&0&1\\0&1&0\end{pmatrix}$	$\begin{pmatrix}0&0&1\\0&1&0\\1&0&0\end{pmatrix}$	$\begin{pmatrix}0&1&0\\1&0&0\\0&0&1\end{pmatrix}$
$D^{(1)}(a)$	1	1	1	1	1	1

Fig. 4.2. Vector diagram invariant under all operations of $\mathscr{C}_{3v}$. It transforms according to the identity REP $D^{(1)}(a)$ of $\mathscr{C}_{3v}$

$$D(a^{-1}) = D^{-1}(a) \ . \tag{4.2.8b}$$

The functions $\{f_i\}$ set up a basis for the *vector* REP $D(\mathscr{G})$. Any function f_j transforms according to the jth row of $D(\mathscr{G})$, see (4.2.6). Matrix REPs which satisfy (4.2.8a) are known as vector REPs. This has to be compared with the projective REPs in Sect. 9.2.1.

In our example in (4.2.4) the functions f_A, f_B, f_C are linearly independent and thus permitted to be taken as a basis. The corresponding matrix REP is defined by (4.2.4). So, using (4.2.6), it obviously follows that

$$D(c_3) = \begin{pmatrix}0&0&1\\1&0&0\\0&1&0\end{pmatrix} .$$

The complete REP of the group $\mathscr{C}_{3v}$ within this basis system is then $D(a)$ as given in Table 4.1. However, if we choose the function $F(x)$ of Fig. 4.1 as a basis, then $\dim \mathscr{L}^{(F)} = 1$. The function remains invariant against all transformations

$$P_a F(x) = F(a^{-1}x) = F(x) \ .$$

The REP matrices consist of (the number) 1 for all the elements of $\mathscr{C}_{3v}$. In Table 4.1 this identity REP is denoted by $D^{(1)}(a)$. It is a (totally) *symmetric* REP which exists for all (finite) groups.

Often the linear space $\mathscr{L}$ is a (displacement) vector space. This is so in the case of vibrations of molecules and lattices where the vectors just represent the displacements of the particles from their equilibrium positions (Sects. 7.1, 10.2). A vector diagram corresponding to the identity REP is given in Fig. 4.2. It is invariant under all operations of $\mathscr{C}_{3v}$.

4.2.2 Equivalent Representations and Characters

We have seen in (4.1.29b) that on changing the basis in $\mathscr{L}$ by a nonsingular matrix S, an operator P in $\mathscr{L}$ transforms according to $P' = S^{-1}PS$. This leads us to the following definition:

Two REPs $D = \{P_a | a \in \mathscr{G}\}$ and $D' = \{P'_a | a \in \mathscr{G}\}$ of a group $\mathscr{G}$ are *equivalent* or *similar*, if for every P_a and P'_a there exists a nonsingular operator S such that $P'_a = S^{-1}P_aS$ for each $a \in \mathscr{G}$. (4.2.9)

Let $\{f'_i\}$ be the new basis. Then in the matrix REP we have, using (4.1.25a) and (4.2.6)

$$f'_i = \sum_j S_{ji} f_j \ ,$$

$$P_a f'_i = \sum_j S_{ji} P_a f_j = \sum_{jk} S_{ji} D_{kj}(a) f_k$$

$$= \sum_{jkl} S_{ji} D_{kj}(a)(S^{-1})_{lk} f'_l = \sum_l (S^{-1}D(a)S)_{li} f'_l \ ,$$

therefore

$$D'(a) = S^{-1}D(a)S \qquad \text{for each } a \in \mathscr{G} \ . \tag{4.2.10}$$

Since this is an equivalence relation, according to (2.2.2, 3) we can separate all the REPs of a group into classes of equivalent REPs and choose *one* representative from each of these classes. Then the only group-theoretical problem we are left with is the search for all the *inequivalent* REPs.

We now have the following theorem:

Any class of equivalent REPs of a finite group contains at least one *unitary* REP, or: Any REP of a finite group is equivalent to a unitary one. (4.2.11a)

Here a unitary REP means a REP by unitary operators as in Sect. 4.1.3. For a proof we consider the positive definite Hermitian matrix $H = \sum_{a \in \mathscr{G}} D(a)D^+(a)$ with $\det H \neq 0$. The matrix H can be diagonalized by a unitary matrix U ($U^+U = 1$),

$$\Lambda = U^{-1}HU = \sum_a D'(a)D'^+(a) \qquad \text{with} \qquad D'(a) = U^{-1}D(a)U \ .$$

Therefore

$$\Lambda_i \delta_{ij} = \sum_{a,k} D'_{ik} D'^*_{jk} \qquad \text{or} \qquad \Lambda_i = \sum_{a,k} |D'_{ik}|^2 > 0 \ ,$$

since $\det \Lambda = \det H \neq 0$. Consequently, $\sqrt{\Lambda_i}$ exists and so does $S = U\Lambda^{1/2}$. The matrix S transforms $D(a)$ into a unitary REP because

$$D^u(a)D^{u+}(a) = \Lambda^{-1/2}U^{-1}D(a)U\Lambda U^{-1}D^+(a)U\Lambda^{-1/2}$$
$$= \Lambda^{-1/2}U^{-1}\sum_{b\in\mathscr{G}} D(a)D(b)D^+(b)D^+(a)U\Lambda^{-1/2}$$
$$= \Lambda^{-1/2}U^{-1}\sum_{b\in\mathscr{G}} D(ab)D^+(ab)U\Lambda^{-1/2}$$
$$= \Lambda^{-1/2}\Lambda\Lambda^{-1/2} = 1 ,$$

since with b, ab also assumes all the elements of $\mathscr{G}$. The transformation matrix S can always be found by construction. In the following we therefore restrict ourselves to unitary REPs (in the case of finite or compact groups, Sect. 11.2).

For the extension of theorem (4.2.11) to continuous groups we have to define two further terms.

We denote as *characters of a* REP the numbers

$$\chi(a) = \mathrm{Tr}\{P_a\} \qquad \text{for each } a \in \mathscr{G} . \tag{4.2.12a}$$

Having chosen a basis $\{e^{(i)}\}$, the character

$$\chi(a) = \mathrm{Tr}\{D(a)\} = \sum_{i=1}^{n} \langle e^{(i)}|P_a|e^{(i)}\rangle = \sum_{i=1}^{n} D_{ii}(a) \tag{4.2.12b}$$

is thus the sum of the diagonal elements of the REP matrices. Obviously

$$\chi(e) = \dim \mathscr{L} = \mathrm{rank}\, D(e) \qquad \text{and} \tag{4.2.12c}$$

$$\chi(b) = \chi(a) \qquad \text{if } b = cac^{-1} , \tag{4.2.12d}$$

i.e. all the elements of a (conjugacy) class have the same character within a definite REP.

We furthermore define a *function on a group* as a unique mapping of a group $\mathscr{G}$ onto a set of complex or real numbers. As an example we may consider the mapping $\varphi\colon \mathscr{G} \to \mathbb{C}$ with $\varphi(a) = \langle x|P_a|y\rangle$, where $|x\rangle$, $|y\rangle$ are vectors of the REP space $\mathscr{L}$ of P_a with $a \in \mathscr{G}$. Another $\varphi(a)$ would be $\varphi(a) = \int f_i(\boldsymbol{x})P_a f_j(\boldsymbol{x})\, d^n\boldsymbol{x}$ with the functions $f_i(\boldsymbol{x})$ of a space $\mathscr{L}$. The character $\chi(a)$ of a REP $D(a)$ is also a function on $\mathscr{G}$. Now, if φ is a function on $\mathscr{G}$, then a linear mapping $M(\varphi)$ of φ onto the complex or real numbers is said to be a *mean value on the group* $\mathscr{G}$ if for all the $a, b \in \mathscr{G}$ it holds that:

(i) if $\varphi(a) > 0$, then also $M(\varphi) > 0$,
(ii) if $\varphi(a) = 1$, then also $M(\varphi) = 1$, (4.2.13)
(iii) if $\varphi_1(a) = \varphi(ab)$ and $\varphi_2(a) = \varphi(ba)$, then also $M(\varphi_1) = M(\varphi_2) = M(\varphi)$.

The most important mean value for finite groups is

$$M(\varphi) = \frac{1}{g}\sum_{a\in\mathscr{G}} \varphi(a) , \tag{4.2.14}$$

in which $\varphi(a)$, for example, can be replaced by the character $\chi(a)$ too. Replacing the sum by an integral, (4.2.14) can be extended to continuous groups (see also Sect. 11.1.2). Now, by means of (4.2.14), we can reformulate theorem (4.2.11a):

> If it is possible to define a mean value on a group $\mathscr{G}$, then every class of equivalent REPs contains a least one unitary REP. (4.2.11b)

The proof is similar to that given for (4.2.11a).

4.2.3 Reducible and Irreducible Representations

A representation $D(\mathscr{G})$ in the REP space $\mathscr{L}$ is said to be *reducible* if $\mathscr{L}$ possesses at least one nontrivial subspace $\mathscr{L}' \subset \mathscr{L}$, i.e. $\mathscr{L}' \neq \{0\}$ and $\mathscr{L}' \neq \mathscr{L}$, which is mapped onto itself by all the operators P_a of the REP and is thus invariant. If there is no such subspace, the REP is said to be *irreducible*. Finally, if $\mathscr{L}$ can be decomposed into a direct sum of invariant subspaces as in (4.1.9a)

$$\mathscr{L} = \sum_{\alpha} \oplus \mathscr{L}^{(\alpha)} , \tag{4.2.15}$$

such that the REP is irreducible on every $\mathscr{L}^{(\alpha)}$, then the REP is said to be completely reducible. These concepts can be illustrated best by a matrix REP $D(a)$. Let the linear space be $\mathscr{L} = \mathscr{L}^{(1)} \oplus \mathscr{L}^{(2)}$, which means that every $\boldsymbol{x} \in \mathscr{L}$ can be written as

$$\boldsymbol{x} = \begin{pmatrix} \boldsymbol{x}^{(1)} \\ \boldsymbol{x}^{(2)} \end{pmatrix} \quad \text{with} \quad \begin{matrix} \boldsymbol{x}^{(1)} \in \mathscr{L}^{(1)} \\ \boldsymbol{x}^{(2)} \in \mathscr{L}^{(2)} \end{matrix} .$$

The matrix REP then is reducible if it can be transformed into the form

$$D(a) = \begin{pmatrix} D^{(1)}(a) & 0 \\ N(a) & D^{(2)}(a) \end{pmatrix} \tag{4.2.16}$$

for each $a \in \mathscr{G}$, where again $D^{(1)}$, $D^{(2)}$ and N are matrices (including one-dimensional ones). If $D^{(1)}$ and $D^{(2)}$ are irreducible, we have the wanted decomposition into invariant subspaces. If not, the procedure has to be continued. Since

$$D(a) \cdot \boldsymbol{x} = \begin{pmatrix} D^{(1)}(a) \cdot \boldsymbol{x}^{(1)} \\ N(a) \cdot \boldsymbol{x}^{(1)} + D^{(2)}(a) \cdot \boldsymbol{x}^{(2)} \end{pmatrix} ,$$

we realize that $D(a)$ transforms the subspace $\mathscr{L}^{(2)}$ into itself: $\mathscr{L}^{(2)}$ is an invariant subspace. Furthermore, if $N(a) = 0$ (matrix) for each $a \in \mathscr{G}$, then both subspaces are invariant under $\mathscr{G}$. For $\mathscr{G} = \mathscr{C}_i$ and $\mathscr{L}$ as above, we have for example

$$D(e) = \begin{pmatrix} 1 & 0 \\ 0 & 1 \end{pmatrix} , \quad D(i) = \begin{pmatrix} 0 & -1 \\ -1 & 0 \end{pmatrix} .$$

A transformation to

$$\begin{pmatrix} x^{(+)} \\ x^{(-)} \end{pmatrix} = \frac{1}{\sqrt{2}} \begin{pmatrix} x^{(1)} + x^{(2)} \\ x^{(1)} - x^{(2)} \end{pmatrix} = S \begin{pmatrix} x^{(1)} \\ x^{(2)} \end{pmatrix} , \quad S = \frac{1}{\sqrt{2}} \begin{pmatrix} 1 & 1 \\ 1 & -1 \end{pmatrix} ,$$

yields the reduced REP $D' = S^{-1}DS$ with

$$D'(e) = \begin{pmatrix} 1 & 0 \\ 0 & 1 \end{pmatrix} , \quad D'(i) = \begin{pmatrix} -1 & 0 \\ 0 & 1 \end{pmatrix} .$$

For all groups possessing a mean value the following theorem can be proved [see (11.2.13)ff.].

Every reducible (finite) REP $D(\mathcal{G})$ can be decomposed completely into irreducible parts (can be reduced completely). (4.2.17)

If $D(\mathcal{G})$ is a unitary REP (4.2.11), then $D(\mathcal{G})$ can be decomposed into unitary irreducible REPs $D^{(\alpha)}(\mathcal{G})$, to which orthogonal irreducible subspaces $\mathcal{L}^{(\alpha)}$ belong, i.e.

$$\mathcal{L} = \sum_{\alpha} \oplus \mathcal{L}^{(\alpha)} , \quad D(\mathcal{G}) = \sum_{\alpha} \oplus D^{(\alpha)}(\mathcal{G}) , \quad D^{(\alpha)} \text{ irreducible.} \tag{4.2.18}$$

The decomposition into irreducible parts can again be explained best by means of a matrix REP. The transformation of one REP

$$D(\mathcal{G}) = \{D(a) | a \in \mathcal{G}; \text{ basis } e^{(i)}\}$$

into another

$$D'(\mathcal{G}) = \{D'(a) = S^{-1}D(a)S | a \in \mathcal{G}; \text{ basis } e'^{(k)}\}$$

is described by a nonsingular matrix S with

$$e'^{(k)} = \sum_i S_{ik} e^{(i)} .$$

We choose S such that $D'(\mathcal{G})$ is completely reduced

$$D'(a) = \begin{pmatrix} D^{(1)}(a) & 0 & \cdots & 0 \\ 0 & D^{(2)}(a) & \cdots & 0 \\ \vdots & \vdots & & \vdots \\ 0 & 0 & \cdots & D^{(m)}(a) \end{pmatrix} \quad \text{for each } a \in \mathcal{G} . \tag{4.2.19}$$

Such an S always exists as a consequence of (4.2.17, 18). The action of P_a is given by the completely reduced matrices $D'(a)$ in the basis $\{e'(k)\}$. Every "block matrix" $D^{(\alpha)}$ in the main diagonal of D' represents the action of P_a in the corresponding invariant subspace $\mathcal{L}^{(\alpha)}$, which is defined by the basis functions (vectors) $\{e_1'^{(\alpha)}, \ldots, e_{d_\alpha}'^{(\alpha)}\}$, with

$$d_\alpha = \dim \mathscr{L}^{(\alpha)} . \tag{4.2.20}$$

A certain irreducible REP $D^{(\alpha)}(\mathscr{G})$ may occur in (4.2.19) repeatedly. If the multiplicity of $D^{(\alpha)}$ in D' or D is m_α, then

$$D(\mathscr{G}) \sim D'(\mathscr{G}) = \sum_\alpha{}_\oplus \, m_\alpha D^{(\alpha)}(\mathscr{G}) , \tag{4.2.21a}$$

and $\mathscr{L}$ contains m_α subspaces of equal quality which are invariant under P_a, which means

$$P_a e_j'^{(\alpha)} = \sum_i D_{ij}^{(\alpha)}(a) \cdot e_i'^{(\alpha)} . \tag{4.2.21b}$$

The inequivalent irreducible REPs or subspaces are of fundamental importance for REP theory. They are the pieces from which *all the* REPs of a group can be constructed. In the following discussions we can restrict ourselves to these *inequivalent, irreducible, unitary* REPs (IR) of a group.

The reduction is illustrated using the group $\mathscr{C}_{3v}$ again. The functions $\{f_A, f_B, f_C\}$ constitute a basis for a REP. Is it possible to decompose this space $\mathscr{L} = \{f_A, f_B, f_C\}$ into irreducible subspaces? We choose new basis functions

$$f_1' = f_A + f_B + f_C , \qquad f_2' = \sqrt{3}(f_B - f_C) , \qquad f_3' = -2f_A + f_B + f_C ,$$

which are evidently linearly independent, too. But now f_1' is invariant under all the operations of $\mathscr{C}_{3v}$, since these merely permute the functions f_A, f_B, f_C, see (4.2.4), which means that f_1' defines an invariant subspace $\mathscr{L}^{(1)}$ belonging to the identity REP $D^{(1)}(\mathscr{C}_{3v}) \equiv A_1$ and in which all elements are mapped onto the identity element (Table 4.1).

This function f_1' has the same properties as F in Fig. 4.1. Obviously it cannot be reduced any further. The functions f_2' and f_3' define a further invariant subspace. Using (4.2.4) we obtain

$$P_{c_3} f_2' = \sqrt{3}(f_C - f_A) \quad = -\frac{1}{2} f_2' + \frac{\sqrt{3}}{2} f_3' ,$$

$$P_{c_3} f_3' = -2f_B + f_C + f_A = -\frac{\sqrt{3}}{2} f_2' - \frac{1}{2} f_3' ,$$

or

$$P_{\sigma_v} f_2' = \sqrt{3}(f_C - f_B) \quad = -f_2' ,$$

$$P_{\sigma_v} f_3' = -2f_A + f_C + f_B = +f_3' .$$

The REP $D^{(3)}(\mathscr{C}_{3v}) \equiv E$ therefore has the structure (2.1.18)

$$D^{(3)}(c_3) = \begin{pmatrix} -1/2 & -\sqrt{3}/2 \\ \sqrt{3}/2 & -1/2 \end{pmatrix}, \qquad D^{(3)}(\sigma_v) = \begin{pmatrix} -1 & 0 \\ 0 & 1 \end{pmatrix}$$

and correspondingly for the other operators. In the invariant subspace with basis $\{f_2', f_3'\}$ the REP matrices have the form (2.1.18). We will show later, that $D^{(3)}$ is irreducible.

Using the basis $\{f_1', f_2', f_3'\}$. $D(\mathscr{C}_{3v})$ is completely reduced: $D'(\mathscr{C}_{3v}) = D^{(1)} \oplus D^{(3)}$. The invariant subspaces are constituted by f_1' and $\{f_2', f_3'\}$, respectively: $\mathscr{L} = \mathscr{L}^{(1)} \oplus \mathscr{L}^{(3)}$. In this case the transformation matrix is

$$S = \begin{pmatrix} 1 & 0 & -2 \\ 1 & \sqrt{3} & 1 \\ 1 & -\sqrt{3} & 1 \end{pmatrix} .$$

4.2.4 Orthogonality Theorems

Some theorems on orthogonality can be derived starting with the two lemma's of Schur. We state these theorems here without proof.

Schur's Lemma I. If A is a linear operator ($\neq 0$) that maps the REP space $\mathscr{L}^{(\alpha)}$ of an IR $D^{(\alpha)}(\mathscr{G})$ onto itself and that commutes with all operators $P_a^{(\alpha)}$ of the IR, $AP_a^{(\alpha)} = P_a^{(\alpha)}A$, then A is a multiple of the unit operator, i.e.

$$A = \lambda e , \qquad \lambda \in \mathbb{C} . \tag{4.2.22}$$

Referring to REP matrices this statement can also be expressed as: Any matrix A_{ik} that commutes with all the matrices of an IR is a multiple of the unit matrix, i.e. $A_{ik} = \lambda\delta_{ik}$. The converse is true as well: If the only matrix that commutes with all the matrices of a REP is equal to $\lambda\delta_{ik}$, then this REP is irreducible.

Schur's Lemma II. Let $D^{(\alpha)}(\mathscr{G})$ and $D^{(\beta)}(\mathscr{G})$ be two nonequivalent (equivalent) IRs of $\mathscr{G}$ with REP spaces $\mathscr{L}^{(\alpha)}$ and $\mathscr{L}^{(\beta)}$ and the corresponding operators $P_a^{(\alpha)}$ and $P_a^{(\beta)}$; let A be a linear operator that maps $\mathscr{L}^{(\beta)}$ onto $\mathscr{L}^{(\alpha)}$. If now $P_a^{(\alpha)}A = AP_a^{(\beta)}$ for each $a \in \mathscr{G}$, then A is the null or zero operator (or non-singular, resp.). (4.2.23)

The set of functions (see end of Sect. 4.2.2) on $\mathscr{G}$

$$\{D_{ij} | D_{ij}: \mathscr{G} \to \mathbb{C} \quad \text{with} \quad D_{ij}(a) = \langle f_i | P_a | f_j \rangle \; ; \quad a \in \mathscr{G}\} \tag{4.2.24}$$

is said to be a *system of coefficients* of the REP which is defined by (4.2.6). The D_{ij} are the matrix elements with respect to the orthonormal basis $\{f_i\}$. The coefficients of two IRs $D^{(\alpha)}$ and $D^{(\beta)}$ of a group $\mathscr{G}$ are connected by the *orthogonality theorem*

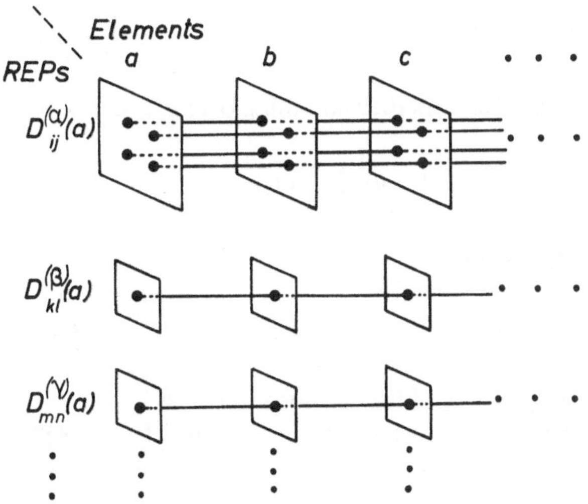

Fig. 4.3. Illustration of the orthogonality theorems (4.2.25a, b). All the vectors constructed from the matrix elements connected by horizontal lines are mutually orthogonal, and so are the vectors constructed from matrix elements belonging to one element $a \in \mathcal{G}$ (columns)

$$\sum_{a \in \mathcal{G}} D_{ij}^{(\alpha)}(a) \cdot D_{kl}^{(\beta)-1}(a) = \frac{g}{d_\alpha} \delta_{il} \delta_{jk} \delta_{\alpha\beta} \tag{4.2.25a}$$

where $g = \operatorname{ord} \mathcal{G}$ and $d_\alpha = \dim \mathcal{L}^{(\alpha)}$ or $D^{(\alpha)}$, as appropriate.[6] In unitary REPs D^{-1} can be replaced equivalently by

$$(D^{-1})_{kl} = (D^+)_{kl} = D^*_{lk} \; .$$

Figure 4.3 illustrates this relation. The coefficients $D_{ij}^{(\alpha)}(a)$ with i, j, α fixed can be looked upon as vectors in a g-dimensional unitary space $\mathcal{L}$ on $\mathbb{C}$ with the components $\{D_{ij}^{(\alpha)}(a), D_{ij}^{(\alpha)}(b), \dots\}$. Because of (4.2.25a), these vectors

$$\{\sqrt{d_\alpha/g} \cdot D_{ij}^{(\alpha)}(a) | \alpha = 1, \dots, r'; i, j = 1, \dots, d_\alpha\}$$

constitute an orthonormal system in $\mathcal{L}$. Since in $\mathcal{L}$ there are at most g mutually orthogonal vectors and since moreover $D_{ij}^{(\alpha)}$ defines $\sum_{\alpha=1}^{r'} d_\alpha^2$ vectors,

$$\sum_{\alpha=1}^{r'} d_\alpha^2 \leqslant g \tag{4.2.26a}$$

must be valid.

This means: *The number of inequivalent* IRs *of a finite group is finite* ($\leqslant g$); *the dimension of these* IRs *is also finite* ($\leqslant \sqrt{g}$). With the help of the regular REP (end of Sect. 4.3.1) it can further be proved that (4.2.26a) holds with the equality sign. Then we have Burnside's theorem

---

[6] It is $\delta_{\alpha\beta} = 0$ if the IRs α and β are inequivalent, $\delta_{\alpha\beta} = 1$ if they are the same. If the IRs α and β are equivalent, but not the same, $\delta_{\alpha\beta}$ has to be replaced by a certain number.

$$\sum_{\alpha=1}^{r'} d_\alpha^2 = g \ . \qquad (4.2.26b)$$

By multiplying (4.2.25a) by $D_{ji}^{(\alpha)-1}(b)$ and then summing over α, i, j, we obtain

$$\sum_{a\in\mathcal{G}} \sum_{ji} \sum_{\alpha=1}^{r'} d_\alpha \cdot D_{ij}^{(\alpha)}(a) \cdot D_{ji}^{(\alpha)-1}(b) \cdot D_{kl}^{(\beta)-1}(a) = gD_{kl}^{(\beta)-1}(b) \ .$$

For this to be valid, it must hold necessarily:

$$\sum_{ij} \sum_{\alpha=1}^{r'} d_\alpha \cdot D_{ij}^{(\alpha)}(a) \cdot D_{ji}^{(\alpha)-1}(b) = g \cdot \delta_{ab} \ . \qquad (4.2.25b)$$

This relation expresses the orthogonality of the columns in Fig. 4.3.

The character of a REP has been defined by (4.2.12b). Because $\mathrm{Tr}\{S^{-1}AS\} = \mathrm{Tr}\{A\}$, it is obvious that

(i) all equivalent REPs of a group and
(ii) all the elements of a (conjugacy) class have the same character. (4.2.12e)

Using (4.2.12b) we derive from (4.2.25a) the orthogonality theorem of characters of IRs.

$$\sum_{a\in\mathcal{G}} \chi^{(\alpha)}(a) \cdot \chi^{(\beta)}(a^{-1}) = g \cdot \delta_{\alpha\beta} \ , \qquad (4.2.27a)$$

where in unitary IRs $\chi^{(\beta)}(a^{-1})$ can be replaced by $\chi^{(\beta)*}(a)$. According to (4.2.12d) the character is a class function, therefore we can write instead of (4.2.27a)

$$\sum_{a=1}^{r} r_a \cdot \chi^{(\alpha)}(K_a) \cdot \chi^{(\beta)*}(K_a) = g \cdot \delta_{\alpha\beta} \ , \qquad (4.2.27b)$$

where r_a is the number of elements in the class K_a containing a, see (2.3.9).

According to (4.2.27b) the vectors $\sqrt{r_a/g} \cdot \chi^{(\alpha)}(K_a)$ constitute an orthonormal system in the r-dimensional space of class functions of $\mathcal{G}$. From this it follows, similarly to the case of (4.2.26a), that the number r' of inequivalent IRs $D^{(\alpha)}(\mathcal{G})$ of a group satisfies

$$r' \leqslant r \ . \qquad (4.2.27c)$$

From (4.2.27b) we can derive by multiplication by $\chi^{(\alpha)*}(K_b)$ and summation over α the corresponding orthogonality theorem [$a, b = 1, 2, \ldots, r$, see (4.2.25b)]

$$\sum_{\alpha=1}^{r'} \chi^{(\alpha)}(K_a) \cdot \chi^{(\alpha)*}(K_b) = \frac{g}{r_a} \cdot \delta_{ab} \ . \qquad (4.2.28a)$$

However, this demands

$$r \leqslant r' \ . \qquad (4.2.28b)$$

Thus (4.2.27c and 28b) constitute the fundamental theorem of representation theory:

The number of inequivalent IRs of a (finite) group is equal to the number of conjugacy classes in this group. (4.2.29)

Equation (4.2.27a) can be used to determine the reducibility of a REP. Obviously the character of a reducible REP is equal to the sum of the characters of the IRs into which it can be decomposed, see (4.2.21a), therefore

$$\chi(a) = \sum_{\alpha} m_\alpha \chi^{(\alpha)}(a) \ . \tag{4.2.30}$$

Using (4.2.27a) for unitary REPs,

$$\sum_{a} \chi(a)\chi^{(\beta)*}(a) = \sum_{\alpha} m_\alpha \sum_{a \in \mathcal{G}} \chi^{(\alpha)}(a) \cdot \chi^{(\beta)*}(a) = g \cdot m_\beta \ ,$$

we have

$$m_\alpha = \frac{1}{g} \sum_{a \in \mathcal{G}} \chi(a) \cdot \chi^{(\alpha)*}(a) \tag{4.2.31a}$$

or, from (4.2.27b),

$$m_\alpha = \frac{1}{g} \sum_{a=1}^{r} r_a \chi(K_a) \cdot \chi^{(\alpha)*}(K_a) \ . \tag{4.2.31b}$$

Equations (4.2.31a, b) yield the multiplicity, with which an IR is contained in a reducible REP. For this to be found, the REP $D(\mathcal{G})$ does not have to be reduced explicitly and the transformation matrix S does not have to be determined.

From (4.2.27a,b, and 30) we also have

$$\sum_{a \in \mathcal{G}} |\chi(a)|^2 = \sum_{a=1}^{r} r_a |\chi(K_a)|^2 = g \cdot \sum_{\alpha} m_\alpha^2 \geqslant g \ , \tag{4.2.32}$$

where the equality sign is valid only if $D(g)$ with character $\chi(a)$ is an IR. In the case that $D(\mathcal{G})$ is reducible, then the "greater than" sign is always valid.

The group $\mathcal{C}_{3v}$ contains 3 classes [see Sect. 2.3, after (2.3.11)]. With the REP $D(a)$ from Table 4.1 we have

$$\begin{aligned} 1 \cdot |\chi(K_e)|^2 + 2 \cdot |\chi(K_3)|^2 + 3 \cdot |\chi(K_\sigma)|^2 &= 1 \cdot 9 + 2 \cdot 0 + 3 \cdot 1 \\ &= 12 > g = 6 \ ; \end{aligned}$$

the REP is reducible. However, the REP $D^{(1)}(a)$ is certainly irreducible and also the REP $D^{(3)}(a)$ (2.1.18):

$$1 \cdot 4 + 2 \cdot 1 + 3 \cdot 0 = 6 = g \ .$$

The multiplicities with which the IRs occur in $D(a)$ result from (4.2.31b)

$$m_1 = \tfrac{1}{6}[1\cdot 3\cdot 1 + 2\cdot 0\cdot 1 + 3\cdot 1\cdot 1] = 1 \ ,$$

$$m_3 = \tfrac{1}{6}[1\cdot 3\cdot 2 + 2\cdot 0\cdot(-1) + 3\cdot 1\cdot 0] = 1 \ .$$

Since $\dim D(a) = 3$, $\dim D^{(1)} = 1$, $\dim D^{(3)} = 2$, we already have the complete decomposition of $D(a)$:

$$D(a) = D^{(1)}(a) \oplus D^{(3)}(a) \ .$$

Besides these IRs $D^{(1)}$ and $D^{(3)}$, according to (4.2.26b) and (4.2.29) there must exist exactly one more 1-dimensional IR of the group $\mathscr{C}_{3v}$:

$$\sum_{\alpha=1}^{3} d_\alpha^2 = 1^2 + 2^2 + d_2^2 = g = 6 \ , \qquad d_2 = 1 \ .$$

With the help of Burnside's theorem (4.2.26b) and (4.2.32), together with (2.3.12), we can determine the characters of all the IRs of a group. Using Schur's Lemma, (2.3.12) yields

$$r_a\cdot r_b\cdot \chi^{(\alpha)}(K_a)\cdot \chi^{(\alpha)}(K_b) = d_\alpha \sum_{c=1}^{r} h_{ab,c}\cdot r_c\cdot \chi^{(\alpha)}(K_c) \ . \tag{4.2.33a}$$

For practical purposes we introduce

$$l_\alpha = \frac{1}{d_\alpha}\sum_{a=1}^{r} r_a\chi^{(\alpha)}(K_a)\cdot y_a \ , \qquad L_{bc} = \sum_{a=1}^{r} h_{ab,c}y_a \ ,$$

with arbitrary vectors y_a which are eliminated after the procedure. Then (4.2.33a) leads to the eigenvalue problem

$$l_\alpha\cdot r_b\chi^{(\alpha)}(K_b) = \sum_{c}^{r} L_{bc}\cdot r_c\chi^{(\alpha)}(K_c) \ . \tag{4.2.33b}$$

The characteristic polynomial is decomposed into linear factors $\det(L_{bc} - l\delta_{bc}) = (l-l_1)(l-l_2)\dots(l-l_r)$. Comparing the coefficients we obtain from

$$l_\alpha = \sum_{a=1}^{r} r_a\frac{\chi^{(\alpha)}(K_a)}{d_\alpha}y_a \tag{4.2.33c}$$

the quotients $\chi^{(\alpha)}(K_a)/d_\alpha$. Equation (4.2.32) can be rewritten as

$$1 = \frac{1}{g}\sum_{a=1}^{r} r_a|\chi^{(\alpha)}(K_a)|^2 = d_\alpha^2\cdot\frac{1}{g}\sum_{a=1}^{r} r_a\cdot\left|\frac{\chi^{(\alpha)}(K_a)}{d_\alpha}\right|^2 \ . \tag{4.2.33d}$$

This last equation, together with (4.2.33c), gives the d_α and then also the $\chi^{(\alpha)}(K_a)$.

Table 4.2. The values of the quotients $\chi^{(\alpha)}(K_a)/d_\alpha$ in (4.2.33c) for the group $\mathscr{C}_{3v}$

	$\chi(K_e)/d$	$\chi(K_3)/d$	$\chi(K_\sigma)/d$	
l_1	1	1	1	$d_1 = 1$
l_2	1	1	-1	$d_2 = 1$
l_3	1	$-1/2$	0	$d_3 = 2$

Table 4.3. Character table of $\mathscr{C}_{3v}$

$D(\mathscr{C}_{3v})$	χ	K_e $\{e\}$	K_3 $\{c_3, c_3^2\}$	K_σ $\{\sigma_v, \sigma_v', \sigma_v''\}$
A_1	$\chi^{(1)}$	1	1	1
A_2	$\chi^{(2)}$	1	1	-1
E	$\chi^{(3)}$	2	-1	0

For the group $\mathscr{C}_{3v}$ we have with Table 2.3

$$L_{bc} = \begin{pmatrix} y_1 & y_2 & y_3 \\ 2y_2 & y_1 + y_2 & 2y_3 \\ 3y_3 & 3y_3 & y_1 + 2y_2 \end{pmatrix} .$$

The roots are

$$\left.\begin{aligned} l_1 &= y_1 + 2y_2 + 3y_3 \\ l_2 &= y_1 + 2y_2 - 3y_3 \\ l_3 &= y_1 - y_2 \end{aligned}\right\} = \sum_{a=1}^{r} r_a \frac{\chi^{(\alpha)}(K_a)}{d_\alpha} y_a .$$

By comparison of the coefficients we have Table 4.2. The dimensions are determined from (4.2.33d). The complete *character table* of $\mathscr{C}_{3v}$ is given in Table 4.3, where the IRs are denoted following R.S. Mulliken [4.3].

The procedure given above allows the determination of characters in any case; but often it is too laborious to be applicable. In many cases the character tables can be found more or less by "guessing". Then use of the orthogonality relations of the rows (4.2.27b) and columns (4.2.28a) is helpful. Table 4.3 illustrates the orthogonality relations once more.

4.2.5 Subduction. Reality of Representations

Several further statements can be made about REPs. We will discuss only two of them, which are important for applications.

Table 4.4. Correlation table for $\mathscr{C}_{3v}$

$\mathscr{C}_{3v}$	$\mathscr{C}_3$	$\mathscr{C}_s$
A_1	A	A'
A_2	A	A''
E	$E' \oplus E''$	$A' \oplus A''$

The symmetry of a system is often reduced by a perturbation. The symmetry group is then only a subgroup $\mathscr{U} \subset \mathscr{G}$. Consequently, the REPs $D(\mathscr{G})$ of $\mathscr{G}$ are REPs of $\mathscr{U}$ too, but in general not IRs. These REPs are said to be subduced REPs $D^{\text{sub}}(\mathscr{U})$ of $\mathscr{G}$ on $\mathscr{U}$:

$$D^{\text{sub}} \to \{D(a)|a \in \mathscr{U} \subset \mathscr{G}\} \ .$$

Obviously the REP space $\mathscr{L}^{(\alpha)}$ of $\mathscr{G}$ decomposes into irreducible subspaces of $\mathscr{U}$:

$$\mathscr{L}^{(\alpha)} = \sum_{\beta} \oplus \, m_{\beta,\alpha} \mathscr{L}^{(\beta)} \ .$$

If $D^{(\alpha)}(a)$ is an IR of $\mathscr{G}$ with character $\chi^{(\alpha)}(a)$ and $\Delta^{(\beta)}(a)$ is an IR of $\mathscr{U}$ with character $\psi^{(\beta)}(a)$ then the *Frobenius theorem* is valid:

$$\chi^{(\alpha)}(a) = \sum_{\beta} m_{\beta,\alpha} \psi^{(\beta)}(a) \qquad \text{for each } a \in \mathscr{U} \ . \tag{4.2.34a}$$

The multiplicities result from (4.2.31a):

$$m_{\beta,\alpha} = \frac{1}{n_u} \sum_{a \in \mathscr{U}} \chi^{(\alpha)}(a) \psi^{(\beta)*}(a) \ , \qquad n_u = \text{ord}\, \mathscr{U} \ . \tag{4.2.34b}$$

The group $\mathscr{C}_{3v}$ has subgroups $\mathscr{C}_3$ and $\mathscr{C}_s$. By means of the character tables we obtain what is known as the correlation table (Table 4.4). In the table, E' and E'' denote one-dimensional complex REPs, which are often combined to form a two-dimensional REP E (see below, type III); A, A' and A'' are other one-dimensional REPs of $\mathscr{C}_3$ and $\mathscr{C}_s$.

The IRs $D^{(\alpha)}$ of a group can be classified into three types. This classification is of importance in physical applications in which time reversal (reversal of motion, $t \to -t$) is an allowed symmetry operation (Sect. 9.4.2). A REP is said to be real, or of type I, if $D^{(\alpha)}$ is real; it is pseudoreal, or of type II, if $D^{(\alpha)*}$ is equivalent to $D^{(\alpha)}$, i.e. $D^{(\alpha)*} = S^{-1}D^{(\alpha)}S$ but $D^{(\alpha)}$ is *not* real; finally it is complex, or of type III, if $D^{(\alpha)*}$ is not equivalent to $D^{(\alpha)}$, i.e. there is another IR $D^{(\beta)} \sim D^{(\alpha)*}$ with $\beta \neq \alpha$. Representations of types I and II have real characters, which can be seen by evaluating the trace. In complex REPs at least one character has to be complex.

If a REP of a group $\mathscr{G}$ is of type III, then there must exist another inequivalent REP among the IRs of the group which has the same dimension and is of type

III. The reality of a REP can be checked with the following criterion:

$$\frac{1}{g}\sum_{a\in\mathscr{G}}\chi^{(\alpha)}(a^2)=\begin{cases}+1 & \text{type I, real}\\ -1 & \text{type II, pseudoreal}\\ 0 & \text{type III, complex}\ .\end{cases} \tag{4.2.35}$$

The IRs of the group $\mathscr{C}_{3v}$ are real.

Exercise 4.3. (a) Attach an arbitrary displacement vector to every point A, B, C of Fig. 4.1. and determine the total REP of this system, e.g. for P_{c_3}. Realize that the total REP has a certain block structure in which a block corresponds to every "1" in Table 4.1. Any block itself describes the rotation of basis vectors under P_{c_3}.
(b) Draw vector diagrams which transform according to the different REPs of $\mathscr{C}_{3v}$ [see Fig. 4.2. and also (2.1.18, 19)], i.e. reduce the total REP.

Exercise 4.4. Show that (4.2.14) possesses the properties of a mean value.

Exercise 4.5. Reformulate Schur's lemma II in the language of (representation) matrices.

Exercise 4.6. Prove (4.2.33a).

4.3 Group Algebra

4.3.1 The Regular Representation

In a group $\mathscr{G}$ there is defined a composition (multiplication) between the elements. By introducing a further (summation) composition law between the elements we can extend the formal group theory according to (2.1.3) to a *group algebra* (Frobenius algebra) $\mathscr{A}(\mathscr{G})$ over the complex numbers $\mathbb{C}$. For this purpose we define a summation of elements $a_i \in \mathscr{G}$ and further a multiplication of the elements with numbers $c_i \in \mathbb{C}$. Then $\mathscr{A}(\mathscr{G})$ is a g-dimensional linear space over $\mathbb{C}$ with the elements $a_i \in \mathscr{G}$ as a basis; $\mathscr{A}(\mathscr{G})$ consists of all the vectors

$$a=\sum_{i=1}^{g}c_i a_i\ , \qquad c_i\in\mathbb{C}\ , \qquad a_i\in\mathscr{G}\ . \tag{4.3.1a}$$

The *summation* in $\mathscr{A}(\mathscr{G})$ is defined by

$$a+a'=\sum_{i=1}^{g}(c_i+c_i')a_i \tag{4.3.1b}$$

and the *multiplication* by

$$a \cdot a' = \sum_{i,j} c_i c_j' a_i a_j = \sum_k c_k'' a_k \;, \qquad c_k'' = \sum_{i,j} c_i c_j' \tag{4.3.1c}$$

with the restriction $a_i a_j = a_k$. Every element $a_i \in \mathscr{G}$ now can be looked upon as a linear operator which acts in the space $\mathscr{A}(\mathscr{G})$. By this procedure we generate analogously to (4.2.6) a REP called the *regular representation.* In the regular REP the operators *and* the basis functions according to (4.2.6) are the group elements $a_i \in \mathscr{G}$. The space $\mathscr{A}(\mathscr{G})$ is invariant under all $a_i \in \mathscr{G}$, the REP matrices are $(g \times g)$ matrices, therefore

$$a_i \cdot a_j = \sum_{k=1}^{g} D_{kj}^{\mathrm{reg}}(a_i) \cdot a_k = a_l(i, j) \tag{4.3.2}$$

because of the group properties. Equation (4.3.2) means that $D_{kj}^{\mathrm{reg}}(a_i) = 0$ if $a_i a_j \neq a_k$ and $D_{kj}^{\mathrm{reg}}(a_i) = 1$ if $a_i a_j = a_k$. Every column and every row of D^{reg} possess only one element which is different from zero (and equal to one). The structure of D^{reg} results from the group multiplication table (Table 2.1). If $a_i = e$, then $D^{\mathrm{reg}}(e)$ has to be the unit matrix. For $a_i = c_3$ we have from Table 2.1 with $a_j = \sigma_v$, $D_{k4}^{\mathrm{reg}}(c_3) = 1$ if $k = 6(\sigma_v'')$ and zero if $k \neq 6$. In this way we obtain for the group $\mathscr{C}_{3v}$ [numbering according to Table 2.1)]

$$D^{\mathrm{reg}}(e) = \begin{pmatrix} 1&0&0&0&0&0\\ 0&1&0&0&0&0\\ 0&0&1&0&0&0\\ 0&0&0&1&0&0\\ 0&0&0&0&1&0\\ 0&0&0&0&0&1 \end{pmatrix}, \qquad D^{\mathrm{reg}}(c_3) = \begin{pmatrix} 0&0&1&0&0&0\\ 1&0&0&0&0&0\\ 0&1&0&0&0&0\\ 0&0&0&0&1&0\\ 0&0&0&0&0&1\\ 0&0&0&1&0&0 \end{pmatrix},$$

$$D^{\mathrm{reg}}(c_3^2) = \begin{pmatrix} 0&1&0&0&0&0\\ 0&0&1&0&0&0\\ 1&0&0&0&0&0\\ 0&0&0&0&0&1\\ 0&0&0&1&0&0\\ 0&0&0&0&1&0 \end{pmatrix}, \qquad D^{\mathrm{reg}}(\sigma_v) = \begin{pmatrix} 0&0&0&1&0&0\\ 0&0&0&0&1&0\\ 0&0&0&0&0&1\\ 1&0&0&0&0&0\\ 0&1&0&0&0&0\\ 0&0&1&0&0&0 \end{pmatrix},$$

$$D^{\mathrm{reg}}(\sigma_v') = \begin{pmatrix} 0&0&0&0&1&0\\ 0&0&0&0&0&1\\ 0&0&0&1&0&0\\ 0&0&1&0&0&0\\ 1&0&0&0&0&0\\ 0&1&0&0&0&0 \end{pmatrix}, \qquad D^{\mathrm{reg}}(\sigma_v') = \begin{pmatrix} 0&0&0&0&0&1\\ 0&0&0&1&0&0\\ 0&0&0&0&1&0\\ 0&1&0&0&0&0\\ 0&0&1&0&0&0\\ 1&0&0&0&0&0 \end{pmatrix}. \tag{4.3.3}$$

These matrices give an exact description of Table 2.1. As seen immediately, the characters satisfy

$$\chi^{\text{reg}}(e) = g \qquad \text{and} \qquad \chi^{\text{reg}}(a_i \neq e) = 0 \ . \tag{4.3.4}$$

The criterion of reducibility (4.2.31) gives

$$m_\alpha^{\text{reg}} = \frac{1}{g}\chi^{\text{reg}}(e)\chi^{(\alpha)*}(e) = d_\alpha \ , \tag{4.3.5}$$

since $\chi^{(\alpha)}(e) = d_\alpha$ according to (4.2.12c). Thus we may formulate: *The regular* REP *contains each* IR $D^{(\alpha)}(\mathscr{G})$ *just as often as the dimension of this* REP *indicates*. All the IRs of a group can be obtained by the reduction of the regular REP. Using (4.3.4, 5) we have from (4.2.30)

$$\chi^{\text{reg}}(e) = g = \sum_{\alpha=1}^{r} m_\alpha^{\text{reg}}\chi^{(\alpha)}(e) = \sum_{\alpha=1}^{r} d_\alpha^2 \ , \tag{4.2.26c}$$

which is again Burnside's theorem.

4.3.2 Projection Operators

A linear space $\mathscr{A}'$ which is contained in $\mathscr{A}(\mathscr{G})$ and which is closed with respect to the multiplication defined in $\mathscr{A}(\mathscr{G})$ is said to be a *subalgebra*:

> If $a, b \in \mathscr{A}'$, then also $a \cdot b$ and $b \cdot a \in \mathscr{A}'$ but in general $a \cdot c \notin \mathscr{A}'$ for $c \in \mathscr{A}$. (4.3.6)

A *left ideal* (and similarly *right ideal*) $\mathscr{S}$ of $\mathscr{A}(\mathscr{G})$ is a subalgebra with

$$a \cdot c \in \mathscr{S} \qquad \text{for} \qquad a \in \mathscr{A}(\mathscr{G}) \ , \qquad c \in \mathscr{S} \tag{4.3.7}$$

If the left ideal is equal to the right ideal, that is, if

$$a \cdot c \text{ and } c \cdot a \in \mathscr{S} \qquad \text{for } a \in \mathscr{A}(\mathscr{G}) \ , \qquad c \in \mathscr{S} \ , \tag{4.3.8}$$

then $\mathscr{S}$ is said to be an *invariant subalgebra* (remember the concepts of subgroup, left, right coset and invariant subgroup in Sect.2.1, 2.2).

We now consider the elements

$$P_{kj}^{(\alpha)} = \frac{d_\alpha}{g}\sum_{a \in \mathscr{G}} D_{jk}^{(\alpha)}(a^{-1}) \cdot P_a \rightarrow \frac{d_\alpha}{g}\sum_{a \in \mathscr{G}} D_{kj}^{(\alpha)*}(a) \cdot P_a \tag{4.3.9a}$$

(for unitary REPs), in which the $D_{kj}^{(\alpha)}$ are the IR matrices of $\mathscr{G}$. The quantities (4.3.9a) are elements of the group algebra represented by the operators P_a which have a one-to-one correspondence to the elements $a \in \mathscr{G}$. The reverse of (4.3.9a) is obtained using (4.2.25b)

$$P_a = \sum_{\alpha=1}^{r} \sum_{k,j=1}^{d_\alpha} D_{kj}^{(\alpha)}(a) \cdot P_{kj}^{(\alpha)} . \tag{4.3.9b}$$

This is an expansion of P_a with respect to the basis $P_{kj}^{(\alpha)}$.

The linear space $\mathscr{A}(\mathscr{G})$ can be decomposed completely into irreducible subspaces, as can be seen from the reduction of the regular REP. But this also means that the group algebra $\mathscr{A}(\mathscr{G})$ can be decomposed completely into a sum of irreducible left ideals $\mathscr{S}_i^{(\alpha)}$:

$$\mathscr{A}(\mathscr{G}) = \sum_{\alpha=1}^{r} \sum_{i=1}^{d_\alpha} \mathscr{S}_i^{(\alpha)} ,$$

the basis of which is generated by the $P_{ki}^{(\alpha)}$ with $k = 1, \ldots, d_\alpha$. In each of these $\mathscr{S}_i^{(\alpha)}$ the P_a are represented by the corresponding IRs $D^{(\alpha)}$. Applying $P_c \in \mathscr{A}(\mathscr{G})$ to (4.3.9a) we have with $ca = b$

$$\begin{aligned} P_c P_{kj}^{(\alpha)} &= \frac{d_\alpha}{g} \sum_{a \in \mathscr{G}} D_{kj}^{(\alpha)*}(a) P_c P_a = \frac{d_\alpha}{g} \sum_{b \in \mathscr{G}} D_{kj}^{(\alpha)*}(c^{-1}b) P_b \\ &= \frac{d_\alpha}{g} \sum_{b \in \mathscr{G}} \sum_i D_{ki}^{(\alpha)*}(c^{-1}) D_{ij}^{(\alpha)*}(b) P_b = \sum_i D_{ik}^{(\alpha)}(c) P_{ij}^{(\alpha)} , \end{aligned} \tag{4.3.10}$$

i.e. the $\{P_{kj}^{(\alpha)} | k = 1, \ldots, d_\alpha, j \text{ fixed}\}$ transform by application of P_c according to the IR $D^{(\alpha)}$ and therefore span the irreducible left ideals $\mathscr{S}_j^{(\alpha)}$. The $\{P_{kj}^{(\alpha)} | k, j = 1, \ldots, d_\alpha; \alpha = 1, \ldots, r\}$ then define a complete basis of the algebra $\mathscr{A}(\mathscr{G})$.

For the group $\mathscr{C}_{3v}$, $\dim \mathscr{A} = 6$ and there are six operators $P_{kj}^{(\alpha)}$: $\alpha = 1, 2$ with $k = j = 1$ and $\alpha = 3$ with $k, j = 1, 2$. With the REPs from (2.1.18) and Tables 4.1, 4.3 we obtain

$$\begin{aligned} 6P_{11}^{(1)} &= P_e + P_{c_3} + P_{c_3^2} + P_{\sigma_v} + P_{\sigma_v'} + P_{\sigma_v''} \\ 6P_{11}^{(2)} &= P_e + P_{c_3} + P_{c_3^2} - P_{\sigma_v} - P_{\sigma_v'} - P_{\sigma_v''} \\ 3P_{11}^{(3)} &= P_e - P_{\sigma_v} + \tfrac{1}{2}(-P_{c_3} - P_{c_3^2} + P_{\sigma_v'} + P_{\sigma_v''}) \\ 3P_{12}^{(3)} &= \frac{\sqrt{3}}{2}(-P_{c_3} + P_{c_3^2} + P_{\sigma_v'} - P_{\sigma_v''}) \\ 3P_{21}^{(3)} &= \frac{\sqrt{3}}{2}(+P_{c_3} - P_{c_3^2} + P_{\sigma_v'} - P_{\sigma_v''}) \\ 3P_{22}^{(3)} &= P_e + P_{\sigma_v} + \tfrac{1}{2}(-P_{c_3} - P_{c_3^2} - P_{\sigma_v'} - P_{\sigma_v''}) . \end{aligned} \tag{4.3.11}$$

The example shows that the $P_{kj}^{(\alpha)}$ can be determined only if the $D^{(\alpha)}$ are already known. They have the following properties (see Exercise 4.7.) when P_a and $D^{(\alpha)}(a)$

are assumed to be unitary:

$$P_{jk}^{(\alpha)+} = P_{kj}^{(\alpha)} , \qquad \text{therefore } P_{jj}^{(\alpha)+} = P_{jj}^{(\alpha)}\text{: Hermitian,}$$
$$P_{kj}^{(\alpha)} \cdot P_{li}^{(\beta)} = \delta_{\alpha\beta}\delta_{jl} \cdot P_{ki}^{(\alpha)} , \qquad \text{therefore } (P_{jj}^{(\alpha)})^2 = P_{jj}^{(\alpha)}\text{: idempotent;} \tag{4.3.12}$$

the $P_{jj}^{(\alpha)}$ are therefore projection operators. They project a function f of the REP space onto the basis functions $f_j^{(\alpha)}$ of the IR $D^{(\alpha)}$, since the functions

$$f_{kj}^{(\alpha)} := P_{kj}^{(\alpha)} f := c_j^{(\alpha)} f_k^{(\alpha)} \qquad \text{for } j \text{ fixed}, \qquad k = 1, \ldots, d_\alpha \tag{4.3.13}$$

transform according to the IR $D^{(\alpha)}$

$$P_a(P_{kj}^{(\alpha)} f) = \sum_{i=1}^{d_\alpha} D_{ik}^{(\alpha)}(a) \cdot (P_{ij}^{(\alpha)} f) . \tag{4.3.14}$$

Thus for $k = j$ we just get the projection of f onto $f_j^{(\alpha)}$, for $k \neq j$ we obtain the other partners of $f_j^{(\alpha)}$. The operator $P_{kj}^{(\alpha)}$ projects out of f the basis functions belonging to the kth row of $D^{(\alpha)}$ or, when j is fixed, we obtain from f all the basis functions $k = 1, \ldots, d_\alpha$ of $D^{(\alpha)}$, except perhaps for a common factor.

Scalar products of basis functions satisfy

$$\begin{aligned} \langle f_k^{(\alpha)} | f_l^{(\beta)} \rangle &= \frac{1}{g} \sum_{a \in \mathcal{G}} \langle P_a f_k^{(\alpha)} | P_a f_l^{(\beta)} \rangle \\ &= \frac{1}{g} \sum_{ij} \sum_{a \in \mathcal{G}} D_{ik}^{(\alpha)*}(a) D_{jl}^{(\beta)}(a) \cdot \langle f_i^{(\alpha)} | f_j^{(\beta)} \rangle \\ &= \delta_{\alpha\beta}\delta_{kl} \cdot \frac{1}{d_\alpha} \sum_i \langle f_i^{(\alpha)} | f_i^{(\alpha)} \rangle \end{aligned} \tag{4.3.15}$$

if the REPs are unitary and (4.2.6, 25a) are taken into account.[7] Basis functions belonging to inequivalent IRs or to different rows of identical IRs are mutually orthogonal. The scalar product of two basis functions belonging to the same row ($k = l$) of an IR ($\alpha = \beta$) is independent of the row number. Equation (4.3.15) is an essential relation for the calculation of matrix elements (expectation values, see Sect. 6.3).

Use is often made of the *character projection operator*, the construction of which is performed with characters only:

$$P^{(\alpha)} = \sum_{k=1}^{d_\alpha} P_{kk}^{(\alpha)} = \frac{d_\alpha}{g} \sum_{a \in \mathcal{G}} \chi^{(\alpha)*}(a) P_a . \tag{4.3.9c}$$

It projects that part of f which lies in the REP space of $D^{(\alpha)}$. For the group $\mathcal{C}_{3v}$, for example, $D = D^{(1)} \oplus D^{(3)}$, see Table 4.1. The basis functions should be those of Sect. 4.2.3. We obtain with (4.2.4) and (4.3.11)

[7] See footnote 6 on p. 67.

$$P_{11}^{(1)} f_A = \tfrac{1}{3}(f_A + f_B + f_C) ,$$

$$P_{12}^{(3)} f_A = \frac{1}{\sqrt{3}}(f_C - f_B) , \qquad (4.3.16)$$

$$P_{22}^{(3)} f_A = \tfrac{1}{3}(2f_A - f_B - f_C) ,$$

which agree with the functions given in Sect. 4.2.3, except for factors of 1/3 and −1/3, respectively. The method, consequently, is simple: We have to apply all the operators $P_{kj}^{(\alpha)}$ with $k, j = 1, \ldots, d_\alpha$ to the function f. Thus, in general, we obtain d_α^2 functions, which implies d_α sets of partner functions $c_j^{(\alpha)} f_k^{(\alpha)}$ of $D^{(\alpha)}$ (j fixed, $k = 1, \ldots, d_\alpha$). These sets can be linearly dependent or even be projected to "zero". The details depend on the choice of the starting function f. If f already possesses certain symmetries with respect to $\mathscr{G}$, then there are "zeros", for example, if we project with operators belonging to "other" IRs. But in any case, with this procedure we can determine the basis functions belonging to $D^{(\alpha)}(\mathscr{G})$.

In applications we often find that the space $\mathscr{S}$ which is invariant under P_a, $a \in \mathscr{G}$ consists of the functions $\{P_e f = f_e;\ P_a f = f_a, \ldots, P_g f = f_g\}$. If these functions are linearly independent, then $\dim \mathscr{S} = g$ and the effect of the P_a is given by the regular REP. In order to see this, we have only to apply (4.3.2) to f. From the reduction of D^{reg} we then have that as a basis in $\mathscr{S}$ we can use all the functions $f_{kj}^{(\alpha)} = P_{kj}^{(\alpha)} f$ ($g = \sum_\alpha^r m_\alpha d_\alpha = \sum_\alpha^r d_\alpha^2$ in number), which are classified according to the IRs. Here $k, j = 1, \ldots, d_\alpha$; α numbers all the IRs. Because the functions are linearly independent, any function of $\mathscr{S}$ can be expanded with respect to them. This is especially simple if the symmetry group of the physical system is $\mathscr{G} = \mathscr{C}_i$. Every function describing the system can be split into an even and an odd function with respect to the inversion i:

$$P_i f_g(\boldsymbol{x}) = f_g(-\boldsymbol{x}) = f_g(\boldsymbol{x}) ,$$

$$P_i f_u(\boldsymbol{x}) = f_u(-\boldsymbol{x}) = -f_u(\boldsymbol{x}) , \qquad (4.3.17)$$

$$f_g = P_{11}^{(1)} f = \tfrac{1}{2}(P_e f + P_i f) = \tfrac{1}{2}[f(\boldsymbol{x}) + f(-\boldsymbol{x})] ,$$

$$f_u = P_{11}^{(1)} f = \tfrac{1}{2}(P_e f - P_i f) = \tfrac{1}{2}[f(\boldsymbol{x}) - f(-\boldsymbol{x})] ,$$

therefore $f = f_g + f_u$ and $P_e = P_{11}^{(1)} + P_{11}^{(2)}$.

In general it can be shown that every function f can be expanded with respect to the basis functions of the IRs of a group $\mathscr{G}$ (expansion theorem). First we realize that the identity P_e in $\mathscr{A}(\mathscr{G})$ decomposes according to (4.3.9b)

$$P_e = \sum_{\alpha=1}^{r} \sum_{j=1}^{d_\alpha} P_{jj}^{(\alpha)} , \qquad (4.3.18)$$

which can be seen directly from (4.3.17) or (4.3.11) for the given examples. Then, also [see (4.3.13)]

$$f = P_e f = \sum_{\alpha=1}^{r} \sum_{j}^{d_\alpha} P_{jj}^{(\alpha)} f := \sum_{\alpha} \sum_{j=1}^{d_\alpha} c_j^{(\alpha)} f_j^{(\alpha)} \;, \tag{4.3.19}$$

where the $P_{jj}^{(\alpha)} f$ are basis functions belonging to the jth row of the IRs $D^{(\alpha)}$ of the group $\mathscr{G}$. Generally the $f_j^{(\alpha)}$ are linear combinations of functions belonging to the jth row of $D^{(\alpha)}$ if the multiplicity of $D^{(\alpha)}$ in the REP space of f is larger than one. The index α numbers the IRs, including the multiplicity with which they occur. The dimension of the REP space is $\sum_{\alpha=1}^{r} m_\alpha \cdot d_\alpha$.

Exercise 4.7. Show the validity of (4.3.12) by using the fact that P_a and $D^{(\alpha)}$ are unitary and that (4.2.25a) holds.

4.4 Direct Products

4.4.1 Representations of Direct Products of Groups

Since multiple direct products can be decomposed into direct products of two factors (2.2.19, 20) it is sufficient to discuss these. By the *direct product* (*Kronecker product*) of two (square) matrices A and B we understand a matrix $A \otimes B$, the elements of which in the row (ij) and column (kl) are given by

$$(A \otimes B)_{ij,kl} := A_{ik} B_{jl} \qquad \text{with } i, k = 1, \ldots, d_A \;, \quad j, l = 1, \ldots, d_B \;. \tag{4.4.1a}$$

For example, with

$$A = \begin{pmatrix} a_{11} & a_{12} \\ a_{21} & a_{22} \end{pmatrix} \quad \text{and} \quad B = \begin{pmatrix} b_{11} & b_{12} \\ b_{21} & b_{22} \end{pmatrix}$$

we have

$$A \otimes B = \begin{pmatrix} a_{11} B & a_{12} B \\ a_{21} B & a_{22} B \end{pmatrix} = \begin{pmatrix} a_{11}b_{11} & a_{11}b_{12} & a_{12}b_{11} & a_{12}b_{12} \\ a_{11}b_{21} & a_{11}b_{22} & \cdot & \cdot \\ \cdot & \cdot & \cdot & \cdot \\ \cdot & \cdot & \cdot & a_{22}b_{22} \end{pmatrix} .$$

Obviously the dimension of $A \otimes B$ is

$$d_{A \otimes B} = d_A \cdot d_B \;. \tag{4.4.1b}$$

The actual arrangement of rows and columns used in the Kronecker product is irrelevant. The arrangement can be changed by a similarity transformation, e.g.

$$A \otimes B \neq B \otimes A \ , \tag{4.4.2a}$$

but on the other hand

$$A \otimes B = P^{-1}(B \otimes A)P \tag{4.4.2b}$$

with a $d_A \cdot d_B$-dimensional permutation matrix P, which is given by

$$P = \begin{pmatrix} 1 & 0 & 0 & 0 \\ 0 & 0 & 1 & 0 \\ 0 & 1 & 0 & 0 \\ 0 & 0 & 0 & 1 \end{pmatrix}$$

in the above example. Because of (4.4.1a, 2b) we have also

$$\begin{aligned} \mathrm{Tr}\{A \otimes B\} &= \mathrm{Tr}\{B \otimes A\} = \sum_{ij} (A \otimes B)_{ij,ij} = \sum_i A_{ii} \sum_j B_{jj} \\ &= \mathrm{Tr}\{A\} \cdot \mathrm{Tr}\{B\} \ . \end{aligned} \tag{4.4.3}$$

If A, C have the same dimension, and likewise B, D, the matrix product of two Kronecker products is given by

$$\begin{aligned} [(A \otimes B) \cdot (C \otimes D)]_{ij,mn} &= \sum_{kl} (A \otimes B)_{ij,kl} (C \otimes D)_{kl,mn} \\ &= \sum_{kl} A_{ik} B_{jl} C_{km} D_{ln} = (AC)_{im} (BD)_{jn} \\ &= [(AC) \otimes (BD)]_{ij,mn} \ . \end{aligned} \tag{4.4.4}$$

Two sets of matrices, e.g. REP matrices of a group or two groups, allow the definition of an inner and an outer Kronecker product. If $\mathcal{M}_A = \{A_1, \ldots, A_n\}$ is a set of d_A-dimensional matrices and $\mathcal{M}_B = \{B_1, \ldots, B_m\}$ a set of d_B-dimensional matrices, then for $n = m$ (!) the *inner Kronecker product* is defined as

$$\mathcal{M}_A \otimes \mathcal{M}_B = \{A_1 \otimes B_1; A_2 \otimes B_2; \ldots; A_n \otimes B_n\} \tag{4.4.5}$$

and for n, m arbitrary (!) the *outer Kronecker Product* as

$$\mathcal{M}_A \times \mathcal{M}_B = \{A_i \otimes B_j | i = 1, \ldots, n; j = 1, \ldots, m\} \ . \tag{4.4.6}$$

In (4.4.5, 6) the rules (4.4.1–4) for the direct matrix products are valid. Let $\mathcal{M}_A \to D^{(\alpha)}$ and $\mathcal{M}_B \to D^{(\beta)}$ be two matrix REPs of two groups $\mathcal{G}_a$ and $\mathcal{G}_b$. Then obviously the outer Kronecker product is a REP of the outer product (2.2.19, 20) of the groups $\mathcal{G}_a$ and $\mathcal{G}_b$, since with $a_i \in \mathcal{G}_a$ and $b_j \in \mathcal{G}_b$ and $c_1 = a_1 b_1$, $c_2 = a_2 b_2$ we have according to (2.2.19) and (4.4.4)

$$
\begin{aligned}
D^{(\alpha)} \otimes D^{(\beta)}(c_1 c_2) &= D^{(\alpha)} \otimes D^{(\beta)}(a_1 b_1 a_2 b_2) = D^{(\alpha)} \otimes D^{(\beta)}(a_1 a_2 b_1 b_2) \\
&= D^{(\alpha)}(a_1 a_2) \otimes D^{(\beta)}(b_1 b_2) \\
&= D^{(\alpha)}(a_1) D^{(\alpha)}(a_2) \otimes D^{(\beta)}(b_1) D^{(\beta)}(b_2) \\
&= [D^{(\alpha)}(a_1) \otimes D^{(\beta)}(b_1)] \cdot [D^{(\alpha)}(a_2) \otimes D^{(\beta)}(b_2)] \\
&= D^{(\alpha)} \otimes D^{(\beta)}(c_1) \cdot D^{(\alpha)} \otimes D^{(\beta)}(c_2) \ .
\end{aligned}
$$

The character of the direct product follows from (4.4.3) with $c = a \cdot b \in \mathscr{G}_a \times \mathscr{G}_b$,

$$
\chi^{(\alpha \times \beta)}(c) = \chi^{(\alpha)}(a) \cdot \chi^{(\beta)}(b) \ . \tag{4.4.7}
$$

Using this and (4.2.32) with the number $r_{ab} = r_a \cdot r_b$ of the elements in the class K_{ab} of $\mathscr{G}_a \times \mathscr{G}_b$,

$$
\begin{aligned}
\sum_a \sum_b r_{ab} |\chi(K_{ab})|^2 &= \sum_a r_a |\chi^{(\alpha)}(K_a)|^2 \cdot \sum_b r_b |\chi^{(\beta)}(K_b)|^2 \\
&\geqslant g_a \cdot g_b = g \ ,
\end{aligned}
$$

where the equality sign is valid if and only if $D^{(\alpha)}$ and $D^{(\beta)}$ are irreducible. This gives the important theorem:

> The REP $D^{(\alpha)} \otimes D^{(\beta)}$ of an outer direct product of the groups $\mathscr{G}_a \times \mathscr{G}_b$ is irreducible if and only if $D^{(\alpha)}$ and $D^{(\beta)}$ are irreducible. (4.4.8)

For groups which can be written as a direct product of two groups, $\mathscr{G} = \mathscr{G}_a \times \mathscr{G}_b$, the elements of $\mathscr{G}_a$ and $\mathscr{G}_b$ commute and the IRs are the Kronecker products of the IRs of $\mathscr{G}_a$ and $\mathscr{G}_b$. The projectors (4.3.9a) of $\mathscr{G}$ are products of the projectors of $\mathscr{G}_a$ and $\mathscr{G}_b$, i.e.

$$
P^{(\alpha \times \beta)}(\mathscr{G}) = P^{(\alpha)}(\mathscr{G}_a) \cdot P^{(\beta)}(\mathscr{G}_b) \ . \tag{4.4.9}
$$

When the symmetry group of a physical system is given by the outer direct product of two or more groups, then we can determine their IRs from those of the single groups. For example, when the symmetry group is $\mathscr{G} \times \mathscr{C}_i$ (a frequently occurring case), the IRs are as in Table 4.5; the group $\mathscr{C}_i$ has only IRs ± 1, equal to the character. The character tables of $\mathscr{G} \times \mathscr{C}_i$ can be constructed analogously to Table 4.5.

Table 4.5. Irreducible representations for the symmetry groups **(a)** $\mathscr{G} \times \mathscr{C}_i$ and **(b)** $\mathscr{C}_i$

$\mathscr{G} \times \mathscr{C}_i$	$a \in \mathscr{G}$	$ia \in i\mathscr{G}$ **(a)**	$\mathscr{C}_i$	e	i **(b)**
$D^{(\alpha,+)}$	$D^{(\alpha)}(a)$	$D^{(\alpha)}(a)$	$D^{(+)}$	1	1
$D^{(\alpha,-)}$	$D^{(\alpha)}(a)$	$-D^{(\alpha)}(a)$	$D^{(-)}$	1	-1

Groups of this kind, e.g. $\mathscr{D}_{nh} = \mathscr{D}_n \times \mathscr{C}_s (\mathscr{C}_s \cong \mathscr{C}_i)$ or $\mathcal{O}(3) = \mathscr{S}\mathcal{O}(3) \times \mathscr{C}_i$, have two REPs for every $D^{(\alpha)}$ of $\mathscr{D}_n$ or $\mathscr{S}\mathcal{O}(3)$, which are denoted by $D^{(\alpha,+)} = D^{(\alpha,g)}$ and $D^{(\alpha,-)} = D^{(\alpha,u)}$, respectively. The pure proper rotations are represented in both the REPs $D^{(\alpha,\pm)}$ by $D^{(\alpha)}$.

4.4.2 The Inner Direct Product of Representations of a Group. Clebsch-Gordan Expansion

Let $D^{(\alpha)}$ and $D^{(\beta)}$ be two matrix REPs of *one* group $\mathscr{G}$. Then, obviously, the inner Kronecker product $D^{(\alpha\times\beta)} := \{D^{(\alpha)}(a) \otimes D^{(\beta)}(a)\}$ is also a REP of $\mathscr{G}$. In the formulation that is independent of the choice of basis, there exists a linear operator $P_a^{(\alpha\times\beta)} = P_a^{(\alpha)} \otimes P_a^{(\beta)}$ with $P_a^{(\alpha\times\beta)}: \mathscr{L}^{(\alpha)} \otimes \mathscr{L}^{(\beta)} \to \mathscr{L}^{(\alpha)} \otimes \mathscr{L}^{(\beta)}$, which corresponds to $P_a^{(\alpha)}: \mathscr{L}^{(\alpha)} \to \mathscr{L}^{(\alpha)}$ and $P_a^{(\beta)}: \mathscr{L}^{(\beta)} \to \mathscr{L}^{(\beta)}$ with $a \in \mathscr{G}$. The new operator also maps the unitary product space (tensor space) onto itself according to (4.1.10); $P_a^{(\alpha\times\beta)}$ is said to be the tensor product of $P_a^{(\alpha)}$ and $P_a^{(\beta)}$. Correspondingly, the matrix REP $D^{(\alpha\times\beta)}(a)$ of $P_a^{(\alpha\times\beta)}$ is related to the basis

$$\{e_{ij}^{(\alpha\beta)}\} = \{e_i^{(\alpha)} e_j^{(\beta)} | i = 1, \dots, d_\alpha = \dim \mathscr{L}^{(\alpha)};\ j = 1, \dots, d_\beta\} \tag{4.4.10}$$

if $\{e_i^{(\alpha)} | i = 1, \dots, d_\alpha\}$ and $\{e_i^{(\beta)} | i = 1, \dots, d_\beta\}$ are the bases of $D^{(\alpha)}$ and $D^{(\beta)}$, respectively. Then

$$\begin{aligned} P_a^{(\alpha\times\beta)} e_i^{(\alpha)} e_k^{(\beta)} &:= P_a e_i^{(\alpha)} \otimes P_a e_k^{(\beta)} = \sum_{jl} D_{ji}^{(\alpha)}(a) D_{lk}^{(\beta)}(a) e_j^{(\alpha)} e_l^{(\beta)} \\ &= \sum_{jl} [D^{(\alpha)}(a) \otimes D^{(\beta)}(a)]_{jl,ik} e_j^{(\alpha)} e_l^{(\beta)} = \sum_{jl} D_{jl,ik}^{(\alpha\times\beta)}(a) e_j^{(\alpha)} e_l^{(\beta)} \ , \end{aligned} \tag{4.4.11}$$

and for the characters

$$\chi^{(\alpha\times\beta)}(a) = \chi^{(\alpha)}(a) \cdot \chi^{(\beta)}(a) \ . \tag{4.4.12}$$

If $D^{(\alpha)}$ and $D^{(\beta)}$ are IRs then $D^{(\alpha\times\beta)}$ is in general a reducible REP of $\mathscr{G}$, which, if required, can in turn be decomposed into IRs $D^{(\gamma)}$. This decomposition is the *Clebsch-Gordan expansion:*

$$D^{(\alpha\times\beta)} = \sum_\gamma{}_\oplus (\alpha\beta|\gamma) D^{(\gamma)} \ ; \tag{4.4.13a}$$

$$\mathscr{L}^{(\alpha)} \otimes \mathscr{L}^{(\beta)} = \sum_\gamma{}_\oplus (\alpha\beta|\gamma) \mathscr{L}^{(\gamma)} \ . \tag{4.4.13b}$$

The $(\alpha\beta|\gamma) = (\beta\alpha|\gamma)$ are called *reduction coefficients.* Using (4.4.12) and (4.2.31a) we obtain

$$\begin{aligned} (\alpha\beta|\gamma) &= \frac{1}{g} \sum_{a\in\mathscr{G}} \chi^{(\alpha)}(a) \cdot \chi^{(\beta)}(a) \cdot \chi^{(\gamma)*}(a) \ , \\ &= \frac{1}{g} \sum_{a=1}^{r} r_a \chi^{(\alpha)}(K_a) \cdot \chi^{(\beta)}(K_a) \cdot \chi^{(\gamma)*}(K_a) \ . \end{aligned} \tag{4.4.14}$$

Table 4.6. Decompositions of the tensor products of $\mathscr{C}_{3v}$

$\otimes$	A_1	A_2	E
A_1	A_1	A_2	E
A_2	A_2	A_1	E
E	E	E	$A_1 + A_2 + E$

This means that for REPs with real characters, $(\alpha\beta|\gamma)$ is totally symmetric. The tensor products of $\mathscr{C}_{3v}$ have the decompositions given in Table 4.6.

An interesting case in connection with quantum-mechanical selection rules occurs if $D^{(\alpha)} = D^{(\beta)}$ and both are related to different bases $e_i^{(\alpha)}$ and $e_j'^{(\alpha)}$. It follows from (4.4.11) that

$$P_a^{(\alpha\times\alpha)}(e_i^{(\alpha)}e_k'^{(\alpha)} \pm e_k^{(\alpha)}e_i'^{(\alpha)}) = \sum_{jl} (D_{ji}^{(\alpha)}D_{lk}^{(\alpha)} \pm D_{jk}^{(\alpha)}D_{li}^{(\alpha)})e_j^{(\alpha)}e_l'^{(\alpha)} \; .$$

Interchanging j and l in the sum and summing corresponding expressions, we obtain the symmetrized form

$$P_a^{(\alpha\times\alpha)}(e_i^{(\alpha)}e_k'^{(\alpha)} \pm e_k^{(\alpha)}e_i'^{(\alpha)})$$

$$= \sum_{jl} \tfrac{1}{2}[D_{ji}^{(\alpha)}(a)D_{lk}^{(\alpha)}(a) \pm D_{jk}^{(\alpha)}(a)D_{li}^{(\alpha)}(a)]\cdot(e_j^{(\alpha)}e_l'^{(\alpha)} \pm e_l^{(\alpha)}e_j'^{(\alpha)}) \; . \tag{4.4.15}$$

The functions $[e_j^{(\alpha)}e_l'^{(\alpha)} \pm e_l^{(\alpha)}e_j'^{(\alpha)}]$ constitute $[d_\alpha(d_\alpha \pm 1)/2]$-dimensional bases for REPs of $\mathscr{G}$, which are the symmetric and antisymmetric tensor products. They are denoted by $[D^{(\alpha)} \otimes D^{(\alpha)}]_+$ and $(D^{(\alpha)} \otimes D^{(\alpha)})_-$, respectively. The characters follow from (4.4.15) and (4.4.12):

$$\chi_\pm^{(\alpha\times\alpha)}(a) = \tfrac{1}{2}\{[\chi^{(\alpha)}(a)]^2 \pm \chi^{(\alpha)}(a^2)\} \; . \tag{4.4.16}$$

The tensor product space $\mathscr{L}^{(\alpha)} \otimes \mathscr{L}^{(\alpha)}$ can therefore be divided into a symmetric space $[\mathscr{L}^{(\alpha)} \otimes \mathscr{L}^{(\alpha)}]_+$ and an antisymmetric one $(\mathscr{L}^{(\alpha)} \otimes \mathscr{L}^{(\alpha)})_-$; obviously

$$D^{(\alpha)} \otimes D^{(\alpha)} = [D^{(\alpha)} \otimes D^{(\alpha)}]_+ \oplus (D^{(\alpha)} \otimes D^{(\alpha)})_- \; . \tag{4.4.17}$$

However, the symmetric and antisymmetric parts are in general reducible. If the bases $\{e_i^{(\alpha)}\}$ and $\{e_k'^{(\alpha)}\}$ are identical then the REP is always symmetric. For $\mathscr{C}_{3v}$ we have the trivial relations

$$[A_1 \otimes A_1]_+ = A_1 \; , \qquad (A_1 \otimes A_1)_- = 0 \; ,$$

$$[A_2 \otimes A_2]_+ = A_1 \; , \qquad (A_2 \otimes A_2)_- = 0 \; ,$$

and, using (4.2.31a) and (4.4.16)

Table 4.7. Character table for $[E \otimes E]_+$ and $(E \otimes E)_-$ for the symmetry group $\mathscr{C}_{3v}$

$\mathscr{C}_{3v}$	e	$2c_3$	$3\sigma_v$
$\chi_+^{E \times E}$	3	0	1
$\chi_-^{E \times E}$	1	1	-1

$$[E \otimes E]_+ = A_1 \oplus E \ , \qquad (E \otimes E)_- = A_2 \ .$$

Table 4.7 gives the character table for these two products.

The reduction of $D^{(\alpha \times \beta)}$ is achieved according to (4.2.19) by a nonsingular matrix $C^{\alpha\beta}$, which transforms the basis $\{e_i^{(\alpha)} e_k^{(\beta)}\}$ into a new basis $\{e_l^{(\gamma,s)} | s = 1, 2, \ldots, (\alpha\beta|\gamma)\}$, which is classified according to the IRs $D^{(\gamma)}$ and their multiplicities $(\alpha\beta|\gamma)$. The matrix elements of $C^{\alpha\beta}$ are denoted by $\left(\begin{smallmatrix}\alpha\beta \\ ik\end{smallmatrix}\middle|\begin{smallmatrix}\gamma s \\ l\end{smallmatrix}\right)$ and called *Clebsch-Gordan* or *Wigner coefficients* (CGC). Then

$$e_l^{(\gamma,s)} = \sum_{ik} \left(\begin{matrix}\alpha\beta \\ ik\end{matrix}\middle|\begin{matrix}\gamma s \\ l\end{matrix}\right) e_i^{(\alpha)} e_k^{(\beta)} \ ,$$

$$(C^{\alpha\beta})^{-1} D^{(\alpha \times \beta)} C^{\alpha\beta} = \left\{\begin{matrix} D^{(\gamma,1)} & & \\ & \ddots & D^{(\gamma,m_\gamma)} \\ & & \ddots \end{matrix}\right\} , \quad \begin{matrix} m_\gamma = (\alpha\beta|\gamma) \ , \\ \gamma = 1, \ldots \ , \end{matrix} \tag{4.4.18a}$$

$$\sum_\gamma (\alpha\beta|\gamma) d_\gamma = d_\alpha d_\beta \ .$$

If the basis systems $\{e_i^{(\alpha)} e_k^{(\beta)}\}$ and $\{e_l^{(\gamma,s)}\}$ are orthonormal, then $C^{\alpha\beta}$ is unitary and the CGC obey

$$\sum_{ik} \left(\begin{matrix}\alpha\beta \\ ik\end{matrix}\middle|\begin{matrix}\gamma' s' \\ l'\end{matrix}\right)^* \left(\begin{matrix}\alpha\beta \\ ik\end{matrix}\middle|\begin{matrix}\gamma s \\ l\end{matrix}\right) = \delta_{\gamma\gamma'} \delta_{ss'} \delta_{ll'} \ , \tag{4.4.19}$$

$$\sum_{\gamma s l} \left(\begin{matrix}\alpha\beta \\ ik\end{matrix}\middle|\begin{matrix}\gamma s \\ l\end{matrix}\right) \left(\begin{matrix}\alpha\beta \\ i'k'\end{matrix}\middle|\begin{matrix}\gamma s \\ l\end{matrix}\right)^* = \delta_{ii'} \delta_{kk'} \ ,$$

and therefore also

$$e_i^{(\alpha)} e_k^{(\beta)} = \sum_{\gamma s l} \left(\begin{matrix}\alpha\beta \\ ik\end{matrix}\middle|\begin{matrix}\gamma s \\ l\end{matrix}\right)^* e_l^{(\gamma,s)} \ , \tag{4.4.18b}$$

The CGC can be chosen real for many groups so that a few relations are simplified.

4.4.3 Simply Reducible Groups

When the reduction coefficients (multiplicities) $(\alpha\beta|\gamma)$ assume only the values 0 or 1, and a^{-1}, $a \in \mathscr{G}$ lie in the same (conjugacy) class, then the groups are said to be *simply reducible*. Most of the point groups, but also $\mathcal{SO}(3)$ and $\mathcal{SU}(2)$ are simply reducible, whereas double groups are not. In simply reducible groups the index s is unnecessary and some further simplifications follow, especially for the CGC, which are uniquely determined apart from a phase factor. If the multiplicities obey $(\alpha\beta|\gamma) > 1$ then the CGC are no longer unique. Using (4.4.13a) we have from (4.4.18a)

$$D^{(\alpha\times\beta)} = C^{\alpha\beta}\sum_{\gamma}(\alpha\beta|\gamma)D^{(\gamma)}(C^{\alpha\beta})^{-1} = C^{\alpha\beta}\sum_{\gamma s}D^{(\gamma,s)}(C^{\alpha\beta})^{-1}$$

or

$$D^{(\alpha\times\beta)}_{ik,jl}(a) = D^{(\alpha)}_{ij}(a)D^{(\beta)}_{kl}(a) = \sum_{\substack{\gamma s\\ mn}}\left(\begin{smallmatrix}\alpha\beta\\ ik\end{smallmatrix}\middle|\begin{smallmatrix}\gamma s\\ m\end{smallmatrix}\right)D^{(\gamma s)}_{mn}(a)\left(\begin{smallmatrix}\alpha\beta\\ jl\end{smallmatrix}\middle|\begin{smallmatrix}\gamma s\\ n\end{smallmatrix}\right)^{*} .$$

Multiplying this by $D^{(\gamma')*}_{m'n'}(a)$ and summing over all $a \in \mathcal{G}$, we obtain with (4.2.25a)

$$\sum_{s}\left(\begin{smallmatrix}\alpha\beta\\ ik\end{smallmatrix}\middle|\begin{smallmatrix}\gamma s\\ m\end{smallmatrix}\right)\left(\begin{smallmatrix}\alpha\beta\\ jl\end{smallmatrix}\middle|\begin{smallmatrix}\gamma s\\ n\end{smallmatrix}\right)^{*} = \frac{d_\gamma}{g}\sum_{a\in\mathcal{G}}D^{(\alpha)}_{ij}(a)D^{(\beta)}_{kl}(a)D^{(\gamma)*}_{mn}(a) , \tag{4.4.20}$$

in which for simply reducible groups the sum over s on the left hand side drops out. In this case we can calculate the CGC easily: we take $i = j$, $k = l$, $m = n$ and obtain (except for a phase) at least one CGC which is different from zero. Then we keep, for example, i, k, m constant and vary j, l, n. In this way we determine the other coefficients together with their relative phases for these fixed i, k, m (in real REPs of simply reducible groups the absolute values of the CGC are totally symmetric).

For simply reducible groups we obtain by means of such a procedure all the CGC and can thus carry out the transformation explicitly. With such groups all related physical problems can be solved by symmetry considerations. Another method of determining the CGC consists in the application of the generators of the group to (4.4.18a), followed by a comparison of the coefficients (compare Exercise 4.11 and Table 8.1). For the example (see Table 4.6)

$$\mathcal{L}^{E\times E} = \mathcal{L}^{A_1} \oplus \mathcal{L}^{A_2} \oplus \mathcal{L}^{E}$$

we obtain for the CGC, according to (4.4.20),

$$\left(\begin{matrix}EE\\ ik\end{matrix}\middle|\begin{matrix}A_1\\ 1\end{matrix}\right) = \frac{1}{\sqrt{2}}\begin{pmatrix}1 & 0\\ 0 & 1\end{pmatrix} , \qquad \left(\begin{matrix}EE\\ ik\end{matrix}\middle|\begin{matrix}A_2\\ 1\end{matrix}\right) = \frac{1}{\sqrt{2}}\begin{pmatrix}0 & 1\\ -1 & 0\end{pmatrix} ,$$

$$\left(\begin{matrix}EE\\ ik\end{matrix}\middle|\begin{matrix}E\\ 1\end{matrix}\right) = \frac{1}{\sqrt{2}}\begin{pmatrix}0 & 1\\ 1 & 0\end{pmatrix} , \qquad \left(\begin{matrix}EE\\ ik\end{matrix}\middle|\begin{matrix}E\\ 2\end{matrix}\right) = \frac{1}{\sqrt{2}}\begin{pmatrix}1 & 0\\ 0 & -1\end{pmatrix} .$$

If f_1, f_2 and g_1, g_2 are basis functions belonging to $\mathcal{L}^{(E)}$, then the basis functions in $\mathcal{L}^{(E\times E)}$ are, from (4.4.18a),

$$e^{(A_1)} = \frac{1}{\sqrt{2}}(f_1 g_1 + f_2 g_2) , \quad e^{(A_2)} = \frac{1}{\sqrt{2}}(f_1 g_2 - f_2 g_1) ,$$
$$e_1^{(E)} = \frac{1}{\sqrt{2}}(f_1 g_2 + f_2 g_1) , \quad e_2^{(E)} = \frac{1}{\sqrt{2}}(f_1 g_1 - f_2 g_2) \tag{4.4.21}$$

apart from phase factors.

Exercise 4.8. Show that if A and B are unitary matrices then $A \otimes B$ is also a unitary matrix: $(A \otimes B)^{+} = (A \otimes B)^{-1}$

Exercise 4.9. Construct the character table of $\mathscr{D}_{2h}$ from that of $\mathscr{D}_2$.

Exercise 4.10. Discuss the addition of angular momenta in quantum mechanics and satisfy yourself that this is just the Clebsch-Gordan expansion of the group $\mathscr{SO}(3)$.

Exercise 4.11. Consider the determination of the CGC with the help of the projectors (4.3.9a) by applying $P_{mn}^{(\gamma)}$ to $e_i^{(\alpha)} e_k^{(\beta)}$ and decomposing the result.

Exercise 4.12. In (4.4.21) put $f_1 = x$, $f_2 = y$ and $g_1 = x^2 - y^2$, $g_2 = -2xy$ and determine the new bases. See Table 5.1.

5. Irreducible Representations of Special Groups

In the preceding chapter we discussed the general properties of representations. For applications we have to specify these REPs for the different groups occurring in physics. In solid-state physics these are the irreducible representations of the point groups, including the double groups and the corepresentations for the magnetic (colour) groups together with the representations of the translation group for the lattices. In many-particle systems the representations of the permutation groups are also of interest. A large part of this chapter is devoted to the representations of the general linear group and its subgroups, which are the essential groups in atomic physics and modern theories of particles.

5.1 Point and Double Point Groups

Once the IRs are known, projectors and other quantities can be given explicitly. In principle, these can be determined from the regular REP (Sect. 4.3.1), which contains every IR. For this, however, we have to transform D^{reg} into block diagonal form, which is not always an easy task. In some cases the IRs can be found by "guessing", e.g. the totally symmetric identity REP, which exists for every (finite) group. The rotation matrices in $\mathbb{R}_3$ also constitute (with properly chosen angles and axes of rotation) REPs of point groups, being sometimes irreducible (for $\mathscr{T}$, $\mathscr{O}$, etc.) and sometimes reducible (see $\mathscr{C}_{3v}$ in Sect. 4.2.1). The latter can be generalized. Let $\mathscr{L}$ be an n-dimensional linear vector space, which is defined by n linearly independent functions or vectors $\{f^{(k)}\}$. Application of the symmetry operators $P_a \in \mathscr{G}$ to the functions of $\mathscr{L}$ induces a REP of $\mathscr{G}$, which is generally reducible (4.2.6). This REP then has to be reduced. In the case that we do not obtain all the IRs in this way we have to start with another space $\mathscr{L}$, which is often of higher dimension (Sect. 4.2.1).

The REP matrices evidently depend on $\mathscr{L}$. But they can be standardized by a similarity transformation (standard REPs). For the point groups that are subgroups of $\mathscr{SO}(3)$ we obtain the standard REPs via the spherical harmonics Y^l_m. These Y^l_m are the basis functions of the IRs $D^{(l)}$ of the rotation group $\mathscr{SO}(3)$ and according to Sect. 4.2.5 constitute a (generally reducible) basis for the REPs of point groups. The reduction yields the IRs of the point groups in standardized form. The spherical harmonics (ϑ, φ: angles of spherical polar coordinates)

$$Y^l_m(\vartheta,\varphi) = \frac{1}{\sqrt{2\pi}}\theta^l_m(\vartheta)e^{im\varphi}\ , \qquad m \geqslant 0\ , \tag{5.1.1a}$$

$$\theta^l_m(\vartheta) = \left(\frac{2l+1}{2}\cdot\frac{(l-m)!}{(l+m)!}\right)^{1/2}\frac{(-1)^m}{2^l l!}\sin^m\vartheta\left(\frac{\partial}{\partial\cos\vartheta}\right)^{l+m}(\cos^2\vartheta-1)^l$$

$$Y^l_{-m} = (-1)^m Y^{l*}_m\ , \qquad -l \leqslant m \leqslant l\ ,$$

constitute a $(2l+1)$-dimensional linear space for the REPs of point groups; often it is useful to choose real functions, thus $(m \neq 0)$

$$Y^l_{m,c} = \frac{1}{\sqrt{2}}[(-1)^m Y^l_m + Y^l_{-m}] = (-1)^m\frac{1}{\sqrt{\pi}}\theta^l_m\cos m\varphi\ ,$$

$$Y^l_{m,s} = \frac{-\mathrm{i}}{\sqrt{2}}[(-1)^m Y^l_m - Y^l_{-m}] = (-1)^m\frac{1}{\sqrt{\pi}}\theta^l_m\sin m\varphi\ . \tag{5.1.1b}$$

For those point groups with one main axis, these real functions are bases for the standard REPs. In the case of cubic point groups we have to choose linear combinations of spherical harmonics adapted to the cubic symmetry: the cubic harmonics are different from the real spherical harmonics for $l \geqslant 3$. The spherical harmonics are denoted as $s-$, $p-$, $d-$, $f-$, ..., or more detailed as p_x, p_y, ..., according to their use in quantum mechanics (see Table 5.1).

In $\mathscr{C}_{3v}$ we choose as a basis e.g. $Y^1_{1c} \sim x$; $Y^1_{1s} \sim y$; $Y^1_0 \sim z$ and again obtain the standard REPs E and A_1 by using P_a [see (2.1.18) and Table 4.1]. Alternatively we could have started with functions belonging to $l = 2$, e.g. $Y^2_{2s} \sim -2xy = f_1$

Table 5.1. Real linear combinations of spherical harmonics, according to (5.1.1), for $l = 0$ to $l = 3$

Y^l_m		Normalization	Spherical coordinates	Cartesian coordinates
Y^0_0	s	$\sqrt{1/4\pi}$	1	1
Y^1_0	p_z	$\sqrt{3/4\pi}$	$\cos\vartheta$	z
Y^1_{1c}	p_x	$\sqrt{3/4\pi}$	$\sin\vartheta\cos\varphi$	x
Y^1_{1s}	p_y	$\sqrt{3/4\pi}$	$\sin\vartheta\sin\varphi$	y
Y^2_0	d_0	$\sqrt{5/16\pi}$	$(3\cos^2\vartheta-1)$	$3z^2-1=2z^2-x^2-y^2$
Y^2_{1c}	d_{1c}	$\sqrt{15/4\pi}$	$\sin\vartheta\cos\vartheta\cos\varphi$	zx
Y^2_{1s}	d_{1s}	$\sqrt{15/4\pi}$	$\sin\vartheta\cos\vartheta\sin\varphi$	zy
Y^2_{2c}	d_{2c}	$\sqrt{15/16\pi}$	$\sin^2\vartheta\cos 2\varphi$	x^2-y^2
Y^2_{2s}	d_{2s}	$\sqrt{15/16\pi}$	$\sin^2\vartheta\sin 2\varphi$	$2xy$
Y^3_0	f_0	$\sqrt{7/16\pi}$	$\cos\vartheta(5\cos^2\vartheta-3)$	$z(5z^2-3)=z(2z^2-3x^2-3y^2)$
Y^3_{1c}	f_{1c}	$\sqrt{21/32\pi}$	$\sin\vartheta(5\cos^2\vartheta-1)\cos\varphi$	$x(5z^2-1)=x(4z^2-x^2-y^2)$
Y^3_{1s}	f_{1s}	$\sqrt{21/32\pi}$	$\sin\vartheta(5\cos^2\vartheta-1)\sin\varphi$	$y(5z^2-1)=y(4z^2-x^2-y^2)$
Y^3_{2c}	f_{2c}	$\sqrt{105/16\pi}$	$\sin^2\vartheta\cos\vartheta\cos 2\varphi$	$z(x^2-y^2)$
Y^3_{2s}	f_{2s}	$\sqrt{105/16\pi}$	$\sin^2\vartheta\cos\vartheta\sin 2\varphi$	$2xyz$
Y^3_{3c}	f_{3c}	$\sqrt{35/32\pi}$	$\sin^3\vartheta\cos 3\varphi$	$x(x^2-3y^2)$
Y^3_{3s}	f_{3s}	$\sqrt{35/32\pi}$	$\sin^3\vartheta\sin 3\varphi$	$y(3x^2-y^2)$

and $Y_{2c}^2 \sim x^2 - y^2 = f_2$ and obtained, using (4.2.3) and Fig. 2.1 (Exercise 5.1),

$$P_{c_3} \cdot (f_1, f_2) = (f_1, f_2) \cdot \begin{pmatrix} -1/2 & -\sqrt{3}/2 \\ +\sqrt{3}/2 & -1/2 \end{pmatrix}, \tag{5.1.2}$$

$$P_{\sigma_v} \cdot (f_1, f_2) = (f_1, f_2) \cdot \begin{pmatrix} -1 & 0 \\ 0 & +1 \end{pmatrix},$$

i.e. f_1 and f_2 also constitute the IR E of $\mathscr{C}_{3v}$, whereas z^2 represents the A_1 REP. In character tables the basis functions belonging to the REPs as well as some components of axial vectors (e.g. angular momentum $\boldsymbol{L}$) are additionally given (Table 5.2).

The double point groups assigned to the point groups are defined by $\mathscr{G}^{\mathrm{D}} = \mathscr{G} + c_0 \mathscr{G}$ (Sect. 3.3), where $\mathscr{G} \subseteq \mathscr{SO}(3)$. Determination of the characters according to (4.2.33) requires knowledge of the class multiplication coefficients, and thus the classes of $\mathscr{G}^{\mathrm{D}}$ must be known. These can be obtained from the multiplication table of $\mathscr{G}^{\mathrm{D}}$, which can be constructed with the help of the matrices $D^{(1/2)}(\alpha, \beta, \gamma)$ according to (3.3.2).

The quotient group $\mathscr{G}^{\mathrm{D}}/\{e, c_0\}$ is isomorphic to $\mathscr{G}$. Thus a number of IRs of $\mathscr{G}^{\mathrm{D}}$ can be obtained directly from the IRs of $\mathscr{G}$. These REPs are "*single-valued*", they have $D(c_0) = D(e)$. The other REPs of $\mathscr{G}^{\mathrm{D}}$ are "*double-valued*" and are called *extra* or *spinor* REPs of $\mathscr{G}$. These have $D(c_0) = -D(e)$.

Some examples may illustrate this.

(i) The cyclic point groups $\mathscr{C}_n^{\mathrm{D}} = \{e, c_n, \ldots, c_n^{n-1}, c_0 = c_n^n, c_0 c_n, \ldots, c_0 c_n^{n-1}\}$ have order $2n$. Every element defines a separate class; there are $2n$ one-dimensional IRs with characters from $\{\chi(c_n)\}^{2n} = 1$, thus $\chi(c_n) = \exp(i\pi m/n)$ with $m = 1, 2, \ldots, 2n$ or $m = 0, 1, \ldots, 2n - 1$, thus we obtain Table 5.3.

Table 5.2. Character table of $\mathscr{C}_{3v}$ with assigned functions

$\mathscr{C}_{3v}$	K_e	K_3	K_σ	s	p	d	$\mathbf{L}$
A_1	1	1	1	1	z	z^2	
A_2	1	1	-1				L_z
E	2	-1	0		x, y	$-2xy, x^2 - y^2; xz, yz$	L_x, L_y

Table 5.3. Character table of $\mathscr{C}_2^{\mathrm{D}}$

$\mathscr{C}_2^{\mathrm{D}}$	e	c_2	c_0	$c_0 c_2$	
A	1	1	1	1	single-valued
B	1	-1	1	-1	
$\bar{A}$	1	i	-1	$-i$	double-valued
$\bar{B}$	1	$-i$	-1	i	extra REP

Table 5.4. The extra REP of $\mathcal{D}_2^{\mathrm{D}}$

$\mathcal{D}_2^{\mathrm{D}}$	e	c_0	$c_2'(x)$	c_0c_2'	$c_2''(y)$	c_0c_2''	$c_2(z)$	c_0c_2
$\bar{E}$	$\begin{pmatrix}1&0\\0&1\end{pmatrix}$	$\begin{pmatrix}-1&0\\0&-1\end{pmatrix}$	$\begin{pmatrix}0&i\\i&0\end{pmatrix}$	$\begin{pmatrix}0&-i\\-i&0\end{pmatrix}$	$\begin{pmatrix}0&1\\-1&0\end{pmatrix}$	$\begin{pmatrix}0&-1\\1&0\end{pmatrix}$	$\begin{pmatrix}i&0\\0&-i\end{pmatrix}$	$\begin{pmatrix}-i&0\\0&i\end{pmatrix}$

(ii) The group $\mathcal{D}_2^{\mathrm{D}}$ (Sect. 3.3) is divided into 5 classes. That means there is one extra REP which is just given by the corresponding matrices of $D^{(1/2)}$, as illustrated in Table 5.4.

(iii) The group $\mathcal{O}^{\mathrm{D}}$ is different from $\mathcal{O}$ in two respects. (1) There are IRs of dimension 4, whereas the crystallographic point groups only have REPs up to dimension 3 (the non-crystallographic[1] group $\mathcal{Y}$ of course has REPs of dimension 5). (2) For the construction of the REP matrices we need the spinor space of rank 1. The elements of this space can be given as products (Sect. 8.4)

$$\Psi(\boldsymbol{x},\sigma)=\psi(\boldsymbol{x})\cdot\chi(\sigma) \tag{5.1.3}$$

with $\sigma=\pm$ (spin variable). The spin function $\chi(\sigma)$ in the basis $\{|+\rangle,|-\rangle\}$ has the REP

$$|\chi\rangle=\langle+|\chi\rangle\cdot|+\rangle+\langle-|\chi\rangle\cdot|-\rangle\ . \tag{5.1.4}$$

The spin function

$$\chi(\sigma):=\langle\sigma|\chi\rangle=\delta_{\sigma+}\langle+|\chi\rangle+\delta_{\sigma-}\langle-|\chi\rangle \tag{5.1.5}$$

is the probability amplitude of finding the electronic spin component $\hbar\sigma/2$ with respect to a given axis. The basis spinors have the REP

$$\begin{aligned}\langle\sigma|+\rangle&:=\alpha(\sigma)=\delta_{\sigma+}=\begin{pmatrix}\langle+|+\rangle\\\langle-|+\rangle\end{pmatrix}=\begin{pmatrix}1\\0\end{pmatrix},\\\langle\sigma|-\rangle&:=\beta(\sigma)=\delta_{\sigma-}=\begin{pmatrix}\langle+|-\rangle\\\langle-|-\rangle\end{pmatrix}=\begin{pmatrix}0\\1\end{pmatrix}.\end{aligned} \tag{5.1.6}$$

They transform according to

$$P_a\{|+\rangle,|-\rangle\}=\{|+\rangle,|-\rangle\}\cdot D^{(1/2)}(\alpha,\beta,\gamma) \tag{5.1.7}$$

for $a\in\mathcal{G}$ and $D^{(1/2)}$, see (3.3.2). On the other hand, the spherical harmonics (5.1.1a) transform according to

$$P_aY_m^l=\sum_{m'=-l}^{l}D^{(l)}_{m'm}(a(\alpha,\beta,\gamma))\,Y_{m'}^l\ , \tag{5.1.8}$$

[1] But see the recent theory of quasi-crystals!

see (7.1.23). Thus the behaviour of product functions (5.1.3) in transformations is described. Let $\{f_i^{(\alpha)}\}$ be a basis for the single-valued REP $D^{(\alpha)}$ of $\mathscr{G}$, e.g. a real spherical harmonic (5.1.1b), and let $D^{(1/2)}$ be the REP subduced onto $\mathscr{G}$ with basis $\{|+>, |->\}$, then $\{|f_i^{(\alpha)}, +>, |f_i^{(\alpha)}, ->\}$ is a basis for the inner Kronecker product

$$D^{(\alpha)\times(1/2)} = D^{(\alpha)} \otimes D^{(1/2)} = \begin{pmatrix} D^{(\alpha)}D_{11}^{(1/2)} & D^{(\alpha)}D_{12}^{(1/2)} \\ D^{(\alpha)}D_{21}^{(1/2)} & D^{(\alpha)}D_{22}^{(1/2)} \end{pmatrix} . \tag{5.1.9}$$

Thus $D^{(\alpha)\times(1/2)}$ is a double-valued (extra) REP of $\mathscr{G}$, which might be reducible. However, often $D^{(\alpha)}$ can be chosen so that $D^{(\alpha)\times(1/2)}$ is a double-valued IR.

For $\mathscr{O}^{\mathrm{D}}$ there are five single-valued IRs: A_1, A_2, E, F_1, F_2 (see Appendix A). With the REP of $D^{(1/2)} \in \mathscr{SU}(2)$ subduced onto $\mathscr{O}$, $A_1 \otimes D^{(1/2)}(a)$, $a \in \mathscr{O}$ is a two-dimensional REP $\bar{E}_1$ (Appendix A). From the criterion (4.2.32) we obtain

$$\sum_{a \in \mathscr{O}^{\mathrm{D}}} |\chi^{\bar{E}_1}(a)|^2 = 48 \ ,$$

i.e. $\bar{E}_1$ is an IR of $\mathscr{O}^{\mathrm{D}}$. Correspondingly we get another two-dimensional REP $A_2 \otimes D^{(1/2)}(a) = \bar{E}_2$. Finally we form $E \otimes D^{(1/2)}(a) = \bar{G}$ and have a four-dimensional REP which proves to be irreducible, too. The IR matrices follow from (5.1.9). Using (4.2.26b) we have for these eight IRs $\sum d_\alpha^2 = 48 = \mathrm{ord}\, \mathscr{O}^{\mathrm{D}}$, i.e. there can be no further IRs of $\mathscr{O}^{\mathrm{D}}$. The products $F_{1,2} \otimes D^{(1/2)}(a)$ thus yield two 6-dimensional reducible REPs, the irreducible parts of which are equivalent to one of the REPs above. Representations of the generators of the point groups are given in Appendix B.

Exercise 5.1. Show the validity of (5.1.2).

Exercise 5.2. Show that the IRs (i.e. characters!) of the groups $\mathscr{C}_n$ are given by $D^{(m)}(c_n) = \chi^{(m)}(c_n) = \exp(-\mathrm{i}2\pi m/n)$ with $m = 1, \ldots, n$ and that Y_m^l are bases of the IRs of $\mathscr{C}_n$.

5.2 Magnetic Point Groups. Time Reversal

In Sect. 3.2 we defined the colour groups which are realized as magnetic groups in physics. The operator for changing colour corresponds to the time reversal operator ϑ (reversal of motion), which is antilinear and antiunitary:

$$\vartheta(c_1 f_1 + c_2 f_2) = c_1^* \vartheta f_1 + c_2^* \vartheta f_2 \ , \tag{5.2.1a}$$

$$<\vartheta f_1 | \vartheta f_2> \ = \ <f_1 | f_2>^* \ = \ <f_2 | f_1> \ , \tag{5.2.1b}$$

$$\vartheta f(\boldsymbol{x}) = f^*(\boldsymbol{x}) \text{ in a direct-space REP } (\boldsymbol{x}\text{-REP}). \tag{5.2.1c}$$

Thus, in a magnetic group (3.2.1-3)

$$\mathcal{M} = \mathcal{G} + r\mathcal{G} \ , \qquad r \notin \mathcal{G} \ , \qquad \mathcal{G}\text{: point group} \ , \tag{5.2.2}$$

only half the elements ($a \in \mathcal{G}$) forming an invariant subgroup with index 2 are unitary. The decomposition into cosets (5.2.2) is always valid, since the product of two antiunitary elements is again a unitary one; r in (5.2.2) may be any element of $r\mathcal{G}$. The magnetic groups $\mathcal{M}_{\text{II}}$ according to (3.2.2) (e.g. paramagnetic systems) contain the time-reversal operator ϑ itself. In this case we can take $r = \vartheta$. But in the groups $\mathcal{M}_{\text{III}}$ (e.g. ferro- and antiferromagnetic systems) ϑ only occurs in combination with a unitary element $a' \notin \mathcal{G}$, e.g. $r = \vartheta a'$. Since ϑ commutes with every $a \in \mathcal{G}$, there exists a group $\mathcal{M}' = \mathcal{G} + a'\mathcal{G}$, which is isomorphic to $\mathcal{M}_{\text{III}}$ in this case [see the construction of $\mathcal{M}_{\text{III}}$ in (3.2.3)]. The occurrence of the antiunitary elements in $r\mathcal{G}$ requires a modification of the REP theory, which is called the theory of *corepresentations* [5.1].

Let $a, b, \ldots,$ be elements of $\mathcal{G}$, and $r, s, t, \ldots$ elements of $r\mathcal{G}$. If Δ is a unitary IR of $\mathcal{G}$ with dimension d and basis $\boldsymbol{f} = \{f_1, \ldots, f_d\}$, using (4.2.6) we have

$$P_a \boldsymbol{f} = \boldsymbol{f}\Delta(a) \ . \tag{5.2.3}$$

Furthermore, let $\boldsymbol{g} = \{g_1, \ldots, g_d\}$ be the functions that result from applying P_r {$r \in r\mathcal{G}$, but fixed) to $\boldsymbol{f}$, thus

$$P_r \boldsymbol{f} = \boldsymbol{g} \ . \tag{5.2.4}$$

Using (5.2.1a,3) we have, since $r^{-1}ar \in \mathcal{G}$,

$$P_a \boldsymbol{g} = P_a P_r \boldsymbol{f} = P_r(P_r^{-1} P_a P_r)\boldsymbol{f} = P_r \boldsymbol{f}\Delta(r^{-1}ar) = \boldsymbol{g} \cdot \Delta^*(r^{-1}ar) \ . \tag{5.2.5}$$

A $2d$-dimensional vector $\boldsymbol{h} = \{\boldsymbol{f}, \boldsymbol{g}\} = \{f_1, \ldots, f_d, g_1, \ldots, g_d\}$ then obeys

$$P_a \boldsymbol{h} = P_a\{\boldsymbol{f}, \boldsymbol{g}\} = \{\boldsymbol{f}, \boldsymbol{g}\} D^M(a) = \boldsymbol{h} \cdot D^M(a)$$

with

$$D^M(a) = \begin{pmatrix} \Delta(a) & 0 \\ 0 & \Delta^*(r^{-1}ar) \end{pmatrix} \text{ for each } a \in \mathcal{G} \ . \tag{5.2.6}$$

Correspondingly, for an element $s = ra \in r\mathcal{G}$

$$\begin{aligned} P_s \boldsymbol{h} &= P_s\{\boldsymbol{f}, \boldsymbol{g}\} = P_r P_a\{\boldsymbol{f}, \boldsymbol{g}\} = \{P_r \boldsymbol{f}\Delta(a); P_s P_r \boldsymbol{f}\} \\ &= \{\boldsymbol{g}\Delta^*(a); \boldsymbol{f}\Delta(sr)\} := \boldsymbol{h} D^M(s) \text{ with} \end{aligned} \tag{5.2.7}$$

$$D^M(s) = \begin{pmatrix} 0 & \Delta(sr) \\ \Delta^*(r^{-1}s) & 0 \end{pmatrix} \text{ for each } s \in r\mathcal{G} \ .$$

The unitary matrices (5.2.6,7) constitute the corepresentation (COR) D^M of $\mathcal{M}$ derived from Δ. The COR does not define a usual homomorphism (4.2.8a), instead it satisfies with (5.2.6,7)

$$D^M(a)\cdot D^M(b) = D^M(ab)\ , \qquad D^M(s)\cdot D^{M*}(a) = D^M(sa)\ ,$$
$$D^M(a)\cdot D^M(s) = D^M(as)\ , \qquad D^M(s)\cdot D^{M*}(t) = D^M(st)\ , \tag{5.2.8}$$

Now let S be a unitary transformation from the basis $\boldsymbol{h}$ to another basis $\boldsymbol{h}'$ with

$$\boldsymbol{h}' = \{f', g'\} = \{f, g\}S = \boldsymbol{h}S\ . \tag{5.2.9}$$

Then the new CORs $D^{M'}$ transform according to

$$D^{M'}(a) = S^{-1}D^M(a)S\ , \qquad \text{for} \quad a \in \mathscr{G}\ ,$$
$$D^{M'}(s) = S^{-1}D^M(s)S^*\ , \qquad \text{for} \quad s \in r\mathscr{G}\ , \tag{5.2.10}$$

Equation (5.2.10) enables us to define the equivalence between two CORs D^M and $D^{M'}$. Likewise the reducibility can be introduced as for common groups: If it is possible to decompose the space with basis $\boldsymbol{h}$ into subspaces in such a way that the latter remain invariant when applying all the elements of $\mathscr{M}$ to $\boldsymbol{h}$, then D^M is reducible. Otherwise D^M is irreducible. The reducibility implies a transformation of the matrices of D^M into block-diagonal form.

The reducibility of D^M depends on the relation between the unitary IRs $\Delta(a)$ and $\bar{\Delta}(a) := \Delta^*(r^{-1}ar)$ of the subgroup $\mathscr{G} \subset \mathscr{M}$. In addition, $\bar{\Delta}(a)$ is a REP of $\mathscr{G}$ because $r^{-1}\, ar \in \mathscr{G}$. Now we have to distinguish between two or three cases.

1) $\Delta(a)$ and $\bar{\Delta}(a)$ are equivalent, i.e. there exists a unitary matrix S with

$$\Delta(a) = S\cdot\Delta^*(r^{-1}ar)\cdot S^{-1} = S\cdot\bar{\Delta}(a).S^{-1} \text{ for each } a \in \mathscr{G}\ , \tag{5.2.11}$$

and further

$$S\cdot S^* = \pm\Delta(r^2) \tag{5.2.12}$$

(for a proof see [5.2]). With

$$\hat{S} = \begin{pmatrix} 1 & 0 \\ 0 & S^{-1} \end{pmatrix}$$

and (5.2.10) we then obtain ($S \to \hat{S}$)

$$D^{M'}(a) = \begin{pmatrix} \Delta(a) & 0 \\ 0 & \Delta(a) \end{pmatrix}, \quad D^{M'}(s) = \begin{pmatrix} 0 & \Delta(sr)S^{*-1} \\ S\Delta^*(r^{-1}s) & 0 \end{pmatrix}. \tag{5.2.13}$$

a) If $S\cdot S^* = +\Delta(r^2)$, then there exists a unitary matrix $\left[\frac{1}{\sqrt{2}}\begin{pmatrix} 1 & -1 \\ 1 & 1 \end{pmatrix}\right]$ which diagonalizes $D^{M'}$, i.e. D^M *is reducible* into

$$D^{M''}(a) = \begin{pmatrix} \Delta(a) & 0 \\ 0 & \Delta(a) \end{pmatrix}, \quad D^{M''}(s) = \begin{pmatrix} \Delta(sr^{-1})\cdot S & 0 \\ 0 & -\Delta(sr^{-1})\cdot S \end{pmatrix}. \tag{5.2.14}$$

b) If $S \cdot S^* = -\Delta(r^2)$, then such a unitary matrix does not exist: This means D^M *is irreducible.*

2) $\Delta(a)$ and $\bar{\Delta}(a) = \Delta^*(r^{-1}ar)$ are inequivalent. Then the COR of $\mathcal{M}$ is *irreducible*, therefore using (5.2.6,7) we get

$$D^M(a) = \begin{pmatrix} \Delta(a) & 0 \\ 0 & \bar{\Delta}(a) \end{pmatrix}, \quad D^M(s) = \begin{pmatrix} 0 & \Delta(sr) \\ \bar{\Delta}(sr^{-1}) & 0 \end{pmatrix}. \tag{5.2.15}$$

In general, it can be shown that all the unitary ICORs of $\mathcal{M}$ can be obtained from the unitary IRs Δ of $\mathcal{G}$ in this way [5.1].

For practical reasons it is essential to know which of the three cases above is realized if $\Delta(a)$ is a given REP. The following criterion can be used. It is similar to that given in (4.2.35):

$$\sum_{s \in r\mathcal{G}} \chi(s^2) = \begin{cases} g & \text{in case 1a)} \\ -g & \text{in case 1b)} \\ 0 & \text{in case 2)} \end{cases} \tag{5.2.16}$$

with $g = \operatorname{ord} \mathcal{G}$ and $\chi(s^2)$ as the character of $\Delta(s^2)$ with $s^2 \in \mathcal{G}$.

Now let us consider the case that ϑ itself is contained in $\mathcal{M}$($\mathcal{M}_{\text{II}}$, paramagnetic systems), then $r = \vartheta$ and $s^2 = \vartheta a \vartheta a = \vartheta^2 a^2$. The eigenvalues of ϑ^2 (8.2.11) are $\lambda = +1$ (systems with an even number of electrons or without spin) and $\lambda = -1$ (systems with an odd number of electrons taking spin into account, see Sect. 8.2). Then it follows from (5.2.16) that

$$\sum_{a \in \mathcal{G}} \chi(a^2) = \begin{cases} \lambda g & \text{in case 1a)} \\ -\lambda g & \text{in case 1b)} \\ 0 & \text{in case 2)} \end{cases} \tag{5.2.17a}$$

with

$$\lambda = \begin{cases} 1 & \text{for an even number of electrons or no spin ,} \\ -1 & \text{for an odd number of electrons with spin .} \end{cases} \tag{5.2.17b}$$

In this case we have in addition $\bar{\Delta}(a) = \Delta^*(\vartheta^{-1}a\vartheta) = \Delta^*(a)$, i.e. (5.2.17a) is just the criterion for reality of IRs (4.2.35). The ICORs of the magnetic point groups can be determined by (5.2.13–16). For this reason perhaps only the matrix S in (5.2.11) has to be determined. It can be represented by

$$S = \frac{1}{g} \sum_{a \in \mathcal{G}} \Delta(a) X \bar{\Delta}(a^{-1}) \tag{5.2.18}$$

(proof in Exercise 5.4). The matrix X must be chosen in such a way that S becomes unitary, but is arbitrary otherwise. Furthermore, (5.2.12) may be helpful in determining S.

All these statements are also valid for the double-valued REPs of the magnetic double groups. The latter must be applied in systems with half-odd-integer spin. The classification of the REPs then obeys (5.2.17) with $\lambda = -1$.

Some physical consequences of this theory of CORs will be briefly pointed out. In a physical system (crystal) with symmetry group $\mathscr{M}$, the occurrence of antiunitary elements $r\mathscr{G} \subset \mathscr{M}$ implies that in cases 1b) and 2) the degeneracy of energy eigenvalues is doubled, when these have previously been classified according to the unitary subgroup $\mathscr{G} \subset \mathscr{M}$ (Sect. 4.3). In case 1a) there is no doubling. If the symmetry is reduced from $\mathscr{M}$ to $\mathscr{G}$ by a perturbation (Sect. 6.4), then in cases 1b) and 2) the energy eigenvalues split, while in case 1a) there is no splitting.

An especially important case of an universal degeneracy caused by a time reversal symmetry is the *Kramers degeneracy*. A quantum-mechanical system with an odd number of electrons (one-electron states in a crystal, theory of band structure) may be described by a Hamiltonian which is symmetric with respect to time reversal, $[H, \vartheta] = 0$. The symmetry (double) group is (Sect. 5.1) $\mathscr{G} = \mathscr{C}_1^{\mathrm{D}} = \{e, c_0\}$. The extra REP of $\mathscr{C}_1^{\mathrm{D}}$ is $\Delta(e) = +1; \Delta(c_0) = -1$, thus $\chi(e^2) + \chi(c_0^2) = 2 = g$. Since in this case $\lambda = -1$ (see above), it follows from (5.2.17) that case 1b) is realized: Every energy eigenvalue is at least doubly degenerate according to time reversal symmetry (except for an accidental degeneracy, see Sect. 6.3).

Exercise 5.3. Prove (5.2.16) by using the theorem of orthogonality (4.2.25a) for $\Delta(a)$.

Exercise 5.4. Show that S in the REP (5.2.18) satisfies (5.2.11).

Exercise 5.5. Determine the characters of the extra REPs of the icosahedral group $\mathscr{Y}$.

Exercise 5.6. Determine the characters of the extra REPs of the group $\mathscr{C}_{\infty v}$. Take into account that the characters for $c(\varphi)$ have to constitute a complete system of (double-valued) functions.

Exercise 5.7. Discuss which are the physical systems where the states have to be classified by magnetic groups.

Exercise 5.8. Discuss the physical relevance of the symmetry reduction from $\mathscr{M}$ to $\mathscr{G}$ (e.g. on switching on a magnetic field) by using simple examples.

5.3 Translation Groups

In Sect. 3.4 we introduced the translation group

$$\mathbb{T} = \{e|\boldsymbol{R}^h\} = \left\{e \,\middle|\, \sum_{i=1}^{3} h_i \boldsymbol{a}^{(i)}\right\} \tag{5.3.1}$$

as an invariant subgroup of a space group $\mathscr{R}$. Thus the REPs of the translation group are essential for space groups, too. Instead of (5.3.1) we can write

$$\{e|\boldsymbol{R}^{\boldsymbol{h}}\} = \{e|\boldsymbol{a}^{(1)}\}^{h_1}\cdot\{e|\boldsymbol{a}^{(2)}\}^{h_2}\cdot\{e|\boldsymbol{a}^{(3)}\}^{h_3}\ , \qquad h_i \in \mathbb{Z} \tag{5.3.2}$$

showing that $\mathbb{T}$ is the outer direct product of translations in the directions $\boldsymbol{a}^{(i)}$:

$$\mathbb{T} = \mathbb{T}_1 \times \mathbb{T}_2 \times \mathbb{T}_3\ . \tag{5.3.3}$$

The Abelian group $\mathbb{T}_i$ is generated by the element $\{e|\boldsymbol{a}^{(i)}\}$. Both $\mathbb{T}$ and $\mathbb{T}_i$ define *infinite* groups which can be made finite only by choice of appropriate boundary conditions (Born–von Kármán). We choose the well-known *periodic boundary conditions* and assume

$$\{e|\boldsymbol{a}^{(i)}\}^G = \{e|G\boldsymbol{a}^{(i)}\} = \{e|0\}\ , \qquad G \in \mathbb{Z} \text{ being of arbitrary size.} \tag{5.3.4}$$

This means geometrically that the space (crystal) is divided into physically and mathematically identical regions with $V = G^3 V_z$, G being the same for each direction $\boldsymbol{a}^{(i)}$. Group theoretically this means the introduction of a "super translation group"

$$\mathbb{T}^s = \left\{e\left|\sum_{i=1}^{3} n_i(G\boldsymbol{a}^{(i)})\right.\right\}\ , \qquad n_i \in \mathbb{Z}\ . \tag{5.3.5}$$

The quotient group $\mathbb{T}/\mathbb{T}^s$ is exactly the finite translation group defined by (5.3.4) which for simplicity also denote by $\mathbb{T}$. All the translations differing by an element of $\mathbb{T}^s$ are looked upon as being equivalent, i.e.

$$\{e|\boldsymbol{R}^{\boldsymbol{h}}\}\cdot\{e|\boldsymbol{R}^{\boldsymbol{h}'}\} = \{e|\boldsymbol{R}^{\boldsymbol{h}} + \boldsymbol{R}^{\boldsymbol{h}'}\} \qquad \operatorname{mod} \mathbb{T}^s\ . \tag{5.3.6}$$

The groups $\mathbb{T}_i$ according to (5.3.4) are cyclic groups of order G. Thus they have only one-dimensional IRs, the number of which is equal to the number of classes (or elements), i.e. equal to G. To an element of order G the Gth unit roots are assigned as the IRs:

$$D^{(m_i)}(e|\boldsymbol{a}^{(i)}) = \exp(-2\pi i m_i/G)\ , \qquad m_i = 1, 2, \ldots, G\ , \tag{5.3.7}$$

where m_i characterizes the G different IRs. The IRs of the other elements of $\mathbb{T}_i$ follow from the powers of (5.3.7) so that all the G^3 IRs of $\mathbb{T}$ are given by

$$D^{(\boldsymbol{m})}(e|\boldsymbol{R}^{\boldsymbol{h}}) = \exp\left(-2\pi i \sum_{i=1}^{3} m_i h_i/G\right)\ . \tag{5.3.8}$$

A convenient form of these REPs can be found by introducing the *reciprocal lattice*. If $\{\boldsymbol{a}^{(i)}\}$ is the basis of a lattice, then the scalar product

$$(\boldsymbol{a}^{(i)}, \boldsymbol{b}^{(j)}) = \delta_{ij}\ , \qquad i, j = 1, 2, 3\ , \tag{5.3.9}$$

defines another basis, the *reciprocal basis* $\{\boldsymbol{b}^{(i)}\}$ with

$$b^{(i)} = \frac{a^{(j)} \times a^{(k)}}{\det A} , \qquad i, j, k \text{ cyclic,} \tag{5.3.10}$$

$$\det A = \det[a^{(i)} \cdot (a^{(j)} \times a^{(k)})] = V_z ,$$

which describes the reciprocal lattice via

$$K^n = \sum_{i=1}^{3} 2\pi n_i b^{(i)} , \qquad n_i \in \mathbb{Z} . \tag{5.3.11}$$

In this reciprocal space defined by $b^{(i)}$, further k-vectors

$$k = \sum_{i=1}^{3} (2\pi/G) m_i b^{(i)} , \qquad m_i = 1, \ldots, G , \tag{5.3.12}$$

can be introduced, in terms of which the REPs (5.3.8) can be written as

$$D^{(k)}(e|R^h) = \exp(-\mathrm{i}k \cdot R^h) . \tag{5.3.13}$$

Since $\exp(-\mathrm{i}K^n R^h) = 1$, we have

$$\exp(-\mathrm{i}k \cdot R^h) = \exp[-\mathrm{i}(k + K^n) \cdot R^h] . \tag{5.3.14}$$

The k-vectors do not have to be chosen from the *unit cell of the reciprocal lattice* defined by $b^{(j)}$, but instead they can be selected in such a way that the complete set of k-vectors is equivalent to those given by (5.3.12), thus $k \bmod K^n$. The most convenient way is to choose the k-vectors from the (first) *Brillouin zone* (BZ), which corresponds to the Wigner-Seitz cell of direct space (see Sect. 3.4.1, Fig. 3.8). These G^3 k-vectors lying in the first BZ are denoted as *reduced k-vectors*. The IRs of the finite translation group $\mathbb{T}$ are now defined by the G^3 k-vectors of the first BZ, where (with even G)

$$m_i = -G/2, -G/2 + 1, \ldots, +G/2 - 1. \tag{5.3.15}$$

Note that k-vectors on opposite surfaces of the BZ should be counted only once!

The one-dimensional spaces $\mathscr{L}^{(k)}$ serving as a basis for the IRs of $\mathbb{T}$ are formed from *Bloch functions* $\psi^{(k)}$. The G^3 Bloch functions obey (Bloch theorem):

$$P(e|R^h)\psi^{(k)}(x) = \psi^{(k)}(x - R^h) = \exp(-\mathrm{i}k \cdot R^h)\psi^{(k)}(x) \tag{5.3.16}$$

(for a proof see Exercise 5.9). The generation of a Bloch function from an arbitrary function is very important in some applications. This can be achieved by the projectors of the translation group:

$$\psi^{(k)} = P^{(k)}\psi \text{ with } P^{(k)} = \frac{1}{N} \sum_{h} \exp(-\mathrm{i}k \cdot R^h) P(e|R^h) , \qquad N = G^3 . \tag{5.3.17}$$

The discussion of interactions and selection rules needs the Clebsch-Gordan decomposition of $\mathbb{T}$. When the functions $\psi^{(k)}$ define the space $\mathscr{L}^{(k)}$ and the $\psi^{(k')}$ the space $\mathscr{L}^{(k')}$, then the product space $\mathscr{L}^{(k\times k')} = \mathscr{L}^{(k)} \otimes \mathscr{L}^{(k')}$ is given by the products $\psi^{(k)} \cdot \psi^{(k')}$. The REP of the inner Kronecker product can be decomposed into (4.4.13)

$$D^{(k\times k')} = \sum_{k''} (kk'|k'')D^{(k'')} \tag{5.3.18}$$

with the reduction coefficients (Exercise 5.10)

$$(kk'|k'') = \delta_{k+k',k''} \ . \tag{5.3.19}$$

Thus

$$D^{(k\times k')} = D^{(k+k')} \ , \qquad \mathscr{L}^{(k\times k')} = \mathscr{L}^{(k+k')} \ . \tag{5.3.20}$$

Exercise 5.9. Prove the Bloch theorem (5.3.16).
Exercise 5.10. Verify (5.3.19) and (5.3.20).
Exercise 5.11. Starting with the theorems of orthogonality (4.2.27a and 28a) for the characters, show that the following relations are valid ($N = G^3$):

$$\text{(a)}\ \frac{1}{N} \sum_{k \in 1\text{ st BZ}} \exp[ik\cdot(R^{h} - R^{h'})] = \delta_{h,h'} \ ;$$

$$\text{(b)}\ \frac{1}{N} \sum_{h} \exp[i(k - k')\cdot R^{h}] = \delta_{k,k'}, \qquad k, k' \in 1\text{ st BZ} \ .$$

5.4 Permutation Groups

In principle the characters of the IRs can be determined according to the method given in Sect. 4.2.4. However, in the case of permutation groups it may be useful to choose another way which makes use of the idempotent projectors $P_{jj}^{(\alpha)}$ given in (4.3.9a). This implies, at least partly, knowledge of the REPs $D^{(\alpha)}$.

If K_a is the class with elements a_i, then $a_i \in K_a \subset \mathscr{G}$, and if r_a is the number of elements in K_a, i.e. $i = 1, \dots, r_a$, then the number

$$z(\alpha, K_a) = \frac{d_\alpha}{g} \sum_{i=1}^{r_a} D_{jj}^{(\alpha)*}(a_i) \quad \text{or} \quad \frac{d_\alpha}{g} \sum_{i=1}^{r_a} D_{jj}^{(\alpha)}(a_i^{-1}) \tag{5.4.1}$$

is *independent* of j and the character of this class is given by

$$\chi^{(\alpha)*}(K_a) = \frac{g}{r_a} \cdot z(\alpha, K_a) \ . \tag{5.4.2}$$

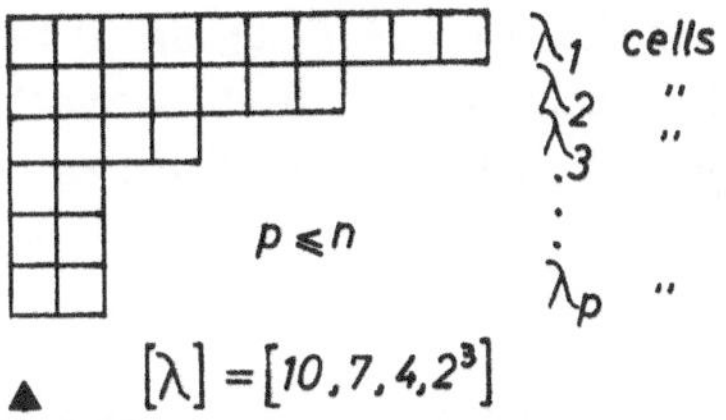

Fig. 5.1. Young diagram of the partition $[\lambda] = [\lambda_1, \lambda_2, \ldots, \lambda_p]$ with $\sum_{i=1}^{n} \lambda_i = n$

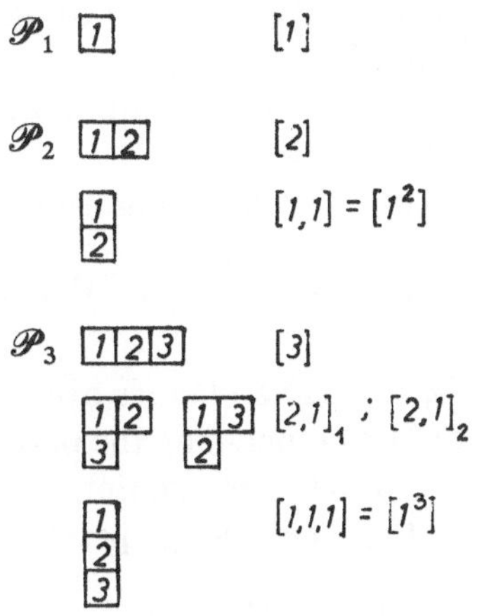

Fig. 5.2. Standard Young tableaux for the groups $\mathscr{P}_n$, $n \leqslant 3$. For non-standard tableaux the numbers may have any sequence, but these do not occur in the theories considered here

Obviously the sum (5.4.1) is also independent of the choice of equivalent IRs entering the definition (4.3.9a), see Sect. 4.2.2. A detailed proof will not be given here. Using the projection operators (4.3.11) and the classes from Table 4.3 we obtain for the group $\mathscr{C}_{3v}$, for example, $z(3, K_e) = 1/3$, $z(3, K_3) = -1/3$, $z(3, K_\sigma) = 0$ and therefore the already familiar characters.

As mentioned above, this method for the calculation of characters is specially adapted for the permutation groups $\mathscr{P}_n$. According to (3.5.11) the conjugacy classes of $\mathscr{P}_n$ are uniquely determined by the partitions $[\lambda]$ of n, (see 3.5.8). Furthermore, according to (4.2.29) the number of IRs is equal to the number of classes of $\mathscr{P}_n$. Now the question arises how to combine an IR of $\mathscr{P}_n$ with every partition $[\lambda]$ of n, or how to connect characters and matrices of the IRs of $\mathscr{P}_n$ with $[\lambda]$.

To illustrate these relations we introduce a graphical representation of the partitions: the *Young diagrams*. (See [5.3,4] and references therein for more details and proofs.) The diagram which is assigned to the partition $[\lambda] := [\lambda_1, \lambda_2, \ldots, \lambda_p]$ with $p \leqslant n$ has λ_1 cells in the first row, λ_2 cells in the second row, etc., and finally λ_p cells in the pth row (Fig. 5.1). If we put the numbers 1, 2, ..., n into the cells we obtain the *Young tableau* $T[\lambda]$. If the numbers increase within a row from left to right, and within a column from top to bottom, we have a *standard Young tableau* (SYT, Fig. 5.2). Instead of numbers, we often choose letters a, b, c and agree on alphabetical (or numerical) order. The SYTs assigned to a Young diagram can be numbered (Fig. 5.2 for $T[2, 1]$).

For the following considerations, three elements of the group algebra $\mathscr{A}(\mathscr{P}_n)$ are of importance. These can be related directly to a definite Young tableau (Sect. 4.3.1,2).

(i) The row-symmetrizer $S[\lambda]$ of $T[\lambda]$:

$$S[\lambda] = \sum_{P \in \text{row}} P \; ; \tag{5.4.3}$$

the summation is over *all* the permutations within the rows of $T[\lambda]$, or, to be more precise, it is a product of the sum of the permutations in the first row with the sum of the permutations in the second row and with that in the third row, etc., see (5.4.6).

(ii) The column-antisymmetrizer $A[\lambda]$ of $T[\lambda]$:

$$A[\lambda] = \sum_{P \in \text{column}} (-1)^p P \; ; \tag{5.4.4}$$

the summation here is over *all* the possible permutations within the columns of $T[\lambda]$, but including the parity of the permutation: $(-1)^p = 1$ for even, $(-1)^p = -1$ for odd permutations.

(iii) The Young operator (sometimes Young symmetrizer) $Y[\lambda]$ of $T[\lambda]$:

$$Y[\lambda] = A[\lambda] \cdot S[\lambda] = \sum_{P \in \text{column}} (-1)^p P \cdot \sum_{P' \in \text{row}} P' \; . \tag{5.4.5}$$

The Young operators assigned to the SYTs [2, 1] of $\mathscr{P}_3$ in Fig. 5.2 are therefore

$$\begin{aligned} Y[2,1]_1 &= [e-(13)]\cdot[e+(12)] = e-(13)+(12)-(13)(12) \\ &= e-(13)+(12)-(123) \; , \\ Y[2,1]_2 &= [e-(12)]\cdot[e+(13)] = e-(12)+(13)-(12)(13) \\ &= e-(12)+(13)-(132). \end{aligned} \tag{5.4.6}$$

Apart from a factor, the Young operators are just the diagonal [$j = k$ in (4.3.9a)] idempotent irreducible projection operators for the group $\mathscr{P}_n$ and thus generate an IR of $\mathscr{P}_n$ as in Sect. 4.3.2. All the Young operators belonging to SYTs with the same diagram (like $[2,1]_1$ and $[2,1]_2$) generate equivalent IRs. All the Young operators belonging to different diagrams generate inequivalent IRs of $\mathscr{P}_n$. Since $[\lambda]$ or the corresponding diagram characterizes the inequivalent IRs we can denote these by $D^{[\lambda]}$. For every Young diagram there exists exactly one IR of $\mathscr{P}_n$. The three inequivalent IRs of $\mathscr{P}_3$ (isomorphic to $\mathscr{C}_{3v}$) are $D^{[3]}$, $D^{[2,1]}$, $D^{[1^3]}$.

By comparison of (5.4.5) with (4.3.9a) we can also determine the factor causing the Young operator to be an irreducible projector. Let $d_{[\lambda]}$ be the dimension of the IR (to be calculated later) $g = n! = \operatorname{ord} \mathscr{P}_n$, then

$$\hat{Y}[\lambda] = \frac{d_{[\lambda]}}{n!} Y[\lambda] \; , \qquad \hat{Y}^2[\lambda] = \hat{Y}[\lambda] \; . \tag{5.4.7}$$

For the permutation groups $\mathscr{P}_n$ the characters (5.4.2) are now given by

$$\chi^{[\lambda]}(\nu_1 \ldots \nu_n) = \frac{n!}{r(\nu_1 \ldots \nu_n)} \cdot z^*([\lambda],(\nu_1 \ldots \nu_n)), \tag{5.4.8}$$

where $(\nu_1 \ldots \nu_n) = (\nu)$ describes the structure of the cycles of the permutation, which on the other hand defines the classes [see (3.5.4) onwards].
Since the dimension of the REPs has not yet been determined, it is useful to rewrite (5.4.8) and to introduce the quantities

$$\hat{z} = \frac{n!}{d_{[\lambda]}} \cdot z \ . \tag{5.4.9}$$

These $\hat{z}$ can be determined by summing up the coefficients of the permutations for the classes $(\nu_1 \ldots \nu_n)$ entering the ("unnormalized") operators $Y[\lambda]$. Then the characters are

$$\chi^{[\lambda]}(\nu) = \frac{d_{[\lambda]}}{r(\nu)} \cdot \hat{z}^*([\lambda], (\nu)) \ . \tag{5.4.10}$$

The dimension $d_{[\lambda]}$ can now be calculated from the criterion of irreducibility (4.2.32)

$$\sum_{(\nu)} r(\nu) |\chi^{[\lambda]}(\nu)|^2 = \sum_{(\nu)} \frac{d^2_{[\lambda]}}{r(\nu)} |\hat{z}([\lambda], (\nu))|^2 = n! \ . \tag{5.4.11}$$

As an example we consider $D^{[2,1]}$ of $\mathscr{P}_3$. The three classes have the following cycle structure [see after (3.5.12)]: (1^3), (3), $(1, 2)$ with $r(1^3) = 1$, $r(3) = 2$, $r(1, 2) = 3$. According to (5.4.9) we have using (5.4.6)

$$\hat{z} = ([2, 1], (1^3)) = 1 \ , \qquad \hat{z}([2, 1], (3)) = -1 \ , \qquad \hat{z}([2, 1], (1, 2)) = 0$$

and therefore from (5.4.11) $d_{[2,1]} = 2$ and finally with (5.4.10)

(5.4.12)

	(1^3)	(3)	$(1, 2)$
$\chi^{[2,1]}$	2	-1	0

Without giving proofs, we quote some important statements about the IRs of $\mathscr{P}_n$ and provide explanations using examples.

(i) The dimension $d_{[\lambda]}$ of $D^{[\lambda]}$ is given by the number of SYTs that can be derived from a Young diagram.

Examples are given in Fig. 5.2 and Exercise 5.13.

(ii) Algebraically the $d_{[\lambda]}$ can be calculated from a formula by Rutherford:

$$d_{[\lambda]} = \frac{n! \prod_{i<j} (h_i - h_j)}{h_1! h_2! \ldots h_m!} \ , \tag{5.4.13}$$

where $h_i = \lambda_i + m - i$, λ_i is the number of cells in the ith row and m is the number of rows of a Young diagram belonging to $[\lambda]$. For $[\lambda] = [2, 1^2]$ of $\mathscr{P}_4$ we have $h_1 = 2 + 3 - 1 = 4$, $h_2 = 1 + 3 - 2 = 2$, $h_3 = 1 + 3 - 3 = 1$, thus $d_{[\lambda]} = 4!(2.3.1)/(4!2!1!) = 3$.

(iii) For $n > 1$, every permutation group $\mathscr{P}_n$ has *exactly two one-dimensional* IRs: the (totally) symmetric one $D^{[n]}$, in which all the elements (permutations) are represented by $+1$ (identity REP), and the (totally) antisymmetric one $D^{[1^n]}$, in which all the even permutations are represented by $+1$, the odd ones by -1. (5.4.14)

(iv) The inner direct product of a (not necessarily irreducible) REP D of $\mathscr{P}_n$ with $D^{[1^n]}$ is said to be the *associate* (or sometimes conjugate or adjoint) REP *of* D

$$\bar{D} = D \otimes D^{[1^n]} \,, \qquad \text{thus} \tag{5.4.15a}$$

$$\begin{aligned} \bar{D}(p) &= +D(p) \,, \qquad \text{for } p \text{ even,} \\ &= -D(p) \,, \qquad \text{for } p \text{ odd;} \end{aligned} \tag{5.4.15b}$$

if D is irreducible, so is $\bar{D}$. If the characters of all the odd permutations $\chi^{(D)}(p_u) = 0$, then D and $\bar{D}$ are equivalent, since all their characters agree. In this case D is said to be *self-associated.* In the other case, i.e. if $\chi^{(D)}(p_u) \neq 0$, D and $\bar{D}$ are inequivalent (see Exercises 5.16 and 5.17).

Graphically we obtain the graph $[\tilde{\lambda}]$ associated with $\bar{D} := D^{[\tilde{\lambda}]} := D^{[\lambda]} \otimes D^{[1^n]}$ by interchanging the rows and columns of $[\lambda]$. The transpose graph $[\tilde{\lambda}]$ thus belongs to the IR $D^{[\tilde{\lambda}]}$ associated with $D^{[\lambda]}$. Examples follow from Exercise 5.13. In $\mathscr{P}_4$, for example, $[\widetilde{3,1}] = [2,1^2]$; $[2,2]$ is self-associated.

For the determination of the REP matrices we will describe two methods, which, however, cannot give unique results because of the equivalence of REPs. Also, the matrices detemined by these approaches are not unitary, but they can be brought into a unitary form (Sect. 4.2.2).

The simplest method is that already described in Sect. 5.1. It starts with a complete orthonormal system and generates the basis functions by applying the projections operators. First, a Young operator $Y^{[\lambda]}$ or $\hat{Y}^{[\lambda]}$ is applied to an arbitrary general n-particle function $\psi(1,\dots,n) = |1,\dots,n\rangle$ and thus a certain linear combination is generated. To this function the permutations $p \in \mathscr{P}_n$ are applied and a maximal set of linearly independent functions is selected. This set can be used as the basis functions of $D^{[\lambda]}$ of $\mathscr{P}_n$ according to (4.3.14) (with $a \to p$ for permutations, $\alpha \to [\lambda]$). Again, we consider the example from (5.4.6) and apply $Y[2,1]$ to $|1,2,3\rangle$:

$$\psi_1^{[2,1]} := \hat{Y}[1,2]\cdot|123\rangle = \tfrac{1}{3}(|123\rangle - |321\rangle + |213\rangle - |231\rangle) \,.$$

Applying the generator $p = (12)$ to this function we get

$$(12)\psi_1^{[2,1]} = \tfrac{1}{3}(|213\rangle - |312\rangle + |123\rangle - |132\rangle) = \psi_2^{[2,1]} \,. \tag{5.4.16a}$$

Since this function is linearly independent of $\psi_1^{[2,1]}$ it can be taken as the second basis function of $D^{[2,1]}$. Obviously also

$$(12)\psi_2^{[2,1]} = \psi_1^{[2,1]} \,; \tag{5.4.16b}$$

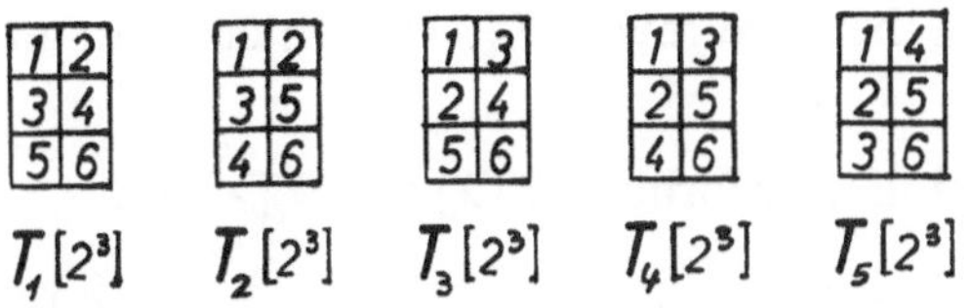

Fig. 5.3. Standard Young tableaux in numerical order for the partition $[2^3]$ of $\mathscr{P}_6$. The tableau $T_1[2^3]$ has the natural order. Note that the numbers increase from left to right and from top to bottom (see text)! The REP $[2^3]$ of $\mathscr{P}_6$ has dimension 5

thus according to (4.3.14) we can construct a REP matrix for

$$p = (12) :$$
$$D^{[2,1]}(12) = \begin{pmatrix} 0 & 1 \\ 1 & 0 \end{pmatrix} . \qquad (5.4.16c)$$

Correspondingly, applying $p = (123)$ to $\psi_{1,2}^{[2,1]}$ we obtain the REP

$$D^{[2,1]}(123) = \begin{pmatrix} -1 & -1 \\ +1 & 0 \end{pmatrix} . \qquad (5.4.16d)$$

Thus we have the matrices of both the generators of $\mathscr{P}_3$ for the IR[2, 1] and can also find the other matrices easily. However, $D^{[2,1]}$ according to (5.4.16) is obviously *not unitary*, although of course, the characters agree with those of the unitary standard REPs.

Another method for the determination of the REP matrices originates from Young and Littlewood who further generalized the Young operators. In Sect. 4.3.2 we saw that the projectors[2] (4.3.9a) constitute a basis of the group algebra, from which unitary IRs of the group can be deduced. For this generalization we have to introduce an order (sequence) of the SYTs. We therefore arrange the SYTs according to the numerical deviation of the numbers in the Young cells from the natural sequence of the numbers in the tableau, reading the tableaux from left to right and from top to bottom (see Fig. 5.3). Thus the SYTs can be uniquely numbered: $T_i[\lambda]$, $i = 1, 2, \ldots$ and the subscript i is well defined.

Now, if the following conditions are met for all the pairs $T_i[\lambda]$ and $T_j[\lambda]$[3]:

1) there exist two numbers a and b in the same row of $T_i[\lambda]$ which occur in the same column of $T_j[\lambda]$;

and *simultaneously*

---

[2] Here and in the following we also use the term "projection operator" for the $P_{kj}^{(\alpha)}$ with $k \neq j$, though this is not completely correct (4.3.12).

[3] They are satisfied for all the SYTs of the groups $\mathscr{P}_3$ and $\mathscr{P}_4$. For $\mathscr{P}_n$ with $n \geqslant 5$ they are sometimes satisfied (Fig. 5.3), but not always. In the latter case the $e_{ij}^{[\lambda]}$ can be redefined for application of a modified method for the determination of the IRs (analogously to the Schmidt procedure of orthogonalization, see [5.4]).

2) there exist two numbers c and d in the same row of $T_j[\lambda]$ which occur in the same column of $T_i[\lambda]$,

and furthermore if p_{ij} is the permutation transforming $T_j[\lambda]$ into $T_i[\lambda]$, i.e.

$$T_i[\lambda] = p_{ij} T_j[\lambda] \ , \tag{5.4.17}$$

then the $d_{[\lambda]}^2$ operators

$$e_{ij}^{[\lambda]} = \frac{d_{[\lambda]}}{n!} A_i[\lambda] p_{ij} S_j[\lambda] \ , \qquad i,j = 1, \ldots, d_{[\lambda]} \tag{5.4.18a}$$

satisfy the relation [see 4.3.12)]

$$e_{ij}^{[\lambda]} e_{kl}^{[\lambda]} = \delta_{jk} e_{il}^{[\lambda]} \ . \tag{5.4.19}$$

The $e_{ij}^{[\lambda]}$ thus constitute a basis of the group algebra $\mathscr{A}(\mathscr{P}_n)$ (Sect. 4.3.2) and generate the IRs of $\mathscr{P}_n$; $A_i[\lambda]$ and $S_i[\lambda]$ are the antisymmetrizer and symmetrizer, respectively, of the ordered SYTs. If $i = j$, (5.4.18a) is identical with the Young operator of the corresponding SYT.

From the $e_{ij}^{[\lambda]}$ the matrices of the IRs can be determined by comparing coefficients (4.3.9a):

$$e_{ij}^{[\lambda]} = \frac{d_{[\lambda]}}{n!} \sum_{p \in \mathscr{P}_n} D_{ji}^{[\lambda]}(p^{-1}) \cdot P_p \ . \tag{5.4.20}$$

The numbering of the matrix elements of $D^{[\lambda]}$ and of the basis functions corresponds to the numbering of the SYTs in Fig. 5.3. Figure 5.2. represents the SYTs of $\mathscr{P}_3$. For the REP [2, 1] the p_{ij} of (5.4.17) is the permutation (23). With this we have as in (5.4.6)

$$\begin{aligned} e_{11}^{[2,1]} &= \tfrac{1}{3}[e - (13) + (12) - (123)] \ , \\ e_{22}^{[2,1]} &= \tfrac{1}{3}[e - (12) + (13) - (132)] \ . \end{aligned} \tag{5.4.21a}$$

Furthermore we obtain

$$\begin{aligned} e_{12}^{[2,1]} &= \tfrac{1}{3} A_1[2,1](23) S_2[2,1] = \tfrac{1}{3}[(23) + (123) - (132) - (12)] \ , \\ e_{21}^{[2,1]} &= \tfrac{1}{3} A_2[2,1](32) S_1[2,1] = \tfrac{1}{3}[(23) + (132) - (123) - (13)] \ . \end{aligned} \tag{5.4.21b}$$

Using (5.4.20) the generators (12) and (123) become

$$\begin{aligned} D^{[2,1]}((12)^{-1}) &= D^{[2,1]}(12) = \begin{pmatrix} 1 & 0 \\ -1 & -1 \end{pmatrix} \ , \\ D^{[2,1]}((132)^{-1}) &= D^{[2,1]}(123) = \begin{pmatrix} 0 & 1 \\ -1 & -1 \end{pmatrix} \ . \end{aligned} \tag{5.4.22}$$

Again, the REPs are not unitary and they are even different from those given in (5.4.16). But, of course, (5.4.16) and (5.4.22) are equivalent to each other.

It is not difficult to convince ourselves that

$$p_{ij}S_j[\lambda] = S_i[\lambda]p_{ij} \ . \tag{5.4.23}$$

Therefore we may also write instead of (5.4.18a)

$$e_{ij}^{[\lambda]} = \frac{d_{[\lambda]}}{n!}A_i[\lambda]S_i[\lambda]p_{ij} = \hat{Y}_i[\lambda]p_{ij} \ ; \tag{5.4.18b}$$

$\hat{Y}_i$ is the Young operator belonging to the ith SYT. Equation (5.4.18b) is especially suited to project basis functions of $D^{[\lambda]}$ out of an arbitrary n-particle function. This is important for many applications. For example, if $\psi_1 := \psi(1, 2, \ldots, n)$ belongs to $T_1[\lambda]$ then we can first generate functions ψ_i by $p_{i1}\psi_1$ with $i = 2, \ldots, d_{[\lambda]}$ from which basis functions ϕ_i of $D^{[\lambda]}$ can then be derived by applying $\hat{Y}_i[\lambda]$. This means, the

$$\phi_{i1}^{[\lambda]} = \hat{Y}_i[\lambda]p_{i1}\psi(1, 2, \ldots, n) \ , \qquad i = 1, \ldots, d_{[\lambda]} \tag{5.4.24}$$

are the basis functions we have been looking for.

In our example we choose $\psi_1 = \psi(1, 2, 3) = |1, 2, 3\rangle$. Applying $e_{i1}^{[\lambda]}$ according to (5.4.21), the basis functions are

$$\begin{aligned} \phi_{11}^{[2,1]} &= \tfrac{1}{3}[|1,2,3\rangle + |2,1,3\rangle - |3,2,1\rangle - |2,3,1\rangle] \ , \\ \phi_{21}^{[2,1]} &= \tfrac{1}{3}[|1,3,2\rangle - |2,3,1\rangle + |3,1,2\rangle - |3,2,1\rangle] \ . \end{aligned} \tag{5.4.25}$$

Applying further the generators (permutations) (12) and (123) to (5.4.25), we obtain linear combinations of (5.4.25), from which the REPs (5.4.22) can again be derived.

The methods just discussed enable us to derive certain rules which allow a determination of the standard REPs, which are unitary and even real in our case (see [5.5]). These IRs are denoted as Young-Yamanouchi REPs. We state these rules without any proof. Since every permutation can be written as a product of transpositions, (see end of Sect. 3.5), it is sufficient to give the REPs of the permutations $p_{i-1,i}$, that is, $p_{12}, p_{23}, \ldots$. All the others can be calculated from these. The $p_{i-1,i}$ constitute a system of generators, but not a minimal one!

The rows and columns of the REPs $D^{[\lambda]}$ are numbered according to the ordered SYTs (Fig. 5.3). We now have the following rules for the *nonvanishing* matrix elements of the IRs $D_{jk}^{[\lambda]}(p_{i-1,i})$; all the other matrix elements vanish.

$$(1) \quad D_{kk}^{[\lambda]}(p_{i-1,i}) = +1 \ , \tag{5.4.26a}$$

if in $T_k[\lambda]$ the numbers $i-1$ and i occur in the same *row*.

$$(2) \quad D_{kk}^{[\lambda]}(p_{i-1,i}) = -1 \ , \tag{5.4.26b}$$

if in $T_k[\lambda]$ the numbers $i-1$ and i occur in the same *column*

$$(3)\quad D^{[\lambda]}_{jk}(p_{i-1,i}) = \begin{matrix} \\ j \\ \\ k \\ \\ \end{matrix}\left\{\begin{matrix} & & j & & k & \\ & & \vdots & & \vdots & \\ \cdots & & -1/s & \cdots & \sqrt{1-1/s^2} & \cdots \\ & & \vdots & & \vdots & \\ \cdots & & \sqrt{1-1/s^2} & \cdots & 1/s & \cdots \\ & & \vdots & & \vdots & \end{matrix}\right\}, \qquad (5.4.26c)$$

if $T_j[\lambda]$ and $T_k[\lambda]$ differ only by a permutation of the numbers $i-1$ and i and furthermore if in $T_j[\lambda]$ the row which contains $i-1$ is above the row which contains i. The parameter s counts the horizontal and vertical steps necessary to advance from the number $i-1$ to i in the tableau $[\lambda]$.

For the group $\mathscr{P}_3$ we have to find the REPs of (12) and (23). We calculate $D^{[2,1]}$ (Fig. 5.2). There are two SYTs, thus $d_{[2,1]} = 2$. According to (5.4.26a, b), $D_{11}(12) = 1$, $D_{22}(12) = -1$ and the other elements vanish. For $D_{jk}(23)$ (5.4.26c) can be used with $s = 2$. Therefore we obtain

$$D^{[2,1]}_{jk}(12) = \begin{pmatrix} 1 & 0 \\ 0 & -1 \end{pmatrix}, \qquad D^{[2,1]}_{jk}(23) = \begin{pmatrix} -1/2 & \sqrt{3}/2 \\ \sqrt{3}/2 & 1/2 \end{pmatrix}; \qquad (5.4.27)$$

the REPs of all the other permutations can be derived from these. For example, with $(12)(23) = (123)$

$$D^{[2,1]}_{jk}(123) = \begin{pmatrix} -1/2 & \sqrt{3}/2 \\ -\sqrt{3}/2 & -1/2 \end{pmatrix}.$$

These REPs are, indeed, equivalent to those of the isomorphic group $\mathscr{C}_{3v}$ [see (2.1.18) and associated REPs; $(12) \leftrightarrow \sigma_v$, $(23) \leftrightarrow \sigma''_v$, $(123) \leftrightarrow c_3^2$]. The matrices of the associated REPs $D^{[\tilde{\lambda}]}$ differ from those of $D^{[\lambda]}$ by the factor $(-1)^p$ denoting the parity of the permutation. Since the $p_{i-1,i}$ have parity -1,

$$D^{[\tilde{\lambda}]}(p_{i-1,i}) = -D^{[\lambda]}(p_{i-1,i}). \qquad (5.4.28)$$

The numbering of rows and columns of the associated REPs is achieved according to the transposed tableaux. This means

$$[\widetilde{2,1}]_1 \leftrightarrow [2,1]_2 \qquad \text{and} \qquad [\widetilde{2,1}]_2 \leftrightarrow [2,1]_1$$

correspond to each other. In $\mathscr{P}_3$ the associated REPs [2, 1] are equivalent to each other. The REPs of $\mathscr{P}_3 - \mathscr{P}_5$ are given in Appendix C.

Because all finite groups are isomorphic to subgroups of permutation groups (3.5.13), we can, in principle, construct all the REPs of finite groups. However, in general the subgroups decompose into classes different from those of $\mathscr{P}_n$, so that the REPs have to be subduced. The alternating group $\mathscr{A}_n$ contains all the even permutations and is a subgroup of $\mathscr{P}_n$. However, its classes are not those of the even elements of $\mathscr{P}_n$; e.g. $\mathscr{P}_5$ has seven classes (four with even elements), but $\mathscr{A}_5$ has five classes (of even elements).

In the case of the permutation groups one often uses instead of the projection operators (4.3.14) or (5.4.20) a standardized, normalized projection operator:

$$P_{ik}^{[\lambda]} = (d_{[\lambda]}/n!)^{1/2} \sum_{p \in \mathscr{P}_n} D_{ik}^{[\lambda]}(p) P_p \,. \tag{5.4.29}$$

It has the advantage that its application to a product of n orthogonal one-particle functions $\prod_{j=1}^{n} \varphi_{\alpha_j}(j)$ generates a function that is normalized to one.

Exercise 5.12. Check the statement that z in (5.4.1) is independent of j by using different projection operators, that is, $P_{11}^{(3)}$ and $P_{22}^{(3)}$, for the determination of z for the group $\mathscr{C}_{3v}$.

Exercise 5.13. Give the Young diagrams and the SYTs for $\mathscr{P}_4$.

Exercise 5.14. Determine $Y^{[2,2]}$ for $\begin{array}{|c|c|}\hline 1&2\\\hline 3&4\\\hline\end{array}$ of $\mathscr{P}_4$ and calculate the characters of the IR [2, 2].

Exercise 5.15. Complete the character table of $\mathscr{P}_3$.

Exericse 5.16. Prove the statements following (5.4.15).

Exercise 5.17. (a) Prove that the characters of the IRs of $\mathscr{P}_n$ are real.

(b) For two IRs $D^{[\lambda]}$ and $D^{[\lambda']}$ of $\mathscr{P}_n$ show:

(i) $D^{[\lambda]} \otimes D^{[\lambda']}$ contains $D^{[n]}$ exactly once, if $D^{[\lambda]}$ and $D^{[\lambda']}$ are equivalent; if they are inequivalent, $D^{[n]}$ is not contained in the direct product; and

(ii) $D^{[\lambda]} \otimes D^{[\lambda']}$ contains $D^{[1^n]}$ exactly once, if $D^{[\lambda]}$ and $D^{[\lambda']}$ are associated to each other, i.e if $D^{[\tilde{\lambda}]} = D^{[\lambda']} = D^{[\lambda]} \otimes D^{[1^n]}$, otherwise $D^{[1^n]}$ is not contained in the direct product.

These statements are very important for systems of identical Bose and Fermi particles, respectively.

5.5 Tensor Representations

In physics we often have to deal with symmetries under an exchange of coordinates or indices or quantum numbers. This symmetry is a permutation symmetry which is closely connected to other symmetry properties of the quantities in question. Thus it is of interest to investigate the connection between transformations in vector spaces and permutations, which will be done in the following sections.

5.5.1 Tensor Transformations. Irreducible Tensors

Let $\{x_1, x_2, \ldots, x_i, \ldots, x_n\}$ be the n (complex) components of an arbitrary vector (point) in an n-dimensional linear space. Then we can transform to new coordinates by a non-singular complex ($n \times n$) matrix, namely

$$x_i' = \sum_{j=1}^{n} a_{ij} x_j , \qquad i = 1, \ldots, n ,$$
$$\det a \neq 0 , \qquad a_{ij} \in \mathbb{C} ; \tag{5.5.1}$$

the transformation has $2n^2$ real parameters. Obviously the a_{ij} constitute a (continuous) group, the *general linear group* $\mathcal{GL}(n, \mathbb{C})$, Appendix D.

The operators P_a, isomorphically assigned to the elements $a \in \mathcal{GL}(n, \mathbb{C})$ act on the basis vectors $\{e_i | i = 1, \ldots, n\}$ of the n-dimensional (complex) linear space $\mathcal{L}$ (Sects. 4.1, 11.1):

$$P_a e_j = \sum_{k=1}^{n} a_{kj} e_k , \qquad j = 1, \ldots, n . \tag{5.5.2}$$

The $\{a_{jk}\}$ thus constitute the fundamental *defining* REP of $\mathcal{GL}(n, \mathbb{C})$; of course, this n-dimensional REP is irreducible. An arbitrary vector $x \in \mathcal{L}$ can be decomposed with respect to this basis $\{e_i\}$:

$$x = \sum_{i=1}^{n} x_i e_i , \tag{5.5.3}$$

then we have according to (5.5.2)

$$x' = P_a x = \sum_i x_i P_a e_i = \sum_{ik} x_i a_{ki} e_k = \sum_k x_k' e_k \tag{5.5.4}$$

with x_k' according to (5.5.1), i.e. x_i and e_k transform with the transposed matrices. The basis vectors e_k are also said to be *basis tensors* of rank 1 with respect to $\mathcal{GL}(n, \mathbb{C})$: the vectors x, correspondingly, are denoted as tensors of rank 1. Analogously, basis tensors of rank m with respect to $\mathcal{GL}(n, \mathbb{C})$ can be defined:

$$\{e_{i_1 \ldots i_m} = e_{i_1} e_{i_2} \ldots e_{i_m} | i_1, \ldots, i_m = 1, \ldots, n\} , \tag{5.5.5}$$

where the n^m basis elements are not necessarily products. From the definition (5.5.2) we have

$$P_a e_{i_1 \ldots i_m} = \sum_{j_1=1}^{n} \ldots \sum_{j_m=1}^{n} a_{j_1 i_1} \ldots a_{j_m i_m} e_{j_1 \ldots j_m} . \tag{5.5.6a}$$

This means, (5.5.5) gives a set of basis vectors (functions) for the m-fold inner direct product of the IRs a_{ij} of $\mathcal{GL}(n, \mathbb{C})$, the so-called *tensor* REP,

$$D^{(\times)}(a) := D(a \otimes a \otimes a \otimes \ldots \otimes a) = a \otimes a \otimes a \otimes \ldots \otimes a . \tag{5.5.6b}$$

Every element T of this linear n^m-dimensional space $\mathcal{L}_n^{(m)}$ defined by (5.5.5) is said to be a tensor of rank m with respect to $\mathcal{GL}(n)$. It has the decomposition into components

$$T = \sum_{i_1 \ldots i_m}^{n} T_{i_1 \ldots i_m} e_{i_1 \ldots i_m} \ . \tag{5.5.7}$$

The components transform according to

$$T'_{i_1 \ldots i_m} = \sum_{j_1 \ldots j_m}^{n} a_{i_1 j_1} \ldots a_{i_m j_m} T_{j_1 \ldots j_m} \ . \tag{5.5.8}$$

If the basis in (5.5.5) is a product, then $\mathscr{L}_n^{(m)} = \mathscr{L}_n^{(1)} \otimes \ldots \otimes \mathscr{L}_n^{(m)}$ is a direct product space (see Sect. 4.1.1 and 4.4.2). Although a_{ij} is an IR, $D^{(\times)}(a)$ is in general reducible; its reduction is an important problem. An example of a physical application of these basis tensors of rank m is given by the n^m products of one-particle functions

$$\Psi_{i_1 \ldots i_m}(1, 2, \ldots, m) = \psi_{i_1}(1) \ldots \psi_{i_m}(m) \ . \tag{5.5.9a}$$

The arguments are any physical variables (like position, spin, etc.). Any factor of (5.5.9a) transforms according to (here P_a is assumed to be a one-particle operator)

$$P_a \psi_i(l) = \sum_{k=1}^{n} a_{ki} \psi_k(l) \ , \qquad i = 1, \ldots, n \ , \tag{5.5.9b}$$

i.e. as a tensor of rank 1, thus (5.5.9a) transforms according to $D^{(\times)}(a)$ of $\mathscr{GL}(n)$.

For the reduction of the basis (5.5.5) into invariant subspaces we use the symmetry "projection operators" of $\mathscr{P}_m$. This is appropriate since the coefficients $a_{i_1 j_1} \ldots a_{i_m j_m}$ show a certain permutation symmetry. Then we obtain the irreducible subspaces of $\mathscr{L}_n^{(m)}$ and the reduction of $D^{(\times)}(a)$ into IRs of $\mathscr{GL}(n)$.

Applying a permutation

$$p = \begin{pmatrix} 1\,2 \ldots & m \\ p_1 p_2 \ldots & p_m \end{pmatrix}$$

to the set of tensor components $\{T_{i_1 \ldots i_m}\}$ we find

$$p T_{i_1 \ldots i_m} := T_{i_{p_1} \ldots i_{p_m}} \ . \tag{5.5.10}$$

The corresponding Tensor is

$$\sum p T_{i_1 \ldots i_m} e_{i_1 \ldots i_m} = \sum T_{i_{p_1} \ldots i_{p_m}} e_{i_1 \ldots i_m} \ .$$

This means *p is defined as acting on the components, not on the basis*! *Simultaneous* application of the same p to *both* the subscripts l of $a_{i_l j_l}$ in the product $a \otimes a \otimes \ldots \otimes a$ gives

$$\begin{aligned} pa_{i_1j_1}\dots a_{i_mj_m} &= a_{i_{p_1}j_{p_1}}\dots a_{i_{p_m}j_{p_m}} \\ &= a_{i_1j_1}\dots a_{i_mj_m} \ , \end{aligned} \tag{5.5.11}$$

i.e. *after rearrangement* we have the original product which is thus invariant under a permutation p. Using (5.5.10,11) and (5.5.8) it follows immediately that

$$p\cdot D^{(\times)}(a) = D^{(\times)}(a)\cdot p \ ; \tag{5.5.12}$$

p commutes with any REP $D^{(\times)}(a)$ of $\mathscr{GL}(n,\mathbb{C})$.

Applying an arbitrary operator $\sum_p c_p p$ of the group algebra $\mathscr{A}(\mathscr{P}_m)$ to the components of T, we have

$$\hat{T}_{i_1\dots i_m} = \sum_p c_p p T_{i_1\dots i_m} \tag{5.5.13a}$$

and with (5.5.12)

$$\begin{aligned} D^{(\times)}(a)\hat{T} = \hat{T}' &= \sum_p c_p p\cdot D^{(\times)}(a) T \\ &= \sum_p c_p p\cdot T' \ . \end{aligned} \tag{5.5.13b}$$

But since $\hat{T}' = 0$ whenever $\hat{T} = 0$, and vice versa, it can be deduced from (5.5.13) that the set of tensor components $\{T\}$ satisfying

$$cT := \sum_p c_p p T = 0 \ , \qquad c \in \mathscr{A}(\mathscr{P}_m) \tag{5.5.14}$$

is invariant against $\mathscr{GL}(n)$. Thus we can assign an invariant subspace of tensors of rank m to every element $c \in \mathscr{A}(\mathscr{P}_m)$. These are just those tensors T whose components satisfy (5.5.14), i.e., which are mapped onto the null space of c, see (5.5.17). The tensor space $\mathscr{L}_n^{(m)}$ can be decomposed into invariant (independent) subspaces with respect to $\mathscr{GL}(n)$ by choosing a complete set of irreducible idempotent projection operators for the c-operators in (5.5.14):

$$P_{kk}^{[\lambda]} \ , \qquad k = 1, \dots, d_{[\lambda]} \text{ from } \mathscr{P}_m \ .$$

Of course, the $P_{kj}^{[\lambda]}$ satisfy (4.3.9a,b and 4.3.12).

Applying (4.3.18) to the tensor components $\{T_{i_1\dots i_m}\}$ we obtain

$$P_e T_{i_1\dots i_m} = T_{i_1\dots i_m} = \sum_\lambda \sum_{k=1}^{d_{[\lambda]}} T_{kk,i_1\dots i_m}^{[\lambda]} \tag{5.5.15}$$

with

$$T_{kk,i_1\dots i_m}^{[\lambda]} = P_{kk}^{[\lambda]} T_{i_1\dots i_m} = \frac{d_{[\lambda]}}{m!} \sum_{p\in\mathscr{P}_m} D_{kk}^{[\lambda]}(p^{-1}) p T_{i_1\dots i_m} \ .$$

The corresponding tensor T then has the expansion

$$T = \sum_{\lambda} \sum_{k} T_{kk}^{[\lambda]} \text{ with } T_{kk}^{[\lambda]} = \sum_{i_1 \ldots i_m} T_{kk,i_1 \ldots i_m}^{[\lambda]} e_{i_1 \ldots i_m} . \tag{5.5.16}$$

Accordingly, the tensor space $\mathscr{L}_n^{(m)}$ decomposes into the direct sum of (independent) invariant subspaces $\mathscr{L}_{n,k}^{[\lambda]}$ with respect to $\mathscr{GL}(n)$:

$$\mathscr{L}_n^{(m)} = \sum_{\lambda,k} \oplus \mathscr{L}_{n,k}^{[\lambda]} , \qquad k = 1, \ldots, d_{[\lambda]} .$$

Using (4.3.12) we have from (5.5.15)

$$P_{jj}^{[\lambda']} T_{kk,i_1 \ldots i_m}^{[\lambda]} = P_{jj}^{[\lambda']} P_{kk}^{[\lambda]} T_{i_1 \ldots i_m} = 0 \qquad \text{for} \qquad \lambda \neq \lambda', j \neq k . \tag{5.5.17}$$

This means the $T_{kk}^{[\lambda]}$ satisfy (5.5.14), lying thus in the null space of each of the projection operators $P_{jj}^{[\lambda']}$, $j \neq k$, $\lambda \neq \lambda'$ and hence also in their intersection. According to (5.5.14), these spaces and their intersection are invariant with respect to $\mathscr{GL}(n)$. The $T_{kk}^{[\lambda]}$ are thus uniquely defined. Furthermore, the subspaces $\mathscr{L}_{n,k}^{[\lambda]}$ of different symmetry are all independent. This may be seen because $\sum_{\lambda,k} T_{kk}^{[\lambda]} = 0$ implies $T_{kk}^{[\lambda]} = 0$. If we assumed $T_{kk}^{[\lambda]} \neq 0$ we could operate with $P_{jj}^{[\lambda']}$ from the left, giving $P_{jj}^{[\lambda']} \cdot \sum_{\lambda,k} T_{kk}^{[\lambda]} = T_{jj}^{[\lambda']} = 0$; but this would be a contradiction.

We omit the proof that the invariant subspaces $\mathscr{L}_{n,k}^{[\lambda]}$ of $\mathscr{L}_n^{(m)}$ are irreducible (see [5.6]). For the following discussions it is important that, in the subspaces $\mathscr{L}_{n,k}^{[\lambda]}$ defined by the $T_{kk}^{[\lambda]}$ from (5.5.16), the REP of the operators P_a belonging to $a \in \mathscr{GL}(n, \mathbb{C})$ is given by an IR $D^{[\lambda]}(n)$ of $\mathscr{GL}(n, \mathbb{C})$. The group $\mathscr{GL}(n, \mathbb{C})$ is the complex extension of $\mathscr{GL}(n, \mathbb{R})$ see (10.4.31). According to Appendix D, $\mathscr{GL}(n, \mathbb{R})$ and $\mathscr{U}(n, \mathbb{C})$ have n^2 parameters, $\mathscr{SL}(n, \mathbb{R})$ and $\mathscr{SU}(n, \mathbb{C})$ have $n^2 - 1$ parameters. Furthermore, $\mathscr{GL}(n, \mathbb{R}) = \mathscr{GL}(1, \mathbb{R}) \times \mathscr{SL}(n, \mathbb{R})$ and $\mathscr{U}(n, \mathbb{C}) = \mathscr{U}(1, \mathbb{C}) \times \mathscr{SU}(n, \mathbb{C})$, where $\mathscr{GL}(1, \mathbb{R})$ consists only of real numbers $a \neq 0$, and $\mathscr{U}(1, \mathbb{C})$ only of phase factors $\exp(i\phi)$. All these groups therefore have essentially the same decompositions of the linear spaces into irreducible subspaces as $\mathscr{GL}(n, \mathbb{R})$ has.

The symmetrized basis tensors $T_{kk}^{[\lambda]}$ of the IRs of $\mathscr{GL}(n, \mathbb{C})$ may be constructed using one-particle functions $\Psi_{i_1 \ldots i_m} \in \mathscr{L}_n^{(m)}$ as starting functions. According to (5.5.16),

$$\sum_{\lambda} \sum_{k=1}^{d_{[\lambda]}} d_{[\lambda]n} = \sum_{\lambda} d_{[\lambda]} d_{[\lambda]n} = n^m , \tag{5.5.18}$$

where $d_{[\lambda]n}$ is the dimension of the IR $D^{[\lambda]}(n)$ of $\mathscr{GL}(n)$. Equation (5.5.18) holds because the IRs of $\mathscr{GL}(n)$ belonging to the spaces $\mathscr{L}_{n,k}^{[\lambda]}$ with the same $[\lambda]$ are equivalent. The *multiplicity* with which an IR $D^{[\lambda]}(n)$ occurs in $D^{(\times)}(a)$ is thus given by $d_{[\lambda]}$, the dimension of the IR $D^{[\lambda]}$ of $\mathscr{P}_m$. In the following we combine the set of subscripts

$$i_1, i_2, \ldots, i_m \to \boldsymbol{i} \tag{5.5.19a}$$

and of permutated subscripts

$$i_{p_1}, i_{p_2}, \ldots, i_{p_m} \rightarrow \boldsymbol{i}_p \ . \qquad (5.5.19b)$$

Each set has m components, equal to the rank of the tensor. Every subscript i_k or i_{p_k} can take the values $1, 2, \ldots, n$ equal to the numbers of rows or columns of the $a_{ij} \in \mathscr{GL}(n)$.

By projection with $P_{kk}^{[\lambda]}$ we generate from the one-particle basis

$$\Psi_{\boldsymbol{i}}(1, 2, \ldots, m) = \psi_{i_1}(1) \ldots \psi_{i_m}(m) \qquad (5.5.20)$$

a set of symmetrized tensors

$$\psi_{k,\boldsymbol{i}}^{[\lambda]} := P_{kk}^{[\lambda]} \Psi_{\boldsymbol{i}} \ ,$$

altogether $\sum_\lambda d_{[\lambda]} \cdot n^m$ functions, which cannot be independent since the dimension of $\mathscr{L}_n^{(m)}$ is only n^m. A maximal set of linearly independent tensors defines the irreducible subspaces $\mathscr{L}_{n,k}^{[\lambda]}$ of $\mathscr{L}_n^{(m)}$ with dimension $d_{[\lambda]n}$, and also the IRs $D^{[\lambda]}(n)$ of $\mathscr{GL}(n)$.

In order to perform the projection we have to investigate the effect of a permutation on the tensor components (5.5.10) and on the basis. We consider a *fixed* basis function $\boldsymbol{i} = \boldsymbol{f} = \{f_1, \ldots, f_m\}$ and a given permutation p. The basis tensor $\Psi_{\boldsymbol{f}}(1, \ldots m) = \psi_{f_1}(1) \ldots \psi_{f_m}(m)$ then has the "coordinate" REP $\Psi_{\boldsymbol{f}} = \sum T_{\boldsymbol{i}}^{(\boldsymbol{f})} \Psi_{\boldsymbol{i}}$ with

$$T_{\boldsymbol{i}}^{(\boldsymbol{f})} = \begin{cases} 1, & \text{if } \boldsymbol{i} = \boldsymbol{f} \\ 0, & \text{otherwise} \end{cases}$$

and according to (5.5.10)

$$p T_{\boldsymbol{i}}^{(\boldsymbol{f})} = T_{\boldsymbol{i}_p}^{(\boldsymbol{f})} = \begin{cases} 1, & \text{if } \boldsymbol{i}_p = \boldsymbol{f} \text{ or } \boldsymbol{i} = \boldsymbol{f}_{p^{-1}} \\ 0, & \text{otherwise.} \end{cases}$$

That means, $p T_{\boldsymbol{i}}^{(\boldsymbol{f})}$ belongs to the basis tensor $\Psi_{\boldsymbol{f}_{p^{-1}}}(1, \ldots, m)$. Consequently we have

$$p \Psi_{\boldsymbol{f}}(1, \ldots, m) = \Psi_{\boldsymbol{f}_{p^{-1}}}(1, \ldots, m) = \psi_{f_{p_1^{-1}}}(1) \ldots \psi_{f_{p_m^{-1}}}(m) \ . \qquad (5.5.21a)$$

Applying p to a basis tensor the sub-subscripts i on f thus have to be interchanged according to the inverse permutation p^{-1}. In turn, an application of p to the labels of the arguments of the basis tensor (one-particle functions) is equivalent to this, thus finally

$$p \Psi_{\boldsymbol{f}}(1, \ldots, m) = \Psi_{\boldsymbol{f}}(p_1, \ldots, p_m) = \psi_{f_1}(p_1) \ldots \psi_{f_m}(p_m) \ . \qquad (5.5.21b)$$

Compare this last statement with the rotation of a coordinate system and the inverse rotation of vectors in a coordinate system!

5.5.2 Induced Representations

In Sect. 4.2.5 we considered subduced REPs as the REPs of subgroups $\mathscr{U} \subset \mathscr{G}$. The "counterparts" to these are the *induced* REPs which we need for the REPs of outer products of permutation groups (Sect. 5.5.4) and of space groups (Sect. 9.1.2).

Let $\Delta^{(\alpha)}$ be an IR of a subgroup $\mathscr{U} \subset \mathscr{G}$ and $\{e_j | j = 1, \ldots, d_\alpha\}$ the basis functions belonging to $\Delta^{(\alpha)}$ so that any operator $P_a \notin \mathscr{U}$, if applied to e_j, leads to a function not contained in the set of the e_j. Accordingly to (2.2.8)

$$\mathscr{G} = a_1\mathscr{U} + a_2\mathscr{U} + \cdots + a_{n_i}\mathscr{U} \tag{5.5.22}$$

is a decomposition of $\mathscr{G}$ into (left) cosets with respect to $\mathscr{U}$. By applying the representative elements a_i of the coset $a_i\mathscr{U}$ to the e_j we obtain new sets of d_α independent functions $a_i e_j$, $i = 1, \ldots, n_i$; $j = 1, \ldots, d_\alpha$, thus altogether $n_i \cdot d_\alpha$ functions which are assigned to the $n_i \cdot d_\alpha$-dimensional REP $D^{(\alpha,\mathrm{ind})}(\mathscr{G})$ induced in $\mathscr{G}$ by the IR $\Delta^{(\alpha)}(\mathscr{U})$. Such an induced REP is in general reducible, just as a subduced REP is.

If $b \in \mathscr{U}$ and $a \in \mathscr{G}$ and if a_i, a_j are the representatives of the decomposition into cosets (5.5.22), then there exists a matrix

$$\sigma_{ij}(a,b) = \begin{cases} 1, & \text{if } a = a_i b a_j^{-1} \qquad i,j = 1,\ldots,n_i \\ 0, & \text{otherwise.} \end{cases} \tag{5.5.23}$$

If different from zero at all, it has exactly one number "one" in every row and every column; therefore the REP induced by $\Delta^{(\alpha)}$ is given by

$$D^{(\alpha,\mathrm{ind})}(a) = \sum_{b \in \mathscr{U}} \sigma(a,b) \otimes \Delta^{(\alpha)}(b) \ . \tag{5.5.24}$$

That $D^{(\alpha,\mathrm{ind})}(a)$ is a coarsened permutation matrix, in which the blocks $\Delta^{(\alpha)}$ occur exactly once in every row and every column, can easily be seen, since for a fixed coarsened row i the relation $a^{-1}a_i = a_j b^{-1}$ in (5.5.23) can only be satisfied by a single pair a_j, b. The proof of (5.5.24) makes use of the properties of REP $D^{(\alpha,\mathrm{ind})}$ (see [5.7]).

If, for example, $\mathscr{U} = \mathscr{C}_s = \{e, \sigma_v\}$ and $\mathscr{G} = \mathscr{C}_{3v}$, then f_A in Sect. 4.2.1 is a basis function of $\mathscr{C}_s$ and $\mathscr{G} = \mathscr{C}_{3v} = e\mathscr{U} + c_3\mathscr{U} + c_3^2\mathscr{U}$. Also, $\Delta^{(\alpha)}(e) = 1$, $\Delta^{(\alpha)}(\sigma_v) = 1$ is the A_g-REP of $\mathscr{C}_s$. Then, $P_e f_A = f_A$, $P_{c_3} f_A = f_B$, $P_{c_3^2} f_A = f_C$, where $\{f_A, f_B, f_C\}$ define the induced REP $D^{(A_g,\mathrm{ind})}(\mathscr{G})$. This is just the REP of Table 4.1 which is reducible into $D^{(A_g,\mathrm{ind})} = A_1 \oplus E$. Furthermore we have for the $\sigma(a,b)$

$$\sigma(e,e) = \begin{pmatrix} 1 & 0 & 0 \\ 0 & 1 & 0 \\ 0 & 0 & 1 \end{pmatrix} , \quad \sigma(c_3,e) = \begin{pmatrix} 0 & 0 & 1 \\ 1 & 0 & 0 \\ 0 & 1 & 0 \end{pmatrix} , \quad \sigma(c_3^2,e) = \begin{pmatrix} 0 & 1 & 0 \\ 0 & 0 & 1 \\ 1 & 0 & 0 \end{pmatrix} ,$$

$$\sigma(\sigma_v,\sigma_v)=\begin{pmatrix}1&0&0\\0&0&1\\0&1&0\end{pmatrix}\ ,\quad \sigma(\sigma'_v,\sigma_v)=\begin{pmatrix}0&0&1\\0&1&0\\1&0&0\end{pmatrix}\ ,\quad \sigma(\sigma''_v,\sigma_v)=\begin{pmatrix}0&1&0\\1&0&0\\0&0&1\end{pmatrix},$$

$$\sigma(\sigma_v,e)=\sigma(\sigma'_v,e)=\sigma(\sigma''_v,e)=\sigma(e,\sigma_v)=\sigma(c_3,\sigma_v)=\sigma(c_3^2,\sigma_v)=0\ .$$

With these matrices we again obtain Table 4.1. Now, if $\Delta^{(\beta)}(\mathscr{U})$ with $\beta=1,\dots,r_u$ are the IRs of $\mathscr{U}\subset\mathscr{G}$ and if $D^{(\alpha)}(\mathscr{G})$ with $\alpha=1,\dots,r_g$ are the IRs of $\mathscr{G}$ and furthermore if $\Delta^{(\beta,\mathrm{sub})}(\mathscr{U})$ is the REP subduced from $D^{(\beta)}(\mathscr{G})$ onto $\mathscr{U}$ and $D^{(\alpha,\mathrm{ind})}(\mathscr{G})$ is the REP induced from $\Delta^{(\alpha)}(\mathscr{U})$ in $\mathscr{G}$, we have according to the foregoing discussion and to Sect. 4.2.5

$$\begin{aligned}\Delta^{(\beta,\mathrm{sub})}(\mathscr{U})&=\sum_{\alpha=1}^{r_u}m_{\beta,\alpha}\Delta^{(\alpha)}(\mathscr{U})\ ,\\ D^{(\alpha,\mathrm{ind})}(\mathscr{G})&=\sum_{\beta=1}^{r_g}m'_{\alpha,\beta}D^{(\beta)}(\mathscr{G})\ .\end{aligned}\tag{5.5.25}$$

The multiplicities $m_{\beta,\alpha}$, $m'_{\alpha,\beta}$ are not independent of each other. The *reciprocity theorem of Frobenius* states that

$$m_{\beta,\alpha}=m'_{\alpha,\beta}\ ;\qquad \alpha=1,\dots,r_u\ ;\qquad \beta=1,\dots,r_g\ .\tag{5.5.26}$$

The multiplicity of $\Delta^{(\alpha)}(\mathscr{U})$ in $\Delta^{(\beta,\mathrm{sub})}(\mathscr{U})$ is equal to the multiplicity of $D^{(\beta)}(\mathscr{G})$ in $D^{(\alpha,\mathrm{ind})}(\mathscr{G})$ if $\mathscr{U}\subset\mathscr{G}$.

For $\mathscr{G}=\mathscr{C}_{3v}$ and $\mathscr{U}=\mathscr{C}_s$,

$\Delta^{(\alpha)}$	e	σ_v
A_g	1	1
A_u	1	-1

The REP subduced from $E(\mathscr{C}_{3v})$, $E^{(\mathrm{sub})}$, is

$$E^{(\mathrm{sub})}(\mathscr{C}_s)=\left\{\begin{pmatrix}1&0\\0&1\end{pmatrix};\begin{pmatrix}1&0\\0&-1\end{pmatrix}\right\}=A_g(\mathscr{C}_s)+A_u(\mathscr{C}_s)\ ,\qquad \text{thus}\tag{5.5.27a}$$

$$\underline{m_{E,g}=1}\ ,\qquad m_{E,u}=1\ .$$

The REP induced from $A_g(\mathscr{C}_s)$ in $\mathscr{C}_{3v}$ is (see above)

$$D^{(\mathrm{ind})}(\mathscr{C}_{3v})=A_1(\mathscr{C}_{3v})\oplus E(\mathscr{C}_{3v})\ ,\qquad \text{thus}\tag{5.5.27b}$$

$$m'_{g,1}=1\ ,\qquad \underline{m'_{g,E}=1}\ .$$

5.5.3 Irreducible Tensor Spaces

In order to specify the symmetry-adapted basis functions (symmetrized tensors) we investigate the set of the projections $P_{kj}^{[\lambda]}\Psi_f:=\Psi_{kj}^{[\lambda]}$ with $k,j=1,\dots,d_{[\lambda]}$ and with a fixed starting function $\Psi_f := \psi_{f_1}(1)\dots\psi_{f_m}(m)$. We arrange them according to the following scheme:

$$\begin{array}{llll} \Psi^{[\lambda]}_{11} & \Psi^{[\lambda]}_{12} & \dots & \Psi^{[\lambda]}_{1d_\lambda} \\ \Psi^{[\lambda]}_{21} & \Psi^{[\lambda]}_{22} & \dots & \Psi^{[\lambda]}_{2d_\lambda} \text{ ---------- row } k=2 \\ \vdots & \vdots & & \vdots \\ \Psi^{[\lambda]}_{d_\lambda 1} & \Psi^{[\lambda]}_{d_\lambda 2} & \dots & \Psi^{[\lambda]}_{d_\lambda d_\lambda} \\ | & & & \\ \text{column } j=1 & & & \end{array} \tag{5.5.28}$$

The functions arranged in columns are the $d_{[\lambda]}$ linearly independent partner functions belonging to the IRs $D^{[\lambda]}$ of $\mathscr{P}_m$ [see (4.2.21b) and Sect. 4.3.2]. Altogether there are $d_{[\lambda]}$ sets of such functions, but these sets cannot be mutually linearly independent; this means, that in general not all of the functions arranged in one row are linearly independent. However the functions in a row belong to the invariant subspaces $\mathscr{L}^{[\lambda]}_{n,k}$ with respect to $\mathscr{GL}(n)$ (see below); we have to select the independent ones of these functions.[4] For this purpose we first consider the case that in the function Ψ_f all the $f_i = f_1, \dots, f_m$ *are different from each other*. In this case the $m!$ functions $\{p\Psi_f\}$ are also linearly independent from each other and constitute a basis for the regular REP of $\mathscr{P}_m$ according to Sect. 4.3.1. In the regular REP every IR $D^{[\lambda]}$ occurs just $d_{[\lambda]}$ times, which means that in this case all the row functions $\Psi^{[\lambda]}_{k1}, \dots, \Psi^{[\lambda]}_{kd_\lambda}$ are linearly independent. Since the f_i take values from 1 to n, *in general*, however, all the f_i are *not* different from each other, i.e. some one-particle functions are equal. But this implies that not all of the $m!$ functions $\{p\Psi_f\}$ are linearly independent. In order to find the number $m_{[\lambda]}$ of linearly independent row functions or sets of partner functions of $D^{[\lambda]}$ we have to determine how often $D^{[\lambda]}$ occurs in the reducible REP of $\mathscr{P}_m$ defined by $\{p\Psi_f\}$. This is possible by means of the reciprocity theorem of Frobenius (5.5.26): If some f_i in Ψ_f are equal, then there exists a subgroup $\mathscr{P}'_l \subset \mathscr{P}_m$ which acts only on the arguments of one-particle functions with equal f_i, i.e. Ψ_f remains invariant under $p' \in \mathscr{P}'_l \subset \mathscr{P}_m$: $p'\Psi_f = 1 \cdot \Psi_f$. Thus Ψ_f belongs to the identity REP of $\mathscr{P}'_l$. Consequently, the set $\{p\Psi_f\}$ contains $m!/l'$ ($l' = \operatorname{ord}\mathscr{P}_l$) linearly independent functions; the REP given by these functions can be looked upon as being a REP $D^{(1,\mathrm{ind})}(\mathscr{P}_m)$ induced by the identity REP of $\mathscr{U} = \mathscr{P}'_l$ in $\mathscr{P}_m$. According to (5.5.26), $D^{(1,\mathrm{ind})}(\mathscr{P}_m)$ contains the REP $D^{[\lambda]}$ as many times as $D^{([\lambda],\mathrm{sub})}(\mathscr{P}'_l)$ contains the identity REP $\Delta^{(1)}$ of $\mathscr{P}'_l$. Thus according to (4.2.34b)

$$m_{[\lambda]} = \frac{1}{l'} \sum_{p' \in \mathscr{P}'_l} \chi^{([\lambda],\mathrm{sub})}(p') \cdot 1 \ . \tag{5.5.29}$$

This is just the number of linearly independent row functions in (5.5.28).

---

[4] One has to distinguish strictly between the IRs $D^{[\lambda]}$ of $\mathscr{P}_m$ and $D^{[\lambda]}(n)$ of $\mathscr{GL}(n)$!

The assignment of functions to the spaces $\mathscr{L}_{n,k}^{[\lambda]}$ can be seen from the relation

$$P_{kk}^{[\lambda]} p = \sum_j D_{kj}^{[\lambda]}(p) P_{kj}^{[\lambda]} \ , \tag{5.5.30}$$

which is analogous to (4.3.10). Hence $P_{kk}^{[\lambda]} p$ with $p \in \mathscr{P}_m$ is a linear combination of the elements $P_{kj}^{[\lambda]}, j = 1, \ldots, d_{[\lambda]}$ with a fixed subscript k for the row. This means the elements $P_{kk}^{[\lambda]} p \Psi_f$ according to (5.5.15) lying in $\mathscr{L}_{n,k}^{[\lambda]}$ are linear combinations of the functions $\Psi_{kj}^{[\lambda]}$ belonging to the kth row of (5.5.28), and thus are in general linear combinations of the $m_{[\lambda]}$ linearly independent functions $\Psi_{ki}^{[\lambda]}$, $i = 1, \ldots, m_{[\lambda]}$. In order to find all the $d_{[\lambda]n}$ linearly independent basis functions of the IRs $D^{[\lambda]}(n)$ of $\mathscr{GL}(n)$, we have only to take all the independent starting functions Ψ_f.

For this is a graphical procedure of Littlewood et al. [5.3,4,8,9], which we will give without proof here. It uses the Young tableaux. If all the f_i are different from each other, then $m_{[\lambda]} = d_{[\lambda]}$. However, the latter is equal to the number of SYTs which we can label numerically (Fig. 5.3). Thus, $m_{[\lambda]}$ is equal to the number of possible ways of inserting the m numbers $f_1, \ldots, f_m$ alphanumerically into the diagrams of $[\lambda]$. In the case that not all the f_i are different, this prescription can be generalized:

The multiplicity $m_{[\lambda]}$ of the REP $D^{[\lambda]}$ in the space of the $\{p\Psi_f\}$ follows from the number of the special tableaux belonging to $[\lambda]$ which we obtain by alphanumerical arrangement of the numbers $f_1, \ldots, f_m$ with a fixed configuration. Starting with the smallest f_i we have to insert the integers in nondecreasing order from left to right in a row and in increasing order from top to bottom in a column; repetition of a number is allowed in a row, but *not* in a column of the tableau belonging to $[\lambda]$. Each of the $m_{[\lambda]}$ tableaux $T_f[\lambda]$ (with $f = \{f_1, \ldots, f_m\}$, f_i: number in the ith cell of the tableau) defined in this way corresponds to a starting function $\Psi_f = \psi_{f_1}(1) \ldots \psi_{f_m}(m)$, in which the one-particle function ψ_{f_i} is assigned to the ith particle. Application of $P_{kk}^{[\lambda]}$ to these $m_{[\lambda]}$ starting functions yield $m_{[\lambda]}$ linearly independent symmetrized basis functions $\Psi_{f,k}^{[\lambda]}$ of $\mathscr{L}_{n,k}^{[\lambda]}$ (Chap. 12)

$$\Psi_{f,k}^{[\lambda]} := P_{kk}^{[\lambda]} \psi_{f_1}(1) \ldots \psi_{f_m}(m) \ . \tag{5.5.31}$$

Finally, in order to get all the $d_{[\lambda]n}$ symmetrized functions, this procedure has to be applied to *all the different* functions (disregarding the order in f) Ψ_f with $f = f_1, \ldots, f_m$; $f_i = 1, \ldots, n$. All the permutation groups $\mathscr{P}_m$ with arbitrary m have to be taken into account in order to get all the corresponding basis functions.

As a result we may summarize: The basis functions Ψ_f constitute a REP space for the two groups of commuting operators

$$p \in \mathscr{P}_m \qquad \text{and} \qquad P_a \in \mathscr{GL}(n, \mathbb{C}) \ . \tag{5.5.32}$$

In this basis the *outer product* $\mathscr{P}_m \times \mathscr{GL}(n, \mathbb{C})$ is realized whose IRs follow from the direct product of the IR $D^{[\lambda]}$ of $\mathscr{P}_m$ with the IR $D^{[\lambda]}(n)$ of $\mathscr{GL}(n, \mathbb{C})$. The IRs $D^{[\lambda]} \times D^{[\lambda]}(n)$ of $\mathscr{P}_m \times \mathscr{GL}(n, \mathbb{C})$ have dimension $d_{[\lambda]} \cdot d_{[\lambda]n}$. The basis functions of the spaces $\mathscr{L}_{n,k}^{[\lambda]}$, $k = 1, \ldots, d_{[\lambda]}$ define these IRs.

We shall illustrate this again with the example $\mathscr{P}_3$. Then functions (subscript f_i) are labelled with $a \leqslant b \leqslant c \leqslant \ldots$, in order to distinguish them from the arguments (particle numbers), which we label with $1, 2, 3, \ldots$. In the case $m = 3$ there are at most three "numbers" a, b, c and $1, 2, 3$, but a, b, c are allowed to take values of numbers between 1 and n. First we ask how many linearly independent row functions can be formed from one starting function $\Psi_{aab} = \psi_a(1)\psi_a(2)\psi_b(3)$. We have $\mathscr{P}'_2 = \{e, (12)\} \subset \mathscr{P}_3$ with $l' = 2$. Hence according to (5.5.29) and Appendix A

$$m_{[3]} = \tfrac{1}{2}(1+1) = 1 \ ; \quad m_{[2,1]} = \tfrac{1}{2}(2+0) = 1 \ ; \quad m_{[1^3]} = \tfrac{1}{2}(1-1) = 0 \ .$$

The graphical rule on the preceding page yields

$$\boxed{a}\boxed{a}\boxed{b} \to m_{[3]} = 1 \ , \qquad \begin{array}{ll}\boxed{a}&\boxed{a}\\ \boxed{b}&\end{array} \ \text{(but } \begin{array}{ll}\boxed{a}&\boxed{b}\\ \boxed{a}&\end{array} \text{ forbidden!)} \to m_{[2,1]} = 1$$

and $\begin{array}{l}\boxed{a}\\ \boxed{a}\\ \boxed{b}\end{array}$ forbidden $\to m_{[1^3]} = 0$, just like above.

The dimensions $d_{[\lambda]n}$ and the bases of $\mathscr{L}^{[\lambda]}_{n,k}$ or the $D^{[\lambda]}(n)$ of $\mathscr{GL}(n)$ for $[\lambda] = [3]$ with tableau $\boxed{}\boxed{}\boxed{}$ are, taking the different starting functions,

$\boxed{a}\boxed{a}\boxed{a} \to$, since $a = 1, \ldots, n$,	n linearly independent functions			$\Psi^{[3]}_{aaa}$
$\boxed{a}\boxed{a}\boxed{b} \to$, since $a < b$,	$\frac{1}{2}n(n-1)$	"		$\Psi^{[3]}_{aab}$
$\boxed{a}\boxed{b}\boxed{b} \to$, since $a < b$,	$\frac{1}{2}n(n-1)$	"		$\Psi^{[3]}_{abb}$
$\boxed{a}\boxed{b}\boxed{c} \to$, since $a < b < c$,	$\frac{1}{3!}n(n-1)(n-2)$	"		$\Psi^{[3]}_{abc}$

That is, there are altogether $d_{[3]n} = \frac{1}{3!}n(n+1)(n+2)$ functions. We acquire the corresponding basis functions by applying the projection operator $P^{[3]}_{11} = \frac{1}{6}\{e + (12) + (23) + (31) + (123) + (132)\}$ to the particle numbers of the "allowed" starting functions $\Psi_{abc} = \psi_a(1)\psi_b(2)\psi_c(3)$, $a \leqslant b \leqslant c$ given above (Exercise 5.19):

$$\begin{aligned}\Psi^{[3]}_{abc} = \tfrac{1}{6}[&\psi_a(1)\psi_b(2)\psi_c(3) + \psi_a(2)\psi_b(1)\psi_c(3) + \psi_a(1)\psi_b(3)\psi_c(2)\\ &+ \psi_a(3)\psi_b(2)\psi_c(1) + \psi_a(2)\psi_b(3)\psi_c(1) + \psi_a(3)\psi_b(1)\psi_c(2)] \ .\end{aligned}$$

In addition to the graphical method for the determination of $d_{[\lambda]n}$ there is an algebraic relation, (Weyl):

$$\begin{aligned} d_{[\lambda]n} &= \frac{\prod_{i<j}(h_i - h_j)}{(n-1)!(n-2)!\ldots 1!} \qquad h_i = \lambda_i + n - i \ ; \\ [\lambda] &= [\lambda_1, \lambda_2, \ldots, \lambda_n] \ ; \qquad i, j = 1, \ldots, n \ ; \end{aligned} \tag{5.5.33}$$

the number z of rows of $[\lambda]$ satisfies $z \leqslant n$, thus $\lambda_i = 0$ if $i > z$ in h_i.

The characters are given by the so-called *Schur function*

$$\chi^{[\lambda]}(n,a)=\frac{d_{[\lambda]}}{m!}\cdot\sum_{(\nu)\in\mathscr{P}_m} r(\nu)\cdot\chi^{[\lambda]}(\nu)\cdot[\chi(a)]^{\nu_1}\ldots[\chi(a^m)]^{\nu_m} \tag{5.5.34a}$$

with $d_{[\lambda]}$ from (5.4.13) and $r(\nu)$ from (3.5.12); $\nu=(\nu_1,\nu_2,\ldots,\nu_m)$ according to (3.5.4) (note the notation and subscripts!). In the special case of the totally symmetric REP, (5.5.34a) simplifies to

$$\chi^{[m]}(n,a)=\sum_{(\nu_1\ldots\nu_m)}\frac{[\chi(a)]^{\nu_1}\ldots[\chi(a^m)]^{\nu_m}}{1^{\nu_1}\cdot\nu_1!\ldots m^{\nu_m}\cdot\nu_m!}\,, \tag{5.5.34b}$$

and for the totally antisymmetric REP[5]

$$\chi^{[1^m]}(n,a)=\sum_{(\nu_1\ldots\nu_m)}(-1)^{\nu_2+\nu_4+\ldots}\frac{\{\chi(a)\}^{\nu_1}\ldots\{\chi(a^m)\}^{\nu_m}}{1^{\nu_1}\cdot\nu_1!\ldots m^{\nu_m}\cdot\nu_m!}\,. \tag{5.5.34c}$$

These REPs are of course especially important in connection with spin functions or corresponding functions describing other internal properties of particles (isospin, hypercharge, flavour, colour etc.). In these cases $\mathscr{GL}(n,\mathbb{C})$ has to be replaced by one of its subgroups mostly $\mathscr{SU}(n)$ with $n=2,3,4,5$ (see Sect. 7.2, 13.1 and Appendix D).

The statements given above remain valid for the subgroups $\mathscr{U}(n)$, $\mathscr{SU}(n)\subset\mathscr{GL}(n)$; however, for $\mathscr{SU}(n)$ certain inequivalent IRs of $\mathscr{GL}(n)$ are equivalent with respect to $\mathscr{SU}(n)$ and similarly for $\mathscr{SL}(n\mathbb{R})$. In order to show this equivalence, we consider the graphs belonging to $[\lambda]=[\lambda_1,\ldots,\lambda_n]$ and $[\lambda']=[\lambda_1+1,\lambda_2+1,\ldots,\lambda_n+1]$ and the corresponding IRs of $\mathscr{GL}(n)$ (Fig. 5.4). If $[\lambda]$ belongs to the space of tensors of rank m, then $[\lambda']$ belongs to the tensors of rank $(m+n)$. From the "counting rule" on p. 116, for the number of the independent symmetrized basis tensors we can see that $D^{[\lambda]}(n)$ and $D^{[\lambda']}(n)$ have the same dimension, because the numbers $1,2,\ldots,n$ can be inserted into the diagram $[\lambda']$ in only one way (if the additional cells are numbered by $m+1,\ldots,m+n$ this consideration still holds). Finally it can be shown that in transformations according to $\mathscr{GL}(n)$, $\mathscr{U}(n)$, $\mathscr{SL}(n)$, the symmetrized tensors belonging to $[\lambda']$ behave just as those which belong to $[\lambda]$ apart from a factor $\det a$ (this is implied by the additional subscripts $1,\ldots,n$). Since for $\mathscr{SL}(n)$, $\det a=+1$, it follows that a successive introduction of additional columns of length n gives no new inequivalent REP:

For every number $z\in\mathbb{Z}$, $z>-\lambda_n$, the IRs of $\mathscr{SU}(n)$ belonging to the graphs $[\lambda_1,\ldots,\lambda_n]$ and $[\lambda_1+z,\ldots,\lambda_n+z]$ are equivalent, see [5.5]. Furthermore, the IRs $[\lambda_1,\ldots,\lambda_n]$ and $[\lambda_1-\lambda_n,\lambda_1-\lambda_{n-1},\ldots,\lambda_1-\lambda_2]$ are mutually equivalent, too (Fig. 5.4). (5.5.35)

[5] $\chi^{[1^m]}(\nu_1,\ldots,\nu_m)=(-1)^{\nu_2+\nu_4+\ldots}$ is the character of the classes represented by $\nu_1,\ldots,\nu_m$ in the REP $[1^m]$. A cycle of length l is even if l is odd, and odd if l is even.

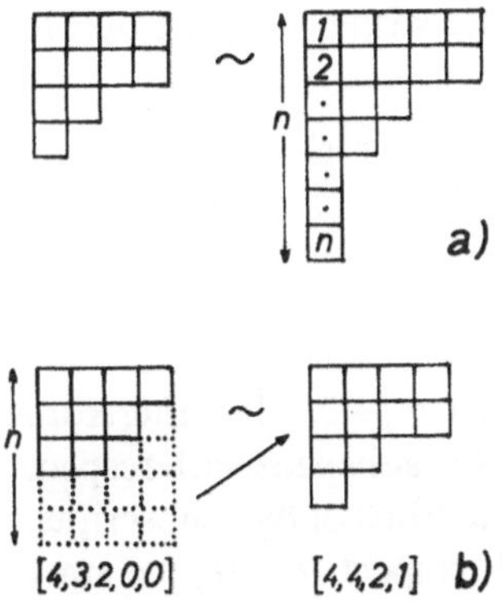

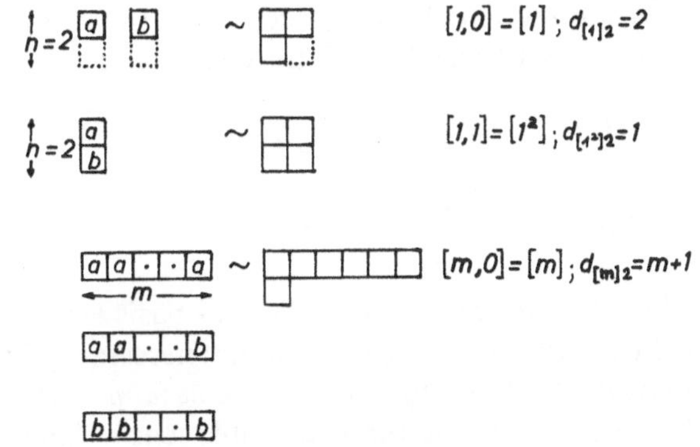

▲

Fig. 5.4a, b. Diagrams belonging to equivalent IRs of $\mathscr{SU}(n)$. **(a)** On adding columns with n cells, graphs belonging to equivalent IRs result. **(b)** The REPs $[\lambda_1, \ldots, \lambda_n]$ and $[\lambda_1 - \lambda_n, \lambda_1 - \lambda_{n-1}, \ldots, \lambda_1 - \lambda_2]$ are equivalent

Fig. 5.5. Diagrams belonging to the IRs of $\mathscr{SU}(2)$; some equivalent REPs are given. The dimensions $d_{[\lambda]n}$ follow from (5.5.33)

For example, for $\mathscr{SU}(n)$ the REPs $[1^\nu]^n$ and $[1^{n-\nu}]^n$, $\nu < n$, are always equivalent to each other. Here the superscript n indicates the assignment to $\mathscr{SU}(n)$.

Figure 5.5 shows the diagrams for the IRs of $\mathscr{SU}(2)$, including some belonging to equivalent REPs. The REP $[\lambda] = [1,0] := [1]$ is the defining two-dimensional IR of $\mathscr{SU}(2)$ [see also (3.3.2) with $\chi(a) = 2\cos(\varphi/2)$, φ: angle of rotation]. For this we have $d_{[1]2} = 2$ and $\chi^{[1]}(2,a) = \chi(a)$. The REP $[\lambda] = [1^2]$ is the identity REP of $\mathscr{SU}(2)$ with $d_{[1^2]2} = 1$ and $\chi^{[1^2]} = [\chi^2(a) - \chi(a^2)]/2 \to 1$, see (5.5.34c). The basis function is the Slater determinant (see Exercise 5.19)

$$\Psi_{ab}^{[1^2]} = \frac{1}{2}\begin{vmatrix} \psi_a^{(1)} & \psi_b^{(1)} \\ \psi_a^{(2)} & \psi_b^{(2)} \end{vmatrix} \quad . \tag{5.5.36}$$

All the other inequivalent IRs have the structure $[m,0] := [m]$ with (consider the counting rule on p. 116) $d_{[m]2} = m + 1$ and

$$\chi^{[m]}(2,a) \to \frac{\sin[(m+1)\varphi/2]}{\sin(\varphi/2)}$$

(see Exercise 5.20 for $m = 2, 3$). For $\mathscr{SU}(2)$ there is exactly one IR with the dimensions 1, 2, 3, ...

For the *orthogonal group* $\mathscr{O}(n, \mathbb{R}) \subset \mathscr{GL}(n, \mathbb{C})$ (see Appendix D) the method of symmetrizing tensors does not lead to irreducible spaces. A further reduction can be achieved by a *contraction* of tensor subscripts. This can be seen from the behaviour of tensors in $\mathscr{O}(n)$ transformations: If $P_o \cong o \in \mathscr{O}(n)$, then according to (5.5.6a)

$$P_o e_{i_1 \ldots i_m} := e'_{i_1 \ldots i_m} = \sum_{j_i}^{n} o_{j_1 i_1} o_{j_2 i_2} \ldots o_{j_m i_m} e_{j_1 \ldots j_m} \quad . \tag{5.5.37}$$

By a contraction of two subscripts (e.g. by taking the trace with $i_1 = i_2$) we obtain with $\mathscr{o}\tilde{\mathscr{o}} = 1$ together with the contraction

$$e(12)_{i_3 \dots i_m} := \sum_{i_1} e_{i_1 i_1 i_3 \dots i_m} , \tag{5.5.38}$$

$$e'(12)_{i_3 \dots i_m} = \sum_{j_i} \mathscr{o}_{j_3 i_3} \dots \mathscr{o}_{j_m i_m} e(12)_{j_3 \dots j_m} . \tag{5.5.39}$$

Thus the contractions transform as basis tensors of rank $m - 2$. They therefore define a separate, invariant, but not necessarily irreducible subspace with respect to $\mathscr{O}(n)$. This holds for all the possible $m(m-1)/2$ contractions of two subscripts. In order to acquire the irreducible tensor spaces with respect to $\mathscr{O}(n)$ we start with tensors whose pairwise contractions (traces) vanish. Then we apply the projectors $P_{kk}^{[\lambda]}$ of $\mathscr{P}_m$ to these and obtain in this way symmetrized tensors with vanishing contractions (zero traces). These define the irreducible spaces with respect to $\mathscr{O}(n)$.

It can be shown that every tensor $T_{i_1 \dots i_m}$ can be decomposed into a tensor $\mathring{T}_{i_1 \dots i_m}$ having zero trace and a tensor

$$\hat{T}_{i_1 \dots i_m} = \delta_{i_1 i_2} \cdot X(12)_{i_3 \dots i_m} + \dots + \delta_{i_\alpha i_\beta} \cdot X(\alpha\beta)_{i_1 \dots i_{\alpha-1} i_{\alpha+1} \dots i_{\beta-1} i_{\beta+1} \dots i_m} \tag{5.5.40}$$

(Exercise 5.21). For $m = 2$ we have e.g.

$$T \to T_{ij} = \mathring{T}_{ij} + \hat{T}_{ij} ; \qquad \hat{T}_{ij} = \delta_{ij} X(12) .$$

From

$$\sum_{i=1}^{n} \mathring{T}_{ii} = 0 = \sum_{i=1}^{n} (T_{ii} - \hat{T}_{ii}) = \mathrm{Tr}\{T\} - nX(12)$$

it follows that

$$T_{ij} = \mathring{T}_{ij} + \frac{1}{n} \delta_{ij} \mathrm{Tr}\{T\} , \qquad \mathrm{Tr}\{\mathring{T}\} = 0 \tag{5.5.41}$$

or, for a product of one-particle functions,

$$\psi_i(1)\psi_j(2) = \mathring{\Psi}_{ij}(12) + \frac{1}{n}\delta_{ij} \sum_k \psi_k(1)\psi_k(2) .$$

Applying the projection opertors $P_{kk}^{[\lambda]}$ of $\mathscr{P}_2$ to the tensors $\mathring{T}$ or $\mathring{\Psi}$ with zero trace, we get

$$\mathring{\Psi}_{ij}^{[2]} = \frac{1}{2}\{\psi_i(1)\psi_j(2) + \psi_i(2)\psi_j(1)\} - \frac{1}{n}\delta_{ij} \sum_k \psi_k(1)\psi_k(2)$$

with $i \leqslant j$, i.e. $n(n+1)/2 - 1$ functions, and

$$\mathring{\psi}_{ij}^{[1^2]} = \frac{1}{2}\{\psi_i(1)\psi_j(2) - \psi_i(2)\psi_j(1)\}$$

with $i < j = 1, \ldots, n$, i.e. $n(n-1)/2$ functions, making a total of $(n^2 - 1)$ functions.

The rule for finding symmetrized tensors with zero trace is the following:

> Tensors with zero trace, belonging to Young diagrams which have in the first two columns a number of cells larger than n, are forbidden (or identically zero). Thus the only diagrams allowed are those in which the sum of the cells (s_1, s_2) in the first two columns is smaller than or equal to n: $s_1 + s_2 \leqslant n$. (5.5.42)

A proof can be found in [5.5]. For the $\mathscr{SO}(n)$ groups, diagrams with s_1 and $n - s_1$ cells in the first column while all other columns have the same length give equivalent REPs if $s_1 < n/2$. If $s_1 = n/2$ for even n then the REP assigned to these diagrams decays into two nonequivalent IRs. For the $\mathscr{O}(n)$ groups the corresponding REPs are only associated or self-associated, i.e. they differ only in sign for the proper and improper elements of $\mathscr{O}(n)$. The identity REP is always represented by $[1^n]$.

Since we started in Sect. 5.5.1 and in this section with tensor transformations in the normal vector space $\mathbb{R}_n$, we obtain in this way only the single-valued vector REPs of $\mathscr{SO}(n)$. The spinor REPs cannot be illustrated by partitions in the above sense. This can clearly be seen with the group $\mathscr{SO}(3)$, for which only the odd-numbered REPs can be obtained from $\mathscr{SL}(3)$ by contractions. Only these seem to be relevant for the physics of $\mathscr{SO}(n)$ symmetries. For the symplectic groups $\mathscr{Sp}(2n)$ there are also rules for a contraction which we will not discuss in detail, but it should be mentioned that for $\mathscr{Sp}(2n)$ only diagrams with $s_1 \leqslant n$ are allowed. The only exception is the identity REP with $[1^{2n}]$.

5.5.4 Direct Products and Their Reduction

The Clebsch-Gordan expansion of the REP of an inner product of two IRs $D^{[\lambda]}$ and $D^{[\lambda']}$ of $\mathscr{P}_m$ is given by (4.4.13):

$$D^{[\lambda]} \otimes D^{[\lambda']} = \sum_{\lambda''}{}_{\oplus} (\lambda\lambda'|\lambda'')D^{[\lambda'']} \qquad \text{with} \tag{5.5.43}$$

$$(\lambda\lambda'|\lambda'') = \frac{1}{m!}\sum_p \chi^{[\lambda]}(p)\chi^{[\lambda']}(p)\chi^{[\lambda'']}(p) \; .$$

This has to be distinguished from the outer product of two REPs of different groups. For example, let $\mathscr{P}_{m_1}$ be the permutation group of the objects $1, \ldots, m_1$ and $\mathscr{P}_{m_2}$ that of the objects $m_1 + 1, \ldots, m_1 + m_2 = m$. The operators $p_1 \in \mathscr{P}_{m_1}$ and $p_2 \in \mathscr{P}_{m_2}$ commute since the objects they operate on are different. According to (2.2.20) we can construct the *outer* product group $\mathscr{P}_{m_1} \times \mathscr{P}_{m_2}$ whose

IRs $D^{[\lambda]\times[\lambda']}(\mathscr{P}_{m_1}\times\mathscr{P}_{m_2})$ can be formed from the two subgroups according to $D^{[\lambda]}(\mathscr{P}_{m_1})\otimes D^{[\lambda']}(\mathscr{P}_{m_2})$; $\mathscr{P}_{m_1}\times\mathscr{P}_{m_2}$ is a subgroup of $\mathscr{P}_m$. We define a *new outer product* $\odot$ of $D^{[\lambda]}(\mathscr{P}_{m_1})$ and $D^{[\lambda']}(\mathscr{P}_{m_2})$ via the REP $D^{[\lambda]\odot[\lambda']}(\mathscr{P}_m)$ *induced* from the IR $D^{[\lambda]\times[\lambda']}(\mathscr{P}_{m_1}\times\mathscr{P}_{m_2})$ in $\mathscr{P}_m$ (Sect. 5.5.2). If the $d_{[\lambda]}\cdot d_{[\lambda']}$ basis functions $\varphi_i(1,\dots,m_1)\cdot\psi_j(m_1+1,\dots,m_1+m_2)$ with $i=1,\dots,d_{[\lambda]}$, $j=1,\dots,d_{[\lambda']}$ of $D^{[\lambda]\times[\lambda']}(\mathscr{P}_{m_1}\times\mathscr{P}_{m_2})$ are given, we obtain a basis of $D^{[\lambda]\odot[\lambda']}$ by applying p_k to the $\varphi_i\psi_j$, i.e. by $\{p_k\varphi_i\psi_j|k=1,\dots,s\}$. Here p_k are the representatives of the decomposition of $\mathscr{P}_m$ into cosets:

$$\mathscr{P}_m=\sum_{k=1}^{s}p_k(\mathscr{P}_{m_1}\times\mathscr{P}_{m_2})\ .$$

The p_k mix the objects in the two sets $\{1,\dots,m_1\}$ and $\{m_1+1,\dots,m_1+m_2\}$, thus $s=m!/m_1!m_2!$. For $D^{[\lambda]\odot[\lambda']}$ there are altogether

$$d_{[\lambda]\odot[\lambda']}=\frac{m!}{m_1!m_2!}\cdot d_{[\lambda]}\cdot d_{[\lambda']} \tag{5.5.44}$$

linearly independent basis functions. Equation (5.5.44) is the dimension of the REP of the new outer product. Generally the induced REP $D^{[\lambda]\odot[\lambda']}$ is reducible and can be decomposed into IRs of $\mathscr{P}_m$ according to the reciprocity theorem of Frobenius (5.5.25). Hence we have for the multiplicity of the IR $D^{[\lambda'']}$ of $\mathscr{P}_m$ in $D^{[\lambda]\odot[\lambda']}$

$$m_{[\lambda]\odot[\lambda'],[\lambda'']}=\frac{1}{m_1!m_2!}\sum_{p_1\in\mathscr{P}_{m_1}}\sum_{p_2\in\mathscr{P}_{m_2}}\chi^{[\lambda]}(p_1)\cdot\chi^{[\lambda']}(p_2)\cdot\chi^{[\lambda'']}(p_1p_2)\ . \tag{5.5.45}$$

For example, if $\varphi(1)$ is a basis of $[\lambda]=[1]$ of $\mathscr{P}_{m_1}$, $m_1=1$ and $\psi(2)$ is a basis of $[\lambda']=[1]$ of $\mathscr{P}_{m_2}$, $m_2=1$, then $\mathscr{P}_m=\mathscr{P}_2=\{e,(12)\}$ and $\{p_k\varphi_i\psi_j\}=\{\varphi(1)\psi(2);\varphi(2)\psi(1)\}$, $d_{[\lambda]\odot[\lambda']}=2$. Now we know that $[\lambda'']=[2]$ has the basis $\varphi(1)\psi(2)+\varphi(2)\psi(1)$ and $[\lambda'']=[1^2]$ has the basis $\varphi(1)\psi(2)-\varphi(2)\psi(1)$, consequently, $[1]\odot[1]=[2]+[1^2]$. The same result follows from (5.5.45):

$$m_{[1]\odot[1],[2]}=m_{[1]\odot[1],[1^2]}=\frac{1}{1\cdot 1}(1\cdot 1\cdot 1)=1\ .$$

As another example we consider $\mathscr{P}_{m_1}=\{e,(12)\}$, $\mathscr{P}_{m_2}=\{e,(34)\}$. Then

$$\mathscr{P}_{m_1}\times\mathscr{P}_{m_2}=\{e,(12),(34),(12)(34)\}\subset\mathscr{P}_4=\mathscr{P}_m\ .$$

For the decomposition of $[1^2]\odot[2]$ into IRs of $\mathscr{P}_4$ we obtain according to (5.5.45) and Appendix A

$$m_{[1^2]\odot[2],[\lambda'']}=\tfrac{1}{4}[\chi^{[\lambda'']}(e)-\chi^{[\lambda'']}(12)+\chi^{[\lambda'']}(34)-\chi^{[\lambda'']}((12)(34))]\ .$$

Thus, see Appendix A,

$[\lambda'']$	[4]	$[1^4]$	$[2^2]$	[3, 1]	$[2, 1^2]$
$m_{[1^2]\odot[2],[\lambda'']}$	0	0	0	1	1

We have $d_{[1^2]\odot[2]} = 6$.

Just as on p. 116, also in this case there is again a graphical method (Littlewood et al. [5.3,4,8,9]) which allows a reduction according to (5.5.45). First we realize that the new outer product $\odot$ is commutative, associative and distributive: Then we have the following rule

> In one of the diagrams $[\lambda]$ or $[\lambda']$ of the product $[\lambda] \odot [\lambda']$, the cells of the first row are labelled with $a \leftrightarrow 1$, those of the second row with $b \leftrightarrow 2$, etc. The cells of this diagram have to be added to those of the other diagram so that there results a new allowed diagram as follows:
> First the "$a \leftrightarrow 1$" cells are added, then the "$b \leftrightarrow 2$" cells, then the "$c \leftrightarrow 3$" cells, etc., such that
> (i) in the same column of the resulting diagram no two cells with the same symbol (number) are allowed to occur, and
> (ii) if the numbers in the added cells of the resulting diagram are read *row by row from right to left and from top to bottom* at any stage there must never be more natural numbers $j + 1$ than j. (5.5.46)

In Fig. 5.6 this procedure is illustrated for the example given above and for $[2, 1] \odot [2, 1]$ of the product $\mathscr{P}_{m_1} \times \mathscr{P}_{m_2} \subset \mathscr{P}_6$, $m_1 = m_2 = 3$. Hence

$$[2,1] \odot [2,1] = [4,2] \oplus [4,1^2] \oplus 2\cdot[3,2,1] \oplus [3^2] \oplus [3,1^3] \oplus [2^3] \oplus [2^2,1^2]$$

and $d_{[2,1]\odot[2,1]} = 80$.

With these tools we can also give the Clebsch-Gordan expansion of an inner direct product of two IRs $D^{[\lambda]}(n)$ and $D^{[\lambda']}(n)$ of $\mathscr{GL}(n)$, $\mathscr{U}(n)$ and $\mathscr{SU}(n)$. For illustration we use the decomposition of the tensor space $\mathscr{L}_n^{(m)}$ into irreducible subspaces of symmetrized tensors $\mathscr{L}_{n,k}^{[\lambda]}$, (5.5.15–17). Hence

$$\mathscr{L}_n^{(2)} = \mathscr{L}_n^{[2]} \oplus \mathscr{L}_n^{[1^2]} \qquad \text{or} \qquad D(a \otimes a) = a \otimes a = D^{[2]}(n) + D^{[1^2]}(n)$$

or (see Exercise 5.19 and the example in Sect. 5.5.3)

$$\mathscr{L}_n^{(3)} = \mathscr{L}_n^{[3]} + \mathscr{L}_{n,1}^{[2,1]} + \mathscr{L}_{n,2}^{[2,1]} + \mathscr{L}_n^{[1^3]} .$$

These examples are illustrated graphically in Fig. 5.7. We see that this reduction obeys the same rules as the reduction of the new product $\odot$ according to (5.5.46). Here, too, we have to consider associativity and distributivity. Thus (Littlewood et al. [5.3,4,8,9]):

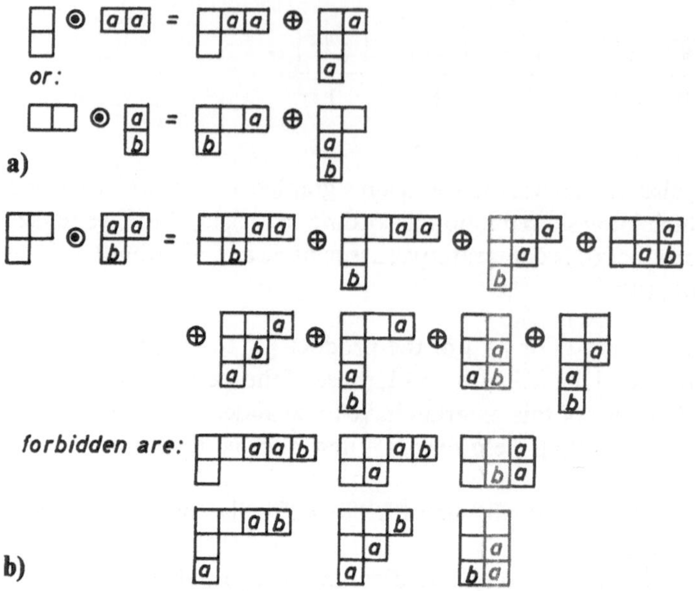

Fig. 5.6. Reduction of the (new outer product according to (5.5.46) for (**a**) $[1^2] \odot [2]$ and (**b**) $[2,1] \odot [2,1]$. It is best to add all the cells with "a" first, then those with "b", etc., in such a way that allowed diagrams result

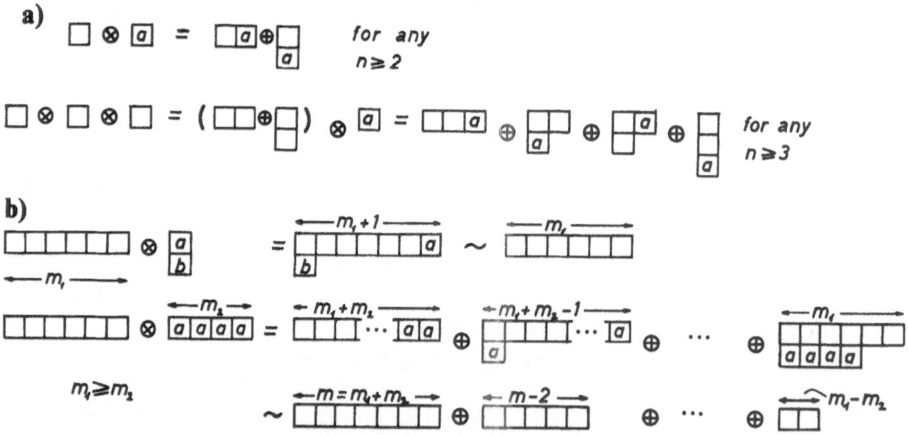

Fig. 5.7a, b. Clebsch-Gordan decomposition of inner products of REPs of $\mathcal{GL}(n)$ according to (5.5.47). (**a**) Decomposition of $a \otimes a$ and $a \otimes a \otimes a$, (**b**) Decomposition of $D^{[\lambda]}(2) \otimes D^{[\lambda]}(2)$ of $\mathcal{SU}(2)$. The dimensions of REPs $[m]$ are $d_{[m]2} = m + 1$

The decomposition of the inner direct product $D^{[\lambda]}(n) \otimes D^{[\lambda']}(n)$ of $\mathscr{GL}(n)$, $\mathscr{U}(n)$ and $\mathscr{SU}(n)$ into IRs $D^{[\lambda'']}(n)$ obeys the rule (5.5.46), provided the resulting allowed diagrams do not have more than n rows. (5.5.47)

The Clebsch-Gordan expansion of $\mathscr{SU}(2)$ thus yields the well-known rules for the coupling of angular momenta (Chaps. 12, 13). With the REPs of Fig. 5.5. and considering the equivalence rules (5.5.35) we obtain the decompositions given in Fig. 5.7b.

The simple rules given in (5.5.46 and 47) are valid for the $\mathscr{SU}(n)$ groups. If they are to be used for the $\mathscr{SO}(n)$ and $\mathscr{Sp}(2n)$ groups, we first have to investigate the assignment of these groups to $\mathscr{SL}(n)$. For details we refer the reader to specialist literature in this field.

Exercise 5.18. Show for $\mathscr{C}_{3v}$ that the induced REPs can also be obtained by the group algebra: The irreducible idempotents $P_{kk}^{(\alpha)}$ of $\mathscr{U}$ generate an induced REP of $\mathscr{G}$ that is in general reducible.

Exercise 5.19. For $\mathscr{P}_3$ and $\mathscr{GL}(n, \mathbb{C})$:

a) Show that $d_{[2,1]n} = n(n^2 - 1)/3$ and $d_{[1^3]n} = n(n-1)(n-2)/6$.

b) Give the linear independent functions $\Psi_{abc,k=1}^{[2,1]}$, $\Psi_{abc,k=2}^{[2,1]}$ and $\Psi_{abc,k=1}^{[1^3]}$; use the projectors $P_{kk}^{[\lambda]}$ from (4.4.6,21) and show that $\Psi_{abc}^{[1^3]}$ can be represented by a Slater determinant.

c) Show that (5.5.18) is satisfied with $m = 3$.

d) For tensors of rank m ($m = 1, 2, 3$, thus T_i, T_{ij}, T_{ijk}) in $\mathscr{L}_n(n = 2, 3)$ determine the tensor components assigned to the IRs of $\mathscr{P}_m$ by using (5.5.15). This also gives the decomposition of $\mathscr{L}_n^{(m)}$ into invariant subspaces with respect to $\mathscr{GL}(n, \mathbb{R})$.

Exercise 5.20. Calculate $\chi^{[m]}(n)$ and $\chi^{[1^m]}(n)$ for $m = 2$ and $m = 3$.

Exercise 5.21. Prove the decomposition (5.5.40) by using

$$\langle \mathring{T} | \hat{T} \rangle := \sum \mathring{T}_{i_1 \dots i_m} \cdot \hat{T}_{i_1 \dots i_m} \equiv 0 \ ,$$

which means that all the traces of $\mathring{T}$ have to vanish.

Exercise 5.22. Discuss the example (5.5.41) for $\mathscr{O}(n = 3)$ with $\psi_1 = x$, $\psi_2 = y$, $\psi_3 = z$, if both particles 1, 2 are identical. What is the scalar that is invariant with respect to $\mathscr{O}(n)$?

Exercise 5.23. Discuss the rule in Sect. 5.5.3, before (5.5.31), as a special case of the rule (5.5.46). Notice that Ψ_f is invariant under $\mathscr{P}_{m_1} \times \mathscr{P}_{m_2} \times \dots \subset \mathscr{P}_m$; $m = m_1 + m_2 + \dots$ if there are m_1 subscripts of the same kind "a", m_2 of "b", etc. among the subscripts $f = \{f_i\}$. The set $\{p\Psi_f\}$ with $p \in \mathscr{P}_m$ is the basis of the direct product of the identity REPs which are induced in $\mathscr{P}_m$ by the REPs of $\mathscr{P}_{m_j}$. Thus one has to consider the new outer product $[m_1] \odot [m_2] \odot \dots$.

6. Tensor Operators and Expectation Values

A large amount of work in physics deals with the calculation of matrix elements and the solution of eigenvalue problems, either exactly or by perturbation methods. The basic ideas of using group theory in simplifying such problems (Wigner-Eckart theorem) will be given in this chapter.

6.1 Tensors and Spinors

Many physical quantities are not invariant under rotations (or general transformations) of the space, but they transform into each other[1]. Examples are vectors, the components of which transform into each other and all tensors (of rank m). A special case is that of scalar quantities, which remain invariant under rotations of the space. For vectors in $\mathbb{R}_3$ their change under rotation is described by (4.1.21), i.e. the components of a vector (operator) $\boldsymbol{T} = \{T_i\}$ obey

$$T_i' = P_d T_i P_d^{-1} = \sum_{k=1}^{3} d_{ik} T_k = \sum_k D_{ik}(d) T_k \ . \tag{6.1.1}$$

The vector components transform according to a (three-dimensional) REP D of the rotation group $\mathcal{SO}(3)$. Tensors of rank 2 transform correspondingly:

$$T_{ij}' = P_d T_{ij} P_d^{-1} = \sum_{k,l=1}^{3} D_{ik}(d) D_{jl}(d) T_{kl} \ , \tag{6.1.2}$$

i.e. according to the inner product (Sect. 4.4.2) of two REPs $D(d) \otimes D(d)$ of the rotation group, which in turn constitutes a (nine-dimensional) REP of $\mathcal{SO}(3)$. In general, tensors of rank m in $\mathbb{R}_3$ transform according to a 3^m-dimensional REP of the rotation group which can be written as an m-fold inner direct product (5.5.6b).

The possible independent tensors now form linear spaces, the tensor spaces $\mathcal{L}_3^{(m)}$ (dimension $d = 3^m$ for tensors of rank m) which can be written as a direct product space of vector spaces:

$$\mathcal{L}_3^{(m)} = \mathbb{R}_3^{(1)} \otimes \mathbb{R}_3^{(2)} \otimes \cdots \otimes \mathbb{R}_3^{(m)} \ . \tag{6.1.3}$$

---

[1] This is a special case of Sect. 5.5, but it is self-contained and can thus be read independently of the more complicated Sect. 5.5. Unlike Sect. 5.5, here the tensors are defined with respect to $\mathcal{SO}(3)$ or $\mathcal{SU}(2)$, instead of more generally with respect to $\mathcal{GL}(n)$. The dimension of the space is n, here $n = 3$.

By rotations the components transform linearly into each other. In each of these spaces a REP of the rotation group $\mathscr{SO}(3)$ is realized. This concept can easily be generalized to vectors or tensors with n components ($n > 3$). The basis space is then $\mathbb{R}_n$ and in the tensor spaces $\mathscr{L}_n^{(m)}$ a REP of $\mathscr{SO}(n)$ is realized. These REPs are said to be tensor REPs (5.5.6b).

Let $\mathbb{R}_n$ be an n-dimensional vector space and D a REP of $\mathscr{SO}(n)$ in $\mathbb{R}_n$, i.e. of the set of linear mappings $\mathbb{R}_n \to \mathbb{R}_n$ with $\det d = \pm 1$, $d\tilde{d} = \tilde{d}d = 1$. Furthermore, let the vectors $e^{(\nu)} = \{e_i^{(\nu)}\}$ constitute an orthonormal basis for the REPs D in the space $\mathbb{R}_n^{(\nu)}$, thus

$$e_i'^{(\nu)} = \sum_{k=1}^{n} D_{ki} e_k^{(\nu)} \ . \tag{6.1.4}$$

The space $\mathscr{L}_n^{(m)}$ of the tensors of rank m in $\mathbb{R}_n$ is then the product space (6.1.3) with $3 \to n$ and is defined analogously to (5.5.5) by the basis:

$$e_{i_1 \ldots i_m} = \{e_{i_1}^{(1)} e_{i_2}^{(2)} \ldots e_{i_m}^{(m)} | i_1, \ldots, i_m = 1, \ldots, n\} \ . \tag{6.1.5}$$

A tensor T of rank m then is an element of the tensor space $\mathscr{L}_n^{(m)}$ with the components $T_{i_1 \ldots i_m}$

$$T = \sum_{i_\nu = 1}^{n} T_{i_1 \ldots i_m} \cdot e_{i_1 \ldots i_m} \ , \qquad \nu = 1, \ldots, m \ , \tag{6.1.6}$$

where the components transform according to (6.1.1, 2) or to the generalization of those equations, see (5.5.8). A rotation D in $\mathbb{R}_n$ thus induces a rotation

$$D_\nu^{(\times)} = D \otimes D \otimes \cdots \otimes D \tag{6.1.7}$$

in the tensor space $\mathscr{L}_n^{(m)}$. The corresponding matrices form a tensor REP (of rank m) of $\mathscr{SO}(n)$. It is an inner direct product of the REPs D and in general reducible, i.e. it can be decomposed into irreducible parts (Sect. 4.4.2). The tensor components or the tensor operators, which are important in quantum mechanics, can be decomposed into corresponding components (Sect. 5.5.1). It is then sufficient to give the behavior in transformations of the different irreducible parts of T, denoted as $T_i^{(\alpha)}$. The total number of independent tensor components (operators) is $\sum_{\alpha=1}^{r} m_\alpha d_\alpha$ and in general

$$T_i'^{(\alpha)} = P_a T_i^{(\alpha)} P_a^{-1} = \sum_{k=1}^{d_\alpha} D_{ki}^{(\alpha)}(a) T_k^{(\alpha)} \tag{6.1.8}$$

is the transformation under any element a of the symmetry group $\mathscr{G}$. If the tensors can be split into invariant tensors of lower rank with respect to $\mathscr{SO}(3)$, e.g. by taking the trace (contraction) [see (5.5.37) onwards], then it is sufficient to give the transformation behaviour of these components. The tensor space thus reduces and a further decomposition into irreducible subspaces can be started from these independent tensor subspaces using the procedure in Sect. 5.5. If for example T_{ij}

is totally symmetric (antisymmetric), it is sufficient to give the behaviour of 6(3) components in the transformation, or of certain linear combinations, such as $1/2(T_{ij} \pm T_{ji})$. In this case $\mathscr{L}_3^{(\pm)} = 6(3)$ if $n = 3$ (see example in Sect. 5.5.3).

A special case occurs for scalar operators like the Hamiltonian H of a physical system. Such operators are invariant with respect to the total symmetry group $\mathscr{G}(H)$ with $[P_a, H] = 0$ or $H' = P_a H P_a^{-1} = H$ (Sect. 6.3), i.e. according to (6.1.8), $D^{(\alpha)}(a) = 1$ for each $a \in \mathscr{G}(H)$.

> Scalar operators such as the Hamiltonian of a physical system transform according to the (one-dimensional) identity REP of the symmetry group of H. (6.1.9)

For a polar vector (e.g. position vector or electric dipole moment) we have in the case of the group $\mathscr{C}_{3v}$ the following decomposition of $\{T_i\}$ into irreducible parts (see also Exercise 6.1):

$$\begin{aligned} &\begin{pmatrix} T_x \\ T_y \end{pmatrix} \Rightarrow T_i^{(E)} , \quad i = 1, 2 ; \quad d_E = 2 ; \quad m_E = 1 ; \\ &T_z \Rightarrow T_i^{(A_1)} , \quad i = 1 ; \quad d_{A_1} = 1 ; \quad m_{A_1} = 1 . \end{aligned} \tag{6.1.10}$$

The above considerations hold for polar vectors and tensors, respectively. Axial vectors or tensor components transform with different signs in reflections and mirror rotations of $\mathscr{O}(3) = \mathscr{C}_i \times \mathscr{SO}(3)$. Then in (6.1.1, 2) or in their generalization, a factor det d has to be added for every axial component. However, we can also look upon axial components as being totally antisymmetric components of a tensor of rank 2.

A spinor of rank m with respect to the rotation group $\mathscr{SO}(3)$ or $\mathscr{SU}(2)$ can be similarly defined:

$$\chi = \sum_{\sigma_v = \pm} \chi_{\sigma_1 \ldots \sigma_m} \cdot e_{\sigma_1}^{(1)} \ldots e_{\sigma_m}^{(m)} , \tag{6.1.11}$$

which is an element of the spinor space that is constituted by the basis[2]

$$\{e_{\sigma_1}^{(1)} \ldots e_{\sigma_m}^{(m)} | \sigma_v = \pm\} ; \tag{6.1.12}$$

the $e_{\sigma_v}^{(v)}$ here transform according to the double-valued REP $D^{(1/2)}(d)$ of the rotation group $\mathscr{SO}(3)$ or according to the defining REP of $\mathscr{SU}(2)$, see (3.3.2). For the components we have

$$\chi'_{\sigma_1 \ldots \sigma_m} = \sum_{\tau_v} D_{\sigma_1 \tau_1}^{(1/2)}(d) \ldots D_{\sigma_m \tau_m}^{(1/2)}(d) \cdot \chi_{\tau_1 \ldots \tau_m} . \tag{6.1.13}$$

They transform according to the m-fold direct product

[2] $\pm$ is used in connection with the spin components $\pm 1/2$ introduced in Sect. 5.1.

$$D_s^{(\times)} = D^{(1/2)} \otimes D^{(1/2)} \otimes \cdots \otimes D^{(1/2)} \; ; \tag{6.1.14}$$

$D^{(1/2)}$ is given in (3.3.2) as a function of the Euler angles.

A spinor field which describes particles with spin 1/2 is a field of two-component spinors of rank 1: $\chi_\sigma(\boldsymbol{x})$. It is transformed by rotations according to

$$\chi'_\sigma(\boldsymbol{x}) = P_d \chi_\sigma(\boldsymbol{x}) = \sum_{\tau=\pm} D^{(1/2)}_{\sigma\tau} \chi_\tau(d^{-1}\boldsymbol{x}) \; . \tag{6.1.15}$$

A one-particle wave function which describes orbital as well as spin angular momentum and which can be written as a product of space and spin functions (5.1.3) transforms according to

$$D^{(l)} \otimes D^{(1/2)} = D^{(l+1/2)} \oplus D^{(l-1/2)} \tag{6.1.16}$$

if the Hamiltonian H is invariant with respect to the rotation group [see also end of Sect. 5.5.4 and (11.4.71)].

Exercise 6.1. Decompose the operator (vector) of angular momentum into its irreducible parts with respect to the symmetry groups $\mathscr{C}_{3v}$ and $\mathscr{D}_3$.

Exercise 6.2. Split the tensor $T_{ij} \neq T_{ji}$ of rank 2 into its irreducible parts with respect to the symmetry groups $\mathscr{C}_{3v}$, $\mathscr{D}_3$, $\mathscr{T}$ and $\mathscr{O}$.

6.2 The Wigner-Eckart Theorem

If a system is invariant with respect to a symmetry group $\mathscr{G}$, the eigenfunctions (basis functions) of the Hamiltonian H can be classified according to the IRs of the group (Sect. 6.3): $|\alpha, i\rangle$; here α denotes the IR and i the row of the REP.

Functions belonging to different α, i are mutually orthogonal. However, there may be several functions $|\alpha, i\rangle$ which transform according to the REP α, since the unitary Hilbert space, in general, may contain the REP α many times. Under a symmetry operation these functions transform according to the corresponding IR (4.3.14)

$$P_a|\alpha, i\rangle = \sum_{k=1}^{d_\alpha} D^{(\alpha)}_{ki}(a)|\alpha, k\rangle \; . \tag{6.2.1}$$

Application of a tensor operator $T_j^{(\beta)}$ to $|\alpha, i\rangle$ then gives a function which transforms according to

$$\begin{aligned} P_a T_j^{(\beta)}|\alpha, i\rangle &= P_a T_j^{(\beta)} P_a^{-1} P_a |\alpha, i\rangle \\ &= \sum_{k=1}^{d_\alpha} \sum_{l=1}^{d_\beta} D^{(\beta)}_{lj}(a) D^{(\alpha)}_{ki}(a) \cdot T_l^{(\beta)}|\alpha, k\rangle \\ &= \sum_{kl} D^{(\beta\times\alpha)}_{lk,ji}(a) \cdot T_l^{(\beta)}|\alpha, k\rangle \; , \end{aligned} \tag{6.2.2}$$

where $D^{(\beta\times\alpha)}$ represents the inner product of the REPs $D^{(\beta)}$ and $D^{(\alpha)}$ of (4.4.5,11). From (4.4.11) it follows that the $T_j^{(\beta)}|\alpha,i\rangle$ transform as the inner product function $|\beta,j\rangle|\alpha,i\rangle$, which can be decomposed into irreducible parts by the Clebsch-Gordan expansion (4.4.13). The reduction coefficient (4.4.14) $(\beta\alpha|\gamma)$ gives the multiplicity of the REP $D^{(\gamma)}$ in $D^{(\beta\times\alpha)}$. Correspondingly we can decompose $T_j^{(\beta)}|\alpha,i\rangle$ according to (4.4.18b) into basis functions of the Hilbert space, where now s_γ numbers the IRs occurring several times; $s_\gamma = 1,\dots,(\alpha\beta|\gamma)$. Thus $T_j^{(\beta)}|\alpha,i\rangle$ transforms as

$$|\beta,j\rangle|\alpha,i\rangle = \sum_{\gamma,s_\gamma,l}\begin{pmatrix}\beta\alpha & \gamma s_\gamma\\ ji & l\end{pmatrix}^* |\gamma,s_\gamma,l\rangle \tag{6.2.3}$$

with the CGC according to (4.4.18a). Multiplying (6.2.3) by $\langle\gamma',l'|$, we obtain on the right hand side

$$\sum_{s_\gamma}\begin{pmatrix}\beta\alpha & \gamma s_\gamma\\ ji & l\end{pmatrix}^* \cdot \bar{C}^\gamma_{s_\gamma} \ .$$

The remaining sum runs only over s_γ because of the orthogonality of basis functions (4.3.15) to different IRs

$$\langle\gamma',l'|\gamma,s_\gamma,l\rangle \sim \delta_{\gamma\gamma'}\delta_{ll'} \ . \tag{6.2.4}$$

The factor $\bar{C}^\gamma_{s_\gamma}$ depends on γ, s_γ, but according to (4.3.15) not on the row subscripts l, l'. Since $T_j^{(\beta)}|\alpha,i\rangle$ transforms in the same way as $|\beta,j\rangle|\alpha,i\rangle$, (6.2.4) must also be valid for the matrix elements (expectation values, transition elements); thus

$$\langle\gamma,l|\,T_j^{(\beta)}|\alpha,i\rangle = \sum_{s_\gamma} C^\gamma_{s_\gamma}\cdot\begin{pmatrix}\beta\alpha & \gamma s_\gamma\\ ji & l\end{pmatrix}^* \ . \tag{6.2.5a}$$

As above, $C^\gamma_{s_\gamma}$ depends on the tensor operator $T_j^{(\beta)}$ and on γ, s_γ, but it is independent of the row indices i, j, l.

Equation (6.2.5a) represents the *Wigner-Eckart theorem* in its generalized form. The $C^\gamma_{s_\gamma}$ are called reduced matrix elements. If the eigenfunctions $|\alpha,i\rangle$ depend on further quantum numbers not related to the symmetry group $\mathcal{G}$, then, of course, the $C^\gamma_{s_\gamma}$ are also dependent on these. A calculation of the $C^\gamma_{s_\gamma}$ requires knowledge of the functions. The importance of the Wigner-Eckart theorem consists in the fact that it allows statements on zero matrix elements, that means, for example, forbidden transitions. This happens always if $D^{(\gamma)}$ does not occur in the reduction of $D^{(\beta\times\alpha)}$. The CGC then vanish.

Our considerations become especially simple for *simply reducible groups* (Sect. 4.4.3) since the reduction coefficients are equal to 0 or 1 for these groups. The subscript s_γ is not necessary in this case. The relative values of the matrix elements for fixed γ are given by the ratios of the CGC alone. The value of $C^\gamma_{s_\gamma}$ can be calculated with the "most convenient" functions $|\alpha,i\rangle$, $|\gamma,l\rangle$. Using (4.4.19) the reduced matrix elements can be written as (Sect. 11.3)

$$C_{s_\gamma}^{\gamma} = \frac{1}{d_\gamma} \sum_{ijl} \begin{pmatrix} \beta\alpha & \gamma s_\gamma \\ ji & l \end{pmatrix} \cdot \langle \gamma, l | T_j^{(\beta)} | \alpha, i \rangle \; . \tag{6.2.5b}$$

For electric dipole transitions with the symmetry $\mathscr{C}_{3v}$ we have according to (4.4.21, 6.2.5) and Table 4.6 the following elements:

$$\langle A_1, 1 | T_j^{(E)} | A_1, 1 \rangle = 0$$

$$\langle A_2, 1 | T_j^{(E)} | A_1, 1 \rangle = 0$$

$$\langle E, l | T_j^{(E)} | A_1, 1 \rangle = C^E \begin{pmatrix} EA_1 & E \\ j1 & l \end{pmatrix}^* = \frac{C^E}{\sqrt{2}} \begin{pmatrix} 1 & 0 \\ 0 & 1 \end{pmatrix}$$

$$\langle A_2, 1 | T_j^{(E)} | A_2, 1 \rangle = 0$$

$$\langle E, l | T_j^{(E)} | A_2, 1 \rangle = C^E \begin{pmatrix} EA_2 & E \\ j1 & l \end{pmatrix}^* = \frac{C^E}{\sqrt{2}} \begin{pmatrix} 0 & 1 \\ -1 & 0 \end{pmatrix}$$

$$\langle E, l | T_j^{(E)} | E, i \rangle = C^E \begin{pmatrix} EE & E \\ ji & l \end{pmatrix}^* = \frac{C^E}{\sqrt{2}} \begin{pmatrix} 0 & 1 \\ 1 & 0 \end{pmatrix} \qquad l = 1$$

$$\frac{C^E}{\sqrt{2}} \begin{pmatrix} 1 & 0 \\ 0 & -1 \end{pmatrix} \qquad l = 2 \tag{6.2.6}$$

$$\langle A_1, 1 | T_1^{A_1} | A_1, 1 \rangle = C^{A_1} \begin{pmatrix} A_1 A_1 & A_1 \\ 1 \;\; 1 & 1 \end{pmatrix} = C^{A_1} \cdot 1$$

$$\langle A_2, 1 | T_1^{A_1} | A_1, 1 \rangle = 0$$

$$\langle E, l | T_1^{A_1} | A_1, 1 \rangle = 0$$

$$\langle A_2, 1 | T_1^{A_1} | A_2, 1 \rangle = c^{A_2} \begin{pmatrix} A_1 A_2 & A_2 \\ 1 \;\; 1 & 1 \end{pmatrix} = C^{A_2} \cdot 1$$

$$\langle E, l | T_1^{A_1} | A_2, 1 \rangle = 0$$

$$\langle E, l | T_1^{A_1} | E, i \rangle = C^E \begin{pmatrix} A_1 E & E \\ 1 \;\; i & l \end{pmatrix} = \frac{C^E}{\sqrt{2}} \begin{pmatrix} 0 & 0 \\ 1 & 1 \end{pmatrix} \; .$$

Consequently, allowed electric dipole transitions are found to be (Table 5.2, column p)

$$A_1 \leftrightarrow E \; ; \qquad A_2 \leftrightarrow E \; ; \qquad E \leftrightarrow E \; , \qquad \text{for } x, y \text{ polarization}$$

$$A_1 \leftrightarrow A_1 \; ; \qquad A_2 \leftrightarrow A_2 \; ; \qquad E \leftrightarrow E \; , \qquad \text{for } z \text{ polarization}$$

where further forbidden transitions are given by the zeros in the above matrices:

For x polarization the transition $A_1 \leftrightarrow E$ couples to the dipole only if the x-component of the E state is different from zero, etc.

Exercise 6.3. Calculate the matrix elements of the angular momentum for the symmetries $\mathscr{C}_{3v}$ and $\mathscr{D}_3$ (see Exercise 6.1).

6.3 Eigenvalue Problems

The problem of calculating eigenvalues of an operator A which is invariant with respect to a symmetry group $\mathscr{G}$, can be simplified considerably by means of the methods discussed above. Let A be such an operator in a vector space $\mathbb{R}_n$ or Hilbert space $\mathscr{H}$, then

$$\begin{aligned} &A|\psi\rangle = \lambda|\psi\rangle \;, \\ &[A, P_a] = AP_a - P_aA = 0 \;, \qquad \text{for all } a \in \mathscr{G} \;. \end{aligned} \tag{6.3.1}$$

If A is equal to the Hamiltonian H, then it follows from $[H, P_a] = 0$ that the (degenerate) eigenfunctions $\{|E,i\rangle; i = 1, \ldots, d_E\}$ of H belonging to the eigenvalue E, and therefore all the eigenfunctions of H, can be classified according to the IRs of $\mathscr{G}$, since

$$P_aH|E,i\rangle = H\cdot P_a|E,i\rangle = E\cdot P_a|E,i\rangle \;.$$

The $\{P_a|E,i\rangle; i = 1, \ldots, d_E\}$ are thus also degenerate eigenfunctions of H with eigenvalue E and consequently must be linear combinations of the $\{|E,i\rangle\}$:

$$P_a|E,i\rangle = \sum_{j=1}^{d_E} D_{ji}^{(E)}(a)\cdot|E,j\rangle \;.$$

The eigenspace $\mathscr{L}^{(E)} := \{|E,i\rangle\}$ is thus a REP space of the symmetry group $\mathscr{G}$. The REP matrices are

$$D_{ji}^{(E)}(a) := \langle E,j|P_a|E,i\rangle \;.$$

The dimension d_E of $D^{(E)}$ is equal to the degeneracy of E. In general, $\mathscr{L}^{(E)}$ can be assumed to be irreducible. This is easily seen, because if $\mathscr{L}^{(E)}$ were reducible into two invariant subspaces $\mathscr{L}_1 \oplus \mathscr{L}_2 = \mathscr{L}^{(E)}$, for example, then the basis functions constituting $\mathscr{L}_1$ and $\mathscr{L}_2$, respectively would not *necessarily* belong to the same eigenvalue. This would be in contradiction to the definition of $\mathscr{L}^{(E)}$.

However, if in a special case $\mathscr{L}^{(E)}$ (or $D^{(E)}$) proves to be reducible, then we have an *accidental* degeneracy. This is mostly due to some special form of the interaction potential. We have a "*hidden symmetry*", which means the originally assumed symmetry group is only a subgroup of the true symmetry group (Sect. 13.4).

The reduction of the eigenvalue problem (6.3.1) can be achieved according to Sect. 4.1.2 by using any complete orthonormal system $\{|e^{(j)}\rangle\}$ of $\mathbb{R}_n$ or $\mathscr{H}$. We write

$$\sum_j \langle e^{(k)}|A|e^{(j)}\rangle\langle e^{(j)}|\psi\rangle = \lambda\langle e^{(k)}|\psi\rangle$$

or

$$\sum_j a_{kj}\psi_j = \lambda\psi_k \ . \tag{6.3.2}$$

The basis system $\{|e^{(j)}\rangle\}$ can then be replaced by a symmetry-adapted system $\{|\mu\rangle := |\alpha, s_\alpha, i\rangle\}$, where α denotes the IR, i the row of the IR and s_α the multiplicity s_α: $1, \ldots, m_\alpha$ with which an IR α occurs in $\mathbb{R}_n$ or $\mathscr{H}$. The determination of the symmetry-adapted basis is performed according to (4.3.14) by the projectors $P_{ik}^{(\alpha)}$. Since $\sum_\mu |\mu\rangle\langle\mu| = 1$, with the unitary matrix

$$S_{\mu j} = \langle\mu|e^{(j)}\rangle = \langle\alpha, s_\alpha, i|e^{(j)}\rangle \tag{6.3.3}$$

(S transforms the old basis into the symmetry-adapted basis), we obtain from (6.3.2)

$$\sum_{\mu'} \langle\mu|A|\mu'\rangle\cdot\langle\mu'|\psi\rangle = \lambda\langle\mu|\psi\rangle$$

$$\sum_{\mu'} A_{\mu\mu'}\psi_{\mu'} = \lambda\psi_\mu \tag{6.3.4}$$

$$A_{\mu\mu'} = \sum_{j,l} S_{\mu j}a_{jl}S^*_{\mu' l} \ , \qquad \psi_\mu = \sum_j S_{\mu j}\psi_j \ .$$

The eigenvalue problem in this basis is thus

$$\det(A_{\mu\mu'} - \lambda\delta_{\mu\mu'}) = 0 \ . \tag{6.3.5}$$

The matrix $A_{\mu\mu'} = \langle\mu|A|\mu'\rangle := \langle\gamma, s_\gamma, l|A|\alpha, s_\alpha, i\rangle$ is a block-diagonal matrix so that the eigenvalue problem is substantially reduced. This statement is an immediate consequence of the Wigner-Eckart theorem, taking into account that A (or H) is a scalar operator which transforms according to the (one-dimensional) identity REP A_1 of the symmetry group $\mathscr{G}$. Since $A_1 \otimes D^{(\alpha)} = D^{(\alpha)}$, we have $(A_1\alpha|\gamma) = \delta_{\alpha\gamma}$. This means the subscript s_γ in the CGC can be dropped and we have the CGC

$$\begin{pmatrix} \beta\alpha & \gamma \\ ji & l \end{pmatrix}^* = \begin{pmatrix} A_1 & \alpha & \gamma \\ 1 & i & l \end{pmatrix}^* = \delta_{\alpha\gamma}\delta_{il} \ . \tag{6.3.6}$$

Thus, using (6.2.5),

$$A_{\mu\mu'} = \langle\gamma, s_\gamma, l|A|\alpha, s_\alpha, i\rangle = \delta_{\alpha\gamma}\delta_{il}\cdot C^\alpha(s_\gamma, s_\alpha) := \delta_{\alpha\gamma}\delta_{il}\cdot C^\alpha(s'_\alpha, s_\alpha) \ , \tag{6.3.7}$$

$$C^\alpha(s'_\alpha, s_\alpha) := \langle s'_\alpha \| A \| s_\alpha \rangle^{(\alpha)} \; ; \tag{6.3.8}$$

$C^\alpha(s'_\alpha, s_\alpha)$ is called the reduced matrix, it has the dimension m_α, i.e. the multiplicity with which $D^{(\alpha)}$ occurs in the total REP of the space $\mathbb{R}_n$ (or $\mathscr{H}$), and is independent of the row index $i, l, \ldots$. Equation (6.3.7) defines the blockdiagonal form of $A_{\mu\mu'}$:

$$A_{\mu\mu'} = \underbrace{\begin{bmatrix} \langle s'_1 \| A \| s_1 \rangle^{(1)} & & & & \\ & \langle s'_1 \| A \| s_1 \rangle^{(1)} & & 0 & \\ & & \ddots & & \\ & 0 & & \langle s'_r \| A \| s_r \rangle^{(r)} & \\ & & & & \langle s'_r \| A \| s_r \rangle^{(r)} \end{bmatrix}}_{d_1 \text{ “blocks” } \qquad d_r \text{ “blocks”}} \tag{6.3.9}$$

Each of the r IRs $D^{(\alpha)}$ leads to d_α identical blocks (d_α: dimension of $D^{(\alpha)}$). The remaining eigenvalue problems have the dimension m_α:

$$\det(\langle s'_\alpha \| A \| s_\alpha \rangle^{(\alpha)} - \lambda^{(\alpha)} \cdot \delta_{s_\alpha s'_\alpha}) = 0 \; . \tag{6.3.10}$$

To every REP α there belong in general m_α *different eigenvalues*, which are all *d_α-fold degenerate*. Hence, the eigenspaces of the operator A (that is, the set of eigenfunctions of A belonging to the same eigenvalue) are given by the REP spaces of the IRs of $\mathscr{G}$. Since this degeneracy is a direct consequence of symmetry it is said to be necessary (natural, symmetry) degeneracy[3]. This has to be distinguished from accidental degeneracy. The latter occurs if eigenvalues belonging to different IRs coincide or if eigenvalues which belong to a reduced part of (6.3.9), are equal. In any case, this degeneracy is determined by the physical parameters of the system, e.g. by the potential $V(r)$. The geometrical symmetry of a central potential is $\mathscr{O}(3)$. The accidental symmetry present when only geometric symmetry is considered can be removed by dynamical symmetries, e.g. $\mathscr{O}(4)$ for a $(1/r)$-potential, $\mathscr{SU}(3)$ for a r^2-potential. This is discussed in Sect. 13.4.1.

Exercise 6.4. Prove that the matrix

$$\begin{pmatrix} A & 0 & C & D & C & -D \\ 0 & B & E & F & -E & F \\ C & E & G & H & K & L \\ D & F & H & G & -L & M \\ C & -E & K & -L & G & -H \\ -D & F & L & M & -H & G \end{pmatrix}$$

possesses $\mathscr{C}_{2v}$ symmetry. Figure 7.1 gives the corresponding geometric reali-

[3] If A is real, i.e. A commutes (in the real space REP of ψ) with the time reversal operator ϑ, additional degeneracies may occur (see end of Sect. 5.2 on two complex conjugated REPs).

zation in a yz-plane. Prove that the symmetry-adapted vectors are $(0, a_1, b_1, c_1, -b_1, c_1)$ and $(a_4, 0, b_4, c_4, b_4, -c_4)$. Give the block-diagonal form.

6.4 Perturbation Calculus

There are only a few cases in which the eigenvalues of H can be calculated exactly. Often a system is perturbed by a (small) external field. Then the Hamiltonian is conveniently split into an unperturbed part H_0 and a perturbation H_1. Let the symmetry group of H_0 be $\mathscr{G}_0$. The solutions of the unperturbed equation $H_0|\psi_0\rangle = E|\psi_0\rangle$ can be classified according to the IRs of $\mathscr{G}_0$ (see Sect. 6.3: $A \to H_0, \lambda \to E_0$). After switching on the perturbation H_1, which is invariant with respect to a symmetry group $\mathscr{G}_1$, the system is described by

$$(H_0 + H_1)|\psi\rangle = E|\psi\rangle \ . \tag{6.4.1}$$

The solutions now have to be classified by the IRs of the group $\mathscr{U} = \mathscr{G}_0 \cap \mathscr{G}_1 \subseteq \mathscr{G}_0$, where in general $\mathscr{G}_1$ is a subgroup of $\mathscr{G}_0$. Hence, $\mathscr{U} = \mathscr{G}_1$. We shall investigate the perturbation of the energy to first order[4] for a d-fold degenerate eigenvalue E_0 of the unperturbed system.

The perturbation calculus for degenerate states yields the secular equation (see also textbooks on quantum mechanics) for the correction of the energy E_1 to first order

$$\det\{\langle\psi_{0,i}|H_1|\psi_{0,k}\rangle - E_1 \cdot \delta_{ik}\} = 0 \tag{6.4.2}$$

with the unperturbed functions $\{|\psi_{0,k}\rangle\}$. The solutions of (6.4.2) give the splitting of degenerate energies E_0 by the perturbation H_1:

$$E = E_0 + \begin{cases} E_{1,1} \\ \vdots \\ E_{1,d} \end{cases} . \tag{6.4.3}$$

Some of the $E_{1,k}$, $k = 1, \ldots, d$ may be equal again (d: degeneracy of E_0). This approximation holds pretty well as long as the splittings are small compared to the energy differences of neighbouring (unperturbed) energy levels (Fig. 6.1).

Concerning the symmetry group $\mathscr{G}_1(H_1)$ we have to distinguish between two cases:

i) $\mathscr{G}_1 \supseteq \mathscr{G}_0$. In this case the group $\mathscr{U}$ of $H = H_0 + H_1$ is equal to $\mathscr{G}_0$ and the states in (6.4.1) are described by the IRs of $\mathscr{G}_0$, as in the unperturbed case. Hence there is no splitting of energy levels but at most a shift.

---

[4] As far as statements on symmetry are concerned, a higher-order perturbation theory gives no essentially new results. The procedure is the same (Sect. 8.3.), therefore we use the first order as an example.

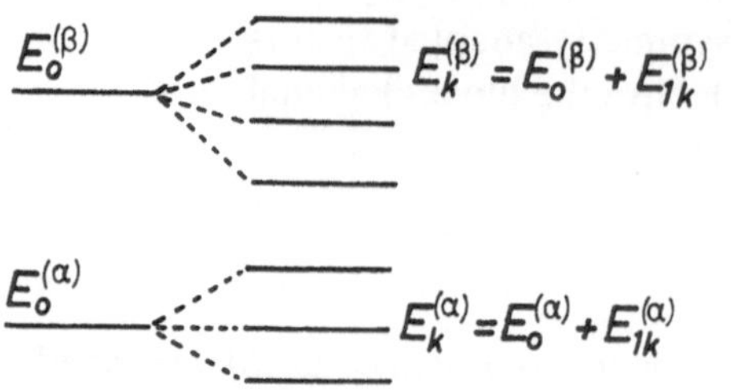

Fig. 6.1. Energy level splitting caused by a perturbation. The first approximation is only useful as long as the splitting is small compared to the separation of neighbouring unperturbed levels: $|E^{(\alpha)}_{1,k}| \ll |E_0^{(\alpha)} - E_0^{(\beta)}|$, $|E^{(\beta)}_{1,k}| \ll |E_0^{(\beta)} - E_0^{(\alpha)}|$

ii) $\mathscr{U} = \mathscr{G}_0 \cap \mathscr{G}_1 \subset \mathscr{G}_0$, usually $\mathscr{U} = \mathscr{G}_1 \subset \mathscr{G}_0$. Then $H = H_0 + H_1$ is invariant under $\mathscr{U}$ ($= \mathscr{G}_1$). Since the dimensions of the IRs of $\mathscr{U}$ are smaller than (or at most equal to) those of $\mathscr{G}_0$ (Sect. 4.2.5), the degeneracy of the eigenvalues is removed at least partly, i.e. in general, the levels are split.

If we intend to make use of the symmetries, we have to change the basis $|\psi_{0,k}\rangle$ to the symmetry-adapted basis $|\alpha, s_\alpha, i\rangle$ [of $\mathscr{U} = \mathscr{G}_0$ in case i), of $\mathscr{G}_1$ in case ii)]. This can be done by using the projection operators $P_{ik}^{(\alpha)}$ (Sects. 6.3 and 4.3.2). Using the Wigner-Eckart theorem we obtain from (6.4.2)

$$\det\{\langle\alpha, s_\alpha, i|H_1|\alpha', s'_\alpha, i'\rangle - E_1\langle\alpha, s_\alpha, i|\alpha', s'_\alpha, i'\rangle\} = 0 \tag{6.4.4}$$

with

$$\langle\alpha, s_\alpha, i|H_1|\alpha', s'_\alpha, i'\rangle = \delta_{\alpha\alpha'}\delta_{ii'}\langle s_\alpha\|H_1\|s'_\alpha\rangle^{(\alpha)}$$

$$\langle\alpha, s_\alpha, i|\alpha', s'_\alpha, i'\rangle = \delta_{\alpha\alpha'}\delta_{ii'}\langle s_\alpha\|s'_\alpha\rangle^{(\alpha)} \tag{6.4.5}$$

For every IR $D^{(\alpha)}$ of $\mathscr{U} = \mathscr{G}_0$ (or $\mathscr{U} = \mathscr{G}_1$) we thus have d_α identical subdeterminants of dimension m_α

$$\det\{\langle s_\alpha\|H_1\|s'_\alpha\rangle^{(\alpha)} - E_1\langle s_\alpha\|s'_\alpha\rangle^{(\alpha)}\} = 0 \ . \tag{6.4.6}$$

The dimension m_α gives the multiplicity of $D^{(\alpha)}$ in the REP space formed by the $\psi_{0,k}$. In our discussion we again have to distinguish the two cases.

i) $\mathscr{U} = \mathscr{G}_0$. The perturbation H_1 is invariant with respect to $\mathscr{G}_0$, i.e. the solutions are determined by the IRs $D^{(\alpha)}$ of $\mathscr{G}_0$. From (6.4.5) together with $m_\alpha = 1$ (the unperturbed degenerate functions are classified according to the IRs of $\mathscr{G}_0$) it follows that all the off-diagonal elements vanish and that all the diagonal elements have the same value. The new eigenvalue is still d_α-fold degenerate, i.e. it does not split, but it is shifted from E_0 by E_1.

If there is an *accidental* degeneracy in the unperturbed system then the degenerate functions $|\psi_{0,k}\rangle$ belonging to E_0 induce the reducible REP D of $\mathscr{U} = \mathscr{G}_0$. A degeneracy of this type may be removed by a perturbation H_1 with the symmetry group $\mathscr{G}_0$. If we decompose D into IRs $D^{(\alpha)}$ of $\mathscr{G}_0$,

Table 6.1. Correlation diagram $\mathscr{SO}(3) \to \mathscr{O}$ (cubic group)

$D^{(l)}$	(d_l)	$\sum_{\beta} \oplus m_{\beta,l}\Delta^{(\beta)}$	Terms	(d_β)
$l = 0\,(s)$	(1)	A_1	A_1	(1)
$1\,(p)$	(3)	F_1	F_1	(3)
$2\,(d)$	(5)	$E \oplus F_2$	E F_2	(2) (3)
$3\,(f)$	(7)	$A_2 \oplus F_1 \oplus F_2$	A_2 F_1 F_2	(1) (3) (3)
$4\,(g)$	(9)	$A_1 \oplus E \oplus F_1 \oplus F_2$	A_1 E F_1 F_2	(1) (2) (3) (3)
$5\,(h)$	(11)	$E \oplus F_1 \oplus F_1 \oplus F_2$	E F_1 F_1 F_2	(2) (3) (3) (3)

$$D = \sum_{\alpha=1}^{m} \oplus D^{(\alpha)} \tag{6.4.7}$$

(a particular $D^{(\alpha)}$ may be repeated), we find using (6.4.4–6) that E_0 can be split into at most m values.

ii) $\mathscr{U} \subset \mathscr{G}_0$ (usually $\mathscr{U} = \mathscr{G}_1$). In this case the REP $D^{(\alpha)}$, which is irreducible with respect to $\mathscr{G}_0$, that characterizes the unperturbed eigenvalues E_0, is in general a reducible REP $D^{(\alpha,\,\mathrm{sub})}(\mathscr{U})$ with respect to $\mathscr{U}$ (Sect. 4.2.5). For a classification of the perturbed eigenvalues this has to be reduced according to the IRs $\Delta^{(\beta)}(\mathscr{U})$ (see the subduction in Sect. 4.2.5):

$$D^{(\alpha,\,\mathrm{sub})}(\mathscr{U}) = \sum_{\beta=1}^{r} \oplus\, m_{\beta,\alpha}\Delta^{(\beta)}(\mathscr{U}) \;, \tag{6.4.8}$$

where $m_{\alpha,\beta}$ are the reduction coefficients of (4.2.34b). This reduction is described by correlation tables like Table 6.1. The eigenvalues split into terms characterized by $\Delta^{(\beta)}$[5]. The symmetry-adapted functions used in (6.4.4–6) are obtained by application of the projectors of $\mathscr{U}$ to the $|\psi_{0,k}\rangle$. The possible splittings are determined by the $m_{\beta,\alpha}$ and the remaining degeneracies by d_β, being the dimension of $\Delta^{(\beta)}$.

In many cases a spherical symmetric system [described by H_0, symmetry $\mathscr{G}_0 = \mathscr{SO}(3)$] e.g. a free atom, is embedded in a system of lower symmetry

[5] If the perturbation H_1 is real, any two eigenvalues belonging to conjugate complex REPs remain degenerate (Sect. 5.2, the end of Sect. 6.3 and the example $\mathscr{C}_{3v}$).

(electrical field in a crystal, see crystal field theory in Sect. 8.4; perturbation H_1, symmetry $\mathcal{G}_1 = \mathcal{O} \subset \mathcal{SO}(3)$; $\mathcal{O}$: octahedral group with cubic symmetry). Then the splittings of the levels of H_0 follow from the reduction of the IR $D^{(l)}$ of $\mathcal{SO}(3)$ with respect to $\Delta^{(\beta)}$ of $\mathcal{O}$. For the first six l-values this reduction is given in Table 6.1. The order (sequence) of the split terms, of course, depends sensitively on the perturbation, i.e. the magnitude of the matrix elements. In this case an accidental degeneracy might be conserved.

Exercise 6.5. Describe the splitting of a term which belongs to the REP E of the group $\mathcal{C}_{3v}$ in the unperturbed system if the perturbation possesses the symmetry (a) $\mathcal{C}_3$ and (b) $\mathcal{C}_s$.

Exercise 6.6. Verify Table 6.1 by using the characters

$$\chi^{(l)}(\varphi) = \frac{\sin(l + 1/2)\varphi}{\sin(\varphi/2)}$$

of the group $\mathcal{SO}(3)$. (φ is the rotation angle of the elements $c(\varphi)$ or c_n.) See (11.4.67) and Appendix A.

7. Molecular Spectra

In molecular physics the vibrational and the electronic states and the interactions between such states are of particular interest. Since (most) molecules possess definite symmetries, the calculation of such states can be reduced to a large extent by using group theoretical methods. This is illustrated in the following, starting with the vibrational states (including infrared absorption and Raman effect), then discussing the properties of one-electron approximations (hybridization, Hückel method, ligand field theory) and finally by considering many-electron problems.

7.1 Molecular Vibrations

7.1.1 Equation of Motion and Symmetry

For an illustration of the symmetry concept it is sufficient to consider only harmonic vibrations. Within this approximation we assume the existence of equilibrium positions for the atoms or ions of a molecule about which they vibrate with small amplitudes. When displaced by u_i^m from their equilibrium positions they exert forces on each other which are proportional to these displacements. If $-\phi_{ij}^{mn}u_j^n$ is the force which acts on the particle m in the direction i due to a displacement of particle n in direction j, then the equation of motion is

$$M^m \ddot{u}_i^m = -\sum_{n,j} \phi_{ij}^{mn} u_j^n \ , \qquad m, n = 1, \ldots, N \ ; \qquad i, j = 1, 2, 3 \ ; \tag{7.1.1}$$

M^m is the mass of particle m. This equation holds for every system of N particles which is able to vibrate harmonically. The ϕ_{ij}^{mn} are known as force constants (FCs). If there exists a potential between the particles we have

$$\phi_{ij}^{mn} = \left. \frac{\partial^2 \phi}{\partial u_i^m \partial u_j^n} \right|_{\text{equilibrium positions}} \ . \tag{7.1.2}$$

The FCs reflect the (geometrical) symmetry of the system. This means the FCs are not completely independent of each other, but there are certain relations between them (Fig. 7.1). For example, the H_2O molecule possesses $\mathscr{C}_{2v}$ symmetry, the FCs between the O atom and the H atoms must have reflection symmetry. However, it might happen that the symmetry of the FCs is different from the *geometrical symmetry* of the system. This would be the case for example in Fig.

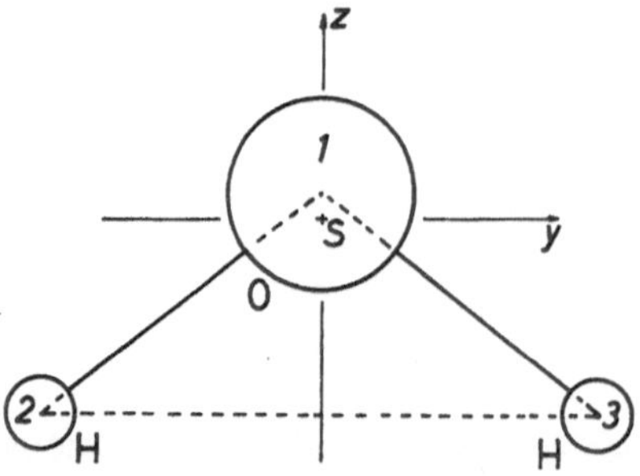

Fig. 7.1. Water molecule with $\mathscr{C}_{2v}$-symmetry. The elements of symmetry are e, c_{2z}, σ_{vx} and σ'_{vy}. O–H distance: 0.956 Å; H–H distance: 1.512 Å; H–O–H angle: 104.5°; S: centre of mass

7.1 if the coupling[1] (interaction) between both the H atoms were just the same as that between the O atom and the H atoms. The interaction would then have $\mathscr{C}_{3v}$ (or $\mathscr{D}_{3h}$) symmetry, that is, a higher symmetry. Such a symmetry of interactions is said to be a *dynamical symmetry* (Sect. 13.4). These are the relevant symmetries of a physical system, not so much the geometrical symmetries. The interaction, however, can of course have a lower symmetry than that indicated by the geometrical configuration, e.g. in Fig. 7.1 if the interaction between the O atom and the H atoms is changed asymmetrically by an external field. But in many cases dynamical and geometrical symmetries coincide.

Our interest lies now in the determination of the eigenfrequencies (eigenvalues) of the vibrations given by (7.1.1). The time dependence can always be eliminated by $u_i^m \sim \exp(-i\omega t)$. Introducing

$$\sqrt{M^m}u_i^m = U_i^m \ , \qquad \phi_{ij}^{mn} \cdot (M^m M^n)^{-1/2} = \Phi_{ij}^{mn} \tag{7.1.3}$$

we obtain (7.1.1) in the mass-symmetrized standard form (see also Sect. 10.2)

$$\omega^2 U_i^m = \sum_{n,j} \Phi_{ij}^{mn} U_j^n \ , \qquad \omega^2 = \lambda \ . \tag{7.1.4}$$

In a group theoretical investigation of the solutions of (7.1.4) we have to perform a series of steps, which we will illustrate with H_2O as an example:

(i) Determination of the (dynamical) symmetry group $\mathscr{G}$. In vibrational problems this is in general the geometrical symmetry, which is seen to be

$$\begin{aligned}\mathscr{C}_{2v} &= \{e, c_{2z}, \sigma_{vx}, \sigma'_{vy}\} \\ &= \{e, c_2, \sigma_v, \sigma'_v\}\end{aligned} \tag{7.1.5}$$

for H_2O if the atoms stay in their equilibrium positions (c_{2z}: rotation about z-axis, etc.)

[1] In the case of molecular vibrations the dynamical symmetry is determined by the mass-symmetrized FCs (see below).

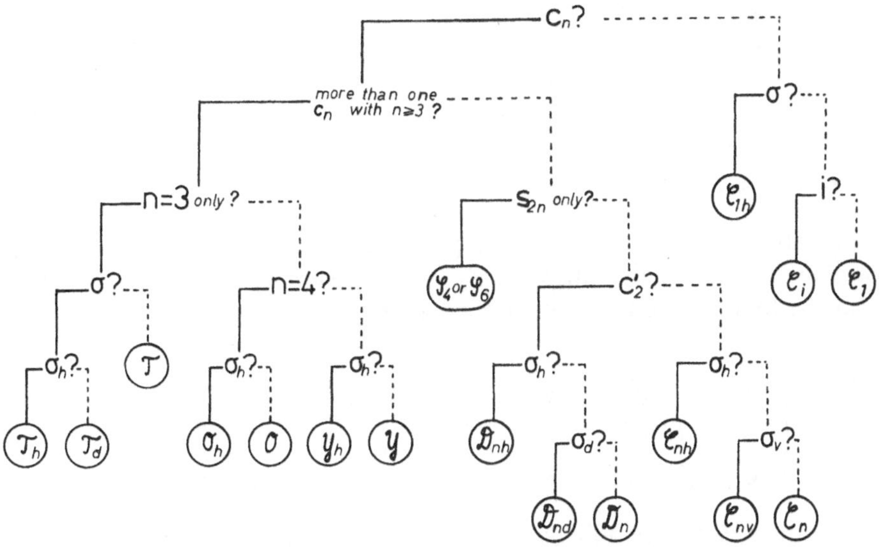

Fig. 7.2. Flow chart method of determining point groups ———: "yes"; - - -: "no"

In most cases the symmetry can be determined in a same simple manner. It is convenient first to look for the axis (axes) of highest-order rotation, then to look for other axes perpendicular to the main axis and finally to look for the mirror planes or the roto-inversion elements. Figure 7.2 shows a flow chart method of determining the point group of any object.

If the dynamical symmetry of the Φ_{ij}^{mn} is different from the geometrical one, its determination requires detailed knowledge of the interaction. In the case that the geometrical symmetry group is a subgroup of the dynamical one, we can use this geometrical group, but then we do not get all the possible symmetry relations.

(ii) Determination of the physical (REP) space $\mathbb{R}_{3N}$. The dimension of the physical space is given by the degrees of freedom of the system. For vibrational problems these degrees of freedom are the $3N$ displacements of the particles from their equilibrium positions. The physical space is constituted by these $3N$ displacements U_i^m:

$$\mathbb{R}_{3N} = \{U_i^m | m = 1, 2, \ldots, N; i = 1, 2, 3\} \ . \tag{7.1.6a}$$

The assignment of the U_i^m to the "directions in space" is arbitrary, but we have to retain the choice once made, e.g.

$$\{U_1^1, U_2^1, U_3^1; U_1^2 U_2^2 U_3^2; \ldots ; U_1^N, U_2^N, U_3^N\}$$
$$= \{\boldsymbol{U}^1; \boldsymbol{U}^2; \ldots ; \boldsymbol{U}^N\} \ . \tag{7.1.6b}$$

(iii) Determination of the REPs of the operators P_a in $\mathbb{R}_{3N}$ which are isomorphically assigned to the elements $a \in \mathcal{G}$. In a rotation c_2 about the z-axis both the H atoms are exchanged. Apart from this, x- and y-components are changed. Thus in $\mathbb{R}_{3N}$ (with $N = 3$)

$$\begin{pmatrix} U'^1 \\ U'^2 \\ U'^3 \end{pmatrix} = P_{c_2} \begin{pmatrix} U^1 \\ U^2 \\ U^3 \end{pmatrix} = \begin{pmatrix} c_2 U^1 \\ c_2 U^3 \\ c_2 U^2 \end{pmatrix} = \begin{pmatrix} c_2 & 0 & 0 \\ 0 & 0 & c_2 \\ 0 & c_2 & 0 \end{pmatrix} \cdot \begin{pmatrix} U^1 \\ U^2 \\ U^3 \end{pmatrix} . \qquad (7.1.7a)$$

The operation P_{c_2} thus describes first a permutation of particles ($2 \leftrightarrow 3$) and then a rotation through π in the subspaces $\mathbb{R}_3$ of the individual particles. Consequently P_{c_2} can be looked upon as a coarsened permutation matrix in $\mathbb{R}_{3N}$ with the rotations

$$c_2 = \begin{pmatrix} -1 & 0 & 0 \\ 0 & -1 & 0 \\ 0 & 0 & 1 \end{pmatrix}$$

as submatrices. Correspondingly

$$P_e = \begin{pmatrix} e & 0 & 0 \\ 0 & e & 0 \\ 0 & 0 & e \end{pmatrix} , \qquad P_\sigma = \begin{pmatrix} \sigma_v & 0 & 0 \\ 0 & \sigma_v & 0 \\ 0 & 0 & \sigma_v \end{pmatrix} , \qquad P_{\sigma'} = \begin{pmatrix} \sigma'_v & 0 & 0 \\ 0 & 0 & \sigma'_v \\ 0 & \sigma'_v & 0 \end{pmatrix} ,$$

$$e = \begin{pmatrix} 1 & 0 & 0 \\ 0 & 1 & 0 \\ 0 & 0 & 1 \end{pmatrix} , \qquad \sigma_v = \begin{pmatrix} -1 & 0 & 0 \\ 0 & 1 & 0 \\ 0 & 0 & 1 \end{pmatrix} , \qquad \sigma'_v = \begin{pmatrix} 1 & 0 & 0 \\ 0 & -1 & 0 \\ 0 & 0 & 1 \end{pmatrix} . \qquad (7.1.7b)$$

In the case of a general system, the P_a in $\mathbb{R}_{3N}$ obviously satisfy

$$U_i'^m = \sum_{nj} (P_a)_{ij}^{mn} U_j^n , \qquad (P_a)_{ij}^{mn} = (\Pi \otimes a)_{ij}^{mn} , \qquad (7.1.8a)$$

where

$$\Pi_{mn} = \begin{cases} 1, & \text{if } n \text{ changes into } m \text{ with } P_a \\ 0, & \text{otherwise ;} \end{cases} \qquad (7.1.8b)$$

i.e. P_a is the direct product of a permutation (of N particles) and a point symmetry operation $a \in \mathcal{G}$.

(iv) Determination of the characters of P_a in the total REP of the $\mathbb{R}_{3N}$. Decomposition of $\mathbb{R}_{3N}$ into invariant subspaces of $\mathcal{G}$. From (7.1.8) it follows immediately that

$$\chi(P_a) = N_a \cdot \chi^{(v)}(a) , \qquad (7.1.9a)$$

where N_a is the number of particles remaining in a fixed position under the

Table 7.1. Decomposition of the total REP of H_2O into IRs

$D^{(\alpha)}$	e	c_2	σ_v	σ_v'	m_α	m_α(int)	m_α(trans)	m_α(rot)
A_1	1	1	1	1	3	2	1	0
A_2	1	1	−1	−1	1	0	0	1
B_1	1	−1	−1	1	2	0	1	1
B_2	1	−1	1	−1	3	1	1	1
$D(H_2O)$	9	−1	3	1				

operation P_a and $\chi^{(v)}(a) = \mathrm{Tr}\{a\}$ is the character of the three-dimensional vector REP of a (rotation, reflection, etc.). For H_2O

$$\begin{array}{l|cccc} & e & c_2 & \sigma_v & \sigma_v' \\ \hline \chi(P_a): & 3\cdot 3 = 9 & 1\cdot(-1) = -1 & 3\cdot 1 = 3 & 1\cdot 1 = 1 \end{array} \quad . \tag{7.1.9b}$$

With this, the total REP $D(P_a)$ can be decomposed into its IRs.

Using (4.2.21a) and (31a) we obtain Table 7.1. Thus

$$D^{\mathrm{tot}}(H_2O) = 3A_1 \oplus A_2 \oplus 2B_1 \oplus 3B_2 \quad . \tag{7.1.10a}$$

Now, every finite system of particles possesses 3 degrees of freedom for translations (trans) (of the centre of mass) and 3 degrees of freedom for rotations (rot) (about the centre of mass, for example). These degrees of freedom have eigenfrequencies equal to zero (motion without any interaction). Thus they are irrelevant for an investigation of the internal motions (int). Their subspaces have to be eliminated

$$D^{(\mathrm{tot})} = D^{(\mathrm{int})} \oplus D^{(\mathrm{trans})} \oplus D^{(\mathrm{rot})} \quad . \tag{7.1.10b}$$

Both these subspaces are three-dimensional. The translation space has a polar vector character (momentum), the rotation space has an axial character (angular momentum). Therefore we have

$$\begin{gathered} \chi^{(\mathrm{tot})}(a) = \chi^{(\mathrm{int})}(a) + \chi^{(\mathrm{trans})}(a) + \chi^{(\mathrm{rot})}(a) \ , \\ \chi^{(\mathrm{trans})}(a) = \chi^{(v)}(a) \ , \qquad \chi^{(\mathrm{rot})}(a) = \det a \cdot \chi^{(v)}(a) \ , \end{gathered} \tag{7.1.10c}$$

thus

$$\chi^{(\mathrm{int})}(a) = [N_a - 1 - \det a] \cdot \chi^{(v)}(a) \tag{7.1.10d}$$

The results are given in Table 7.1 as well. The REP of the internal vibrations is then

$$D^{(\mathrm{int})}(H_2O) = 2A_1 \oplus B_2 \quad . \tag{7.1.10e}$$

Degeneracies arising from symmetry do not occur in this case since all the REPs are one dimensional.

7.1.2 Determination of Eigenvalues and Eigenvectors

The eigenvalue problem (7.1.4) in general leads to $3N$ eigenvalues and the same number of eigenvectors $\hat{e}_k^{(\sigma)}$, $\sigma = 1, \ldots, 3N$, which can be written as column or row vectors as in (7.1.6b). The insignificant degrees of freedom for translations and rotations in Table 7.1 to which frequencies "zero" are assigned have to be eliminated in an appropriate way (see below). Since often the atomic FCs ϕ are not known, or are only roughly known, one has to set up an ansatz for the FCs compatible with the symmetry of the molecules. The magnitude of these FCs is then determined by comparison with the experimentally measured frequencies. Every type of vibration corresponds to a symmetry-adapted subspace, which has to be determined first. The translational and rotational degrees of freedom follow from this procedure as well.

Determination of Subspaces. The projection operator $P_{kk}^{(\alpha)}$ according to (4.3.9a, 13) projects (apart from a factor) the symmetry-adapted vector $|\alpha, s_\alpha, k\rangle$ out of an arbitrary vector of $\mathbb{R}_{3N}$ for a fixed k. We then obtain the $(d_\alpha - 1)$ partner vectors by applying $P_{jk}^{(\alpha)}$, $j \neq k$ to $|\alpha, s_\alpha, k\rangle$. It is convenient to apply the $P_{kk}^{(\alpha)}$ successively to the vectors of an orthonormal basis $|e^{(\rho)}\rangle$ of $\mathbb{R}_{3N}$. Then those components of $\mathbb{R}_{3N}$ are projected which are assigned to a definite REP α. These components transform with P_a according to the kth row of the IR α (4.3.14).

If we denote an arbitrary state of displacements by $|U\rangle$, the displacements of the particles m are the projections of $|U\rangle$ onto the position vectors $|X_i^m\rangle$, thus

$$U_i^m = \langle X_i^m | U\rangle \ . \tag{7.1.11}$$

These displacements can be expanded with respect to the symmetry-adapted basis $|\alpha, s_\alpha, k\rangle$:

$$U_i^m = \langle X_i^m | U\rangle = \sum_{\alpha s k} (S^+)_{mi,\alpha s k} \langle \alpha, s_\alpha, k | U\rangle \ , \qquad s = s_\alpha \tag{7.1.12a}$$

with

$$(S^+)_{mi,\alpha s k} = \langle X_i^m | \alpha, s_\alpha, k\rangle \ . \tag{7.1.12b}$$

The $\langle \alpha, s_\alpha, k | U\rangle = Q_k^{\alpha,s}$ are called symmetry coordinates. They are the projections of an arbitrary state vector $|U\rangle$ onto the basis vectors $|\alpha, s_\alpha, k\rangle$. From (7.1.12b) we see directly that the symmetry-adapted vector $|\alpha, s_\alpha, k\rangle$ gives the $(\alpha s_\alpha k)$ column of the matrix S^+, which is crucial in (6.3.4). We elucidate this method using the example of H_2O, starting with a Cartesian basis

$$\{e^{(\rho)} | \rho = 1, \ldots, 3N\} = \{(0, \ldots, 1, \ldots, 0), \ldots\} \ , \tag{7.1.13}$$

where the "1" takes all the positions in the sequence of (7.1.6b). The projection operators $P_{kk}^{(\alpha)}$ from (4.3.9a) are constructed with the REPs of $\mathscr{C}_{2v}$, where we can use Table 7.1 because all the REPs are one dimensional:

$$
\begin{aligned}
P_{11}^{A_1} &= \tfrac{1}{4}(P_e + P_{c_2} + P_{\sigma_v} + P_{\sigma'_v}) \ , \\
P_{11}^{A_2} &= \tfrac{1}{4}(P_e + P_{c_2} - P_{\sigma_v} - P_{\sigma'_v}) \ , \\
P_{11}^{B_1} &= \tfrac{1}{4}(P_e - P_{c_2} - P_{\sigma_v} + P_{\sigma'_v}) \ , \\
P_{11}^{B_2} &= \tfrac{1}{4}(P_e - P_{c_2} + P_{\sigma_v} - P_{\sigma'_v}) \ .
\end{aligned}
\tag{7.1.14}
$$

Using (7.1.7), we have for example

$$
P_{11}^{A_1} = \frac{1}{2}\begin{pmatrix}
0 & 0 & 0 & 0 & 0 & 0 & 0 & 0 & 0 \\
0 & 0 & 0 & 0 & 0 & 0 & 0 & 0 & 0 \\
0 & 0 & 2 & 0 & 0 & 0 & 0 & 0 & 0 \\
0 & 0 & 0 & 0 & 0 & 0 & 0 & 0 & 0 \\
0 & 0 & 0 & 0 & 1 & 0 & 0 & -1 & 0 \\
0 & 0 & 0 & 0 & 0 & 1 & 0 & 0 & 1 \\
0 & 0 & 0 & 0 & 0 & 0 & 0 & 0 & 0 \\
0 & 0 & 0 & 0 & -1 & 0 & 0 & 1 & 0 \\
0 & 0 & 0 & 0 & 0 & 1 & 0 & 0 & 1
\end{pmatrix} .
\tag{7.1.15}
$$

Applying this successively to the basis vectors (7.1.13) we obtain 5 projections that are different from zero, namely (including normalization)

$$
\begin{aligned}
P_{11}^{A_1} e^{(3)} &= \{0,0,1;0,0,0;0,0,0\} = |A_1,1,1\rangle \ , \\
P_{11}^{A_1} e^{(5)} &= \frac{1}{\sqrt{2}}\{0,0,0;0,1,0;0,-1,0\} = |A_1,2,1\rangle \ , \\
P_{11}^{A_1} e^{(6)} &= \frac{1}{\sqrt{2}}\{0,0,0;0,0,1;0,0,1\} = |A_1,3,1\rangle \ , \\
P_{11}^{A_1} e^{(8)} &= \frac{1}{\sqrt{2}}\{0,0,0;0,-1,0;0,1,0\} = -|A_1,2,1\rangle \ , \\
P_{11}^{A_1} e^{(9)} &= \frac{1}{\sqrt{2}}\{0,0,0;0,0,1;0,0,1\} = |A_1,3,1\rangle \ .
\end{aligned}
\tag{7.1.16a}
$$

However, two of these are linearly dependent, i.e. there are 3 basis vectors belonging to A_1, as there has to be according to Table 7.1. The vibrational problem thus has to be solved in the subspace defined by (7.1.16a).

We can eliminate the translation contained in A_1 by taking appropriate linear combinations of (7.1.16a):

$$\frac{1}{\sqrt{3}}\{|A_1,1,1\rangle + \sqrt{2}\cdot|A_1,3,1\rangle\}\ , \text{ translation in } z \text{ direction,}$$

$$\frac{1}{\sqrt{3}}\{|A_1,1,1\rangle - \sqrt{3/2}\cdot|A_1,2,1\rangle - 1/\sqrt{2}\cdot|A_1,3,1\rangle\}\ , \text{ vibration,}$$

$$\frac{1}{\sqrt{3}}\{|A_1,1,1\rangle + \sqrt{3/2}\cdot|A_1,2,1\rangle - 1/\sqrt{2}\cdot|A_1,3,1\rangle\}\ , \text{ vibration.} \quad (7.1.16b)$$

Any other linear combination is possible. For the A_2 type we have correspondingly

$$P_{11}^{A_2}e^{(4)} = \frac{1}{\sqrt{2}}\{0,0,0;1,0,0;-1,0,0\} = |A_2,1,1\rangle \quad (7.1.16c)$$

as the only linearly independent function. This is a rotation about the z-axis [cf. (7.1.10a,e)]. The B_1 type has the projections

$$P_{11}^{B_1}e^{(1)} = \{1,0,0;0,0,0;0,0,0\} = |B_1,1,1\rangle\ ,$$

$$P_{11}^{B_1}e^{(4)} = \frac{1}{\sqrt{2}}\{0,0,0;1,0,0;1,0,0\} = |B_1,2,1\rangle\ . \quad (7.1.16d)$$

The linear combinations $|1\rangle \pm \sqrt{2}|2\rangle$ of these represent a translation (in the x direction) and a rotation (about the y-axis). Finally, for the B_2 type we have

$$P_{11}^{B_2}e^{(2)} = \{0,1,0;0,0,0;0,0,0\} = |B_2,1,1\rangle\ ,$$

$$P_{11}^{B_2}e^{(5)} = \frac{1}{\sqrt{2}}\{0,0,0;0,1,0;0,1,0\} = |B_2,2,1\rangle\ , \quad (7.1.16e)$$

$$P_{11}^{B_2}e^{(6)} = \frac{1}{\sqrt{2}}\{0,0,0;0,0,1;0,0,-1\} = |B_2,3,1\rangle\ .$$

Here appropriate combinations are

$$\frac{1}{\sqrt{3}}\{|B_2,1,1\rangle + \sqrt{2}|B_2,2,1\rangle\}\ , \quad \text{translation in } y \text{ direction,}$$

$$\frac{1}{\sqrt{3}}\{|B_2,1,1\rangle - 1/\sqrt{2}|B_2,2,1\rangle + \sqrt{3/2}|B_2,3,1\rangle\}\ , \quad \text{rotation about } x\text{-axis,}$$

$$\frac{1}{\sqrt{3}}\{|B_2,1,1\rangle - 1/\sqrt{2}|B_2,2,1\rangle - \sqrt{3/2}|B_2,3,1\rangle\}\ , \quad \text{vibration.} \quad (7.1.16f)$$

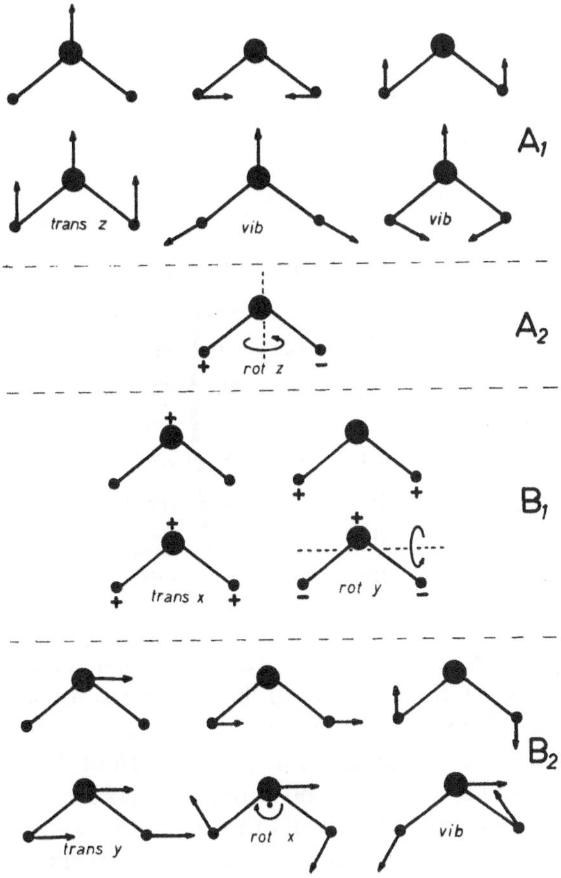

Fig. 7.3. Symmetry modes of a H_2O molecule. Projected basis vectors and linear combinations are given in order to show translations and rotations. Motions perpendicular to the plane of the paper are indicated by +, −

The different situations are illustrated by the "arrow diagrams" in Fig. 7.3. These figures allow one to prove easily that the diagram belonging to the REP α transforms just in the way prescribed by this REP if P_a, $a \in \mathscr{G}$, is applied. The transformation matrices according to (7.1.12) are

$$S = \frac{1}{\sqrt{2}} \cdot \begin{pmatrix} 0 & 0 & \sqrt{2} & 0 & 0 & 0 & 0 & 0 & 0 \\ 0 & 0 & 0 & 0 & 1 & 0 & 0 & -1 & 0 \\ 0 & 0 & 0 & 0 & 0 & 1 & 0 & 0 & 1 \\ 0 & 0 & 0 & 1 & 0 & 0 & -1 & 0 & 0 \\ \sqrt{2} & 0 & 0 & 0 & 0 & 0 & 0 & 0 & 0 \\ 0 & 0 & 0 & 1 & 0 & 0 & 1 & 0 & 0 \\ 0 & \sqrt{2} & 0 & 0 & 0 & 0 & 0 & 0 & 0 \\ 0 & 0 & 0 & 0 & 1 & 0 & 0 & 1 & 0 \\ 0 & 0 & 0 & 0 & 0 & 1 & 0 & 0 & -1 \end{pmatrix} \begin{matrix} A_1 \\ A_1 \\ A_1 \\ A_2 \\ B_1 \\ B_1 \\ B_2 \\ B_2 \\ B_2 \end{matrix} ,$$

Table 7.2. Symmetry-adapted subspaces of H_2O vibrations

A_1	0	0	a_1	0	b_1	c_1	0	$-b_1$	c_1
A_2	0	0	0	a_2	0	0	$-a_2$	0	0
B_1	a_3	0	0	b_3	0	0	b_3	0	0
B_2	0	a_4	0	0	b_4	c_4	0	b_4	$-c_4$

$$\hat{S} = \frac{1}{\sqrt{3}} \cdot \begin{pmatrix} 0 & 0 & 1 & 0 & 0 & 1 & 0 & 0 & 1 \\ 0 & 0 & 1 & 0 & -\varepsilon & -\omega & 0 & \varepsilon & -\omega \\ 0 & 0 & 1 & 0 & \varepsilon & -\omega & 0 & -\varepsilon & -\omega \\ 0 & 0 & 0 & \eta & 0 & 0 & -\eta & 0 & 0 \\ 1 & 0 & 0 & 1 & 0 & 0 & 1 & 0 & 0 \\ 1 & 0 & 0 & -1 & 0 & 0 & -1 & 0 & 0 \\ 0 & 1 & 0 & 0 & 1 & 0 & 0 & 1 & 0 \\ 0 & 1 & 0 & 0 & -\omega & \varepsilon & 0 & -\omega & -\varepsilon \\ 0 & 1 & 0 & 0 & -\omega & -\varepsilon & 0 & -\omega & \varepsilon \end{pmatrix} \begin{matrix} \text{trans}_z \\ \text{vib} \\ \text{vib} \\ \text{rot}_z \\ \text{trans}_x \\ \text{rot}_y \\ \text{trans}_y \\ \text{rot}_x \\ \text{vib} \end{matrix} . \tag{7.1.17}$$

Here $\varepsilon = \sqrt{3}/2$, $\omega = 1/2$; $\eta = \sqrt{3/2}$ and $S^+S = \hat{S}^+\hat{S} = 1$; $\hat{S}$ will be used later.

It is often convenient to specify the basis vectors of the different subspaces not completely but only by general linear combinations of the basis vectors assigned to a definite IR. Then we have the subspaces displayed in Table 7.2. In many cases such symmetry-adapted subspaces can be found by guessing, if we study the behaviour of an arbitrary vector of $\mathbb{R}_{3N}$ under the operations $a \in \mathcal{G}$.

To solve the eigenvalue problem we have to know the FCs in (7.1.1) explicitly. For this reason one conveniently makes an ansatz for the ϕ_{ij}^{mn} compatible with the symmetry of the problem where some general symmetry conditions have to be satisfied:

$$\sum_n \phi_{ij}^{mn} = \sum_m \phi_{ij}^{mn} = 0 \ , \qquad \sum_n \phi_{ij}^{mn} X_k^n = \sum_n \phi_{ik}^{mn} X_j^n \ . \tag{7.1.18}$$

The first of these equations represents the translational condition (conservation of momentum), the second one the rotational condition [conservation of angular momentum, see (7.1.22)]. The FCs compatible with symmetry are, in the case of H_2O (Sect. 10.2),

$$\phi^{12} = \begin{pmatrix} \alpha' & 0 & 0 \\ 0 & \beta & \zeta + \zeta' \\ 0 & \zeta - \zeta' & \gamma \end{pmatrix} , \quad \phi^{13} = \begin{pmatrix} \alpha' & 0 & 0 \\ 0 & \beta & -\zeta - \zeta' \\ 0 & -\zeta + \zeta' & \gamma \end{pmatrix} ,$$

$$\phi^{23} = \begin{pmatrix} \alpha'' & 0 & 0 \\ 0 & \beta' & \zeta' \\ 0 & -\zeta' & \gamma' \end{pmatrix} . \tag{7.1.19a}$$

The translational condition yields the FCs ϕ_{ij}^{mm}, the rotational condition yields (see also Fig. 7.4 for the angle α)

$$\alpha' = \alpha'' = 0 \ ,$$

$$\beta\cos^2(\alpha/2) = (\zeta + \zeta')\sin(\alpha/2)\cos(\alpha/2) = (\gamma + 2\gamma')\sin^2(\alpha/2) \qquad (7.1.19b)$$

so that there are only four independent FCs for H_2O[2]. Then the vibrational frequencies according to (7.1.4) can be determined by using the eigenvectors which have been calculated group theoretically (6.3.4,5). In our case we have for the three proper vibrational frequencies the equations ($\lambda = \omega^2$)

$$A_1: \lambda^2 + \lambda\left(\frac{2\gamma}{M_1} + \frac{\beta + \gamma + 2\beta'}{M_2}\right) + \frac{M_1 + 2M_2}{M_1 M_2^2}[\gamma\beta + 2\gamma\beta' - (\zeta - \zeta')^2] = 0 \ ,$$

$$B_2: \lambda = \frac{2\beta}{M_1} - \frac{\beta + \gamma + 2\gamma'}{M_2} \ . \qquad (7.1.20)$$

By comparison with the (three) measured frequencies we can derive statements about the (four) FCs. The reader should do the necessary calculations as an exercise.

The foregoing method always works. However, instead of discussing it in more detail, we will describe another method which uses only the internal coordinates, i.e. those of proper vibrations. The degrees of freedom (translations, rotations having zero frequencies) that are not of interest then have first to be eliminated.

For this purpose we transform to new coordinates w_j^n by a matrix $\hat{S}$[3]:

$$w_j^n = \sum_{mi} \hat{S}_{ji}^{nm} u_i^m \ , \qquad \text{or, more briefly, } w = \hat{S}u \ . \qquad (7.1.21)$$

The w_j^n then contain a subspace which describes only the 3 translations and 3 rotations of the total system and another one which contains the $3N - 6$ internal vibrations denoted by $\bar{w}_s$, $s = 1, \ldots, 3N - 6$. Here s replaces ${}^n_j$ in (7.1.21). The conditions for the proper vibration vectors follow from the conservation laws of momentum and angular momentum[4]:

$$\text{trans}_i := \sum_{n=1}^{N} M^n u_i^n = \sum_{n=1}^{N} \sqrt{M^n} U_i^n = 0 \qquad \text{for } i = 1, 2, 3 \ , \qquad (7.1.22a)$$

[2] The general construction of such FCs taking into account all the existing symmetry conditions can be learned from books on lattice dynamics, e.g. [7.1–3].

[3] The reverse, i.e. the solution of the inhomogenous linear system of equations (7.1.21), has the form $u = \hat{R}w + u'$, where $\langle u'|w\rangle = 0$, i.e. the inhomogeneity w has to be orthogonal to the nontrivial solution u' (Fredholm's alternative [7.4]); u' describes a common translation or rotation of the system.

[4] In infinitely extended systems (lattices!) (7.1.22b) has to be dropped. Then the sum over n does not converge.

$$\mathrm{rot}_i := \sum_{n=1}^{N} M^n(X^n \times u^n)_i = \sum_{n=1}^{N} \sqrt{M^n}(X^n \times U^n)_i = 0 \qquad \text{for } i = 1, 2, 3 \; ; \tag{7.1.22b}$$

$X^n = \{X_i^n\}$ are the equilibrium positions of the particles. Usually, one can easily find the subspaces for the translations and rotations. The vibrational vectors have to be orthogonal to these. In this way the w_j^n or $\hat{S}_{ij}^{mn}$ can be constructed. Formally we obtain from (7.1.4) using (7.1.21)

$$\omega^2 w = GFw \tag{7.1.23}$$

with (M being diagonal)

$$G = \hat{S}\frac{1}{M}\hat{S}^+ \qquad \text{and} \qquad F = \hat{R}^+\phi\hat{R} \; . \tag{7.1.24}$$

The advantage of this equation is that we do not have to determine the complete $\hat{S}$, but only that part which belongs to the internal vibrations. Then, however, we have to make an ansatz for F instead of ϕ.

The part belonging to the internal coordinates is according to (7.1.21)

$$\bar{w}_s = \sum_{mi} \hat{S}_{si}^m \cdot u_i^m = \sum_{m=1}^{N} \langle v_s^m | u^m \rangle \tag{7.1.25a}$$

with the vectors

$$v_s^m = \sum_{i=1}^{3} \hat{S}_{si}^m \cdot e_i \; , \qquad \langle e_i | u^m \rangle = u_i^m \; ; \tag{7.1.25b}$$

e_i projects the i-component out of a general displacement vector. The vectors e_i may be an orthogonal Cartesian basis, but it does not have to be. The only essential property is that e_i is a complete basis (Fig. 7.4). The projection of a displacement u^m onto v_s^m (for a fixed m) gives that part of u^m, which contributes to the internal coordinate $\bar{w}_s$. If u^m is parallel to v_s^m, then we have the maximal contribution to $\bar{w}_s$. Thus v_s^m denotes the direction in which a displacement u^m of the particle m gives the maximal contribution to $\bar{w}_s$. The absolute value of v_s^m is equal to the contribution to $\bar{w}_s$ which is provided by a "unit" displacement of m in the optimal direction. We will illustrate this again with the example H_2O for which we first have to define the internal coordinates (Fig. 7.4) and springs.

We have to distinguish between valence (ν) vibrations (motions "parallel" to the bonding direction) and angle bending (δ, deformation) vibrations (in this case the angle H-O-H). Torsion (τ) vibrations do not occur in this case. The first two coordinates are denoted by $\bar{w}_1, \bar{w}_2$, the latter by $\bar{w}_3 = a \cdot \delta\alpha$, which is the change of the angle multiplied by the equilibrium distance O-H, in order to have the same dimensions as $\bar{w}_1, \bar{w}_2$. For the coupling matrix F of the internal motions we make the ansatz, see (7.1.24):

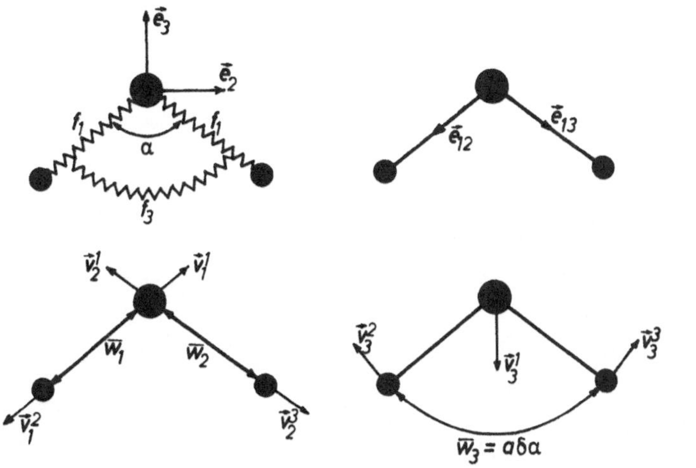

Fig. 7.4. Internal coordinates $\bar{w}_s$, springs f_i and basis e_i for the vibrations of H_2O. The nonorthogonal basis $\{e_{12}, e_{13}\}$ may be used instead of a Cartesian basis $\{e_2, e_3\}$, but then the transformation $\hat{S}$ and the FC have to be changed appropriately since their definition depends on the basis chosen. A nonorthogonal basis is sometimes more convenient

Table 7.3. The elements $\hat{S}_{si}^m$ of the transformation from displacements u_i^m to internal coordinates $\bar{w}_s$

	i:	2			3		
s	m:	1	2	3	1	2	3
1		$\sin(\alpha/2)$	$-\sin(\alpha/2)$	0	$\cos(\alpha/2)$	$-\cos(\alpha/2)$	0
2		$-\sin(\alpha/2)$	0	$\sin(\alpha/2)$	$\cos(\alpha/2)$	0	$-\cos(\alpha/2)$
3		0	$-\cos(\alpha/2)$	$\cos(\alpha/2)$	$-2\sin(\alpha/2)$	$\sin(\alpha/2)$	$\sin(\alpha/2)$

$$F_{st} = \begin{pmatrix} f_1 & f_2 & f_4 \\ f_2 & f_1 & f_4 \\ f_4 & f_4 & f_3 \end{pmatrix} , \tag{7.1.26}$$

which corresponds to the symmetry $\mathscr{C}_{2v}$. Here f_1 and f_3 describe the pure valence and deformation interactions, respectively, whereas f_2 and f_4 describe the coupling between the valence motions and the valence and deformation motions, respectively.

For a solution of the eigenvalue problem we have to calculate G or $\hat{S}$. As a basis e_i we choose the Cartesian vectors e_2, e_3 (Fig. 7.4). After a simple geometric calculation we obtain the elements $\hat{S}_{si}^m$ given in Table 7.3. Equation (7.1.22) is satisfied if the $\boldsymbol{v}_s^m$ (or $\hat{S}_{si}^m$) obey

$$\sum_m \boldsymbol{v}_s^m = 0 , \qquad \sum_m \boldsymbol{X}^m \times \boldsymbol{v}_s^m = 0 . \tag{7.1.27}$$

Table 7.4. Scheme for the application of the symmetry operators P_a to the internal coordinates $\bar{w}_t$ for the symmetry $\mathscr{C}_{2v}$

P_a / $\bar{w}_t$	e	c_{2z}	σ_{vx}	σ'_{vy}
$\bar{w}_1$	$\bar{w}_1$	$\bar{w}_2$	$\bar{w}_1$	$\bar{w}_2$
$\bar{w}_2$	$\bar{w}_2$	$\bar{w}_1$	$\bar{w}_2$	$\bar{w}_1$
$\bar{w}_3$	$\bar{w}_3$	$\bar{w}_3$	$\bar{w}_3$	$\bar{w}_3$

With

$$G_{st} = \sum_{m,i} \hat{S}_{si}^{m} \frac{1}{M^m} \hat{S}_{ti}^{*m}$$

it follows that

$$G_{st} = \begin{pmatrix} g_1 & g_2 & g_4 \\ g_2 & g_1 & g_4 \\ g_4 & g_4 & g_3 \end{pmatrix} \quad \text{with} \tag{7.1.28}$$

$$g_1 = \frac{1}{M_1} + \frac{1}{M_2} , \qquad g_2 = \frac{1}{M_1}[\cos^2(\alpha/2) - \sin^2(\alpha/2)] ,$$

$$g_3 = \frac{4}{M_1}\sin^2(\alpha/2) + \frac{2}{M_2} , \qquad g_4 = -\frac{2}{M_1}\sin(\alpha/2)\cos(\alpha/2) .$$

Finally, the internal coordinates $\bar{w}_t$ have to be transformed into symmetry-adapted coordinates, which can be done analogously to (7.1.12)

$$Q_k^{as_\alpha} = \sum_t S_{ask,t}\bar{w}_t . \tag{7.1.12c}$$

For the application of the symmetry operators P_a the scheme of Table 7.4 holds.

From the general considerations (7.1.10e) we know that there are three vibrational states: two A_1 and one B_2. The projection operators satisfy

$$\begin{aligned} P^{A_1}\bar{w}_1 &= P^{A_1}\bar{w}_2 = \tfrac{1}{2}(\bar{w}_1 + \bar{w}_2) = Q^{A_1 1} : \nu(\pi) \\ P^{A_1}\bar{w}_3 &= \bar{w}_3 = Q^{A_1 2} : \delta(\pi) \\ P^{B_2}\bar{w}_1 &= -P^{B_2}\bar{w}_2 = \tfrac{1}{2}(\bar{w}_1 - \bar{w}_2) = Q^{B_2} : \nu(\sigma) \\ P^{B_2}\bar{w}_3 &= 0 . \end{aligned} \tag{7.1.29}$$

Thus the transformation matrix is, see (6.3.4),

$$S = \frac{1}{\sqrt{2}} \begin{pmatrix} 1 & 1 & 0 \\ 0 & 0 & \sqrt{2} \\ 1 & -1 & 0 \end{pmatrix} \quad \text{and therefore} \tag{7.1.30}$$

$$SFS^+ = \begin{pmatrix} f_1 + f_2 & 2f_4 & 0 \\ 2f_4 & f_3 & 0 \\ 0 & 0 & f_1 - f_2 \end{pmatrix}, \quad SGS^+ = \begin{pmatrix} g_1 + g_2 & 2g_4 & 0 \\ 2g_4 & g_3 & 0 \\ 0 & 0 & g_1 - g_2 \end{pmatrix}.$$

With these matrices the eigenvalues can be calculated according to (7.1.23). From the measured frequencies we can derive the FCs ($10^5\ \text{dyn}\cdot\text{cm}^{-1} = 1\ \text{Nm}^{-1}$)

$$\begin{aligned} &A_1, \nu(\pi): \omega = 6.8785 \times 10^{14}\ \text{s}^{-1}, && f_1 = +73.81 \times 10^4\ \text{dyn}\cdot\text{cm}^{-1}, \\ &A_1, \delta(\pi): \omega = 3.0044 \times 10^{14}\ \text{s}^{-1}, && f_2 = -4.89 \times 10^4\ \text{dyn}\cdot\text{cm}^{-1}, \\ &B_2, \nu(\sigma): \omega = 7.0746 \times 10^{14}\ \text{s}^{-1}, && f_3 = +9.43 \times 10^4\ \text{dyn}\cdot\text{cm}^{-1}, \\ & && f_4 = -7.30 \times 10^4\ \text{dyn}\cdot\text{cm}^{-1}, \end{aligned} \tag{7.1.31}$$

Comparing this result with (7.1.20) we also find the connection between the FCs according to (7.1.19) and (7.1.26):

$$\begin{aligned} \beta &= -(f_1 - f_2)\sin^2(\alpha/2) = -49.17 \times 10^4\ \text{dyn}\cdot\text{cm}^{-1}, \\ \gamma &= -(f_1 + f_2)\cos^2(\alpha/2) - 2f_3\sin^2(\alpha/2) + 2f_4\sin\alpha \\ &= -51.78 \times 10^4\ \text{dyn}\cdot\text{cm}^{-1}, \\ \beta' &= -f_2\sin^2(\alpha/2) - f_3\cos^2(\alpha/2) - f_4\sin\alpha = +6.59 \times 10^4\ \text{dyn}\cdot\text{cm}^{-1}, \\ \gamma' &= +f_2\cos^2(\alpha/2) + f_3\sin^2(\alpha/2) - f_4\sin\alpha = +11.12 \times 10^4\ \text{dyn}\cdot\text{cm}^{-1}, \\ \zeta &= -f_1\sin(\alpha/2)\cos(\alpha/2) + f_3\sin(\alpha/2)\cos(\alpha/2) - f_4\cos\alpha \\ &= -33.00 \times 10^4\ \text{dyn}\cdot\text{cm}^{-1}, \\ \zeta' &= +f_2\sin(\alpha/2)\cos(\alpha/2) - f_3\sin(\alpha/2)\cos(\alpha/2) + f_4\cos\alpha \\ &= -5.11 \times 10^4\ \text{dyn}\cdot\text{cm}^{-1}. \end{aligned} \tag{7.1.32}$$

The potential energy can also be expressed in terms of the symmetry-adapted coordinates. According to (7.1.23, 25) the potential energy is

$$\phi = \tfrac{1}{2}\sum_{st} \bar{w}_s F_{st} \bar{w}_t, \tag{7.1.33a}$$

and using (7.1.12) it then follows that

$$\phi = \tfrac{1}{2} \sum Q_k^{\alpha s *} \langle \alpha s k | SFS^+ | \beta s' i \rangle Q_i^{\beta s'} \ . \tag{7.1.33b}$$

But SFS^+ is diagonal with respect to α, β and independent of k, i, which follows from (6.3.7). Therefore

$$\phi = \tfrac{1}{2} \sum_{\alpha s s'} \langle s_\alpha \| SFS^+ \| s'_\alpha \rangle^{(\alpha)} \cdot \sum_k Q_k^{\alpha s *} Q_k^{\alpha s'} \tag{7.1.33c}$$

is represented by a sum of symmetrized coordinates. There is no connection (coupling) between different IRs and between the rows of the IRs. A coupling exists only between those symmetry coordinates which belong to the same partner of the same IR (s_α, s'_α). Corresponding statements hold for the kinetic energy, which is in Cartesian coordinates diagonal also in s_α, s'_α.

The number of independent FCs can be seen directly from (7.1.33c). If the dimension of the block of the REP α is m_α, then this block contains m_α^2 elements. But SFS^+ is Hermitian, i.e. it is symmetric if real symmetry coordinates are used, and thus there are $\tfrac{1}{2} m_\alpha (m_\alpha + 1)$ independent FCs in one block, that is, the total number of independent FCs is given by

$$\tfrac{1}{2} \sum_{\alpha=1}^{r} m_\alpha (m_\alpha + 1) \ . \tag{7.1.34}$$

In the H_2O example, $m_{A_1} = 2$, $m_{B_2} = 1$, thus there are four independent FCs.

7.1.3 Selection Rules

Optical investigations of molecular vibrations are performed using infrared and Raman spectroscopy. In *infrared absorption* the electromagnetic radiation interacts with the dipole moment (polar vector) of the molecules. We distinguish between fundamental processes and those of higher order according as whether there is an excitation of one or of more than one vibrational quantum of the molecule. The higher-order processes are furthermore divided into combination processes (two or more vibrational states belonging to different IRs are excited) and overtones (two or more vibrational states belonging to the same IR are excited). Here transitions between different vibrational states have to be formed with the tensor (vector) operator T_i^v of the dipole moment, which is a polar vector.

The *Raman effect* is due to inelastic scattering of electromagnetic radiation (photons) at the molecular vibrations and these are excited in this process. It is described by the tensor of polarizability (symmetrical tensor of rank 2); the tensor operator is the symmetrized tensor product T_+^{vv} of rank 2. Here we can also distinguish between fundamental processes and those of higher order.

The Wigner-Eckart theorem (6.2.5a) allows for statements on the possible processes. In general it is even sufficient to determine the reduction coefficients (4.4.14). They give the multiplicity with which an IR is contained in the direct product of two IRs or of a tensor operator with one IR (Sects. 6.1, 2). To find

Table 7.5. Character table, see (7.1.35), for the group $\mathscr{C}_{2v}$

	e	c_{2z}	σ_{vx}	σ'_{vy}
χ^v	3	-1	1	1
χ^s_+	6	2	2	2

out which matrix elements $\langle\gamma|T^\beta|\alpha\rangle$ are different from zero, we just have to decompose the direct product $D^{(\beta\times\alpha)}$. All those states γ contained in $D^{(\beta\times\alpha)}$ can be reached from α via an interaction T^β (α: initial state, γ: final state). For this, according to (4.4.14) we only need the characters of the REPs of the tensor operators, e.g. from Exercise 6.6 with $l = 1$ and (4.4.16),

$$\chi^v(\phi) = \pm 1 + 2\cos\phi\ , \qquad (+\text{: proper rotation; } -\text{: improper rotation})$$

$$\chi^s_+(a) = \tfrac{1}{2}[\chi^v(a)]^2 + \tfrac{1}{2}\chi^v(a^2)\ . \tag{7.1.35}$$

Thus for the group $\mathscr{C}_{2v}$ we have Table 7.5.

a) Infrared Absorption

According to (4.4.14) we have for H_2O the following decompositions for $T_i^v|\alpha\rangle$ depending on the initial state α:

$$\begin{aligned} T_i^v|A_1\rangle &\mathrel{\hat=} A_1 \oplus B_1 \oplus B_2\ ,\\ T_i^v|B_2\rangle &\mathrel{\hat=} A_1 \oplus A_2 \oplus B_2\ , \end{aligned} \tag{7.1.36}$$

i.e. in the fundamental process[5], transitions between all the proper vibrational states (A_1, B_2) are possible (Sect. 7.1.2). In the combination processes the final state γ itself consists of two different IRs:

$$\begin{aligned} &A_1\otimes A_2 = A_2\ , \quad A_1\otimes B_1 = B_1\ , \quad A_1\otimes B_2 = B_2\ ,\\ &A_2\otimes B_1 = B_2\ , \quad A_2\otimes B_2 = B_1\ , \quad B_1\otimes B_2 = A_2\ . \end{aligned} \tag{7.1.37}$$

There are possible transitions from A_1 as well as from B_2 into the combination states $A_1\otimes B_2$ and $A_2\otimes B_1$, but the latter are not proper vibrations. The transitions from A_1 to $A_1\otimes A_2$ and $B_1\otimes B_2$ and from B_2 to $A_1\otimes B_1$ and $A_2\otimes B_2$ are forbidden. In the overtones the final state γ is the symmetrized product $D_+^{(\alpha\times\alpha)}$ because the states for vibrations (phonons, bosons) are symmetric. The character to be used is thus (4.4.16)

$$\chi_+^{(\alpha\times\alpha)}(a) = \tfrac{1}{2}[\chi^{(\alpha)}(a)]^2 + \tfrac{1}{2}[\chi^{(\alpha)}(a^2)]\ ,$$

[5] Note that B_1, A_2 are not proper vibrational states. They are of interest only if the molecule as a whole is bound to another complex in such a way that it can have quasi-free translations or rotations.

which corresponds to that of the A_1 REP for all the IRs of the group $\mathscr{C}_{2v}$, thus

$$D_+^{(\alpha\times\alpha)} = A_1 \qquad \text{for } \alpha = A_1, A_2, B_1, B_2 \ . \tag{7.1.38}$$

Since A_1 according to (7.1.36) can be reached from A_1 as well as from B_2 all the overtones can be excited.

b) Raman Effect

Using (4.4.14) together with Table 7.5 we find for $T_+^{vv}|\alpha\rangle$ the decompositions

$$\begin{aligned} T_+^{vv}|A_1\rangle &\triangleq 3A_1 \oplus A_2 \oplus B_1 \oplus B_2 \ , \\ T_+^{vv}|B_2\rangle &\triangleq A_1 \oplus A_2 \oplus B_1 \oplus 3B_2 \ . \end{aligned} \tag{7.1.39}$$

Since in these decompositions all the IRs of $\mathscr{C}_{2v}$ occur, all the Raman processes are allowed. This is also valid for the combination processes and the overtones.

This analysis of the vibrations can of course be used for a determination of the structure of molecules. For this all the geometrical structures compatible with the chemical formula are investigated with respect to their vibrational properties and the calculated values are compared with the experimental results. Definite models may be used in such calculations (Sect. 7.1.2). However, in general the direct methods for a determination of structures, like x-ray, neutron or electron diffraction, are superior to such an analysis.

Exercise 7.1. Discuss the NH_3 molecule with respect to its vibrational properties, in complete analogy to H_2O. The NH_3 molecule possesses $\mathscr{C}_{3v}$ symmetry (triangular pyramid) and has six internal vibrational degrees of freedom.

7.2 Electron Functions and Spectra

7.2.1 Symmetry in Many-Particle Systems

The Hamiltonian of a system of nuclei and electrons can be written as

$$H = H_K + H_e \ ; \tag{7.2.1}$$

H_K contains only nuclear coordinates, H_e, nuclear as well as electronic coordinates:

$$H_K = -\frac{1}{2}\hbar^2 \sum_{k=1}^{N_K} \frac{\Delta_k}{M_k} + \frac{e^2}{2} \sum_{k \neq k'} \frac{Z^k Z^{k'}}{|\boldsymbol{R}^k - \boldsymbol{R}^{k'}|} \ , \tag{7.2.2}$$

$$H_e = -\frac{1}{2}\hbar^2 \sum_{j=1}^{N_e} \frac{\Delta_j}{m_e} - e^2 \sum_{jk} \frac{Z^k}{|\boldsymbol{R}^k - \boldsymbol{r}_j|} + \frac{e^2}{2} \sum_{i \neq j} \frac{1}{|\boldsymbol{r}_i - \boldsymbol{r}_j|} \ . \tag{7.2.3}$$

In the Born-Oppenheimer approximation for molecules and crystals the eigenvalue problem of the electronic states is set up assuming *fixed nuclear positions*; i.e. it is discussed using (7.2.3)

$$H_e \Psi(\boldsymbol{r}_1, \ldots, \boldsymbol{r}_{N_e}) = E(\ldots, \boldsymbol{R}^k, \ldots) \Psi(\boldsymbol{r}_1, \ldots, \boldsymbol{r}_{N_e}) , \tag{7.2.4}$$

where the $\boldsymbol{R}^k$ are taken to be fixed parameters which occur in the arguments of Ψ as well.

Equations (7.2.3, 4) also describe the atomic problem (one nucleus only) if we put $\boldsymbol{R}^k = 0$. Equation (7.2.2) as well as (7.2.3) may be extended if there is a spin interaction (spin-orbit coupling, spins in a magnetic field, etc.). In Russell-Saunders (*LS*) coupling, spin terms are neglected in the Hamiltonian but taken into account in the total wave function.

The atomic problem has the symmetry group of H_e which is $\mathcal{O}(3) \times \mathcal{P}_m$ with $m = N_e$. The operations of $\mathcal{O}(3)$ are to be taken with respect to the nucleus[6]. The eigenfunctions of H_e can be classified according to the IRs of $\mathcal{O}(3)$ and $\mathcal{P}_m$, since the IRs of $\mathcal{O}(3) \times \mathcal{P}_m$ are the direct products of the IRs of both (Sect. 13.2). Neglecting spin (and Pauli's principle) for the moment, every approximation for $\Psi(\boldsymbol{r}_1, \ldots, \boldsymbol{r}_m)$ should be a basis function of the IR $D^{[\lambda]}$ of $\mathcal{P}_m$. Thus it can be generated by the Young operators $e_{ij}^{[\lambda]}$, see (5.4.18a). Since (7.2.4) cannot in general be solved exactly the interaction is replaced by a mean effective potential $V(\boldsymbol{r})$, in which every electron moves independently of the others ("shell" model, one-particle approximation):

$$H_0 = \sum_j h_j(\boldsymbol{r}_j) , \qquad h_j = -\frac{\hbar^2}{2m_e} \Delta_j + V(\boldsymbol{r}_j) . \tag{7.2.5}$$

If at all, the difference $H_e - H_0 = W$ is treated as a perturbation. The solutions of (7.2.5) are products of type (5.5.9), where the atomic orbitals (AO) ψ_i satisfy the equation

$$h\psi_i(\boldsymbol{r}) = \varepsilon_i \psi_i(\boldsymbol{r}) . \tag{7.2.6}$$

The product function $\psi_{i_1 \ldots i_m}(1, \ldots, m)$ describes an electronic state with $E = \varepsilon_{i_1} + \cdots + \varepsilon_{i_m}$. This state is highly degenerate. Every ψ_i possesses $\mathcal{O}(3)$ degeneracy, the product possesses additionally $\mathcal{P}_m$ degeneracy. Both H_0 and H_e are invariant under $\mathcal{P}_m$: we can always start with the basis functions of the IRs of $\mathcal{P}_m$, i.e. the $\psi_k^{[\lambda]} \sim e_{kj}^{[\lambda]} \psi(1, \ldots, m)$ according to (5.4.24).

For molecules, the same approximations are made (using the Born-Oppenheimer approximation), but the atomic orbitals have to be replaced by molecular orbitals (MO). The latter are often represented as linear combinations of AOs (LCAO-MO method) located at the different nuclei. The extension of this

[6] In $\mathcal{O}(3)$ all the coordinates $\boldsymbol{r}_i$ have to be rotated or reflected simultaneously. Without any interaction between the electrons $\mathcal{O}(3)$ has to be replaced by the m-fold outer product $\mathcal{G} \cong \mathcal{O}(3) \times \mathcal{O}(3) \times \cdots \times \mathcal{O}(3)$, in which every electron is moved independently. An interaction breaks this m-fold symmetry.

method to crystals is also called the tight-binding method (Sects. 7.2.3 and 10.3). The symmetry group in molecules (and in crystals) is $\mathcal{G} \times \mathcal{P}_m$, $\mathcal{G}$ being the corresponding point group (space group).

Because of Pauli's principle, the spin is an essential quantity, even if it does not occur explicitly in the Hamiltonian. The spatial part of the total wave function also depends on the spin state. The total wave function, which can be written as $\phi = \sum_{ij} \psi_i(\boldsymbol{r}_1 \dots \boldsymbol{r}_m)\varphi_j(\sigma_1 \dots \sigma_m)$ with the spin coordinates σ_j, has to be a totally antisymmetric function with respect to interchanges of position and spin coordinates. The functions $\{\psi_i\}$ are a set of degenerate eigenfunctions of H_e, φ_j is an arbitrary spin function. According to Exercise 5.17, $D^{[\lambda]} \otimes D^{[\lambda']}$ contains the antisymmetric REP $D^{[1^m]}$ just once if $[\lambda'] = [\tilde{\lambda}]$. Thus $[\lambda']$ has to be the associated REP of $[\lambda]$. We get just one totally antisymmetric function by linear combination of functions $\psi_i^{[\lambda]}\varphi_j^{[\tilde{\lambda}]}$ with $i, j = 1, \dots, d_{[\lambda]}$, which can be generated by the corresponding Young operators. Since $D^{[1^m]}$ occurs in $D^{[\lambda]} \otimes D^{[\tilde{\lambda}]}$ just once, we can start with any $\psi_1^{[\lambda]} \cdot \varphi_1^{[\tilde{\lambda}]}$ and then apply the Young operator $\hat{Y}[1^m]$ to this function according to (5.4.4, 7):

$$\phi^{[1^m]} = \frac{1}{m!} \sum_{p \in \mathcal{P}_m} (-1)^p p[\psi_1^{[\lambda]}(\boldsymbol{r}_1 \dots \boldsymbol{r}_m)\phi_1^{[\tilde{\lambda}]}(\sigma_1 \dots \sigma_m)] \ . \tag{7.2.7}$$

The symmetry of the spin functions $\varphi^{[\tilde{\lambda}]}$ in (7.2.7) implies that of the spatial functions. Since these spin functions are basis functions of $\mathcal{SU}(2)$, the corresponding Young diagrams have at most two rows, and therefore the Young diagrams corresponding to spatial functions only have at most two columns (Sect. 5.5.4). The possible spin functions of an m-electron problem constitute a 2^m-dimensional space $\mathcal{L}_2^{(m)}$, which we can define by the 2^m product functions

$$\varphi_{i_1 \dots i_m} := \varphi_{i_1}(\sigma_1)\varphi_{i_2}(\sigma_2) \dots \varphi_{i_m}(\sigma_m) \ , \qquad i_1, \dots = 1, 2 \ . \tag{7.2.8}$$

In the scheme (5.5.28) the column functions belong to the IRs of $\mathcal{P}_m$, the row functions in this case belong to the IRs of $\mathcal{SU}$ $(n = 2)$. Often $\varphi_1(\sigma)$ and $\varphi_2(\sigma)$ are taken as $\alpha(\sigma)$ and $\beta(\sigma)$, respectively. The spin functions are eigenfunctions of the z-(3-) component of the spin operator S_z. With

$$S_z = \sum_{j=1}^{m} s_z(j) \ , \quad s_z(j)\varphi_1(j) = \tfrac{1}{2}\varphi_1(j) \ , \quad s_z(j)\varphi_2(j) = -\tfrac{1}{2}\varphi_2(j) \tag{7.2.9}$$

we have

$$S_z\varphi_{i_1 \dots i_m} = \left(\sum_{k=1}^{m} s_{i_k}\right)\varphi_{i_1 \dots i_m} \qquad \text{with} \qquad s_{i_k} = \begin{cases} 1/2 & i_k = 1 \\ -1/2 & i_k = 2 \ . \end{cases}$$

The s_{i_k} correspond to the Λ^{f_i} in Chap. 12.

A permutation $p \in \mathcal{P}_m$ interchanges the particle numbers in the arguments, but does not influence the quantum numbers $i_1 \dots i_m$. Therefore the symmetrized tensors according to (5.5.31)

$$\varphi_{i,k}^{[\lambda]} = P_{kk}^{[\lambda]} \varphi_{i_1 \ldots i_m} \tag{7.2.10}$$

as well as the $\varphi_{i_1 \ldots i_m}$ are eigenfunctions of S_z with the same eigenvalue. We can assign an eigenvalue of S_z—and an IR of $\mathscr{SU}(2)$ with a corresponding Young diagram $[\lambda]$—to every $\varphi_{i,k}^{[\lambda]}$. The possible eigenvalues which are correlated to the different IRs of $\mathscr{SU}(2)$ are said to be a *weight system* (Chap. 12). Figure 5.5 shows the IRs of $\mathscr{SU}(2)$. The identity REP $[1, 1]$ has one allowed tableau with $\sum s_{i_k} = s_1 + s_2 = 1/2 - 1/2 = 0$. Apart from equivalences, the other REPs are of the form $[m, 0]$ with dimension $d_{[m]2} = m + 1$ and with the $m + 1$ eigenvalues (weight system) $\sum_k s_{i_k} = rs_1 + (m - r)s_2 = r - m/2$ with $r = 0, \ldots, m$, thus $\sum_k s_{i_k} = -m/2, -m/2 + 1, \ldots, m/2 - 1, m/2 = S_z$. The weight system (distribution of S_z) characterizes the IRs uniquely, since equivalent REPs according to Fig. 5.5 only give contributions to $\sum_k s_{i_k}$ of the form $1/2 - 1/2$.

According to (5.5.31) the basis functions are

$$\varphi_{111\ldots,222\ldots}^{[m]} = \frac{1}{m!} \sum_{p \in \mathscr{P}_m} p\varphi_1(1) \ldots \varphi_1(r)\varphi_2(r+1) \ldots \varphi_2(m) \; .$$

Because of the S_z values given above, the maximal value (total spin) $m/2$ can be uniquely assigned to the REP $[m, 0] = [m]$, or more generally: the total spin $S = (\lambda_1 - \lambda_2)/2$ has to be assigned to a REP $[\lambda] = [\lambda_1, \lambda_2]$. In an m-electron system with total spin S the spin function has the permutation symmetry $[\lambda_1, \lambda_2]$ with

$$\lambda_1 = \tfrac{1}{2}m + S \qquad \text{and} \qquad \lambda_2 = \tfrac{1}{2}m - S \; ; \qquad m = \lambda_1 + \lambda_2 \; . \tag{7.2.11}$$

7.2.2 Symmetry-Adapted Atomic and Molecular Orbitals

According to Sects. 6.3 and 6.4, the treatment of eigenvalue problems needs symmetry-adapted functions, which we may obtain by the projection operator method of Sect. 4.3.2. The point groups $\mathscr{G}$ of interest are subgroups of $\mathscr{O}(3)$, so that the functions will be definite linear combinations of spherical harmonics (Sect. 5.1). According to (4.3.9a, 13) we get the symmetry-adapted functions $\psi_{ik}^{(\alpha)}$ belonging to the IR $D^{(\alpha)}(\mathscr{G})$ from

$$\psi_{ik}^{(\alpha)} = P_{ik}^{(\alpha)} \cdot Y_m^l(\vartheta, \varphi) = \frac{d_\alpha}{g} \sum_{a \in \mathscr{G}} D_{ik}^{(\alpha)*}(a) \cdot P_a \cdot Y_m^l(\vartheta, \varphi) \; , \tag{7.2.12}$$

where

$$P_a \cdot Y_m^l(\vartheta, \varphi) = \sum_{m'=-l}^{+l} D_{m'm}^{(l)}(a) \cdot Y_{m'}^l(\vartheta, \varphi) \; . \tag{7.2.13a}$$

The rotation element a is conveniently expressed by the Euler angles α, β, γ according to (3.3.1). The REP $D^{(l)}$ is an IR of $\mathscr{SO}(3)$ of dimension $2l + 1$ and can be written as (for proper rotations)

$$D^{(l)}_{m,m}(a) = c_{m'm} \cdot \mathrm{e}^{-im'\gamma} \cdot d^l_{m'm}(\beta) \cdot \mathrm{e}^{-im\alpha} \qquad \text{with} \qquad c_{m'm} = \mathrm{i}^{|m'|-|m|+m'-m} ,$$

$$d^l_{m'm}(\beta) = \sum_k \frac{(-1)^{k-m+m'}\sqrt{(l+m')!(l+m)!(l-m')!(l-m)!}}{(l-m'-k)!(l+m-k)!k!(k-m+m')!} \times \cos^{2l+m-m'-2k}(\beta/2) \cdot \sin^{2k+m'-m}(\beta/2) \; ; \tag{7.2.14}$$

see (11.4.66) and [7.5, 6]. The summation runs from $k = \max(0, m - m')$ to $k = \min(l - m', l + m)$. The $a \in \mathscr{G}$ in (7.2.12) may also be a mirror rotation (roto-inversion). Then $a = ia'$. In this case we have to take into account in (7.2.12) that $\mathrm{i} \cdot Y^l_m = (-1)^l Y^l_m$. The symmetry-adapted spherical harmonics for most of the point groups are tabulated using (7.2.12, 14) [7.6].

According to (5.1.7, 8) double point groups have an extended basis. With $|s\rangle \in \{|+\rangle, |-\rangle\}$ we have instead of (7.2.13a)

$$P_a Y^l_m |s\rangle = \sum_{m'=-l}^{l} D^{(l)}_{m'm}(a) \sum_{s=\pm} D^{(1/2)}_{s's}(a) Y^l_{m'} |s'\rangle \; ; \tag{7.2.13b}$$

$D^{(1/2)}$ is the REP subduced from (3.3.2) for the corresponding point group. The REPs $D^{(\alpha)}$ in (7.2.12) are then of course the IRs of the double groups, thus especially the extra REPs of Sect. 5.1, too. In order to obtain the symmetry-adapted basis of double groups we can also reduce the product basis $f^{(\alpha)}_{ik}|s\rangle$ of $D^{(\alpha)} \otimes D^{(1/2)}$, provided the $f^{(\alpha)}_{ik}$ of the simple groups are known.

The basis functions can also be acquired by a slightly modified method. The functions $r^l Y^l_m$ are polynomials of degree l in Cartesian coordinates, which means we can construct a polynomial basis for the point groups from these functions. This will be explained using the example $\mathscr{C}_{3v}$. As a test function we take the basis $r \cdot \{Y^1_{-1}, Y^1_1, Y^1_0\}$ or $\{x, y, z\}$. We have

$$D^{(l=1)}(\mathscr{C}_{3v}) = A_1 \oplus E \; .$$

This means for example that a state $l = 1$ which is 3-fold degenerate in a central field splits into a (single) A_1 and a 2-fold degenerate E state in a field of $\mathscr{C}_{3v}$ symmetry. Using the P^E_{ik} projection operator we have from the basis $\{x, y, z\}$

$$\Psi^E_{11} = P^E_{11} x = x \; , \qquad \Psi^E_{21} = P^E_{21} x = y \; ,$$

where $P_a x_i = P_a \langle e_i | x \rangle = \langle e_i | a^{-1} x \rangle$ has to be taken into account. Thus $\{x, y\}$ is a basis of the E REP, and correspondingly $\{z\}$ is a basis of the A_1 REP.

Once a set of basis functions is known, then there is a simple method to obtain another set of basis functions belonging to the same IR. If $\{\Psi^{(\alpha)}_{ik}\}$ is the known basis, then a new basis $\{\varphi^{(\alpha)}_{ji}\}$ follows from (see Exercise 7.4)

$$\sum_j \Psi^{(\alpha)*}_{jk} \varphi^{(\alpha)}_{ji} = \frac{d_\alpha}{g} \sum_a P_a (\Psi^{(\alpha)}_{ik})^+ \varphi \tag{7.2.15}$$

by comparison of the coefficients of $\Psi_{jk}^{(\alpha)}$. Here φ is an arbitrary test function. The matrices of the IRs do not enter (7.2.15) explicitly. In our example $\{\Psi_{jk}^{(\alpha)}\} = \{x, y\}$. Choosing $\varphi = X^2$ (using capital letters to distinguish it from $\Psi^{(\alpha)}$) as a test function we cannot derive any statement from application of

$$\sum_a P_a = (P_e + P_\sigma)(P_e + P_c + P_{c_3^2}) \text{ on } X^2, \text{ since } \sum_a P_a x X^2 = 0 \ .$$

However,

$$\tfrac{1}{3}\sum_a P_a y X^2 = \tfrac{1}{2}x \cdot 2XY + \tfrac{1}{2}y \cdot (X^2 - Y^2)$$

gives a new (quadratic) basis of the E REP, namely $\{2XY, X^2 - Y^2\}$. Similarly, we obtain basis functions of higher order from higher-order test functions.

In molecules the functions are localized at different centres, some of which may be equivalent. These interchange under symmetry operations. The set of AOs, of which the MOs $\{\psi_i\}$ are composed, has to have identical AOs at atoms of the same kind. Therefore the MOs can be constructed according to the following scheme.

(i) The AOs are arranged into equivalent sets, which contain all those AOs that transform into each other under all the operations P_a with $a \in \mathcal{G}$, but which are localized at different centres.

(ii) Each of these sets realizes a REP of $\mathcal{G}$, the dimension of which is smaller than or equal to the dimension of the regular REP; thus $m_\alpha \leqslant d_\alpha$ (Sect. 4.3.1)

(iii) Using the projection operators $P_{kj}^{(\alpha)}$ of $\mathcal{G}$ we get a set of symmetrized MOs $\{\Psi_k^{\alpha s}\}$ from every set of equivalent AOs if we apply $P_{kj}^{(\alpha)}$ to the $\{\psi_i\}$ as long as we obtain $m_\alpha \leqslant d_\alpha$ functions $\Psi_k^{\alpha s}$, $s = 1, \ldots, m_\alpha$.

If the AOs constitute an orthonormal system (ONS) then these projections just define a unitary transformation between the ψ_i and the $\Psi_k^{\alpha s}$. The operators $P_a \in \mathcal{G}$ act on the AOs $\psi_i(\boldsymbol{x} - \boldsymbol{X}^n)$ which are localized at the atoms $\boldsymbol{X}^n$ in the molecule just as the operators in (7.1.8). This means P_a transforms the centre of the AO from $\boldsymbol{X}^n$ to $\boldsymbol{X}^k$ and simultaneously the function to $P_a\psi_i(\boldsymbol{x}) = \psi_i(a^{-1}\boldsymbol{x})$. If the AOs $\psi_i(\boldsymbol{x}) \sim Y_m^l(\vartheta, \varphi)$, then they transform according to (7.2.13a), thus with $\boldsymbol{X}^k = a\boldsymbol{X}^n$

$$\begin{aligned} P_a\psi_{lm}(\boldsymbol{x} - \boldsymbol{X}^n) &= \psi_{lm}(a^{-1}\boldsymbol{x} - \boldsymbol{X}^n) = \psi_{lm}(a^{-1}(\boldsymbol{x} - a\boldsymbol{X}^n)) \\ &= \sum_{m'=-l}^{l} D_{m'm}^{(l)}(a)\psi_{lm'}(\boldsymbol{x} - a\boldsymbol{X}^n) = \sum_{m'=-l}^{l} D_{m'm}^{(l)}(a)\psi_{lm'}(\boldsymbol{x} - \boldsymbol{X}^k) \ . \end{aligned}$$

In this case the REP has dimension $(2l + 1)N$, where N denotes the number of equivalent centres. If the AOs are polynomials in Cartesian coordinates we have to use the vector REP $D^{(v)}$ or its tensorial extension. It is convenient to introduce local coordinate systems at the equivalent atoms which are orientated in such a

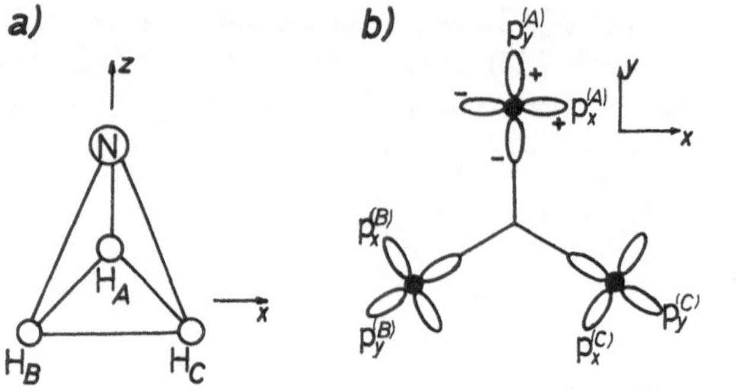

Fig. 7.5. (a) NH_3 molecule, which has $\mathscr{C}_{3v}$ symmetry. (b) p functions localized at the H atoms in a local coordinate system

way that a maximal number of AOs maps onto each other under the transformations.

The basis functions $\Psi_k^{\alpha s}$ for the NH_3 molecule which has $\mathscr{C}_{3v}$ symmetry will be determined starting with a set of given s and p functions. The AOs arranged in equivalent sets $\{\phi_i\}$ are (Fig. 7.5)

$$\{\phi_i\} = \{\phi_1 = \{s^N\}; \phi_2 = \{p_x^N, p_y^N\}; \phi_3 = \{p_z^N\}; \phi_4 = \{s^A, s^B, s^C\},$$

$$\phi_5 = \{p_x^A, p_y^A, \ldots, p_x^C, p_y^C\}; \phi_6 = \{p_z^A, p_z^B, p_z^C\}\} \ . \tag{7.2.16}$$

We have six sets of equivalent functions for the two atomic species (N, H = A, B, C). The $\{\phi_i\}$ constitute a REP $(\Pi \otimes D(a))_{ij}^{mn}$ just as in (7.1.8). Reduction yields the multiplicities determining how often the IRs of $\mathscr{C}_{3v}$ occur in this REP. The characters for the sets of equivalent functions are given by

$$\chi^{(\phi_i)}(a) = N_a \cdot \chi^{(D)}(a) \ , \tag{7.2.17}$$

as in (7.1.9a)[7]; (D) indicates the identity REP for s functions, the vector REP for p functions. For (7.2.16) we have

$$(\Pi \otimes D)_{ij}^{mn} = 5A_1 \oplus A_2 \oplus 5E \tag{7.2.18}$$

(see Exercise 7.5). With the projection operators from (4.3.11) we obtain for example for $\{\phi_4\}$ and the E REP (including normalizing factors)

$$P_{22}^E s^A = \frac{1}{3}(2s^A - s^B - s^C) = \sqrt{2/3}\, \Psi_2^E \ ,$$

$$P_{12}^E s^A = \frac{1}{2\sqrt{3}}(-s^B + s^C) = -\frac{1}{\sqrt{6}} \Psi_1^E \ .$$

We assume that the $\{\phi_i\}$ form an ONS. Then by successive applications of the projection operators we finally get all the symmetry-adapted functions;

[7] Translations and rotations of the system, as discussed in (7.1.10), do not play any role here, of course.

Table 7.6. The unitary transformation U which maps the N and 3H s, p functions onto symmetry-adapted functions ($\alpha = 1/\sqrt{3}$, $\beta = 1/\sqrt{6}$, $\gamma = 1/\sqrt{2}$)

	N				H											
					A	B	C	A			B			C		
	s	p_x	p_y	p_z	s	s	s	p_x	p_y	p_z	p_x	p_y	p_z	p_x	p_y	p_z
A_1 1	1															
A_1 2				1												
A_1 3					α	α	α									
A_1 4									α			α			α	
A_1 5										α			α			α
A_2								α			α			α		
E_1 1		1														
E_2 1			1													
E_1 2					2β	$-\beta$	$-\beta$									
E_2 2					0	γ	$-\gamma$									
E_1 3									2β			$-\beta$			$-\beta$	
E_2 3									0			γ			$-\gamma$	
E_1 4								2β			$-\beta$			$-\beta$		
E_2 4								0			γ			$-\gamma$		
E_1 5										2β			$-\beta$			$-\beta$
E_2 5										0			γ			$-\gamma$

consequently the unitary transformation U which maps ψ onto $\Psi = U\psi$ is as in Table 7.6.

In chemical bonding, directed valences are of importance. For this it is convenient first to construct symmetry orbitals (hybrids) from the AOs (s, p, d, etc.). These hybrids have to be adapted to the bonding directions. In general they constitute a reducible REP of the point group of the molecule. The electrons of the orbitals localized at different atoms may interact with the electrons of neighbouring atoms and thus form (covalent) bonds. The overlap of these orbitals is a measure of the strength of the bonding. The hybrids are classified according to the nodal planes of the functions: σ orbitals have no nodal plane in the bonding direction and possess rotational symmetry about the bonding axis while π orbitals are orthogonal to this axis and have a nodal plane through it. Thus the latter wave functions are antisymmetric with respect to the σ-bonding plane, as shown in Fig. 7.6. Further, δ orbitals have two nodal planes, etc.

The three H atoms in Fig. 7.5, 6 have $\mathscr{D}_{3h}$ symmetry. Their hybrids, which correspond to the equivalent functions earlier in this section, have characters according to (7.2.17) and thus the decomposition in Table 7.7. The form of the hybrids depends on the AOs used for the construction. Important for chemical bonding are the sp^2 hybrids, which we will discuss as an example. We choose the functions which are symmetrized with respect to the IRs of $\mathscr{D}_{3h}$ as the AO basis ψ^{AO}. With $\psi^{AO} = \{s, p_x, p_y\}$ we have for a single atom in the H plane according to (7.2.12)

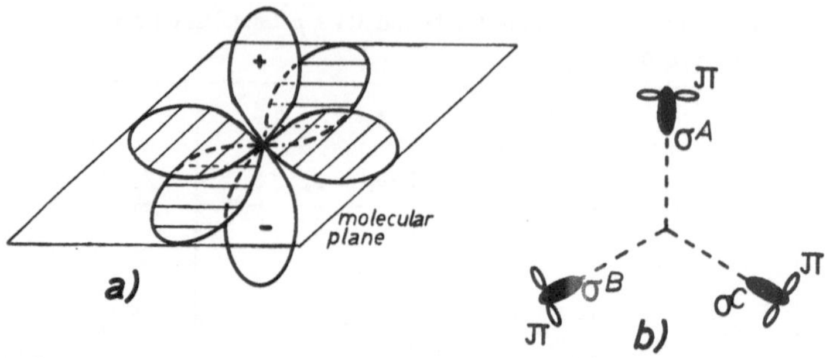

Fig. 7.6. **(a)** Plan of a molecule having σ bonds to four neighbours (*hatched*) and π bonds perpendicular to them (+, −). **(b)** H plane of the NH_3 molecule, which has σ functions in that plane and π functions in the plane (π_{xy}) as well as perpendicular to the plane (π_z)

Table 7.7. The characters and decomposition of the hybrids of the three H atoms in Fig. 7.5, 6

$\mathscr{D}_{3h}$	e	$2c_3$	$3c_2$	σ_h	$2s_3$	$3\sigma_v$	Decomposition
σ	3	0	1	3	0	1	$A_1' + E'$
π_{xy}	3	0	−1	3	0	−1	$A_2' + E'$
π_z	3	0	−1	−3	0	1	$A_2'' + E''$
π	6	0	−2	0	0	0	$A_2' + A_2'' + E' + E''$.

$$\psi^{A_1'} = s \quad \text{and} \quad \psi^{E'}_{1,2} = \{p_x, p_y\} \ . \tag{7.2.19a}$$

Applying for example the projection operators (4.3.9a) of $\mathscr{D}_{3h}$ to a hybrid σ^A of $\sigma = \{\sigma^A, \sigma^B, \sigma^C\}$, we obtain

$$\begin{aligned} \psi^{A_1'} &= \frac{1}{\sqrt{3}}(\sigma^A + \sigma^B + \sigma^C) \sim P^{A_1'}\sigma^A \\ \psi^{E'}_{1,2} &\begin{cases} = \dfrac{1}{\sqrt{6}}(2\sigma^A - \sigma^B - \sigma^C) \sim P^{E'}_{22}\sigma^A \\ = \dfrac{1}{\sqrt{2}}(\sigma^B - \sigma^C) \sim P^{E'}_{12}\sigma^A \ , \end{cases} \end{aligned} \tag{7.2.19b}$$

thus

$$\psi^{AO} = \begin{pmatrix} s \\ p_y \\ p_x \end{pmatrix} = \begin{pmatrix} 1/\sqrt{3} & 1/\sqrt{3} & 1/\sqrt{3} \\ 2/\sqrt{6} & -1/\sqrt{6} & -1/\sqrt{6} \\ 0 & 1/\sqrt{2} & -1/\sqrt{2} \end{pmatrix} \cdot \begin{pmatrix} \sigma^A \\ \sigma^B \\ \sigma^C \end{pmatrix} ,$$

$$\psi^{AO} = U\sigma \ , \qquad \sigma = U^+\psi^{AO} \ .$$

The reverse defines the σ orbitals in terms of the AOs

$$\sigma^A = \frac{1}{\sqrt{3}}(s + \sqrt{2}p_y) , \qquad \sigma^{B,C} = \frac{1}{\sqrt{6}}(\sqrt{2}s - p_y \mp \sqrt{3}p_x) .$$

The π hybrids are divided into weak and strong ones; the first contain functions which also occur in σ hybrids, the latter do not contain such functions. In Table 7.7, the π_{xy} hybrids are weak, the π_z ones are strong.

As in (7.2.19) we obtain the π_z (pd^2) hybrids, if we also allow for d-electrons (see also Table 5.1). From

$$\psi^{A_2''} = p_z , \qquad \psi^{E''}_{1,2} = \{d_{1c}, d_{1s}\}$$

we get with projection operators of $\mathscr{D}_{3h}$:

$$\pi^A = \frac{1}{\sqrt{3}}(p_z + \sqrt{2}d_{1s}) , \qquad \pi^{B,C} = \frac{1}{\sqrt{6}}(\sqrt{2}p_z - d_{1s} \mp \sqrt{3}d_{1c}) . \qquad (7.2.20)$$

7.2.3 The Hückel Method and Ligand Field Theory

Some further applications will be discussed here, mainly with respect to the group theoretical aspects. We assume that the reader is familiar with the quantum-mechanical background.

The *Hückel method* deals with the calculation of binding energies of molecules by means of hybrid functions. We illustrate this with the example of benzene (C_6H_6) which is sketched in Fig. 7.7. The four valence electrons of the carbon are hybridized into three σ orbitals (in directions separated by 120° in the C plane) and one π_z orbital. The σ electrons are assumed to be strongly localized (at the C atoms), so that they do not contribute to the overlap and thus to the bonding. Only the π_z electrons participate in the overlap (bonding). Inner (s) electrons are even more localized and can be neglected. The LCAO function is

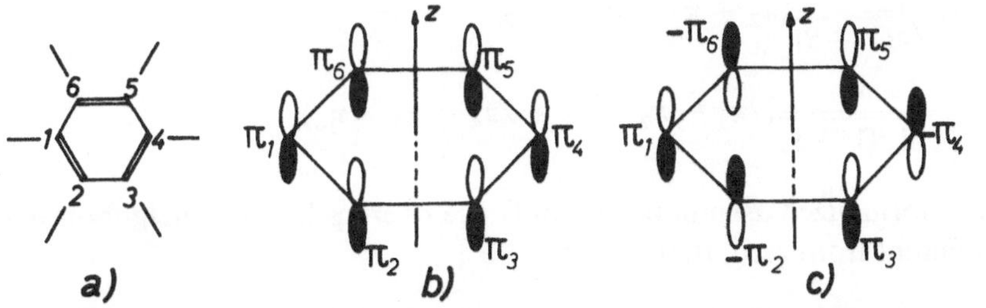

Fig. 7.7. (**a**) Benzene structure after Kekulé. The C atoms are numbered 1 to 6. Symmetry is $\mathscr{D}_{6h} = \mathscr{C}_i \times \mathscr{D}_6$. (**b**) Localized π orbitals contribute most to the binding. If combined as indicated, the A_{2u} state in (7.2.23) is obtained. The overlap between positive and negative contributions is maximal: bonding state. (**c**) The distribution corresponds to the B_{2g} state. Positive and negative contributions partly cancel so the overlap is small: antibonding state. ⬮ : wave function positive; () : wave function negative

Table 7.8. Characters of the representation constituted by the π_n for benzene

$\mathscr{D}_{6h}$	e	$2c_6$	$2c_3$	c_2	$3c_2'$	$3c_2''$	i	$2s_3$	$2s_6$	σ_h	$3\sigma_d$	$3\sigma_v$
D^{π_n}	6	0	0	0	−2	0	0	0	0	−6	0	2

then[8]

$$\psi(x) = \sum_{n=1}^{6} a_n \pi_n(x - X^n) , \qquad \pi_n = p_z ; \tag{7.2.21}$$

$|a_n|^2$ is the probability of finding an electron at the atom n. The characters of the REP constituted by the π_n are given by Table 7.8 (see also Appendix A, so that the REP space has the decomposition

$$D^{\pi} = B_{2g} \oplus A_{2u} \oplus E_{1g} \oplus E_{2u} . \tag{7.2.22}$$

The projection operators can be obtained by means of the REPs (Appendix B). They can be written as $P^{(\alpha)}(\mathscr{D}_{6h}) = P^{(\alpha')}(\mathscr{C}_i) \cdot P^{(\alpha'')}(\mathscr{D}_6)$ and give after application to (7.2.21)

$$\psi^{B_{2g}} = \frac{1}{\sqrt{6(1-2S)}}(\pi_1 - \pi_2 + \pi_3 - \pi_4 + \pi_5 - \pi_6) \sim P^{B_{2g}}(\mathscr{D}_{6h})\pi_1$$

$$\psi^{A_{2u}} = \frac{1}{\sqrt{6(1+2S)}}(\pi_1 + \pi_2 + \pi_3 + \pi_4 + \pi_5 + \pi_6)$$

$$\psi_{1,2}^{E_{1g}} = \begin{cases} \dfrac{1}{\sqrt{4(1+S)}}(\pi_2 + \pi_3 - \pi_5 - \pi_6) \\ \dfrac{1}{\sqrt{12(1+S)}}(2\pi_1 + \pi_2 - \pi_3 - 2\pi_4 - \pi_5 + \pi_6) \end{cases} \tag{7.2.23}$$

$$\psi_{1,2}^{E_{2u}} = \begin{cases} \dfrac{1}{\sqrt{4(1-S)}}(\pi_2 - \pi_3 + \pi_5 - \pi_6) \\ \dfrac{1}{\sqrt{12(1-S)}}(2\pi_1 - \pi_2 - \pi_3 + 2\pi_4 - \pi_5 - \pi_6) . \end{cases}$$

The $\psi^{(\alpha)}$ are normalized assuming that only the overlap between neighbouring atoms is different from zero, that means

$$\langle \pi_i | \pi_j \rangle = S_{ij} = \begin{cases} 1 & \text{for } i = j \\ S > 0 & \text{for } i = j \pm 1, \quad \text{mod } 6 \\ 0 & \text{otherwise.} \end{cases} \tag{7.2.24}$$

[8] A numerical fit gives e.g. $p_z = 1.71 r \cos \vartheta e^{-1.56r}$.

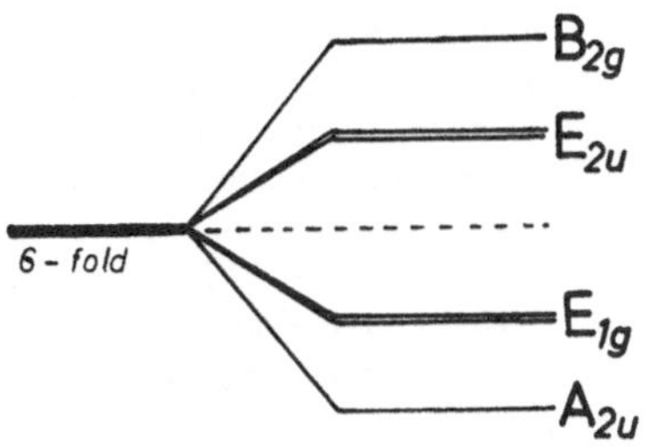

Fig. 7.8. Splitting of sixfold-degenerate atomic levels in a molecular field having $\mathscr{D}_{6h}$ symmetry (π orbitals)

The functions (7.2.21) [or (7.2.23)] can now be used in a simple variational or perturbation method for the calculation of the eigenvalues from (Sects. 6.3, 4)

$$\sum_n (H_{in} - ES_{in})a_n = 0 \ . \tag{7.2.25}$$

This can be simplified using the symmetry-adapted functions. Because of (7.2.22) there are two nondegenerate and two two-fold-degenerate states. The different IRs occur only once, which means the symmetry adapted MOs are already the "correct" eigenfunctions of (7.2.25). Quantitatively[9] we obtain after using (7.2.23) in (6.3.10), with

$$H_{ij} = \langle \pi_i | H | \pi_j \rangle = \begin{cases} C > 0 & \text{for } i = j\text{: Coulomb integral} \\ R < 0 & \text{for } i = j \pm 1\text{: hopping integral} \ , \\ 0 & \text{otherwise} \end{cases}$$

$$\begin{aligned} E^{B_{2g}} &= \frac{C - 2R}{1 - 2S} \ , \qquad E^{A_{2u}} = \frac{C + 2R}{1 + 2S} \ , \\ E^{E_{1g}} &= \frac{C + R}{1 + S} \ , \qquad E^{E_{2u}} = \frac{C - R}{1 - S} \ , \qquad \text{each twofold} \ . \end{aligned} \tag{7.2.26}$$

See Fig. 7.8. Considering the twofold degeneracy due to the spin, the six electrons occupy the three lowest states A_{2u}, E_{1g}. The binding energy is the difference between this energy and the unperturbed energy, that is, the energy if all the six electrons are in their p_z-orbitals:

$$E_{\text{bind}} = 4\frac{(R - CS)(2 + 3S)}{(1 + S)(1 + 2S)} \ . \tag{7.2.27}$$

In a better calculation, of course, we can include the σ orbitals having a smaller overlap. Then we have to construct 42 symmetry-adapted functions from the 42 AOs H(1s) and C(1s, 2s, 2p), and put the 42 electrons in the lowest energy levels in order to calculate the binding energy.

In *ligand field theory* the foregoing method is applied to ion complexes to determine the electronic structure. These complexes consist of a central atom

[9] The signs of C and R can only be determined by quantitative calculations.

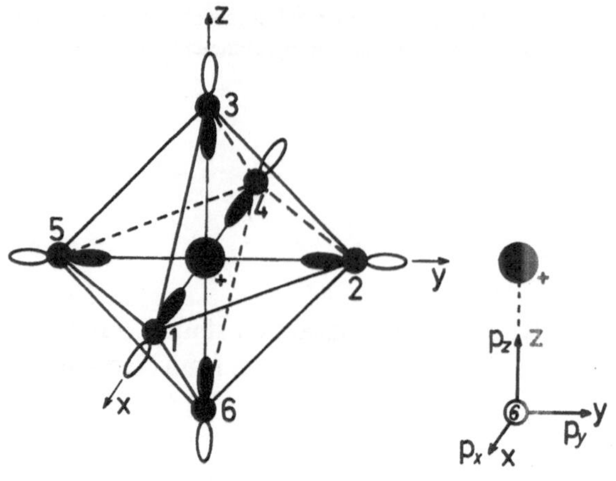

Fig. 7.9. ML_6 complex with $\mathcal{O}_h$ symmetry. The atomic p_z orbitals of the ligands in a local coordinate system are indicated. This distribution corresponds to the bonding A_{1g} state in (7.2.29). The other p-states not indicated only contribute to the π hybrids. ● +; ○ −

Table 7.9. Character table for the hybrid representations for complexes of the type ML_6

$\mathcal{O}_h$	e	$8c_3$	$3c_2$	$6c_4$	$6c_2'$	i	$8s_6$	$3\sigma_h$	$6s_4$	$6\sigma_d$	Decomposition
$D^{M\sigma}$	6	0	2	2	0	0	0	4	0	2	$A_{1g} \oplus E_g \oplus F_{1u}$
$D^{L\sigma}$	6	0	2	2	0	0	0	4	0	2	$A_{1g} \oplus E_g \oplus F_{1u}$
$D^{L\pi}$	12	0	−4	0	0	0	0	0	0	0	$F_{1g} \oplus F_{2g} \oplus F_{1u} \oplus F_{2u}$

(ion) and several ligands which again may be neighbouring atoms in a crystal. We consider complexes of the type ML_6, where M denotes a metal ion (Fe, Co etc.), L a ligand (CN, NH_3 etc.) These complexes have octahedral ($\mathcal{O}_h$) symmetry. For the bonding we assign s, p and d AOs to the central ion, which are assumed to form six identical σ hybrids in the direction of the ligands. These hybrids overlap with the hybrids of the ligands which may be of σ or π type. We assume that the ligands contribute p electrons to the bonding: p_z states (in a local coordinate system) contribute to the σ hybrids, and p_x, p_y states form π hybrids (Fig. 7.9).

(i) We construct the hybrids of the central ion, and decompose these with respect to the IRs of $\mathcal{O}_h$.
(ii) We construct the σ and π hybrids of the ligands, also decomposing them with respect to the IRs.
(iii) We form the symmetry-adapted MOs from the hybrids of (i) and (ii).

In the case of ML_6 the characters of the hybrid REPs according to (7.2.17) are given in Table 7.9.

In the following we investigate only the σ hybrids, which give the main contribution to the bonding. As at the end of Sect. 7.2.2, using the projection operators $P^{(\alpha)}(\mathcal{O}_h) = P^{(\alpha')}(\mathcal{O}) \cdot P^{(\alpha'')}(\mathcal{C}_i)$ we get for the hybrid functions of the central ion ($\sigma = U^+ \psi^{AO}$)

$$\sigma_{1,4}^M = \frac{1}{\sqrt{12}}(\sqrt{2}s \pm \sqrt{6}p_x - d_0 + \sqrt{3}d_{2c}) ,$$

$$\sigma_{2,5}^M = \frac{1}{\sqrt{12}}(\sqrt{2}s \pm \sqrt{6}p_y - d_0 - \sqrt{3}d_{2c}) , \qquad (7.2.28)$$

$$\sigma_{3,6}^M = \frac{1}{\sqrt{6}}(s \pm \sqrt{3}p_z + \sqrt{2}d_0) .$$

Here the AOs s belong to A_{1g}; p_x, p_y, p_z to F_{1u} and d_0, d_{2c} to E_g. Correspondingly we obtain the symmetrized MOs of the ligands $\psi_k^{(\alpha)} = Up_{z,i}$ belonging to the σ hybrids $p_{z,i}$ $(i = 1, \dots, 6)$, where U^+ can be taken from (7.2.28) since the reductions in Table 7.9 are the same for $D^{M\sigma}$ and $D^{L\sigma}$:

$$\psi^{A_{1g}} = \frac{1}{\sqrt{6}}(p_1 + p_2 + p_3 + p_4 + p_5 + p_6) ,$$

$$\psi_1^{E_g} = \frac{1}{\sqrt{12}}(p_1 + p_2 - 2p_3 + p_4 + p_5 - 2p_6) ,$$

$$\psi_2^{E_g} = \frac{1}{2}(p_1 - p_2 + p_4 - p_5) ,$$

$$\psi_1^{F_{1u}} = \frac{1}{\sqrt{2}}(p_1 - p_4) ,$$

$$\psi_2^{F_{1u}} = \frac{1}{\sqrt{2}}(p_2 - p_5) ,$$

$$\psi_3^{F_{1u}} = \frac{1}{\sqrt{2}}(p_3 - p_6) . \qquad (7.2.29)$$

The variational function corresponding to (7.2.25) belongs to the REP $D^\sigma := D^{M\sigma} \oplus D^{L\sigma} = 2(A_{1g} \oplus E_g \oplus F_{1u})$ and reads in this case (σ bonding)

$$\Psi = c_1^{A_{1g}}s + c_2^{A_{1g}}\psi^{A_{1g}} + c_{11}^{E_g}d_0 + c_{12}^{E_g}\psi_1^{E_g} + c_{21}^{E_g}d_{2c} + c_{22}^{E_g}\psi_2^{E_g}$$
$$+ c_{11}^{F_{1u}}p_x + c_{12}^{F_{1u}}\psi_1^{F_{1u}} + c_{21}^{F_{1u}}p_y + c_{22}^{F_{1u}}\psi_2^{F_{1u}} + c_{31}^{F_{1u}}p_z + c_{32}^{F_{1u}}\psi_3^{F_{1u}} .$$

According to (6.3.9) the symmetrized eigenvalue problem consists of six blocks, A_{1g}, $2E_g$, $3F_{1u}$, each of them being two-dimensional. The multiplicities and degeneracies can be seen from this decomposition. There are three quadratic equations, for example the one for the A_{1g} states is

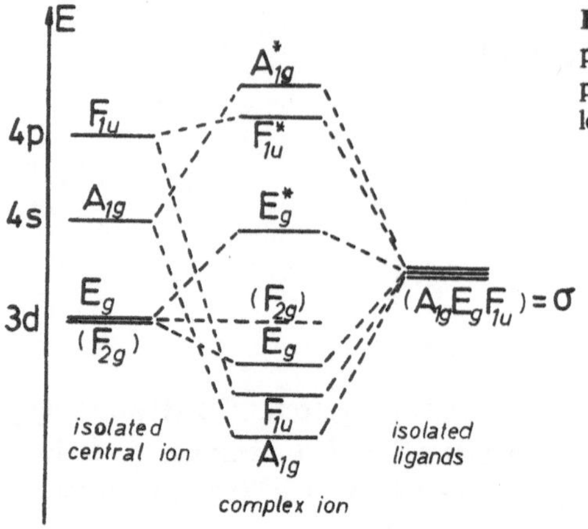

Fig. 7.10. Qualitative spectrum of a ML_6 complex. The quantitative position of the levels depends sensitively on the matrix elements. The F_{2g} level does not change in a σ approximation

$$\begin{vmatrix} H_{11}-E & H_{12}-ES_{12} \\ H_{12}^*-ES_{12}^* & H_{22}-E \end{vmatrix} = 0 \quad \text{with} \tag{7.2.30}$$

$$H_{11} = \langle s|H|s\rangle \ , \qquad H_{12} = \langle s|H|\psi^{A_{1g}}\rangle \ ,$$

$$H_{22} = \langle \psi^{A_{1g}}|H|\psi^{A_{1g}}\rangle \ , \qquad S_{12} = \langle s|\psi^{A_{1g}}\rangle \ .$$

Thus

$$E_{1,2} = a \pm \sqrt{a^2-b} \ , \qquad \text{where}$$

$$a = \frac{H_{11}+H_{22}-H_{12}S_{12}^*-H_{12}^*S_{12}}{2(1-S_{12}S_{12}^*)} \ ,$$

$$b = \frac{H_{11}H_{22}-H_{12}H_{12}^*}{1-S_{12}S_{12}^*} \ .$$

The higher state (with respect to energy) is an antibonding state A_{1g}^*, the lower one a bonding state A_{1g}. The function is sketched in Fig. 7.9. Correspondingly we obtain the other energy levels which are shown in Fig. 7.10 (for a transition metal ion as the central one with σ coupling).

As an F_{2g} basis, d_{1c}, d_{1s}, d_{2s} functions cannot couple to σ ligand hybrids because of the decomposition of states given in Table 7.9. They can only interact with π ligands. In this approximation the ground state is defined by filling up the energetically lowest states with the electrons, including the spinstates. A quantitative calculation of the energies is difficult even in this approximation, because the integrals in (7.2.30) are two-or-more-centre integrals. Therefore the integrals are often parametrized or expressed by parametrized overlap integrals S_{jk} (Mulliken). Their values are then fitted to experimental data.

The methods given (Hückel method, ligand field theory) can also be applied in other branches of physics, e.g. in nuclear physics for the discussion of the α-particle model of light nuclei (Exercise 7.10).

Exercise 7.2. Give the possible permutation symmetries for an m-electron system ($m = 1, 2, 3, 4, 5$).

Exercise 7.3. Give the spin functions of a three-electron system by particularizing Exercise 5.19 to $n = 2$. Assign the corresponding S_z values. The basis functions belonging to $D^{[3]}$ and $D^{[2,1]}$ of $\mathscr{P}_3$ should be given in α, β notation.

Exercise 7.4. Prove (7.2.15).

Exercise 7.5. Show the validity of (7.2.18).

Exercise 7.6. Discuss sp^3 hybridization for $\mathscr{T}_d$-symmetry and determine the σ functions.

Exercise 7.7. Illustrate (graphically) the spherical harmonics given in Table 5.1, especially the s, p, d functions.

Exercise 7.8. Give the transformation matrix U for the transformation (7.2.23) between the π orbitals and the symmetry-adapted functions $\psi^{(\alpha)}$ and determine the block-diagonal form of the eigenvalue problem (7.2.25). [See (6.3.9).]

Exercise 7.9. Sketch the symmetrized MO states E_g and F_{1u} as in Fig. 7.9.

Exercise 7.10. According to Hartree-Fock calculations for the ^{12}C nucleus, the particle density can be assumed to be concentrated in three α particles, located at the vertices of an equilateral triangle ($\mathscr{C}_{3v}$). Calculate the ground state energy of the ^{12}C nucleus with this model, ψ^A, ψ^B, ψ^C being the α-particle functions. Solve the eigenvalue problem in a symmetrized form using a model Hamiltonian

$$H = E_0 \sum_{i=1}^{3} \hat{\psi}_i^+ \hat{\psi}_i + \tfrac{1}{2} h \sum_{i=j} \hat{\psi}_i^+ \hat{\psi}_j \ ;$$

E_0 is the unperturbed energy of an α particle, h is a transition probability element between α particles. The operator $\hat{\psi}_i^+$ generates one particle in state i if applied to a vacuum state.

7.3 Many-Electron Problems

7.3.1 Permutation Symmetry

The eigenvalue problem in Sect. 6.3 is based on the orthogonality of the basis functions, or, put another way, it uses unitary IRs. On the other hand, in Sect. 5.4 we used nonunitary REPs for the Young operators. This is more convenient in many cases. Then we have to formulate the eigenvalue problem in terms of such REPs, too. This can be achieved by introducing a dual Young operator, compare (5.4.18a) defined by

$$\bar{e}_{ij}^{[\lambda]} = \frac{d_{[\lambda]}}{n!} S_i[\lambda] p_{ij} A_j[\lambda] = \frac{d_{[\lambda]}}{n!} S_i[\lambda] A_i[\lambda] p_{ij} \ . \tag{7.3.1}$$

With its help we can also generate a basis dual to $\Psi_{ij}^{[\lambda]}$ (5.4.24)

$$\bar{\Psi}_{ij}^{[\lambda]} := \bar{e}_{ij}^{[\lambda]}\psi(1,\dots,m) , \qquad \Psi_{ij}^{[\lambda]} := e_{ij}^{[\lambda]}\psi(1,\dots,m) . \tag{7.3.2}$$

It can be shown that for nonorthogonal bases we have instead of (6.3.7)

$$\langle \bar{\Psi}_{ik}^{[\lambda']} | H | \Psi_{jl}^{[\lambda]} \rangle = \delta_{\lambda\lambda'}\delta_{ij} H_{kl}^{[\lambda]} , \qquad \text{and} \tag{7.3.3a}$$

$$\langle \bar{\Psi}_{ik}^{[\lambda']} | \Psi_{jl}^{[\lambda]} \rangle = \delta_{\lambda\lambda'}\delta_{ij} S_{kl}^{[\lambda]} , \tag{7.3.3b}$$

provided that $[H,p] = 0$ for each $p \in \mathscr{P}_m$. In (7.3.3) the energy $H_{kl}^{[\lambda]}$ and overlap matrix elements $S_{kl}^{[\lambda]}$ are independent of the row index i. See (6.3.9).

The m-electron Hamiltonian (7.2.3) satisfies $[H,p] = 0$ and has the form

$$H = H(\boldsymbol{x}_1 \dots \boldsymbol{x}_m) := \sum_{i=1}^{m} h(i) + \tfrac{1}{2} \sum_{i,j=1}^{m}{}' g(i,j) . \tag{7.3.4}$$

Its eigenvalues can best be determined by a variational method for which we choose one-particle product functions as a basis symmetrized with respect to $\mathscr{P}_m$ (5.5.20). Using (7.3.2) we have

$$\phi = \sum_{[\lambda]} \sum_{i,j}^{d_{[\lambda]}} c_{ij}^{[\lambda]} \Psi_{ij}^{[\lambda]} = \sum_{[\lambda]} \sum_{i,j}^{d_{[\lambda]}} c_{ij}^{[\lambda]} e_{ij}^{[\lambda]} \psi_{i_1}(1) \dots \psi_{i_m}(m) . \tag{7.3.5}$$

First we consider the case that all the $\boldsymbol{i} = \{i_1,\dots,i_m\}$ *are different from each other*. According to the discussion following (5.5.28), the $\Psi_{ij}^{[\lambda]}$ then constitute the regular REP of $\mathscr{P}_m$, which means every $D^{[\lambda]}$ occurs $m_{[\lambda]} = d_{[\lambda]}$ times in (7.3.5). Using (7.3.3 and 5) the eigenvalue problem for H reduces to the secular equations (6.3.10)

$$\det(H_{kl}^{[\lambda]} - E S_{kl}^{[\lambda]}) = 0 \tag{7.3.6}$$

of dimension $d_{[\lambda]} = m_{[\lambda]}$. Since the dimension of the REP $D^{[\lambda]}$ is also equal to $d_{[\lambda]}$, (7.3.6) comprises $d_{[\lambda]}$ identical subproblems. The degeneracy of the energy is then $d_{[\lambda]}$. The eigenfunctions belonging to the energy $E_k^{[\lambda]}$, $k = 1, \dots, d_{[\lambda]}$ are

$$\phi_{i,k}^{[\lambda]} = \sum_{j=1}^{d_{[\lambda]}} c_{ij,k} \Psi_{ij}^{[\lambda]} , \qquad i = 1, \dots, d_{[\lambda]} . \tag{7.3.7}$$

The functions $\{\phi_{i,k}^{[\lambda]}\}$ are the degenerate basis functions of the IR $D^{[\lambda]}$, row i. These statements also hold without introducing one-particle functions into (7.3.5). But for product functions we obtain using (7.3.2) and (5.4.19)

$$S_{kl}^{[\lambda]} = \frac{d_{[\lambda]}}{m!} \sum_{p \in \mathscr{P}_m} D_{lk}^{[\lambda]}(p^{-1}) \langle \psi | p\psi \rangle , \tag{7.3.8a}$$

$$H_{kl}^{[\lambda]} = \frac{d_{[\lambda]}}{m!} \sum_{p \in \mathscr{P}_m} D_{lk}^{[\lambda]}(p^{-1}) \langle \psi | Hp\psi \rangle , \tag{7.3.8b}$$

where ψ denotes the product function in (7.3.5) Assuming the one-particle functions in (7.3.5) to be an ONS, i.e. $\langle\psi_{i_1}|\psi_{i_2}\rangle = \delta_{i_1 i_2}$, etc. then

$$\langle\psi|p\psi\rangle = \begin{cases}\langle\psi|\psi\rangle = 1 & \text{for } p = e\\ 0 & \text{otherwise}\end{cases},$$

thus

$$S_{kl}^{[\lambda]} = \frac{d_{[\lambda]}}{m!}\delta_{kl} \; . \tag{7.3.8c}$$

For the first term in (7.3.4) we obtain with one-particle functions

$$\langle\psi|h(j)|p\psi\rangle = \begin{cases}\langle\psi|h(j)|\psi\rangle = \langle\psi_{a_j}|h(j)|\psi_{a_j}\rangle = \langle h(j)\rangle & \text{for } p = e\\ 0 & \text{otherwise}\end{cases}, \tag{7.3.8d}$$

and for the second one

$$\langle\psi|g(j,l)|p\psi\rangle = \begin{cases}\left.\begin{array}{l}\langle\psi|g(j,l)|\psi\rangle = C(j,l)\\ = \langle\psi_{a_j}(j)\psi_{a_l}(l)|g(j,l)|\psi_{a_j}(j)\psi_{a_l}(l)\rangle\end{array}\right\} & \text{for } p = e\\ \left.\begin{array}{l}\langle\psi|g(j,l)|\psi\rangle = A(j,l)\\ = \langle\psi_{a_j}(j)\psi_{a_l}(l)|g(j,l)|\psi_{a_j}(l)\psi_{a_l}(j)\rangle\end{array}\right\} & \text{for } p = (jl) = p^{-1}\\ 0 & \text{otherwise}\end{cases}, \tag{7.3.8e}$$

where $C(j,l)$ is the Coulomb integral and $A(j,l)$ the exchange integral, in which the arguments j, l of the one-particle functions in $p\psi$ are exchanged. Using (7.3.8c–e) we obtain

$$H_{kl}^{[\lambda]} = \frac{d_{[\lambda]}}{m!}\left\{\left[\sum_i \langle h(i)\rangle + \frac{1}{2}\sum_{i,j}{}' C(i,j)\right]\delta_{kl} + \frac{1}{2}\sum_{i,j}{}' D_{kl}^{[\lambda]}((ij))\cdot A(i,j)\right\} \tag{7.3.8f}$$

and thus the secular equation

$$\det\left(\frac{1}{2}\sum_{i,j}{}' D_{kl}^{[\lambda]}((ij))\cdot A(i,j) - \left\{E - \left[\sum_i \langle h(i)\rangle + \frac{1}{2}\sum_{i,j}{}' C(i,j)\right]\right\}\delta_{kl}\right) = 0 \; . \tag{7.3.9}$$

Neglecting the two-particle interaction, $A(i,j) = C(i,j) = 0$, the energy in (7.3.9) is just the sum of one-particle energies. The Coulomb energy just means a shift of the energy due to a two-particle interaction, whereas the exchange energy in the off-diagonal elements means a splitting of the energy levels.

If we let an $m = 3$ electron system have the one-particle functions $\psi_{a_1}(1)\psi_{a_2}(2)\psi_{a_3}(3)$, then $[\lambda] = [2,1]$ is a two-dimensional REP whose nonunitary matrices are given in (5.4.22). The possible permutations are $\{(ij)\} = \{(12),(13),(23)\}$. Thus, $\Delta E := E - \langle h(1)\rangle - \langle h(2)\rangle - \langle h(3)\rangle - C(1,2) - C(1,3) - C(2,3)$ and

$$\begin{vmatrix} A(1,2) - A(1,3) - \Delta E & A(2,3) - A(1,3) \\ A(2,3) - A(1,2) & A(1,3) - A(1,2) - \Delta E \end{vmatrix} = 0 \ .$$

The resulting energies $E_{k=1}^{[2,1]}$, $E_{k=2}^{[2,1]}$ are doubly degenerate with eigenfunctions $\phi_{1,k}^{[2,1]}$ and $\phi_{2,k}^{[2,1]}$, $k = 1, 2$. Finally, in order to get totally antisymmetric functions when spin is included (Pauli's principle), according to (7.2.7) the spatial functions have to be combined with the spin functions of the associated REP

$$\phi_{kj}^{[1^3]} = \frac{1}{3!} \sum_{p \in \mathscr{P}_3} (-1)^p p[\phi_{1,k}^{[2,1]}(\boldsymbol{r}_1, \boldsymbol{r}_2, \boldsymbol{r}_3)\varphi_{1,j}^{[2,1]}(\sigma_1, \sigma_2, \sigma_3)] \ , \quad j = 1, 2 \ . \tag{7.3.10}$$

The spin functions $\varphi_{1,j}^{[2,1]}$ can be chosen to be both the sets of partner functions in Exercise 7.3.

If not all the a_l in $\Psi_{a_1 \ldots a_m}$ are different from each other we have to go back to Sect. 5.5.3. The multiplicity $m_{[\lambda]}$ with which a REP $D^{[\lambda]}$ occurs in the space constituted by the $m!$ functions $\{p\Psi_i\}$ with $p \in \mathscr{P}_m$ is then $m_{[\lambda]} \leqslant d_{[\lambda]}$. The secular equation has dimension $m_{[\lambda]}$ only. As described in Sect. 5.5.3, we obtain the basis functions $\Psi_{ij}^{[\lambda]}$, $i = 1, \ldots, d_{[\lambda]}$; $j = 1, \ldots, m_{[\lambda]}$ by application of $P_{kk}^{[\lambda]} := e_{kk}^{[\lambda]}$ to starting functions which belong to allowed tableaux $T_i[\lambda]$. But now, the one-particle matrix elements $\langle\psi|h(i)|p\psi\rangle$, $\langle\psi|p\psi\rangle$ in $H_{kl}^{[\lambda]}$ and $S_{kl}^{[\lambda]}$ do not vanish for all the $p \in \mathscr{P}_{m'} \subset \mathscr{P}_m$. Here $\mathscr{P}_{m'}$ is that permutation group which leaves the set $a_1, \ldots, a_m$ invariant if some of the a_l are equal. Correspondingly $\langle\psi|g(i,j)|p\psi\rangle \neq 0$ for all $p \in \mathscr{P}_{m'}$ *and* $p \in (ij)\mathscr{P}_{m'}$, the latter being a coset of $\mathscr{P}_{m'}$. Again we consider the REP $D^{[2,1]}$ of $\mathscr{P}_3$ with the starting function $\psi_a(1)\psi_a(2)\psi_b(3)$. The only allowed tableau is (Sect. 5.5.3)

$$\begin{array}{|c|c|}\hline a & a \\ \hline b \\ \cline{1-1}\end{array} \ , \quad \text{thus } m_{[2,1]} = 1 \ .$$

Then the energy is given by

$$E^{[2,1]} = \frac{\langle\bar{\Psi}^{[2,1]}|H|\Psi^{[2,1]}\rangle}{\langle\bar{\Psi}^{[2,1]}|\Psi^{[2,1]}\rangle} \tag{7.3.11}$$

Choosing $\psi_a := a = 1s$ (function), $\psi_b := b = 2s$ (function), meaning we have an electronic configuration $(1s)^2 2s$ as in lithium, then (7.3.11) represents the ground state energy of a Li atom (without spin interaction). According to Exercise 5.19, the spatial function is

$$\Psi_{aab}^{[2,1]} = \tfrac{2}{3}[a(1)a(2)b(3) - a(2)a(3)b(1)] \tag{7.3.12a}$$

and the dual function, see (7.3.1),

$$\bar{\Psi}^{[2,1]}_{aab} = \tfrac{1}{3}[2 \cdot a(1)a(2)b(3) - a(2)a(3)b(1) - a(1)a(3)b(2)] \ . \tag{7.3.12b}$$

Therefore we have $\langle \bar{\Psi} | \Psi \rangle = 2/3$ and

$$E^{[2,1]} = 2\varepsilon_a + \varepsilon_b + C_{aa} + 2C_{ab} - A_{ab} \ , \tag{7.3.13}$$

$$\varepsilon_a = \langle a(1) | h(1) | a(1) \rangle \ ,$$

$$C_{ab} = \left\langle a(1)b(2) \left| \frac{e^2}{|\boldsymbol{r}_1 - \boldsymbol{r}_2|} \right| a(1)b(2) \right\rangle \ ,$$

$$A_{ab} = \left\langle a(1)b(2) \left| \frac{e^2}{|\boldsymbol{r}_1 - \boldsymbol{r}_2|} \right| a(2)b(1) \right\rangle \ .$$

The spin function associated with (7.3.12a) is according to Exercise 7.3

$$\varphi^{\widetilde{[2,1]}} \sim [\alpha(2)\alpha(3)\beta(1) - \alpha(1)\alpha(3)\beta(2)] \tag{7.3.14}$$

and corresponds to total spin 1/2. The totally antisymmetric function has to be constructed according to (7.3.10) using (7.3.12a) and (7.3.14). It can be written as a Slater determinant

$$\phi^{[1^3]} = \frac{1}{3!} \begin{vmatrix} a(1)\alpha(1) & a(1)\beta(1) & b(1)\alpha(1) \\ a(2)\alpha(2) & a(2)\beta(2) & b(2)\alpha(2) \\ a(3)\alpha(3) & a(3)\beta(3) & b(3)\alpha(3) \end{vmatrix} \tag{7.3.15}$$

If we calculate expectation values of H with the function (7.3.15) we obtain (7.3.13), of course. In the last column of (7.3.15) α can be replaced by β. The unpaired electron can have either spin state.

7.3.2 Point and Permutation Symmetry. Molecular States

In the calculation of molecular states as well as of atomic states one usually starts with a model in which the electrons are looked upon as being independent particles. This has been discussed in Sect. 7.2.2 and 7.2.3. If the total number of electrons is m, such a state consists of a product of z one-particle functions, in which a spatial one-particle state is allowed to occur atmost twice because of the Pauli principle:

$$\psi_{i_1}(1) \dots \psi_{i_z}(z) \ , \qquad z \geqslant m/2 \text{ indices different from each other, } \sum_{k=1}^{z} i_k = m \ . \tag{7.3.16}$$

Expression (7.3.16) is said to be a *configuration*; if there is a spin function α or β assigned to the spatial functions $\psi_{i_1}(1), \dots$ we speak of *spin orbitals*. The number of the possible totally antisymmetric functions $\phi^{[1^m]}$ which can be built up from

z spatial orbitals together with the spin functions α and β is equal to the number of possible ways of distributing m electrous among the $2z$ spin orbitals, thus

$$Z(z,m) = \binom{2z}{m} = \frac{(2z)!}{(2z-m)!m!} \tag{7.3.17}$$

is in general a large number.

In a variational method for the determination of the energy we would have to represent all possible functions $\phi^{[1^m]}$ by Slater determinants and to use a linear combination of all these determinants as the starting function for a variation. This method is said to be the (unsymmetrized) *configuration interaction method* (CI). For molecules we distinguish two variants of the CI method: the valence-bond method, in which the ψ_i consist of AOs, and the MO method, in which the ψ_i consist of molecular orbitals, e.g. in LCAO form (Sect. 7.2). When we use the same basis set of functions and all the possible configurations both methods provide the same result for the energy. Nonetheless, in these methods, the symmetry of the problem, the permutation as well as the geometrical symmetry, is not taken into account. Consequently the secular equations have a very high dimension (see Exercise 7.11).

However, if we take into account the symmetries we are led to a decomposition of the eigenvalue problem as in Sect. 6.3. In Sect. 7.3.1 a factorization of the secular determinant with respect to the IRs $D^{[\lambda]}$ of $\mathscr{P}_m$ has been made. By the unique assignment $D^{[\lambda]} \leftrightarrow D^{[\tilde{\lambda}]} \leftrightarrow$ spin S according to Sect. 7.2.1, we obtain a secular equation for every S whose dimension is equal to the number of ways $R(m, S)$ of realizing the spin S in an m-electron system. In the case that the molecules have a geometrical symmetry, the projection operator method Sect. (4.3.2) can be used for a further symmetrization which leads to wave functions that also transform according to the IRs $D^{(\alpha)}$ of the point group $\mathscr{G}$. With these functions we finally obtain reduced eigenvalue problems for the molecular multiplets ${}^{2S+1}D^{(\alpha)}$, determined by spin S and $D^{(\alpha)}$. They have the dimension $m_{\alpha,S}$ which is the multiplicity with which a state ${}^{2S+1}D^{(\alpha)}$ occurs in the total REP of $\mathscr{P}_m \times \mathscr{G}$ originally defined by the $\binom{2z}{m}$ basis configurations. The valence bond method furthermore distinguishes between *covalent configurations*

$$\psi_{i1}(1)\ldots\psi_{i_m}(m)\ , \qquad \text{all the } i_1, \ldots, i_m \text{ being different from each other,} \tag{7.3.18}$$

(i.e. every AO of an outer electron which participates in the bonding is occupied at most once) and *ionic configurations*

$$\psi_{i_1}(1)\psi_{i_1}(2)\ldots\psi_{i_l}(k)\psi_{i_l}(k+1)\psi_{i_{l+1}}(k+2)\ldots\psi_{i_{m-2l}}(m)\ , \tag{7.3.19}$$

i.e. l AOs are doubly occupied. This also means that bonding electrons can pass from one atom of a molecule to another. Usually both kinds of bonding are present and have to be considered.

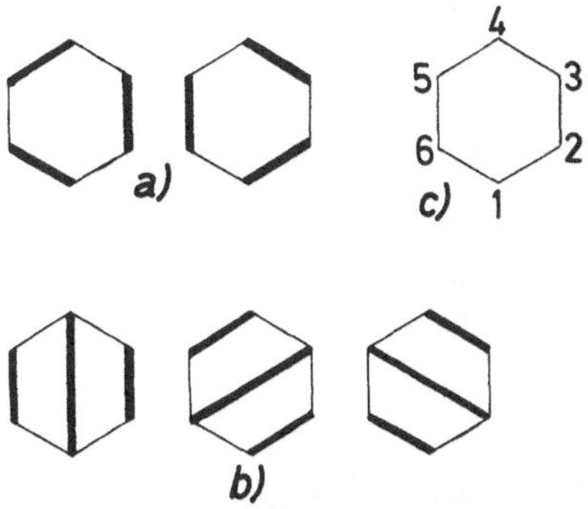

Fig. 7.11. Kekulé (a) and Dewar (b) structures of benzene. The heavy lines indicate the pair-wise saturation of the electronic spins. (c) Numbers used for the hypothetical H_6 example

Since all the i_l are different from each other, the number of covalent structures $R(m, S)$ with spin S is given by the multiplicity with which $D^{[\lambda]}$ of $\mathscr{P}_m$ occurs in the regular REP constituted by $p(\psi_{i_1}(1)\ldots\psi_{i_m}(m))$, thus, using (5.4.13) and (7.2.11)

$$R(m, S) = d_{[\lambda]} = d_{[\tilde{\lambda}]} = \frac{m!(2S+1)}{(m/2+S+1)!(m/2-S)!} \quad . \tag{7.3.20}$$

(In covalent structures the number of electrons contributing to the binding is always even.) The wave functions assigned to the covalent structures with $S = (\tilde{\lambda}_1 - \tilde{\lambda}_2)/2$, $m = \tilde{\lambda}_1 + \tilde{\lambda}_2$ are given by (7.2.7).

For the benzene model in Sect. 7.2.3, $S = 0$ and $m = 6$, thus $R(m, S) = 5$. The five covalent structures are sketched in Fig. 7.11. The *rule of Rumer* allows us to find the possible covalent structures: The valence orbitals $\psi(i)$ are symbolically arranged on a circle and connected in pairs (saturation of spins) by internal lines, so that there is no crossing of lines (Fig. 7.11).[10]

A further reduction of the eigenvalue problem is possible by means of the geometrical symmetry. We have now to investigate which IRs $D^{(\alpha)}$ of $\mathscr{G}$ occur in $D^{[\lambda]}$. Since the AOs are localized at the atoms, to every $a \in \mathscr{G}$ there corresponds a permutation $p \in \mathscr{P}'_m \subset \mathscr{P}_m$. Hence $\mathscr{G}$ is isomorphic to a subgroup $\mathscr{P}'_m \subset \mathscr{P}_m$, see (3.5.13), and we can solve this problem by the subduction (Sect. 4.2.5)

$$D^{[\lambda]}(\mathscr{P}_m) \to D^{[\lambda]\mathrm{sub}}(\mathscr{P}'_m) \cong \sum_{\alpha}{}_{\oplus}\, m_{\alpha,S} D^{(\alpha)}(\mathscr{G}) \ . \tag{7.3.21}$$

In this way, to every $D^{(\alpha)}$ there is assigned a definite spin S: $D^{[\tilde{\lambda}]}$. The character of $a \in \mathscr{G}$ can easily be given if in the molecular models we restrict ourselves to orbitals of the same type ($\sigma, \pi, \ldots$ electrons only), which transform into each other under $a \in \mathscr{G}$. Then no further operation of $\mathscr{G}$ in the space of the orbitals is necessary. In this case the character which is assigned to the permutation symmetry $[\lambda]$ is given by

[10] If the lines for the saturation of spins were to cross, the spatial wave function would have such a configuration that no binding could arise.

Table 7.10. Number $R(6,S)$ of possible covalent structures with spin S of the hypothetical molecule H_6

$[\tilde{\lambda}] = [\tilde{\lambda}_1, \tilde{\lambda}_2]$	[6,0]	[5,1]	[4,2]	[3,3]
S	3	2	1	0
$R(6,S)$	1	5	9	5
$[\lambda]$	$[1^6]$	$[2,1^4]$	$[2^2,1^2]$	$[2^3]$

Table 7.11. Assignment of the elements of $\mathscr{P}_6'$ to $\mathscr{D}_6$, together with the characters and decompositions

$\mathscr{P}_6' \cong \mathscr{D}_6$	e e (1^6)	$2c_6$ (123456) (6)	$2c_3$ (135)(246) (3^2)	c_2 (14)(25)(36) (2^3)	$3c_2'$ (26)(35) $(1^2,2^2)$	$3c_2''$ (12)(36)(45) (2^3)	Decomposition
$\chi^{[1^6]\,\mathrm{sub}}$	1	-1	1	-1	1	-1	${}^7B_1$
$\chi^{[2,1^4]\,\mathrm{sub}}$	5	1	-1	1	1	1	${}^5A_1 \oplus {}^5E_1 \oplus {}^5E_2$
$\chi^{[2^2,1^2]\,\mathrm{sub}}$	9	0	0	-3	1	-3	${}^3A_2 \oplus 2\cdot{}^3B_1 \oplus 2\cdot{}^3E_1 \oplus {}^3E_2$
$\chi^{[2^3]\,\mathrm{sub}}$	5	0	2	3	1	3	$2\cdot{}^1A_1 \oplus {}^1B_2 \oplus {}^1E_2$

$$\chi^{[\lambda]\,\mathrm{sub}}(a) = \chi^{[\lambda]}(p)\ , \qquad p \in \mathscr{P}_m'\ , \qquad a \in \mathscr{G}\ . \tag{7.3.22}$$

Using this, (7.3.21) can be reduced.

As an example we consider a hypothetical molecule with $\mathscr{D}_{6h}$ symmetry and six s electrons (H_6 or C_6H_6 with six π-electrons only; the H_6-model is unstable: $H_6 \to 3H_2$). According to (7.3.17), $Z = 924$ and with (7.3.20) we have the covalent structures with possible spins S given in Table 7.10; altogether 20 covalent structures. The other 904 structures have ionic contributions. We consider only the covalent ones. The geometrical symmetry can be taken as being $\mathscr{D}_6$, since the s functions are invariant with respect to σ_h. The assignment of the elements from $\mathscr{P}_6' \subset \mathscr{P}_6$ to $\mathscr{D}_6$, together with the characters, can be taken from Table 7.11, which also includes the decompositions. The decompositions also include the multiplet indices, so we can see directly the multiplet structure ${}^{2S+1}D^{(\alpha)}$. The dimension $m_{\alpha,s}$ of the resulting secular equations is at most 2, that means the originally 20-dimensional problem of the covalent structures reduces to 8 one- and 3 two-dimensional ones, some of which are twofold degenerate (E_1, E_2). The molecule multiplet functions in the valence-bond approximation symmetrized according to $\mathscr{G} \times \mathscr{P}_m$ are obtained by using the projection operators again. Restricting ourselves again to the covalent configuration $\psi_1(1)\ldots\psi_m(m)$, we get the functions $\Psi_{jl}^{[\lambda]}$ by projection with $P_{jl}^{[\lambda]}$ of $\mathscr{P}_m$. Furthermore, from (7.3.21) we can learn which projection operators $P_{ik}^{(\alpha)}$ of the point group $\mathscr{G}$ have to be applied to the $\Psi_{jl}^{[\lambda]}$ in order to get the symmetry-adapted functions $\Psi_{jl,ik}^{[\lambda](\alpha)}$ of $\mathscr{G} \times \mathscr{P}_m$.

$$\Psi_{jl,ik}^{[\lambda](\alpha)} \sim P_{ik}^{(\alpha)}\Psi_{jl}^{[\lambda]} = \sum_{a\in\mathscr{G}} D_{ik}^{(\alpha)*}(a)P_a\Psi_{jl}^{[\lambda]}$$

$$= \sum_{a\in\mathscr{G}}\sum_{n} D_{ik}^{(\alpha)*}(a)D_{nl}^{[\lambda]}(p)\Psi_{jn}^{[\lambda]} , \qquad (7.3.23)$$

where P_a has been replaced by the corresponding permutation p of the AOs (see Exercise 7.12), and

$$p\Psi_{jl}^{[\lambda]} = \sum_n D_{nl}^{[\lambda]}(p)\Psi_{jn}^{[\lambda]}$$

has been used. The functions in (7.3.23) are determined only to within a normalization constant.

7.3.3 The H_2 Molecule

The hydrogen molecule is a well-known example which we will nevertheless discuss in more detail here, too (Fig. 7.12). The H_2 molecule has the symmetry groups $\mathscr{G} = \mathscr{D}_{\infty h} = \mathscr{C}_{\infty v} \times \mathscr{C}_i$ and $\mathscr{P}_m = \mathscr{P}_2$. For the ground state we only consider the 1*s* functions localized at the H atoms *A* and *B*. The AOs are s^A and s^B in the valence-bond method. The projection operators of $\mathscr{P}_2$ are $P^{[2]} = (1/\sqrt{2})[e + (12)]$ and $P^{[1^2]} = (1/\sqrt{2})[e - (12)]$; those of $\mathscr{D}_{\infty h}$ are $P^{(\alpha)}(\mathscr{C}_i)P^{(\beta)}(\mathscr{C}_{\infty v})$. Thus the *covalent configuration* $s^A(1)\cdot s^B(2)$ has the normalized functions $\Psi_{jl}^{[\lambda]}$ $(j = l = 1)$

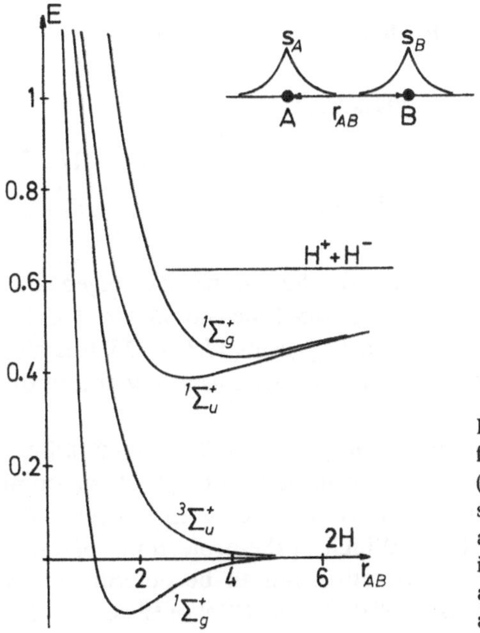

Fig. 7.12. Σ states of the H_2 molecule as a function of nuclear distance r_{AB}. Three states $(2\,^1\Sigma_g^+, {}^1\Sigma_u^+)$ are bonding states. The lower two states are mainly covalent, the upper two states are mainly ionic. The covalent states dissociate into two (neutral) H atoms, the ionic ones into an H^- ion and a proton (H^+). Units of E and r_{AB} are e^2/a_B and a_B, respectively

$$\Psi^{[2]} = \frac{1}{\sqrt{2(1+S^2)}}[s^A(1)s^B(2) + s^B(1)s^A(2)]$$

$$\leftrightarrow [\tilde{2}] \sim [1^2] \rightarrow S = 0 \ , \qquad \text{singlet} \ ,$$

$$\Psi^{[1^2]} = \frac{1}{\sqrt{2(1+S^2)}}[s^A(1)s^B(2) - s^B(1)s^A(2)] \qquad (7.3.24a)$$

$$\leftrightarrow [\widetilde{1^2}] \sim [2] \rightarrow S = 1 \ , \qquad \text{triplet} \ .$$

Here the spin states represented by the associated tableaux are also given; $S = \langle s^A | s^B \rangle$ is the overlap integral. Symmetrizing these functions with respect to $\mathscr{D}_{\infty h}$ according to (7.3.23) gives the functions $\Psi_{ij}^{[\lambda](\alpha)}$ belonging to $\mathscr{D}_{\infty h} \times \mathscr{P}_2$. Since the s functions already belong to the $\Sigma^+(A_1)$ REP of $\mathscr{C}_{\infty v}$ (Table A.16), we only have to apply the projectors of $\mathscr{C}_i$ to (7.3.24). Then the functions reproduce themselves, so that

$$\Psi^{[2]}(\Sigma_g^+) := \Psi(^1\Sigma_g^+) = \Psi^{[2]} \ ,$$

$$\Psi^{[1^2]}(\Sigma_u^+) := \Psi(^3\Sigma_u^+) = \Psi^{[1^2]} \ . \qquad (7.3.24b)$$

The *ionic configuration* only allows for $s^A(1)s^A(2)$ and $s^B(1)s^B(2)$. The spin function then belongs to $[\tilde{\lambda}] = [1^2] = [\tilde{2}]$ with $S = 0$ (singlet), thus the spatial function belongs to $[\lambda] = [2]$. The symmetrizing procedure first gives

$$\Psi^{[2]}(A) = s^A(1)s^A(2) \ , \qquad \Psi^{[2]}(B) = s^B(1)s^B(2) \ ,$$

and from this with the projection operators of $\mathscr{C}_i$

$$\Psi^{[2]}(^1\Sigma_g^+) = \frac{1}{\sqrt{2(1+S^2)}}[s^A(1)s^A(2) + s^B(1)s^B(2)]$$

$$\Psi^{[2]}(^1\Sigma_u^+) = \frac{1}{\sqrt{2(1+S^2)}}[s^A(1)s^A(2) - s^B(1)s^B(2)] \qquad (7.3.25)$$

(both the starting functions result in the same set). The symmetrized eigenvalue problem according to the CI method leads to a two-dimensional problem for $^1\Sigma_g^+$ and two one-dimensional ones for $^3\Sigma_u^+$ and $^1\Sigma_u^+$ using (7.3.24, 25) (see also Sect. 6.3). The resulting eigenvalues are shown in Fig. 7.1.2 as a function of the distance AB.

With this example we can also discuss the MO method mentioned in Sect. 7.3.2. Here it is convenient to apply the projection operator $P_{ik}^{(\alpha)}$ of the point group $\mathscr{G}$ to a product of MOs which is *not* yet symmetrized with respect to $\mathscr{P}_m$ but already symmetrized with respect to $\mathscr{G}$. Which of the projection operators $P_{ik}^{(\alpha)}$ of $\mathscr{G}$ have to be used for the symmetrization has to be determined by reduction of the (inner direct) product REP $D^{(\alpha_1)} \otimes \cdots \otimes D^{(\alpha_m)}$ of $\mathscr{G}$ constituted by the MOs $\{\psi_i^{(\alpha)}\} := \{\psi^{\alpha_1}(1)\psi^{\alpha_2}(2)\ldots\psi^{\alpha_m}(m)\}$. Finally, the projection operators $P_{jl}^{[\lambda]}$ of $\mathscr{P}_m$ are applied. For the H_2 molecule the s functions are already basis functions belonging to $\Sigma^+ = A_1$ of $\mathscr{C}_{\infty v}$. The $\mathscr{D}_{\infty h}$ symmetrization thus gives

$$\psi(\Sigma_g^+) \sim P^{\Sigma_g^+}(\mathscr{D}_{\infty h})s_A \sim P^{(g)}(\mathscr{C}_i)\cdot P^{(A_1)}(\mathscr{C}_{\infty v})s^A$$
$$\sim P^{(g)}(\mathscr{C}_i)s^A = \frac{1}{\sqrt{2(1+S)}}(s^A + s^B)$$

and correspondingly

$$\psi(\Sigma_u^+) = \frac{1}{\sqrt{2(1-S)}}(s^A - s^B) \ . \tag{7.3.26}$$

From these MOs we are able to form the configurations

$$[\psi(\Sigma_g^+)]^2 \ , \qquad [\psi(\Sigma_u^+)]^2 \ , \qquad \psi(\Sigma_g^+)\psi(\Sigma_u^+) \ . \tag{7.3.27}$$

The assigned product REPs have the reductions

$$\Sigma_g^+ \otimes \Sigma_g^+ = \Sigma_g^+ \ , \qquad \Sigma_u^+ \otimes \Sigma_u^+ = \Sigma_g^+ \ , \qquad \Sigma_g^+ \otimes \Sigma_u^+ = \Sigma_u^+ \ , \tag{7.3.28}$$

i.e. we can use the projection operators $P^{\Sigma_g^+}$ and $P^{\Sigma_u^+}$ of $\mathscr{D}_{\infty h}$ and then again we obtain the functions (7.3.27) because of (7.3.28). The Young diagrams for the spatial and spin states assigned to the configurations (7.3.27) are

$$[\psi(\Sigma_g^+)]^2 : [\lambda] = [2] \leftrightarrow [\tilde{\lambda}] = [1^2] \rightarrow S = 0 \ ,$$
$$[\psi(\Sigma_u^+)]^2 : [\lambda] = [2] \leftrightarrow [\tilde{\lambda}] = [1^2] \rightarrow S = 0 \ , \tag{7.3.29}$$
$$\psi(\Sigma_g^+)\psi(\Sigma_u^+) : [\lambda] = [2] \ , \qquad [1^2] \leftrightarrow [\tilde{\lambda}] = [1^2] \ , \qquad [2] \rightarrow S = 0, 1 \ .$$

Thus we have three singlet functions and one triplet MO function when applying the projection operators of $\mathscr{P}_2$. Specifically, using the same order as in (7.3.27–29) and $s^A = A$, $s^B = B$,

$$\Psi(^1\Sigma_g^+) = P^{[2]}\psi(^1\Sigma_g^+, 1)\psi(^1\Sigma_g^+, 2)$$
$$= \frac{1}{2(1+S)}[A(1)A(2) + B(1)B(2) + A(1)B(2) + A(2)B(1)] \ ,$$

$$\Psi(^1\Sigma_g^+) = \frac{1}{2(1-S)}[A(1)A(2) + B(1)B(2) - A(1)B(2) - A(2)B(1)] \ ,$$

$$\Psi(^1\Sigma_u^+) = \frac{1}{\sqrt{2(1-S^2)}}[A(1)A(2) - B(1)B(2)] \ , \tag{7.3.30}$$

$$\Psi(^3\Sigma_u^+) = \frac{1}{\sqrt{2(1-S^2)}}[A(1)B(2) - B(1)A(2)] \ .$$

The energy scheme resulting from this method is completely identical with that of the valence bond method, which again emphasizes the equivalence of both methods if one starts with the same set of one-particle basis functions (Sect. 7.2.2). The transformation between (7.3.30) and (7.3.24a, 25) can be seen directly (Exercise 7.13).

Exercise 7.11. Discuss how many functions in the form of a determinant can be formed from 2, 6, 10 AOs and MOs, respectively, if there are $m = 2, 3, 4, 5, 6$ electrons in the system.

Exercise 7.12. Prove (7.3.23) by taking into account that the permutation of quantum numbers (p^{-1}) is inverse to that of the particle numbers (p); see also (5.5.21).

Exercise 7.13. Give the transformation between valence bond functions (7.3.24a, 25) and MO functions (7.3.30) directly.

8. Selection Rules and Matrix Elements

The basis for the investigation of matrix elements and selection rules lies in the application of the Wigner-Eckart theorem. This is illustrated in this chapter for tensor (scalar, vector) operators. Examples are connected with the Jahn-Teller effect, including spin and time reversal symmetry, radiative transitions between different energy levels, the splitting of energy levels in a crystalline field and the Stark and Zeeman effects. Section 8.5 deals with the determination of independent components of a material tensor (a tensor describing physical properties).

8.1 Selection Rules of Tensor Operators

In Sects. 6.3 and 6.4 we used a special case of the Wigner-Eckart theorem in which the operators are invariant quantities. In this case only matrix elements between states of the same IR are different from zero. In general we have to consider matrix elements of tensor operators T with states $|\gamma, l\rangle$ and $|\alpha, i\rangle$ (Sect. 6.2). The components of T transform according to a reducible REP of $\mathcal{G}$ which we can look upon as being decomposed into the IRs $D^{(\beta)}$ with the basis elements $T_j^{(\beta)}, j = 1, \ldots, d_\beta$ (Sects. 5.5, 6.1, 6.2). Thus we have to investigate the matrix elements

$$\langle\gamma, l|\, T_j^{(\beta)}\, |\alpha, i\rangle \tag{8.1.1}$$

from (6.2.5), where α, β, γ are IRs of $\mathcal{G}$. The selection rules tell us when (8.1.1) will be different from zero. The Wigner-Eckart theorem (6.2.5) provides a necessary condition: Expression (8.1.1) is different from zero only if the IR $D^{(\gamma)}$ occurs in the reduction of the inner direct product

$$D^{(\beta)} \otimes D^{(\alpha)} = \sum_{\mu \oplus} (\beta\alpha|\mu) \cdot D^{(\mu)} \ ,$$

where, see (4.4.14),

$$(\beta\alpha|\mu) = \frac{1}{g} \sum_{a \in \mathcal{G}} \chi^{(\beta)}(a)\chi^{(\alpha)}(a)\chi^{(\mu)*}(a) \ . \tag{8.1.2}$$

Sometimes it is convenient to use the equivalent formulation: The expression (8.1.1) is different from zero only if the inner direct product $D^{(\gamma)*} \otimes D^{(\beta)} \otimes D^{(\alpha)}$

contains the unit REP $D^{(1)}$ at least once. Often $D^{(1)}$ is denoted as A_1, A_{1g}, etc., and for permutation groups $\mathscr{P}_n$, $[\lambda] = [n]$. In the following we investigate some matrix elements relevant to several physical problems.

8.2 The Jahn-Teller Theorem

8.2.1 Spinless States

The *static*[1] *Jahn-Teller theorem* tells us whether a physical system (molecule, ionic complex, crystal) is stable under static displacements of the particles (atoms, nuclei) or not. It also states with which changes of symmetry such displacements are compatible. The displacements are caused by the coupling between electronic and vibrational states of the system.

As in Sect. 7.1 we denote the displacements of the nuclei (ion cores) from the equilibrium positions $\boldsymbol{X} = \{\dots X_i^n \dots\}$ by $\boldsymbol{u} = \{\dots u_i^n \dots\}$, thus

$$H = H_0(\boldsymbol{x}, \boldsymbol{X}) + H_1(\boldsymbol{x}, \boldsymbol{u}) \ ;$$

$$H_1(\boldsymbol{x}, \boldsymbol{u}) = \sum_{n,i} \left.\frac{\partial H}{\partial u_i^n}\right|_X u_i^n + \cdots \tag{8.2.1a}$$

is the Hamiltonian of an electron (coordinate $\boldsymbol{x}$) in the field of the ion cores at the positions $\boldsymbol{X} + \boldsymbol{u}$. The term H_1 couples the motion of electron and ion cores. Transforming to symmetry coordinates according to (7.1.12), we have

$$H_1(\boldsymbol{x}, Q_k^{\alpha s}) = \sum_{\alpha s k} \left.\frac{\partial H}{\partial Q_k^{\alpha s}}\right|_0 Q_k^{\alpha s} \ . \tag{8.2.1b}$$

Higher terms in the expansion of H_1 will be neglected. Since the $Q_k^{\alpha s}$ transform according to the IR $D^{(\alpha)}$ of $\mathscr{G}$, and H according to $D^{(1)}$ of $\mathscr{G}$, $\partial H/\partial Q_k^{\alpha s}$ has to transform according to the complex conjugate REP $D^{(\alpha)*}$ of $\mathscr{G}$.

Now, let E_0 be an eigenvalue of H_0 (ground state) and $E_0 + \Delta E$ an eigenvalue of $H_0 + H_1$, then the system is stable against a displacement and the symmetry change belonging to it if and only if $\Delta E = 0$. Within perturbation theory (Sect. 6.4) this means that the matrix elements

$$\langle \gamma, s_\gamma, l | \partial H/\partial Q_k^{\beta s_\beta} | \alpha, s_\alpha, i \rangle \tag{8.2.2}$$

vanish ($|\alpha, s_\alpha, i\rangle$ are eigenfunctions of H_0). If (8.2.2) vanishes for *all* the symmetry coordinates $Q_k^{\alpha s}$, and thus $\Delta E = 0$, then the system is stable against an arbitrary

[1] In the *dynamical* Jahn-Teller effect a change of symmetry is due to the motion of the system between equivalent stable positions, in which the electronic states are *nondegenerate*. This case will not be considered here.

displacement of the particles. We therefore have to investigate whether or not

$$D^{(\gamma)} \text{ is contained in } D^{(\beta)} \otimes D^{(\alpha)}, \text{ or } D^{(1)} \text{ in } D^{(\gamma)*} \otimes D^{(\beta)} \otimes D^{(\alpha)} \ . \tag{8.2.3}$$

We will distinguish two cases:

(i) E_0 is *nondegenerate*. Then $D^{(\gamma)} = D^{(\alpha)} = D^{(1)}$ and $D^{(\gamma)*} \otimes D^{(\beta)} \otimes D^{(\alpha)} = D^{(\gamma)*} \otimes D^{(\gamma)} \otimes D^{(\beta)} = D^{(\beta)}$. Thus $\Delta E = 0$ for all the displacements $Q_k^{\beta s}$, with the exception of the total-symmetric displacements Q_k^{1s} belonging to the identity REP. The displacement pattern of the particles for Q_k^{1s} has the total symmetry of the group $\mathscr{G}$. This means for a nondegenerate energy level (ground state) that no displacement of the particles can lead to a reduction of the symmetry of the structure.

(ii) E_0 is *degenerate*. Then according to (6.4.5) the matrix element (8.2.2) is diagonal with respect to α, i.e. $\langle \alpha, s_\alpha, l | \partial H/\partial Q_k^{\beta s} \beta | \alpha, s'_\alpha, i \rangle$. Again we have to distinguish between several cases. Let, for example, $D^{(\alpha)}$ be real and let the matrix element be symmetric with respect to i, l. Then the selection rule is: The matrix element is different from zero if $D^{(\beta)}$ occurs in the symmetrized product (4.4.17), $[D^{(\alpha)} \otimes D^{(\alpha)}]_+$. Correspondingly, if there is an antisymmetry in i, l, we have to look for the occurrence of $D^{(\beta)}$ in $[D^{(\alpha)} \otimes D^{(\alpha)}]_-$. Which of these cases is present depends on the approximations used in (8.2.1, 2), (see also the next section). However, if we do not consider any spin dependence, i.e. E_0 has *only spatial degeneracy*, then the symmetric case is present since the perturbation $\partial H/\partial Q_k^{\beta s}$ is Hermitian and time reversal invariant. Consequently we have to ask whether $D^{(1)}$ occurs in $D^{(\beta)} \otimes [D^{(\alpha)} \otimes D^{(\alpha)}]_+$. According to Jahn and Teller [8.1] there exists [except for linear molecules (systems) and for Kramers degeneracy, see end of Sect. 5.2 and Sect. 8.4.2] an asymmetric displacement $Q_k^{\beta s}$ so that the molecule (system) is unstable in the more symmetric structure and changes with $Q_k^{\beta s}$ into the asymmetric structure. Thus the degeneracy is lifted.

As an example we consider the complex ML_6 of Fig. 7.9. With the inclusion of the d-electrons in the $\mathscr{O}_h$ symmetry we have the 3-fold spatially degenerate $^2F_{2g}$ ground state (see also Sect. 8.4.2). In order to find the displacements which raise the degeneracy of the unstable $^2F_{2g}$ ground state and reduce (lower) the $\mathscr{O}_h$ symmetry, we reduce the symmetrized product (Fig. 8.1)

$$[F_{2g} \otimes F_{2g}]_+ = A_{1g} \oplus E_g \oplus F_{2g} \ . \tag{8.2.4}$$

Both E_g and F_{2g} can cause a Jahn-Teller displacement. The E_g displacement for example leads to a stretched or compressed octahedron with $\mathscr{D}_{4h}$ symmetry. In order to get the splitting of the F_{2g} levels, we have to perform the subduction $\mathscr{O}_h$ into $\mathscr{D}_{4h}$, thus

$$F_{2g}^{\text{sub}}(\mathscr{O}_h) = B_{2g}(\mathscr{D}_{4h}) \oplus E_g(\mathscr{D}_{4h}) \ . \tag{8.2.5}$$

Whether this displacement or that of F_{2g} occurs depends on the perturbation. If

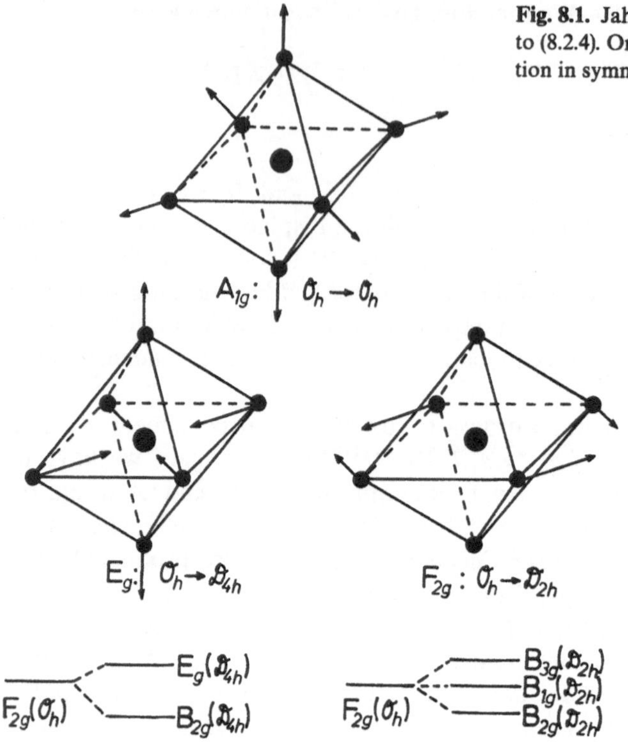

Fig. 8.1. Jahn-Teller distortions according to (8.2.4). Only E_g and F_{2g} lead to a reduction in symmetry

we take into account the electronic spin we have to consider time reversal symmetry, which leads to statements different from the spinless case.

8.2.2 Time Reversal Symmetry

From the discussion in Sect. 5.2 it is clear that time reversal symmetry implies an extension of the selection rules. In classical physics, time reversal (or reversal of motion) means that, with $\boldsymbol{x}(t)$ and $\boldsymbol{p}(t)$, the inversion of the motion, $\boldsymbol{x}(-t)$ and $-\boldsymbol{p}(-t)$, is also a solution of the equation of motion. This is satisfied if the Hamiltonian is an even function of $\boldsymbol{p}$:

$$H(\boldsymbol{x},\boldsymbol{p}) = H(\boldsymbol{x},-\boldsymbol{p}) \ . \tag{8.2.6a}$$

In quantum theory the corresponding condition is in the $\boldsymbol{x}$- (*real-space, Schrödinger*) *representation* with $\boldsymbol{p} = -\mathrm{i}\hbar\partial/\partial\boldsymbol{x}$

$$H\left(x, \frac{\hbar}{\mathrm{i}}\frac{\partial}{\partial x}\right) = H\left(x, -\frac{\hbar}{\mathrm{i}}\frac{\partial}{\partial x}\right) \ ; \tag{8.2.6b}$$

here we can take the operator ϑ_0 of complex conjugation as that for the reversal of momentum

$$\vartheta_0 \frac{\hbar}{\mathrm{i}} \frac{\partial}{\partial x} \vartheta_0^{-1} = -\frac{\hbar}{\mathrm{i}} \frac{\partial}{\partial x} \ . \tag{8.2.7a}$$

Let ψ be an eigenfunction of H with (real) eigenvalue E, then $\vartheta_0 \psi \equiv \psi^*$ is an eigenfunction of H with eigenvalue E, too. This holds since H in the $\boldsymbol{x}$-REP is real because of (8.2.6), i.e. $H\vartheta_0 - \vartheta_0 H = 0$. We can also define the time reversal operator by

$$\vartheta_0 \boldsymbol{x} \vartheta_0^{-1} = \boldsymbol{x} \ , \qquad \vartheta_0 \boldsymbol{p} \vartheta_0^{-1} = -\boldsymbol{p} \ , \qquad \vartheta_0^2 = 1 \tag{8.2.7b}$$

independently of any REP. The angular momentum then satisfies

$$\vartheta_0 \boldsymbol{L} \vartheta_0^{-1} = -\boldsymbol{L} \ . \tag{8.2.7c}$$

We can finally transfer this definition to systems with spin $\boldsymbol{\sigma}$ and describe the time reversal operator ϑ by

$$\vartheta \boldsymbol{x} \vartheta^{-1} = \boldsymbol{x} \ , \qquad \vartheta \boldsymbol{p} \vartheta^{-1} = -\boldsymbol{p} \ , \qquad \vartheta \boldsymbol{\sigma} \vartheta^{-1} = -\boldsymbol{\sigma} \ . \tag{8.2.8}$$

The Hamiltonian of a system with spin-orbit coupling, for example,

$$H = \frac{p^2}{2m} + V(\boldsymbol{x}) + \frac{\hbar}{8m^2c^2} \frac{1}{r} \frac{dV}{dr} (\boldsymbol{L} \cdot \boldsymbol{\sigma}) \tag{8.2.9}$$

is then invariant under time reversal, i.e. $H\vartheta - \vartheta H = 0$. On switching on an external magnetic field this invariance is destroyed by the term $\alpha \boldsymbol{\sigma} \cdot \boldsymbol{B}$ if the direction of $\boldsymbol{B}$ is fixed.

The general operator ϑ can be written as a product of a (unitary) spin operator $\mathscr{U}$ with the (antiunitary, antilinear) operator ϑ_0 [8.2]:

$$\vartheta = U\vartheta_0 \ , \qquad U = \prod_{l=1}^{m} \sigma_y(l) \ , \qquad \vartheta^2 = (-1)^m \ . \tag{8.2.10}$$

Again, m is the number of particles with spin 1/2 (electrons) in the system. Because of the antiunitarity (5.2.1b) it follows for a tensor matrix element T (an arbitrary reducible or irreducible tensor component) that

$$\langle \psi_i^\gamma | T | \psi_i^\alpha \rangle = \langle \vartheta \psi_i^\alpha | \vartheta (T^+ \psi_i^\gamma) \rangle = \langle \vartheta \psi_i^\alpha | \hat{T} | \vartheta \psi_i^\gamma \rangle \qquad \text{with } \hat{T} := \vartheta T^+ \vartheta^{-1} \ . \tag{8.2.11}$$

We obtain different selection rules according to whether T is symmetric or antisymmetric with respect to ϑ, i.e. according to the sign in

$$\hat{T} = \vartheta T^+ \vartheta^{-1} = \pm T \ , \ T\text{: Hermitean} \ . \tag{8.2.12}$$

Neglecting any spin interaction, e.g. the spin-orbit coupling, we can use ϑ_0 instead of ϑ. In (8.2.11) we consider states belonging to the same IR $D^{(\alpha)}$, e.g. energetically degenerate states. Therefore we take $\psi_l^\gamma = \vartheta_0\psi_l^\alpha$. Using (8.2.11, 12) we then have[2]

$$\langle \vartheta_0\psi_l^\alpha | T | \psi_i^\alpha \rangle = \langle \vartheta_0\psi_i^\alpha | \pm T | \vartheta_0\vartheta_0\psi_l^\alpha \rangle = \pm \langle \vartheta_0\psi_i^\alpha | T | \psi_l^\alpha \rangle \ , \tag{8.2.13a}$$

i.e. depending upon the sign, we have symmetry or antisymmetry in i and l. Thus investigating the selection rules we have to use the REPs $[D^{(\alpha)} \otimes D^{(\alpha)}]_\pm$ or in the corresponding reductions the characters (4.4.16)

$$\chi_\pm^{(\alpha\times\alpha)}(a) = \tfrac{1}{2}\{[\chi^{(\alpha)}(a)]^2 \pm \chi^{(\alpha)}(a^2)\} \ . \tag{8.2.14a}$$

Taking into account a spin interaction we have to use $\psi_l^\gamma = \vartheta\psi_l^\alpha$, and then in analogy to (8.2.13a) we obtain

$$\langle \vartheta\psi_l^\alpha | T | \psi_i^\alpha \rangle = \pm \vartheta^2 \langle \vartheta\psi_i^\alpha | T | \psi_l^\alpha \rangle \ , \tag{8.2.13b}$$

where ϑ^2 is given by (8.2.10). The symmetry in i and l is then determined by $\pm(-1)^m$, where the sign $(\pm)$ is fixed by the behaviour of T with time reversal according to (8.2.12).

In the case of the Jahn-Teller effect, T is equal to $\partial H/\partial Q_k^{\beta s}$ (see Sect. 8.2.1) and thus we have to take the + sign in (8.2.12) and

$$\chi_\pm^{(\alpha\times\alpha)}(a) = \tfrac{1}{2}\{[\chi^{(\alpha)}(a)]^2 + (-1)^m \chi^{(\alpha)}(a^2)\} \tag{8.2.14b}$$

instead of (8.2.14a). These considerations also hold for crystals if the corresponding symmetry coordinates are used; this is true for localized electronic states as well as for Bloch states.

Exercise 8.1. Discuss the Jahn-Teller effect for a spatially degenerate E state in $\mathscr{C}_{3v}$ symmetry.

8.3 Radiative Transitions

The transition probability from a state $|a\rangle$ into a state $|e\rangle$ of the unperturbed system is given to first approximation by

$$W_{a\to e} = \frac{2\pi}{\hbar} |\langle e | H_1 | a \rangle|^2 \delta(E_a - E_e) \ ; \tag{8.3.1}$$

H_1 describes the interaction of the system with a (electromagnetic) field $\boldsymbol{A}(\boldsymbol{x}, t)$ which can be assumed to be

$$H_1 = -\frac{e}{c} \int \boldsymbol{j}(\boldsymbol{x}) \cdot \boldsymbol{A}(\boldsymbol{x}, t)\, d\boldsymbol{x} \tag{8.3.2a}$$

---

[2] According to Sect. 5.2 the energy degeneracy doubles when time reversal symmetry is considered, in cases (1b) and (2), whereas in case (1a) there is no doubling. Thus in case (1a), ψ_i^α and $\vartheta_0\psi_i^\alpha$ are linearly dependent and instead of $\langle \psi_l^\alpha | T | \psi_i^\alpha \rangle$ we can investigate whether $\langle \vartheta_0\psi_l^\alpha | T | \psi_i^\alpha \rangle$ is zero.

with the current operator

$$j(x) = \frac{1}{2m_e} \sum_{i=1}^{m} [p_i \delta(x - x_i) + \delta(x - x_i) p_i] , \tag{8.3.2b}$$

where m is the number of electrons in the system. In general $A(x, t)$ and $j(x, t)$ are decomposed into their Fourier components, so that the relevant transition matrix elements have the form

$$e_\sigma \cdot \langle e | j_k | a \rangle := e_\sigma \cdot \int e^{ik \cdot x} \langle e | j(x) | a \rangle \, dx \tag{8.3.3a}$$

with k as the wave number and e_σ as the (transverse) polarization vector. We are interested in the cases in which the matrix elements are different from zero. It is usual to expand $\exp(ik \cdot x)$ (about the centre of an atom) and to consider one term only, since $k \cdot x \ll 1$ for $r_{at} \ll \lambda$ in the area of integration. The zero-order term

$$e_\sigma \cdot \langle e | j_0 | a \rangle \sim e_\sigma \cdot \langle e | \sum_{i=1}^{m} x_i | a \rangle \tag{8.3.3b}$$

describes electric dipole $E1$ transitions of the system, the next term linear in $k \cdot x$ describes magnetic dipole $M1$ and electric quadrupole $E2$ transitions, and so on. In the $M1$ case the matrix element contains the magnetic moment $M = (e/2mc)L$ of the spatial movement, in the $E2$ case the quadrupole tensor $\sum_i x_i y_i$, $\sum_i y_i z_i$, $\ldots, \sum_i x_i x_i, \ldots$:

$$M1\colon (k \times e_\sigma) \cdot \langle e | L | a \rangle , \qquad E2\colon \langle e | \sum_i^m (k \cdot x_i)(e_\sigma \cdot x_i) | a \rangle .$$

As an example we consider transitions with $\mathscr{C}_{3v} \cong \mathscr{D}_3$ symmetry (Sect. 6.2). The decomposition of products of REPs according to Table 4.6 is

$$\begin{aligned} &A_1 \otimes A_1 = A_1 , && A_2 \otimes A_2 = A_1 , \\ &A_1 \otimes A_2 = A_2 , && A_2 \otimes E = E , \\ &A_1 \otimes E = E , && E \otimes E = A_1 \oplus A_2 \oplus E . \end{aligned} \tag{8.3.4}$$

The electric dipole operator transforms according to the polar vector REP $D^{(v)}$, and the magnetic dipole operator according to the axial vector REP $D^{(A)}$. These are given for $\mathscr{C}_{3v}$ symmetry by

$$D^{(v)}(\mathscr{C}_{3v}) = A_1 \oplus E , \qquad D^{(A)}(\mathscr{C}_{3v}) = A_2 \oplus E . \tag{8.3.5a}$$

The trace of the quadrupole tensor is zero and the tensor transforms according to the IR $D^{(l=2)}$ of $\mathscr{SO}(3)$, for which

$$D^{(l=2)}(\mathscr{C}_{3v}) = A_1 \oplus E \oplus E \tag{8.3.5b}$$

(see Sect. 5.5.3 and Exercise 8.2).

*E*1 *transitions.* According to (8.3.5a) we have to reduce $A_1 \otimes D^{(\alpha)}$ and $E \otimes D^{(\alpha)}$ and look to see whether $D^{(\gamma)}$ is contained in that decomposition (α, γ are IRs of $\mathscr{C}_{3v}$). Using (8.3.4) and Table 5.2 the allowed transitions are: a) radiation polarized parallel to z (A_1): $A_1 \leftrightarrow A_1$, $A_2 \leftrightarrow A_2$, $E \leftrightarrow E$; b) radiation polarized parallel to the xy-plane (E): $A_1 \leftrightarrow E$, $A_2 \leftrightarrow E$, $E \leftrightarrow E$. The transition $A_1 \leftrightarrow A_2$ is the only forbidden one.

*M*1 *transitions.* We have to reduce $A_2 \otimes D^{(\alpha)}$ and $E \otimes D^{(\alpha)}$. Allowed transitions are: a) radiation polarized parallel to z (A_2): $A_1 \leftrightarrow A_2$, $E \leftrightarrow E$; b) radiation polarized parallel to the xy-plane (E): $A_1 \leftrightarrow E$, $A_2 \leftrightarrow E$, $E \leftrightarrow E$. Forbidden are $A_1 \leftrightarrow A_1$ and $A_2 \leftrightarrow A_2$. For groups which do not contain any mirror element, polar and axial vectors cannot be distinguished. Then the selection rules are the same for *E*1 and *M*1 transitions.

*E*2 *radiation.* Because of (8.3.5b), for $\mathscr{C}_{3v}$ symmetry the selection rules for *E*1 and *E*2 transitions are the same, apart from the intensities.

Higher multipoles, that is terms with $(\boldsymbol{k} \cdot \boldsymbol{x})^{\nu > 1}$ in (8.3.3a), can be treated analogously. In the case of a 2^l-multipole one has first to decompose the REP $D^{(l)}$ of $\mathscr{SO}(3)$ into the IRs of the corresponding point group and then to proceed as above. In many-particle problems the Hamiltonian H, the momentum $\boldsymbol{P}$ and the angular momentum $\boldsymbol{L}$ are invariant under particle permutations. This leads to a *selection rule* according to this *permutation symmetry.* The operators H, $\boldsymbol{P}$, $\boldsymbol{L}$ transform according to the identity REP $D^{[m]}$ of $\mathscr{P}_m$; the matrix elements have the form

$$\langle \psi_i^{[\lambda_1]} | A^{[m]} | \psi_l^{[\lambda_2]} \rangle \propto \delta_{\lambda_1 \lambda_2} \delta_{il} \ , \tag{8.3.6}$$

i.e. a symmetric operator $A^{[m]}$ only induces transitions between states of the same permutation symmetry. Therefore transitions between, for example, symmetric (bosons, $\Psi^{[m]}$) and antisymmetric (fermions, $\Psi^{[1^m]}$) states are forbidden. This means that the Hilbert space can be separated into two subspaces between which there are no transitions.

All these statements concern transitions in the first perturbation approximation (8.3.1). To second order the transition probability is given by

$$W_{a \to e} = \frac{2\pi}{\hbar} \left| \sum_z \frac{\langle e | H_1 | z \rangle \langle z | H_1 | a \rangle}{E_a - E_z + i\hbar\varepsilon} \right|^2 \delta(E_e - E_a) \ , \qquad \varepsilon \to 0 \ . \tag{8.3.7}$$

In order to get the selection rules we have to discuss the products of matrix elements $\langle e | H_1 | z \rangle$ and $\langle z | H_1 | a \rangle$ simultaneously. Let the wave functions transform according to $D^{(e)}$, $D^{(a)}$, $D^{(z)}$, and the operator H_1 according to the tensor operator $D^{(T)}$, then

$$\begin{aligned} &\langle z | H_1 | a \rangle \neq 0, \quad \text{if } D^{(z)} \text{ occurs in } D^{(T)} \otimes D^{(a)} \quad \text{and} \\ &\langle e | H_1 | z \rangle \neq 0, \quad \text{if } D^{(e)} \text{ occurs in } D^{(T)} \otimes D^{(z)} \end{aligned} \tag{8.3.8}$$

A necessary condition for the product of matrix elements to remain nonzero is that they satisfy (8.3.8) simultaneously, which means that

$$D^{(e)} \text{ has to occur in } D^{(T)} \otimes D^{(T)} \otimes D^{(a)} . \tag{8.3.9}$$

These findings can be extended to transitions of higher (νth) order:

$$D^{(e)} \text{ has to occur in } \underbrace{D^{(T)} \otimes D^{(T)} \ldots \otimes D^{(T)}}_{\nu \text{ times}} \otimes D^{(a)} \tag{8.3.10}$$

if the ν-fold product of matrix elements is to be different from zero.

Exercise 8.2. Prove (8.3.5a,b) by using the character table of $\mathscr{C}_{3v}$ and (8.4.1)

8.4 Crystal Field Theory

8.4.1 Crystal Field Splitting of Energy Levels

The effective potential V_{cr} for an atomic electron in a crystal possesses either the symmetry of the lattice or the local site symmetry of the atom in the crystal. This symmetry is described by the point group $\mathscr{G}$. Compared to the potential of a free atom, which has $\mathscr{O}(3)$ symmetry in general, there is a reduction of symmetry. Eigenfunctions and levels have to be classified according to $\mathscr{G} \subset \mathscr{O}(3)$. We obtain the splitting by subduction (Sect. 4.2.5) of the corresponding IR of $\mathscr{O}(3)$ into $\mathscr{G}$; a rotation $c(\varphi) \in \mathscr{O}(3)$ has the characters

$$\chi^{(j)}(\varphi) = \frac{\sin[(2j+1)\varphi/2]}{[\sin \varphi/2]} \qquad \text{or} \qquad \chi_{\pm}^{(j)}(ic(\varphi)) = \pm\chi^{(j)}(\varphi) \tag{8.4.1}$$

if it is a roto-inversion $ic(\varphi)$. See also (11.4.67). The details of the subduction depend on the relative strength of V_{cr} in relation to the level splitting of the atomic multiplets, which is given by the electron-electron interaction ΔE and the spin-orbit coupling E_{LS}. We have to distinguish between three cases.

(i) Weak crystal field, $V_{cr} \ll E_{LS}$. Here V_{cr} can be treated by perturbation theory, starting with the atomic states with fixed $\boldsymbol{J} = \boldsymbol{L} + \boldsymbol{S}$ (total angular momentum). This happens to be the case in rare-earth metals.

(ii) Intermediate crystal field, $E_{LS} \ll V_{cr} \ll \Delta E$. Again V_{cr} is treated by perturbation theory. However, the initial states are characterized by $\boldsymbol{L}$ *and* $\boldsymbol{S}$ separately. This case often occurs in transition metals.

(iii) Strong crystal field, $V_{cr} \gg E_{LS}$, $V_{cr} \gg \Delta E$. Then the multiplet splitting has to be treated as a perturbation compared to the crystal field states determined by $\mathscr{G}$.

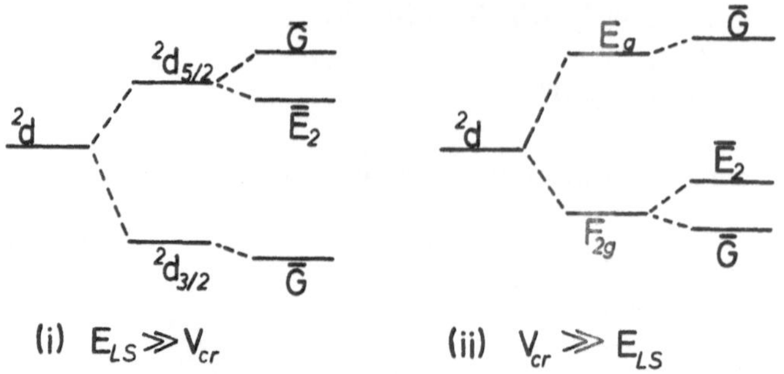

Fig. 8.2. Ligand field splitting of a ^2d level in the cases $E_{LS} \gg V_{cr}$ and $V_{cr} \gg E_{LS}$

Here we will discuss the cases (i) and (ii) for a d-electron ($l = 2$, spin 1/2) in a cubic $\mathcal{O}_h$ crystal field. In the one-electron approximation the free atom has the electronic states $^2d_{5/2}$ and $^2d_{3/2}$ with total angular momentum $j = 5/2$ and $j = 3/2$ [see also Sect. 12.2]. Using (8.4.1) and Appendix A[3], the subduction into $\mathcal{O}$ or $\mathcal{O}_h$ symmetry leads to

$$\begin{aligned} D^{(5/2)} &= \bar{E}_2 \oplus \bar{G} = \Gamma_7 \oplus \Gamma_8 \; , \\ D^{(3/2)} &= \bar{G} = \Gamma_8 \; . \end{aligned} \tag{8.4.2}$$

However, if the spin-orbit splitting is small compared to the crystal field splitting, we first have to determine the latter. For $l = 2$ we have the decomposition $D^{(2)} = E_g \oplus F_{2g} = \Gamma_3 \oplus \Gamma_5 = \Gamma_{12} \oplus \Gamma_{25}$. Then the spatial state has to be coupled to the spin, i.e. we have to reduce the products

$$\begin{aligned} E_g \otimes D^{(1/2)} &= \bar{G} = \Gamma_8 \; , \\ F_{2g} \otimes D^{(1/2)} &= \bar{E}_2 \oplus \bar{G} = \Gamma_7 \oplus \Gamma_8 \; , \end{aligned} \tag{8.4.3}$$

considering $\mathcal{O}_h$ symmetry. The resulting splitting (Fig. 8.2) agrees with that of (8.4.2) qualitatively.

In a solid with *magnetic order* the crystal field has the symmetry of a magnetic group $\mathcal{M}_{\text{III}}$ (Sect. 3.2). The crystal field is composed of a normal electrostatic field and a (magnetic) exchange or anisotropy field. The classification of the energy levels has to be carried out according to the ICORs D^M of the magnetic groups $\mathcal{M} = \mathcal{G} + r\mathcal{G}$ (Sect. 5.2). Since according to (5.2.6, 7) the ICORs of $\mathcal{M}$ are completely and uniquely given by the unitary IRs of $\mathcal{G}$, we can perform all the necessary subductions and reductions in the subgroup $\mathcal{G} \subset \mathcal{M}$ and finally determine the corresponding ICORs of $\mathcal{M}$ for the cases (1a), (1b) and (2) of Sect. 5.2. In doing this we have to observe that the levels assigned to the ICORs may have degeneracies different from those assigned to the IRs of $\mathcal{G}$. In the following scheme the possible cases according to Sect. 5.2 are shown:

[3] There are different notations originating from Mulliken, Bethe, and Bouckart, Smoluchowsky and Wigner [8.3–5]. The parity ($\pm$ or g, u) is irrelevant in this context.

$$\text{(1a)}\quad d_j;\ \Delta^{(j)};\ \phi_k^{(j)} \text{-------} d_j;\ D^{M(j)};\ \phi_k^{(j)}$$

$$\text{(1b)}\ \left\{\begin{array}{l} d_j;\ \Delta^{(j)};\ \phi_k^{(j)} \\ d_j;\ \Delta^{(j)};\ \psi_k^{(j)} \end{array}\right. \text{---}> 2d_j;\ D^{M(j)};\ \{\phi_k^{(j)}, \psi_k^{(j)}\} \tag{8.4.4}$$

$$\text{(2)}\quad \left\{\begin{array}{l} d_j;\ \Delta^{(j)};\qquad\quad \phi_k^{(j)} \\ d_j;\ \Delta^{(i)} \equiv \bar{\Delta}^{(j)};\ \phi_k^{(i)} \end{array}\right. \text{---}> 2d_j;\ D^{M(j)};\quad \begin{array}{c} \{\phi_k^{(j)}, \psi_k^{(j)}\} \\ \text{or} \\ \{\phi_k^{(i)}, \psi_k^{(i)}\} \end{array}$$

We have used the notation:

$\Delta^{(j)}$: IR of $\mathscr{G}$ with dimension d_j,
$\phi_k^{(j)}$: basis function assigned to IR,
$D^{M(j)}$: ICOR assigned to $\Delta^{(j)}$,
$\psi_k^{(j)} = P_r\phi_k^{(j)}$: basis function assigned to $D^{M(j)}$ according to (5.2.4).

If there is a magnetic symmetry the states of the free atom also have to be classified according to the ICORs of the group $\mathcal{O}(3) \times \{e, r\}$. These REPs have to be used for the subduction in case (ii), i.e. $V_{cr} \gg E_{LS}$. According to the above, we outline the scheme

$$\begin{array}{ccccc} & \text{Subduction} & & (1a, b; 2) & \\ \mathcal{O}(3) \times \{e, r\} \leftrightarrow & \mathcal{O}(3) \quad \leftrightarrow & \mathscr{G} & \leftrightarrow & \mathscr{M} \end{array}$$

or

$$D^{M(j)}(\mathcal{O}(3)) \leftrightarrow D^{(j)}(\mathcal{O}(3)) \leftrightarrow D^{(j)\,\mathrm{sub}}(\mathscr{G}) = \sum\nolimits_{\oplus} \Delta^{(j)}(\mathscr{G}) \leftrightarrow \sum\nolimits_{\oplus} D^{M(j)}\ . \tag{8.4.5}$$

The spin-orbit splitting then has to be calculated by reduction of the product $D^{M(j)} \otimes D^{M(1/2)}$ of $\mathscr{M}$[4]. To achieve this, $\Delta^{(j)} \otimes \Delta^{(1/2)}$ of $\mathscr{G}$ has to be reduced and then the ICORs of $\mathscr{M}$ are assigned to the resulting IRs of $\mathscr{G}$.

For a strong spin-orbit coupling $E_{LS} \gg V_{cr}$, first the REP $D^{M(l \times 1/2)}(\mathcal{O}(3))$ has to be reduced, which is done using $D^{(l)} \otimes D^{(1/2)} = D^{(l+1/2)} \oplus D^{(l-1/2)}$:

$$D^{M(l \times 1/2)}(\mathcal{O}(3)) = D^{M(l+1/2)} \oplus D^{M(l-1/2)}\ . \tag{8.4.6}$$

Then the subduction has to be performed as in (8.4.5) (see Exercise 8.4).

8.4.2 Calculation of Splitting

For a quantitative calculation of the crystal field splitting, the potential V_{cr} is looked upon as a perturbation of the potential of the free atom. A determination

---

[4] In many-electron systems with angular momentum L and spin S we have to take $D^{(L)}(\mathcal{O}(3))$ instead of $D^{(j)}(\mathcal{O}(3))$ and $D^{(S)}$ instead of $D^{(1/2)}$.

of V_{cr} is almost impossible. However, a Poisson equation $\Delta V_{cr} = -\rho/\varepsilon$ has to be satisfied, with ρ being the charge density due to the ligands. To a good approximation ρ can be neglected (at the position of the electron, point charge model) and consequently the perturbation has only to satisfy $\Delta V_{cr} = 0$, which means that there exists a multipole expansion

$$V_{cr}(r, \vartheta, \varphi) = \sum_{l,m} b_{lm} r^l Y_m^l(\vartheta, \varphi) \ . \tag{8.4.7a}$$

As V_{cr} has to be invariant with respect to the point group of the crystal field, (8.4.7a) may only contain spherical harmonics, which are basis functions belonging to the identity REP of $\mathscr{G}$. For $\mathcal{O}_h$ symmetry, restricting ourselves to $l \leqslant 4$, this gives

$$V_{cr}(r, \vartheta, \varphi) = b_{00} Y_0^0 + b_{40} r^4 [Y_0^4 + \sqrt{5/14}(Y_4^4 + Y_{-4}^4)] \tag{8.4.7b}$$

(see also Exercise 8.5).

The coefficients b_{lm} are treated as free parameters, although they can also be derived in several models. The zero-order functions needed for a calculation of matrix elements are the atomic functions in a symmetry-adapted form:

$$\psi_{ik}^{(\alpha)} \sim R_{nl}(r) P_{ik}^{(\alpha)} Y_m^l = \sum_{nlm} a_{ik,nlm}^{(\alpha)} R_{nl}(r) Y_m^l \ . \tag{8.4.8}$$

Using (8.4.7b) and (8.4.8) the matrix elements can be determined in a first-order approximation. We will accomplish this for the example of a d-electron in $\mathcal{O}_h$ symmetry (8.4.2, 3).

Let H_0 be the unperturbed atomic Hamiltonian, and $H_1 = -V_{cr}$ the perturbation, which is assumed to be attractive. The solutions of the atomic problem ($l = 2$) are $\psi_{n,2,m} = R_{n2} \cdot Y_m^2$ with $m = -2, \ldots, +2$. We obtain a fivefold spatial degeneracy, which means a tenfold degeneracy when spin is taken into account (without spin-orbit coupling). With the spin functions α and β, respectively, we have ten zero-order functions and therefore a ten- (or five-dimensional) secular problem. By means of symmetry-adapted functions, however, the secular equation can be block-diagonalized as in (6.4.5). According to (8.4.2, 3) we obtain by reduction a twofold degenerate E_g and a threefold degenerate F_{2g} problem (each one dimensional). The basis functions are

$$\begin{aligned}
\psi_1^{E_g} &= R_{n2} Y_0^2 = \psi_{n20} \ , & \psi_1^{F_{2g}} &= \frac{-i}{\sqrt{2}}(\psi_{n22} - \psi_{n2-2}) \ , \\
\psi_2^{E_g} &= R_{n2} Y_{2c}^2 = \frac{1}{\sqrt{2}}(\psi_{n22} + \psi_{n2-2}) \ , & \psi_2^{F_{2g}} &= \frac{1}{\sqrt{2}}(-\psi_{n21} + \psi_{n2-1}) \ , \\
& & \psi_3^{F_{2g}} &= \frac{+i}{\sqrt{2}}(\psi_{n21} + \psi_{n2-1}) \ .
\end{aligned} \tag{8.4.9}$$

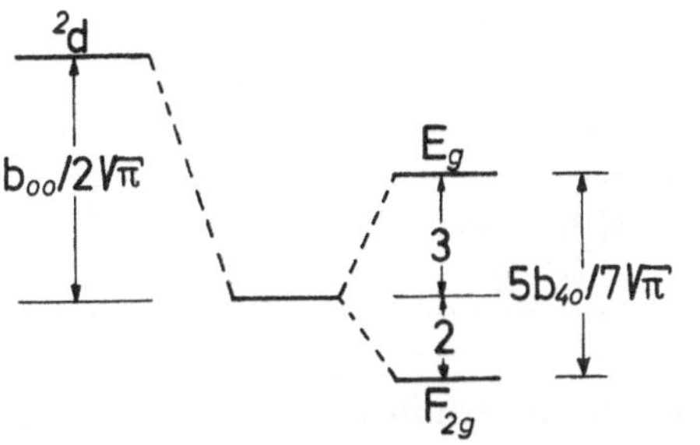

Fig. 8.3. Level splitting in a ligand field with $\mathcal{O}_h$ symmetry according to (8.4.7b, 10a); the expectation values $\langle r\rangle$ and $\langle r^4\rangle$ have been omitted

According to (6.4.6) we then have

$$E^{E_g} = \langle\psi_1^{E_g}|H_1|\psi_1^{E_g}\rangle = -\int \psi_{n20}^* V_{cr}\psi_{n20}\, d^3x$$

$$= -\frac{1}{2\sqrt{\pi}}b_{00}\langle r\rangle - \frac{3}{7\sqrt{\pi}}b_{40}\langle r^4\rangle\ , \qquad (8.4.10a)$$

and

$$E^{F_{2g}} = -\frac{1}{2\sqrt{\pi}}b_{00}\langle r\rangle + \frac{2}{7\sqrt{\pi}}b_{40}\langle r^4\rangle \qquad \text{with} \qquad (8.4.10b)$$

$$\langle r^\nu\rangle = \int_0^\infty R_{n2}^*(r)r^\nu R_{n2}(r)r^2\, dr\ .$$

The angle-independent term b_{00} in V_{cr} thus means a lowering of energy that is of equal magnitude for both states, whereas b_{40} describes a splitting which has the F_{2g} state as the lower one if the potential is attractive (Fig. 8.3). Furthermore, the spin-orbit splitting ought to be considered (Fig. 8.2) as given by (8.4.3). This will not be treated here.

In *many-electron systems* H_1 is a sum over one-electron operators, $H_1 = \sum_i V(x_i)$, where a single $V(x_i)$ represents the contribution of all the ligands to the field of the ith electron. We have the case of a strong field if

$$H_1 \gg \sum_{i,j} e^2/2r_{ij} \gg \sum_i H_{LS}(i)\ .$$

As starting functions in a perturbation calculation we then have to use many-electron product functions in a symmetry-adapted basis. These functions have to transform according to the IRs of the group $\mathcal{G}$ of the field of ligands.

As an example we consider a two-electron system in a $\mathcal{D}_3 \cong \mathcal{C}_{3v}$ symmetry. The starting basis can be chosen to be $\psi_{ik}^{\alpha\beta}(x_1, x_2) = \psi_i^{(\alpha)}(x_1)\psi_k^{(\beta)}(x_2)$ where the $\psi^{(\alpha)}$ and $\psi^{(\beta)}$ transform according to the IRs $D^{(\alpha)}$ and $D^{(\beta)}$, respectively, of $\mathcal{G} = \mathcal{D}_3$. Since we are looking for two-electron functions with well-defined behaviour with respect to $\mathcal{D}_3$ we use a Clebsch-Gordan expansion according to (4.4.18):

$$\psi_j^{\gamma s}(x_1, x_2) = \sum_{ik}\begin{pmatrix}\alpha\beta & \gamma s\\ ik & j\end{pmatrix}\psi_i^\alpha\psi_k^\beta\ . \qquad (8.4.11)$$

Table 8.1. Linear combinations of functions according to (8.4.11)

	ψ^{A_1}	$\psi_1^E\psi_1^E$	$\psi_1^E\psi_2^E$	$\psi_2^E\psi_2^E$
$c_2'\psi$	ψ^{A_1}	$\psi_1^E\psi_1^E$	$-\psi_1^E\psi_2^E$	$\psi_2^E\psi_2^E$
$c_3\psi$	ψ^{A_1}	$\frac{1}{4}(\psi_1^E\psi_1^E - 2\sqrt{3}\psi_1^E\psi_2^E + 3\psi_2^E\psi_2^E)$	$\frac{1}{4}(\sqrt{3}\psi_1^E\psi_1^E - 2\psi_1^E\psi_2^E - \sqrt{3}\psi_2^E\psi_2^E)$	$\frac{1}{4}(3\psi_1^E\psi_1^E + 2\sqrt{3}\psi_1^E\psi_2^E + \psi_2^E\psi_2^E)$

If the CGCs are known, (8.4.11) is completely determined. Otherwise we can evaluate the CGCs according to Sect. 4.4.3. To illustrate this technique once more, we apply the generators of the group (e.g. c_3, c_2') to (8.4.11) for fixed γ, α, β (e.g. $\gamma = A_1$, $\alpha = \beta = E$). By comparison of the coefficients we then obtain the CGC.

In the example

$$D^{(E)}(c_2') = \begin{pmatrix} -1 & 0 \\ 0 & 1 \end{pmatrix} , \qquad D^{(E)}(c_3) = \begin{pmatrix} -1/2 & -\sqrt{3}/2 \\ \sqrt{3}/2 & -1/2 \end{pmatrix} .$$

What we obtain for the various functions occurring in (8.4.11) is given in Table 8.1. By comparison of the coefficients, including normalisation, we then obtain (Sect. 4.4.3)

$$\left(\begin{matrix} EE \\ 11 \end{matrix}\middle| A_1\right) = \left(\begin{matrix} EE \\ 22 \end{matrix}\middle| A_1\right) = \pm\frac{1}{\sqrt{2}} , \qquad \left(\begin{matrix} EE \\ 12 \end{matrix}\middle| A_1\right) = \left(\begin{matrix} EE \\ 21 \end{matrix}\middle| A_1\right) = 0 .$$

Starting with the configuration $\psi_i^E(1)\psi_k^E(2)$, we can use the functions from (4.4.21) with

$$\{f_1, f_2\} = \{\psi_1^E(1), \psi_2^E(1)\} \qquad \text{and} \qquad \{g_1, g_2\} = \{\psi_1^E(2), \psi_2^E(2)\}$$

as the symmetry-adapted two-electron functions. Finally, because of the Pauli principle we have to symmetrize the functions according to the IRs $D^{[\lambda]}$ of $\mathscr{P}_2$ and to combine them with the associated spin functions of $D^{[\tilde{\lambda}]}$, see (7.2.7). In this special case the functions (4.4.21) are already assigned to the REPs $D^{[\lambda]}$ of $\mathscr{P}_2$:

$$A_1, E \Rightarrow D^{[\lambda]} = D^{[2]} \curvearrowright D^{[\tilde{\lambda}]} = D^{[1^2]} \Rightarrow S = 0 ,$$

$$A_2 \Rightarrow D^{[\lambda]} = D^{[1^2]} \curvearrowright D^{[\tilde{\lambda}]} = D^{[2]} \Rightarrow S = 1 .$$

The spin functions are

$$\varphi^{[1^2]} = \frac{1}{\sqrt{2}}[\alpha(1)\beta(2) - \alpha(2)\beta(1)] \qquad \text{singlet} ,$$

$$\varphi^{[2]} = \begin{cases} \alpha(1)\alpha(2) \\ \dfrac{1}{\sqrt{2}}[\alpha(1)\beta(2) + \alpha(2)\beta(1)] \\ \beta(1)\beta(2) \end{cases} \qquad \text{triplet} , \qquad (8.4.12)$$

thus allowing the multiplets $^3A_2$, $^1A_1$, ^1E. Using these functions, the two-electron problem for the ligands is then diagonal.

The influence of external fields can be described within perturbation theory by matrix elements, too. External fields break the spherical symmetry of the $\mathcal{O}_3$, that means there is a reduction of symmetry. A homogeneous electric field $\boldsymbol{E}$ (polar vector) induces a $\mathscr{C}_{\infty v}$ symmetry, and a homogeneous magnetic field $\boldsymbol{B}$ (axial vector) an $\mathcal{O}(2) \cong \mathscr{C}_\infty \times \mathscr{C}_i \cong \mathscr{C}_{\infty h}$ symmetry. In an electric field (*Stark effect*) the Hamiltonian can be written as $H = H_0 + H_1$ with $H = -e\boldsymbol{x} \cdot \boldsymbol{E}$. Whereas H_0 is invariant against $\mathcal{O}(3)$, H_1 is invariant only against $\mathscr{C}_{\infty v}$. The reduction of the REP $D^{(l)}$ of $\mathcal{O}(3)$ subduced into $\mathscr{C}_{\infty v}$ results in a splitting into l doubly degenerate levels $E_m(\pm m)$ and one nondegenerate level A_1 $(m = 0)$.

In a homogeneous magnetic field

$$H_1 = -(e/2mc)\boldsymbol{B} \cdot \boldsymbol{L} + (e^2/8mc^2)(\boldsymbol{x} \times \boldsymbol{B})^2 \; .$$

The symmetry group of H_1 is the Abelian group $\mathcal{O}(2)$ which has only one-dimensional REPs. On subducing $\mathcal{O}(3) \to \mathcal{O}(2)$ there arise $2l + 1$ nondegenerate levels (normal *Zeeman effect*, no spin-orbit coupling). In the case of the Zeeman effect in a crystal field we have to pay attention to the symmetry of H_1 as well as to that of the crystal. In a crystal with local $\mathscr{C}_{3v}$ symmetry ($\boldsymbol{B} \| z$) the axial vector $\boldsymbol{L}$ in $\boldsymbol{B} \cdot \boldsymbol{L}$ has the decomposition $D^{(L)} = A_2 \oplus E$ (8.3.5a). Thus we have to investigate matrix elements of the form $\langle \psi_i^\alpha | L^{A_2} | \psi_k^\beta \rangle$ and $\langle \psi_i^\alpha | L_j^E | \psi_k^\beta \rangle$ with $L^{A_2} = L_z$, $L_j^E = \{L_x, L_y\}$ (Sect. 6.2). From (8.3.4) we learn, for example, that the matrix element $\langle \psi_i^E | L_j^E | \psi_k^E \rangle$ is different from zero, and furthermore, using the Wigner-Eckart theorem in (6.2.5, 6; 6.3.8),

$$\langle \psi_i^E | L_j^E | \psi_k^E \rangle = \begin{cases} \begin{pmatrix} EE & E \\ jk & i=1 \end{pmatrix} \langle 1 \| L_j^E \| 1 \rangle^{(E)} = \dfrac{1}{\sqrt{2}} \begin{pmatrix} 0 & 1 \\ 1 & 0 \end{pmatrix} \langle 1 \| L_j^E \| 1 \rangle^{(E)} \\ \begin{pmatrix} EE & E \\ jk & i=2 \end{pmatrix} \langle 1 \| L_j^E \| 1 \rangle^{(E)} = \dfrac{1}{\sqrt{2}} \begin{pmatrix} 1 & 0 \\ 0 & -1 \end{pmatrix} \langle 1 \| L_j^E \| 1 \rangle^{(E)} \; . \end{cases} \tag{8.4.13}$$

Equation (8.4.13) again shows the advantages of the Wigner-Eckart theorem: (1) the reduced matrix element $\langle \| L^E \| \rangle$ does not depend on the basis indices i, j, k and can be determined with the "most convenient" choice of i, j, k; all the other factors are given by the CGCs alone. (2) Statements on the relative magnitude of the matrix elements can already be found from (8.4.13). The statements, once obtained for a tensor operator $T_j^{(\alpha)}$ (here L_j^E), are also valid for all the tensor operators $\hat{T}_j^{(\alpha)}$ with the same behaviour in transformations (e.g. $x_j^E = \{x, y\}$ for the Stark effect).

Exercise 8.3. Carry out the decompositions (8.4.2, 3) with the help of Appendix A.

Exercise 8.4. Calculate the crystal field splitting of a $^2d_{5/2}$ or $^2d_{3/2}$ level for the magnetic symmetry $\mathscr{M}_{\mathrm{III}} = \mathscr{G} + r\mathscr{G}$ with

a) $\mathscr{G} = \mathscr{C}_{3v}, r = \mathrm{i}\vartheta, \mathscr{G} + \mathrm{i}\mathscr{G} = \mathscr{D}_{3d}$;
b) $\mathscr{G} = \mathscr{D}_{2h}, r = c_{4z}\vartheta, \mathscr{G} + c_{4z}\mathscr{G} = \mathscr{D}_{4h}$.

Exercise 8.5. Calculate (8.4.7b) by applying the generators $a \in \mathcal{O}_h$ to V_{cr} and utilizing the invariance condition $aV_{cr} = V_{cr}$.

8.5 Independent Components of Material Tensors

In Sects. 5.5 and 6.1 we discussed how tensors of rank m behave under a group of transformations. The tensor REP $D^{(\times)}_{V,A}$ in $\mathbb{R}_3$, according to (6.1.7) or (5.5.6b), is the inner product of m (polar) vector REPs $D^{(v)} \cong D^{(l=1)}_-$ or (axial) pseudovector REPs $D^{(A)} \cong D^{(l=1)}_+$ with the characters

$$\chi^{(V,A)}(c(\varphi)) = 1 + 2\cos\varphi \ , \qquad \chi^{(V,A)}(\mathrm{i}c(\varphi)) = \mp(1 + 2\cos\varphi) \ . \tag{8.5.1}$$

If we consider physical quantities in systems which are invariant under $a \in \mathscr{G}$ then the material tensors connecting them must not change either, i.e. besides (5.5.8), the invariance condition

$$\begin{aligned} P_a T_{i'_1\dots i'_m} := T'_{i'_1\dots i'_m} &= \sum_{i_1\dots i_m} a_{i'_1 i_1}\dots a_{i'_m, i_m} T_{i_1\dots i_m} \\ &= T_{i'_1\dots i'_m} \end{aligned} \tag{8.5.2}$$

has to be valid for each $a \in \mathscr{G}$. If we apply the projection operator

$$P^{(1)} = \frac{1}{g}\sum_{a\in\mathscr{G}} P_a \tag{8.5.3a}$$

of the identity REP of $\mathscr{G}$ to the tensor components, then because of (8.5.2) we can express all the tensor components by their invariant combinations $\{P^{(1)}T_{i_1\dots i_m}\}$, as

$$P^{(1)}T_{i_1\dots i_m} = \frac{1}{g}\sum_{a\in\mathscr{G}}\sum_{i'_j} a_{i_1 i'_1}\dots a_{i_m i'_m} T_{i'_1\dots i'_m} = T_{i_1\dots i_m} \ . \tag{8.5.3b}$$

Thus we obtain N_0 linearly independent tensor components by applying the projection operator $P^{(1)}$ to the initial components $T_{i_1\dots i_m}$. The number N_0 of independent components is equal to the multiplicity with which the identity REP $D^{(1)}$ occurs in the tensor REP

$$D^{(\times)}_V(a) = D^{(V)}(a) \otimes D^{(V)}(a) \otimes \dots \otimes D^{(V)}(a) \ , \tag{8.5.4}$$

where the superscript may be V or A:

$$N_0 = \frac{1}{g}\sum_{a\in\mathscr{G}} 1\chi^{(\times)}_V(a) \ . \tag{8.5.5}$$

On the other hand, the independent tensor components can be derived directly from (8.5.2) if we write down the corresponding relations. It is sufficient to choose the generators from $\mathscr{G}$ for the a because the other elements do not lead to new relations.

In general, the physical tensors still have *internal* symmetries, like permutation symmetries of indices. For example, the dielectric constant ε_{ij} and susceptibility χ_{ij} are symmetric in their subscripts. For transport properties, such as electric conductivity σ_{ij} and thermal conductivity κ_{ij}, we have because of the Onsager relations (time reversal symmetry)

$$\sigma_{ij}(\boldsymbol{B}) = \sigma_{ji}(-\boldsymbol{B}) \ , \qquad \kappa_{ij}(\boldsymbol{B}) = \kappa_{ji}(-\boldsymbol{B}) \ , \tag{8.5.6}$$

where we have allowed for a magnetic field dependence. Such symmetries have to be taken into account additionally, by discussing the symmetrized (or antisymmetrized) product of two tensors $[D^{(V)}(a) \otimes D^{(V)}(a)]_{\pm}$ (see Exercises 8.6 and 8.9).

Macroscopic physical properties described by tensors, of course, are invariant against lattice translations. This implies that they transform according to the identity REP $D^{(1)} = D^{(k=0)}$ of the translation group (5.3.13). The REPs of the space group $\mathscr{R}$ in this case agree with those of the point group $\mathscr{G}$ (see the first example in Sect. 9.2.2). The invariance under $\mathscr{R}$ is equal to that under $\mathscr{G}$.

Some examples for point groups of the type $\mathscr{M}_1 = \mathscr{G}$ will be discussed here.

(i) *First rank tensors.* Pyroelectric coefficients, spontaneous polarization ($\boldsymbol{P}$) and spontaneous magnetization ($\boldsymbol{M}$, axial vector) are first rank tensors. According to (8.5.1 and 5),

$$\begin{aligned} &\text{for } \mathscr{C}_3,\ N_0 = \tfrac{1}{3}(3+0+0) = 1 \qquad \text{for } \boldsymbol{P} \text{ and } \boldsymbol{M} \ , \\ &\text{and for } \mathscr{C}_{3v}\text{: } N_0 = \begin{cases} \tfrac{1}{6}(3 + 2\cdot 0 + 3\cdot 1) = 1 & \text{for } \boldsymbol{P} \ , \\ \tfrac{1}{6}(3 + 2\cdot 0 - 3\cdot 1) = 0 & \text{for } \boldsymbol{M} \ . \end{cases} \end{aligned} \tag{8.5.7}$$

Thus for $\mathscr{C}_3$ symmetry spontaneous polarization (ferroelectric) and magnetization (ferromagnetic) are possible, while for $\mathscr{C}_{3v}$ symmetry only spontaneous polarization is possible. From (8.5.2) using the generator $c_3 \in \mathscr{C}_3$ we obtain the independent tensor components $c_3\boldsymbol{P} = \boldsymbol{P}$ and $c_3\boldsymbol{M} = \boldsymbol{M}$. This means that $\boldsymbol{P}$ or $\boldsymbol{M}$ have to be parallel to the main axis (z-axis), therefore $\boldsymbol{P} = \{0, 0, P_z\}$. The same result follows from (8.5.3b) with the projection operator method:

$$P^{(1)}\{P_x, P_y, P_z\} = \tfrac{1}{3}(P_e + P_{c_3} + P_{c_3^2})\{P_x, P_y, P_z\} = \tfrac{1}{3}\{0, 0, 3P_z\} \ .$$

(ii) *Second rank tensors.* Examples of second rank tensors are the (symmetric) polar tensors electric (σ_{ij}) and thermal (κ_{ij}) conductivity, dielectric constants ε_{ij} and the (symmetric, doubly) axial tensor magnetic susceptibility χ_{ij}. However the behavior of $D^{(V)} \otimes D^{(V)}$ and $D^{(A)} \otimes D^{(A)}$ in transformations is the same because $D^{(1)}_{\pm}(a) = D^{(1)}(a)$ and $D^{(1)}_{\pm}(ia) = \pm D^{(1)}(a)$. This, correspondingly, holds for all *pure* polar and axial tensors of even rank. Thus it is not possible to ascertain by a measurement of tensor properties of tensors of even rank whether a crystal does or does not have an inversion centre. The number of independent tensor components in the cases above is given by the symmetrized tensor product. Thus, according to (8.5.5), see (4.4.16),

$$N_0 = \frac{1}{g} \sum_{a \in \mathscr{G}} \frac{1}{2}\{[\chi^{(V)}(a)]^2 + \chi^{(V)}(a^2)\} \; ; \qquad (8.5.8)$$

for $\mathscr{C}_3$ symmetry (trigonal system) we have

$$N_0 = \tfrac{1}{6}(9 + 3 + 0 + 0 + 0 + 0) = 2 \; .$$

From (8.5.2) we obtain with $T_{ij} := ij;\, i, j = x, y, z,$

$$P_{c_3} xx = \left(-\frac{x}{2} - \frac{\sqrt{3}}{2} y\right)\left(-\frac{x}{2} - \frac{\sqrt{3}}{2} y\right) = \frac{1}{4} xx + \frac{\sqrt{3}}{2} xy + \frac{3}{4} yy = xx$$

$$P_{c_3} xy = \left(-\frac{x}{2} - \frac{\sqrt{3}}{2} y\right)\left(\frac{\sqrt{3}}{2} x - \frac{1}{2} y\right) = -\frac{\sqrt{3}}{4} xx + \frac{\sqrt{3}}{4} yy = xy \; ,$$

from which it follows that $xx = yy$ and $xy = 0$. Correspondingly, $xz = yz = 0$. Thus T_{ij} has the form

$$T_{ij} = \begin{pmatrix} T_{xx} & 0 & 0 \\ 0 & T_{xx} & 0 \\ 0 & 0 & T_{zz} \end{pmatrix} \; . \qquad (8.5.9)$$

Using (8.5.3b) we have $P^{(1)} xx = \frac{1}{2}(xx + yy) = P^{(1)} yy$, $P^{(1)} zz = zz$, $P^{(1)} ij = 0$ otherwise, and thus again (8.5.9).

(iii) *Third rank tensors.* The tensor of the piezoelectric constants (moduli) $h_{k,ij}$ with three polar components and the Hall tensor $a_{ij,k}$ with two polar (ij) components and one axial (k) component are third rank tensors. In piezoelectric crystals a strain[5] $v_{ij} = v_{ji}$ induces an electric field $E_k = -\sum_{i,j} h_{k,ij} v_{ij}$, which means that $h_{k,ij}$ is symmetric in i and j and transforms according to $D_V^{(\times)} = [D^{(V)} \otimes D^{(V)}]_+ \otimes D^{(V)}$. The Hall tensor is defined by

$$E_i = \sum_{jk} a_{ij,k} j_j B_k := \sum_j \rho_{ij}(\boldsymbol{B}) j_j \; ,$$

where because of (8.5.6), $a_{ij,k} = -a_{ji,k}$. Therefore it transforms according to $D_V^{(\times)} = [D^{(V)} \otimes D^{(V)}]_- \otimes D^{(A)}$. Since $[D^{(V)} \otimes D^{(V)}]_-$ transforms as $D^{(A)}$ (see Exercise 8.8), the Hall tensor is often represented as a nonsymmetric tensor of second

[5] We shall not take into account the difference between infinitesimal and finite strains, although this difference may be important in symmetry considerations. For further details see special books on this topic.

rank with two axial components. The number of independent components for $h_{k,ij}$ is thus given by

$$N_0 = \frac{1}{g} \sum_{a \in \mathscr{G}} \frac{1}{2}\{[\chi^{(V)}(a)]^2 + \chi^{(V)}(a^2)\}\chi^{(V)}(a) \to 4 \text{ for } \mathscr{C}_{3v} , \tag{8.5.10}$$

but for $a_{ij,k}$ by

$$N_0 = \frac{1}{g} \sum_{a \in \mathscr{G}} [\chi^{(A)}(a)]^2 \to 2 \text{ for } \mathscr{C}_{3v} .$$

(iv) *Fourth rank tensors.* We shall only consider the tensor of the second-order elastic constants, which is defined by $\sigma_{ij} = \sum_k C_{ij,kl} v_{kl}$ (σ_{ij}: stress). The subscripts have the permutation symmetry

$$C_{ij,kl} = C_{ji,kl} = C_{ij,lk} = C_{kl,ij} . \tag{8.5.11}$$

All the possible permutations of subscripts of a tensor of rank m which leave this tensor invariant form a group $\mathscr{P}$, which is a subgroup of $\mathscr{P}_4$. In this case the permutations are

$$\mathscr{P} = \{e, (ij), (kl), (ik)(jl), (ij)(kl), (il)(jk), (ikjl), (iljk)\} \subset \mathscr{P}_4 . \tag{8.5.12}$$

Consequently, we have to use the character of the correspondingly symmetrized tensor product in (8.5.5). This is given by [see Exercise 8.9 and (5.5.34a)]

$$\chi_p(a) = \frac{1}{n_p} \sum_{p \in \mathscr{P}} \{\chi(a)\}^{\nu_1}\{\chi(a^2)\}^{\nu_2} \dots \{\chi(a^m)\}^{\nu_m} , \tag{8.5.13}$$

where $n_p = \operatorname{ord} \mathscr{P}$ and ν_i is the length of the cycles in the permutation $p \in \mathscr{P}$. Then (8.5.5) has to be replaced by

$$N_0 = \frac{1}{g} \sum_{a \in \mathscr{G}} \frac{1}{n_p} \sum_{p \in \mathscr{P}} \{\chi(a)\}^{\nu_1}\{\chi(a^2)\}^{\nu_2} \dots \{\chi(a^m)\}^{\nu_m} . \tag{8.5.14a}$$

In the case that a tensor is antisymmetric with respect to a permutation of subscripts the corresponding permutation in (8.5.13) has to be taken with a negative sign. For example, if a tensor is antisymmetric under all the odd permutations of $\mathscr{P}$, then

$$N_0 = \frac{1}{g} \sum_{a \in \mathscr{G}} \frac{1}{n_p} \sum_{p \in \mathscr{P}} (-1)^p \{\chi(a)\}^{\nu_1}\{\chi(a^2)\}^{\nu_2} \dots \{\chi(a^m)\}^{\nu_m} . \tag{8.5.14b}$$

Consequently, for the elastic constants (8.5.11,12)

$$N_0 = \frac{1}{8g} \sum_{a \in \mathscr{G}} \{[\chi(a)]^4 + 2[\chi(a)]^2\chi(a^2) + 3[\chi(a^2)]^2 + 2\chi(a^4)\} \to 6 \text{ for } \mathscr{C}_{3v} . \tag{8.5.15}$$

If the form of the tensor is to be determined explicitly it is best to proceed as for nonsymmetric tensors according to (8.5.2, 3) and then to symmetrize the tensor.

In the case of an *isotropic medium* (material) the total REP is decomposed according to Exercise 8.8. Then we have to investigate whether the identity REP $D_+^{(0)}$ occurs in it. For the piezoelectric tensor we have finally

$$D_V^{(\times)} = [D_+^{(0)} \otimes D_+^{(0)}]_+ \otimes D_-^{(1)} = 2D_-^{(1)} \oplus D_-^{(2)} \oplus D_-^{(3)} \qquad (8.5.16a)$$

and for the Hall tensor

$$D_V^{(\times)} = [D_-^{(1)} \otimes D_-^{(1)}]_- \otimes D_+^{(1)} = D_+^{(1)} \otimes D_+^{(1)} = D_+^{(0)} \oplus D_+^{(1)} \oplus D_+^{(2)} \ , \qquad (8.5.16b)$$

i.e. there is no piezoelectric constant and one Hall constant $a_{ij,k} = -\varepsilon_{ijk} \cdot a_H$ or $\boldsymbol{E} = -a_H(\boldsymbol{j} \times \boldsymbol{B})$. For the elastic constants the corresponding decomposition is

$$D_V^{(\times)} = [D_-^{(1)} \otimes D_-^{(1)} \otimes D_-^{(1)} \otimes D_-^{(1)}]_{\text{sym}} = 2D_+^{(0)} \oplus 2D_+^{(2)} \oplus D_+^{(4)} \ ; \qquad (8.5.17)$$

there are therefore two elastic constants for an isotropic material.

In crystals with magnetic symmetry (type $\mathscr{M}_{\text{III}}$) the above procedure has to be extended. Let $\mathscr{M} = \mathscr{G} + r\mathscr{G}$ and let the independent components with respect to $\mathscr{G}$ be determined according to the above methods, then it is only necessary to investigate the effect of the (antiunitary) generator $r = \vartheta a'$ (Sect. 5.2) on the tensors in order to obtain the independent components with respect to the total $\mathscr{M}$. In this technique one has to take into account that $\vartheta \notin \mathscr{M}_{\text{III}}$, but $\vartheta^2 \in \mathscr{M}_{\text{III}}$ in these $\mathscr{M}_{\text{III}}$ groups. We have

$$\vartheta^2 T_{i_1 i_2 \dots} = T_{i_1 i_2 \dots} \ , \qquad \vartheta T_{i_1 i_2 \dots} = \pm T_{i_1 i_2 \dots} \ , \qquad (8.5.18)$$

i.e. ϑ causes a multiplication of all the tensor components by $+1$ [for invariant tensors (i-tensors), e.g. $\boldsymbol{P}$] or by -1 [for change sign tensors (c-tensors), e.g. $\boldsymbol{M}$]. For i-tensors we have to use (8.5.2) with P_a and $P_{a'}$ ($a \in \mathscr{G}$, $\vartheta a' \in r\mathscr{G}$), while for c-tensors we have to add a minus sign to the $P_{a'}$ operators in (8.5.2).

Exercise 8.6. Demonstrate the symmetry $\varepsilon_{ij} = \varepsilon_{ji}$ by using the existence of an energy density.

Exercise 8.7. Calculate the components different from zero for the vectors $\boldsymbol{P}$ and $\boldsymbol{M}$ in $\mathscr{C}_{3v}$ symmetry from (8.5.2) *and* (8.5.3b).

Exercise 8.8. Show with the help of (8.5.1), (4.4.16) and Exercise 6.6 that

$$[D_\pm^{(l=1)} \otimes D_\pm^{(l=1)}]_+ = D_+^{(0)} + D_+^{(2)} \qquad \text{and}$$

$$[D_\pm^{(l=1)} \otimes D_\pm^{(l=1)}]_- = D_+^{(1)} \ .$$

Exercise 8.9. Prove (8.5.13) by applying the permutations $p \in \mathscr{P}$ to the subscripts in (8.5.2) and then taking appropriate traces; give reasons for their use in (8.5.5).

9. Representations of Space Groups

Electronic and vibrational states of crystals are classified according to the IRs of space groups. Thus, as a first step we have to establish these IRs, which is done in this chapter for the ordinary as well as the magnetic and double space groups. In any case the basic group to be discussed is the little group. Its IRs can be obtained from the projective REPs as well as from ordinary vector REPs. The discussion of the magnetic space groups needs the CORs. In addition, the projection operators for the construction of the symmetry adapted basis functions are given.

9.1 Representations of Normal Space Groups

9.1.1 Decompositions into Cosets

In Sect 2.3 we defined an element conjugate (with respect) to b by means of $a^{-1}ba$. Correspondingly we define a REP of an invariant subgroup $\mathcal{N} \subset \mathcal{G}$ *conjugate* to $D(b)$ by

$$D_a(b) := D(a^{-1}ba) \ , \qquad a \in \mathcal{G} \ , \qquad b \in \mathcal{N} \ , \tag{9.1.1}$$

where D is a REP of $\mathcal{N}$. In the case of space groups $\mathcal{R}$, $\mathcal{N} = \mathbb{T}$ (Sect. 3.4.1) and according to (5.3.13) the conjugate REPs of an IR $D^{(\boldsymbol{k})}$ of $\mathbb{T}$ are, using (3.4.8) and $\boldsymbol{k} \cdot d^{-1}\boldsymbol{R}^h = d\boldsymbol{k} \cdot \boldsymbol{R}^h$,

$$D^{(\boldsymbol{k})}_{\{d|t\}}(e|\boldsymbol{R}^h) := D^{(\boldsymbol{k}')}(e|\boldsymbol{R}^h) = D^{(\boldsymbol{k})}(e|d^{-1}\boldsymbol{R}^h) = \exp(-\mathrm{i}\, d\boldsymbol{k} \cdot \boldsymbol{R}^h) \tag{9.1.2}$$

where with (5.3.14)[1]

$$\boldsymbol{k}' = d\boldsymbol{k} + \boldsymbol{K}_m \ . \tag{9.1.3a}$$

A *maximal set* of inequivalent conjugate IRs of an invariant subgroup $\mathbb{T}$ is said to be an *orbit* of $\mathbb{T}$ with respect to $\mathcal{R}$. With $D^{(\boldsymbol{k})}$, all the IRs for which (9.1.3a) is satisfied belong to the same orbit (see also Sect. 5.3). The set of $\boldsymbol{k}$-vectors belonging to one orbit according to (9.1.3a) is said to be the *star of* $\boldsymbol{k}$:

[1] Equation (9.1.3a) is an equivalence relation on the set of IRs of $\mathbb{T}$; the stars (orbits) can thus be divided into classes (Sect. 2.2).

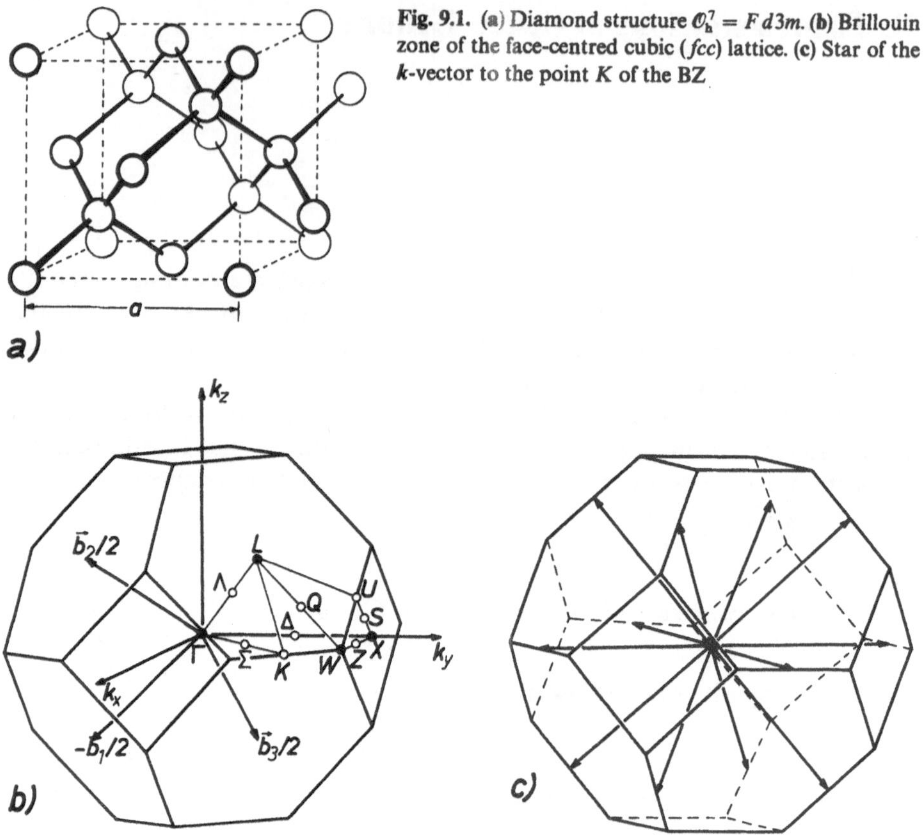

Fig. 9.1. (a) Diamond structure $\mathcal{O}_h^7 = Fd3m$. (b) Brillouin zone of the face-centred cubic (*fcc*) lattice. (c) Star of the $\boldsymbol{k}$-vector to the point K of the BZ

$$*\boldsymbol{k} = \{\boldsymbol{k}' | \boldsymbol{k}' = d\boldsymbol{k} + \boldsymbol{K};\, d \in \mathcal{G};\, \boldsymbol{K} \in \{\boldsymbol{K}_m\}\} \ . \tag{9.1.4}$$

The order of a star (equal to the number of different $\boldsymbol{k}$-vectors) can at most be equal to the order g of the point group $\mathcal{G}$ assigned to the space group $\mathcal{R}$. In Fig. 9.1 the BZ of a fcc lattice ($\mathcal{O}_h$) is shown together with the star of the K point of the BZ. The star of $\boldsymbol{k} = 0$ (Γ point) only consists of this single vector (one point). Of special importance for the REP theory of $\mathcal{R}$ is the *little group* $\mathcal{G}_{\boldsymbol{k}}$ (second kind). It contains all those elements of $\{d|t\} \in \mathcal{R}$ whose rotation part $d = \{d|0\}$ maps the vector $\boldsymbol{k}$ modulo $\boldsymbol{K}$ onto itself:

$$\mathcal{G}_{\boldsymbol{k}} := \{\{d|\boldsymbol{t}\} | d\boldsymbol{k} = \boldsymbol{k} + \boldsymbol{K};\, \{d|\boldsymbol{t}\} \in \mathcal{R}\} \ . \tag{9.1.5}$$

According to (9.1.2, 3a) the elements of $\mathcal{G}_{\boldsymbol{k}}$ yield a number of equivalent (self-conjugate) IRs assigned to $D^{(\boldsymbol{k})}$ of $\mathbb{T}$. Using $\mathcal{G}_{\boldsymbol{k}}$ a decomposition of $\mathcal{R}$ into cosets can be given:

$$\mathcal{R} = \mathcal{G}_{\boldsymbol{k}} + \{d_2|\boldsymbol{s}_2\}\mathcal{G}_{\boldsymbol{k}} + \cdots + \{d_s|\boldsymbol{s}_s\}\mathcal{G}_{\boldsymbol{k}} \ ; \tag{9.1.6}$$

the representatives $\{d_\nu|\boldsymbol{s}_\nu\}$ of the cosets just generate the star of $\boldsymbol{k}$:

$$*\boldsymbol{k} = \{\boldsymbol{k}_1 = \boldsymbol{k}, \boldsymbol{k}_2 = d_2\boldsymbol{k}, \ldots, \boldsymbol{k}_s = d_s\boldsymbol{k}\} \ . \qquad (9.1.7)$$

The nonprimitive translations $\boldsymbol{s}_\nu$ do not enter here (Sect. 3.4.1).

For points in the interior of the BZ $\mathscr{G}_{\boldsymbol{k}}$ consists only of elements which leave $\boldsymbol{k}$ invariant ($\boldsymbol{K} \equiv 0$). For $\boldsymbol{k}$-vectors at the surface of the first BZ $\mathscr{G}_{\boldsymbol{k}}$ also contains elements which map $\boldsymbol{k}$ onto equivalent vectors (modulo $\boldsymbol{K}$). Since $\mathscr{G}_{\boldsymbol{k}}$ contains the translation group $\mathbb{T}$ as an invariant subgroup there is a decomposition of $\mathscr{G}_{\boldsymbol{k}}$ into cosets

$$\mathscr{G}_{\boldsymbol{k}} = \mathbb{T} + \{d'_2|s'_2\}\mathbb{T} + \cdots + \{d'_n|s'_n\}\mathbb{T} \ . \qquad (9.1.8)$$

The quotient group $\mathscr{G}_{0\boldsymbol{k}} := \mathscr{G}_{\boldsymbol{k}}/\mathbb{T}$ contains (apart from isomorphism) all the elements d' of the point group $\mathscr{G}$ of $\mathscr{R}$, for which

$$\boldsymbol{k} = d'\boldsymbol{k} + \boldsymbol{K}_m \ . \qquad (9.1.3\text{b})$$

Since $d' \in \mathscr{G}$ we can also decompose $\mathscr{G}$ into cosets of $\mathscr{G}_{0\boldsymbol{k}}$

$$\mathscr{G} \cong \mathscr{G}_{0\boldsymbol{k}} + d_2\mathscr{G}_{0\boldsymbol{k}} + \cdots + d_s\mathscr{G}_{0\boldsymbol{k}} \ . \qquad (9.1.9)$$

The number s of points of $*\boldsymbol{k}$ is equal to the number of cosets in (9.1.9), i.e. equal to the index of $\mathscr{G}_{0\boldsymbol{k}}$ in $\mathscr{G}$. Thus

$$g = sn_0 \ , \qquad n_0 = \operatorname{ord}\mathscr{G}_{0\boldsymbol{k}} \ . \qquad (9.1.10)$$

We shall elucidate this notation for the cases of the nonsymmorphic diamond structure ($\mathcal{O}_\mathrm{h}^7 = Fd3m$, Fig. 9.1) and the symmorphic NaCl structure ($\mathcal{O}_\mathrm{h}^5 = Fm3m$, Fig. 9.2). Both of them have the fcc translation group with the basis translations $\boldsymbol{a}^{(1)} = (0, a/2, a/2)$; $\boldsymbol{a}^{(2)} = (a/2, 0, a/2)$; $\boldsymbol{a}^{(3)} = (a/2, a/2, 0)$.

According to (5.3.10) the reciprocal basis is $\boldsymbol{b}^{(1)} = (-1, 1, 1)\ 2\pi/a$; $\boldsymbol{b}^{(2)} = (1, -1, 1)\ 2\pi/a$; $\boldsymbol{b}^{(3)} = (1, 1, -1)\ 2\pi/a$ yielding the BZ in Fig. 9.1b. The decomposition into cosets in these cases is

$$\mathcal{O}_\mathrm{h}^5 = \{e|0\}\mathbb{T} + \cdots + \{d_{48}|0\}\mathbb{T} \ , \qquad (9.1.11\text{a})$$

where the elements $\{e, \ldots, d_{48}\} = \mathcal{O}_\mathrm{h}$ constitute the point group $\mathscr{G}$ and

$$\mathcal{O}_\mathrm{h}^7 = \{e|0\}\mathbb{T} + \cdots + \{d_{24}|0\}\mathbb{T} + \{i|\boldsymbol{s}\}\mathbb{T} + \cdots + \{id_{24}|\boldsymbol{s}\}\mathbb{T} \ , \qquad (9.1.11\text{b})$$

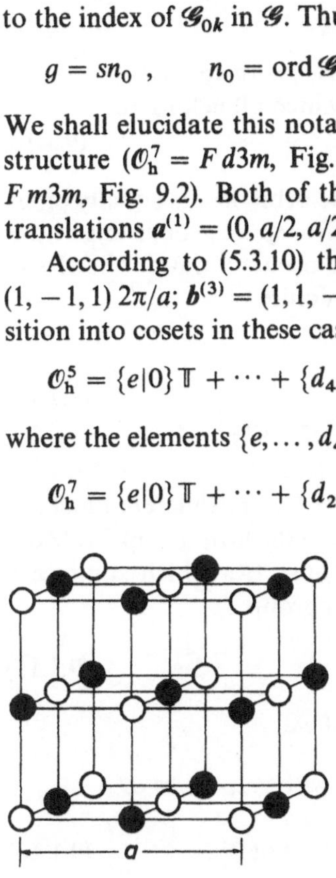

Fig. 9.2. NaCl structure $\mathcal{O}_\mathrm{h}^5 = Fm3m$

where the elements $\{e,\dots,d_{24}\} = \mathscr{T}_d \subset \mathscr{O}_h$ form a point group; $\mathscr{T}_d \times \mathscr{C}_i$ is isomorphic to $\mathscr{O}_h = \mathscr{O} \times \mathscr{C}_i$. The nonprimitive translation is $\boldsymbol{s} = (1,1,1)a/4$.

As an example for the decomposition (9.1.6) we choose the point X (Fig. 9.1b) with the star $*X = \{X_1, X_2, X_3\}$ and $X_1 = (1,0,0)2\pi/a$, cyclic. Then for $X = X_1$

$$\mathscr{O}_h^5 = \mathscr{G}_X + \{c_3|0\}\mathscr{G}_X + \{c_3^{-1}|0\}\mathscr{G}_X \ , \qquad c_3 = \text{rotation about } x = y = z \ . \tag{9.1.12}$$

The quotient group $\mathscr{G}_{0X} = \mathscr{G}_X/\mathbb{T}$ is isomorphic to $\mathscr{D}_{4h}$. The representatives of the cosets according to (9.1.8) are the elements of $\mathscr{D}_{4h}$, since $\mathscr{O}_h^5$ is symmorphic. For $\mathscr{O}_h^7$, (9.1.12) is valid, too, but the representatives in (9.1.8) contain non-primitive translations $\boldsymbol{s}$. The representatives are elements of the form $\{d|0\}$: $\{e|0\}$, $\{c_{2x}|0\}$, $\{c_{2y}|0\}$, $\{c_{2z}|0\}$, $\{s_{4x}|0\}$, $\{s_{4x}^3|0\}$, $\{\sigma_{yz}|0\}$, $\{\sigma_{y\bar{z}}|0\}$ and the elements $\{id|\boldsymbol{s}\}$. Equation (9.1.10) is satisfied under all circumstances.

9.1.2 Induction of the Representations of $\mathscr{R}$

For the description of the IRs of $\mathscr{R}$ we start with the following theorem which will not be proved here (see, for example, [9.1]):

> Every REP D^{sub} subduced from an IR D of $\mathscr{R}$ by an invariant subgroup $\mathbb{T}$ decomposes into irreducible parts which all belong to the same orbit. (9.1.13)

In the subduced REP every IR and thus the complete orbit occurs equally often, i.e. α times (α: multiplicity of the orbit in the REP D). For space groups this means that their IRs can be characterized by $\boldsymbol{k}$. Every star occurs α times. The IRs of $\mathscr{R}$ in the subduction onto $\mathbb{T}$ thus have the block structure

$$D^{(*\boldsymbol{k},\alpha)\text{sub}}(e|\boldsymbol{R}^h) = \begin{pmatrix} \mathbf{1}_\alpha \exp(-i\boldsymbol{k}_1\cdot\boldsymbol{R}^h) & & \\ & \ddots & \\ & & \mathbf{1}_\alpha \exp(-i\boldsymbol{k}_s\cdot\boldsymbol{R}^h) \end{pmatrix} \tag{9.1.14}$$

with $\boldsymbol{k}_\nu$ from (9.1.7) and the α-dimensional unit matrix $\mathbf{1}_\alpha$. The different blocks in (9.1.14) are *allowable* IRs $D^{(\boldsymbol{k},\alpha)}$ of the translation part of the little group $\mathscr{G}_{\boldsymbol{k}}$. More generally, allowable IRs of a little group $\mathscr{G}_{\boldsymbol{k}}$ are those IRs which subduce a multiple of $D^{(\boldsymbol{k})} = \exp(-i\boldsymbol{k}\cdot\boldsymbol{R}^h)$ onto the translation group $\mathbb{T}$

$$D^{(\boldsymbol{k},\alpha)}(e|\boldsymbol{R}^h) = \mathbf{1}_\alpha \exp(-i\boldsymbol{k}\cdot\boldsymbol{R}^h) \ , \qquad d_\alpha := \alpha \ . \tag{9.1.15}$$

All the IRs $D^{(*\boldsymbol{k},\alpha)}$ of a space group $\mathscr{R}$ are obtained if we

1) take just one IR $D^{(\boldsymbol{k})}$ out of every orbit of $\mathbb{T}$ with respect to $\mathscr{R}$ and constitute the assigned little group $\mathscr{G}_{\boldsymbol{k}}$, and
2) induce an IR of $\mathscr{R}$ from every allowable IR $D^{(\boldsymbol{k},\alpha)}$ of $\mathscr{G}_{\boldsymbol{k}}$. (9.1.16)

The form of the induced REP is given by the block structure (coarsened permutation matrix) defined by (5.5.23). The induction process will be performed once more in this context (Sect. 5.5.2). Let

$$\mathscr{L}_\alpha^{(k)} \equiv \{\psi_1^{(k,\alpha)},\ldots,\psi_\alpha^{(k,\alpha)}\}\ ,\qquad d_\alpha := \alpha\ . \tag{9.1.17}$$

be the irreducible space of $D^{(k,\alpha)}$ constituted by the α Bloch functions $\psi_i^{(k,\alpha)}$. By means of the representatives $\{d_\nu|s_\nu\}$ of the cosets in (9.1.6) we form the $s\alpha$ independent functions $\{d_\nu|s_\nu\}\psi_i^{(k,\alpha)}$, which are basis functions of the IR $D^{(*k,\alpha)}$ with dimension $d = s\alpha$ induced from $D^{(k,\alpha)}$. These basis functions belong to the points $k_\nu = d_\nu \boldsymbol{k} + \boldsymbol{K}$ of $*\boldsymbol{k}$, as can easily be seen by application of $P(e|\boldsymbol{R}^h) := \{e|\boldsymbol{R}^h\}$:

$$\begin{aligned}\{e|\boldsymbol{R}^h\}\{d_\nu|s_\nu\}\psi_i^{(k,\alpha)} &= \{d_\nu|s_\nu\}\{e|d_\nu^{-1}\boldsymbol{R}^h\}\psi_i^{(k,\alpha)}\\ &= \exp(-\mathrm{i}d_\nu \boldsymbol{k}\cdot\boldsymbol{R}^h)\cdot\{d_\nu|s_\nu\}\psi_i^{(k,\alpha)}\ .\end{aligned}$$

The irreducible space $\mathscr{L}_\alpha^{(*k)}$ belonging to $D^{(*k,\alpha)}$ is then given by

$$\mathscr{L}_\alpha^{(*k)} = \{\{d_\nu|s_\nu\}\psi_i^{(k,\alpha)}|\nu = 1,\ldots,s;\, i = 1,\ldots,d_\alpha\} \tag{9.1.18a}$$

or

$$\mathscr{L}_\alpha^{(*k)} = \mathscr{L}_\alpha^{(k)} \oplus \mathscr{L}_\alpha^{(k_1)} \oplus \cdots \oplus \mathscr{L}_\alpha^{(k_s)} \tag{9.1.18b}$$

The matrix elements of $D^{(*k,\alpha)}$ can be expressed by those of $D^{(k,\alpha)}$, see (5.5.24). Let $b \in \mathscr{G}_k$ and let $a = \{d|t\} \in \mathscr{R}$ be arbitrary, but in such a way that it maps $\boldsymbol{k}_\nu$ onto $\boldsymbol{k}_\mu$, i.e.

$$a\boldsymbol{k}_\nu = \boldsymbol{k}_\mu\ . \tag{9.1.19}$$

Furthermore let $a_\nu = \{d_\nu|s_\nu\}$ and $a_\mu = \{d_\mu|s_\mu\}$ be representatives of cosets from (9.1.6) with

$$a_\nu \boldsymbol{k} = \boldsymbol{k}_\nu\ ,\qquad a_\mu \boldsymbol{k} = \boldsymbol{k}_\mu\ ,$$

then we can write every $b \in \mathscr{G}_k$ as

$$b = a_\mu^{-1} a a_\nu\ . \tag{9.1.20}$$

Applying a to $\psi_i^{(k_\nu)}$ it follows that

$$\begin{aligned}a\psi_i^{(k_\nu,\alpha)} &= a_\mu b a_\nu^{-1}\psi_i^{(k_\nu,\alpha)} = a_\mu b\psi_i^{(k,\alpha)} = a_\mu \sum_{j=1}^{d_\alpha} D_{ji}^{(k,\alpha)}(b)\psi_j^{(k,\alpha)}\\ &= \sum_{j=1}^{d_\alpha} D_{ji}^{(k,\alpha)}(a_\mu^{-1} a a_\nu)\psi_j^{(k_\mu,\alpha)}\ .\end{aligned} \tag{9.1.21}$$

From (9.1.21) we can see the matrix REP of the elements $a \in \mathscr{R}$

$$D^{(*k,\alpha)}_{\mu j,\nu i}(a) = \begin{cases} D^{(k,\alpha)}_{ji}(b) & \text{if } b = a_\mu^{-1} a a_\nu \in \mathscr{G}_k \,, \\ 0 & \text{otherwise,} \end{cases} \tag{9.1.22}$$

which means that the REP matrices have block structure with respect to μ, ν. In every (coarsened) row and column there is exactly one matrix $D^{(k,\alpha)}_{ji}$. With the definition

$$\mathring{D}^{(k,\alpha)}(b) = \begin{cases} D^{(k,\alpha)}(b) & \text{for } b \in \mathscr{G}_k \\ 0 & \text{for } b \notin \mathscr{G}_k \end{cases} \tag{9.1.23}$$

we can write also for (9.1.22)

$$D^{(*k,\alpha)}_{\mu,\nu}(a) = \mathring{D}^{(k,\alpha)}(a_\mu^{-1} a a_\nu) \,, \qquad a_\mu = \{d_\mu | s_\mu\}, \dots . \tag{9.1.24}$$

The character of this REP is

$$\chi^{(*k,\alpha)}(a) = \sum_{\mu=1}^{s} \mathring{\chi}^{(k,\alpha)}(a_\mu^{-1} a a_\mu) \,. \tag{9.1.25}$$

With this, the determination of the IRs $D^{(*k,\alpha)}$ of $\mathscr{R}$ becomes the determination of the allowable IRs $D^{(k,\alpha)}$ of the little group $\mathscr{G}_k$. Two methods of calculating the latter will be discussed in Sect. 9.2.

Exercise 9.1. Induce by means of (9.1.23, 24) the IRs $D^{(*k,\alpha)}$ of $\mathscr{R}$ from the $D^{(k,\alpha)}$ of $\mathscr{G}_k$ for the following two cases:
a) $\mathscr{R}$ symmorphic, k in a general position, i.e. $\mathscr{G}_k \equiv \mathbb{T}$
b) $\mathscr{R}$ nonsymmorphic, k in a special position, i.e. $\mathscr{G}_k \neq \mathbb{T}$.
Give the block structure of $D^{(*k,\alpha)}$ explicitly.

Exercise 9.2. Show with the help of the character of $D^{(*k,\alpha)}$ in (9.1.25) that an IR of $\mathscr{R}$ contains a complete star of k with an integer multiplicity.

9.2 Allowable Irreducible Representations of the Little Group $\mathscr{G}_k$

9.2.1 Projective Representations. Representations with a Factor System for $\mathscr{G}_{0k} = \mathscr{G}_k/\mathbb{T}$

For the multiplication of two representatives of cosets in (9.1.8) we have according to (3.4.5)

$$\{d'_\mu | s'_\mu\}\{d'_\nu | s'_\nu\} = \{d'_\mu d'_\nu | s'_\mu + d'_\mu s'_\nu\} = \{e | R_{\mu\nu}\}\{d'_{\mu\nu} | s'_{\mu\nu}\} \tag{9.2.1}$$

with

$$R_{\mu\nu} := s'_\mu + d'_\mu s'_\nu - s'_{\mu\nu} \in \{R^k\} = \mathbb{T} \,.$$

Thus for an allowable IR $D^{(k,\alpha)}$ of $\mathcal{G}_k$ to exist ($b_\mu := \{d'_\mu | s'_\mu\}$),

$$D^{(k,\alpha)}(b_\mu) \cdot D^{(k,\alpha)}(b_\nu) = \exp(-\mathrm{i}\boldsymbol{k} \cdot \boldsymbol{R}_{\mu\nu}) \mathbf{1}_\alpha D^{(k,\alpha)}(b_{\mu\nu}) \tag{9.2.2}$$

must hold. From the isomorphism $\mathcal{G}_{0k} = \mathcal{G}_k/\mathbb{T} \cong \{e, d'_2, \ldots, d'_n\}$ we see that the matrices $D^{(k,\alpha)}$ in (9.2.2), being images of the elements d'_ν, constitute a *projective* REP $D^{(0k,\alpha)}$ with the factor system

$$\omega^{(k)}(\mu, \nu) := \exp(-\mathrm{i}\boldsymbol{k} \cdot \boldsymbol{R}_{\mu\nu}) \ . \tag{9.2.3}$$

In general a *projective* REP (PR) (which is also denoted as a ray or multiplier REP, see [9.2–6]) with the *factor system* $\omega(b_1, b_2)$ of a group $\mathcal{G}$ is defined by

$$D(b_1) \cdot D(b_2) := \omega(b_1, b_2) \cdot D(b_1 b_2) \ , \quad |\omega(b_1, b_2)| = 1 \ , \quad b_1, b_2 \in \mathcal{G} \ . \tag{9.2.4}$$

The factor system $\omega(b_1, b_2)$ consists of g^2 numbers, which are not independent of each other because of the associativity law (2.1.3). From (9.2.4) we thus have the identity

$$\omega(b_1, b_2 b_3) \cdot \omega(b_2, b_3) = \omega(b_1 b_2, b_3) \cdot \omega(b_1, b_2) \ . \tag{9.2.5}$$

Any system of g^2 numbers satisfying (9.2.4, 5) can be used as a factor system[2]. Thus, it is necessary to introduce a further idea, i.e. that of a projective-equivalent (p-equivalent) factor system. Let D be a PR according to (9.2.4), then also

$$\hat{D}(b) = \frac{1}{f(b)} D(b) \ , \qquad |f(b)| = 1 \ , \qquad f(b)\text{: function on } \mathcal{G} \ , \tag{9.2.6a}$$

is such a PR with the factor system

$$\hat{\omega}(b_1, b_2) = \frac{\omega(b_1, b_2) \cdot f(b_1 b_2)}{f(b_1) f(b_2)} \ . \tag{9.2.6b}$$

Equation (9.2.6) satisfies (9.2.5). By different choices of $f(b)$ we can generate an arbitrary number of factor systems which are said to be p-equivalent if they satisfy (9.2.6a, b). The set of all p-equivalent factor systems constitutes an equivalence class (multiplier class). Projective-equivalent factor systems are identical only if $f(b_1 b_2) = f(b_1) \cdot f(b_2)$; then $f(b_i)$ defines an ordinary one-dimensional vector REP on $\mathcal{G}$ according to (4.2.8a).

By means of (9.2.6b) we cannot obtain all the possible factor systems of a group $\mathcal{G}$, i.e. a group may have different multiplier classes of factor systems. Schur has shown that a group of finite order possesses a finite number of classes of factor systems (see Exercise 9.3).

---

[2] For the production of tables of the PRs of groups, a standard factor system would be helpful. Since this does not exist, different authors use different systems. Thus one has to be careful when using these tables!

The "natural" factor system (9.2.3) can be converted into a simple p-equivalent system $\hat{\omega}^{(k)}(\mu, \nu)$ according to (9.2.6), as is used for example in the tables of *Kovalev* [9.5]. The p-equivalent REP according to (9.2.6) is by definition

$$\hat{D}^{(k,\alpha)}(b_\mu) := \exp(\mathrm{i}\boldsymbol{k} \cdot \boldsymbol{s}'_\mu) D^{(k,\alpha)}(b_\mu) \;, \qquad b_\mu = \{d'_\mu | \boldsymbol{s}'_\mu\} \tag{9.2.7a}$$

with the factor system

$$\hat{\omega}^{(k)}(\mu, \nu) = \omega^{(k)}(\mu, \nu) \frac{\exp(-\mathrm{i}\boldsymbol{k} \cdot \boldsymbol{s}'_{\mu\nu})}{\exp[-\mathrm{i}\boldsymbol{k} \cdot (\boldsymbol{s}'_\mu + \boldsymbol{s}'_\nu)]} = \exp[\mathrm{i}(\boldsymbol{k} - d'^{-1}_\mu \boldsymbol{k}) \cdot \boldsymbol{s}'_\nu] \;. \tag{9.2.7b}$$

We now define the inequivalent allowable IRs $D^{(k,\alpha)}$ of $\mathscr{G}_k$ by the inequivalent IPRs

$$\hat{D}^{(0k,\alpha)}(d'_\mu) = \exp(\mathrm{i}\boldsymbol{k} \cdot \boldsymbol{s}'_\mu) D^{(0k,\alpha)}(d'_\mu)$$

of the group $\{e, d'_2, \ldots\}$ isomorphic to $\mathscr{G}_{0k}$ by

$$D^{(k,\alpha)}(\{e|\boldsymbol{R}^h\} b_\mu) = \exp(-\mathrm{i}\boldsymbol{k} \cdot \boldsymbol{R}^h) \mathbf{1}_\alpha D^{(k,\alpha)}(b_\mu) \;, \tag{9.2.8a}$$

$$D^{(k,\alpha)}(b_\mu) = \exp(-\mathrm{i}\boldsymbol{k} \cdot \boldsymbol{s}'_\mu) \hat{D}^{(0k,\alpha)}(d'_\mu) \;, \qquad b_\mu = \{d'_\mu | \boldsymbol{s}'_\mu\} \;. \tag{9.2.8b}$$

In order to assure ourselves that the allowable IRs of $\mathscr{G}_k$ can indeed be written in this way, we have to check the REP properties. Since the REPs in (9.2.8a) are allowable, it is sufficient to consider (9.2.8b). On one hand we have

$$D^{(k,\alpha)}(b_\mu) \cdot D^{(k,\alpha)}(b_\nu) = \exp[-\mathrm{i}\boldsymbol{k} \cdot (\boldsymbol{s}'_\mu + \boldsymbol{s}'_\nu)] \cdot \hat{D}^{(0k,\alpha)}(d'_\mu) \cdot \hat{D}^{(0k,\alpha)}(d'_\nu)$$

and on the other

$$D^{(k,\alpha)}(b_\mu \cdot b_\nu) = \exp[-\mathrm{i}\boldsymbol{k}(d'_\mu \boldsymbol{s}'_\nu + \boldsymbol{s}'_\mu)] \cdot \hat{D}^{(0k,\alpha)}(d'_\mu d'_\nu) \;.$$

Using the factor system (9.2.7b) in $D^{(0k,\alpha)}$, we see from these equations that

$$\hat{D}^{(0k,\alpha)}(d'_\mu) \cdot \hat{D}^{(0k,\alpha)}(d'_\nu) = \exp[\mathrm{i}(\boldsymbol{k} - d'^{-1}_\mu \boldsymbol{k}) \cdot \boldsymbol{s}'_\nu] \cdot \hat{D}^{(0k,\alpha)}(d'_\mu d'_\nu) \;. \tag{9.2.9}$$

Thus the REP property is demonstrated. Using (9.2.8, 9) the allowable IRs of $\mathscr{G}_k$ are related to the IPRs of the assigned point group $\mathscr{G}_{0k}$.

However, for most $\boldsymbol{k}$-vectors of the BZ it is not necessary to consider the factor system; namely, whenever the exponent in (9.2.9) vanishes: This happens if 1) $\boldsymbol{k}$ lies *in the interior* of the BZ, i.e. $\boldsymbol{k} - d'^{-1}_\mu \boldsymbol{k} = 0$ for each d'_μ, or 2) $\mathscr{G}_k$ is a symmorphic space group, i.e. all the $\{\boldsymbol{s}'_\nu\}$ vanish. Then the $\hat{D}^{(0k,\alpha)}$ according to (9.2.9) represent a normal (vector) REP of $\mathscr{G}_{0k}$ as discussed in Chaps. 4, 5. Equations (9.2.8) then give the allowable IRs of $\mathscr{G}_k$. In the case of symmorphic space groups, $\mathscr{R}$, as well as $\mathscr{G}_k$, can be written as a semidirect product, see (3.4.10),

$$\mathscr{R} = \mathbb{T} \circledS \mathscr{G} \;, \qquad \mathscr{G}_k = \mathbb{T} \circledS \mathscr{G}_{0k} \;. \tag{9.2.10}$$

Statement 2 above then says that for semidirect products the (vector) REPs of $\mathcal{G}_{0k}$ can be directly extended to $\mathcal{G}_k$ (9.2.8). Only for *nonsymmorphic* space groups where the $\boldsymbol{k}$-vector additionally ends *on the surface* of the BZ, do we need the PRs explicitly. They will be constructed in the following.

The multiplier group $\mathbb{M}(\mathcal{G})$ assigned to a group $\mathcal{G}$ has by definition the multiplier classes $\{K_0, K_1, \ldots, K_{m-1}\}$ of the p-equivalent factor systems of $\mathcal{G}$ as its elements. The multiplication in $\mathbb{M}$ is defined by the one-to-one assignment

$$K_p \Leftrightarrow \omega^{(p)}(b_i, b_j) \qquad \text{with} \tag{9.2.11}$$

$$K_r = K_p \cdot K_s = K_s \cdot K_p \leftrightarrow \omega^{(r)}(b_i, b_j) = \omega^{(p)}(b_i, b_j) \cdot \omega^{(s)}(b_i, b_j) \ ;$$

$\mathbb{M}(\mathcal{G})$ is Abelian. The unit element is K_0 with the factor system $\omega^{(0)}(b_i, b_j) \equiv 1$, i.e. K_0 belongs to the normal (vector) REP of $\mathcal{G}$. Equivalence and irreducibility of PRs are defined in the same way as for vector REPs. In particular, unitary equivalent REPs belong to the same factor system because

$$D'(b_1) \cdot D'(b_2) \equiv U^+ D(b_1) U U^+ D(b_2) U = \omega(b_1, b_2) D'(b_1 b_2) \ .$$

In *Schur's method* for constructing the PRs of $\mathcal{G}$ this group is extended by $\mathbb{M}(\mathcal{G})$, the resulting group being $\mathcal{G}^*$. It has the order gm and is built up in such a way that $\mathbb{M}(\mathcal{G})$ is isomorphic to the centre of $\mathcal{G}^*$, i.e.

$$\mathcal{G}^*/\mathbb{M}(\mathcal{G}) \cong \mathcal{G} \tag{9.2.12}$$

(Sect. 2.3). According to Schur we then obtain all the inequivalent IPRs of $\mathcal{G}$ by taking all the inequivalent vector IRs of $\mathcal{G}^*$ and restricting these REPs to the elements of $\mathcal{G}$.

Determination of $\mathbb{M}(\mathcal{G})$. The group $\mathcal{G} = \{e, a, b, c, \ldots\}$ may be defined by the ν relations

$$a^{n_i} b^{l_i} c^{m_i} \ldots = e \ , \qquad i = 1, \ldots, \nu \tag{9.2.13}$$

between their generators $a, b, c, \ldots$. If now $A \equiv D(a), \ldots$ are the REP matrices of an IR of $\mathcal{G}$, then obviously we have

$$\alpha_i A^{n_i} B^{l_i} C^{m_i} \ldots = \mathbf{1} \ ; \qquad i = 1, \ldots, \nu \ . \tag{9.2.14}$$

In vector REPs, $\alpha_i \equiv 1$, while in PRs $\alpha_i \neq 1$, but $|\alpha_i| = 1$. According to (9.2.4, 5)

$$\alpha_i^{-1} = [\omega(a^{n_i}, b^{l_i} c^{m_i} \ldots)\omega(b^{l_i}, c^{m_i} \ldots)\omega(c^{m_i}, \ldots)\ldots]\omega_{n_i}(a)\omega_{l_i}(b)\omega_{m_i}(c)\ldots \tag{9.2.15}$$

with

$$\omega_{n_i}(a) = \omega(a, a)\omega(a, a^2)\ldots\omega(a, a^{n_i - 1})\ldots \ .$$

Obviously this factor system is very complicated. Consequently one tries to find a transformation $u(a)$ such that there is a large number of factors equal to one. We put

$$\hat{A} = A/u(a)\ , \qquad \hat{B} = B/u(b)\ ,\ldots \tag{9.2.16a}$$

and get p-equivalent REPs with

$$\hat{\alpha}_i \hat{A}^{n_i} \hat{B}^{l_i} \hat{C}^{m_i} \ldots = 1\ , \tag{9.2.16b}$$

$$\hat{\alpha}_i = \alpha_i \cdot u^{n_i}(a) \cdot u^{l_i}(b) \cdot u^{m_i}(c) \ldots .$$

From the new $\hat{\alpha}_i$ there may be $\nu' \leqslant \nu$ independent ones different from 1, whereas $\nu - \nu'$ can be chosen to be equal to 1. By varying the function u together with (9.2.16) it can be shown [9.6] for the 32 point groups that the $\alpha_i \neq 1$ can be written as

$$\hat{\alpha}_i^2 = 1\ ; \qquad i = 1, \ldots, \nu' \tag{9.2.17}$$

(see Example 2 at the end of this section). Every solution of this equation, i.e.

$$\{\hat{\alpha}_i^{(k)}\} := \{\hat{\alpha}_1, \ldots, \hat{\alpha}_{\nu'}\}^{(k)}\ , \qquad \text{with } \hat{\alpha}_i = \pm 1\ , \qquad k = 1, 2, \ldots, 2^{\nu'}\ , \tag{9.2.18}$$

defines a multiplier class of factor systems, from which there are altogether $2^{\nu'} = m$. The order of $\mathbb{M}(\mathscr{G})$ is thus $2^{\nu'}$. The solutions $\{\hat{\alpha}_i^{(k)}\}$ and $\{\hat{\alpha}_i^{(l)}\}$ and their product $\{\hat{\alpha}_i^{(m)}\} = \{\hat{\alpha}_i^{(k)} \cdot \hat{\alpha}_i^{(l)}\}$ correspond one-to-one to the classes K_k, K_l and K_m, respectively and thus define multiplication and structure in the Abelian group $\mathbb{M}(\mathscr{G})$.

Finally we introduce the generators $\tilde{\alpha}_i$, $i = 1, \ldots, \nu'$ where the $\tilde{\alpha}_i$ are abstract elements assigned to the numbers $\hat{\alpha}_i$, which define a group $\mathbb{H}_M$ of order $2^{\nu'}$ with the elements

$$h = \prod_{i=1,\ldots,\nu'} \tilde{\alpha}_j^{m_i} = \tilde{\alpha}_1^{m_i} \cdot \tilde{\alpha}_2^{m_i} \cdot \ldots \cdot \tilde{\alpha}_{\nu'}^{m_i}\ , \qquad m_i = 0, 1\ . \tag{9.2.19a}$$

The group $\mathbb{H}_M$ is *isomorphic* to $\mathbb{M}(\mathscr{G})$ if the $\tilde{\alpha}_i$ satisfy the defining relations

$$\tilde{\alpha}_i^2 = e\ ; \qquad \tilde{\alpha}_i \tilde{\alpha}_j = \tilde{\alpha}_j \tilde{\alpha}_i\ ; \qquad i, j = 1, \ldots, \nu'\ , \tag{9.2.19b}$$

which correspond to those of the $\hat{\alpha}_i$.

Definition of $\mathscr{G}^*$. The extended group $\mathscr{G}^*$ now contains the generators $\{a, b, c, \ldots; \tilde{\alpha}_1, \ldots, \tilde{\alpha}_{\nu'}\}$ with the defining relations

$$\tilde{\alpha}_i a^{n_i} b^{l_i} c^{m_i} \ldots = e\ , \qquad \tilde{\alpha}_i^2 = e\ , \qquad \tilde{\alpha}_i \tilde{\alpha}_j = \tilde{\alpha}_j \tilde{\alpha}_i\ , \qquad \tilde{\alpha}_i a = a \tilde{\alpha}_i\ , \qquad \text{etc.} \tag{9.2.20}$$

Thus $\mathbb{H}_M$ lies in the centre of $\mathscr{G}^*$ and we have

$$\mathcal{G}^*/\mathbb{H}_M \cong \mathcal{G} \text{ or } \mathcal{G}^* = \mathbb{H}_M + a\mathbb{H}_M + \ldots + a_g\mathbb{H}_M \; ; \qquad a, \ldots, a_g \in \mathcal{G} \; . \tag{9.2.21}$$

The isomorphism is established by[3] $a\mathbb{H}_M \leftrightarrow a$. Every element $a^* \in \mathcal{G}^*$ can be represented by $a^* = h \cdot a = a \cdot h$ with $h \in \mathbb{H}_M$, $a \in \mathcal{G}$. This means that the product of the two elements $a_i a_j \in \mathcal{G}$ generally contains in $\mathcal{G}^*$ a further element $h_{ij} \in \mathbb{H}_M$, i.e.

$$a_i a_j = h_{ij} a_k \; . \tag{9.2.22}$$

Now, if $D^{(\gamma)}$ is a vector REP of $\mathcal{G}^*$ then

$$D^{(\gamma)}(a_i) \cdot D^{(\gamma)}(a_j) = D^{(\gamma)}(h_{ij} a_k) = D^{(\gamma)}(h_{ij}) \cdot D^{(\gamma)}(a_k) \; .$$

Since h_{ij} commutes with each $a^* \in \mathcal{G}^*$, this is also valid for the corresponding REP matrices. Consequently, according to Schur's lemma (Sect. 4.2.4), $D^{(\gamma)}(h_{ij}) = \omega^{(\gamma)}(i,j) \cdot \mathbf{1}$, thus

$$D^{(\gamma)}(a_i) \cdot D^{(\gamma)}(a_j) = \omega^{(\gamma)}(i,j) \cdot D^{(\gamma)}(a_i a_j) \; . \tag{9.2.23}$$

With respect to $\mathcal{G}$, this defines just one PR with factor system $\omega^{(\gamma)}(i,j)$. The set of all inequivalent vector REPs of $\mathcal{G}^*$ yields all inequivalent IPRs of $\mathcal{G}$ (Schur) belonging to all the possible classes of factor systems.

The determination of the PRs of $\mathcal{G}$ is thus reduced to the calculation of the vector REPs of $\mathcal{G}^*$ according to Chaps. 4, 5. Analogously to the case of vector REPs there are orthogonality relations for the PRs. For unitary IPRs with the same or different factor systems (see Exercise 9.4)

$$\sum_{a \in \mathcal{G}} D_{ik}^{(\gamma)}(a) D_{jl}^{(\beta)*}(a) = \frac{g}{d_\gamma} \delta_{ij} \delta_{kl} \delta_{\beta\gamma} \; . \tag{9.2.24}$$

With the *same* factor system we also have

$$\sum_{\gamma, jl} d_\gamma D_{jl}^{(\gamma)}(a) D_{jl}^{(\gamma)*}(b) = g \cdot \delta_{ab} \; . \tag{9.2.25}$$

The reduction of a reducible PR into irreducible parts obeys

$$D = \sum_{\gamma} {}_{\oplus}\, m_\gamma D^{(\gamma)} \qquad \text{with} \qquad m_\gamma = \frac{1}{g} \sum_{a \in \mathcal{G}} \chi(a) \chi^{(\gamma)*}(a) \tag{9.2.26}$$

if the factor system is the same.

Example 1: Cyclic groups of order $g = n$ ($\mathcal{C}_n, \mathcal{S}_n$, n even; $\mathcal{S}_2 = \mathcal{C}_i$; $\mathcal{C}_s$; $\mathcal{S}_3 = \mathcal{C}_{3h}$, order $g = 6$). There is only one generator with $a^g = e$. From (9.2.14, 15)

---

[3] $a\mathbb{H}_M \cdot b\mathbb{H}_M = ab \cdot \mathbb{H}_M \leftrightarrow a \cdot b = ab$.

$$A^g = \omega_g(a)\mathbf{1} \qquad \text{with} \qquad \omega_g(a) = \omega(a,a)\ldots\omega(a,a^{g-1}) \ .$$

We choose $u(a) = [\omega_g(a)]^{1/g}$ and obtain using (9.2.16a)

$$\hat{A} = \hat{D}(a) = A\cdot[\omega_g(a)]^{-1/g} \qquad \text{or} \qquad \hat{A}^g = A^g/\omega_g(a) = 1 \ .$$

The PRs of cyclic groups are thus p-equivalent to vector REPs. The multiplier group $\mathbb{M}(\mathscr{G})$ consists of only the identity element K_0 [see below (9.2.11)]. The p-equivalent PRs are one dimensional, as are the vector REPs. According to (9.2.6b) the function $f(a^k) = [\omega_g(a)]^{k/g}/\omega_k(a)$ transforms the matrices of the PR $D(a^k)$ into the p-equivalent matrices $\hat{D} = D/f$ according to (9.2.6a) because

$$\hat{\omega}(a,a^k) = \frac{\omega(a,a^k)\cdot[\omega_g(a)]^{(k+1)/g}\cdot\omega_k(a)\cdot\omega_1(a)}{\omega_{k+1}(a)\cdot[\omega_g(a)]^{k/g}\cdot[\omega_g(a)]^{1/g}} = 1 \ . \tag{9.2.27}$$

Example 2: Groups of two (commuting) generators, e.g. $\mathscr{C}_{nh}$ with $a = c_n$, $b = \sigma_h$, $a^n = b^m = e$, $ab = ba$. With (9.2.14–16) then

$$A^n = \omega_n(a)\cdot 1 \ , \qquad B^m = \omega_m(b)\cdot 1 \ , \qquad AB = \frac{\omega(a,b)}{\omega(b,a)}\cdot BA = \alpha BA \ .$$

The transformation to p-equivalent REPs with $u(a) = [\omega_n(a)]^{1/n}$ and $u(b) = [\omega_m(b)]^{1/m}$ gives then, using (9.2.16a),

$$\hat{A}^n = 1 \ , \tag{9.2.28a}$$

$$\hat{B}^m = 1 \ , \tag{9.2.28b}$$

$$\hat{A}\hat{B} = \alpha\hat{B}\hat{A} \ , \tag{9.2.28c}$$

i.e. there is $v' = 1$ independent $\hat{\alpha}_i(\rightarrow\alpha)$ in the sense of (9.2.17, 18). We now have to determine the possible values (± 1) of α, in order to get the multiplier group and the extended group according to (9.2.20, 21). Repeated multiplication of (9.2.28c) with $\hat{A}$ gives $\hat{A}^n\hat{B} = \hat{A}^{n-1}\alpha\hat{B}\hat{A} = \cdots = \alpha^n\hat{B}\hat{A}^n$, thus $\alpha^n = 1$. Correspondingly, by multiplication with $\hat{B}$ we also have $\alpha^m = 1$. Thus we can write

$$\alpha^t = 1 \ , \qquad t \text{ largest common divisor of } \{m,n\} \ . \tag{9.2.29}$$

If m and n are "prime" numbers then $t = 1$, $\alpha = +1$ (e.g. $\mathscr{C}_{3h} \cong \mathscr{S}_3$). In this case all the PRs are p-equivalent to vector REPs (Example 1). For even $n = 2k$, thus for the groups $\mathscr{C}_{2k,h}$, $t = 2$, therefore $\alpha = \pm 1$. The group $\mathbb{M}(\mathscr{G})$ consists of two classes $\{K_0, K_1\}$, which are uniquely assigned to the solutions $\alpha = \{+1, -1\}$. Since the groups $\mathscr{C}_{nh}$ are Abelian, all the vector IRs and thus all the PRs belonging to K_0 are *one dimensional*. The REPs belonging to K_1 obey (9.2.28c) with $\hat{A}\hat{B} = -\hat{B}\hat{A}$, that means these PRs cannot be one dimensional[4]. To find out their

---

[4] $\hat{A}, \hat{B}$ are REP *matrices* of the elements a, b; however, one-dimensional matrices commute and cannot satisfy $\hat{A}\hat{B} = -\hat{B}\hat{A}$.

type we define the group $\mathcal{G}^*(\mathcal{C}_{nh}^*)$ according to (9.2.20) by

$$a^n = b^m = \tilde{\alpha}^2 = e \; ; \qquad a\tilde{\alpha} = \tilde{\alpha}a \; , \qquad b\tilde{\alpha} = \tilde{\alpha}b \; ; \qquad ab = \tilde{\alpha}ba \; . \tag{9.2.30}$$

Using (9.2.22, 23) we can get the factor system $\hat{\omega}$ of the group $\mathcal{G}$ from (9.2.30). Let $a_{i'} = a^i b^k$ and $a_{j'} = a^j b^l$, then $a_{i'} a_{j'} = a^i b^k a^j b^l = \tilde{\alpha}^{kj} a^{i+j} b^{k+l}$, i.e. $h_{i'j'} = \tilde{\alpha}^{kj}$ and

$$\hat{\omega}(a^i b^k, a^j b^l) = \alpha^{k \cdot j} \qquad \text{with} \qquad \alpha = \pm 1 \; . \tag{9.2.31}$$

This is the *standard* factor system of $\hat{D}$. For $\mathcal{C}_{2h}$ ($\cong \mathcal{C}_{2v} \cong \mathcal{D}_2$), $g = 4$ and for the REPs assigned to the class K_1 according to (4.2.26b)[5] $d_j^2 = 2^2 = 4$, i.e. the dimension of the IPR is $d_j = 2$. According to (9.2.28), $\hat{A}^2 = \hat{B}^2 = 1$, $\hat{A}\hat{B} = -\hat{B}\hat{A}$. We can choose two of the Pauli matrices as REP matrices for $\hat{A}$ and $\hat{B}$, for example

$$\hat{A} = \sigma_x \; , \qquad \hat{B} = \sigma_y \Rightarrow \hat{A}\hat{B} = -\hat{B}\hat{A} = \mathrm{i}\sigma_z \; .$$

This example illustrates how to get the IPRs of the 32 point groups, at least in principle. These can be used as the IPRs of $\mathcal{G}_{0k}$. In general, however, a gauge transformation has to be performed between the standard factor system of the method just explained and the p-equivalent factor systems used in the literature [Kovalev [9.5] and (9.2.7b)]. With the IPRs of $\mathcal{G}_{0k}$ finally all the allowable IRs of $\mathcal{G}_k$ according to (9.2.8, 9) and the IRs of $\mathcal{R}$ according to (9.1.22, 24) can be constructed.

9.2.2 Vector Representations of the Group $\mathcal{S}_k = \mathcal{G}_k/\mathbb{T}_k$

In (9.1.15) the allowable IRs of $\mathcal{G}_k$ were defined. For those translations $\{e|\boldsymbol{R}(\boldsymbol{k})\}$ obeying

$$\boldsymbol{k} \cdot \boldsymbol{R}(\boldsymbol{k}) = 2\pi n \; , \qquad n \in \mathbb{Z} \; , \qquad \boldsymbol{R}(\boldsymbol{k}) \in \{\boldsymbol{R}^{\boldsymbol{k}}\} \; , \tag{9.2.32}$$

this means that the projections of them onto $\boldsymbol{k}$ are equal to a multiple of the wavelength $\lambda = 2\pi/k$, and the allowable IRs according to (9.1.15) just are the unit matrices $\mathbf{1}_\alpha$. The translations defined by (9.2.32) obviously constitute a subgroup $\mathbb{T}_k \subset \mathbb{T}$. The quotient group $\mathbb{T}/\mathbb{T}_k$ is defined by the decomposition into cosets

$$\mathbb{T} = \mathbb{T}_k + \{e|\boldsymbol{R}_2(\boldsymbol{k}')\}\mathbb{T}_k + \cdots + \{e|\boldsymbol{R}_f(\boldsymbol{k}')\}\mathbb{T}_k \; , \tag{9.2.33}$$

where $\boldsymbol{k}'$ indicates that $\{e|\boldsymbol{R}_i(\boldsymbol{k}')\} \notin \mathbb{T}_k$. With (9.1.8) and (9.2.33) we can give further a decomposition of $\mathcal{G}_k$ into cosets with respect to $\mathbb{T}_k$:

---

[5] Burnside's theorem is valid for $\mathcal{G}^*$ as well as for $\mathbb{M}(\mathcal{G})$ and $\mathcal{G}$ in (9.2.12) and consequently can be used here.

$$\begin{aligned}\mathcal{G}_k = \mathbb{T}_k &+ \{e|\boldsymbol{R}_2(\boldsymbol{k}')\}\mathbb{T}_k + \cdots + \{e|\boldsymbol{R}_f(\boldsymbol{k}')\}\mathbb{T}_k \\ &+ \{d_2'|\boldsymbol{s}_2'\}\mathbb{T}_k + \cdots + \{d_2'|\boldsymbol{s}_2' + \boldsymbol{R}_f(\boldsymbol{k}')\}\mathbb{T}_k + \cdots \\ &+ \{d_n'|\boldsymbol{s}_n' + \boldsymbol{R}_f(\boldsymbol{k}')\}\mathbb{T}_k \ . \end{aligned} \tag{9.2.34}$$

The quotient group $\mathcal{S}_k = \mathcal{G}_k/\mathbb{T}_k$ is thus defined by the $n \cdot f$ cosets

$$\{d_\nu'|\boldsymbol{s}_\nu' + \boldsymbol{R}_i(\boldsymbol{k}')\}\mathbb{T}_k \ ; \qquad \nu = 1, \dots, n \ ; \qquad i = 1, \dots, f \ . \tag{9.2.35}$$

We consider $\mathcal{S}_k$ as an abstract group of the order $n \cdot f$, the IRs $D(\mathcal{S}_k)$ of which can be determined by the standard methods explained in Chaps. 4, 5. With the homomorphisms $\mathcal{G}_k \to \mathcal{G}_k/\mathbb{T}_k = \mathcal{S}_k \to D(\mathcal{S}_k)$ the IRs $D(\mathcal{S}_k)$ are also IRs of $\mathcal{G}_k$. Using the character equation

$$\chi^{(\boldsymbol{k},\alpha)}(e|\boldsymbol{R}_i(\boldsymbol{k}')) = d_\alpha \exp[-\mathrm{i}\boldsymbol{k} \cdot \boldsymbol{R}_i(\boldsymbol{k}')] \tag{9.2.36}$$

the allowable IRs $D^{(\boldsymbol{k},\alpha)}$ of $\mathcal{G}_k$ have to be selected from the possible IRs $D(\mathcal{S}_k)$. With this method care has to be taken that the $n \cdot f$ representatives of the cosets are closed and isomorphic to the abstract group $\mathcal{S}_k$. According to (3.4.5) the multiplication is defined by

$$\begin{aligned}\{d_\nu'|\boldsymbol{s}_\nu' + \boldsymbol{R}_i(\boldsymbol{k}')\}\{d_\mu'|\boldsymbol{s}_\mu' + \boldsymbol{R}_j(\boldsymbol{k}')\} &= \{d_\nu' d_\mu'|\boldsymbol{s}_{\nu\mu}' + \boldsymbol{R}_{ij}(\boldsymbol{k}')\} \\ &= \{e|\boldsymbol{R}_{ij}(\boldsymbol{k}')\} \cdot \{d_\nu' d_\mu'|\boldsymbol{s}_{\nu\mu}'\} \end{aligned} \tag{9.2.37}$$

with

$$\boldsymbol{s}_{\nu\mu}' + \boldsymbol{R}_{ij}(\boldsymbol{k}') = d_\nu'[\boldsymbol{s}_\mu' + \boldsymbol{R}_j(\boldsymbol{k}')] + \boldsymbol{s}_\nu' + \boldsymbol{R}_i(\boldsymbol{k}') + \boldsymbol{R}(\boldsymbol{k}) \ .$$

Here a vector $\{e|\boldsymbol{R}(\boldsymbol{k})\} \in \mathbb{T}_k$ has been added in order to cause the multiplication to be closed. This is always possible since the elements of $\mathbb{T}_k$ are mapped onto the identity.

Examples: Γ and L points of the BZ of $\mathcal{O}_\mathrm{h}^5$ (Fig. 9.1b). For the *Γ point*, $\mathcal{G}_\Gamma = \mathcal{R}$, i.e. $\mathcal{G}_\Gamma$ is equal to the total space group. Correspondingly we have $\mathbb{T}_\Gamma = \mathbb{T}$, equal to the total translation group. Then $\mathcal{S}_\Gamma = \mathcal{G}_\Gamma/\mathbb{T}_\Gamma = \mathcal{R}/\mathbb{T} \cong \mathcal{G}_{0\Gamma} = \mathcal{O}_\mathrm{h}$, thus $\mathcal{S}_\Gamma$ is isomorphic to the total point group of $\mathcal{O}_\mathrm{h}^5$. Since here all the translations $\{e|\boldsymbol{R}^{\boldsymbol{h}}\}$ are represented by unit matrices $\mathbf{1}_\alpha$, we obtain, in the case of the Γ point, by means of the character table of $\mathcal{O}_\mathrm{h}$, the character table of $\mathcal{R}$ for the IR $D^{(*\Gamma,m)}$, if we replace the elements $a \in \mathcal{O}_\mathrm{h}$ by the cosets $\{d|0\}\mathbb{T} \in \mathcal{R}$ (See Sect. 8.5 for a discussion of direction symmetries).

As an L point we choose the point $\boldsymbol{L} = (1,1,1)\pi/a$ (Fig. 9.1b); it represents $*L$. According to (9.2.32) $\mathbb{T}_L$ is given by $\boldsymbol{L} \cdot \boldsymbol{R}(\boldsymbol{L}) = 2\pi n$ or $h_1 + h_2 + h_3 = 2n$ with

$$\boldsymbol{R}^{\boldsymbol{h}} = \sum_{i=1}^{3} h_i \boldsymbol{a}^{(i)} \ , \qquad n \in \mathbb{Z} \ ; \tag{9.2.38}$$

in $\mathcal{G}_L$ all these translations are mapped onto $\mathbf{1}_\alpha$. The elements $\{e|R_i(L')\}$ from (9.2.33) then satisfy the condition

$$h_1 + h_2 + h_3 = 2n + 1 \ , \qquad n \in \mathbb{Z} \ , \tag{9.2.39}$$

because of (9.2.38). Since the h_i are integers, other points L' cannot exist, i.e. according to (9.2.33)

$$\mathbb{T} = \mathbb{T}_L + \{e|\boldsymbol{R}_2(L')\}\cdot\mathbb{T}_L \ ,$$

where $\{e|\boldsymbol{R}_2(L')\} = \{e|\boldsymbol{a}^{(1)}\}$ can be chosen since $\boldsymbol{L}\cdot\boldsymbol{a}^{(1)} = \pi$. Thus, in $\mathcal{G}_L$ the elements of the coset are mapped onto $-\mathbf{1}_\alpha$. The representatives of the cosets in (9.2.34, 35) are the elements $\{d'_\nu|0\}$ and $\{d'_\nu|\boldsymbol{a}^{(1)}\}$, where the d'_ν have to be determined in such a way that they leave $\boldsymbol{L}$ invariant. In this case these are (Fig. 9.1b) $\{e, c_{3xyz}, c^2_{3xyz}, c_{2x\bar{y}}, c_{2y\bar{z}}, c_{2z\bar{x}}, i, s_{6xyz}, s^5_{6xyz}, \sigma_{x\bar{y}}, \sigma_{y\bar{z}}, \sigma_{z\bar{x}}\} \cong \mathcal{D}_{3d}$. The abstract group $\mathcal{S}_L = \mathcal{G}_L/\mathbb{T}_L$ consists of $2\times 12 = 24$ elements. The restriction to allowable IRs $D^{(\boldsymbol{k},\alpha)}$ of $\mathcal{G}_L$ provides a relation between the elements $\{d'_\nu|0\}$ and $\{d'|\boldsymbol{a}^{(1)}\}$, namely

$$D^{(L,\alpha)}(d'_\nu|\boldsymbol{a}^{(1)}) = D^{(L,\alpha)}(e|\boldsymbol{a}^{(1)})\cdot D^{(L,\alpha)}(d'_\nu|0) = -\mathbf{1}_\alpha\cdot D^{(L,\alpha)}(d'_\nu|0) \ . \tag{9.2.40}$$

From this we see that $\mathcal{S}_L$ is isomorphic to the direct product $\mathcal{C}_i \times \mathcal{D}_{3d}$, the REPs of which then have to be used.

The given method for the determination of the IRs of $\mathcal{R}$ via the vector REPs of $\mathcal{S}_{\boldsymbol{k}}$ is of course not independent of the method of PRs of $\mathcal{G}_{0\boldsymbol{k}}$ presented in Sect. 9.2.1. This can be seen if we investigate the multiplication of the elements $\{d'_\nu|s'_\nu\}$ occurring in both groups. According to (9.2.37) we have

$$\{d'_\nu|s'_\nu\}\{d'_\mu|s'_\mu\} = \{e|\boldsymbol{R}(\boldsymbol{k}')\}\cdot\{d'_\nu d'_\mu|s'_{\nu\mu}\} \qquad \text{with} \tag{9.2.41}$$

$$\boldsymbol{R}(\boldsymbol{k}') = d'_\nu s'_\mu + s'_\nu - s'_{\nu\mu} \notin \mathbb{T}_{\boldsymbol{k}} \ .$$

Since $D^{(\boldsymbol{k},\alpha)}$ is an allowable IR of $\mathcal{G}_{\boldsymbol{k}}$ it follows that

$$D^{(\boldsymbol{k},\alpha)}(e|\boldsymbol{R}(\boldsymbol{k}')) = \mathbf{1}_\alpha\cdot\exp[-i\boldsymbol{k}\cdot\boldsymbol{R}(\boldsymbol{k}')] \ . \tag{9.2.42}$$

Furthermore

$$\{e|\boldsymbol{R}(\boldsymbol{k}')\}\cdot\{d'_\nu d'_\mu|s'_{\nu\mu}\} = \{d'_\nu d'_\mu|s'_{\nu\mu}\}\cdot\{e|d'^{-1}_\mu d'^{-1}_\nu \boldsymbol{R}(\boldsymbol{k}')\} \ .$$

So, finally, we have the chain of equations

$$\begin{aligned} D^{(\boldsymbol{k},\alpha)}(d'_\nu|s'_\nu)\cdot D^{(\boldsymbol{k},\alpha)}(d'_\mu|s'_\mu) &= D^{(\boldsymbol{k},\alpha)}(e|\boldsymbol{R}(\boldsymbol{k}'))\cdot D^{(\boldsymbol{k},\alpha)}(d'_\nu d'_\mu|s'_{\nu\mu}) \\ &= D^{(\boldsymbol{k},\alpha)}(d'_\nu d'_\mu|s'_{\nu\mu})\cdot D^{(\boldsymbol{k},\alpha)}(e|d'^{-1}_\mu d'^{-1}_\nu \boldsymbol{R}(\boldsymbol{k}')) \\ &= D^{(\boldsymbol{k},\alpha)}(d'_\nu d'_\mu|s'_{\nu\mu})\cdot D^{(\boldsymbol{k},\alpha)}(e|\boldsymbol{R}(\boldsymbol{k}')) \ . \end{aligned} \tag{9.2.43}$$

This shows that the REPs constitute a system which is closed, if the allowable IRs are given by (9.2.42). From (9.2.43) we can also see that the allowable IRs of $\mathcal{G}_k$ or $\mathcal{S}_k$ with respect to $\mathcal{G}_{0k}$ just are the PRs with the natural factor system $\omega^{(k)}(\mu, \nu) = \exp[-i\boldsymbol{k} \cdot \boldsymbol{R}(k')]$. This follows from (9.2.1–3). *Zak* [9.7] gives another method (induction method) for the determination of the allowable IRs of a nonsymmorphic group $\mathcal{G}_k$. They are induced from the allowable IRs of a symmorphic invariant subgroup of index 2 or 3, the IRs of which are known from (9.2.8). We will not go into greater detail here.

9.2.3 Representations of Double Space Groups. Spinor Representations

In Sect. 5.1 we saw that the IRs of a double point group with respect to the assigned point group divide into single- and double-valued IRs. We are especially interested in the double-valued IRs. The elements of the double space groups $\mathcal{R}^{\mathrm{D}}$ can be represented by[6] (Sect. 3.4.3)

$$\{d^{\pm}|\boldsymbol{t}\} \equiv \pm D^{(1/2)}(d) \cdot \{d|\boldsymbol{t}\} \tag{9.2.44a}$$

or using the element $c_0 = c(2\pi) = \{e^-|0\}$ as in (3.3.4, 3.4.15) by[7]

$$\{d^{\pm}|\boldsymbol{t}\} := \{\{d|\boldsymbol{t}\}; c_0\{d|\boldsymbol{t}\} = \{c_0 d|\boldsymbol{t}\} := \{\bar{d}|\boldsymbol{t}\}\} \ . \tag{9.2.44b}$$

The decompositions of $\mathcal{R}^{\mathrm{D}}$ into cosets being analogous to those of the normal space groups are thus obvious, e.g.

$$\mathbb{T}^{\mathrm{D}} = \mathbb{T}^+ + \{e^-|0\} \cdot \mathbb{T}^+ \tag{9.2.45a}$$

and with (9.1.8)

$$\begin{aligned} \mathcal{G}_k^{\mathrm{D}} = {} & \mathbb{T}^+ + \{e^-|0\}\mathbb{T}^+ + \{d_2'|s_2'\}\mathbb{T}^+ + \{\bar{d}_2'|s_2'\}\mathbb{T}^+ + \cdots \\ & + \{d_n'|s_n'\}\mathbb{T}^+ + \{\bar{d}_n'|s_n'\}\mathbb{T}^+ \ . \end{aligned} \tag{9.2.45b}$$

Because of (3.4.15) there are twice as many IRs of $\mathbb{T}^{\mathrm{D}}$ as of $\mathbb{T}^+$. See Table 9.1.

Table 9.1. Representations of the double translation groups

$\mathbb{T}^{\mathrm{D}}$	$\mathbb{T}^+$	$\{e^-\|0\} \cdot \mathbb{T}^+$
$D^{(k)}$	$\exp(-i\boldsymbol{k} \cdot \boldsymbol{R}^{\boldsymbol{\iota}})$	$\exp(-i\boldsymbol{k} \cdot \boldsymbol{R}^{\boldsymbol{\iota}})$
$\bar{D}^{(k)}$	$\exp(-i\boldsymbol{k} \cdot \boldsymbol{R}^{\boldsymbol{\iota}})$	$-\exp(-i\boldsymbol{k} \cdot \boldsymbol{R}^{\boldsymbol{\iota}})$

[6] We do not always distinguish between the elements $\{d|\boldsymbol{t}\} \in \mathcal{R}$ and the assigned operators $P(d|\boldsymbol{t})$. Since these are isomorphic, this leads to no difficulties (Sect. 3.4.3).

[7] Products $d_i \bar{d}_j$ multiply according to the multiplication of double point groups, i.e. as $D^{(1/2)}(d_i) \cdot (-1) \cdot D^{(1/2)}(d_j)$. Besides this, $\bar{d}_j \boldsymbol{t} = d_j \boldsymbol{t}$, i.e. d_j and $\bar{d}_j$ have the same effect on translations $\boldsymbol{t}$.

The REPs $D^{(k)}$ and $\bar{D}^{(k)}$ can never belong to the same orbit. To every star $\boldsymbol{k}$ there belongs an orbit of $D^{(k)}$ REPs and an orbit of $\bar{D}^{(k)}$ REPs. If the allowable IRs $D^{(\boldsymbol{k},\alpha)}$ and $\bar{D}^{(\boldsymbol{k},\alpha)}$ of $\mathcal{G}_{\boldsymbol{k}}^{\mathrm{D}}$ are known then we can induce the IRs $D^{(*\boldsymbol{k},\alpha)}$ and $\bar{D}^{(*\boldsymbol{k},\alpha)}$ of the double space groups in the same way as for the simple space groups (Sect. 9.1.2). For the determination of $D^{(\boldsymbol{k},\alpha)}$, $\bar{D}^{(\boldsymbol{k},\alpha)}$ of $\mathcal{G}_{\boldsymbol{k}}^{\mathrm{D}}$ we distinguish the following cases.

1a) The $D^{(\boldsymbol{k},\alpha)}$ belonging to $D^{(k)}$ are directly obtained from the $D^{(\boldsymbol{k},\alpha)}$ of $\mathcal{G}_{\boldsymbol{k}}$. The same REP matrix is assigned to the elements $\{d|t\}$ and $\{\bar{d}|t\}$:

$$D^{(\boldsymbol{k},\alpha)}(\bar{d}|\boldsymbol{t}) = D^{(\boldsymbol{k},\alpha)}(d|\boldsymbol{t}) \ . \tag{9.2.46a}$$

1b) The more important double-valued (spinor) IRs are those which are assigned to the $\bar{D}^{(k)}$. For these

$$\bar{D}^{(\boldsymbol{k},\alpha)}(\bar{d}|\boldsymbol{t}) = -\bar{D}^{(\boldsymbol{k},\alpha)}(d|\boldsymbol{t}) \tag{9.2.46b}$$

is always valid.

If $\mathcal{G}^{\mathrm{D}}$ is *symmorphic or* $\boldsymbol{k}$ is a point of the *interior of the* BZ, the allowable IRs of $\mathcal{G}_{\boldsymbol{k}}^{\mathrm{D}}$ can be traced back to the IRs of the double point group $\mathcal{G}_{0\boldsymbol{k}}^{\mathrm{D}} = \{d'_\nu, \bar{d}'_\nu\}$. We obtain (see Exercise 9.7)

$$\bar{D}^{(\boldsymbol{k},\alpha)}(a_\nu|\boldsymbol{t}) = \mathrm{e}^{-\mathrm{i}\boldsymbol{k}\cdot\boldsymbol{t}} \cdot \bar{D}^{(\boldsymbol{k},\alpha)}(a_\nu) \ , \qquad a_\nu = d'_\nu \text{ or } \bar{d}'_\nu \ . \tag{9.2.46c}$$

The $\bar{D}^{(\boldsymbol{k},\alpha)}(a_\nu)$ are the double-valued IR matrices of $\mathcal{G}_{0\boldsymbol{k}}^{\mathrm{D}}$ for which, according to (9.2.46b), we have

$$\bar{D}^{(\boldsymbol{k},\alpha)}(\bar{d}'_\nu) = -\bar{D}^{(\boldsymbol{k},\alpha)}(d'_\nu) \ . \tag{9.2.46d}$$

2) Representation theory becomes more complicated for some points at the surface of the BZ in nonsymmorphic double space groups than for the normal space groups. This will be briefly discussed with the help of the PRs in Sect. 9.2.1. First we realize that the double-valued (spinor) REPs of the point groups $\mathcal{G}$ are also just the PRs of $\mathcal{G}$ with a certain factor system. According to Exercise 3.8

$$D^{(1/2)}(\boldsymbol{n}, \varphi) = (\cos\varphi/2)\cdot\mathbf{1} - \mathrm{i}(\sin\varphi/2)(\boldsymbol{n}\cdot\boldsymbol{\sigma}) \ , \tag{9.2.47}$$

$$\boldsymbol{\sigma} = \{\sigma_x, \sigma_y, \sigma_z\}\text{: Pauli-matrices} \ .$$

Due to the multiplication rule we furthermore have to distinguish between rotations through φ and $\varphi + 2\pi$.

If $a_i \in \mathcal{G}$ are the elements of the point group, then[8]

$$D^{(1/2)}(a_i)\cdot D^{(1/2)}(a_j) = \pm\mathbf{1}\cdot D^{(1/2)}(a_k) \ , \qquad \text{e.g.} \tag{9.2.48}$$

$$D^{(1/2)}(c_2)\cdot D^{(1/2)}(c_2) = -\mathbf{1} \ .$$

[8] Improper rotations a'_i can be taken into account by $D^{(1/2)}(a'_i) = D^{(1/2)}(a_i)$ because $D^{(1/2)}(a)P_a\psi$ and $\lambda D^{(1/2)}(a)P_a\psi$ describe the same state. Factors occurring by chance can be dropped.

By this relation the factor system $\omega^{(1/2)} = \{+1, -1\}$ of $D^{(1/2)}$ is established. Combining this factor system with that of the PRs of the point groups given in Sect. 9.2.1 we obtain the factor system for the projective spinor REPs $\bar{D}(a_i)$ of the point groups:

$$\bar{\omega}(a_i, a_j) := \omega^{(1/2)}(a_i, a_j) \cdot \omega(a_i, a_j) \ . \tag{9.2.49}$$

Sometimes it is necessary that (9.2.49) is subject to a gauge transformation as in (9.2.6a) in order to fix the class of the factor system $\bar{\omega}$. The only multiplier classes possible are those which are compatible with

$$\bar{D}(\bar{a}_i) = -\bar{D}(a_i) \ . \tag{9.2.50a}$$

These are only two classes (usually denoted by K_0, K_1). Using (9.2.50a) the spinor REP of $\mathscr{G}$ can be extended to form the complete vector REP of the double group $\mathscr{G}^{\mathrm{D}}$, i.e. the spinor REPs of $\mathscr{G}$ give the vector REPs of $\mathscr{G}^{\mathrm{D}} = \mathscr{G} + c_0\mathscr{G}$. This construction provides the IRs of double space groups $\mathscr{R}^{\mathrm{D}}$. The factor system $\hat{\bar{\omega}}^{(\boldsymbol{k})}$ of a spinor REP of $\mathscr{G}_{0\boldsymbol{k}}$ is given by $(d'_\nu \to \nu)$

$$\hat{\bar{\omega}}^{(\boldsymbol{k})}(\mu, \nu) := \omega^{(1/2)}(\mu, \nu) \cdot \hat{\omega}^{(\boldsymbol{k})}(\mu, \nu) \ , \tag{9.2.50b}$$

where the $\hat{\omega}^{(\boldsymbol{k})}$ can be seen from (9.2.7) (factor system according to *Kovalev* [9.5]). Using

$$D^{(\boldsymbol{k},\alpha)}(a_\nu | s_\nu + \boldsymbol{R}^h) = \exp(-\mathrm{i}\boldsymbol{k} \cdot \boldsymbol{R}^h) \cdot D^{(\boldsymbol{k},\alpha)}(a_\nu) \ , \tag{9.2.51}$$

$$a_\nu \to d'_\nu \text{ or } \bar{d}'_\nu \ ; \qquad D \to D \text{ or } \bar{D}$$

and

$$\bar{D}^{(\boldsymbol{k},\alpha)}(\bar{d}'_\nu) = -\bar{D}^{(\boldsymbol{k},\alpha)}(d'_\nu) \ , \qquad D^{(\boldsymbol{k},\alpha)}(\bar{d}'_\nu) = +D^{(\boldsymbol{k},\alpha)}(d'_\nu)$$

we are able to extend the PRs of $\mathscr{G}_{0\boldsymbol{k}}$ (single-valued $D^{(\boldsymbol{k},\alpha)}$, double-valued $\bar{D}^{(\boldsymbol{k},\alpha)}$ of $\mathscr{G}^{\mathrm{D}}_{0\boldsymbol{k}}$) to the allowable vector IPRs of $\mathscr{G}^{\mathrm{D}}_{\boldsymbol{k}}$ and finally transfer to IPRs $D^{(*\boldsymbol{k},\alpha)}$ or $\bar{D}^{(*\boldsymbol{k},\alpha)}$ of the double space groups $\mathscr{R}^{\mathrm{D}}$ by induction.

Exercise 9.3. Show that if ω and $\hat{\omega}$ are two factor systems and if for a pair of commuting elements $a, b \in \mathscr{G}$ it holds that $[\hat{\omega}(a,b)/\hat{\omega}(b,a)] \neq [\omega(a,b)/\omega(b,a)]$, then these factor systems belong to different (multiplier) classes. The transformation (9.2.6b) leaves $\omega(a,b)/\omega(b,a)$ invariant.

Exercise 9.4. Prove the relations (9.2.24–26) by starting with the theorems of orthogonality (4.2.25–27) for the vector REPs $D^{(\gamma)}(a^*)$, $a^* \in \mathscr{G}^*$.

Exercise 9.5. Determine a PR from (9.2.30, 31) explicitly for $\mathscr{C}_{4\mathrm{h}}$.

Exercise 9.6. Using (9.1.6) and (9.1.25) determine the characters $\chi^{(*L,\alpha)}(d | \boldsymbol{R}^h)$ for the point $\boldsymbol{k} = \boldsymbol{L}$ in $\mathscr{O}_{\mathrm{h}}^5$.

Exercise 9.7. Show that the matrices in (9.2.46) have the properties of a REP.

9.3 Projection Operators and Basis Functions

The linear operators which are assigned isomorphically to the space group [double space group] elements $\{d|t\} = \{d|s(d) + R^h\}$ and which act on scalar functions $\psi(x)$ [two-component spinor functions $\phi(x)$] we denote by $P(d|t)$[$P^s(d|t)$ (Sect. 6.1)]. Using (3.4.6) and (4.2.3),

$$P(d|t)\psi(x) \equiv \psi(\{d|t\}^{-1}x) = \psi(d^{-1}(x-t)) \ , \tag{9.3.1a}$$

$$P^s(d|t)\phi(x) = D^{(1/2)}(d)\phi(d^{-1}(x-t)) \ . \tag{9.3.1b}$$

The matrix $D^{(1/2)}(d)$ describes the rotation d in the spinor REP (9.2.47). The projection operators for the little group $\mathscr{G}_k$ of order g_k corresponding to (4.3.9a) are

$$P_{ij}^{(k,\alpha)} = \frac{d_\alpha}{g_k} \sum_{(d|t)\in\mathscr{G}_k} D_{ij}^{(k,\alpha)*}(d|t)P(d|t) \ . \tag{9.3.2}$$

Since $D^{(k,\alpha)}$ is an allowable IR of $\mathscr{G}_k$ and since

$$\{d|t\} = \{d|s(d) + R^h\} = \{d|s(d)\}\{e|d^{-1}R^h\} \ , \tag{9.3.3}$$

which is also valid for the operators isomorphically assigned,

$$P_{ij}^{(k,\alpha)} = \frac{d_\alpha}{g_k} \sum_{d\in\mathscr{G}_k} D_{ij}^{(k,\alpha)*}(d|s)P(d|s) \sum_h \exp(+ik\cdot R^h)P(e|d^{-1}R^h) \tag{9.3.4a}$$

or

$$P_{ij}^{(k,\alpha)} = \frac{d_\alpha}{g_{0k}} \sum_d D_{ij}^{(k,\alpha)*}(d|s)P(d|s)P^{(k)}$$

with the projection operator

$$P^{(k)} = \frac{1}{N} \sum_h \exp(ik\cdot R^h)P(e|R^h) \ , \qquad R^h \in \mathbb{T} \text{ and } g_k = g_{0k}N \tag{9.3.4b}$$

of the translation group $\mathbb{T}$; ord $\mathbb{T} = N$, see (5.3.17). Equation (9.3.4a) shows that the basis functions of the IRs of $\mathscr{G}_k$, and by induction those of $\mathscr{R}$, can be constructed without extending the IPRs of $\mathscr{G}_{0k}$ to the IRs of $\mathscr{G}_k$. Applying $P_{ij}^{(k,\alpha)}$ according to (9.3.4) to a function $\psi(x)$ we obtain the function

$$\psi_{ij}^{(k,\alpha)} = \frac{d_\alpha}{g_{0k}} \sum_d D_{ij}^{(k,\alpha)*}(d|s)P(d|s)\psi^{(k)}(x) \ , \qquad d \in \mathscr{G}_{0k} \tag{9.3.5a}$$

symmetrized according to $\mathcal{G}_k$. Here $\psi^{(k)}(x)$ is the *Bloch* function [see (5.3.16)]

$$\psi^{(k)}(x) = \frac{1}{N}\sum_h \exp(\mathrm{i}k\cdot R^h)P(e|R^h)\psi(x) = \frac{1}{N}\sum_h \exp(\mathrm{i}k\cdot R^h)\psi(x - R^h) \; . \tag{9.3.5b}$$

generated by $P^{(k)}$ from $\psi(x)$.

Consequently, if the starting functions are already Bloch functions (5.3.16) i.e. transform as IRs of $\mathbb{T}$ (plane waves, OPW, APW, Bloch sums of LCAOs, etc.) then according to (9.3.5a) it is sufficient to use the IPRs of the point groups for symmetry adaptation. Since the matrices of the IPRs also satisfy orthogonality relations (9.2.24, 25) (see Exercise 9.4) the functions obtained with (9.3.5a) can be used in the symmetrization of secular equations and in selection rules (Sects. 6.2–4). Finally, because of (9.2.51), which shows how to extend the IPRs of $\mathcal{G}_{0k}$ to the allowable IRs of $\mathcal{G}_k^{\mathrm{D}}$, the basis functions of $\mathcal{G}_k^{\mathrm{D}}$, and after induction also those of $\mathcal{R}^{\mathrm{D}}$, can be obtained in the same way. Of course, here the scalar functions $\psi(x)$ and $\psi^{(k)}(x)$ have to be replaced by the two-component spinor functions $\phi(x)$ and $\phi^{(k)}(x)$, respectively; (9.3.1b) has to be taken into account.

9.4 Representations of Magnetic Space Groups

9.4.1 Corepresentations of Magnetic Space Groups

According to (3.2.2 and 5.2.2), every finite magnetic group can be written as

$$\mathcal{M} = \mathcal{G} + r\mathcal{G} \; , \qquad \operatorname{ind}\mathcal{G} = 2 \; , \qquad \text{unitary;} \tag{9.4.1}$$

if $r = \vartheta$ (ϑ: time reversal), then $\mathcal{M}$ is of type II, if $r = \vartheta a'$ ($a' \notin \mathcal{G}$ unitary), then $\mathcal{M}$ is of type III. Type and form of an ICOR of $\mathcal{M}$, assigned to an IR of $\mathcal{G}$ (here $\mathcal{G} \to \mathcal{R}$, $\mathcal{R}^{\mathrm{D}}$) can be determined from (5.2.13–16) if we replace Δ by the single- or double-valued IRs of the groups of type I, thus by $D^{(*k,\alpha)}$ or $D^{(*k,\alpha)}$ and $\bar{D}^{(*k,\alpha)}$, respectively, or if we use the corresponding characters. However, although criterion (5.2.16) is useful for point groups, it cannot be used for space groups in the given form. Since the IRs Δ of $\mathcal{R}$ are determined by the allowable IRs of $\mathcal{G}_k$ and these by the IPRs of $\mathcal{G}_{0k}$, it is to be expected that (5.2.16) can be reformulated into a criterion for space groups containing only the characters $\chi^{(k,\alpha)}$ of $\mathcal{G}_k$. In the decomposition into cosets with respect to $\mathbb{T}$, $\mathcal{M}$ can be written as

$$\mathcal{M} = \sum_d \{d|s\}\mathbb{T} + \sum_{d'} \vartheta\{d'|s'\}\mathbb{T} \; , \tag{9.4.2}$$

where in $\mathcal{M}_{\mathrm{II}}$ groups $\{d|s\} = \{d'|s'\}$ but in $\mathcal{M}_{\mathrm{III}}$ groups all the elements $\{d'|s'\}$ are different from the elements $\{d|s\}$. In any case, $\{d'|s'\}^2$ is contained in the set $\{d|s\}$. The criterion which corresponds to (5.2.16) is then (for a proof see [9.8])

$$\sum_{d'} \chi^{(k,\alpha)}(\{d'|s'\}^2) = \begin{cases} \lambda g_{0k} & \text{in case 1a)} \\ -\lambda g_{0k} & \text{in case 1b)} \\ 0 & \text{in case 2)} \end{cases} \tag{9.4.3}$$

where g_{0k} is the order of $\mathscr{G}_{0k}$; λ has to be taken according to (5.2.17b) thus $\lambda = 1$ for vector REPs (no spin, or an even number of electrons) and $\lambda = -1$ for spinor REPs (odd number of electrons). In (9.4.3) $\chi^{(k,\alpha)}$ is the character of the REP $D^{(k,\alpha)}$ of $\mathscr{G}_k$. The elements d' in the sum (9.4.3) have to be chosen in such a way that $\vartheta\{d'|s'\} \in \mathscr{M}$ and that

$$D^{(k)}_{\vartheta\{d'|s'\}}(e|R^h) = D^{(-k)}(e|R^h) \ , \tag{9.4.4a}$$

for the conjugated REP according to (9.1.2). This is satisfied for

$$d'k = -k + K \ . \tag{9.4.4b}$$

Thus the determination of the ICORs of $\mathscr{M}$ can be described by the scheme

$$\text{allowable IR of } \mathscr{G}_k \xrightarrow{\text{induction}} \text{IR of } \mathscr{R} \xrightarrow{(9.4.3)} \text{ICOR of } \mathscr{M} \ . \tag{9.4.5}$$

Besides this method there is another one, which we shall also illustrate briefly. As for normal space groups, it starts with the *magnetic little group* $\mathscr{M}_k$ or the assigned point group $\mathscr{M}_{0k}$ which we have to define first. By comparing (5.2.16, 17) and (9.4.3) we expect the definition [see (9.4.4)]

$$\mathscr{M}_{0k} = \{d \in \mathscr{G}_{0k}\} \ \textit{and} \ \{d'|d'k = -k + K\} \ . \tag{9.4.6}$$

Then[9]

$$\mathscr{M}_k = \sum_d \{d|s\} \cdot \mathbb{T} + \sum_{d'} \vartheta\{d'|s'\} \cdot \mathbb{T} \ , \qquad d \in \mathscr{G}_{0k} \ , \qquad d' \in \mathscr{M}_{0k} \backslash \mathscr{G}_{0k} \ ; \tag{9.4.7}$$

$\mathscr{M}_k$ is a magnetic space group, $\mathscr{G}_k$ its unitary subgroup of index 2. In this case the criterion (5.2.16) reads

$$\sum_b \chi^{(k,\alpha)}(b^2) = \begin{cases} \lambda g_k & \text{case 1a)} \\ -\lambda g_k & \text{case 1b)} \ , \\ 0 & \text{case 2)} \end{cases} \qquad b \in \mathscr{M}_k \backslash \mathscr{G}_k \ . \tag{9.4.8}$$

With this the ICORs $D^{M(k,\alpha)}$ of $\mathscr{M}_k$ assigned to the $D^{(k,\alpha)}$ can be determined. By summing (9.4.8) over the elements of the translation group $\mathbb{T}$ we again get (9.4.3). Thus the definition (9.4.6, 7) is justified, too. With the decomposition

$$\mathscr{M} = \sum_{\nu=1}^{s} \{d_\nu|s_\nu\} \mathscr{M}_k \tag{9.4.9}$$

---

[9] The set $\mathscr{M}_{0k} \backslash \mathscr{G}_{0k}$ contains all the elements which *do not* occur in $\mathscr{G}_{0k}$.

analogous to that in (9.1.6) we are able to induce the ICORs of $\mathcal{M}$ from those of $\mathcal{M}_k$ according to the procedure in Sect. 9.1.2. The scheme corresponding to that of (9.4.5) is thus

$$\text{allowable IR of } \mathcal{G}_k \xrightarrow{(9.4.3)} \text{allowable ICOR of } \mathcal{M}_k \xrightarrow{\text{induction}} \text{ICOR of } \mathcal{M} \,. \tag{9.4.10}$$

The details of, among other things, the reduction of $D^{M(k,\alpha)}$ to the IPCORs of $\mathcal{M}_{0k}$ will not be given here.

9.4.2 Time Reversal Symmetry in $\mathcal{M}_{\mathrm{II}}$ Groups

For the grey space groups (type II) with $\mathcal{M}_{\mathrm{II}} = \mathcal{R} + \vartheta\mathcal{R}$, (9.4.3) reduces to the Herring criterion for the degeneracy with respect to time reversal symmetry in nonmagnetic crystals (Sect. 3.4.2):

$$\sum_{d} \chi^{(k,\alpha)}((d|s)^2) = \begin{cases} \lambda g_{0k} & \text{case 1a)} \\ -\lambda g_{0k} & \text{case 1b)} \\ 0 & \text{case 2)} \end{cases} , \tag{9.4.11}$$

where the sum runs over all the $\{d|s\}$ of the decomposition (3.4.9) of $\mathcal{R}$ with respect to $\mathbb{T}$, for which $d\boldsymbol{k} = -\boldsymbol{k} + \boldsymbol{K}$. Equation (9.4.11) is a direct consequence of (9.4.3) because

$$\vartheta\{d'|s'\} = \vartheta\{d|s\} \,, \qquad d \in \mathcal{G} \,, \tag{9.4.12}$$

for $\mathcal{M}_{\mathrm{II}}$ groups. The classification according to (9.4.11) allows determination of the additional degeneracy caused by the time reversal symmetry ϑ according to (5.2.13–15). The spectra $E(\boldsymbol{k})$ of particles and quasi-particles become identical at $\boldsymbol{k}$ and $-\boldsymbol{k}$ if time reversal symmetry is present (*Kramer's rule*). This statement can be verified by discussing separately the three cases

$$\begin{aligned} &\text{(i)} && \mathcal{M}_k = \mathcal{G}_k \,, \\ &\text{(ii)} && \mathcal{M}_k = \mathcal{G}_k + \vartheta\mathcal{G}_k \,, \\ &\text{(iii)} && \mathcal{M}_k = \mathcal{G}_k + \vartheta a'\mathcal{G}_k \,. \end{aligned} \tag{9.4.13}$$

(i) In this case according to (9.4.7) there are no elements d' which map $\boldsymbol{k}$ onto $-\boldsymbol{k} + \boldsymbol{K}$. Because of (9.4.2) and (9.4.12) such elements do not exist even in the complete $\mathcal{R}$, i.e. $-\boldsymbol{k} \notin *\boldsymbol{k}$. Thus the inversion i cannot be contained in $\mathcal{G}$. Since $\mathcal{M}_k = \mathcal{G}_k$ there cannot exist any additional degeneracy due to time reversal symmetry as far as the REPs $D^{(k,\alpha)}$ of $\mathcal{G}_k$ are concerned. But since the set $\mathcal{M}_k \backslash \mathcal{G}_k$ is empty, according to (9.4.3) we have case 2). This means that according to (5.2.15–17) $D^{M(*k,\alpha)}$ contains $D^{(*k,\alpha)}$ as well as $D^{(*k,\alpha)*}$. But now $D^{(*k,\alpha)*}$ is induced by $D^{(k,\alpha)*}$ which is an IR of $\mathscr{G}_{-k}$. The additional degeneracy of spectra at $\boldsymbol{k}$ and $-\boldsymbol{k}$ thus also exists if $-\boldsymbol{k} \notin *\boldsymbol{k}$.

(ii) According to (9.4.6, 7) in this case the elements of $\mathscr{G}_k$ just map $\boldsymbol{k}$ onto $-\boldsymbol{k}+\boldsymbol{K}$. But since to agree with (9.1.3, 7) these elements also have to map $\boldsymbol{k}$ onto $+\boldsymbol{k}+\boldsymbol{K}$ the relation

$$-\boldsymbol{k} = +\boldsymbol{k} + \boldsymbol{K} \qquad \text{i.e.} \qquad \boldsymbol{k} = -\boldsymbol{K}/2 \text{ or } 0 \tag{9.4.14}$$

must hold. Then of course, we have $-\boldsymbol{k} \in *\boldsymbol{k}$, which means that even without time reversal symmetry $E(\boldsymbol{k}) = E(-\boldsymbol{k})$. In this case there are elements $d \in \mathscr{G}$, namely

$$d \in \{d_\nu \mathscr{G}_{0k}\} \qquad \text{with} \qquad d_\nu \boldsymbol{k} = -\boldsymbol{k} + \boldsymbol{K} \ ,$$

which map $\boldsymbol{k}$ onto $-\boldsymbol{k}+\boldsymbol{K}$. Besides, for these $d \in \mathscr{G}$ also $E(\boldsymbol{k}) = E(d\boldsymbol{k})$, see (10.1.7).

(iii) As in (ii), here we again have $-\boldsymbol{k} \in *\boldsymbol{k}$ so that $E(\boldsymbol{k}) = E(-\boldsymbol{k})$. But $-\boldsymbol{k} \neq \boldsymbol{k}+\boldsymbol{K}$, since $d' \notin \mathscr{G}_k$ as in case (ii).

By using (9.4.3) we can determine in cases (ii) and (iii), where $-\boldsymbol{k} \in *\boldsymbol{k}$, a further possible degeneracy due to time reversal by the classification of the ICORs of $\mathscr{M}_k$ or $\mathscr{M}$, according to 1a), 1b) and 2).

In more general cases we have to include the spin when discussing Kramer's rule. Thus for a $\boldsymbol{k}$-point in a general position in the BZ (the REPs are then $D^{(k,\alpha)} \equiv 1$) with time reversal symmetry we always have $E(\boldsymbol{k}) = E(-\boldsymbol{k})$. Now, which are the (Bloch) wave functions assigned to these energies? If the *spin does not occur explicitly* we can use ϑ_0 as the time reversal operator (8.2.7). For Bloch functions then, apart from a phase factor, $\Psi^{(k)}$ and $\vartheta_0 \Psi^{(k)} = \Psi^{(k)*} = \Psi^{(-k)}$ are degenerate, i.e.

$$E(\boldsymbol{k}) = E(-\boldsymbol{k}) \ . \tag{9.4.15}$$

If the *spin occurs explicitly* (in the Hamiltonian) ϑ has to be represented by (8.2.10), again perhaps with a different phase factor. Then for one electron we have

$$\vartheta\begin{pmatrix}\psi^{(k)}\\0\end{pmatrix} = \sigma_y\begin{pmatrix}\psi^{(k)*}\\0\end{pmatrix} = \mathrm{i}\begin{pmatrix}0\\\psi^{(k)*}\end{pmatrix} = \mathrm{i}\begin{pmatrix}0\\\psi^{(-k)}\end{pmatrix} ,$$

$$\vartheta\begin{pmatrix}0\\\psi^{(k)}\end{pmatrix} = \sigma_y\begin{pmatrix}0\\\psi^{(k)*}\end{pmatrix} = -\mathrm{i}\begin{pmatrix}\psi^{(k)*}\\0\end{pmatrix} = -\mathrm{i}\begin{pmatrix}\psi^{(-k)}\\0\end{pmatrix} ,$$

which means $\psi^{(k)}(\boldsymbol{x},\uparrow)$ and $\psi^{(-k)}(\boldsymbol{x},\downarrow)$ are degenerate, thus

$$E(\boldsymbol{k},\uparrow) = E(-\boldsymbol{k},\downarrow) \ . \tag{9.4.16}$$

For crystals containing the inversion in their point group [cases (ii), (iii), e.g. diamond structure O_h^7] and for $\boldsymbol{k}$ at a general position in the BZ, $d = i$ in (9.4.11)

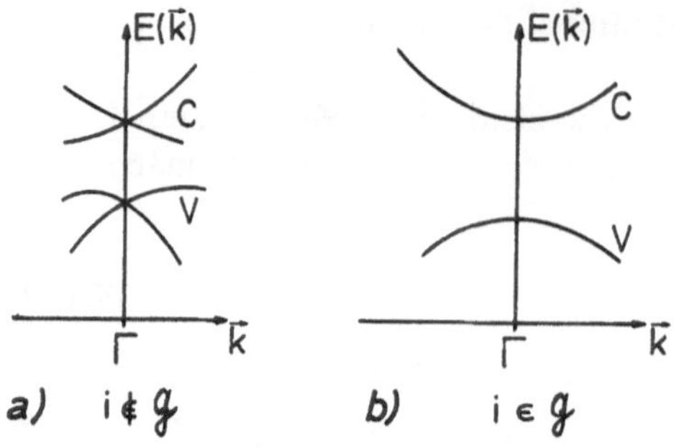

Fig. 9.3. Energy curves in the neighbourhood of the Γ point for a general $\boldsymbol{k}$-direction. For the diamond structure (**b**) valence (V) and conduction (C) bands are at least twofold spin degenerate. For the zinc-blende structure (**a**) the missing inversion symmetry lifts the degeneracy for $\boldsymbol{k} \neq 0$

(the sum over d is only one term!). Consequently we have

$$\{d|s\}^2 = \{e|0\} , \qquad \sum_d \chi^{(\boldsymbol{k},\alpha)}(e|0) = 1 .$$

As now $\lambda = -1$ and $g_{0k} = 1$, according to (9.4.11) we have case 1b). Taking spin into account, all bands are (because of the spin) at least doubly degenerate

$$E(\boldsymbol{k}, \uparrow) = E(-\boldsymbol{k}, \downarrow) = E(\boldsymbol{k}, \downarrow) . \tag{9.4.17a}$$

However, if a crystal does not have i in its point group (zinc-blende structure[10] $\mathscr{T}_d^2$) and in case (i) ($\boldsymbol{k}$ in a general position $\mathscr{C}_1$) then

$$E(\boldsymbol{k}, \uparrow) = E(-\boldsymbol{k}, \downarrow) \tag{9.4.17b}$$

still holds but the additional degeneracy (9.4.17a) is lifted (see Fig. 9.3).

We shall discuss further details with the use of special $\boldsymbol{k}$-vectors in a symmetry position. We choose an example which also allows for case (i) in (9.4.13), i.e. for $-\boldsymbol{k} \notin *\boldsymbol{k}$ and $i \notin \mathscr{G}$. This case occurs for the Γ-Λ-L direction of the $\mathscr{T}_d^2$ space group (Fig. 9.1). We take into account spin-orbit coupling and thus we have to use the double space group $\mathscr{G}_{0k}^{\mathrm{D}}$ in our discussion. Since the spin is equal to 1/2, only the double-valued spinor REPs of $\mathscr{G}_{0k}^{\mathrm{D}}$ from (9.2.46) are essential, for which

$$\chi^{(\boldsymbol{k},\alpha)}(\bar{d}_\nu) = -\chi^{(\boldsymbol{k},\alpha)}(d_\nu) . \tag{9.4.18a}$$

It is therefore sufficient to know the characters of d_ν. For double space groups the sum in (9.4.11) is over all d and $\bar{d}$ of $\mathscr{G}^{\mathrm{D}}$ with $d\boldsymbol{k} = -\boldsymbol{k} + \boldsymbol{K}$ and $\bar{d}\boldsymbol{k} =$

---

[10] Zinc-blende structure is similar to diamond structure (Fig. 9.1a). The two positions of the basis have only to be occupied by two different kinds of ions. The point group is $\mathscr{T}_d$ and leaves the lattice (fcc) invariant, i.e. $\mathscr{T}_d^2$ is symmorphic. The BZ is that of Fig. 9.1b, but the $\boldsymbol{k}$-points partly have a lower symmetry, as given in Appendix E, since $i \notin \mathscr{T}_d$ and correspondingly other elements of $\mathscr{O}_h$ are not contained in $\mathscr{T}_d$. Examples for the space group $\mathscr{T}_d^2$ are the III-V compounds such as GaAs and InSb.

Table 9.2. Character table for the d_v, d_v^2 belonging to the Γ point of ZnS structures

d_v / d_v^2	e / e	$8c_3$ / $8c_3^2$	$6\sigma_d$ / $6c_0$	$6s_4$ / $6c_2$	$3c_2$ / $3c_0$	$\sum_d$	Case corresponding to (9.4.11)	(9.4.13)
Γ_6	2	−1	−2	0	−2	−24	1a)	(ii)
Γ_7	2	−1	−2	0	−2	−24	1a)	(ii)
Γ_8	4	1	−4	0	−4	−24	1a)	(ii)

Table 9.3. Character table for the d_v, d_v^2 belonging to the L point of ZnS structures

d_v / d_v^2	e / e	$2c_3$ / $2c_3^2$	$3\sigma_v$ / $3c_0$	$\sum_d$	Case corresponding to (9.4.11)	(9.4.13)
L_4	1	1	−1	0	2	(ii)
L_5	1	1	−1	0	2	(ii)
L_6	2	−1	−2	−6	1a	(ii)

$-\boldsymbol{k} + \boldsymbol{K}$. But since

$$\overline{\chi}^{(\boldsymbol{k},\alpha)}(\bar{d}^2) = \overline{\chi}^{(\boldsymbol{k},\alpha)}(d^2) , \qquad (9.4.18b)$$

it suffices to carry out a summation over d in (9.4.11).

Γ Point. The sum runs over all $\{d_v, \bar{d}_v\} \in \mathscr{T}_d^D$, thus $d_v \in \mathscr{T}_d$. We have to calculate all the d_v^2 and to take the characters from Appendix A for the extra REPs $\Gamma_6 - \Gamma_8$ of $\mathscr{T}_d^D$; d_v and d_v^2 are contained in Table 9.2. For c_3 we have to take into account that in $\mathscr{T}_d^D$, c_3 and c_3^2 belong to different classes. For $l = 1/2$, from (8.4.1), $\chi = 2\cos(\varphi/2)$ and thus $\chi^{(1/2)}(c_3) = +1$ and $\chi^{(1/2)}(c_3^2) = -1$, i.e. $\{c_3, \bar{c}_3^2\}$ and $\{c_3^2, \bar{c}_3\}$ form a class [(see 3.3.6) and Appendix A]. Table 9.2 shows that there is no additional degeneracy due to time reversal symmetry for the REPs $\Gamma_6 - \Gamma_8$.

Λ and L Points. In both cases $\mathscr{G}_{0\boldsymbol{k}} \cong \mathscr{C}_{3v}$, since $i \notin \mathscr{G}_{0\boldsymbol{k}}$ (contrary to the $\mathscr{O}_h$ groups in Appendix E). For Λ points then $-\boldsymbol{k} \notin *\boldsymbol{k}$; we have case (i) from (9.4.13) thus case (2) from (9.4.11). There is no additional degeneracy due to time reversal symmetry [see discussion following (9.4.13)], i.e. in general Λ_4, Λ_5, split. At the L point $\boldsymbol{k} = \boldsymbol{K}/2 \in *\boldsymbol{k}$, i.e. we have case (ii). In (9.4.11) the summation has to be carried out over all $d_v \in \mathscr{C}_{3v}$ (for Λ points this sum is empty). Using Appendix A we obtain Table 9.3. Since $L_4^* = L_5$ (Sect. 5.2), L_4 and L_5 are always doubly degenerate, whereas L_6 has no additional degeneracy. For zinc-blende structure this is shown in Fig. 9.4, which also shows the $E(\boldsymbol{k})$ curves for diamond structure for comparison.

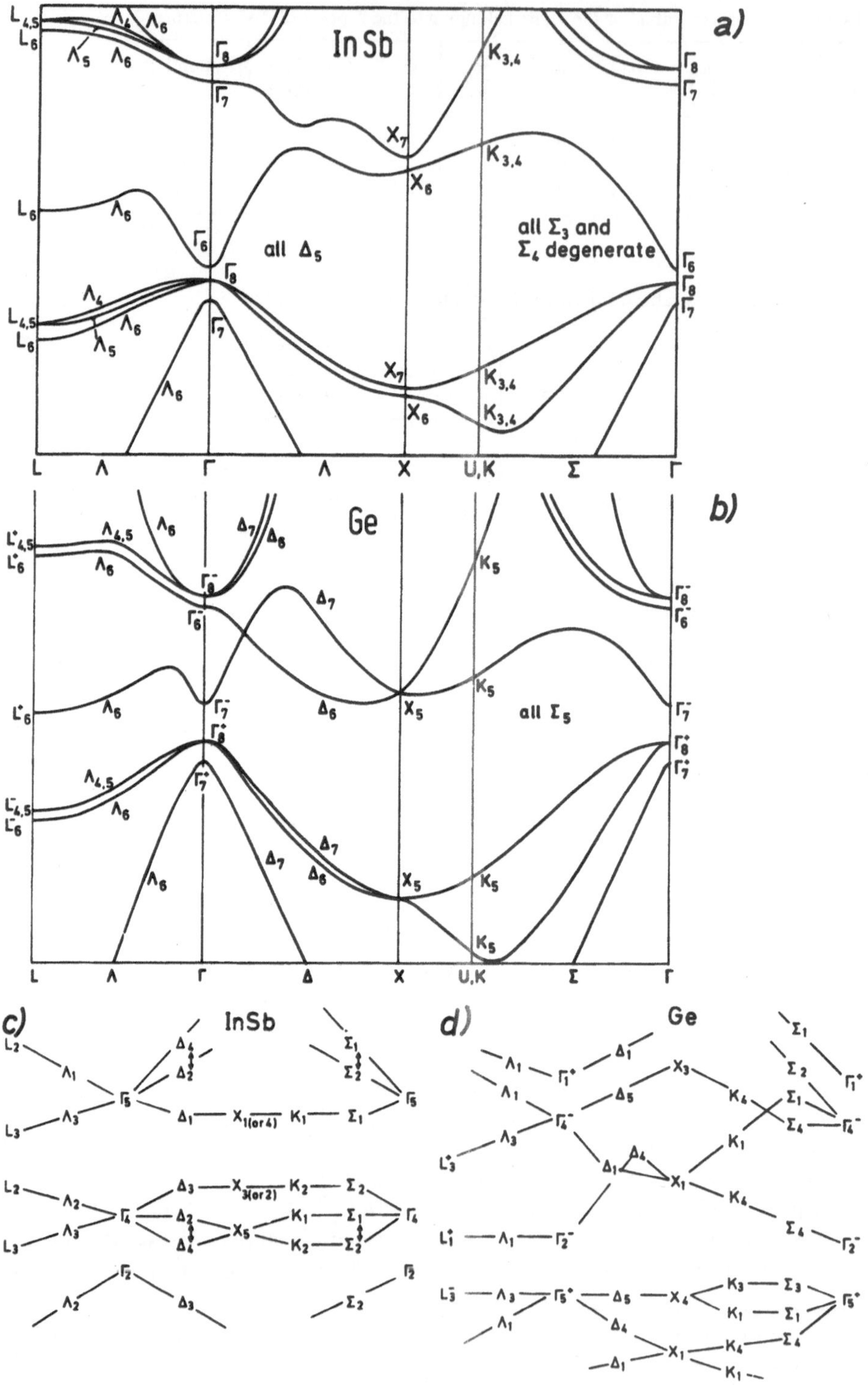
a)
InSb
all Δ5
all Σ3 and Σ4 degenerate
L Λ Γ Δ X U,K Σ Γ
b)
Ge
all Σ5
L Λ Γ Δ X U,K Σ Γ
c)
InSb
d)
Ge

Exercise 9.8. Discuss the Γ-Λ-L energy curves for diamond structure $\mathcal{O}_h^7$, for which $i \in \mathcal{G}$, and give the classification of the single-valued as well as the double-valued IRs with respect to time reversal symmetry.

◀ **Fig. 9.4.** Energy curves (qualitative), including spin, for different directions. For zinc-blende structure (**a**) Λ_4 and Λ_5, for example, are split, sometimes not to be found in the literature. For diamond structure (**b**) Λ_4 and Λ_5 are degenerate. For zinc-blende structure spin-orbit coupling must be included if the splitting is to be calculated quantitatively. It is largest in InSb. The connections Γ-Λ-L follow from the compatibility relations (10.1.16): $\Gamma_6 \rightarrow \Lambda_6 \rightarrow L_6$; $\Gamma_7 \rightarrow \Lambda_6 \rightarrow L_6$; $\Gamma_8 \rightarrow \Lambda_4 \oplus \Lambda_5 \oplus \Lambda_6 \rightarrow L_4 \oplus L_5 \oplus L_6$, and similarly for the other directions. Without spin, some bands coincide; then they have to be denoted according to one-valued REPs, as shown for InSb (**c**) and Ge (**d**). The notation corresponds to that in Appendix A. (Different notations are also given in Appendix A) Γ_7 has $s = 1/2$, Γ_8 has $s = 3/2$ (Sect. 8.3)

10. Excitation Spectra and Selection Rules in Crystals

In the preceding section we developed the methods necessary for a discussion of space group symmetries in crystals. In the following we apply these methods to an investigation of the elementary excitations in crystals, which have many aspects in common. Among these there are phonons (lattice vibrations) and electronic excitations for which we discuss the symmetrized eigenvalue problem. Finally we study the selection rules for the electron-phonon, electron-photon and phonon-photon interactions in crystals, including infrared absorption and the Raman effect.

10.1 Spectra—Some General Statements

10.1.1 Bands and Branches

The Hamiltonians H of crystals, and thus those of the electrons, of the atomic or ionic motions, of the spin interactions, etc., commute with all the symmetry operators $P(d|\boldsymbol{t})$ of the space group[1]:

$$[H, P(d|\boldsymbol{t})] = 0 \qquad \text{for each } \{d|t\} \in \mathscr{R} \ . \tag{10.1.1}$$

According to Sect. 6.3 the eigenspaces of H, i.e. the spaces defined by the eigenfunctions belonging to a fixed eigenvalue E of H, are described by basis functions, which are assigned to the IRs $D^{(*\boldsymbol{k},\alpha)}$ of the corresponding symmetry group $\mathscr{R}$, see (9.1.18). The eigenfunctions degenerate with respect to energy are thus

$$P(\dot{d}_\nu|\boldsymbol{t}_\nu)\psi_i^{(\boldsymbol{k},\alpha)} := \{d_\nu|\boldsymbol{t}_\nu\}\psi_i^{(\boldsymbol{k},\alpha)} := \Psi_{i,\nu}^{(\boldsymbol{k},\alpha)} \ ; \tag{10.1.2}$$

$$i = 1, \ldots, d_\alpha = \alpha \ ; \qquad \nu = 1, \ldots, s \ .$$

According to (5.3.16) we have for the translational part of (10.1.2)

$$P(e|\boldsymbol{R}^h)\Psi_{i,\nu}^{(\boldsymbol{k},\alpha)}(\boldsymbol{x}) = \exp(-\mathrm{i}\boldsymbol{k}_\nu \cdot \boldsymbol{R}^h)\Psi_{i,\nu}^{(\boldsymbol{k},\alpha)}(\boldsymbol{x}) = \Psi_{i,\nu}^{(\boldsymbol{k},\alpha)}(\boldsymbol{x} - \boldsymbol{R}^h) \ ,$$

---

[1] The space group is one of the groups $\mathscr{M}_{\mathrm{I}}$, $\mathscr{M}_{\mathrm{II}}$ or $\mathscr{M}_{\mathrm{III}}$, where we usually restrict ourselves to problems with $\mathscr{M}_{\mathrm{I}}$ or $\mathscr{M}_{\mathrm{II}}$ symmetry.

therefore also

$$\exp[-\mathrm{i}\boldsymbol{k}_\nu\cdot(\boldsymbol{x}-\boldsymbol{R}^h)]\,\Psi_{i,\nu}^{(\boldsymbol{k},\alpha)}(\boldsymbol{x}-\boldsymbol{R}^h)=\exp(-\mathrm{i}\boldsymbol{k}_\nu\cdot\boldsymbol{x})\,\Psi_{i,\nu}^{(\boldsymbol{k},\alpha)}(\boldsymbol{x}):=u_{i,\nu}^{(\boldsymbol{k},\alpha)}(\boldsymbol{x})\ ; \tag{10.1.3}$$

the right hand side of (10.1.3) is thus a *lattice-periodic function*, so that any eigenfunction of a crystal Hamilton can be written as

$$\Psi_{i,\nu}^{(\boldsymbol{k},\alpha)}(\boldsymbol{x})=\exp(\mathrm{i}\boldsymbol{k}_\nu\cdot\boldsymbol{x})u_{i,\nu}^{(\boldsymbol{k},\alpha)}(\boldsymbol{x})\quad\text{with}\quad u(\boldsymbol{x}-\boldsymbol{R}^h)=u(\boldsymbol{x})\ \text{arbitrary}\ . \tag{10.1.4}$$

This is the Bloch theorem.

Given the functions $\Psi_\sigma^{(\boldsymbol{k})}$ the eigenvalue problem[2]

$$H\Psi_\sigma^{(\boldsymbol{k})}=E_\sigma(\boldsymbol{k})\Psi_\sigma^{(\boldsymbol{k})} \tag{10.1.5}$$

for a fixed σ leads to $E_\sigma(\boldsymbol{k})$ being a continuous function of $\boldsymbol{k}$ in the BZ: (*energy*) *dispersion curve, energy band, branch*. The complete ensemble of the eigenvalues $E_\sigma(\boldsymbol{k})$ forms the band structure. For practical reasons, $E_\sigma(\boldsymbol{k})$ is usually calculated in certain definite directions of $\boldsymbol{k}$ only, or energy surfaces $E_\sigma(\boldsymbol{k})=$ constant are determined in $\boldsymbol{k}$ space. First we shall discuss the possible degeneracies:

$$E_\sigma(\boldsymbol{k})=E_{\sigma'}(\boldsymbol{k}')\ . \tag{10.1.6}$$

(i) *Band degeneracy*. The degeneracy of a band $E_\sigma(\boldsymbol{k})$ is given by the dimension of the allowed IR $D^{(\boldsymbol{k},\alpha)}$ of $\mathscr{G}_{\boldsymbol{k}}$.

(ii) *Star degeneracy*. Since the eigenspaces (9.1.18) of H are given by $\mathscr{L}_\alpha^{(*\boldsymbol{k})}$,

$$E_\sigma(\boldsymbol{k})=E_\sigma(d\boldsymbol{k})\qquad\text{with}\qquad d\in\mathscr{G} \tag{10.1.7}$$

is always valid; $\mathscr{G}$ is the point group of the crystal. Thus it is sufficient to determine the energies for one point of a star (9.1.4), i.e. in the so-called irreducible region of the BZ. Using (10.1.7), $E_\sigma(\boldsymbol{k})$ can be extended all over the BZ. The volume of the basis region is $1/g$ of the total volume of the BZ (Fig. 9.1b).

(iii) *Band overlap*. By this we understand an accidental degeneracy which is not a consequence of symmetry. One, two or even more bands overlap as is indicated in Fig. 10.1a: $E_\sigma(\boldsymbol{k})=E_{\sigma'}(\boldsymbol{k}')$ for $\sigma=\sigma'$ or $\sigma\neq\sigma'$, too, but in any case $\boldsymbol{k}\neq\boldsymbol{k}'$. According to Sect. 6.3, the REP belonging to an accidentally degenerate eigenvalue is reducible. We assume it to be decomposed into IRs, e.g. $D=D^{(*\boldsymbol{k},\alpha)}\oplus D^{(*\boldsymbol{k}',\beta)}$. Band overlap occurs if $*\boldsymbol{k}\neq *\boldsymbol{k}'$.

(iv) *Level crossing* (*band crossing*). If we have $*\boldsymbol{k}=*\boldsymbol{k}'$ then bands belonging to $D^{(\boldsymbol{k},\alpha)}$ and to $D^{(\boldsymbol{k},\beta)}$ may approach or cross each other (Fig. 10.1b). At the intersection the energies $E_\sigma(\boldsymbol{k})$ are accidentally degenerate. The IR for the lowest

---

[2] Notation: For a fixed $\boldsymbol{k}$, in general there are (infinitely) many eigenspaces $\mathscr{L}_\alpha^{(*\boldsymbol{k})}$ of H that contain this $\boldsymbol{k}$. If these spaces are ordered, e.g. with increasing energy $E_\sigma(\boldsymbol{k})$; $E_1(\boldsymbol{k})\leqslant E_2(\boldsymbol{k})\leqslant\ldots$, we can write for the different eigenfunctions $\Psi_{i,\nu}^{(\boldsymbol{k},\alpha)}$ with $E_i(\boldsymbol{k})$, $i=1,\ldots,\alpha$; just $\Psi_\sigma^{(\boldsymbol{k})}$, $\sigma=1,\ldots,\alpha,\alpha+1,\ldots,\infty$, where σ renumbers all the functions.

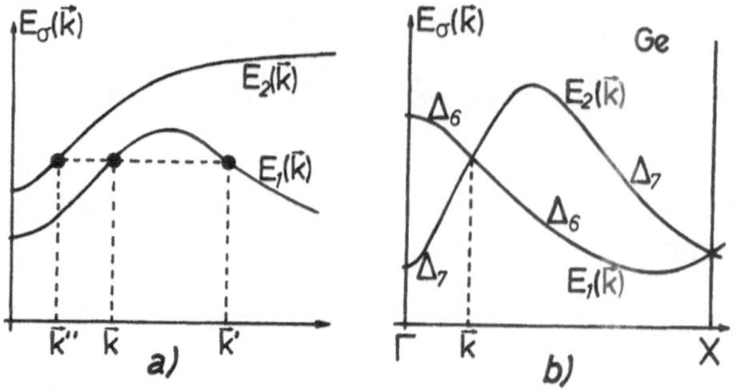

Fig. 10.1. Band overlap and crossing. In (**a**) there is degeneracy between $E_1(\boldsymbol{k})$ and $E_1(\boldsymbol{k}')$ and also $E_2(\boldsymbol{k}'')$, $\boldsymbol{k} \neq \boldsymbol{k}' \neq \boldsymbol{k}''$. In (**b**) two bands with the REPs $D^{(\boldsymbol{k},\alpha)} = \Delta_6$ and $D^{(\boldsymbol{k},\beta)} = \Delta_7$ cross. If we always denote the lowest energy by $E_1(\boldsymbol{k})$, then the REP changes on passing along $E_1(\boldsymbol{k})$ or $E_2(\boldsymbol{k})$

energy changes from Δ_6 into Δ_7 or vice versa when going in a symmetry direction (in this case in the Δ direction).

Let $\boldsymbol{k}$ lie in a general position, then there is only one allowable IR $D^{(\boldsymbol{k})}$ for every $\boldsymbol{k}$, since $\mathscr{G}_{\boldsymbol{k}} = \mathbb{T}$. The converse statement is: if two REPs $D^{(\boldsymbol{k},\alpha)}$ and $D^{(\boldsymbol{k},\beta)}$ are *not* equivalent, $\boldsymbol{k}$ cannot be in a general position, but has to be located in a symmetry position. In Fig. 10.1b, $\Delta_6 \not\sim \Delta_7$, $\boldsymbol{k}$ lies in the Δ direction.

These statements are valid for all elementary excitations in crystals, e.g. lattice vibrations (phonons), spin waves (magnons) and so on. For phonons, the $E_\sigma(\boldsymbol{k})$ have to be replaced by the frequencies $\omega_\sigma(\boldsymbol{k})$ or the squared frequencies $\omega_\sigma^2(\boldsymbol{k})$. The subscript σ in this case takes only $3r$ values (*branches*) if there are r basis atoms in the unit (elementary) cell. The statements on symmetries and degeneracies of $\omega_\sigma(\boldsymbol{k})$ correspond to those on $E_\sigma(\boldsymbol{k})$. Because time reversal symmetry (grey groups, no spin) is generally present for lattice vibrations, we always have according to (9.4.15)

$$\omega_\sigma(\boldsymbol{k}) = \omega_\sigma(-\boldsymbol{k}) \ . \tag{10.1.9}$$

Examples will be considered in Sect. 10.2.

10.1.2 Compatibility Relations

Let $\boldsymbol{k}$ take all the values of the first BZ; for a fixed band index σ the IRs $D^{(\boldsymbol{k},\alpha)}$ of $\mathscr{G}_{\boldsymbol{k}}$ can be given for every energy $E_\sigma(\boldsymbol{k})$. The degenerate eigenfunctions $\Psi_i^{(\boldsymbol{k},\alpha)}$, $i = 1, \ldots, d_\alpha$ transform according to this IR. For the line L, Λ, Γ, Δ, X of Fig. 9.1b such a band can be seen in Fig. 9.4b. The interrelation of different IRs assigned to the $\boldsymbol{k}$-points of this line, is determined by the symmetries of these $\boldsymbol{k}$-points. When advancing, for example, from a point $\boldsymbol{k}$ (say Γ) to a neighbouring point $\boldsymbol{k}' = \boldsymbol{k} + \boldsymbol{\kappa}$ (say Δ) of lower symmetry ($\mathscr{G}_{\boldsymbol{k}'} \subset \mathscr{G}_{\boldsymbol{k}}$), an IR $D^{(\boldsymbol{k},\alpha)}$ in $\mathscr{G}_{\boldsymbol{k}}$ may be reducible with respect to $\mathscr{G}_{\boldsymbol{k}'}$. This means we have to form the REP $D^{(\boldsymbol{k},\alpha)\text{sub}}(\mathscr{G}_{\boldsymbol{k}'})$ in the limit $\boldsymbol{\kappa} \to 0$, which is subduced from $\mathscr{G}_{\boldsymbol{k}}$

on $\mathscr{G}_{k'}$. If necessary it should be reduced in order to get the splitting from $\boldsymbol{k}$ to $\boldsymbol{k}'$ (Γ to Δ). This splitting is determined by the IRs $D^{(\boldsymbol{k}',\beta)}(\mathscr{G}_{k'})$ compatible with $D^{(\boldsymbol{k},\alpha)}(\mathscr{G}_{k})$. Thus we have to carry out the decomposition

$$D^{(\boldsymbol{k},\alpha)\mathrm{sub}}(\mathscr{G}_{k'}) = \sum_{\beta} m_{\beta,\alpha}^{\boldsymbol{k}',\boldsymbol{k}} D^{(\boldsymbol{k}',\beta)}(\mathscr{G}_{k'}) \tag{10.1.10}$$

in the limit $\boldsymbol{k}' \to \boldsymbol{k}$. According to (4.2.34), using the orthogonality-theorem (4.2.27a), we obtain from the corresponding character relation

$$m_{\beta,\alpha}^{\boldsymbol{k}',\boldsymbol{k}} = \lim_{\boldsymbol{k}'\to\boldsymbol{k}} \left(\frac{1}{g_{\boldsymbol{k}'}} \sum_{a\in\mathscr{G}_{k'}} \chi^{(\boldsymbol{k}',\beta)*}(a)\chi^{(\boldsymbol{k},\alpha)\mathrm{sub}}(a) \right) , \tag{10.1.11}$$

where the limit has to be taken *after* summing over a. Setting $a = \{d_\nu | s_\nu + \boldsymbol{R}^h\} \in \mathscr{G}_{k'}$ and using (9.2.8) we obtain

$$m_{\beta,\alpha}^{\boldsymbol{k}',\boldsymbol{k}} = \lim_{\boldsymbol{k}'\to\boldsymbol{k}} \left(\frac{1}{g_{0\boldsymbol{k}'}} \sum_{d_\nu} \chi^{(\boldsymbol{k}',\beta)*}(d_\nu|s_\nu)\chi^{(\boldsymbol{k},\alpha)\mathrm{sub}}(d_\nu|s_\nu) \right) , \qquad d_\nu \in \mathscr{G}_{0\boldsymbol{k}'} , \tag{10.1.12}$$

which according to (4.2.31b) can also be written as a sum over corresponding classes.

For *symmorphic* space groups, or if $\boldsymbol{k}$ *and* $\boldsymbol{k}'$ *lie in the interior* of the BZ, (10.1.12) can easily be evaluated since according to Sect. 9.2.1 only characters of ordinary vector REPs of the point groups $\mathscr{G}_{0\boldsymbol{k}}$ or $\mathscr{G}_{0\boldsymbol{k}'}$ occur.

For *nonsymmorphic space groups* we have to use the characters of the IPRs for the $\boldsymbol{k}$-vectors lying at the surface of the BZ. Evaluation of the compatibility relations is simplified by the following theorem:

single-valued REPs of a group $\mathscr{G}$ are never compatible with the double-valued (extra, spinor) REPs. (10.1.13)

If $X^{(\beta)}$ are the characters of the IRs of the double group (3.3.5) $\mathscr{G}^D (\triangleq \mathscr{G}_{k'}^D)$ and if $\chi^{(\alpha)}$ is the character of any reducible REP of $\mathscr{G}^D$, e.g. corresponding to $\chi^{(\boldsymbol{k},\alpha)\mathrm{sub}}(a)$ with $a \in \mathscr{G}_{k'}^D$, then analogously to (10.1.12) we have

$$m_{\beta,\alpha} = \frac{1}{g^D} \sum_{a\in\mathscr{G}} [X^{(\beta)*}(a)\chi^{(\alpha)}(a) + X^{(\beta)*}(\bar{a})\chi^{(\alpha)}(\bar{a})] \tag{10.1.14a}$$

with

$$\chi^{(\alpha)}(\bar{a}) = \pm\chi^{(\alpha)}(a) \qquad \text{and} \qquad X^{(\beta)}(\bar{a}) = \pm X^{(\beta)}(a) ,$$

where the upper sign holds for single-valued, the lower sign for double-valued REPs of $\mathscr{G}$. If now $D^{(\alpha)}$ and $D^{(\beta)}$ are different types of representations then (10.1.14a) vanishes since then $X^{(\beta)*}(\bar{a})\chi^{(\alpha)}(\bar{a}) = -X^{(\beta)*}(a)\chi^{(\alpha)}(a)$, which proves the above theorem. However, if the REPs are of the same type, (10.1.14a) simplifies to

$$m_{\alpha,\beta} = \frac{1}{g} \sum_{a \in \mathcal{G}} \chi^{(\beta)*}(a)\chi^{(\alpha)}(a) \ , \tag{10.1.14b}$$

i.e. we have only to sum over the elements a, but using the character tables of the double group.

As an example we investigate the Γ-Λ compatibility relations for the symmorphic space group $\mathcal{T}_{\mathrm{d}}^{2}$ (see end of Sect. 9.4.2). We have

$$\mathcal{G}_{0\boldsymbol{k}} \equiv \mathcal{G}_{0\Gamma} \cong \mathcal{T}_{\mathrm{d}} \ , \qquad \mathcal{G}_{0\boldsymbol{k}'} \equiv \mathcal{G}_{0\Lambda} \cong \mathcal{C}_{3\mathrm{v}} \ , \tag{10.1.15a}$$

thus with (10.1.12) and (10.1.14b)

$$m_{\beta,\alpha}^{\Lambda\to\Gamma} = \tfrac{1}{6} \sum_{a \in \mathcal{C}_{3\mathrm{v}}} \chi^{(\Lambda,\beta)*}(a)\chi^{(\Gamma,\alpha)\mathrm{sub}}(a) \ . \tag{10.1.15b}$$

To evaluate (10.1.15b) one conveniently writes the corresponding character tables from Appendix A under each other and then performs the corresponding multiplications and summations. It should be noted that the notation is not unique, and that care is needed when comparing with other publications. The notation is explained and commented on in Appendix A. For the m-matrix of (10.1.15b) then

$$m_{\beta,\alpha}^{\Lambda\to\Gamma} = \begin{array}{c|ccccc:ccc} {}_{\Lambda}\diagdown^{\Gamma}\ \beta = & \alpha = 1 & 2 & 3 & 4 & 5 & 6 & 7 & 8 \\ \hline 1 & 1 & 0 & 0 & 0 & 1 & & & \\ 2 & 0 & 1 & 0 & 1 & 0 & & \text{incompatible} & \\ 3 & 0 & 0 & 1 & 1 & 1 & & & \\ \hdashline 4 & & & & & & 0 & 0 & 1 \\ 5 & & \text{according to} & & & & 0 & 0 & 1 \\ 6 & & \text{(10.1.13)} & & & & 1 & 1 & 1 \end{array} \ , \tag{10.1.16a}$$

$$\begin{aligned} &\Gamma_1^{\mathrm{sub}} \Rightarrow \Lambda_1 \ , \qquad \Gamma_2^{\mathrm{sub}} \Rightarrow \Lambda_2 \ , \qquad \Gamma_3^{\mathrm{sub}} \Rightarrow \Lambda_3 \ , \\ &\Gamma_4^{\mathrm{sub}} \Rightarrow \Lambda_2 \oplus \Lambda_3 \ , \qquad \Gamma_5^{\mathrm{sub}} \Rightarrow \Lambda_1 \oplus \Lambda_3 \ , \\ &\Gamma_6^{\mathrm{sub}} \Rightarrow \Lambda_6 \ , \qquad \Gamma_7^{\mathrm{sub}} \Rightarrow \Lambda_6 \ , \qquad \Gamma_8^{\mathrm{sub}} \Rightarrow \Lambda_4 \oplus \Lambda_5 \oplus \Lambda_6 \ . \end{aligned} \tag{10.1.16b}$$

An at least qualitative understanding of the band structures, e.g. as given in Fig. 9.4, is achieved by taking the classification scheme (9.4.11) with time reversal symmetry, the compatibility relations (10.1.12) or (10.1.14) and considering in addition the splitting of the levels or bands by the spin-orbit coupling (Sect. 8.4.1) for $\mathcal{T}_{\mathrm{d}}$ symmetry:

$$\Gamma_4 \otimes D^{(1/2)} = \Gamma_6 \oplus \Gamma_8 \ ,$$
$$\Gamma_5 \otimes D^{(1/2)} = \Gamma_7 \oplus \Gamma_8 \ , \qquad (10.1.17)$$
$$\Gamma_2 \otimes D^{(1/2)} = \Gamma_7 \ .$$

Exercise 10.1. Give the compatibility relations for diamond structure $\mathcal{O}_h^7$ for the $\boldsymbol{k}$-vectors $\Delta \to \Gamma$:

$$\mathcal{G}_{0\boldsymbol{k}} = \mathcal{G}_{0\Gamma} \cong \mathcal{O}_h \ , \qquad \mathcal{G}_{0\boldsymbol{k}'} = \mathcal{G}_{0\Delta} \cong \mathcal{C}_{4v} \ .$$

10.2 Lattice Vibrations

10.2.1 Equation of Motion and Symmetry Properties

Vibrations of crystals can be treated in a similar way to those of molecules. Contrary to the symmetries of molecules described by point groups, ideal infinite crystals are invariant under transformations of the space group $\mathcal{R}$. The vibrations may be described by bands (branches) as explained in Sect. 10.1. The equation of motion in the harmonic approximation is (7.1.1).

$$M_\mu \ddot{u}_i^{\overset{\boldsymbol{m}}{\mu}} = -\sum_{\boldsymbol{n}\nu j} \phi_{ij}^{\overset{\boldsymbol{mn}}{\mu\nu}} \cdot u_j^{\overset{\boldsymbol{n}}{\nu}} \ , \qquad i,j = 1, 2, 3 \ , \qquad \mu, \nu = 1, 2, \dots, r \ . \qquad (10.2.1)$$

Here $\boldsymbol{m}, \boldsymbol{n}$ describe the positions of the lattice (elementary) cells as in (3.4.3), while μ, ν number the r atoms in the basis of the lattice. Introducing a periodicity volume according to (5.3.4), $\boldsymbol{m}, \boldsymbol{n}$ label $G^3 = N$ cells. The $u_i^{\overset{\boldsymbol{m}}{\mu}}$ are the displacements of the particles from their rest (equilibrium) positions $X_i^{\overset{\boldsymbol{m}}{\mu}}$, so that the instantaneous positions are $\boldsymbol{R}^{\overset{\boldsymbol{m}}{\mu}} + \boldsymbol{u}^{\overset{\boldsymbol{m}}{\mu}}$. The Hamiltonian assigned to (10.2.1) is

$$H = \frac{1}{2} \sum_{\boldsymbol{m}\mu i} \frac{1}{M_\mu} (p_i^{\overset{\boldsymbol{m}}{\mu}})^2 + \frac{1}{2} \sum_{\substack{\boldsymbol{mn} \\ \mu\nu \\ ij}} u_i^{\overset{\boldsymbol{m}}{\mu}} \phi_{ij}^{\overset{\boldsymbol{mn}}{\mu\nu}} u_j^{\overset{\boldsymbol{n}}{\nu}} \ , \qquad p_i^{\overset{\boldsymbol{m}}{\mu}} = M_\mu \dot{u}_i^{\overset{\boldsymbol{m}}{\mu}} \ . \qquad (10.2.2)$$

Similarly to in (7.1.3, 4) we obtain with

$$u_i^{\overset{\boldsymbol{m}}{\mu}} = \frac{1}{\sqrt{M_\mu}} U_i^{\overset{\boldsymbol{m}}{\mu}} \mathrm{e}^{-\mathrm{i}\omega t} \qquad (10.2.3)$$

the $3rN$-dimensional secular equation

$$\omega^2 U_i^{\overset{\boldsymbol{m}}{\mu}} = \sum_{\boldsymbol{n}\nu j} F_{ij}^{\overset{\boldsymbol{mn}}{\mu\nu}} U_j^{\overset{\boldsymbol{n}}{\nu}} \ , \qquad F_{ij}^{\overset{\boldsymbol{mn}}{\mu\nu}} = \frac{1}{\sqrt{M_\mu M_\nu}} \phi_{ij}^{\overset{\boldsymbol{mn}}{\mu\nu}} \ , \qquad (10.2.4)$$

which is hard to solve without using symmetry properties. The fundamental symmetry is that of translation, which is reflected in the commutativity of the $\overset{mn}{F}{}^{\mu\nu}_{ij}$ with the translation operator and is expressed by

$$\overset{mn}{\phi}{}^{\mu\nu}_{ij} = \overset{m+h\,n+h}{\phi}{}^{\mu\;\;\nu}_{i\;\;j} = \overset{m-n\,0}{\phi}{}^{\mu\;\;\nu}_{i\;\;j} = \overset{0\,n-m}{\phi}{}^{\mu\;\;\nu}_{i\;\;j} \tag{10.2.5}$$

for infinite lattices. The symmetries are similar to those described in Sect. 7.1.

Since $\mathbb{T}$ is an invariant subgroup of the space group $\mathscr{R}$, we can extract all the statements which originate from the translation symmetry, in other words: the eigenvectors $\overset{m}{e}{}^{\mu}_{i}(\boldsymbol{k},\sigma)$ ($m_i = 1,\dots,N$; $\mu = 1,\dots,r$) of (10.2.4) according to (5.3.16, 17) and (10.1.3, 4) have Bloch form. Therefore we may set $\boldsymbol{m} = 0$ and get[3]

$$e^{\mu}_{i}(\boldsymbol{k},\sigma) := P^{(\boldsymbol{k})}\overset{0}{e}{}^{\mu}_{i}(\sigma) = \frac{1}{\sqrt{N}}\sum_{\boldsymbol{h}} \exp(-i\boldsymbol{k}\cdot\boldsymbol{R}^{\boldsymbol{h}})\overset{\boldsymbol{h}}{e}{}^{\mu}_{i}(\sigma) \; ; \qquad \sigma = 1,\dots,3r \; . \tag{10.2.6a}$$

$$P^{(\boldsymbol{k})} = \frac{1}{\sqrt{N}}\sum_{\boldsymbol{h}} \exp(i\boldsymbol{k}\cdot\boldsymbol{R}^{\boldsymbol{h}})P(e|\boldsymbol{R}^{\boldsymbol{h}}) \tag{10.2.6b}$$

is the normalized projection operator of the translation group in the REP $\boldsymbol{k}$. From the inverse of (10.2.6a)

$$\overset{\boldsymbol{m}}{e}{}^{\mu}_{i}(\sigma) = \frac{1}{\sqrt{N}}\sum_{\boldsymbol{k}} \exp(i\boldsymbol{k}\cdot\boldsymbol{R}^{\boldsymbol{m}})e^{\mu}_{i}(\boldsymbol{k},\sigma) \tag{10.2.6c}$$

(see Exercise 5.11a) we obtain from (10.2.4) the secular equation

$$\sum_{\nu j} F^{\mu\nu}_{ij}(\boldsymbol{k})e^{\nu}_{j}(\boldsymbol{k},\sigma) = \omega^2_{\sigma}(\boldsymbol{k})e^{\mu}_{i}(\boldsymbol{k},\sigma) \; , \qquad \sigma = 1,\dots,3r \; , \tag{10.2.7a}$$

which is only $3r$-dimensional.

$$F^{\mu\nu}_{ij}(\boldsymbol{k}) = \sum_{\boldsymbol{h}} \overset{0\boldsymbol{h}}{F}{}^{\mu\nu}_{ij}\exp(i\boldsymbol{k}\cdot\boldsymbol{R}^{\boldsymbol{h}}) \tag{10.2.7b}$$

is the *dynamical matrix* (frequency tensor)[4].

For a further discussion of (10.2.7) the properties of the representatives of the little group $\mathscr{G}_{\boldsymbol{k}}$ are essential. We shall now construct the REPs $D^{(\boldsymbol{k},t)}$ of $\mathscr{G}_{\boldsymbol{k}}$

---

[3] Note that the state vectors $\overset{\boldsymbol{h}}{e}{}^{\mu}_{i}$, in contrast to the wave functions $\psi(x)$, are only defined at discrete positions.

[4] Often it is appropriate to consider a secular equation derived from (10.2.7) by a unitary transformation $S = \exp(-i\boldsymbol{k}\cdot\boldsymbol{R}_\mu)\cdot 1$:

$$e^{\mu}_{i} \to \bar{e}^{\mu}_{i} = e^{\mu}_{i}\exp(-i\boldsymbol{k}\cdot\boldsymbol{R}_\mu) \; ; \qquad F^{\mu\nu}_{ij} \to \bar{F}^{\mu\nu}_{ij} = \sum \overset{0\boldsymbol{h}}{F}{}^{\mu\nu}_{ij}\exp[i\boldsymbol{k}\cdot(\boldsymbol{R}^{\boldsymbol{h}} - \boldsymbol{R}_\mu + \boldsymbol{R}_\nu)] \; .$$

generated by the eigenvectors $e_i^\mu(\boldsymbol{k},\sigma)$. We consider the behaviour of the e_i^μ in the transformations $\{d|s\} \in \mathcal{G}_{\boldsymbol{k}}/\mathbb{T} \cong \mathcal{G}_{0\boldsymbol{k}}$:

$$P(d|s)e_i^\mu(\boldsymbol{k},\sigma) = P(d|s)P^{(\boldsymbol{k})}\overset{0}{e}{}_i^\mu(\sigma) = P^{(\boldsymbol{k})}P(d|s)\overset{0}{e}{}_i^\mu(\sigma) \ . \tag{10.2.8}$$

Since $\overset{0}{e}{}_i^\mu$ is a vector field defined at discrete positions $\boldsymbol{R}_\mu^0 = \boldsymbol{R}_\mu$, according to (4.2.2, 3) we have

$$P(d|s)\overset{0}{e}{}_i^\mu = \sum_j d_{ij}\overset{0}{e}{}_j^{\bar{\mu}}(\sigma) \qquad \text{with}$$
$$\boldsymbol{R}_{\bar{\mu}} = \{d|s\}^{-1}\boldsymbol{R}_\mu = d^{-1}\boldsymbol{R}_\mu - d^{-1}s := \boldsymbol{R}_\nu^n = \boldsymbol{R}^n + \boldsymbol{R}_\nu \ . \tag{10.2.9}$$

Again, $\boldsymbol{R}_{\bar{\mu}}$ is a lattice point of the crystal. It is that lattice point into which $\boldsymbol{R}_\mu$ is transformed by $\{d|s\}^{-1}$. Thus we also have

$$e_j^{\bar{\mu}} = \overset{n}{e}{}_j^\nu = P(e|-\boldsymbol{R}^n)e_j^\nu = D^{(\boldsymbol{k})}(e|-\boldsymbol{R}^n)e_j^\nu$$
$$= \exp(\mathrm{i}\boldsymbol{k}\cdot\boldsymbol{R}^n)e_j^\nu = \exp[\mathrm{i}\boldsymbol{k}\cdot(\{d|s\}^{-1}\boldsymbol{R}_\mu - \boldsymbol{R}_\nu)]e_j^\nu \ . \tag{10.2.10}$$

From (10.2.6, 8, 9) we first obtain

$$P(d|s)e_i^\mu(\boldsymbol{k},\sigma) = \sum_j d_{ij}e_j^{\bar{\mu}}(\boldsymbol{k},\sigma) \tag{10.2.11}$$

and finally, using (10.2.10), the matrix REP

$$P(d|s)e_i^\mu(\boldsymbol{k},\sigma) := \sum_{\nu j} D_{\mu i,\nu j}^{(\boldsymbol{k},t)}(d|s)e_j^\nu(\boldsymbol{k},\sigma) \ ,$$
$$D_{\mu i,\nu j}^{(\boldsymbol{k},t)}(d|s) = d_{ij}\exp[\mathrm{i}\boldsymbol{k}\cdot(\{d|s\}^{-1}\boldsymbol{R}_\mu - \boldsymbol{R}_\nu)]\delta_{\mu\nu}(d) \ , \tag{10.2.12}$$

where

$$\delta_{\mu\nu}(d) = \begin{cases} 1 & \text{if } \boldsymbol{R}_\mu \text{ transforms into } \boldsymbol{R}_\nu^n \text{ under } \{d|s\}^{-1} \ , \\ 0 & \text{otherwise.} \end{cases}$$

Multiplying $D^{(\boldsymbol{k},t)}$ by $\exp(\mathrm{i}\boldsymbol{k}\cdot s)$ we get another gauge for the REPs so that we finally have

$$\hat{D}_{\mu i,\nu j}^{(\boldsymbol{k},t)}(d|s) = \exp(\mathrm{i}\boldsymbol{k}\cdot s)\mathbf{1}D_{\mu i,\nu j}^{(\boldsymbol{k},t)}(d|s)$$
$$\hat{D}_{\mu i,\nu j}^{(\boldsymbol{k},t)}(d|s) = d_{ij}\exp[\mathrm{i}\boldsymbol{k}\cdot(\boldsymbol{R}_\mu - d\boldsymbol{R}_\nu)]\delta_{\mu\nu}(d) \ , \qquad \{d|s\} \in \mathcal{G}_{\boldsymbol{k}}/\mathbb{T} = \mathcal{G}_{0\boldsymbol{k}} \ . \tag{10.2.13}$$

The $\hat{D}^{(\boldsymbol{k},t)}$ constitute a PR of $\mathcal{G}_{0\boldsymbol{k}}$ with the factor system $\hat{\omega}$ from (9.2.7–9) and commute with the dynamical matrix $F(\boldsymbol{k})$ in (10.2.7), thus

$$\hat{D}^{(\boldsymbol{k},t)}(d|s)\hat{D}^{(\boldsymbol{k},t)}(d'|s') = \exp[\mathrm{i}(\boldsymbol{k} - d^{-1}\boldsymbol{k})\cdot s']\hat{D}^{(\boldsymbol{k},t)}(dd'|s + ds') \tag{10.2.14a}$$

and

$$\hat{D}^{(k,t)}F(k) = F(k)\hat{D}^{(k,t)} . \tag{10.2.14b}$$

The proof is somewhat lengthy and will not be given here (see e.g. [10.1]).

In the sense of the Wigner-Eckart theorem in Sects. 6.2 and 6.3, $F(k)$ is a scalar operator, the diagonalization of which is conveniently performed in a symmetry-adapted basis[5] $e_{ij}^{(\alpha,s_\alpha)}(k) = |\alpha, s_\alpha, ij\rangle$. The basis has to be constructed with the help of the projection operators of $\mathscr{G}_{0k}$ from Sect. 9.3. Consequently we have to determine the transformation S between the old Cartesian basis, in which the e_j^μ from (10.2.7a) are represented, and the symmetry-adapted basis for every k according to (6.3.3, 4)

$$S = \langle\alpha, s_\alpha, i|e_j^\mu\rangle , \tag{10.2.15}$$

i.e. the symmetry-adapted vector $e_i^{(\alpha,s_\alpha)}(k)$ yields the (α, s_α, i) row of S or column of S^+, see (7.1.16b, 20). Since the $e_i^\mu(k,\sigma)$ from (10.2.6a) already have Bloch form, we obtain with (9.3.5a)

$$\{e_{ij}^{(\alpha,s_\alpha)}(k)\} \sim \sum_{\mathscr{G}_{0k}} \hat{D}_{ij}^{(k,\alpha)*}(d|s)P(d|s)\{e^{(l)}\} . \tag{10.2.16}$$

The $\hat{D}^{(k,\alpha)}$ are IPRs of $\mathscr{G}_{0k}$ in the Kovalev gauge (9.2.7a). The notation $\{e^{(l)}\}$ represents a $3r$-dimensional Cartesian basis, e.g. $\{(1, 0, 0, \ldots); (0, 1, 0, \ldots); \ldots; (0, \ldots, 0, 1)\}$. The symbols $\{e_{ij}^{(\alpha,s_\alpha)}\}$ and $\{e^{(l)}\}$ mean that the projection operators have in general to be applied successively to several elements of the basis $\{e^{(l)}\}$, in order to take into account the multiplicities $s_\alpha = 1, \ldots, m_\alpha$ in the new symmetry-adapted basis. The operators $P(d|s)$ in (10.2.16) are represented by the matrices defined in (10.2.13).

The multiplicities m_α with which an IPR $\hat{D}^{(k,\alpha)}$ occurs in the reducible PR $\hat{D}^{(k,t)}$ of $\mathscr{G}_{0k}$ give the dimensions of the symmetrized eigenvalue problems to be solved finally. The dimensions d_α of $\hat{D}^{(k,\alpha)}$ give the degeneracies of the vibrations of type $e_i^{(\alpha)}(k)$ with frequencies $\omega_\alpha(k)$ $(i = 1, \ldots, d_\alpha)$. The multiplicities m_α have to be calculated from

$$m_\alpha = \frac{1}{g_{0k}} \sum_{\mathscr{G}_{0k}} \hat{\chi}^{(k,t)}(d|s)\hat{\chi}^{(k,\alpha)*}(d|s) . \tag{10.2.17}$$

Here, according to (10.2.13),

$$\begin{aligned} &\hat{\chi}^{(k,t)}(d|s) = \mathrm{Tr}\{\hat{D}^{(k,t)}(d|s)\} = \mathrm{Tr}\{d\} \sum_\mu \exp[\mathrm{i}k\cdot(R_\mu - dR_\mu)]\delta_{\mu\mu}(d) ; \\ &\mathrm{Tr}\{d\} = \pm(1 + 2\cos\varphi_d) ; \qquad \varphi_d\text{: angle of rotation in } d . \end{aligned} \tag{10.2.18a}$$

[5] In general the second subscript of i, j is not needed and can thus be dropped. See also (4.3.13, 14) and [10.1], especially for the elastic limit.

If $\boldsymbol{k}$ lies in the interior of a BZ, then according to Sect. 9.2.1, (10.2.18a) simplifies to

$$\hat{\chi}^{(\boldsymbol{k},t)}(d|\boldsymbol{s}) = \pm(1 + 2\cos\varphi_d)\sum_{\mu}\delta_{\mu\mu}(d) \ . \tag{10.2.18b}$$

10.2.2 Vibrations of the Diamond Lattice

The general statements of the previous section can be illustrated using the vibrations of the nonsymmorphic diamond structure. As a special case we shall investigate the Δ direction (Figs. 9.1 and 10.3). The discussion will be carried out using (10.2.16–18).

Γ Point. Here we have $\mathscr{G}_{0\Gamma} \cong \mathscr{O}_\mathrm{h}$. The elements $\{d|0\}$ correspond to those of $\mathscr{T}_\mathrm{d}$, while the others have the form $\{id|\boldsymbol{s}\}$ with $\boldsymbol{s} = (1,1,1)\,a/4$. In the operations with $\boldsymbol{s} \neq 0$ the basis atoms at $\boldsymbol{R}_1$ and $\boldsymbol{R}_2$ are interchanged, i.e. according to (10.2.18), for $\boldsymbol{s} \neq 0$, $\hat{\chi}^{(\Gamma,t)} = 0$. For the operations with $\boldsymbol{s} = 0$, $\sum_{\mu=1}^{2}\delta_{\mu\mu}(d) = 2$. With this we obtain the characters of Table 10.1 and with (10.2.17) the multiplicities. Thus (see Appendix A)

$$\hat{D}^{(\Gamma,t)} = \Gamma_4^- \oplus \Gamma_5^+ = F_{1u} \oplus F_{2g} = \Gamma_{15} \oplus \Gamma_{25}' \ . \tag{10.2.19}$$

The vibrations of the Γ point ($\boldsymbol{k} = 0$) decompose into a 3-fold degenerate acoustic

Table 10.1. Characters of the representations of the Γ, Δ, and X points; their multiplicities and compatibilities and the REPs of Δ_5

	e	$8c_3$	$6\sigma_d$	$6S_4$	$3c_2$	$24\{d\|s\}$			$m_\alpha(\Gamma)$
$\hat{\chi}^{(\Gamma,t)}$	6	0	2	-2	-2	0			
Γ_4^-	3	0	1	-1	-1	–			1
Γ_5^+	3	0	1	-1	-1	–			1
	e	c_{2x}	σ_{yz}	$\sigma_{y\bar{z}}$	$\{c_{4x}\|s\}$	$\{c_{4x}^2\|s\}$	$\{\sigma_y\|s\}$	$\{\sigma_z\|s\}$	$m_\alpha(\Delta)$
$\hat{\chi}^{(\Delta,t)}$	6	-2	2	2	0	0	0	0	
Δ_1	1	1	1	1	1	1	1	1	1
Δ_2	1	1	-1	-1	1	1	-1	-1	0
Δ_3	1	1	-1	-1	-1	-1	1	1	0
Δ_4	1	1	1	1	-1	-1	-1	-1	1
$\hat{\chi}(\Delta_5)$	2	-2	0	0	0	0	0	0	2
Δ_5	e	$-e$	σ_y	$-\sigma_y$	$\mathrm{i}\sigma_z$	$-\mathrm{i}\sigma_z$	σ_x	$-\sigma_x$	–
Γ_4^-	3	-1	1	1	1	1	1	1	$\Delta_1 \oplus \Delta_5$
Γ_5^+	3	-1	1	1	-1	-1	-1	-1	$\Delta_4 \oplus \Delta_5$
X_1	2	2	2	2	0	0	0	0	$\Delta_1 \oplus \Delta_4$
X_2	2	2	-2	-2	0	0	0	0	$\Delta_2 \oplus \Delta_3$
X_3	2	-2	0	0	0	0	0	0	Δ_5
X_4	2	-2	0	0	0	0	0	0	Δ_5

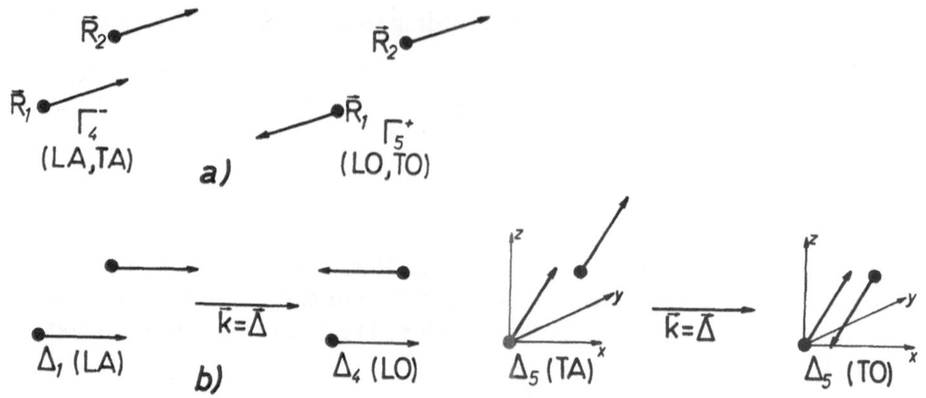

Fig. 10.2. (a) $\boldsymbol{k}=0$ or Γ-vibrations in a cubic crystal with diamond structure (two particles per cell); Γ_4^- is the acoustic, Γ_5^+ the optical mode. The displacements may take place in three arbitrary directions perpendicular to each other. (b) Δ-vibrations. For the Δ_5-vibrations only one special polarization is given. The displacements of the atoms at $\boldsymbol{R}_1$ and $\boldsymbol{R}_2$ contain a phase factor $\exp(i\pi/2)$

mode (Γ_4^-) and a 3-fold degenerate optical mode (Γ_5^+) for diamond structure. This again reflects the cubic symmetry (Fig. 10.2). For lower symmetry of $\mathcal{G}_{0\boldsymbol{k}}$ the three $\boldsymbol{k}=0$ acoustic modes are only twofold degenerate or even nondegenerate based on $\mathcal{G}_{0\boldsymbol{k}}$, but nevertheless there are three frequencies with $\omega_\Gamma = 0$ because of the translation symmetry of the whole crystal.

Δ Direction. We choose $\Delta = (2\pi/a)\ (\xi, 0, 0)$, $0 < \xi < 1$ with $\mathcal{G}_{0\Delta} \cong \mathcal{C}_{4v}$. The characters according to (10.2.18) and those of the IRs are given in Table 10.1, together with the characters of the Γ and X point subduced on Δ. Thus we obtain the multiplicities and compatibilities given on the right hand side, e.g.

$$\hat{D}^{(\Delta,t)} = \Delta_1 \oplus \Delta_4 \oplus 2\Delta_5 = A_1 \oplus B_2 \oplus 2E \ . \tag{10.2.20}$$

With this, the band structure in Fig. 10.3 can be completely understood (see also the bands in Fig. 9.4d).

For a quantitative calculation of the dispersion curves we need the dynamical matrix or the force constants (FCs) explicitly. The FCs are not independent of each other. They reflect the (dynamical) symmetry, i.e. in general the geometrical symmetry of the crystal (Sect. 7.1). The potential energy in (10.2.2) is then invariant under the transformations of the space group. That is, the transformation law for the FCs reads for the elements $\{d|\boldsymbol{t}\}$ (Exercise 10.2)

$$\overset{mn}{\phi}{}^{\mu\nu}_{ij} = \sum_{i',j'} D_{ii'} D_{jj'} \overset{m'n'}{\phi}{}^{\mu'\nu'}_{i'j'} \ . \tag{10.2.21}$$

Primed and unprimed superscripts m, n, μ, ν are connected by $\{d|\boldsymbol{t}\} \in \mathcal{R}$. For the special elements $\{e|\boldsymbol{R}^h\} \in \mathrm{T}$, (10.2.5) is valid. Restricting ourselves to interactions between nearest and next-nearest neighbours in the lattice, symmetry considera-

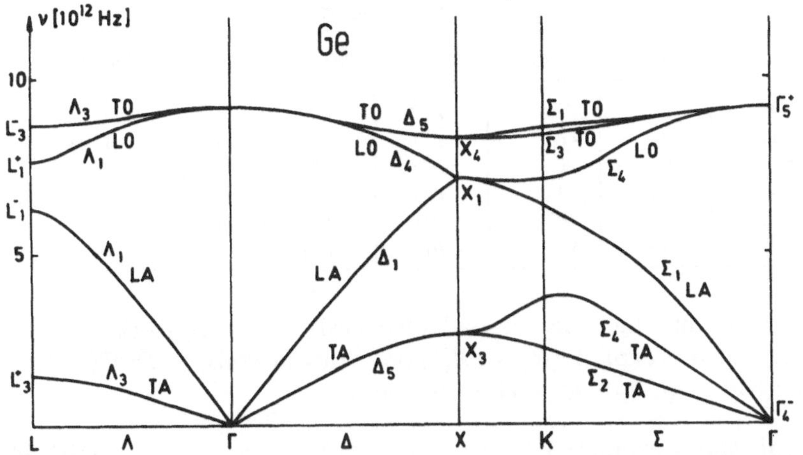

Fig. 10.3. Dispersion curves for germanium according to the model discussed. Symmetry directions Δ, Σ, Λ are shown. Notation from Appendix A. The Σ_1 and Σ_4 modes do not have a "definite" polarization (L, T) but rather have contributions of the "second" component

tions allow the following representative FC tensors for diamond structure:

$$\text{n.n.}\ \phi^{\overset{00}{12}}_{ij} = -\begin{pmatrix} \alpha & \beta & \beta \\ \beta & \alpha & \beta \\ \beta & \beta & \alpha \end{pmatrix}, \qquad \text{n.n.n.}\ \phi^{\overset{0(110)}{1\quad 1}}_{i\quad j} = -\begin{pmatrix} \mu & \nu & 0 \\ \nu & \mu & 0 \\ 0 & 0 & \lambda \end{pmatrix} \tag{10.2.22}$$

and

$$\phi^{\overset{00}{11}}_{ij} = \phi^{\overset{00}{22}}_{ij} = 4(\alpha + \lambda + 2\mu)\delta_{ij} \ .$$

All other FCs are determined by these ones. The model has five independent parameters which can be fitted by comparison with measured frequencies.

Using (10.2.22) we get the frequency tensor (10.2.7b) for the $\Delta = (2\pi/a) \times (\xi, 0, 0)$ direction in the form described by footnote 4:

$$M \cdot F^{\mu\nu}_{ij}(\Delta) = \begin{pmatrix} A & 0 & 0 & C & 0 & 0 \\ 0 & B & 0 & 0 & C & -\mathrm{i}D \\ 0 & 0 & B & 0 & -\mathrm{i}D & C \\ C & 0 & 0 & A & 0 & 0 \\ 0 & C & \mathrm{i}D & 0 & B & 0 \\ 0 & \mathrm{i}D & C & 0 & 0 & B \end{pmatrix} = \begin{pmatrix} R & T \\ T^* & R \end{pmatrix} \tag{10.2.23}$$

with

$$A = 4\alpha + 16\mu \sin^2(\pi\xi/2) \ , \qquad B = 4\alpha + 8(\lambda + \mu)\sin^2(\pi\xi/2) \ ,$$

$$C = -4\alpha\cos(\pi\xi/2) \ , \qquad D = 4\beta\sin(\pi\xi/2) \ , \qquad M = M_1 = M_2 \ .$$

Table 10.2. Vectors resulting from application of $\hat{D}^{(\Delta,t)}$ to a general vector

$\hat{D}(e)$	$\hat{D}(c_{2x})$	$\hat{D}(\sigma_{yz})$	$\hat{D}(\sigma_{y\bar{z}})$	$\hat{D}(c_{4x})$	$\hat{D}(c_{4x}^3)$	$\hat{D}(\sigma_y)$	$\hat{D}(\sigma_z)$
x_1	x_1	x_1	x_1	$\rho^* x_2$	$\rho^* x_2$	$\rho^* x_2$	$\rho^* x_2$
y_1	$-y_1$	$-z_1$	z_1	$-\rho^* z_2$	$\rho^* z_2$	$-\rho^* y_2$	$\rho^* y_2$
z_1	$-z_1$	$-y_1$	y_1	$\rho^* y_2$	$-\rho^* y_2$	$\rho^* z_2$	$-\rho^* z_2$
x_2	x_2	x_2	x_2	ρx_1	ρx_1	ρx_1	ρx_1
y_2	$-y_2$	$-z_2$	z_2	$-\rho z_1$	ρz_1	$-\rho y_1$	ρy_1
z_2	$-z_2$	$-y_2$	y_2	ρy_1	$-\rho y_1$	ρz_1	$-\rho z_1$

In order to determine the symmetry-adapted basis according to (10.2.16) we need the phase factors $\exp[\mathrm{i}\boldsymbol{\Delta}\cdot(\boldsymbol{R}_\mu - d\boldsymbol{R}_\nu)]$ occurring in $P(d|s) := \hat{D}^{(\Delta,t)}(d|s)$, see (10.2.13). With $\boldsymbol{R}_1 = (0,0,0)$, $\boldsymbol{R}_2 = (1,1,1)a/4$ and

$$\rho = \exp(\mathrm{i}k_x a/4) = \exp(\mathrm{i}\pi\xi/2) \tag{10.2.24}$$

we obtain the following assignment of the phase-factors:

$$e,\, c_{2x},\, \sigma_{yz},\, \sigma_{y\bar{z}} \leftrightarrow 1 \qquad \text{for } (\mu,\nu) = (1,1) \text{ and } (2,2)$$

$$c_{4x},\, c_{4x}^3,\, \sigma_y,\, \sigma_z \leftrightarrow \rho^* \qquad \text{for } (\mu,\nu) = (1,2)$$

$$c_{4x},\, c_{4x}^3,\, \sigma_y,\, \sigma_z \leftrightarrow \rho \qquad \text{for } (\mu,\nu) = (2,1)$$

No others occur. We obtain the effect of $\hat{D}^{(\Delta,t)}$ in REP space when we apply $\hat{D}$ to a column vector $(x_1, y_1, z_1, x_2, y_2, z_2)$ of this space, where the phase factors have to be taken into account. The results are given in the *operator table* (Table 10.2). From this we see for the matrix $\hat{D}(c_{4x})$ for example

$$\hat{D}(c_{4x}) = \begin{pmatrix} 0 & \rho^* c_{4x} \\ \rho c_{4x} & 0 \end{pmatrix} \quad \text{with} \quad c_{4x} = \begin{pmatrix} 1 & 0 & 0 \\ 0 & 0 & -1 \\ 0 & 1 & 0 \end{pmatrix}, \ldots . \tag{10.2.25}$$

Together with the IRs $\hat{D}^{(\Delta,\alpha)}$ from Table 10.1 we are able to construct the projection operators according to (10.2.16) and to apply them to a Cartesian basis $(e^{(l)}\}$. Using the Δ_1-REP and applying the projection operator to $\{e^{(l)}\} = \{x_1 = 1, y_1 = 0, z_1 = 0, x_2 = 0, y_2 = 0, z_2 = 0\}$ we obtain $(4,0,0,4\rho,0,0)$ or, normalized,

$$e_1^{(\Delta_1,1)}(\Delta) = \frac{1}{\sqrt{2}}(1,0,0;\rho,0,0) . \tag{10.2.26a}$$

Using the form of footnote 4 for $\bar{F}$ or (10.2.23), we have to multiply this vector by the phase factor $\exp(-\mathrm{i}\boldsymbol{\Delta}\cdot\boldsymbol{R}_\mu)$, which equals 1 for $\mu = 1$ and ρ^* for $\mu = 2$. Thus

$$\bar{e}_1^{(\Delta_1,1)}(\Delta) = \frac{1}{\sqrt{2}}(1,0,0;1,0,0) . \tag{10.2.26b}$$

In this way we finally obtain the transformation matrix which corresponds to (7.1.16) and (6.3.3, 4)

$$S = \frac{1}{\sqrt{2}} \begin{pmatrix} 1 & 0 & 0 & 1 & 0 & 0 \\ 1 & 0 & 0 & -1 & 0 & 0 \\ 0 & 0 & 1 & 0 & -i & 0 \\ 0 & 1 & 0 & 0 & 0 & i \\ 0 & 0 & 1 & 0 & i & 0 \\ 0 & 1 & 0 & 0 & 0 & -i \end{pmatrix} \begin{matrix} \Delta_1 \\ \Delta_4 \\ \left.\begin{matrix} \\ \end{matrix}\right\} \Delta_5 \\ \left.\begin{matrix} \\ \end{matrix}\right\} \Delta_5 \end{matrix} , \tag{10.2.27}$$

and thus from (10.2.23)

$$S(MF)S^{+} = \begin{pmatrix} A+C & 0 & 0 & 0 & 0 & 0 \\ 0 & A-C & 0 & 0 & 0 & 0 \\ 0 & 0 & B+D & -iC & 0 & 0 \\ 0 & 0 & iC & B-D & 0 & 0 \\ 0 & 0 & 0 & 0 & B-D & iC \\ 0 & 0 & 0 & 0 & -iC & B+D \end{pmatrix} \begin{matrix} \Delta_1 \\ \Delta_4 \\ \left.\begin{matrix} \\ \end{matrix}\right\} \Delta_5 \\ \left.\begin{matrix} \\ \end{matrix}\right\} \Delta_5 \end{matrix} . \tag{10.2.28}$$

Hence the vibration problem for the Δ-direction is reduced:

(i) Δ_1-vibration, nondegenerate.

$$M\omega^2(\Delta_1) = A + C , \qquad e = \frac{1}{\sqrt{2}}(1,0,0;1,0,0) ,$$

i.e. a longitudinal acoustic (LA) mode.

(ii) Δ_4-vibration, nondegenerate.

$$M\omega^2(\Delta_4) = A - C , \qquad e = \frac{1}{\sqrt{2}}(1,0,0;-1,0,0) ,$$

i.e. a longitudinal optical (LO) mode.

(iii) Δ_5-vibration, 2-fold degenerate.

$$M\omega^2(\Delta_5) = B \pm \sqrt{D^2 + C^2} , \qquad e = (0,b,a;0,ia,-ib) ,$$

$$e = (0,b,a;0,-ia,ib) ,$$

i.e. two degenerate transverse optical (TO, upper sign) modes and two degenerate transverse acoustic (TA, lower sign) modes.

The vibration vectors are shown in Fig. 10.2, and the dispersion curves for a special model in Fig. 10.3.

Exercise 10.2. a) Prove (10.2.21) by using the invariance of the potential energy.
b) Determine the FCs in (10.2.22) by investigating the effect of c_{3xyz}, c_{3xyz}^2, $\sigma_{x\bar{y}}$, etc.
c) Using (10.2.22) evaluate the form of the other FCs for nearest and next-nearest neighbours of $\binom{0}{1}$.

Exercise 10.3. Determine the frequency branches for vibrations with $\boldsymbol{k}$-vectors in Λ and Σ directions for the diamond lattice (Fig. 10.3).

10.3 Electron Energy Bands

10.3.1 Symmetrization of Plane Waves

Calculation of the electronic band structure (in the one-electron approximation) requires an expansion of the wave function Ψ of the electron with respect to an appropriate complete system of functions $\{\psi_i\}$:

$$\Psi = \sum_{i=1}^{\infty} a_i \psi_i \ . \tag{10.3.1a}$$

For practical purposes the series (10.3.1a) is restricted to a suitable number of terms.

a) In many cases the ψ_i are chosen to have the Bloch form (10.1.4). The periodic function $u_\sigma^{(\boldsymbol{k})}$ then takes into account that the $\psi_\sigma^{(\boldsymbol{k})}$ approximate an atomic wave function in the vincinity of the ionic cores. We therefore set

$$\Psi^{(\boldsymbol{k})} = \sum_\sigma a_\sigma^{(\boldsymbol{k})} \psi_\sigma^{(\boldsymbol{k})} \ . \tag{10.3.1b}$$

The associated eigenvalue problem then follows from the Schrödinger equation:

$$\det\left[\langle \psi_\rho^{(\boldsymbol{k})} | H | \psi_\tau^{(\boldsymbol{k})} \rangle - E(\boldsymbol{k}) \langle \psi_\sigma^{(\boldsymbol{k})} | \psi_\tau^{(\boldsymbol{k})} \rangle\right] = 0 \ , \tag{10.3.2}$$

where H is the one-electron crystal Hamiltonian. The various methods differ in the choice of the system of $\psi_\sigma^{(\boldsymbol{k})}$.

(i) The LCAO (*or tight binding*) method uses

$$\psi_{\mu,\sigma}^{(\boldsymbol{k})}(\boldsymbol{x}) = \frac{1}{\sqrt{N}} \sum_{\boldsymbol{h}} \exp(\mathrm{i}\boldsymbol{k} \cdot \boldsymbol{R}_{\mu}^{\boldsymbol{h}}) u_{\mu,\sigma}(\boldsymbol{x} - \boldsymbol{R}_{\mu}^{\boldsymbol{h}}) \ , \tag{10.3.3}$$

where σ denotes atomic quantum numbers (n, l, m, s) and $u_{\mu,\sigma} \sim R_{nl} \cdot Y_m^l \cdot \chi_s$ are atomic functions. This method describes the limit of strong binding, and is thus appropriate for the calculation of valence bands.

(ii) In the method of *plane waves* (PW) only plane waves reduced to the first BZ are used:

$$\psi_K^{(k)}(x) = \frac{1}{\sqrt{V}} \exp[i(k + K)\cdot x] \to |k + K\rangle \ . \tag{10.3.4}$$

The sum in (10.3.1b) runs over a selection of (or over all) the K-vectors of reciprocal space. This ansatz is appropriate in the case of weak binding, e.g. for the description of conduction bands.

(iii) The OPW *method* improves (10.3.4):

$$\tilde{\psi}_K^{(k)}(x) = \frac{1}{\sqrt{V}} \exp[i(k + K)\cdot x] - \sum_{\sigma=1}^{\sigma_0} b_{K,\sigma}^{(k)} \tilde{\psi}_\sigma^{(k)}(x) \ . \tag{10.3.5a}$$

The $\tilde{\psi}_\sigma^{(k)}$ are σ_0 Bloch functions of the low filled bands. The coefficients $b_{K\sigma}^{k}$ have to be chosen such that the total function (10.3.5) is orthogonal to all the low-energy core functions:

$$b_{K\sigma}^{(k)} := \langle k, \sigma | k + K\rangle = \frac{1}{\sqrt{V}} \int \tilde{\psi}_\sigma^{(k)*}(x) \exp[i(k + K)\cdot x]\, dx \ . \tag{10.3.5b}$$

With the projection operator

$$P_{\sigma_0}^{(k)} = \sum_{\sigma=1}^{\sigma_0} |k\sigma\rangle\langle k\sigma| \tag{10.3.6}$$

acting on the core states, (10.3.5a) can also be written as

$$|\tilde{\psi}_K^{(k)}\rangle = (1 - P_{\sigma_0}^{(k)})|k + K\rangle \ . \tag{10.3.7}$$

This method is used very often. The band structures of semiconductors, for example, are described quite well.

For details of all these methods see textbooks on solid state theory.

b) There are also other cases where (10.3.1a) is expanded in terms of atomic wave functions $\{R_{nl} Y_m^l\}$, and the Schrödinger equation is only solved for a (spherically symmetric) potential in the interior of a Wigner-Seitz cell (*cellular method*). However, boundary conditions between neighbouring cells have to be observed. The solution is then extended over the whole lattice in Bloch form. For practical applications one has to distinguish between the simple cellular method, the method of Green's functions (KKR) and the APW method.

In all these cases one starts with

$$\psi^{(k)}(x) = \sum_l a_l(k) \cdot \sum_m c_{lm}(k) Y_m^l(\vartheta, \varphi) \cdot R_l(E, x) \ , \tag{10.3.8}$$

where $R_l(E, x)$ is a solution of the radial Schrödinger equation with $V(r)$ in a Wigner-Seitz cell. Group theory now yields the linear combinations of the Y_m^l with fixed l which are assigned to the IRs $D^{(k,\alpha)}$ of $\mathscr{G}_k$, e.g. the cubic harmonics in cubic lattices (see Table 5.1). Symmetry adaption is then accomplished by applying the group projection operators to the starting functions $Y_m^l(\vartheta, \varphi)$. Thus, for symmorphic space groups

$$Y_i^{(\boldsymbol{k},\alpha;l)} = \frac{d_\alpha}{g_{0\boldsymbol{k}}} \sum_{\mathcal{G}_{0\boldsymbol{k}}} D_{ii}^{(\boldsymbol{k},\alpha)*}(d|0) \cdot P(d|0) \cdot Y_m^l \tag{10.3.9}$$

with $P(d|0) \cdot Y_m^l$ from (7.2.13a, 14). For nonsymmorphic space groups the localization of the Y_m^l at the positions $\boldsymbol{s}_\nu$ has additionally to be taken into account. Because

$$P(d|\boldsymbol{s}) \cdot \psi(\boldsymbol{x}) = \psi(d^{-1}(\boldsymbol{x} - \boldsymbol{s})) = P(d|0) \cdot \psi(\boldsymbol{x} - \boldsymbol{s}) \ , \tag{10.3.10}$$

we have to replace Y_m^l in (10.3.9) by $Y_m^l(\boldsymbol{s})$, where $Y_m^l(\boldsymbol{s})$ differs from Y_m^l only by localization at $\boldsymbol{x} = \boldsymbol{s}$ instead of $\boldsymbol{x} = 0$. Furthermore $D_{ii}^{(\boldsymbol{k},\alpha)*}(d|\boldsymbol{s})$ has to be used in (10.3.9).

According to Sect. 6.3, the crystal Hamiltonian H ($\leftarrow A$) has vanishing matrix elements for basis functions of different types of symmetry (6.3.7), i.e in (10.3.1b) we only need functions that are basis functions for allowable IRs $D_{ij}^{(\boldsymbol{k},\alpha)}$ of $\mathcal{G}_{\boldsymbol{k}}$ with fixed row index i. Hence the eigenvalue problem (10.3.2) is given in the symmetrized form (6.3.10). The symmetry adaption itself is achieved by applying the projection operators (9.3.4) to the functions ψ_i in the expansions. This method will be illustrated by means of plane waves (PW), but the results can be transferred immediately to the OPW and pseudo-potential methods (end of this section).

The behaviour of plane waves in transformations with $P(d|\boldsymbol{s}) \in \mathcal{G}_{\boldsymbol{k}}$ defines the REP spaces (Sect. 5.1), which are in general reducible, constituted by these functions. We have

$$\begin{aligned} P(d|\boldsymbol{t}) \exp[\mathrm{i}(\boldsymbol{k} + \boldsymbol{K}) \cdot \boldsymbol{x}] &= \exp[\mathrm{i}(\boldsymbol{k} + \boldsymbol{K}) \cdot d^{-1}(\boldsymbol{x} - \boldsymbol{t})] \\ &= \exp[-\mathrm{i}d(\boldsymbol{k} + \boldsymbol{K}) \cdot \boldsymbol{t}] \exp[\mathrm{i}d(\boldsymbol{k} + \boldsymbol{K}) \cdot \boldsymbol{x}] \ , \end{aligned} \tag{10.3.11}$$

i.e. the waves with

$$d(\boldsymbol{k} + \boldsymbol{K}) = \boldsymbol{k} + \boldsymbol{K}' \sim \boldsymbol{k} \ , \qquad d \in \mathcal{G}_{0\boldsymbol{k}} \ ,$$

constitute the REP space. The REP matrices have only one non-vanishing matrix element $\langle \boldsymbol{k} + \boldsymbol{K}' | P(d|\boldsymbol{t}) | \boldsymbol{k} + \boldsymbol{K} \rangle$ in every row and in every column, if $\boldsymbol{k} + \boldsymbol{K}' = d(\boldsymbol{k} + \boldsymbol{K})$. According to (10.3.11) the character is

$$\chi^{(\boldsymbol{k}+\boldsymbol{K})}(d|\boldsymbol{t}) = \sum \exp[-\mathrm{i}(\boldsymbol{k} + \boldsymbol{K}') \cdot \boldsymbol{t}] \ , \tag{10.3.12}$$

where the sum is taken over all $\boldsymbol{k} + \boldsymbol{K}' \in \{d(\boldsymbol{k} + \boldsymbol{K}) | d \in \mathcal{G}_{0\boldsymbol{k}}\}$, $\boldsymbol{k}$, $\boldsymbol{K}$ fixed for which $d(\boldsymbol{k} + \boldsymbol{K}') = \boldsymbol{k} + \boldsymbol{K}'$, which are thus invariant under d. Here $d(\boldsymbol{k} + \boldsymbol{K})$ does not have to be a vector of the first BZ (see the definition in Sect. 5.3). The reduction of this REP yields the corresponding symmetrized PWs.

Table 10.3. Characters of $\chi^{(k+K)}$ for the Γ-point of $\mathcal{O}_h^7$

$\mathcal{O}_h$	K	e	$8c_3$	$6\{c_2'\|s\}$	$6\{c_4\|s\}$	$3c_2$	$\{i\|s\}$	$8\{s_6\|s\}$	$6\sigma_d$	$6s_4$	$3\{\sigma_h\|s\}$
χ^K	0	1	1	1	1	1	1	1	1	1	1
χ^K	(1, 1, 1)	8	2	0	0	0	0	0	4	0	0
χ^K	(2, 0, 0)	6	0	0	−2	2	0	0	2	0	−4

As an example we again consider diamond structure ($\mathcal{O}_h^7$) at the Γ point but now restrict ourselves to the "lowest" PWs according to (10.3.4): $\boldsymbol{K} = 0$, $\boldsymbol{K} = (2\pi/a)(1,1,1)$, $\boldsymbol{K} = (2\pi/a)(2,0,0)$ and equivalent ones. As $\mathcal{G}_{0\Gamma} \cong \mathcal{O}_h$ and with (10.3.12)

$$\chi^{(K)}(d|t) = \sum \mathrm{e}^{-\mathrm{i}\boldsymbol{K}'\cdot\boldsymbol{s}} \qquad \text{with} \qquad \boldsymbol{s} = \frac{a}{4}(1,1,1) \ . \tag{10.3.13}$$

Obviously the identity REP is assigned to $\boldsymbol{K} = 0$. The vectors $\{d\boldsymbol{K}, \boldsymbol{K} \sim (1,1,1), d \in \mathcal{O}_h\} = \{\pm 1, \pm 1, \pm 1\}$ remain invariant only under the elements of $\mathcal{G}_{0\Lambda} \cong \mathcal{C}_{3v}$; for example, c_{3xyz} leaves invariant $(1,1,1)$ and $(\bar{1},\bar{1},\bar{1})$; $\sigma_{x\bar{y}}$ leaves invariant $(1,1,1)$, $(1,1,\bar{1})$, $(\bar{1},\bar{1},1)$ and $(\bar{1},\bar{1},\bar{1})$. Since c_3 and $\sigma_d \in \mathcal{T}_d \subset \mathcal{O}_h$, $\boldsymbol{s} = 0$, thus $\exp(-\mathrm{i}\boldsymbol{K}\cdot\boldsymbol{s}) = 1$, from which Table 10.3 follows. With the character table (Appendix A) we then have the decompositions for $\mathcal{O}_h$:

$$
\begin{aligned}
D^{(0)} &= \Gamma_1^+ \ , \\
D^{(1,1,1)} &= \Gamma_1^+ \oplus \Gamma_2^- \oplus \Gamma_4^- \oplus \Gamma_5^+ \ , \\
D^{(2,0,0)} &= \Gamma_2^- \oplus \Gamma_3^- \oplus \Gamma_5^+ \ .
\end{aligned} \tag{10.3.14}
$$

This means for the symmetrized eigenvalue problem that the problem [originally 9-dimensional with $\boldsymbol{K} = 0$ and $\boldsymbol{K} \sim (1,1,1)$ or 15-dimensional with $\boldsymbol{K} \sim (2,0,0)$ included] decomposes into a 2×2 problem (Γ_1^+) and 3 simple problems (Γ_2^- single, Γ_4^- and Γ_5^+ each 3-fold) or into three 2×2 problems ($\Gamma_1^+, \Gamma_2^-, \Gamma_5^+$) and two simple problems (Γ_3^-, Γ_4^-).

The explicit form of the symmetrized PWs follows by means of the projection operator method (9.3.4, 5). As a starting function we choose $\psi_K^{(\Gamma)} = \psi_0$ with $\boldsymbol{K} = (2\pi/a)(1,1,1)$ and determine the effect of $P(d|\boldsymbol{s})$ on ψ_0. We choose the origin of the coordinate to be in the centre of the basis, thus at $\boldsymbol{x} = (a/8)(1,1,1) = \boldsymbol{s}/2$. Then $\{i|0\}$ belongs to $\mathcal{R}$ and to $\mathcal{G}_\Gamma$, $\mathcal{G}_{0\Gamma}$. Consequently, the transformation laws simplify considerably. The decomposition of $\mathcal{O}_h$ into cosets with respect to the invariance group $\mathcal{C}_{3v}$ of $\boldsymbol{K} \sim (1,1,1)$ is

$$
\begin{aligned}
\mathcal{O}_h &= \mathcal{T}_d + i\mathcal{T}_d \ , \\
\mathcal{T}_d &= \mathcal{C}_{3v} + c_{2x}\mathcal{C}_{3v} + c_{2y}\mathcal{C}_{3v} + c_{2z}\mathcal{C}_{3v} := K_0 + K_1 + K_2 + K_3 \\
\mathcal{C}_{3v} &= \{e, c_{3xyz}^{\pm 1}, \sigma_{x\bar{y}}, \sigma_{y\bar{z}}, \sigma_{z\bar{x}}\} \ .
\end{aligned} \tag{10.3.15}
$$

The K_j are the cosets in $\mathscr{T}_d$, and $iK_j = K_{\bar{j}}$ are those in $i\mathscr{T}_d$. Then we have $[s = s(d)]$

$$P(d|s)\psi_0 = \frac{1}{\sqrt{V}}\exp[\mathrm{i}\boldsymbol{K}\cdot d^{-1}(\boldsymbol{x}-\boldsymbol{s})] = \frac{1}{\sqrt{V}}\exp(-\mathrm{i}d\boldsymbol{K}\cdot\boldsymbol{s})\exp(\mathrm{i}d\boldsymbol{K}\cdot\boldsymbol{x})\ .$$

We now take

$$\exp(\mathrm{i}d\boldsymbol{K}\cdot\boldsymbol{x}) = \begin{cases}\psi_j, & \text{if } d \in K_j,\quad j = 0, 1, 2, 3\\ \psi_{\bar{j}}, & \text{if } d \in K_{\bar{j}}\ .\end{cases} \tag{10.3.16}$$

According to Exercise 3.11, if we define the origin at the centre of the basis, we obtain new nonprimitive displacements $\boldsymbol{s}'(d) = (d\boldsymbol{s} - \boldsymbol{s})/2$ for $d \in \mathscr{T}_d$ and $\boldsymbol{s}'(d') = -\boldsymbol{s}'(d)$ for $d' \in i\mathscr{T}_d$, and thus

$$\begin{aligned}\exp(-\mathrm{i}d\boldsymbol{K}\cdot\boldsymbol{s}') &= \begin{cases}\ \ 1 & \text{for } d \in K_0\\ -1 & \text{for } d \in K_1, K_2, K_3\end{cases}\\ P(d|\boldsymbol{s}')\psi_0 &= \begin{cases}\psi_0, \bar{\psi}_0 & \text{for } d \in K_0, K_{\bar{0}}\\ -\psi_j, -\bar{\psi}_j & \text{for } d \in K_j, K_{\bar{j}}, j = 1, 2, 3\ .\end{cases}\end{aligned} \tag{10.3.17}$$

The projections onto the waves symmetrized with respect to $\mathscr{G}_{0\Gamma} \cong \mathscr{O}_h$ can now be performed by using $P^{(\alpha)}(\mathscr{O}_h) \sim P^{\gamma(\alpha)}(\mathscr{C}_i)P^{\beta(\alpha)}(\mathscr{T}_d)$, and hence we obtain the symmetrized PWs $\psi_i^{\Gamma} \sim P_{i1}^{\Gamma}\psi_0$ given in Table 10.4 from ψ_0 using (10.3.17). For $\Gamma_2^-, \Gamma_4^-, \Gamma_5^+$ the energies are given by

$$E(\Gamma) = \langle\psi_i(\Gamma)|H|\psi_i(\Gamma)\rangle\ .$$

For Γ_1^+ the function $\psi(\Gamma_1^+, 1) = 1/\sqrt{V}$ belonging to $\boldsymbol{K} = 0$ has to be taken into account besides the function $\psi(\Gamma_1^+, 2)$ from Table 10.4. Then the equation

$$\begin{vmatrix}\langle\psi(\Gamma_1^+,1)|H|\psi(\Gamma_1^+,1)\rangle - E(\Gamma_1^+) & \langle\psi(\Gamma_1^+,1)|H|\psi(\Gamma_1^+,2)\rangle\\ \langle\psi(\Gamma_1^+,2)|H|\psi(\Gamma_1^+,1)\rangle & \langle\psi(\Gamma_1^+,2)|H|\psi(\Gamma_1^+,2)\rangle - E(\Gamma_1^+)\end{vmatrix} = 0$$

has to be solved (see Fig. 10.4 for free electrons).

Table 10.4. Transformation of plane waves to symmetry-adapted functions

$$\begin{pmatrix}\psi(\Gamma_1^+)\\ \psi(\Gamma_2^-)\\ \psi_1(\Gamma_5^+)\\ \psi_2(\Gamma_5^+)\\ \psi_3(\Gamma_5^+)\\ \psi_1(\Gamma_4^-)\\ \psi_2(\Gamma_4^-)\\ \psi_3(\Gamma_4^-)\end{pmatrix} \sim \begin{pmatrix}1 & -1 & -1 & -1 & 1 & -1 & -1 & -1\\ 1 & -1 & -1 & -1 & -1 & 1 & 1 & 1\\ 1 & -1 & 1 & 1 & 1 & -1 & 1 & 1\\ 1 & 1 & -1 & 1 & 1 & 1 & -1 & 1\\ 1 & 1 & 1 & -1 & 1 & 1 & 1 & -1\\ 1 & -1 & 1 & 1 & -1 & 1 & -1 & -1\\ 1 & 1 & -1 & 1 & -1 & -1 & 1 & -1\\ 1 & 1 & 1 & -1 & -1 & -1 & -1 & 1\end{pmatrix} \cdot \begin{pmatrix}\psi_0\\ \psi_1\\ \psi_2\\ \psi_3\\ \psi_{\bar{0}}\\ \psi_{\bar{1}}\\ \psi_{\bar{2}}\\ \psi_{\bar{3}}\end{pmatrix}$$

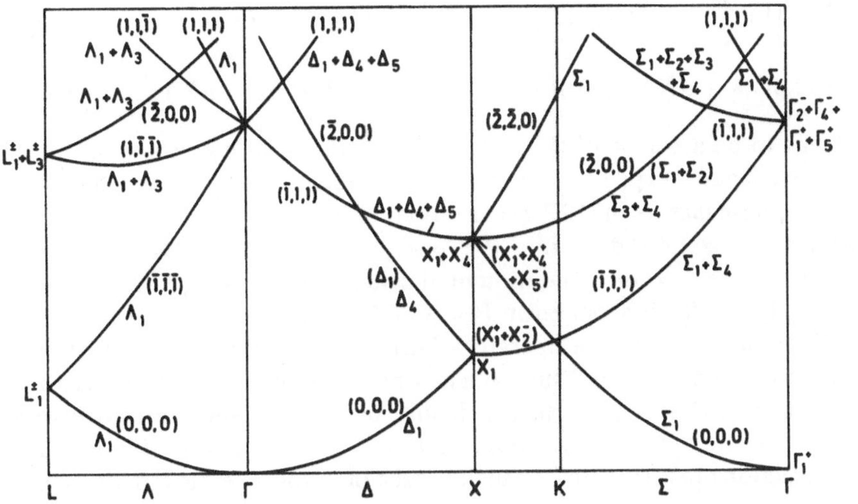

Fig. 10.4. Energy bands in the free-electron approximation for the crystal type O_h^7. The scheme is also valid for a fcc lattice but with different notation (in brackets) for some bands

We now consider the eigenvalue problem in a symmetrized PW basis according to (6.3.10) in its general form. In this case it reads

$$\det\{\langle\psi_i^{(\boldsymbol{k},\alpha,s)}|H|\psi_i^{(\boldsymbol{k},\alpha,s')}\rangle - E_i^{(\alpha)}(\boldsymbol{k})\delta_{ss'}\} = 0 \ , \tag{10.3.18}$$

since

$$\langle\psi_i^{(\boldsymbol{k},\alpha,s)}|\psi_i^{(\boldsymbol{k},\alpha,s')}\rangle = \delta_{ss'} \qquad \text{and} \qquad {}_i^{(\alpha,s)} \mathrel{\hat{=}} \sigma$$

in (10.3.2). According to (9.3.5, 10.3.11) the symmetrized functions are

$$\langle\boldsymbol{x}|\psi_i^{(\boldsymbol{k},\alpha,s)}\rangle := \frac{c_s}{\sqrt{V}} \sum_{\mathscr{G}_{0k}} D_{ii}^{(\boldsymbol{k},\alpha)*}(d|\boldsymbol{s})\exp[-\mathrm{i}d(\boldsymbol{k}+\boldsymbol{K}_s)\cdot\boldsymbol{s}]\exp[\mathrm{i}d(\boldsymbol{k}+\boldsymbol{K}_s)\cdot\boldsymbol{x}] \tag{10.3.19}$$

with a normalizing constant c_s; s gives the multiplicities of the IRs occurring in the reduction of the space defined by the $\exp[\mathrm{i}(\boldsymbol{k}+\boldsymbol{K}_s)\cdot\boldsymbol{x}]$. With $H = T + V$, $T = -\hbar^2\Delta/2m$ and (10.3.19), and using $[d(\boldsymbol{k}+\boldsymbol{K}_s)]^2 = (\boldsymbol{k}+\boldsymbol{K}_s)^2$, (10.3.18) can be written as

$$\langle\psi_i^{(\boldsymbol{k},\alpha,s)}|T|\psi_i^{(\boldsymbol{k},\alpha,s')}\rangle = \frac{\hbar^2}{2m}(\boldsymbol{k}+\boldsymbol{K}_s)^2\delta_{ss'} \ , \tag{10.3.20}$$

and with

$$\tilde{V}(\boldsymbol{K}_{s'}-\boldsymbol{K}_s) = \frac{1}{V_z}\int_{V_z}\exp[\mathrm{i}(\boldsymbol{K}_{s'}-\boldsymbol{K}_s)\cdot\boldsymbol{x}]V(\boldsymbol{x})d^3\boldsymbol{x} \ ,$$

$$\langle\psi_i^{(k,\alpha,s)}|V|\psi_i^{(k,\alpha,s')}\rangle = c_{s'}c_s^* \sum_d \sum_{d' \in \mathscr{G}_{0k}} D_{ii}^{(k,\alpha)}(d|s)D_{ii}^{(k,\alpha)*}(d'|s')$$

$$\times \exp[\mathrm{i}(k + dK_s)\cdot s]\exp[-\mathrm{i}(k + d'K_{s'})\cdot s']\tilde{V}(d'K_{s'} - dK_s) \; . \qquad (10.3.21)$$

Here k has to be a vector of the interior of the BZ. If it is a vector of the BZ surface then factors $\exp(\pm \mathrm{i}dK\cdot x)$ have to be added. For symmorphic space groups the phase factors in (10.3.21) are equal to one.

The symmetrization, done in a PW basis, is also valid for OPW functions and thus it can be used in pseudopotential calculations as well. For a proof we show that the projection operator $P = \sum_k P_{\sigma_0}^{(k)}$ acting on the core states, see (10.3.6, 7), commutes with the operators $P(d|s)$, because then we can first symmetrize the PW functions and subsequently perform the core projections. Since the wave functions $|k,\sigma\rangle$ in $P_{\sigma_0}^{(k)}$ have a definite behaviour under transformations defined by the IRs of the space group, see (9.1.21), we decompose P into a sum of sub-projection operators $\hat{P}(E_k)$ which consist of the degenerate functions $|k,i\rangle$ belonging to a fixed energy E_k. With

$$P = \sum \hat{P}(E_k) \; , \qquad \hat{P}(E_k) = \sum_{k \in *k} \sum_{i=1}^{d_{k\alpha}} |ki\rangle\langle ki| \; , \qquad (10.3.22)$$

we have for an arbitrary function $|\phi\rangle$ [see (9.1.21, 24); the subscripts k', k'' of D correspond to μ, ν in (9.1.22)]

$$\begin{aligned} P(d|s)\hat{P}|\phi\rangle &= \sum_{ki} P(d|s)|ki\rangle\langle ki|\phi\rangle \\ &= \sum_{ki} P(d|s)|ki\rangle\langle P(d|s)ki|P(d|s)\phi\rangle \\ &= \sum_{ki}\sum_{k'j,k''l} D_{k'j,ki}^{(*k,\alpha)}(d|s)D_{k''l,ki}^{(*k,\alpha)*}(d|s)|k'j\rangle\langle k''l|P(d|s)\phi\rangle \\ &= \sum_{k'j} |k'j\rangle\langle k'j|P(d|s)\phi\rangle = \hat{P}P(d|s)|\phi\rangle \; . \end{aligned} \qquad (10.3.23)$$

In this case we have assumed that $P(d|s)$ and $D^{(*k,\alpha)}$ are unitary. Thus we may always perform the symmetrization for the PWs first, e.g. for $|\phi\rangle = |k + K\rangle$.

10.3.2 Energy Bands and Atomic Levels

Bands grow from atomic levels if we bring free atoms close together to form well-defined structures. The tight-binding method allows one to visualize this development of bands. We rewrite functions (10.3.3) as

$$\begin{aligned} \psi_{\mu,\sigma}^{(k)}(x) &= \mathrm{e}^{\mathrm{i}k\cdot x}U_{\mu,\sigma}^{(k)}(x) \; , \\ U_{\mu,\sigma}^{(k)}(x) &= \frac{1}{\sqrt{N}}\sum_h \exp[-\mathrm{i}k\cdot(x - \overset{h}{R^\mu})]u_{\mu,\sigma}(x - \overset{h}{R^\mu}) \; . \end{aligned} \qquad (10.3.24)$$

Under $P(d|t) \in \mathscr{R}$, (10.3.24) transforms according to

$$P(d|t)\psi_{\mu,\sigma}^{(k)}(x) = P(d|t)e^{ik\cdot x}P(d|t)U_{\mu,\sigma}^{(k)}(x) \ . \qquad (10.3.25)$$

Because

$$P(e|R^h)U_{\mu,\sigma}^{(k)}(x) \equiv U_{\mu,\sigma}^{(k)}(x - R^h) = 1 \cdot U_{\mu,\sigma}^{(k)}(x) \ ,$$

the function $U_{\mu,\sigma}^{(k)}$ has to belong to the $k = 0$-(Γ-)REP $D^{(\Gamma,\alpha)}$ of the space group $\mathscr{R} = \mathscr{G}_\Gamma$. The type ($\alpha$) of the IR is determined by the rotation symmetry of the atomic functions $u_{\mu,\sigma}(x)$, thus in general by s-, p-, d-functions or their hybrids (Sect. 7.2). Then we have to investigate the subductions of the IRs $D^{(l)}$, $l = 0, 1, 2, \dots$ of $\mathscr{SO}(3)$ onto the IRs $D^{(\Gamma,\alpha)}$ of the point group of the crystal (Sect. 8.4). The symmetry of $\psi_{\mu,\sigma}^{(k)}(x)$ is that of the direct product in (10.3.25). Here we additionally have to know the PW REP $D^{(k)}(PW)$ of $P(d|t)$ defined by $\exp(ik \cdot x)$; i.e., according to (10.3.11, 12),

$$P(d|t)e^{ik\cdot x} = e^{idk\cdot t}e^{idk\cdot x} \ . \qquad (10.3.26)$$

Altogether we have to reduce the inner direct product $D^{(k)}(PW) \otimes D^{(\Gamma,\alpha)}$ with respect to IRs $D^{(k,\beta)}$ of $\mathscr{G}_k$. Using (10.3.26) for $k = \Gamma$ we get the identity REP for $D^{(k)}(PW)$. But this means that at the Γ point we may classify the bands by the IR $D^{(\Gamma,\alpha)}$ of $\mathscr{G}$, i.e. by atomic levels. This holds for symmorphic and nonsymmorphic space groups since for $k = 0(\Gamma)$ the nonprimitive translations drop out.

Figure 10.5 shows a correlation diagram, which illustrates a possible connection between quasi-free and atomic states. For the discussion we have used a fcc lattice (e.g. Al), the outer electronic configuration being $3s^2 3p$. The free-electron energies may be taken from Fig. 10.4 when the decomposition (10.3.14) is taken

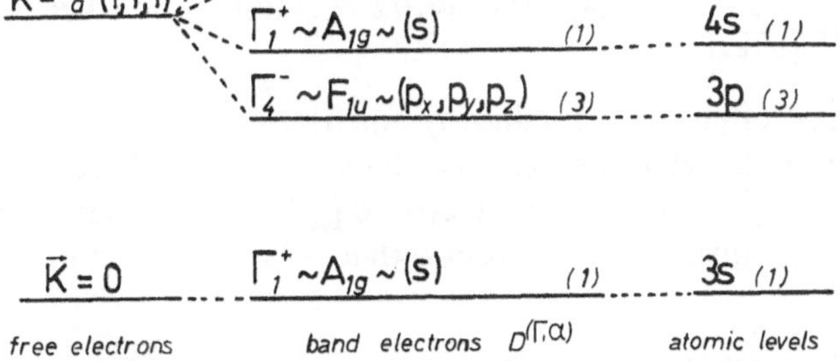

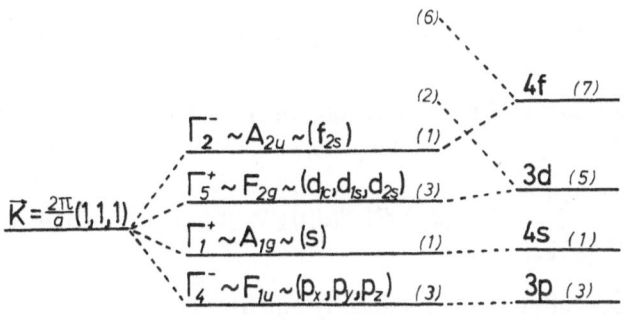

Fig. 10.5. Schematic representation of the correlation between energy bands in the free-electron, tight-binding (Γ point) approximations and atomic levels. Atomic level notation from Table 4.1

into account. To get a quantitative connection we need a free-electron calculation on one hand and a tight-binding calculation on the other. If the electronic spin is considered we only have to realize that the tight-binding functions $U_{\mu,\sigma}^{(k)}$ transform according to a double-valued IR.

Exercise 10.4. Reduce the PWREPs with $K = 0$, $(2\pi/a)(1,1,1)$, $(2\pi/a)(2,0,0)$ for the Δ, Σ, Λ directions in a diamond lattice ($\mathcal{O}_{\mathrm{h}}^{7}$) and a fcc lattice ($\mathcal{O}_{\mathrm{h}}^{5}$) and thus verify Fig. 10.4.

10.4 Selection Rules for Interactions in Crystals

10.4.1 Determination of Reduction Coefficients

In Chap. 8 we discussed the selection rules for tensor operators in general. We will now turn to the application to space groups. The matrix elements to be investigated have the form

$$\langle \boldsymbol{k}', \gamma, l | T_j^{(\boldsymbol{q},\beta)} | \boldsymbol{k}, \alpha, i\rangle \; ; \qquad \boldsymbol{k} \in *\boldsymbol{k} \; , \qquad \boldsymbol{q} \in *\boldsymbol{q} \; , \qquad \boldsymbol{k}' \in *\boldsymbol{k}' \; , \tag{10.4.1}$$

where the wave functions $|\boldsymbol{k}, \alpha, i\rangle, \ldots$ and tensor operators T transform according to the IRs $D^{(*\boldsymbol{k},\alpha)}$, $D^{(*\boldsymbol{k}',\gamma)}$ or $D^{(*\boldsymbol{q},\beta)}$ of the space group. Now, the direct product of two IRs, which, depending on the problem, may be a normal, symmetrized or antisymmetrized product, has first to be reduced. Then we have to investigate whether the third IR is contained in this product. Thus according to (4.4.13a)

$$D^{(*\boldsymbol{q},\beta)} \otimes D^{(*\boldsymbol{k},\alpha)} := D^{(*\boldsymbol{q}\times *\boldsymbol{k};\beta\times\alpha)} = \sum_{*\boldsymbol{k}'\gamma}\!{}^{\oplus} \, (*\boldsymbol{q}, \beta; *\boldsymbol{k}, \alpha | *\boldsymbol{k}', \gamma) D^{(*\boldsymbol{k}',\gamma)} \; . \tag{10.4.2}$$

By taking the trace, we obtain the relation for the characters

$$\chi^{(*\boldsymbol{q},\beta)}\chi^{(*\boldsymbol{k},\alpha)} := \chi^{(*\boldsymbol{q}\times *\boldsymbol{k};\beta\times\alpha)} = \sum_{*\boldsymbol{k}'\gamma}\!{}^{\oplus} \, (*\boldsymbol{q}, \beta; *\boldsymbol{k}, \alpha | *\boldsymbol{k}', \gamma) \chi^{(*\boldsymbol{k}',\gamma)} \; . \tag{10.4.3a}$$

From this equation we can now evaluate the reduction coefficients by means of the orthogonality relation

$$(*\boldsymbol{q}, \beta; *\boldsymbol{k}, \alpha | *\boldsymbol{k}', \gamma) = \frac{1}{n_R} \sum_{\{d|\boldsymbol{t}\} \in \mathcal{R}} \chi^{(*\boldsymbol{q},\beta)}(d|\boldsymbol{t}) \chi^{(*\boldsymbol{k},\alpha)}(d|\boldsymbol{t}) \chi^{(*\boldsymbol{k}',\gamma)*}(d|\boldsymbol{t}) \; . \tag{10.4.3b}$$

As already mentioned, we often have to reduce symmetrized products instead of normal ones. In this case, in the relations (10.4.3) we have to take the characters of the pth symmetrized $[\chi^{(*\boldsymbol{q},\beta)}]^{[p]}$ or pth antisymmetrized $[\chi^{(*\boldsymbol{q},\beta)}]^{[1^p]}$ power. The corresponding expressions are given by (5.5.34b, c) with $a \to \{d|\boldsymbol{t}\}$. According to Exercise 5.20, for $p = 2$

$$[\chi^{(*\boldsymbol{k},\alpha)}(d|\boldsymbol{t})]_{\pm} = [\chi^{(*\boldsymbol{k},\alpha)}(d|\boldsymbol{t})]^{[2],[1^2]} = \tfrac{1}{2}\{[\chi^{(*\boldsymbol{k},\alpha)}(d|\boldsymbol{t})]^2 \pm \chi^{(*\boldsymbol{k},\alpha)}(\{d|\boldsymbol{t}\}^2)\} \; . \tag{10.4.4}$$

In (10.4.2, 3a) the summation has to be carried out over all the stars $\boldsymbol{k}'$, which might be very laborious. However, since always only certain stars occur, the introduction of star reduction coefficients $(*\boldsymbol{q}; *\boldsymbol{k}|*\boldsymbol{k}')$ allows a certain simplification. The direct product space $\mathscr{L}^{(*\boldsymbol{q},\beta)} \otimes \mathscr{L}^{(*\boldsymbol{k},\alpha)}$ is constituted by the Bloch functions

$$\{\phi_j^{(\boldsymbol{q},\beta)}\psi_i^{(\boldsymbol{k},\alpha)}|\boldsymbol{q} \in *\boldsymbol{q}; \boldsymbol{k} \in *\boldsymbol{k}; i = 1\ldots d_{k\alpha}; j = 1\ldots d_{q\beta}\} \ . \tag{10.4.5}$$

According to (5.3.20) the product of the wave functions has the wave vector

$$\boldsymbol{q} + \boldsymbol{k} = \boldsymbol{k}' \in *\boldsymbol{k}' \ , \tag{10.4.6}$$

so that we may symbolically write

$$(*\boldsymbol{q}) \otimes (*\boldsymbol{k}) = \sum_{*\boldsymbol{k}'} (*\boldsymbol{q}; *\boldsymbol{k}|*\boldsymbol{k}')(*\boldsymbol{k}') \ . \tag{10.4.7a}$$

With the number of points s_q of $*\boldsymbol{q}$ then

$$s_q s_k = \sum_{*\boldsymbol{k}'} (*\boldsymbol{q}; *\boldsymbol{k}|*\boldsymbol{k}')s_{k'} \ . \tag{10.4.7b}$$

Equation (10.4.3a) can be looked upon as a linear system of equations for the determination of the reduction coefficients (RC). The number of unknown coefficients (which equals the dimension of the system of equations) is

$$\dim(RC) = \sum_{k'=1}^{A} z_{k'} \ , \tag{10.4.8}$$

where $z_{k'}$ is the number of allowable IRs for $*\boldsymbol{k}'$ and A is the number of non-equivalent stars in (10.4.2, 3). In (10.4.3a) we have to take as many space group elements $\{d|\boldsymbol{t}\}$ (one from each conjugacy class) as there are unknown reduction coefficients. In order to solve the system of equations we can make use of an auxiliary condition, which restricts the manifold of solutions. If $d_{*k',\gamma}$ is the dimension of the allowable IR $D^{(*\boldsymbol{k},\gamma)}$, then according to (10.4.2 and 7)

$$\sum_{\gamma=1}^{z_{k'}} (*\boldsymbol{q}, \beta; *\boldsymbol{k}, \alpha|*\boldsymbol{k}', \gamma)d_{*k',\gamma} = (*\boldsymbol{q}; *\boldsymbol{k}|*\boldsymbol{k}')s_{k'}d_{q\beta}d_{k\alpha} \ . \tag{10.4.9}$$

As an illustration we calculate the reduction coefficients $(*X, \beta; *X, \alpha|*\boldsymbol{k}', \gamma)$ for the space group $\mathcal{O}_h^5$ (fcc; NaCl structure) and choose the point $X_1 = (2\pi/a) \times (1, 0, 0)$ as the representative of the star X. The decomposition into cosets is given in (9.1.12) and is $\mathscr{G}_{X_1}/\mathbb{T} = \mathscr{G}_{0X_1} \cong \mathscr{D}_{4h}$. The representatives a_ν of the cosets used for the evaluation of the characters are c_3 and $c_3^2 = c_3^{-1}$ about the axis $x = y = z$. The elements d, the conjugate elements $c_3 d c_3^{-1}, \ldots$ and the characters are given in Table 10.5. Since $\mathcal{O}_h^5$ is symmorphic, the characters according to (9.1.25) then follow from

Table 10.5. Characters of $X_1 = (2\pi/a)(1,0,0)$ and conjuagate elements for $\mathcal{O}_h^5$ symmetry

$\mathcal{G}_{0X_1}$	e	c_{2x}	c_{2y}, c_{2z}	c_{4x}, c_{4x}^3	$c_{2yz}, c_{2y\bar{z}}$	i	σ_x	σ_y, σ_z	s_{4x}, s_{4x}^3	$\sigma_{yz}, \sigma_{y\bar{z}}$
$c_3 d c_3^2$	e	c_{2y}	c_{2z}, c_{2x}	c_{4y}, c_{4y}^3	$c_{2zx}, c_{2z\bar{x}}$	i	σ_y	σ_z, σ_x	s_{4y}, s_{4y}^3	$\sigma_{zx}, \sigma_{z\bar{x}}$
$c_3^2 d c_3$	e	c_{2z}	c_{2x}, c_{2y}	c_{4z}, c_{4z}^3	$c_{2xy}, c_{2x\bar{y}}$	i	σ_z	σ_x, σ_y	s_{4z}, s_{4z}^3	$\sigma_{xy}, \sigma_{x\bar{y}}$
$\alpha = 1^\pm$	1	1	1	1	1	±1	±1	±1	±1	±1
$2^\pm$	1	1	-1	1	-1	±1	±1	∓1	±1	∓1
$3^\pm$	1	1	1	-1	-1	±1	±1	±1	∓1	∓1
$4^\pm$	1	1	-1	-1	1	±1	±1	∓1	∓1	±1
$5^\pm$	2	-2	0	0	0	±2	∓2	0	0	0

Table 10.6. Characters of $(*X_1\alpha)$ for $\mathcal{O}_h^5$

$\chi^{(*X,\alpha)}$	e	$8c_3$	$3c_2$	$6c_4$	$6c_2'$	i	$8s_6$	$3\sigma_h$	$6s_4$	$6\sigma_d$	$\{c_{2x}\|t_1\}$	$\{c_{2z}\|t_3\}$	$\{c_{4x}\|t_1\}$	$\{c_{2xy}\|t_1\}$	$\{c_{2y}\|t_1\}$
$1^\pm$	3	0	3	1	1	±3	0	±3	±1	±1	-1	-1	-1	-1	-1
$2^\pm$	3	0	-1	1	-1	±3	0	∓1	±1	∓1	-1	3	-1	$+1$	-1
$3^\pm$	3	0	3	-1	-1	±3	0	±3	∓1	∓1	-1	-1	$+1$	$+1$	-1
$4^\pm$	3	0	-1	-1	1	±3	0	∓1	∓1	±1	-1	3	$+1$	-1	-1
$5^\pm$	6	0	-2	0	0	±6	0	∓2	0	0	2	-2	0	0	2
$2^- \otimes 5^-$	18	0	2	0	0	18	0	2	0	0	-2	-6	0	0	-2

$$\chi^{(*X,\alpha)}(d|\boldsymbol{R}^h) = \sum_{\nu=1}^{s} \exp(-\mathrm{i}\boldsymbol{X}_\nu \cdot \boldsymbol{R}^h)\mathring{\chi}^{(X_1,\alpha)}(d_\nu^{-1} d d_\nu|0) \qquad \text{with} \qquad \boldsymbol{X}_\nu = d_\nu \boldsymbol{X}_1 \ . \tag{10.4.10}$$

In Table 10.6 the characters of the point group elements ($\cong \mathcal{O}_h, \boldsymbol{R}^h = 0$) and of some elements with $\boldsymbol{t} = \boldsymbol{R}^h \neq 0$ are given. [$\boldsymbol{t}_1 = (a/2, a/2, 0)$; $\boldsymbol{t}_3 = (0, a/2, a/2)$] Using (10.4.10) and Table 10.5 we have for example, for $\{\sigma_x|0\}$

$$\chi^{(*X,\alpha)}(\sigma_x|0) = \chi^{(X_1,\alpha)}(\sigma_x|0) + \chi^{(X_1,\alpha)}(\sigma_y, 0) + \chi^{(X_1,\alpha)}(\sigma_z, 0)$$

$$= \chi^{(X_1,\alpha)}(\sigma_x|0) + 2\chi^{(X_1,\alpha)}(\sigma_y, 0)$$

and for $\{c_{2x}, \boldsymbol{t}_1\}$

$$\chi^{(*X,\alpha)}(c_{2x}|t_1) = -1 \cdot \chi^{(X_1,\alpha)}(c_{2x}|0) - \underbrace{1 \cdot \chi^{(X_1,\alpha)}(c_{2y}|0) + 1 \cdot \chi^{(X_1,\alpha)}(c_{2z}|0)}_{=\,0\text{, since they belong to the same class of } \mathcal{G}_{0X_1}}$$

$$= -\chi^{(X_1,\alpha)}(c_{2x}|0)$$

The star-reduction coefficients in (10.4.7a) have to be calculated with (10.4.6) and with the reciprocal basis given in Fig. 9.1b. For example,

$$*X \otimes *X = 3*\Gamma \oplus 2*X \ , \qquad \text{thus}$$

$$(*X; *X|*\Gamma) = 3 \ ; \qquad (*X; *X|*X) = 2 \ ; \qquad \text{all others are zero.} \tag{10.4.11}$$

With the help of these relations, $D^{(*X,2^-)} \otimes D^{(*X,5^-)}$ will now be reduced, as an example. According to (10.4.3a), we have to solve the inhomogeneous system of equations

$$\sum_{*k'\gamma}\!{}_{\oplus}\, (*X,2^-;\, *X,5^-\,|\,*k',\gamma)\chi^{(*k',\gamma)}(d\,|\,t) = \chi^{(*X,2^-)}(d\,|\,t)\chi^{(*X,5^-)}(d\,|\,t) \tag{10.4.12}$$

with $*k' = \{*\Gamma, *X\}$. In the case of $*\Gamma$, γ denotes the ten IRs of $\mathcal{O}_h$, while in the case of $*X$, γ denotes the ten IRs of Table 10.6. Altogether there are 20 coefficients which have to be determined from (10.4.12). Relation (10.4.9) now reads

$$\begin{aligned}
&\sum_{\gamma} (*X,2^-;\, *X,5^-\,|\,*\Gamma,\gamma) d_{*\Gamma,\gamma} = 3\cdot 1\cdot 1\cdot 2 = 6 \ ,\\
&\sum_{\gamma} (*X,2^-;\, *X,5^-\,|\,*X,\gamma) d_{*X,\gamma} = 2\cdot 3\cdot 1\cdot 2 = 12 \ .
\end{aligned} \tag{10.4.13}$$

The IRs of $*\Gamma$ and $*X$ decompose into even $(+)$ and odd $(-)$ ones. Consequently in (10.4.12) only the even REPs yield coefficients different from zero, since the right hand side is a product of two odd REPs, thus being even itself. We furthermore learn from (10.4.12) that those five classes which contain improper rotations yield no independent conditions. Thus, in order to determine the remaining ten coefficients, we have to take the five elements $\{d\,|\,0\}$ and five further ones $\{d\,|\,t\}$ to close the system of equations. Using Table 10.6, the character table of $\mathcal{O}_h$ and (10.4.13) we are then able to determine the coefficients. The decomposition finally reads [10.2]

$$\begin{aligned}
&D^{(*X,2^-)} \otimes D^{(*X,5^-)} = D^{(*\Gamma,4^+)} \oplus D^{(*\Gamma,5^+)} \oplus D^{(*X,1^+)} \oplus D^{(*X,3^+)} \oplus D^{(*X,5^+)} \ ,\\
&\text{Dim: } 3 \ \cdot \ 6 \ = \ 3 \ + \ 3 \ + \ 3 \ + \ 3 \ + \ 6 \ ,
\end{aligned} \tag{10.4.14}$$

from which the reduction coefficients can be seen. The coefficients can also be determined directly from (10.4.3b), which is not easier. A further way of obtaining selection rules starts directly with (10.4.1). Here the multiplicity with which the unit REP occurs in the threefold product REP defined by (10.4.1) is evaluated (see also Sect. 8.1) The multiplicity gives the number of linearly independent matrix elements.

In analogy to (4.4.18a) the CGCs of the space groups are defined by

$$\psi_l^{(k',\gamma,s)} = \sum_{\substack{ij\\kq}} \left(\begin{smallmatrix}k,\alpha\\ i\end{smallmatrix};\begin{smallmatrix}q,\beta\\ j\end{smallmatrix}\middle|\begin{smallmatrix}k',\gamma,s\\ l\end{smallmatrix}\right)\psi_i^{(k,\alpha)}\psi_j^{(q,\beta)} \ ; \quad k\in *k,\ q\in *q,\ k'\in *k' \ . \tag{10.4.15}$$

They have to be determined according to (4.4.20). The summation in (4.4.20), however, extends over the whole space group $\mathcal{R}$! The allowable IRs of the elements of the translation group have a simple diagonal form. In the sum (4.4.20) the translation group can thus be separated, and it yields the *conservation of quasi-momentum* (*pseudomomentum*) as a selection rule for the wave numbers:

The CGCs vanish if

$$\boldsymbol{k} + \boldsymbol{q} - \boldsymbol{k}' = 0 \bmod \boldsymbol{K} \tag{10.4.16}$$

is *not* satisfied, see (10.4.17).

10.4.2 General Selection Rules

The consequences of the translation symmetry of the crystal are valid for all interactions. The functions in (10.4.1) are always Bloch functions and $T^{(\boldsymbol{q})}$ is an ireducible tensor operator with respect to the translation group. Thus

$$\begin{aligned}\langle \phi^{(\boldsymbol{k}')} | T^{(\boldsymbol{q})} | \psi^{(\boldsymbol{k})} \rangle &= \langle P(e|\boldsymbol{R}^h)\phi^{(\boldsymbol{k}')} | P(e|\boldsymbol{R}^h) T^{(\boldsymbol{q})} P^{+}(e|\boldsymbol{R}^h) | P(e|\boldsymbol{R}^h)\psi^{(\boldsymbol{k})} \rangle \\ &= \exp[-\mathrm{i}(\boldsymbol{k} + \boldsymbol{q} - \boldsymbol{k}')\cdot \boldsymbol{R}^h] \langle \phi^{(\boldsymbol{k}')} | T^{(\boldsymbol{q})} | \psi^{(\boldsymbol{k})} \rangle \ ,\end{aligned}$$

or

$$\langle \phi^{(\boldsymbol{k}')} | T^{(\boldsymbol{q})} | \psi^{(\boldsymbol{k})} \rangle = \frac{1}{n_T} \sum_{h} \exp[-\mathrm{i}(\boldsymbol{k} + \boldsymbol{q} - \boldsymbol{k}')\cdot \boldsymbol{R}^h] \langle \phi^{(\boldsymbol{k}')} | T^{(\boldsymbol{q})} | \psi^{(\boldsymbol{k})} \rangle$$

and according to Exercise 5.11b

$$\langle \phi^{(\boldsymbol{k}')} | T^{(\boldsymbol{q})} | \psi^{(\boldsymbol{k})} \rangle = \sum_{\boldsymbol{K}} \delta(\boldsymbol{k} + \boldsymbol{q} - \boldsymbol{k}' - \boldsymbol{K}) \langle \phi^{(\boldsymbol{k}')} | T^{(\boldsymbol{q})} | \psi^{(\boldsymbol{k})} \rangle \ . \tag{10.4.17}$$

This means, the matrix elements are different from zero if and only if the conservation of quasi-momentum (10.4.16) is satisfied.

A further general statement is related to time reversal symmetry. According to Sects. 5.2 or 9.4.2, in case (1a), see (9.4.3), the functions ϕ and $\vartheta\phi$ are linearly dependent; ϑ is the time reversal operator (8.2.10). Instead of the matrix elements $\langle \phi_i^{(\boldsymbol{k}',\alpha)} | T_j^{(\boldsymbol{q},\beta)} | \psi_l^{(\boldsymbol{k},\alpha)} \rangle$, the elements $\langle \vartheta\phi_i^{(\boldsymbol{k}',\alpha)} | T_j^{(\boldsymbol{q},\beta)} | \psi_l^{(\boldsymbol{k},\alpha)} \rangle$ can be investigated. This leads to an equation which corresponds to (8.2.13b), i.e. the matrix elements (10.4.1) are either symmetric or antisymmetric under an interchange of the row indices $(\boldsymbol{k}', i) \leftrightarrow (\boldsymbol{k}, l)$, depending on whether $\pm\vartheta^2 = +1$ or -1 [see (8.2.10, 13b)]. In investigating which matrix elements are zero we thus have to reduce $D^{(*\boldsymbol{k},\alpha)[2]}$ or $D^{(*\boldsymbol{k},\alpha)[1^2]}$ (symmetrized and antisymmetrized, respectively) and to check whether $D^{(*\boldsymbol{q},\beta)}$ is contained or not.

10.4.3 Electron-Phonon Interaction

In an elementary interaction of electrons and phonons in the matrix elements (10.4.1) we have electronic wave functions in addition to those of phonons. The Hamiltonian of the (free) phonons is

$$H_p = \sum_{\boldsymbol{q}\sigma} \hbar\omega(\boldsymbol{q},\sigma)\{b_{\boldsymbol{q}\sigma}^{+} b_{\boldsymbol{q}\sigma} + 1/2\} \ , \tag{10.4.18}$$

where we have used the particle number REP[6]; the phonon wave functions are

$$|\ldots n(\boldsymbol{q},\sigma)\ldots, n(\boldsymbol{q}',\sigma')\ldots\rangle = \prod_{q\sigma} \frac{1}{\sqrt{n_{q\sigma}!}}(b^+_{q\sigma})^{n_{q\sigma}}|0\rangle \; ; \tag{10.4.19}$$

$b^+_{\boldsymbol{q},\sigma}$, $b_{\boldsymbol{q},\sigma}$ are the creation and annihilation (ladder) operators for phonons with eigenfrequencies $\omega(\boldsymbol{q},\sigma)$ from Sect. 10.2, $|0\rangle$ is the phonon vacuum state and $n(\boldsymbol{q},\sigma)$ the occupation number of the mode $\boldsymbol{e}(\boldsymbol{q},\sigma)$. According to Sect. 10.2, the modes may be classified by the IRs of the little group $\mathscr{G}_{\boldsymbol{q}}$ or of the space group $\mathscr{R}$, $(\boldsymbol{q},\sigma \rightarrow \boldsymbol{q},\alpha,s_\alpha,i)$.

A *one-phonon state* $|n(\boldsymbol{q},\sigma)=1\rangle$ transforms according to an IR $D^{(*\boldsymbol{q},\sigma)}$ of the space group.

For *two-phonon states* we have to distinguish between two cases (see also Sect. 7.1.3):

(i) $\omega(\boldsymbol{q},\sigma) \neq \omega(\boldsymbol{q}',\sigma')$. The two-phonon state $|n(\boldsymbol{q},\sigma)=1;\; n(\boldsymbol{q}',\sigma')=1\rangle = |n(\boldsymbol{q},\sigma)\rangle|n(\boldsymbol{q}',\sigma')\rangle$ then transforms according to the REP of the inner direct product

$$D^{(*\boldsymbol{q},\sigma)} \otimes D^{(*\boldsymbol{q}',\sigma')} \text{ with energy } E = \hbar\omega(\boldsymbol{q},\sigma) + \hbar\omega(\boldsymbol{q}',\sigma') \; . \tag{10.4.20}$$

The states are said to be *combination modes.*

(ii) $\omega(\boldsymbol{q},\sigma) = \omega(\boldsymbol{q}',\sigma')$. Then $\boldsymbol{q}$ and $\boldsymbol{q}'$ necessarily belong to the same star. The two-phonon state transforms according to the symmetrized product

$$[D^{(*\boldsymbol{q},\sigma)}]^{[2]} = [D^{(*\boldsymbol{q},\sigma)}]_+ \qquad \text{with} \qquad E = 2\hbar\omega(\boldsymbol{q},\sigma) \tag{10.4.21}$$

and is said to be an *overtone*, see (10.4.4). In general a phonon state (10.4.19) transforms according to

$$\ldots \otimes [D^{(*\boldsymbol{q},\sigma)}]^{[n]} \ldots \otimes [D^{(*\boldsymbol{q}',\sigma')}]^{[n']} \otimes \ldots , \tag{10.4.22}$$

where $[D^{(*\boldsymbol{q},\sigma)}]^{[n]}$ is the $n(\boldsymbol{q},\sigma)$th symmetrized power of $D^{(*\boldsymbol{q},\sigma)}$.

In the following we denote the electronic state by $|\boldsymbol{k},\nu\rangle$ (see footnote 2) and the phonon state by $|n(\boldsymbol{q},\sigma)\rangle$. The elementary processes of the electron-phonon interaction are illustrated in Fig. 10.6. When the electronic state changes, a

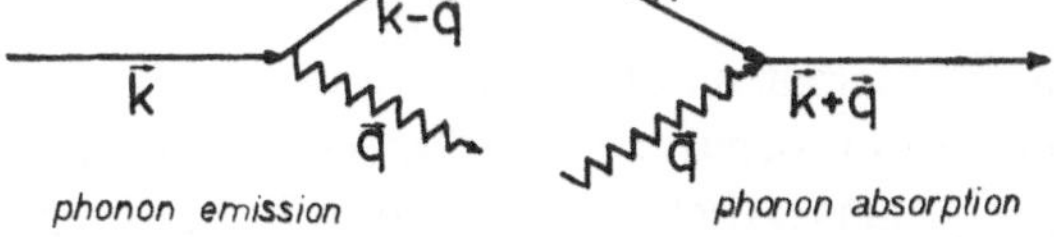

Fig. 10.6. Interaction of electrons $\boldsymbol{k}$ with phonons $\boldsymbol{q}$; scattering with emission or absorption of a phonon

[6] See textbooks on solid-state physics. In the following we denote the phonon wave numbers by $\boldsymbol{q}$. The occupation numbers would be better written as $n(*\boldsymbol{q},\sigma)$, where $n(*\boldsymbol{q},\sigma)$ is the number of phonons that are distributed over all the partner functions of the IR $D^{(*\boldsymbol{q},\sigma)}$.

phonon is emitted or absorbed:

$$|\boldsymbol{k},\nu\rangle|0\rangle \to |\boldsymbol{k}-\boldsymbol{q}+\boldsymbol{K},\nu'\rangle|n(\boldsymbol{q},\sigma)\rangle;\ |\boldsymbol{k},\nu\rangle|n(\boldsymbol{q},\sigma)\rangle \to |\boldsymbol{k}+\boldsymbol{q}+\boldsymbol{K},\nu'\rangle|0\rangle\ . \tag{10.4.23}$$

The electron-phonon interaction H_{EP} is now of such a form (the explicit form is not needed in the following) that in the first-order perturbation approximation only one-phonon processes are possible. The transition matrices $\langle \text{final}\,|H_{\mathrm{EP}}|\,\text{init.}\rangle$ contain the states with $n_{\boldsymbol{q}\sigma} = 1$ defined in (10.4.23). Considering absorption, the initial state transforms according to $D^{(*\boldsymbol{q},\sigma)} \otimes D^{(*\boldsymbol{k},\nu)}$, and the final state according to $D^{(*\boldsymbol{k}',\nu')}$ with $\boldsymbol{k}' = \boldsymbol{k} + \boldsymbol{q} \bmod \boldsymbol{K}$. The scalar interaction H_{EP} transforms according to the identity REP so that the following selection rules and conservation laws apply:

(i) $\langle \text{final}\,|H_{\mathrm{EP}}|\,\text{init.}\rangle \neq 0$ if $D^{(*\boldsymbol{k}',\nu')}$ in $D^{(*\boldsymbol{q},\sigma)} \otimes D^{(*\boldsymbol{k},\nu)}$,
(ii) $\boldsymbol{k}' = \boldsymbol{k} + \boldsymbol{q} \bmod \boldsymbol{K}$, conservation of quasi-momentum; $\boldsymbol{K} = 0$: normal processes; $\boldsymbol{K} \neq 0$: umklapp processes
(iii) $E_\nu(\boldsymbol{k}) + \hbar\omega(\boldsymbol{q},\sigma) = E_{\nu'}(\boldsymbol{k}')$, conservation of energy. (10.4.24)

They have to be satisfied simultaneously.

As an example we consider inter-valley scattering in germanium. By this we understand the transition of an electron between different equivalent (relative) minima of the conduction band induced by H_{EP}. Such minima occur in Ge at the L point (four-fold star generate). They belong to the IR L_1^+ (without spin, see Fig. 9.4d). The electrons may be scattered from one minimum to another. Since all the possibilities are completely equal, we consider the scattering from $\boldsymbol{k} = (1,1,1)(\pi/a)$ into $\boldsymbol{k}' = c_{2x}\cdot\boldsymbol{k} = (1,\bar{1},\bar{1})(\pi/a)$. Conservation of quasi-momentum requires $\boldsymbol{k}' - \boldsymbol{k} = (0,\bar{1},\bar{1})(2\pi/a) = \boldsymbol{q} + \boldsymbol{K}$, thus $\boldsymbol{q} = (1,0,0)(2\pi/a)$; $\boldsymbol{K} = (\bar{1},\bar{1},\bar{1}) \times (2\pi/a)$. Therefore this is an umklapp-type process in which an X-phonon participates (Fig. 9.1b). The interesting REPs are $D^{(*\boldsymbol{k}',\nu')} = D^{(*L,1^+)}$, $D^{(*\boldsymbol{k},\nu)} = D^{(*L,1^+)}$ and $D^{(*\boldsymbol{q},\sigma)} = D^{(*X,\sigma)}$ with $\sigma = 1,3,4$ according to Fig. 10.3. According to (10.4.24) (i) we have to investigate whether or not $D^{(*L,1^+)}$ occurs in $D^{(*L,1^+)} \otimes D^{(*X,\sigma)}$. The reduction[7] yields, according to Sect. 10.4.1 [we set $D^{(*X,\sigma)} := (*X,\sigma)$, etc.],

$$\begin{aligned}
(*X,1)\otimes(*L,1^+) &= \underline{(*L,1^+)}\oplus(*L,1^-)\oplus(*L,3^+)\oplus(*L,3^-)\ ,\\
(*X,3)\otimes(*L,1^+) &= \underline{(*L,1^+)}\oplus(*L,2^-)\oplus(*L,3^+)\oplus(*L,3^-)\ ,\\
(*X,4)\otimes(*L,1^+) &= (*L,2^+)\oplus(*L,1^-)\oplus(*L,3^+)\oplus(*L,3^-)\ .
\end{aligned} \tag{10.4.25}$$

Thus scattering is possible with the participation of X_1-(LA, LO) and X_3-(TA) phonons whereas X_4-(TO) phonons do not occur. In addition, conservation of energy has to be satisfied, in this special case $\Delta E_\nu(L) = \hbar\omega(X,\sigma) \approx 29.3$ and 10.7 meV, respectively. The selection rules discussed here, for absorption are also valid for emission.

---

[7] Anyone not interested in carrying out these reductions may consult tables, e.g. [10.2].

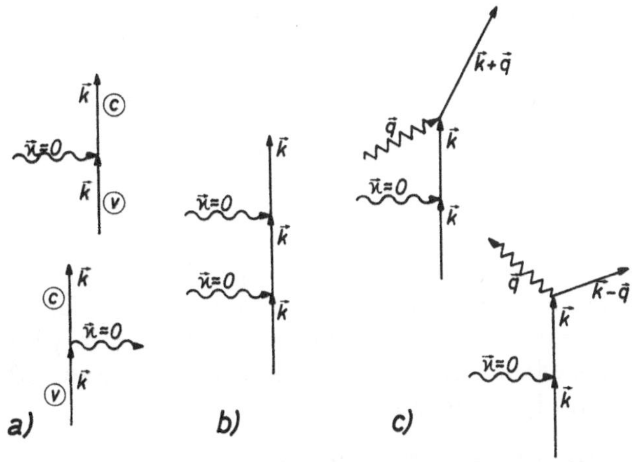

Fig. 10.7a–c. Interactions of electrons $\boldsymbol{k}$ with photons κ. (a) Direct transition between valence (v) and conduction (c) bands with one-photon absorption or emission. (b) Direct two-photon absorption. (c) Indirect transitions. Photon absorption with simultaneous phonon absorption or emission

10.4.4 Electron-Photon Interaction: Optical Transitions

Due to the electron-photon interaction, electrons are scattered into different states with mutual absorption or emission of photons. There are *direct* and *indirect* transitions. Direct means that only electrons and photons participate in the interaction ($\boldsymbol{k} \approx \boldsymbol{k}'$), whereas in indirect transitions other elementary excitations, e.g. phonons (Fig. 10.7), participate.

Direct Transitions. The operator inducing the transition is the interaction operator H_1 discussed in Sect. 8.3, so that the relevant matrix elements have the form

$$M_{\nu\nu'}(\boldsymbol{k}) \cong \boldsymbol{e}_\sigma \cdot \langle \boldsymbol{k}', \nu' | \mathrm{e}^{\mathrm{i}\boldsymbol{\kappa}\cdot\boldsymbol{x}} \boldsymbol{p}_\mathrm{E} | \boldsymbol{k}, \nu \rangle \ , \tag{10.4.26}$$

where in the (electric) dipole approximation the momentum of the photon $\boldsymbol{\kappa}$ can be set equal to zero ($\boldsymbol{k}' \approx \boldsymbol{k}$). Thus we are left with the discussion of the matrix element with the electron momentum $\boldsymbol{p}_\mathrm{E}$. Criteria (10.4.24) have to be checked:

(i) $\langle \boldsymbol{k}', \nu' | \boldsymbol{p}_\mathrm{E} | \boldsymbol{k}, \nu \rangle \neq 0$ if $D^{(*\boldsymbol{k}', \nu')}$ is contained in $D^{(v)} \otimes D^{(*\boldsymbol{k}, \nu)}$ since $\boldsymbol{p}_\mathrm{E}$ is a polar vector.

(ii) Conservation of quasi-momentum is satisfied with $\boldsymbol{k}' \approx \boldsymbol{k}$, $\boldsymbol{K} = 0$, $\boldsymbol{k} \approx 0$.

(iii) $E_{\nu'}(\boldsymbol{k}) = E_\nu(\boldsymbol{k}) \pm \hbar\omega$ for emission/absorption. (10.4.27)

The REP $D^{(v)}$ is the vector REP according to which $\boldsymbol{p}_\mathrm{E}$ transforms; in the cubic group $\mathcal{O}_\mathrm{h}$, $D^{(v)} = D^{(*\Gamma, 4^-)} = \Gamma_4^-$. In Ge the valence band edge according to Figs. 9.4b, d or 10.8 has the symmetry Γ_5^+ (excluding spin), thus

$$(*\Gamma, 4^-) \otimes (*\Gamma, 5^+) = \Gamma_4^- \otimes \Gamma_5^+ = \underline{\Gamma_2^-} \oplus \Gamma_3^- \oplus \Gamma_4^- \oplus \Gamma_5^- \ , \tag{10.4.28a}$$

which means that the direct transition to the point Γ_2^- is possible. The conduction band edge in Ge has $(*L, 1^+)$ symmetry, the corresponding valence band state has $(*L, 3^-)$ symmetry. Because

$$(*\Gamma, 4^-) \otimes (*L, 3^-) = \underline{(*L, 1^+)} \oplus (*L, 2^+) \oplus 2 \cdot (*L, 3^+) \ , \qquad (10.4.28b)$$

this direct transition is possible, as well.

Indirect Transitions. From the above discussion it follows that the direct transition $\Gamma_5^+ \to L_1^+$ cannot take place. If such transitions are to be "forced", further elementary excitations are necessary, e.g. *phonons*. In a perturbation theory this implies the occurrence of intermediate states, which means that at least a second-order approximation with transition probability (8.3.7)

$$w_{kv \to k'v'} = \frac{2\pi}{\hbar} \left| \sum_{\substack{z \\ k''v''}} \frac{\langle k'v', e|H_1|k''v'', z\rangle \langle k''v'', z|H_1|kv, a\rangle}{E_{v,a}(k) - E_{v'',z}(k'')} \right|^2 \times \delta(E_{v,a}(k) - E_{v',e}(k')) \qquad (10.4.29)$$

is necessary. The unperturbed Hamiltonian contains all the occurring elementary excitations, while the perturbation includes all the interactions, i.e.

$$H_0 = H_E + H_\gamma + H_P \ ; \qquad H_1 = H_{E_\gamma} + H_{EP} + \dots \ . \qquad (10.4.30)$$

Thus four different processes contribute in (10.4.29): a two-photon process, two photon-phonon processes and a two-phonon process, of which we shall discuss only the photon-phonon processes I and II in Fig. 10.8. They are the most

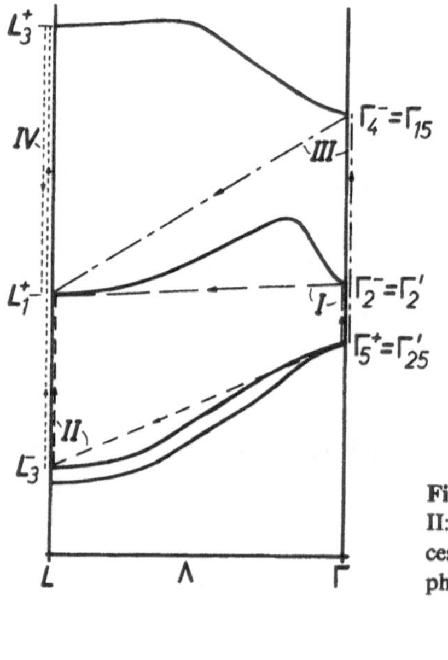

Fig. 10.8. Phonon-induced electronic transitions in Ge. I, II: phonon-photon processes; III: phonon-photon processes involving a "higher" conduction band; IV: two-photon processes involving a "higher" conduction band

probable ones since in these processes the energy difference in the denominator of (10.4.29) is the smallest. Two-photon processes (IV) and other processes (III) entering via "higher" conduction bands are less probable. In the processes I and II, because $\boldsymbol{\kappa} \approx 0$, conservation of quasi-momentum requires $\boldsymbol{k}' = \boldsymbol{k}$ in "vertical" processes and $\boldsymbol{k}' = \boldsymbol{k} + \boldsymbol{q} + \boldsymbol{K}$ in phonon-induced "horizontal" processes (the phonon energy is always "small"). Furthermore the possible processes are restricted by the following reductions:

I. $D^{(*\boldsymbol{k}'',\nu'')}$ in $D^{(v)} \otimes D^{(*\boldsymbol{k},\nu)}$, $\quad \boldsymbol{k}'' = \boldsymbol{k}$, $\quad$ photon absorption;

and

$D^{(*\boldsymbol{k}',\nu')}$ in $D^{(*\boldsymbol{q},\sigma)} \otimes D^{(*\boldsymbol{k}'',\nu'')}$, $\quad \boldsymbol{k}' = \boldsymbol{k}'' + \boldsymbol{q} + \boldsymbol{K}$, $\quad$ phonon process;

II. $D^{(*\boldsymbol{k}'',\nu'')}$ in $D^{(*\boldsymbol{q},\sigma)} \otimes D^{(*\boldsymbol{k},\nu)}$, $\quad \boldsymbol{k}'' = \boldsymbol{k} + \boldsymbol{q} + \boldsymbol{K}$, $\quad$ phonon process;

and

$D^{(*\boldsymbol{k}',\nu')}$ in $D^{(v)} \otimes D^{(*\boldsymbol{k}'',\nu'')}$, $\quad \boldsymbol{k}' = \boldsymbol{k}''$, $\quad$ photon absorption.

For the transition $\Gamma_5^+ \to L_1^+$ in Ge shown in Fig. 10.8 we then have in detail:

I. $(*\Gamma, 2^-)$ in $(*\Gamma, 4^-) \otimes (*\Gamma, 5^+)$ is satisfied according to (10.4.28a), but $(*L, 1^+)$ in $(*L, \sigma) \otimes (*\Gamma, 2^-)$ can only be satisfied for $\sigma = 1^-$, i.e. the transition is only possible with L_1^--(LA) phonons (Fig. 10.3). Of course, the reduction has to be performed in detail. We have used the notation of Appendix A; sometimes other notations are used.

II. $(*L, 3^-)$ in $(*L, \sigma) \otimes (*\Gamma, 5^+)$ yields, since σ^+-phonons are not possible:

$$
\begin{aligned}
&(*L, 1^-) \otimes (*\Gamma, 5^+) = (*L, 1^-) \oplus (*L, 3^-) \\
&(*L, 2^-) \otimes (*\Gamma, 5^+) = (*L, 2^-) \oplus (*L, 3^-) \\
&(*L, 3^-) \otimes (*\Gamma, 5^+) = (*L, 1^-) \oplus (*L, 2^-) \oplus 2 \cdot (*L, 3^-) \ ,
\end{aligned}
\tag{10.4.31}
$$

that means, this process is possible for all the $(*L, \sigma^-)$ phonons. Further we have to check whether $(*L, 1^+)$ is contained in $(*\Gamma, 4^-) \otimes (*L, 3^-)$. This holds because of (10.4.28b). Consequently, in Ge the indirect transition II becomes possible, since L_1^--(LA) and L_3^--(TO) phonons exist.

10.4.5 Phonon-Photon Interaction

We have already discussed this interaction in Sect. 7.1.3 for molecular vibrations. An extension to lattice vibrations within this framework can be achieved easily.

1) Infrared Absorption

Infrared absorption is governed by the matrix elements of the lattice dipole moment, which transforms according to the vector REP $D^{(v)}$ (e.g. Γ_4^- in cubic crystals). The dipole moment is the sum over the moments of the individual particles and can be expanded with respect to the displacements $\overset{n}{u}{}_j^{\nu}$ of the ions (Sect. 10.2.1) or with respect to the corresponding normal coordinates in (7.1.12)

$$M(\kappa) = M(\kappa|0) + \sum_{q,\alpha\ldots} M(\kappa|q, {}^{\alpha s}_{i})Q_i^{(\alpha,s)}(q) + \tfrac{1}{2}\sum_{q,q'\ldots} M(\kappa|q, {}^{\alpha s}_{i}; q', {}^{\alpha' s'}_{j})Q_i^{(\alpha,s)}(q)Q_j^{(\alpha',s')}(q') + \ldots \; . \tag{10.4.32}$$

The normal coordinates $Q_i^{(\alpha,s)}(q)$ transform as the symmetry-adapted eigenvectors $e_{ij}^{(\alpha,s)}(q)$ in (10.2.16) according to the allowable IRs $D^{(q,\alpha)}$ of $\mathscr{G}_q$, thus

$$P(d|t)Q_i^{(\alpha,s)}(q) = \sum_{j=1}^{d_\alpha} D_{ji}^{(q,\alpha)}(d|t)Q_j^{(\alpha,s)}(q) \; . \tag{10.4.33}$$

Expansion (10.4.32) corresponds to a separation into zero-, one-, two-, ... phonon processes. The expansion coefficients in general depend on the electronic state of the system which, however, is not changed in infrared absorption. The relevant matrix elements for absorption are

$$\langle n'|M(\kappa)|n\rangle \qquad \text{with} \qquad \kappa \to 0 \; . \tag{10.4.34}$$

In this process the phonon system changes from $|n(q,\alpha)\rangle = |n\rangle$ to $|n'(q',\sigma')\rangle = |n'\rangle$. Because $\kappa \to 0$, it follows that the total moment in (10.4.32) transforms according to Γ-REPs and furthermore that in the expansion coefficients of first order, $q \to 0$ mod K, in those of second order $q + q' \to 0$ mod K, etc. These statements are a consequence of the conservation of energy or quasi-momentum (10.4.16). The matrix elements (10.4.34) are different from zero if $D^{(n')}$ is contained in $D^{(v)} \otimes D^{(n)}$, where $D^{(n)}$ denotes the REP according to which the phonon wave function transforms (10.4.22).

The matrix element relevant for *one-phonon processes*, is

$$\langle n'|M(\kappa|q, {}^{\alpha s}_{i})Q_i^{(\alpha,s)}(q)|n\rangle \; , \qquad \kappa \approx 0 \; ; \qquad q = 0 \; . \tag{10.4.35}$$

The only coefficients $M(\kappa|q, {}^{\alpha s}_{i})$ that can be different from zero are those for which $Q_i^{(\alpha,s)}(q)$ transforms according to $D^{(v)}$, since the total dipole moment transforms according to $D^{(v)}$. The initial state in the infrared absorption in general is the phonon vacuum $|0\rangle$, which transforms according to the identity REP Γ_1 or Γ_1^+. The selection rule is thus

$$\langle n'|Q_i^{(\alpha,s)}(q)|0\rangle \neq 0 \; , \tag{10.4.36}$$

if $D^{(n')}$ is contained in $D^{(v)} \otimes \Gamma_1^+ = D^{(v)}$, which means that the final state itself has to be correlated to the REP $D^{(v)}$. In cubic crystals $D^{(v)} = \Gamma_4^-$, i.e. only Γ_4^- states are infrared active (in one-phonon processes). In Ge there are only Γ_5^--phonons at the Γ point with frequencies different from zero. Consequently there is no one-phonon infrared absorption. The relevant matrix element in *two-phonon processes* is

$$\langle n'|M(\boldsymbol{\kappa}|\boldsymbol{q},{}^{\alpha s}_{i};\boldsymbol{q}',{}^{\alpha's'}_{j})Q_i^{(\alpha,s)}(\boldsymbol{q})Q_j^{(\alpha',s')}(\boldsymbol{q}')|0\rangle \ , \tag{10.4.37}$$

$$\boldsymbol{\kappa} \approx 0 \ , \qquad \boldsymbol{q}+\boldsymbol{q}' = 0 \ ,$$

as the initial state is again the vacuum. Expression (10.4.37) is different from zero only if

$$\begin{aligned} &D^{(n')} \text{ is contained in } D^{(*\boldsymbol{q},\alpha)} \otimes D^{(*\boldsymbol{q}',\alpha')} \otimes \Gamma_1^+ = D^{(*\boldsymbol{q},\alpha)} \otimes D^{(*\boldsymbol{q}',\alpha')} \\ &\textit{and } D^{(v)} \text{ is contained in } D^{(*\boldsymbol{q},\alpha)} \otimes D^{(*\boldsymbol{q}',\alpha')} \ , \end{aligned} \tag{10.4.38a}$$

i.e. the reduction coefficients have to satisfy

$$(*\boldsymbol{q},\alpha;*\boldsymbol{q}',\alpha'|v) \neq 0 \qquad \textit{and} \qquad (*\boldsymbol{q},\alpha;*\boldsymbol{q}',\alpha'|n') \neq 0 \tag{10.4.38b}$$

Here again we have to distinguish between combination processes transforming according to $D^{(*\boldsymbol{q},\alpha)} \otimes D^{(*\boldsymbol{q}',\alpha')}$ and overtones transforming according to $[D^{(*\boldsymbol{q},\alpha)}]_+$ [see (10.4.21) and Sect. 7.1.3).

a) Ge, Γ point. Figure 10.3 shows that there are only Γ_5^+-phonons with $\omega \neq 0$, that means, $|n'\rangle$ can only have overtones (and no combination processes since only *one* REP occurs). It is

$$[*\Gamma_5^+]_+ = (*\Gamma,1^+) \oplus (*\Gamma,3^+) \oplus (*\Gamma,5^+) \ . \tag{10.4.39}$$

This does not contain $D^{(v)} = \Gamma_4^-$, thus there are no two-phonon processes of this type in Ge.

b) Ge, X point. Figure 10.3 shows X_1-, X_3-, X_4-phonons. The following reductions hold for the combination processes [10.2]:

$$\begin{aligned} (*X,1) \otimes (*X,3) &= (*\Gamma,4^+) \oplus \underline{(*\Gamma,4^-)} \oplus (*\Gamma,5^+) \oplus (*\Gamma,5^-) \\ &\quad \oplus (*X,1) \oplus (*X,2) \oplus (*X,3) \oplus (*X,4) \ , \\ (*X,1) \otimes (*X,4) &= (*X,1) \otimes (*X,3) \ , \\ (*X,3) \otimes (*X,4) &= (*\Gamma,2^+) \oplus (*\Gamma,2^-) \oplus (*\Gamma,3^+) \oplus (*\Gamma,3^-) \\ &\quad \oplus (*\Gamma,4^+) \oplus \underline{(*\Gamma,4^-)} \oplus (*X,1) \oplus (*X,2) \\ &\quad \oplus (*X,3) \oplus (*X,4) \ , \end{aligned} \tag{10.4.40}$$

i.e. all combination processes with $*X$ in Ge are infrared active. The reductions for the overtones are

$$
\begin{aligned}
[(*X,1)]_+ &= (*\Gamma,1^+)\oplus(*\Gamma,2^-)\oplus(*\Gamma,3^+)\oplus(*\Gamma,3^-)\oplus(*\Gamma,5^+)\\
&\quad\oplus(*X,1)\oplus(*X,4)\\
[(*X,3)]_+ &= (*\Gamma,1^+)\oplus(*\Gamma,1^-)\oplus(*\Gamma,3^+)\oplus(*\Gamma,3^-)\oplus(*\Gamma,5^+)\\
&\quad\oplus(*X,1)\oplus(*X,4)\\
[(*X,4)]_+ &= [(*X,3)]_+ \ ,
\end{aligned}
\tag{10.4.41}
$$

which means that $(*\Gamma,4^-)=\Gamma_4^-$ does not occur; the overtones with $*X$ are not infrared active in Ge.

2) Raman Effect

In Sect. 7.1.3 it was explained that in the case of the Raman effect the dipole moment has to be replaced by the polarizability tensor (symmetric second-rank tensor), transforming according to $[D^{(v)}]_+ \to T_+^{vv}$ for $\boldsymbol{\kappa}=0$. The relevant matrix element is thus

$$\langle n'|T_+^{vv}|n\rangle \ , \qquad |n\rangle\to|0\rangle \ . \tag{10.4.42}$$

We now have to investigate the same questions as in the case of infrared absorption. In the same way, T_+^{vv} can be expanded with respect to normal coordinates analogously to (10.4.32). For cubic crystals we have

$$D^{(v)}=\Gamma_4^- \ , \qquad T_+^{vv}\sim[D^{(v)}]_+=[\Gamma_4^-]_+=\Gamma_1^+ + \Gamma_3^+ + \Gamma_5^+ \ . \tag{10.4.43}$$

The question for *one-phonon processes* is thus whether $D^{(n')}$ occurs in $[D^{(v)})]_+\otimes\Gamma_1^+$. In a diamond lattice (Ge) the Γ_5^+-modes are the only ones with $\omega\neq0$ (for $\boldsymbol{q}=0$); these occur in (10.4.43) and are thus Raman active.

Two-phonon processes have to be treated analogously to (10.4.38) where $D^{(v)}$ has to be replaced by $[D^{(v)}]_+$.

a) Ge, Γ point. The decomposition (10.4.43) agrees with (10.4.39), i.e. the Γ_5^+-overtones are Raman active: $[D^{(v)}]_+$ is contained in $[\Gamma_5^+]_+$.

b) Ge, X point. Each of the decompositions in (10.4.40) contains just *one* term of (10.4.43). Thus the combination modes are Raman active. Corresponding results follow for the overtones, since each of the decompositions in (10.4.41) contains all terms of (10.4.43). This also means that the overtones are more intense than the combination modes.

Exercise 10.5. Investigate the two-phonon infrared and Raman processes for Ge at points other than the Γ and X points, for example, L point and Δ, Σ, Λ directions.

Exercise 10.6. a) By using (10.2.21) derive a corresponding transformation law for the dynamical matrix (10.2.7b) under $\{d|\boldsymbol{t}\} \in \mathscr{R}$; use $\bar{F}_{ij}^{\mu\nu}$ of footnote 4.
b) Show that $\bar{F}(\boldsymbol{k})$ is Hermitian and that $\bar{F}(\boldsymbol{k}) = \bar{F}(-\boldsymbol{k})$.
c) Investigate the influence of time reversal symmetry on the dynamical matrix. Take into account that the vibrations can be classified according to the grey space group $\mathscr{M}_{\mathrm{II}}$ and that case (iii) of (9.4.13) is present.
d) Try to find a unitary transformation which makes $\bar{F}(\boldsymbol{k})$ real.
e) Give the transformation of d) explicitly for diamond structure.

11. Lie Groups and Lie Algebras

Whereas discrete groups mainly describe the symmetries of regular geometric structures (crystals), continuous groups are essential in discussing the properties of particles, fields (atoms and all the more elementary particles) and conservation laws. We restrict the investigation here to Lie groups and the Lie algebras connected with them. First we discuss the fundamental notions and relations for these groups, which are the generators, the (unitary) REPs, the invariants, connected spaces, covering and compact groups, simple and semisimple groups, etc. Most of the ideas are illustrated with the $\mathscr{SU}(n)$ groups. The central notions like symmetry projection operators, Clebsch-Gordan decomposition and the Wigner-Eckart theorem known from finite groups are generalized to continuous ones. For the investigation of semisimple groups and their IRs the knowledge of their weight and root systems is extremely useful.

11.1 General Foundations

11.1.1 Infinitesimal Generators and Defining Relations

In Sect. 5.5.1 we defined a continuous group, the general linear group $\mathscr{GL}(n, \mathbb{C})$. The elements a of such a group $\mathscr{G}$ may be characterized by a set $\{\alpha_1, \dots, \alpha_m\}$ of independent real parameters. At least one of these parameters must be defined on a continuous field (of scalars). The elements $a(\alpha_1, \dots, \alpha_m)$ thus can be uniquely identified with points $R(a) = \{\alpha_1(a), \dots, \alpha_m(a)\} \in \mathbb{R}_m$. The subset of $\mathbb{R}_m$ spanned by the parameters $\{\alpha_1, \dots, \alpha_m\}$ is said to be the *parameter space* and the number of parameters $m = n^2$ is said to be the dimension of the continuous group.

Because of the unique (one-to-one) assignment of points $R(a) \in \mathbb{R}_m$ and elements $a \in \mathscr{G}$ in $\mathscr{G}$ a topology is determined by the topology of $\mathbb{R}_m$. A *neighbourhood* $U_\varepsilon(R(a))$ (open sphere[1]) of $R(a)$ is the set of all points $R' \in \mathbb{R}_m$ with $\|R' - R(a)\| < \varepsilon$ and $\varepsilon > 0$, real. The $R' \in U_\varepsilon$ are the images of those elements which lie in a neighbourhood $Z_\varepsilon(a)$ of $a \in \mathscr{G}$.

A group multiplication $a_1 \cdot a_2 = a_3$ is said to be continuous if after choice of $\varepsilon > 0$ there is a number $\delta(\varepsilon) > 0$ so that the following holds for all norms (distances):

---

[1] An open sphere in $\mathbb{R}_m$ with centre 0 and radius ε is a set $\mathbb{K}_\varepsilon := \left\{x \in \mathbb{R}_m \middle| d(0, x) = \left[\sum_{i=1}^{m} (0_i - x_i)^2\right]^{1/2} < \varepsilon\right\}$.

$$\text{If } \|R(a) - R(a_2)\| < \delta(\varepsilon) \text{ then also } \|R(a_1 a) - R(a_3)\| < \varepsilon \ . \tag{11.1.1a}$$

Correspondingly the inversion $a_1 \cdot a_1^{-1} = e$ has to fulfil:

$$\text{If } \|R(a) - R(a_1)\| < \delta(\varepsilon) \text{ then also } \|R(a^{-1}) - R(a_1^{-1})\| < \varepsilon \ . \tag{11.1.1b}$$

Provided these conditions are satisfied by each $a \in \mathcal{G}$, then $\mathcal{G}$ is said to be *topological.*

The product of two elements $a(\alpha_1, \ldots, \alpha_m)$ and $b(\beta_1, \ldots, \beta_m)$ is an element $a \cdot b = c(\gamma_1, \ldots, \gamma_m)$, where in topological groups the $\{\gamma_i\}$ form a set of continuous functions of $\{\alpha_i\}$ and $\{\beta_i\}$:

$$\gamma_i = \gamma_i(\alpha_1 \cdots \alpha_m, \beta_1 \cdots \beta_m) \ , \qquad i = 1, \ldots, m \ . \tag{11.1.2a}$$

These functions γ_i determine the multiplication in $\mathcal{G}$ just as the group table (Table 2.1) does for finite groups. A simple example is given by the group of two-dimensional nonsingular real matrices when we identify the parameters with the matrix elements:

$$\begin{pmatrix} \gamma_1 & \gamma_2 \\ \gamma_3 & \gamma_4 \end{pmatrix} = \begin{pmatrix} \alpha_1 & \alpha_2 \\ \alpha_3 & \alpha_4 \end{pmatrix} \cdot \begin{pmatrix} \beta_1 & \beta_2 \\ \beta_3 & \beta_4 \end{pmatrix} = \begin{pmatrix} \alpha_1\beta_1 + \alpha_2\beta_3 & \alpha_1\beta_2 + \alpha_2\beta_4 \\ \alpha_3\beta_1 + \alpha_4\beta_3 & \alpha_3\beta_2 + \alpha_4\beta_4 \end{pmatrix} \ . \tag{11.1.2b}$$

Examples are given in Tables 11.1, 2. Denoting by $\{\alpha_i(e)\}$ the parameter set belonging to e, and by $\{\alpha_i'\}$ that belonging to a^{-1}, the γ_i satisfy the group axioms (2.1.3)

$$\gamma_i(\ldots \alpha_j \ldots, \ldots \alpha_k(e) \ldots) = \gamma_i(\ldots \alpha_k(e) \ldots, \ldots \alpha_j \ldots) = \alpha_i(a) \ , \tag{11.1.3a}$$

$$\gamma_i(\ldots \alpha_j \ldots, \ldots \alpha_k' \ldots) = \gamma_i(\ldots \alpha_k' \ldots, \ldots \alpha_j \ldots) = \alpha_i(e) \ . \tag{11.1.3b}$$

We furthermore require that the inverse of (11.1.3b) always exists, i.e. each parameter α_i' is to be represented by a function $\alpha_i'(\ldots \alpha_j \ldots, \ldots \alpha_k(e) \ldots)$, and,

Table 11.1. Lie algebras

	Rank	Dimension	Order of the Casimir operator
A_l	l	$l(l+2)$	$2, 3, \ldots, l+1$
B_l	l	$l(2l+1)$	$2, 4, \ldots, 2l$
C_l	l	$l(2l+1)$	$2, 4, \ldots, 2l$
D_l	l	$l(2l-1)$	$2, 4, \ldots, 2(l-1); l$
G_2	2	14	2, 6
F_4	4	52	2, 6, 8, 12
E_6	6	78	2, 5, 6, 8, 9, 12
E_7	7	133	2, 6, 8, 10, 12, 14, 18
E_8	8	248	2, 8, 12, 14, 18, 20, 24, 30

Table 11.2. Special Lie groups of rank 1–4

Dim.	Group	Algebra root system	Weyl group	Casimir operators	p from (11.4.28)
Rank 1					
1	$\mathscr{SO}(2)$	D_1	No root	C_1	–
3	$\mathscr{SU}(2)$; $\mathscr{SL}(2)$; $\mathscr{SO}(3)$; $\mathscr{Sp}(2)$	$A_1 \sim B_1 \sim C_1$	$\mathscr{C}_s$	C_2	0
Rank 2	↓local				
6	$\mathscr{SO}(4) \sim \mathscr{SU}(2) \times \mathscr{SU}(2)$; $\bar{\mathscr{L}}_z^+(4)$	$D_2 \sim A_1 \times A_1$	$\mathscr{C}_{2v}$	C_2, C_2'	0
8	$\mathscr{SU}(3)$; $\mathscr{SL}(3)$	A_2	$\mathscr{C}_{3v}$	C_2, C_3	1
10	$\mathscr{SO}(5)$; $\mathscr{Sp}(4)$	$B_2 \sim C_2$	$\mathscr{C}_{4v}$	C_2, C_4	2
14	Subalgebra of $\mathscr{SO}(7)$	G_2	$\mathscr{C}_{6v}$	C_2, C_6	4
Rank 3					
9	$\mathscr{SU}(2) \times \mathscr{SU}(2) \times \mathscr{SU}(2)$	$D_2 \times A_1$	$\mathscr{D}_{2h}$	C_2, C_2', C_2''	0
11	$\mathscr{SU}(3) \times \mathscr{SU}(2)$	$A_2 \times A_1$	$\mathscr{D}_{3h}$	C_2, C_2', C_3	1
13	$\mathscr{SO}(5) \times \mathscr{SU}(2)$	$B_2 \times A_1$	$\mathscr{D}_{4h}$	C_2, C_2', C_4	2
15	$\mathscr{SU}(4)$; $\mathscr{SL}(4)$; $\mathscr{SO}(6)$	$A_3 \sim D_3$	$\mathscr{T}_d$	C_2, C_3, C_4	3
17	Subalgebra of $\mathscr{SO}(10)$	$G_2 \times A_1$	$\mathscr{D}_{6h}$	C_2, C_2', C_6	4
21	$\mathscr{SO}(7)$	B_3	$\mathscr{O}_h$	C_2, C_4, C_6	6
21	$\mathscr{Sp}(6)$	C_3	$\mathscr{O}_h$	C_2, C_4, C_6	6
Rank 4					
24	$\mathscr{SU}(5)$; $\mathscr{SL}(5)$	A_4		C_2, C_3, C_4, C_5	6
28	$\mathscr{SO}(8)$	D_4		C_2, C_4, C_4', C_6	8
36	$\mathscr{SO}(9)$	B_4		C_2, C_4, C_6, C_8	12
36	$\mathscr{Sp}(8)$	C_4		C_2, C_4, C_6, C_8	12
52		F_4		C_2, C_6, C_8, C_{12}	20
and direct products of groups of lower rank					

moreover, the functions γ_i, α_i' should be (arbitrarily often) continuously differentiable functions of all the parameters. Provided all these requirements are satisfied, $\mathscr{G}$ is said to be a *Lie group*[2].

Let $D(a)$ with $a \in \mathscr{G}$ be a matrix REP of $\mathscr{G}$ in the REP space $\mathscr{L}$ spanned by a basis $\{e_i\}$, then $D(a)$ is said to be *continuous* if the coefficients (see Sect. 4.2.1)

$$D_{kl}(a) := D_{kl}(\ldots \alpha_i(a) \ldots) := \langle e_k | P_a | e_l \rangle \tag{11.1.4}$$

are continuous functions on $\mathscr{G}$, i.e. each $|\psi\rangle \in \mathscr{L}$ satisfies

$$\lim_{a \to b} \| \langle \psi | P_a - P_b | \psi \rangle \| \to 0 \ . \tag{11.1.5}$$

For example, for $\mathscr{G} = \mathscr{SO}(2)$ and $\mathscr{L} = \mathbb{R}_2$ we have $e_1 = (1,0)$, $e_2 = (0,1)$ and

[2] A somewhat less restrictive definition with respect to differentiability may also be sufficient (see [11.1] Wybourne).

$$D_{kl}(\varphi(a)) := \langle e_k | P_a | e_l \rangle = \begin{pmatrix} \cos\varphi & -\sin\varphi \\ \sin\varphi & \cos\varphi \end{pmatrix} ,$$

which is obviously continuous.

In the following, only continuous REPs of Lie groups according to (11.1.4) will be considered. Without lack of generality the parametrization can be chosen in such a way that the identity (unit) element e according to (11.1.3) has the values $\alpha_i(e) = 0$ for the parameters. Then the partial derivatives

$$J_i := \left. \frac{\partial a(\ldots \alpha_i \ldots)}{\partial \alpha_i} \right|_{\alpha_i = 0} \tag{11.1.6}$$

exist. Here a is looked upon as the isomorphic operator P_a, too. The J_i are the *infinitesimal generators* of the continuous Lie group. With them each element of the group in an infinitesimal neighbourhood of the identity can be written as

$$a(\ldots \alpha_i \ldots) = e(\ldots 0 \ldots) + \sum_{i=1}^{m} J_i \alpha_i + O(\ldots \alpha_i^2 \ldots) , \tag{11.1.7}$$

where now α_i is the infinitesimal "deviation from zero".

If a is represented by matrices as in (11.1.4) then we have accordingly

$$J_{i,kl} := \left. \frac{\partial D_{kl}(\ldots \alpha_i(a) \ldots)}{\partial \alpha_i(a)} \right|_{\alpha_i = 0} ; \tag{11.1.8}$$

$J_{i,kl}$ are infinitesimal matrix operators of the REP $D(a)$. The relation corresponding to (11.1.7) is

$$D_{kl}(\alpha) = \delta_{kl} + \sum_{i=1}^{m} J_{i,kl} \alpha_i + O(\ldots \alpha_i^2 \ldots) . \tag{11.1.9}$$

Elements of a finite neighbourhood of e or δ_{kl} may be obtained by successive continuation of (11.1.7, 9). Thus, from investigations in infinitesimal regions, conclusions can be drawn with regard to general elements of the group. Let α/N be an infinitesimal rotation angle about a fixed axis [$\mathscr{SO}(2)$]. Then

$$\begin{aligned} a(\alpha) &= a\left(\frac{\alpha}{N} + \cdots + \frac{\alpha}{N}\right) = a\left(\frac{\alpha}{N}\right) \cdot a\left(\frac{\alpha}{N}\right) \cdots a\left(\frac{\alpha}{N}\right) \\ &= \left[a\left(\frac{\alpha}{N}\right) \right]^N \Rightarrow (e + J\alpha/N)^N \qquad \text{for } N \to \infty , \end{aligned}$$

$$\curvearrowright \exp(J\alpha) = a(\alpha) \qquad \text{for } N \to \infty .$$

Another example is the four-parameter group $\mathscr{GL}(2, \mathbb{R})$ described by (11.1.2b). In infinitesimal form an element a is given by

$$a = e + \begin{pmatrix} \alpha_1 & \alpha_2 \\ \alpha_3 & \alpha_4 \end{pmatrix} = \begin{pmatrix} 1+\alpha_1 & \alpha_2 \\ \alpha_3 & 1+\alpha_4 \end{pmatrix}, \quad a^{-1} = e - \begin{pmatrix} \alpha_1 & \alpha_2 \\ \alpha_3 & \alpha_4 \end{pmatrix}. \qquad (11.1.10)$$

The generating matrix operators are

$$J_1 = \begin{pmatrix} 1 & 0 \\ 0 & 0 \end{pmatrix}, \quad J_2 = \begin{pmatrix} 0 & 1 \\ 0 & 0 \end{pmatrix}, \quad J_3 = \begin{pmatrix} 0 & 0 \\ 1 & 0 \end{pmatrix}, \quad J_4 = \begin{pmatrix} 0 & 0 \\ 0 & 1 \end{pmatrix}. \qquad (11.1.11a)$$

The operator group isomorphic to (11.1.10) has elements P_a, which are defined by their action on a function $f(\boldsymbol{x})$ according to (4.2.3)

$$P_a f(\boldsymbol{x}) = f(a^{-1}\boldsymbol{x}), \quad \begin{pmatrix} x_1' \\ x_2' \end{pmatrix} = \begin{pmatrix} 1-\alpha_1 & -\alpha_2 \\ -\alpha_3 & 1-\alpha_4 \end{pmatrix} \begin{pmatrix} x_1 \\ x_2 \end{pmatrix}. \qquad (11.1.12)$$

The change of f is then

$$\Delta f = P_a f - P_e f = \sum_i J_i \alpha_i \cdot f \qquad (11.1.13)$$

$$J_i f = \lim_{\alpha_i \to 0} \frac{P_a(\ldots\alpha_i\ldots)f - P_e f}{\alpha_i}, \; \alpha_j = \text{const for } j \neq i \, .$$

Using (11.1.12) we have

$$\begin{aligned} P_a f = f(a^{-1}\boldsymbol{x}) &= f(x_1 - \alpha_1 x_1 - \alpha_2 x_2; x_2 - \alpha_3 x_1 - \alpha_4 x_2) \\ &\approx f(\boldsymbol{x}) - (\alpha_1 x_1 + \alpha_2 x_2)\frac{\partial f}{\partial x_1} - (\alpha_3 x_1 + \alpha_4 x_2)\frac{\partial f}{\partial x_2} \, . \end{aligned}$$

By comparison with (11.1.13) we can see that the infinitesimal generators are

$$J_1 = -x_1 \frac{\partial}{\partial x_1}, \quad J_2 = -x_2 \frac{\partial}{\partial x_1}, \quad J_3 = -x_1 \frac{\partial}{\partial x_2}, \quad J_4 = -x_2 \frac{\partial}{\partial x_2} \, . \qquad (11.1.11b)$$

From (11.1.11a, b) we obtain the *commutators*

$$\begin{aligned} &[J_1, J_2] = [J_2, J_4] = J_2 \, , \qquad [J_1, J_3] = [J_3, J_4] = -J_3 \, , \\ &[J_2, J_3] = J_1 - J_4 \, , \qquad [J_1, J_4] = 0 \, . \end{aligned} \qquad (11.1.14)$$

It can generally be shown that the infinitesimal generators J_i of an m-parameter Lie group or of their REPs satisfy commutator relations of the kind (for a proof see Exercise 11.2)

$$J_i J_j - J_j J_i := [J_i, J_j] = \sum_{k=1}^{m} c_{ij}^k J_k \, , \qquad i, j = 1, \ldots, m \, . \qquad (11.1.15a)$$

The commutators themselves are linear in the generators. The real numbers occurring in this procedure are the *structure constants* of the group. Provided the parametrization is the same, mutually isomorphic groups satisfy (11.1.15) with the same structure constants. For the commutators we have

$$[J_i, J_j] = -[J_j, J_i] \ ,$$

$$[\alpha J_i + \beta J_j, J_k] = \alpha[J_i, J_k] + \beta[J_j, J_k] \ , \qquad (11.1.15b)$$

$$[J_i, [J_j, J_k]] + [J_k, [J_i, J_j]] + [J_j, [J_k, J_i]] = 0 \quad \text{(Jacobi identity)} \ .$$

A *Lie algebra* (LA) $\mathscr{A}$ is a vector space over a field $\mathbb{K}$ with a linear composition [,], so that with J_i, $J_j \in \mathscr{A}$ also $[J_i, J_j] \in \mathscr{A}$ and furthermore equations (11.1.15b) are valid with $\alpha, \beta \in \mathbb{K}$, (11.1.16)

If the J_i are operators, then [,] is the commutator. The dimension of the vector space is the dimension of the LA. The infinitesimal generators of an m-parameter Lie group $\mathscr{G}$ define a LA $\mathscr{A}(\mathscr{G})$ of dimension m. The m infinitesimal generators are the basis of the LA. Because of (11.1.15b) the structure constants satisfy

$$c_{ij}^k = -c_{ji}^k \ ,$$

$$\sum_{s=1}^{m} (c_{ik}^s c_{sl}^j + c_{kl}^s c_{si}^j + c_{li}^s c_{sk}^j) = 0 \ . \qquad (11.1.17)$$

The *rank* of a LA (or Lie group) is the maximal number of mutually commuting infinitesimal generators. Thus $\mathscr{SO}(2, \mathbb{R})$ and $\mathscr{SO}(3, \mathbb{R})$ have rank 1, $\mathscr{GL}(2, \mathbb{R})$ has rank 2, see (11.1.14).

The infinitesimal operators of a Lie group determine the Lie group. Thus a Lie group $\mathscr{G}$ is assigned to every LA $\mathscr{A}(\mathscr{G})$. This, together with the following statements, also holds for REPs by operators or matrices. The corresponding REP spaces are usually Hilbert spaces.

When we start with a certain Lie group, determine its LA and then after this evaluate a Lie group assigned to this LA, this last group is not necessarily the Lie group we started with. The reason lies in the topological features of the parameter space (Sect. 11.1.2).

We illustrate the construction of a Lie group from a LA by means of an example. We should mention that every element $a \in \mathscr{G}$ is either an element of a one-parameter subgroup $\mathscr{U} \subset \mathscr{G}$ or can be written as a product of such elements. Let $a(\gamma)$ be such an element. The parametrization should be chosen in such a way that

$$a(\gamma_1) \cdot a(\gamma_2) = a(\gamma_1 + \gamma_2) \ , \qquad a(0) = e \ .$$

After differentiation with respect to γ_1, at the point $\gamma_1 = 0$ we obtain ($\gamma_2 \to \gamma$)

$$Ja(\gamma) = \frac{\partial a}{\partial \gamma} , \qquad J = \left.\frac{\partial a}{\partial \gamma}\right|_{\gamma=0} \tag{11.1.18}$$

This is a differential equation for $a(\gamma)$ with the solution (see before 11.1.10)

$$a(\gamma) = \exp(J\gamma) = e + J\gamma + \frac{1}{2!}(J\gamma)^2 + \cdots . \tag{11.1.19}$$

Changing the parametrization from α_i to $\beta_j(\ldots\alpha_i\ldots)$ we then obtain from (11.1.7), with

$$\begin{aligned} \alpha_j &= \sum_i \left.\frac{\partial \alpha_j}{\partial \beta_i}\right|_{\beta=0} \beta_i := \sum_i d_{ji}\beta_i , \quad J_i' = \sum_j J_j d_{ji} , \\ a(\ldots\beta_i\ldots) &= e + \sum_{i,j} J_j d_{ji}\beta_i := e + \sum_i J_i'\beta_i . \end{aligned} \tag{11.1.20}$$

In our case we thus have, with a parametrization of the elements of the one-parameter subgroup $\alpha_i = \alpha_i(\gamma)$,

$$\begin{aligned} J &= \left.\frac{\partial a}{\partial \gamma}\right|_{\gamma=0} = \sum_{j=1}^{m} J_j \left.\frac{\partial \alpha_j}{\partial \gamma}\right|_{\gamma=0} \quad \text{with} \quad J_j = \frac{\partial a(\alpha_i(\gamma))}{\partial \alpha_j} , \\ a(\gamma) &= \exp\left(\sum_{j=1}^{m} J_j \frac{\partial \alpha_j}{\partial \gamma}\gamma\right) . \end{aligned} \tag{11.1.21}$$

The elements $a(\gamma)$ constitute a one-parameter subgroup of $\mathcal{SO}(3, \mathbb{R})$ (see Exercise 11.1) as well. Thus these elements of $\mathcal{SO}(3, \mathbb{R})$ can be represented as

$$a(\gamma) = a(\gamma_1, \gamma_2, \gamma_3) = \exp(J_1\gamma_1 + J_2\gamma_2 + J_3\gamma_3) . \tag{11.1.22a}$$

For small angles of rotation in $\mathbb{R}_3$ we then have

$$x' = a(\gamma)x \approx (e + \sum_i J_i\gamma_i)x . \tag{11.1.22b}$$

In order to determine IRs of $\mathcal{G}$ we can restrict ourselves to finding sets of (irreducible) matrices which satisfy (11.1.15a) with the structure constants being characteristic for the group. This follows from the above considerations. In other words, in Lie groups the determination of all REPs (or their infinitesimal operators) is reduced to the problem of finding all possible systems of operators or matrices satisfying (11.1.15a) with the corresponding c_{ij}^k. If there are unitary REPs then the operators have to fulfil

$$J_i^+ = -J_i , \qquad i = 1, \ldots, m , \tag{11.1.23}$$

which means they have to be anti-Hermitian (Exercise 11.2). Equation (11.1.23) is compatible with (11.1.15a) when the structure constants are real.

11.1.2 Algebra and Parameter Space

In the following we shall need some further algebraic fundamentals. They are given below:

i) $\mathscr{A}'$ is a *subalgebra*, if $\mathscr{A}' \subset \mathscr{A}$ and $\mathscr{A}'$ itself is a LA.
ii) $\mathscr{S} = \mathscr{A}'$ is an invariant subalgebra or ideal if $[J_i, J_j] \in \mathscr{S}$ for each $J_i \in \mathscr{S}$ and $J_j \in \mathscr{A}$.
iii) An algebra $\mathscr{A}$ or subalgebra $\mathscr{S}$ is said to be *Abelian*, if $[J_i, J_j] = 0$ for each $J_i, J_j \in \mathscr{A}$ or $\mathscr{S}$.
iv) An algebra $\mathscr{A}$ is said to be *simple*, if it has no invariant subalgebras besides $\mathscr{A}$ itself and $\{0\}$.
v) An algebra $\mathscr{A}$ is said to be *semisimple* if it does not contain an invariant Abelian subalgebra.

These terms correspond to those of groups, and because of the relations (Sect. 11.1.1) between LAs and Lie groups they can be transferred correspondingly. Simple and semisimple algebras can be treated and classified together. Also they are those algebras which are most important in applications (see Chapst. 13, 14). The properties of $\mathscr{A}$ and $\mathscr{S}$ may be seen immediately from the structure constants: In Abelian LAs it follows from (iii) that

$$c_{ij}^k \equiv 0 \qquad \text{for each } i, j, k \ . \tag{11.1.24}$$

Let $\mathscr{S} \subset \mathscr{A}$ be an *ideal* with the basis $J_1, \dots, J_p$, $p < m$ then

$$[J_i, J_j] = \sum_{k=1}^{p} c_{ij}^k J_k \text{ with } i \leqslant p,\ j \text{ arbitrary, thus}$$

$$c_{ij}^k = 0 \qquad \text{for } i \leqslant p,\ k > p,\ j \text{ arbitrary.} \tag{11.1.25}$$

If $\mathscr{S} \subset \mathscr{A}$ is an *Abelian ideal* we have correspondingly

$$c_{ij}^k = 0 \text{ for } i \leqslant p,\ k > p \textit{ and } i, j \leqslant p,\ k \text{ arbitrary.} \tag{11.1.26}$$

Let $\mathscr{A}_1$ and $\mathscr{A}_2$ be two LAs with bases $\{J_i | i = 1, \dots, m\}$ and $\{Y_j | j = 1, \dots, n\}$, then the *direct sum* ($\oplus$) of these LAs is that linear space which is spanned by the J_i, Y_j with commutators as the composition, in addition they are assumed to satisfy $[J_i, Y_j] = 0$ for each i, j; thus

$$\mathscr{A} = \mathscr{A}_1 \oplus \mathscr{A}_2 \qquad \text{with } \mathscr{A}_1 \cap \mathscr{A}_2 = \{0\} \ . \tag{11.1.27a}$$

Correspondingly, the direct sum of several LAs is defined as

$$\mathscr{A} = \mathscr{A}_1 \oplus \dots \oplus \mathscr{A}_s \quad \text{with } \mathscr{A}_i \cap \mathscr{A}_j = \{0\} \quad \text{for } i, j = 1, \dots, s; \quad i \neq j \tag{11.1.27b}$$

A LA is semisimple if and only if it can be written as a direct sum of simple LAs.

For this theorem a criterion can be given which makes use of the properties of the *Killing form* (which is the metric of the LA):

$$g_{ij} := \sum_{k,l} c_{ik}^l c_{jl}^k = g_{ji} \ . \tag{11.1.28}$$

The theorem of Cartan then says:

A LA is semisimple if and only if the determinant of the Killing form is different from zero:

$$\det(g_{ij}) \neq 0 \ . \tag{11.1.29}$$

For the LA of $\mathscr{SO}(3, \mathbb{R})$ according to Exercise 11.1, $c_{ij}^k = \varepsilon_{ijk}$, thus $g_{ij} = -2\delta_{ij}$; $\det(g_{ij}) = -8$. The LA of $\mathscr{SO}(3, \mathbb{R})$ denoted by B_1 (Table 11.2) is semisimple. In fact it is even simple, because it cannot be represented as a direct sum.

The group $\mathscr{SO}(4, \mathbb{R})$ gives an example of a semisimple LA which can be decomposed into the sum of two simple LAs; $\mathscr{SO}(4, \mathbb{R})$ has six parameters and describes the transformation of a vector $\boldsymbol{x} = \{x_i | i = 1 \dots 4\} = (x, y, z, t)$ in $\mathbb{R}_4$. Analogously to Exercise 11.1 we get the infinitesimal generators

$$\begin{aligned} A_1 &= z\frac{\partial}{\partial y} - y\frac{\partial}{\partial z} \ , \qquad A_2, A_3 \text{ cyclic in } x, y, z \\ B_1 &= x\frac{\partial}{\partial t} - t\frac{\partial}{\partial x} \ ; \qquad B_2, B_3 \text{ cyclic with} \end{aligned} \tag{11.1.30a}$$

$$[A_i, A_j] = [B_i, B_j] = \sum_k \varepsilon_{ijk} A_k \ , \qquad [A_i, B_j] = \sum_k \varepsilon_{ijk} B_k \ .$$

After a transformation (change of parametrization) to new generators

$$J_i = \tfrac{1}{2}(A_i + B_i) \ , \qquad K_i = \tfrac{1}{2}(A_i - B_i) \tag{11.1.30b}$$

we obtain the somewhat simpler commutators

$$[J_i, J_j] = \sum_k \varepsilon_{ijk} J_k \ ; \qquad [K_i, K_j] = \sum_k \varepsilon_{ijk} K_k \ ; \qquad [J_i, K_j] = 0 \ .$$

These relations satisfy (11.1.27a), i.e. the LA of $\mathscr{SO}(4, \mathbb{R})$, denoted by D_2, is the direct sum of two simple LAs B_1 of $\mathscr{SO}(3, \mathbb{R})$, thus

$$D_2 = B_1 \oplus B_1 \ . \tag{11.1.31}$$

The decomposition of a LA into a direct sum is connected with a *local isomorphism* of the Lie groups $\mathscr{SO}(4, \mathbb{R})$ and $\mathscr{SO}(3, \mathbb{R}) \times \mathscr{SO}(3, \mathbb{R})$. In general we have:

If a LA of a Lie group $\mathscr{G}$ can be decomposed into the direct sum of two LAs of the simple Lie groups $\mathscr{G}_1$ and $\mathscr{G}_2$, then $\mathscr{G}$ can be represented locally as the outer direct product

$$\mathscr{G} = \mathscr{G}_1 \times \mathscr{G}_2 \ . \tag{11.1.32}$$

In addition to the remarks in Sect. 11.1.1, we shall illustrate the concept of local isomorphism by considering the groups $\mathscr{SO}(3,\mathbb{R})$ and $\mathscr{SU}(2,\mathbb{C})$. The LA of $\mathscr{SU}(2,\mathbb{C})$ denoted by A_1 may be derived from the transformations

$$\begin{pmatrix} u' \\ v' \end{pmatrix} = U \begin{pmatrix} u \\ v \end{pmatrix} \text{ with } U = \begin{pmatrix} a & b \\ -b^* & a^* \end{pmatrix} , \quad \det U = |a|^2 + |b|^2 = 1 \ . \tag{11.1.33a}$$

$$\text{Infinitesimal: } U \to \begin{pmatrix} 1 & 0 \\ 0 & 1 \end{pmatrix} - \frac{\mathrm{i}}{2} \begin{pmatrix} \alpha_3 & \alpha_1 - i\alpha_2 \\ \alpha_1 + i\alpha_2 & -\alpha_3 \end{pmatrix} \ . \tag{11.1.33b}$$

According to (11.1.13), using a function $f(u,v)$ we obtain the three generators

$$J_1 = \frac{\mathrm{i}}{2}\left(v\frac{\partial}{\partial u} + u\frac{\partial}{\partial v}\right) , \quad J_2 = \frac{1}{2}\left(v\frac{\partial}{\partial u} - u\frac{\partial}{\partial v}\right) , \quad J_3 = \frac{\mathrm{i}}{2}\left(u\frac{\partial}{\partial u} - v\frac{\partial}{\partial v}\right) ,$$

which satisfy

$$[J_i, J_j] = \sum_k \varepsilon_{ijk} J_k \ , \tag{11.1.34}$$

i.e. the LA A_1 of $\mathscr{SU}(2,\mathbb{C})$ is isomorphic to the LA B_1 of $\mathscr{SO}(3,\mathbb{R})$. The Lie groups themselves, however, are *not* isomorphic. Let $\mathscr{L}_3$ be a three-dimensional real unitary space being spanned by the three Pauli matrices σ_i [see (3.6.2, 4), (4.1.6), (11.4.7)], i.e. $\boldsymbol{x} = \sum_i x_i \sigma_i$ for each $\boldsymbol{x} \in \mathscr{L}_3$. Then $\mathscr{U} \in \mathscr{SU}(2,\mathbb{C})$ induces an orthogonal transformation in $\mathscr{L}_3$ (with $\sum_i x_i'^2 = \sum_i x_i^2$). If we choose $U = D^{(1/2)}(\alpha_i)$, then to each of these matrices there corresponds a matrix $D^{(1)}(\alpha_i) \in \mathscr{SO}(3,\mathbb{R})$ (Sect. 3.3). Because $D^{(1/2)}(\alpha_i + 2\pi) = -D^{(1/2)}(\alpha_i)$, there exists a two-to-one homomorphism of $\mathscr{SU}(2,\mathbb{C})$ onto $\mathscr{SO}(3,\mathbb{R})$ with kernel $\{e, c_0 = -e\}$, i.e. every single-valued REP of $\mathscr{SU}(2)$ can generate a double-valued REP of $\mathscr{SO}(3)$. Conversely, every REP of $\mathscr{SO}(3)$ is a single-valued REP of $\mathscr{SU}(2)$. The group $\mathscr{SU}(2)$ is said to be the *covering group* of $\mathscr{SO}(3)$. Representations of covering groups are always single valued. All those groups that we derive directly from a LA are covering groups. For $\mathscr{SO}(n,\mathbb{R})$ these are the double groups $\mathscr{SO}^{\mathrm{D}}(n,\mathbb{R})$; for example, the LA D_2 generates the double group $\mathscr{SO}^{\mathrm{D}}(4,\mathbb{R})$, which is isomorphic to $\mathscr{SU}(2,\mathbb{C}) \times \mathscr{SU}(2,\mathbb{C})$ because of (11.1.31) and $A_1 \cong B_1$. This isomorphism is not only local but also global.

The non-uniqueness of the REPs of certain Lie groups is directly related to the connectivity of the parameter space. This can easily be understood geometrically. A group $\mathscr{G}$ is said to be *connected* if for every two elements $a_1, a_2 \in \mathscr{G}$ there is a continuous path (in the parameter space) which connects them. Consequently this space is connected.

$$\text{The group } \mathcal{O}(3, \mathbb{R}) \to \begin{pmatrix} a_0 & -\alpha_3 & \alpha_2 \\ \alpha_3 & a_0 & -\alpha_1 \\ -\alpha_2 & \alpha_1 & a_0 \end{pmatrix} \quad \text{with } a_0 = \pm 1, \alpha_i \text{ infinitesimal,} \tag{11.1.35}$$

has four parameters. The parameter space of $\mathcal{O}(3)$ consists of two (disconnected) regions (with $a_0 = \det a = \pm 1$) in the $\mathbb{R}_4$. However, the parameter space of $\mathcal{SO}(3, \mathbb{R})$ according to Exercise 11.1 is a connected sphere K_π in $\mathbb{R}_3$, since only $a_0 = \det a = +1$ is allowed. Groups with disconnected parameter spaces are not called Lie groups. But often (not always) they can be represented as direct products with Lie groups, e.g. $\mathcal{O}(3) = \mathcal{C}_i \otimes \mathcal{SO}(3)$ (see the Lorentz group instead).

The connected groups are separated into simply and multiply connected ones. A parameter space (and thus the corresponding group) is said to be *n-fold connected* if there are just n nonequivalent paths connecting two arbitrary points of this space. We consider paths to be equivalent if they can be transformed into each other by a continuous "deformation" without leaving the space (Fig. 11.1). In simply connected spaces there is only one independent path; in other words, every closed path (curve) can be deformed continuously into a single point.

(i) $\mathcal{SU}(2, \mathbb{C})$; parameter space is a subspace of $\mathbb{R}_3$, simply connected.
(ii) $\mathcal{SO}(2, \mathbb{R})$; parameter space is $[\![0^+, 2\pi]\!]$, with a possible extension (Fig. 11.1c); there are $n \to \infty$ independent paths, thus it is $n(\to\infty)$-fold connected.
(iii) $\mathcal{SO}(3, \mathbb{R})$; parameter space is K_π, twofold connected.

In the case of multiple connectedness it may happen that functions (e.g. REPs of

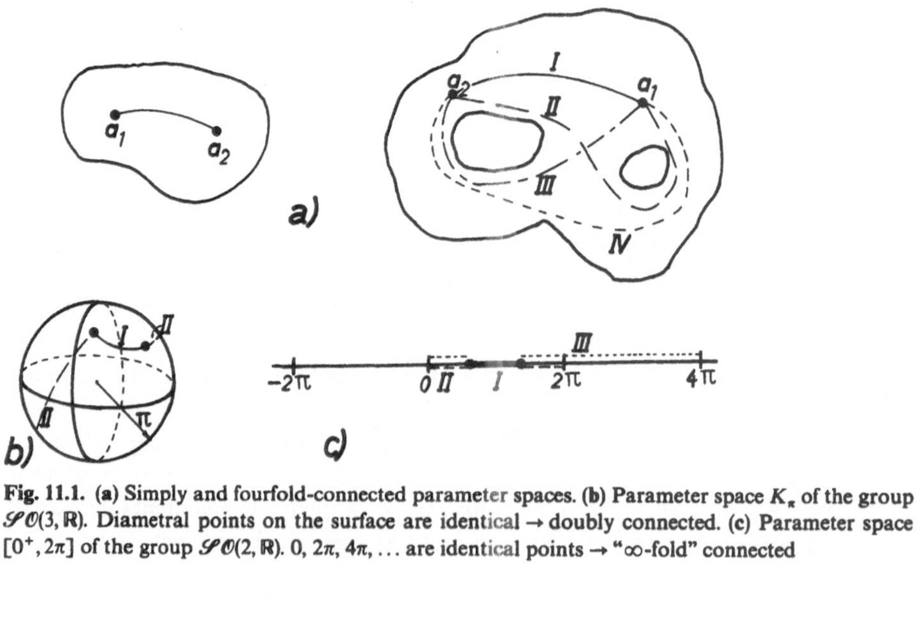

Fig. 11.1. (a) Simply and fourfold-connected parameter spaces. (b) Parameter space K_π of the group $\mathcal{SO}(3, \mathbb{R})$. Diametral points on the surface are identical → doubly connected. (c) Parameter space $[0^+, 2\pi]$ of the group $\mathcal{SO}(2, \mathbb{R})$. 0, 2π, 4π, ... are identical points → "∞-fold" connected

the group) defined on the parameter space, are many valued (i.e. path dependent). Now for every multiply connected group $\mathscr{G}$ there exists a simply connected group $\bar{\mathscr{G}}$ which is n-to-1 homomorphic onto $\mathscr{G}(\bar{\mathscr{G}} \rightarrow \mathscr{G})$ and where none of its simply connected subgroups is homomorphic onto $\mathscr{G}$; $\bar{\mathscr{G}}$ is said to be the *universal covering* group of $\mathscr{G}$. Since $\bar{\mathscr{G}}$ is simply connected, its REPs are single valued. In order to find the REPs of $\mathscr{G}$ it is sufficient to determine the single-valued REPs of $\bar{\mathscr{G}}$.

Examples are (see Table 11.3):

a) $\mathscr{SO}^D(n, \mathbb{R}) \rightarrow \mathscr{SO}(n, \mathbb{R})$; $\mathscr{SU}(2, \mathbb{C}) \rightarrow \mathscr{SO}(3, \mathbb{R})$ double point group $\rightarrow$ point group;

b) If $\mathscr{G} = \{c_n | c_n^n = e\}$ is a cyclic group of order n, and if $c_{2\pi j}$, $j = 1, \ldots, m$ are m further elements with $c_{2\pi m} = e$, then the group

$$\bar{\mathscr{G}} = \mathscr{G} + c_{2\pi}\mathscr{G} + c_{4\pi}\mathscr{G} + \cdots + c_{2\pi(m-1)}\mathscr{G}$$

is m-to-1 homomorphic onto $\mathscr{G}$; in $\bar{\mathscr{G}}$ we have $c_n^{nj} = c_{2\pi j}$, $c_n^{nm} = c_{2\pi m} = e$. The single valued REPs of $\bar{\mathscr{G}}$ give single-, double-, $\ldots$, m-valued REPs of $\mathscr{G}$ (see Fig. 11.1c).

11.1.3 Casimir Operators

From the examples in (11.1.11, 14, 30, 34) it follows that the infinitesimal generators may be represented in very different ways: by operators such as matrix operators, differential operators, etc. Correspondingly, their physical meaning or interpretation is varied: rotations in $\mathbb{R}_n$, operators in any Hilbert space, annihilation and creation operators for quasi-particles, etc. In all applications related to these operators the question arises whether there are certain operators which commute with all the generators J_i of a Lie algebra. In fact, this is the case.

For semisimple LAs we define a metric g^{ij} that is the reciprocal of (11.1.28) by

$$\sum_j g^{ij} g_{jk} = \sum_j g_{kj} g^{ji} = \delta_{ik} . \tag{11.1.36}$$

Then the quadratic form (*Casimir operator*)

$$C_2 = \sum_{ij} g^{ij} J_i J_j \tag{11.1.37}$$

commutes with all the infinitesimal generators J_i of the LA; C_2 is an invariant operator. (There is one exception: the LA D_1 of $\mathscr{SO}(2, \mathbb{R})$, see Table 11.2 and also Exercise 11.3.) The physical meaning of these operators follows from that of the J_i. For the groups $\mathscr{SO}(3, \mathbb{R})$ (Exercise 11.1) and $\mathscr{SU}(2, \mathbb{C})$, (11.1.34),

$$g_{jk} = -2\delta_{jk} , \qquad g^{ij} = -\tfrac{1}{2}\delta_{ij} \qquad \text{thus}$$

$$C_2 = -\tfrac{1}{2}\sum_i J_i^2 = -\tfrac{1}{2}(J_1^2 + J_2^2 + J_3^2) ; \tag{11.1.38}$$

Table 11.3. Isomorphisms

Group	Algebra	Covering group	Isomorphism	Local isomorphisms of the associated algebras
$\mathcal{SO}(2)$	D_1	$\mathbb{R}$	$\mathcal{SO}(2) \cong \mathbb{R}/\mathbb{Z}_\infty$	
$\mathcal{SO}(1,1)$	D_1	$\mathbb{R}$		
$\mathcal{SO}(3)$	$B_1 \sim A_1 \sim C_1$	$\mathcal{SU}(2)$	$\mathcal{SO}(3) \cong \mathcal{SU}(2)/\mathbb{Z}_2$	$\mathcal{SU}(2) \cong \mathcal{SO}(3) \cong \mathcal{Sp}(2) \cong \mathcal{SU}^*(2)$
$\mathcal{SO}(2,1)$	$B_1 \sim A_1 \sim C_1$	$\mathcal{SU}(1,1)$	$\mathcal{SO}(2,1) \cong \mathcal{SU}(1,1)/\mathbb{Z}_2$	$\mathcal{SU}(1,1) \cong \mathcal{SO}(2,1) \cong \mathcal{Sp}(2,\mathbb{R}) \cong \mathcal{SL}(2,\mathbb{R})$
$\mathcal{SO}(4)$	$D_2 \sim A_1 \oplus A_1$	$\mathcal{SU}(2) \times \mathcal{SU}(2)$	$\mathcal{SO}(4) \cong \mathcal{SU}(2) \times \mathcal{SU}(2)/\mathbb{Z}_2$	$\mathcal{SO}(4) \cong \mathcal{SU}(2) \times \mathcal{SU}(2) \cong \mathcal{SO}(3) \times \mathcal{SO}(3) \cong \mathcal{Sp}(2) \times \mathcal{Sp}(2)$
				$\mathcal{SO}^*(4) \cong \mathcal{SU}(2) \times \mathcal{SL}(2,\mathbb{R})$
$\mathcal{SO}(3,1)$	$D_2 \sim A_1 \oplus A_1$	$\mathcal{SL}(2,\mathbb{C})$	$\mathcal{SO}(3,1) \cong \mathcal{SL}(2,\mathbb{C})/\mathbb{Z}_2$	$\mathcal{SO}(3,1) \cong \mathcal{SL}(2,\mathbb{C})$
$\mathcal{SO}(2,2)$	$D_2 \sim A_1 \oplus A_1$	$\mathcal{SU}(1,1) \times \mathcal{SU}(1,1)$	$\mathcal{SO}(2,2) \cong \mathcal{SU}(1,1) \times \mathcal{SU}(1,1)/\mathbb{Z}_2$	$\mathcal{SO}(2,2) \cong \mathcal{SL}(2,\mathbb{R}) \times \mathcal{SL}(2,\mathbb{R})$
$\mathcal{SO}(5)$	$B_2 \sim C_2$	$\mathcal{Sp}(4)$	$\mathcal{SO}(5) \cong \mathcal{Sp}(4)/\mathbb{Z}_2$	$\mathcal{SO}(5) \cong \mathcal{Sp}(4)$
$\mathcal{SO}(4,1)$	$B_2 \sim C_2$	$\mathcal{Sp}(2,2)$	$\mathcal{SO}(4,1) \cong \mathcal{Sp}(2,2)/\mathbb{Z}_2$	$\mathcal{SO}(4,1) \cong \mathcal{Sp}(2,2)$
$\mathcal{SO}(3,2)$	$B_2 \sim C_2$	$\mathcal{Sp}(4,\mathbb{R})$	$\mathcal{SO}(3,2) \cong \mathcal{Sp}(4,\mathbb{R})/\mathbb{Z}_2$	$\mathcal{SO}(3,2) \cong \mathcal{Sp}(4,\mathbb{R})$
$\mathcal{SO}(6)$	$D_3 \sim A_3$	$\mathcal{SU}(4)$	$\mathcal{SO}(6) \cong \mathcal{SU}(4)/\mathbb{Z}_2$	$\mathcal{SO}(6) \cong \mathcal{SU}(4)$
				$\mathcal{SO}^*(6) \cong \mathcal{SU}(3,1)$
				$\mathcal{SO}(5,1) \cong \mathcal{SU}^*(4)$
$\mathcal{SO}(3,3)$	$D_3 \sim A_3$	$\mathcal{SL}(4,\mathbb{R})$	$\mathcal{SO}(3,3) \cong \mathcal{SL}(4,\mathbb{R})/\mathbb{Z}_2$	$\mathcal{SO}(3,3) \cong \mathcal{SL}(4,\mathbb{R})$
$\mathcal{SO}(4,2)$	$D_3 \sim A_3$	$\mathcal{SU}(2,2)$	$\mathcal{SO}(4,2) \cong \mathcal{SU}(2,2)/\mathbb{Z}_2$	$\mathcal{SO}(4,2) \cong \mathcal{SU}(2,2)$

the J_i correspond to the three components of the angular momentum (spin), C_2 corresponds to the square of the angular momentum (spin).

Racah [11.2] has shown that for every semisimple LA (Lie group) of rank r there exist just r independent *Casimir invariants*, the eigenvalues of which completely specify the IRs of semisimple LAs and thus characterize the multiplets. This holds because, according to Schur's lemma, operators commuting with all the elements of a group are multiples of the unit operator and thus behave as numbers. According to Racah the invariants can be represented by the structure constants and the metric

$$C_n = \sum_{\dots i_l, k_l \dots} c_{i_1 k_1}^{k_2} c_{i_2 k_2}^{k_3} \dots c_{i_n k_n}^{k_1} \cdot J^{i_1} J^{i_2} \dots J^{i_n} \ , \qquad J^i = \sum_j g^{ij} J_j \ . \tag{11.1.39}$$

Here n takes all positive integers. However, in this way we arbitrarily often obtain every Casimir operator, or linear combinations of them. Thus we have to choose r independent values for n (as small as possible). For $n = 2$, (11.1.39) leads again to (11.1.37), if (11.1.28) is used.

For non-semisimple Lie groups ($\det g_{ij} = 0$) Casimir invariants can be given too. But their construction is more complicated. Beltrametti and Blasi [11.3] have shown that in such groups there exist $m - r$ invariants (m: number of parameters α_k in $\mathscr{G}$, r is the rank of the matrix $\sum_{k=1}^{m} c_{ij}^k \alpha_k$).

Exercise 11.1. Using (11.1.13), determine the three infinitesimal generators of $\mathscr{SO}(3, \mathbb{R})$ by investigating the rotations about three mutually orthogonal axes with rotation angles $\varphi_1, \varphi_2, \varphi_3$. Give J_x, J_y, J_z in differential and matrix form and prove $[J_i, J_j] = \sum \varepsilon_{ijk} J_k$. Further give the parameter space (subspace of $\mathbb{R}_3$) for $\mathscr{SO}(3, \mathbb{R})$ and discuss the assignment of points of this space to rotations.

Exercise 11.2. a) Prove (11.1.15a) by expanding the commutators $a^{-1}(\alpha)\, a^{-1}(\beta)\, a(\alpha)\, a(\beta)$ up to quadratic terms in α_i, β_j and by realizing that the result can also be represented as an expansion.
b) Show that by changing the parametrization $\alpha_i \to \beta_j$ according to (11.1.20) the structure constants transform according to $c_{ij}^{\prime k} = \sum_{lrs} d_{li}\, d_{rj} (d^{-1})_{ks} c_{lr}^s$.
c) Prove that (11.1.23) is valid for unitary REPs.

Exercise 11.3. Prove that the Casimir operator (11.1.37) commutes with all the infinitesimal generators of a semisimple LA.

11.2 Unitary Representations of Lie Groups

According to (4.2.11b) the existence of *unitary* REPs of a group in a class of equivalent REPs depends on whether a mean value (4.2.12, 13) can be defined or not. For finite groups, according to (4.2.14) and (4.2.13) the mean value satisfies

$$\sum_{a \in \mathscr{G}} \varphi(a) = \sum_{a \in \mathscr{G}} \varphi(ba) = \sum_{a \in \mathscr{G}} \varphi(ab) = \sum_{c \in \mathscr{G}} \varphi(c) \ , \qquad b \in \mathscr{G} \ . \tag{11.2.1}$$

The theorems on unitary REPs and orthogonality can be transferred to continuous groups if it is possible to generalize (11.2.1) appropriately for the latter. Of course, the summation has to be replaced by an integration over the parameter space, i.e. over $d^m\alpha = d\alpha_1\, d\alpha_2 \ldots d\alpha_m$; here a *weight function* $\rho(\ldots\alpha_i\ldots) := \rho(a(\ldots\alpha_i\ldots)) := \rho(a)$ may have to be taken into account for the volume elements. Formally, (11.2.1) has to be replaced by

$$\int \varphi(a(\alpha_i))\cdot\rho(\alpha_i)\, d^m\alpha = \int \varphi(b(\beta_j)a(\alpha_i))\cdot\rho(\alpha_i)\, d^m\alpha$$
$$= \int \varphi(a(\alpha_i)b(\beta_j))\cdot\rho(\alpha_i)\, d^m\alpha = \int \varphi(c(\gamma_i))\cdot\rho(\gamma_i)\, d^m\gamma \ . \tag{11.2.2}$$

This equation, of course, assumes the existence of the integrals. Furthermore, the shape of the weight function has to be such that the volume elements assigned to the elements a and ab agree (11.1.2a):

$$dV(a) = \rho(\alpha_i)\, d^m\alpha \quad \text{and} \quad dV(ab) = dV(c) = \rho(\gamma_i)\, d^m\gamma \ . \tag{11.2.3}$$

From this, the weight function $\rho(\alpha_i)$ can be determined. In the neighbourhood of the identity, i.e. $\alpha_i \to 0$, $\rho(0)$ may be chosen arbitrarily. Furthermore, according to (11.1.2), $d\gamma_j$ and $d\alpha_i$ satisfy

$$d\gamma_j = \sum_{i=1}^{m} \left.\frac{\partial\gamma_j(\ldots\alpha_k\ldots,\ldots\beta_k\ldots)}{\partial\alpha_i}\right|_{\alpha=0} d\alpha_i \ , \tag{11.2.4}$$

thus

$$d^m\gamma = \left(\det \left.\frac{\partial\gamma_j(\ldots\alpha_k\ldots,\ldots\beta_k\ldots)}{\partial\alpha_i}\right|_{\alpha=0}\right) d^m\alpha := W(\gamma_k)\, d^m\alpha \ . \tag{11.2.5}$$

With the functional determinant $W(\beta_i)$ we obtain for the weight function

$$\rho(\beta_i) = W^{-1}(\beta_i)\rho(0) \ . \tag{11.2.6}$$

Using (11.1.2b), i.e. for $\mathscr{GL}(2,\mathbb{R})$[3],

$$W(\beta_i) = (\beta_1\beta_4 - \beta_2\beta_3)^2 = (\det b)^2 \ .$$

For $\mathscr{SO}(2,\mathbb{R})$, $\gamma = \alpha + \beta$ (Sect. 11.1.1). Then $W(\beta)$ and $\rho(\beta)$ are constant and the mean values are given by

$$M(\varphi) = \frac{1}{2\pi}\int_0^{2\pi} \varphi(\alpha)\, d\alpha \ . \tag{11.2.7}$$

For $\mathscr{SU}(2,\mathbb{C})$ with a parametrization according to (11.1.33a) $a = \alpha_0 - i\alpha_3$, $b = \alpha_2 - i\alpha_1$, $\alpha_0^2 = 1 - (\alpha_1^2 + \alpha_2^2 + \alpha_3^2)$, the functional determinant is

---

[3] For $\mathscr{GL}(n,\mathbb{R})$ we find $W(\ldots\beta_i\ldots) = (\det b)^n$.

$$W(\alpha_1, \alpha_2, \alpha_3) = \alpha_0 = (1 - \alpha_1^2 - \alpha_2^2 - \alpha_3^2)^{1/2} . \tag{11.2.8a}$$

With a change of the parametrization the functional determinant changes, too. This change is given by the functional determinant of the transformation between the sets of parameters. Using the Euler angles α, β, γ of (3.3.2) as parameters, we have

$$W(\alpha, \beta, \gamma) = \frac{\partial(\alpha, \beta, \gamma)}{\partial(\alpha_1, \alpha_2, \alpha_3)} W(\alpha_1, \alpha_2, \alpha_3) , \tag{11.2.9}$$

and

$$\frac{\partial(\alpha_1, \alpha_2, \alpha_3)}{\partial(\alpha, \beta, \gamma)} = -\frac{1}{8}\cos[(\alpha + \gamma)/2] \cos(\beta/2) \sin\beta = -\frac{1}{8}\alpha_0 \sin\beta ,$$

so that

$$W(\alpha, \beta, \gamma) = (-)\frac{8}{\sin\beta} . \tag{11.2.8b}$$

With $\rho(0) = 1$ (often other normalizations are used) we obtain for $\mathscr{SU}(2, \mathbb{C})$

$$\rho(\alpha, \beta, \gamma)\, d\alpha\, d\beta\, d\gamma = \tfrac{1}{8} \sin\beta\, d\alpha\, d\beta\, d\gamma . \tag{11.2.8c}$$

For $\mathscr{SO}(3, \mathbb{R})$ a possible parametrization is

$$\begin{pmatrix} 1 - 2(\alpha_2^2 + \alpha_3^2) & 2(\alpha_1\alpha_2 - \alpha_0\alpha_3) & 2(\alpha_1\alpha_3 + \alpha_0\alpha_2) \\ 2(\alpha_1\alpha_2 + \alpha_0\alpha_3) & 1 - 2(\alpha_1^2 + \alpha_3^2) & 2(\alpha_2\alpha_3 - \alpha_0\alpha_1) \\ 2(\alpha_1\alpha_3 - \alpha_0\alpha_2) & 2(\alpha_2\alpha_3 + \alpha_0\alpha_1) & 1 - 2(\alpha_1^2 + \alpha_2^2) \end{pmatrix} \tag{11.2.10a}$$

with $\alpha_0^2 + \alpha_1^2 + \alpha_2^2 + \alpha_3^2 = 1$, thus having three independent parameters. Multiplication of $a(\alpha)$ by $b(\beta)$ gives $\gamma_0 = \alpha_0\beta_0 - \alpha_1\beta_1 - \alpha_2\beta_2 - \alpha_3\beta_3$, $\gamma_1 = \alpha_0\beta_1 + \alpha_1\beta_0 + \alpha_2\beta_3 - \alpha_3\beta_2$, which is cyclic in 1, 2, 3 are thus according to (11.2.5)

$$W(\alpha_1, \alpha_2, \alpha_3) = \alpha_0 = (1 - \alpha_1^2 - \alpha_2^2 - \alpha_3^2)^{1/2} . \tag{11.2.10b}$$

A more appropriate parametrization uses the rotation axis $\boldsymbol{n} = \{\sin\vartheta\cos\varphi, \sin\vartheta\sin\varphi, \cos\vartheta\} = \{n_1, n_2, n_3\}$ and the rotation angle α (about this axis). Then

$$\alpha_0 = \cos(\alpha/2) , \qquad \alpha_i = n_i \sin(\alpha/2) . \tag{11.2.10c}$$

We get the functional determinant

$$\frac{\partial(\alpha_1, \alpha_2, \alpha_3)}{\partial(\vartheta, \varphi, \alpha)} = \frac{1}{2}\sin\vartheta \sin^2\left(\frac{\alpha}{2}\right)\cos\left(\frac{\alpha}{2}\right) \tag{11.2.10d}$$

and thus

$$W(\vartheta, \varphi, \alpha) = \frac{2}{\sin\vartheta \sin^2(\alpha/2)} \; . \tag{11.2.11a}$$

Apart from normalizing factors

$$\rho(\vartheta, \varphi, \alpha)\, d\vartheta\, d\varphi\, d\alpha = \frac{1}{2}\sin^2\frac{\alpha}{2}\sin\vartheta\, d\vartheta\, d\varphi\, d\alpha = \frac{1}{2}\sin^2\left(\frac{\alpha}{2}\right) d\Omega\, d\alpha \tag{11.2.11b}$$

with the solid angle element $d\Omega = \sin\vartheta\, d\vartheta\, d\varphi$. When the integration is over class functions depending only on the rotation angle α, the solid angle integration can be performed. We may then use the weight function

$$\hat{\rho}(\alpha)\, d\alpha = 2\pi \sin^2(\alpha/2)\, d\alpha \; . \tag{11.2.11c}$$

A Lie group is said to be *compact* if the images $R(a)$ of all the elements $a \in \mathscr{G}$ form a closed and bounded set of points, i.e. the parameter space is closed and bounded. It is *locally compact* if an image $R(a)$ has a compact neighbourhood of points that are images of elements of $\mathscr{G}$ (Sect. 11.1.1). A criterion for compact groups is:

$$g_{ij} \text{ is negative-definite} \; , \tag{11.2.12}$$

i.e. *compact groups are necessarily semisimple* (11.1.29). Every compact group *possesses a mean value* and thus according to (4.2.11b) has at least one unitary REP in any class of equivalent REPs. The relations concerning mean values can then be taken from Sec. 4.2.2, replacing everywhere

$$\frac{1}{g}\sum_{a\in\mathscr{G}} \cdots \quad \text{by}\ \frac{1}{V}\int \ldots \rho(\ldots\alpha_i\ldots)\, d^m\alpha \qquad \text{with}\ V = \int \rho(\ldots\alpha_i\ldots)\, d^m\alpha \tag{11.2.13}$$

and understanding the group elements as being functions of the parameters α_i. In particular, this is valid for the important theorems on orthogonality (4.2.25, 27) as well as (4.2.31, 32). The criterion of irreducibility (4.2.32) now reads explicitly

$$\frac{1}{V}\int |\chi^{(k)}(a(\ldots\alpha_i\ldots))|^2 \cdot \rho(\ldots\alpha_i\ldots)\, d^m\alpha = 1 \; , \tag{11.2.14}$$

where k denotes the IR.

In the Abelian group $\mathscr{SO}(2, \mathbb{R})$

$$D^{(k)}(\alpha) = D^{(k)}(\alpha + 2\pi) = e^{ik\alpha} \; ,$$

$$D^{(k)}(\alpha_1)\cdot D^{(k)}(\alpha_2) = D^{(k)}(\alpha_1 + \alpha_2) \; ,$$

$$\rho(\alpha) = 1 \; , \qquad V = \int_0^{2\pi} \rho(\alpha)\, d\alpha = 2\pi \; , \qquad k \in \mathbb{Z} \; .$$

Thus (11.2.14) reads

$$\frac{1}{2\pi}\int e^{-ik\alpha}e^{ik'\alpha}\,d\alpha = \delta_{kk'} \to 1 \qquad \text{for } k = k' \ , \tag{11.2.15}$$

which of course indicates irreducibility. Equation (11.2.15) is simply an expression of the orthogonality of the Fourier basis. With (11.2.11c), Appendix A and (11.4.67) we have for $\mathscr{SO}(3,\mathbb{R})$

$$V = \pi\int_0^\pi (1-\cos\alpha)\,d\alpha = 2\pi\int_0^\pi \sin^2\left(\frac{\alpha}{2}\right)d\alpha = \pi^2 \ ,$$

$$\chi^{(k)}(\alpha) = \frac{\sin[\alpha(k+1/2)]}{\sin(\alpha/2)} \ , \qquad k\in\mathbb{Z}_+ \ , \tag{11.2.16}$$

$$\frac{2\pi}{\pi^2}\int_0^\pi \chi^{(k)}(\alpha)\chi^{(k')}(\alpha)\sin^2\left(\frac{\alpha}{2}\right)d\alpha = \delta_{kk'} \ ,$$

which means these REPs are irreducible too.

Reducible REPs of *compact* Lie groups are completely reducible and *the unitary* IRs *are finite*. For the projection operators there is a relation analogous to (4.3.9a)

$$P_{ij}^{(k)} = \frac{d_k}{V}\int D_{ij}^{(k)*}(a(\ldots\alpha_i\ldots))P_{a(\ldots\alpha_i\ldots)}\rho(\ldots\alpha_i\ldots)\,d^m\alpha \ . \tag{11.2.17}$$

With these projection operators (4.3.12–15) are valid too, especially the fundamental relations for the orthogonality of basis functions. By normalizing them in such a way that

$$|f_i^{(k)}\rangle := \frac{1}{N_k}P_{ij}^{(k)}|f\rangle \qquad \text{with} \qquad N_k^2 = \langle f|P_{ii}^{(k)}|f\rangle \ , \tag{11.2.18}$$

it follows that

$$\begin{aligned}\langle f_{i'}^{(k')}|f_i^{(k)}\rangle &= \frac{1}{N_{k'}N_k}\langle P_{i'j}^{(k')}f|P_{ij}^{(k)}f\rangle \\ &= \frac{1}{N_{k'}N_k}\langle f|P_{i'j}^{(k')+}P_{ij}^{(k)}|f\rangle = \frac{1}{N_{k'}N_k}\langle f|P_{ji'}^{(k')}P_{ij}^{(k)}|f\rangle \\ &= \frac{1}{N_k^2}\delta_{k'k}\delta_{i'i}\langle f|P_{jj}^{(k)}|f\rangle = \delta_{k'k}\delta_{i'i} \ .\end{aligned} \tag{11.2.19}$$

The symmetry group of a linear molecule AB_2 (e.g. CO_2) is $\mathscr{D}_{\infty h} = \mathscr{C}_{\infty v}\times\mathscr{C}_i$, the REPs of which may be taken from Appendix A. In $\mathscr{C}_\infty := \mathscr{SO}(2)$, according to

(11.2.15), $V = 2\pi$. Since in $\mathscr{C}_{\infty v}$ the elements $c(\alpha)$ and $\sigma_v(\alpha)$ are assigned to every point α of the parameter space, the volume for $\mathscr{C}_{\infty v}$ is $V = 4\pi$ and for $\mathscr{D}_{\infty h}$ thus $V = 8\pi$. Using (11.2.17), $\rho(\alpha) = 1$ and Appendix A we obtain

$$P^{(\Sigma^+_{g,u})} = \frac{1}{8\pi} \int_0^{2\pi} (P_{c(\alpha)} + P_{\sigma_v(\alpha)} \pm P_{ic(\alpha)} \pm P_{i\sigma_v(\alpha)})\, d\alpha \; . \tag{11.2.20}$$

The functions $|f\rangle$ of the REP space according to (11.2.18) are chosen to be s orbitals: $|s_i\rangle, i = A, B_1, B_2$, concentrated at the corresponding particle. The $\{|s_i\rangle\}$ span a basis which according to (4.2.6) defines the REP

$$D(c(\alpha)) = D(\sigma_v(\alpha)) = \begin{pmatrix} 1 & 0 & 0 \\ 0 & 1 & 0 \\ 0 & 0 & 1 \end{pmatrix} , \qquad D(ic(\alpha)) = D(i\sigma_v(\alpha)) = \begin{pmatrix} 1 & 0 & 0 \\ 0 & 0 & 1 \\ 0 & 1 & 0 \end{pmatrix} .$$

By reduction we have

$$D = 2\Sigma^+_g + \Sigma^+_u \; . \tag{11.2.21}$$

The symmetry-adapted orbitals are

$$|f^{(\Sigma^+_g,1)}\rangle := P^{(\Sigma^+_g)}|s_A\rangle = |s_A\rangle \; ,$$

$$|f^{(\Sigma^+_g,2)}\rangle := P^{(\Sigma^+_g)}|s_{B1}\rangle = \tfrac{1}{2}(|s_{B1}\rangle + |s_{B2}\rangle) \; ,$$

$$|f^{(\Sigma^+_u)}|s_{B1}\rangle := P^{(\Sigma^+_u)}|s_{B1}\rangle = \tfrac{1}{2}(|s_{B1}\rangle - |s_{B2}\rangle) \; , \qquad \text{etc.}$$

Exercise 11.4. Give the parameters (11.2.10) in terms of the Euler angles (3.3.1); one has to take into account that the (three) Euler angles describe three successive rotations. Then calculate $\rho(\alpha, \beta, \gamma)$ according to (11.2.9, 10).

Exercise 11.5. Give the projection operators for $\Sigma^-_{g,u}$ and $\Pi_{g,u}$ similar to that given in (11.2.20).

11.3 Clebsch-Gordan Coefficients and the Wigner-Eckart Theorem

The calculation of the CGC for compact Lie groups is similar to that for finite groups. We start with the projection operator method described in Exercise 4.11. By application of the projection operators $P^{(k)}_{ij}$ to a product basis

$$|k, i; k', i'\rangle = |f^{(k)}_i\rangle |f^{(k')}_{i'}\rangle \tag{11.3.1}$$

this is decomposed according to $|f^{(k, s_K)}_j\rangle$, as in Sect. 4.4.2. The assigned tensor product space is defined in (4.4.13b). In *simply reducible* groups (Sect. 4.4.3) the index of multiplicity may be dropped. Using (11.2.18) we obtain for these groups

$$|\hat{f}_j^{(k)}\rangle = \frac{1}{K} P_{jj'}^{(k)} |f_i^{(k')}\rangle |f_l^{(k'')}\rangle \ , \qquad i, j', l \text{ fixed} \ , \tag{11.3.2a}$$

with a normalizing constant K, which is determined using $\langle \hat{f}_j^{(k)} | \hat{f}_j^{(k)} \rangle = 1$ and (4.3.12):

$$K = \langle k', i; k'', l | k, j' \rangle^* := \left(\begin{smallmatrix} k' k'' \\ i \ l \end{smallmatrix} \middle| \begin{smallmatrix} k \\ j' \end{smallmatrix}\right)^* \ , \tag{11.3.2b}$$

and thus depends on all the indices. With (11.2.17) we have furthermore

$$P_{jj'}^{(k)} |f_i^{(k')}\rangle |f_l^{(k'')}\rangle = \frac{d_k}{V} \int D_{jj'}^{(k)*}(a) P_a |f_i^{(k')}\rangle |f_l^{(k'')}\rangle \rho(\alpha)\, d^m\alpha \tag{11.3.3}$$

and

$$P_a |f_i^{(k')}\rangle |f_l^{(k'')}\rangle = \sum_{i'l'} D_{i'i}^{(k')}(a) D_{l'l}^{(k'')}(a) |f_{i'}^{(k')}\rangle |f_{l'}^{(k'')}\rangle \ .$$

Forming the scalar product of (11.3.2a) with $\langle f_{i'}^{(k')} | \langle f_{l'}^{(k'')} |$ and inserting (11.3.2) and (11.3.3) we obtain for the CGC $\langle k', i; k'', l | k, j' \rangle := \left(\begin{smallmatrix} k' k'' \\ i \ l \end{smallmatrix} \middle| \begin{smallmatrix} k \\ j' \end{smallmatrix}\right)$

$$\left(\begin{smallmatrix} k' k'' \\ i' \ l' \end{smallmatrix} \middle| \begin{smallmatrix} k \\ j' \end{smallmatrix}\right) \left(\begin{smallmatrix} k' k'' \\ i \ l \end{smallmatrix} \middle| \begin{smallmatrix} k \\ j \end{smallmatrix}\right)^* = \frac{d_k}{V} \int D_{j'j}^{(k)*}(a) D_{i'i}^{(k')}(a) D_{l'l}^{(k'')}(a) \rho(\alpha)\, d^m\alpha \ , \tag{11.3.4}$$

which corresponds to (4.4.20). In non-simply reducible groups the product space $\mathscr{L} = \mathscr{L}^{(k')} \otimes \mathscr{L}^{(k'')}$ has first to be decomposed into irreducible subspaces of the form $\mathscr{L} = \sum_\oplus m_k \mathscr{L}^{(k)}$. This is done by applying $P_{jj}^{(k)}$ with fixed j to $\mathscr{L}$ (Sect. 4.3.2). Then $P_{jj}^{(k)} \mathscr{L} = \mathscr{L}_j^{(k)}$ is a subspace of dimension m_k (the multiplicity). In $\mathscr{L}_j^{(k)}$ we may choose an orthonormal basis $\{|f_s^{(k)}\rangle\}$, $s = 1, \ldots, m_k$, and from this we can construct m_k sets of basis functions for the IRs $D^{(k)}$ by using the projections $P_{ij}^{(k)} |e_s^{(k)}\rangle \sim |f_i^{(k,s)}\rangle$, $i = 1, \ldots, d_k$. With these $|f_i^{(k,s)}\rangle$ the CGC can be constructed similarly to in (11.3.2–4) or Sects. 4.4.2, 3.

In (6.1.8) we defined a general irreducible tensor operator with respect to a group $\mathscr{G}$:

$$P_a T_i^{(k)} P_a^{-1} = \sum_{j=1}^{d_k} D_{ji}^{(k)}(a) \cdot T_j^{(k)} \ , \quad a \in \mathscr{G} \ . \tag{11.3.5}$$

This transformation for the components $\{T_i^{(k)} | i = 1, \ldots, d_k\}$ is valid for Lie groups, too, and may correspondingly be transferred to the LA. According to (11.1.18, 21) the infinitesimal generator of a one-parameter subgroup of a Lie group or its REP satisfies

$$J = \left.\frac{da(\gamma)}{d\gamma}\right|_{\gamma=0} = \sum_{j=1}^{m} J_j \left.\frac{\partial \alpha_j}{\partial \gamma}\right|_{\gamma=0} := \sum_{j=1}^{m} J_j e_j \ . \tag{11.3.6}$$

With $P_a \approx e + J\gamma$ and the REP $D_{ji}^{(k)}(a) = \langle k, j | e + J\gamma | k, i \rangle$ with respect to any irreducible basis we obtain from (11.3.5) the transformation of the tensor components by the generators:

$$[J, T_i^{(k)}] = \sum_j \langle k, j | J | k, i \rangle \cdot T_j^{(k)} . \qquad (11.3.7)$$

Using the generators of $\mathscr{SO}(3, \mathbb{R})$ (see Exercises 11.1, 6) we get for a vector operator $\boldsymbol{x} = \{x_i\} \in \mathbb{R}_3$ (first rank tensor)

$$[J_i, x_j] = \sum_k \varepsilon_{ijk} x_k . \qquad (11.3.8)$$

After these remarks we are able to transfer the Wigner-Eckart theorem (Sect. 6.2) to Lie groups: Let $D^{(k')}$ and $D^{(k'')}$ be unitary IRs of a compact simply reducible Lie group $\mathscr{G}$ with the REP spaces (Hilbert spaces) $\mathscr{L}^{(k')}$ and $\mathscr{L}^{(k'')}$ having orthonormal bases $|k', i\rangle$ and $|k'', j\rangle$, respectively. Furthermore let $T_l^{(k)}$ be an irreducible tensor operator with respect to $\mathscr{G}$. Then

$$\langle k'', j | T_l^{(k)} | k', i \rangle = C^{k''} \left({}^{kk'}_{l\,i} \middle| {}^{k''}_{j} \right)^* = C^{k''} \left({}^{k''}_{j} \middle| {}^{kk'}_{l\,i} \right) \qquad (11.3.9a)$$

with the reduced matrix element (inversion)

$$C^{k''} = \frac{1}{d_{k''}} \sum_{ijl} \left({}^{kk'}_{l\,i} \middle| {}^{k''}_{j} \right) \langle k'', j | T_l^{(k)} | k', i \rangle . \qquad (11.3.9b)$$

These relations follow from (11.3.5) if we express each $a \in \mathscr{G}$ by the operators P_a, or their unitary REPs, and set up the equation

$$\begin{aligned} \langle k'', j | T_l^{(k)} | k', i \rangle &= \langle P_a (k'', j) | P_a T_l^{(k)} P_a^{-1} P_a | k', i \rangle \\ &= \sum_{i'j'l'} D_{j',j}^{(k'')*}(a) D_{l'l}^{(k)}(a) D_{i'i}^{(k')}(a) \langle k'', j' | T_{l'}^{(k)} | k', i' \rangle . \end{aligned}$$

Integration of this equation over $\mathscr{G}$ according to (11.2.13) yields

$$\begin{aligned} V \langle k'', j | T_l^{(k)} | k', i \rangle = &\sum_{i'j'l'} \langle k'', j | T_{l'}^{(k)} | k', i' \rangle \\ &\times \int D_{j'j}^{(k'')*}(a) D_{l'l}^{(k)}(a) D_{i'i}^{(k')}(a) \rho(\ldots \alpha_i \ldots) \, d^m\alpha , \end{aligned}$$

which together with (11.3.4) finally leads to (11.3.9).

The importance of the Wigner-Eckart theorem lies in the fact that it enables us to determine the relative magnitude of matrix elements and consequently to arrive at statements about the intensities in quantum mechanical transitions, scattering processes, reactions, etc. For this purpose, k, k', k'' are kept fixed and the quotients

$$\frac{\langle k'', j | T_l^{(k)} | k', i \rangle}{\langle k'', j' | T_{l'}^{(k)} | k', i' \rangle} = \frac{\left({}^{k''}_{j} \middle| {}^{kk'}_{l\,i} \right)}{\left({}^{k''}_{j'} \middle| {}^{kk'}_{l'i'} \right)} \qquad (11.3.10)$$

are discussed in terms of (11.3.9a). For this, knowledge of the CGC, that is, the symmetry of the problem, is sufficient. Furthermore, we can determine matrix elements from (11.3.10) provided some "simple" ones are known. Finally, the CGC vanish if $D^{(k'')}$ does not occur in $D^{(k)} \otimes D^{(k')}$, i.e. the corresponding matrix elements vanish.

Exercise 11.6. According to Exercise 11.1, for an infinitesimal rotation about the z-axis, $J_\gamma = J_z$, i.e. $a(\gamma) = \exp(-J_z\gamma)$. Investigate the behaviour of (11.3.5) for infinitesimal γ.

11.4 The Cartan-Weyl Basis for Semisimple Lie Algebras

11.4.1 The Lie Group $\mathcal{SU}(n, \mathbb{C})$ and the Lie Algebra A_{n-1}

For the general linear group and some of its relevant subsets (not necessarily subgroups, see Sect. 3.3) we have the following hierarchy:

$$\mathcal{GL}(n,\mathbb{C}) \supset \mathcal{U}(n,\mathbb{C}) \supset \left\{ \begin{matrix} \mathcal{O}(n,\mathbb{R}) \\ \mathcal{SU}(n,\mathbb{C}) \end{matrix} \right\} \supset \mathcal{SO}(n,\mathbb{R}) \tag{11.4.1}$$

and

$$\mathcal{SO}(3,\mathbb{R}) \supset \text{proper point groups} \ .$$

Defining relations and other characteristic data may be seen from Table 11.2 and Appendix D. The groups $\mathcal{SU}(n, \mathbb{C})$, to be used as an example in the following, have $m = n^2 - 1$ real parameters, the space of which is bounded and closed because of the condition $\sum_j |a_{ij}|^2 = 1$; that means (Appendix D) $\mathcal{SU}(n, \mathbb{C})$ is compact, simple and thus also semisimple [see (11.4.2) and (11.1.29, 34) for $n = 2$]. The $n \times n$-matrices $a_{ij} \in \mathcal{SU}(n, \mathbb{C})$ define an irreducible REP of $\mathcal{SU}(n, \mathbb{C})$. From this we may construct a basis for the LA A_{n-1} of $\mathcal{SU}(n, \mathbb{C})$. The commutators of the infinitesimal generators and thus the structure constants are then valid for all the other (irreducible) REPs of $\mathcal{SU}(n, \mathbb{C})$. They also hold for the generators of all the groups that are isomorphic to $\mathcal{SU}(n, \mathbb{C})$ (e.g. operator instead of matrix REPs), if the same parametrization is always used. The $n-1$ generators $\{H_l | l = 1 \dots n-1\}$ and the $n(n-1)$ generators $\{E_{ij} | 1 \leqslant i \neq j \leqslant n\}$ constitute a basis (Cartan-Weyl basis) of the LA A_{n-1} if

$$[H_i, H_l] = 0 \qquad \text{for all } i, l = 1, \dots, n-1 \ ; \tag{11.4.2a}$$

$$[H_l, E_{ij}] = r_{l,ij} E_{ij} \qquad \text{for all } l \text{ and } i \neq j = 1, \dots, n \ ; \tag{11.4.2b}$$

$$[E_{ij}, E_{i'j'}] = \delta_{i'j} E_{ij'} - \delta_{j'i} E_{i'j} \ ; \qquad \text{for } (i,j) \neq (j', i') \ ; \tag{11.4.2c}$$

$$[E_{ij}, E_{ji}] = \text{linear combinations of the } H_l \ . \tag{11.4.2d}$$

To prove this statement it is sufficient to show the validity of these relations for the *defining* REP of the group, which is given by the $\{a_{ij}\}$. Thus with infinitesimal $\bar{a}_{ij}$ we set

$$a_{ij} = \delta_{ij} + \bar{a}_{ij} \quad \text{with} \quad \bar{a}_{ij} = -\bar{a}_{ji}^{*} = -(\bar{a}^{+})_{ij} \quad \text{and} \quad \mathrm{Tr}\{\bar{a}\} = \sum_i \bar{a}_{ii} = 0 \ , \tag{11.4.3}$$

which follows for $|\bar{a}_{ij}| \ll 1$ because $a^{+}a = aa^{+} = 1$ and $\det a = 1$. Thus we may write

$$a = \mathbf{1} + \sum_{i=1}^{m} J_i \alpha_i \ , \qquad m = n^2 - 1 \ , \qquad \alpha_i \text{ infinitesimal} \tag{11.4.4a}$$

where the generators J_i are anti-Hermitian $n \times n$-matrices with zero trace; they span the LA A_{n-1}. In physical applications, for example in the algebra of angular momentum, Hermitian matrices are preferred, i.e., one uses $\hat{J}_j = \mathrm{i}J_j$, $j = 1, \ldots, n^2 - 1$, thus having

$$a = \mathbf{1} - i \sum_{i}^{m} \hat{J}_i \alpha_i \ , \qquad \hat{J}_i = \hat{J}_i^{+} \ , \qquad \mathrm{Tr}\{\hat{J}_i\} = 0 \ . \tag{11.4.4b}$$

The maximal set of those $\hat{J}_i$ which commute must also simultaneously be diagonalizable with real diagonal elements. However, there are only $n - 1$ linearly independent diagonal Hermitian $n \times n$-matrices with zero trace. Consequently the rank of $\mathcal{SU}(n, \mathbb{C})$ is equal to $n - 1$. Apart from the normalizing constants c_l, which may be chosen arbitrarily, see (11.4.30b), a possible choice of the $H_l \in \{\hat{J}_i\}$ is

$$H_1 = c_1 \begin{pmatrix} 1 & & & & \\ & -1 & & & \mathbf{0} \\ & & 0 & & \\ & & & 0 & \\ & \mathbf{0} & & & \ddots \\ & & & & & 0 \end{pmatrix} , \qquad H_2 = c_2 \begin{pmatrix} 1 & & & & \\ & 1 & & & \mathbf{0} \\ & & -2 & & \\ & & & 0 & \\ & \mathbf{0} & & & \ddots \\ & & & & & 0 \end{pmatrix} , \ldots ,$$

$$H_l = c_l \begin{pmatrix} \left.\begin{matrix} 1 & & & \\ & 1 & & \\ & & \ddots & \\ & & & 1 \end{matrix}\right\} l \text{ times} & & & \mathbf{0} \\ & -l & & \\ & & 0 & \\ \mathbf{0} & & & \ddots \\ & & & & 0 \end{pmatrix} , \ldots ,$$

$$H_{n-1} = c_{n-1}\begin{pmatrix} 1 & & & & & \\ & 1 & & & 0 & \\ & & 1 & & & \\ & & & \ddots & & \\ & 0 & & & 1 & \\ & & & & & -n+1 \end{pmatrix}. \tag{11.4.5}$$

We may now define the remaining $n(n-1)$ generators E_{ij} (non-Hermitian) in such a way that E_{ij} has zeros everywhere except for the element in the ith row and jth column, thus

$$(E_{ij})_{mn} = \delta_{im}\delta_{jn} \qquad \text{for } i \neq j \tag{11.4.6a}$$

(*ladder operators*); a Hermitian choice would be $(i < j)$

$$E^R_{ij} = c_{ij}(E_{ij} + E_{ji}) \ , \qquad E^I_{ij} = ic_{ij}(E_{ji} - E_{ij}) \tag{11.4.6b}$$

with real normalizing constants c_{ij}. With this choice for example we get the well-known generators of the LA A_1 of $\mathscr{SU}(2, \mathbb{C})$ with $c_1 = c_{12} = 1/2$:

$$\begin{aligned} H_1 &= \tfrac{1}{2}\begin{pmatrix} 1 & 0 \\ 0 & -1 \end{pmatrix} = \tfrac{1}{2}\sigma_z \ , \\ E_{12} &= \begin{pmatrix} 0 & 1 \\ 0 & 0 \end{pmatrix} = \sigma_+ \ , \qquad E_{21} = \begin{pmatrix} 0 & 0 \\ 1 & 0 \end{pmatrix} = \sigma_- \ , \\ E^R_{12} &= \tfrac{1}{2}\begin{pmatrix} 0 & 1 \\ 1 & 0 \end{pmatrix} = \tfrac{1}{2}\sigma_x = \tfrac{1}{2}(\sigma_+ + \sigma_-) \ , \\ E^I_{12} &= \tfrac{1}{2}\begin{pmatrix} 0 & -i \\ i & 0 \end{pmatrix} = \tfrac{1}{2}\sigma_y = \tfrac{i}{2}(\sigma_- - \sigma_+) \ . \end{aligned} \tag{11.4.7}$$

Similarly, the generators of the LA A_2 of $\mathscr{SU}(3, \mathbb{C})$ choosing $c_1 = c_{ij} = 1/2$, $c_2 = 1/(2\sqrt{3})$ are

$$H_1 = \frac{1}{2}\begin{pmatrix} 1 & 0 & 0 \\ 0 & -1 & 0 \\ 0 & 0 & 0 \end{pmatrix} , \qquad H_2 = \frac{1}{2\sqrt{3}}\begin{pmatrix} 1 & 0 & 0 \\ 0 & 1 & 0 \\ 0 & 0 & -2 \end{pmatrix} ,$$

$$E^R_{12} = \frac{1}{2}\begin{pmatrix} 0 & 1 & 0 \\ 1 & 0 & 0 \\ 0 & 0 & 0 \end{pmatrix} , \qquad E^R_{13} = \frac{1}{2}\begin{pmatrix} 0 & 0 & 1 \\ 0 & 0 & 0 \\ 1 & 0 & 0 \end{pmatrix} , \qquad E^R_{23} = \frac{1}{2}\begin{pmatrix} 0 & 0 & 0 \\ 0 & 0 & 1 \\ 0 & 1 & 0 \end{pmatrix} ,$$

$$E^I_{12} = \frac{1}{2}\begin{pmatrix} 0 & -i & 0 \\ i & 0 & 0 \\ 0 & 0 & 0 \end{pmatrix} , \quad E^I_{13} = \frac{1}{2}\begin{pmatrix} 0 & 0 & -i \\ 0 & 0 & 0 \\ i & 0 & 0 \end{pmatrix} , \quad E^I_{23} = \frac{1}{2}\begin{pmatrix} 0 & 0 & 0 \\ 0 & 0 & -i \\ 0 & i & 0 \end{pmatrix} . \tag{11.4.8}$$

By comparing (11.4.7) and (11.4.8) we recognize that H_1, E_{12}^R and E_{12}^I generate the algebra A_1 of $\mathscr{SU}(2, \mathbb{C})$. Within the Gell-Mann scheme (Sect. 13.1.2) the three components of the isospin are assigned to these three operators, while H_2 is connected with the hypercharge (see also other examples in Chaps. 13, 14).

The matrices (11.4, 5, 6a or 6b) span the LA A_{n-1} of $\mathscr{SU}(n)$. Their commutation relations determine the structure constants (Exercise 11.7). In the following we shall use the basis (11.4.5, 6a) which corresponds to the Cartan-Weyl basis in Sect. 11.4.2. Utilizing this basis we show that the relations (11.4.2) are valid.

(a) $[H_i, H_l] = 0$ for all i, l is satisfied by construction.
(b) Let Λ_l^i be the ith element on the diagonal of H_l, thus $(H_l)_{mn} = \Lambda_l^m \delta_{mn}$; then

$$[H_l, E_{ij}]_{mn} = (\Lambda_l^i - \Lambda_l^j)\delta_{im}\delta_{jn} = (\Lambda_l^i - \Lambda_l^j)(E_{ij})_{mn} \ ,$$

i.e. (11.4.2b) is satisfied, too, if

$$r_{l,ij} = \Lambda_l^i - \Lambda_l^j \ , \quad i \neq j \ . \tag{11.4.9}$$

(c) Because of the definition (11.4.6a), (11.4.2c) is satisfied directly.
(d) $[E_{ij}, E_{ji}] = E_{ii} - E_{jj}$; here E_{ii} is defined analogously to (11.4.6a). However, $E_{ii} - E_{jj}$ is a Hermitian diagonal matrix with zero trace, thus it can be expressed in terms of the H_l and (11.4.2d) is satisfied too.

For $\mathscr{SU}(3)$ we have with $c_l = 1$

$$\begin{aligned} [E_{12}, E_{21}] &= E_{11} - E_{22} = H_1 \ , \\ [E_{13}, E_{31}] &= E_{11} - E_{33} = \tfrac{1}{2}(H_1 + H_2) \ , \\ [E_{23}, E_{32}] &= E_{22} - E_{33} = \tfrac{1}{2}(H_2 - H_1) \ . \end{aligned} \tag{11.4.10}$$

In order to investigate the behaviour of basis functions belonging to IRs of $\mathscr{SU}(n)$ in transformations with the operators H_l and E_{ij}, it is useful to define *weights* (*weight vectors*). With their help, the IRs of $\mathscr{SU}(n)$ may be classified. The weights are generated (very similarly to a finite lattice) by translations of basis vectors, the *roots* (*root vectors*). Let $\{|f_\nu\rangle | \nu = 1, \ldots, d\}$ be the basis functions of an IR of $\mathscr{SU}(n)$, then the operators, H_l and E_{ij} in this basis are represented by d-dimensional matrices $\langle f_\nu | H_l | f_\mu \rangle, \ldots$[4]. According to (11.1.9) these REPs then determine the IRs of the elements of $\mathscr{SU}(n)$ too. The $\{H_l\}$ commute; thus the $\{|f_\nu\rangle\}$ can be chosen such that they are simultaneous eigenfunctions of these $n-1$ operators H_l:

$$H_l|f_\nu\rangle = w_l|f_\nu\rangle \ , \quad l = 1, \ldots, n-1 \ ; \tag{11.4.11}$$

the $n-1$ real numbers (eigenvalues) w_l can be looked upon as components of a

---

[4] We do not distinguish between the operators H_l, E_{ij} and their REPs in some basis because the meaning is always clear.

weight vector $\boldsymbol{w}$ in the $\mathbb{R}_{n-1}$:

$$\boldsymbol{w} = \{w_1, \ldots, w_l, \ldots, w_{n-1}) \ . \tag{11.4.12}$$

Of course, the weights may be degenerate, which happens in general, and there may be several linearly independent basis functions belonging to each weight. Therefore we number the basis functions with $\boldsymbol{w}$ and the index μ of the degeneracy $[\mu = \mu(\boldsymbol{w})] : |f_\nu\rangle \to |\boldsymbol{w}, \mu\rangle$. The set of different weights, including their multiplicity (degeneracy), assigned to the basis functions thus characterizes the IRs of $\mathcal{SU}(n)$.

Analyzing the weight system we employ the ladder operators E_{ij} defined in (11.4.2, 6). From (11.4.2b) and (11.4.11) we have, making use of (11.4.9),

$$\begin{aligned} H_l E_{ij}|\boldsymbol{w}, \mu\rangle &= E_{ij} H_l|\boldsymbol{w}, \mu\rangle + r_{l,ij} E_{ij}|\boldsymbol{w}, \mu\rangle \\ &= (w_l + r_{l,ij}) E_{ij}|\boldsymbol{w}, \mu\rangle \ . \end{aligned} \tag{11.4.13}$$

By this method $n(n-1)$ further vectors, the roots (root vectors)

$$\boldsymbol{r}_{ij} = \{r_{1,ij}, \ldots, r_{l,ij}, \ldots, r_{n-1,ij}\} \ ; \qquad i, j = 1, \ldots, n \ ; \qquad i \neq j \tag{11.4.14}$$

are defined. From (11.4.13) we now find:

(i) If $E_{ij}|\boldsymbol{w}, \mu\rangle \neq 0$, then if $\boldsymbol{w}$ is a weight (eigenvalue) so is $\boldsymbol{w} + \boldsymbol{r}_{ij}$.
(ii) $E_{ij}|\boldsymbol{w}, \mu\rangle$ lies in the subspace of the degenerate functions belonging to the eigenvalue $\boldsymbol{w} + \boldsymbol{r}_{ij}$, i.e.

$$E_{ij}|\boldsymbol{w}, \mu\rangle = \sum_{\mu'} c_{\mu'}|\boldsymbol{w} + \boldsymbol{r}_{ij}, \mu'\rangle \ . \tag{11.4.15}$$

For example, for $\mathcal{SU}(2)$ according to (11.4.7) and (11.4.2b) we have

$$[H_1, \sigma_\pm] = \pm\sigma_\pm \ , \qquad \text{thus } r_{1,\pm} = \pm 1 \ . \tag{11.4.16}$$

The weight only has one component $w_1 = w$, thus according to (11.4.13)

$$H_1 \sigma_\pm|w\rangle = (w \pm 1)\sigma_\pm|w\rangle \ , \quad \sigma_\pm|w\rangle \sim |w \pm 1\rangle \ . \tag{11.4.17}$$

However, this just defines the well-known algebra of angular momentum, in which H_1 corresponds to the z-component $J_z(L_z)$ and $\sigma_\pm$ to the ladder operators $J_\pm(L_\pm)$; $w \to m$ corresponds to the magnetic quantum number, the maximal value of w is the quantum number j of the total angular momentum. The ladder operators thus map a basis function with weight $\boldsymbol{w}$ into a basis function with weight $\boldsymbol{w} + \boldsymbol{r}_{ij}$. Successive application of E_{ij}, that means $(E_{ij})^n|\boldsymbol{w}, \mu\rangle$ with $n = 1, 2, \ldots$, gives a series of functions, which belong to the weights (eigenvalues) $\boldsymbol{w} + n\boldsymbol{r}_{ij}$. However, according to Sect. 11.2, the unitary REPs of $\mathcal{SU}(n, \mathbb{C})$ are finite; upper and lower bounds for the w_l have to exist, which means there must be a maximal $\bar{n}$ such that

$$(E_{ij})^{\bar{n}}|\boldsymbol{w}, \mu\rangle = 0 \ . \tag{11.4.18}$$

By successive application of the E_{ij} to any basis function $|f_\nu\rangle$ of the IR considered, we obtain the functions $|\boldsymbol{w}, \mu\rangle$ belonging to all the weights which are assigned to this IR. Starting with a weight $\boldsymbol{w}_1$ we are able to get all the other weights by translations with the roots $\boldsymbol{r}_{ij}$, thus

$$\boldsymbol{w}_2 - \boldsymbol{w}_1 = \sum_{ij} c_{ij}\boldsymbol{r}_{ij} \; ; \qquad c_{ij} \in \mathbb{Z} \; . \tag{11.4.19}$$

This allows a geometrical interpretation: all the weight systems of the IRs of $\mathscr{SU}(n)$ are obtained by the $n(n-1)$ roots; every weight system of an IR is a finite sublattice of the $(n-1)$-dimensional root lattice (11.4.14), which has weights as lattice points. For $n = 2, 3, 4$ this corresponds to a 1-, 2-, 3-dimensional (crystal) lattice [see (3.4) and Figs. 11.2, 3). The point group of a root (weight) lattice is said to be the *Weyl group* $\mathscr{W}$ (Sect. 11.4.2).

We now consider the fundamental weight system of the defining IR $\{H_l; E_{ij} | l = 1, \dots, n-1;\; i \neq j = 1, \dots, n\}$ of $\mathscr{SU}(n)$ according to (11.4.5, 6). Let $\{|f_i\rangle | i = 1, \dots, n\}$ be a set of basis functions chosen in such a way that $\langle f_i | \dots | f_j\rangle$ is the defining REP; with (11.4.5, 6) then

$$\begin{aligned} H_l|f_j\rangle &= A_l^j|f_j\rangle \qquad \text{or} \qquad \langle f_i|H_l|f_j\rangle = A_l^j\delta_{ij} \; , \\ E_{ij}|f_{j'}\rangle &= \delta_{jj'}|f_i\rangle \qquad \text{or} \qquad \langle f_{i'}|E_{ij}|f_{j'}\rangle = \delta_{ii'}\delta_{jj'} \; . \end{aligned} \tag{11.4.20}$$

By comparison of (11.4.20) with (11.4.11) we recognize that $|f_j\rangle$ belongs to the weight

$$\boldsymbol{w}^{(j)} := (A_1^j, \dots, A_l^j, \dots, A_{n-1}^j) \; , \qquad j = 1, \dots, n \; . \tag{11.4.21}$$

These n weights are said to be the *fundamental weights.* The $n(n-1)$ roots assigned to them are given according to (11.4.9) by

$$\boldsymbol{r}_{ij} = \boldsymbol{w}^{(i)} - \boldsymbol{w}^{(j)} := (A_1^i - A_1^j, \dots, A_{n-1}^i - A_{n-1}^j) \; , \qquad i \neq j \; . \tag{11.4.22}$$

The fundamental weights can be specified explicitly using (11.4.5)

$$\begin{aligned} &\boldsymbol{w}^{(1)} = (c_1, c_2, \dots, c_{n-1}) \; ; \qquad \boldsymbol{w}^{(2)} = (-c_1, c_2, c_3, \dots, c_{n-1}) \; ; \\ &\boldsymbol{w}^{(3)} = (0, -2c_2, c_3, \dots, c_{n-1}) \; ; \qquad \boldsymbol{w}^{(4)} = (0, 0, -3c_3, c_4, \dots, c_{n-1}) \; ; \dots \; ; \\ &\boldsymbol{w}^{(j)} = (\underbrace{0, \dots, 0}_{(j-2)\,\text{times}}, -(j-1)c_{j-1}, c_j, \dots, c_{n-1}) \; ; \dots \; ; \\ &\boldsymbol{w}^{(n)} = (0, \dots, 0, -(n-1)c_{n-1}) \; . \end{aligned} \tag{11.4.23}$$

Since $\mathrm{Tr}\{H_l\} = 0$, we obviously have $\sum_{j=1}^n \boldsymbol{w}^{(j)} = 0$. For $\mathscr{SU}(2)$, with $c_1 = 1/2$, see (11.4.7),

$$\boldsymbol{w}^{(1)} = \tfrac{1}{2} \; , \qquad \boldsymbol{w}^{(2)} = -\tfrac{1}{2} \; , \qquad \boldsymbol{r}_{12} = -\boldsymbol{r}_{21} = 1 \; . \tag{11.4.24}$$

For $\mathscr{SU}(3)$ it follows from (11.4.8), taking $c_1 = 1/2$, $c_2 = 1/(2\sqrt{3})$, that

$$\boldsymbol{w}^{(1)} = \frac{1}{2}\left(1, \frac{1}{\sqrt{3}}\right), \quad \boldsymbol{w}^{(2)} = \frac{1}{2}\left(-1, \frac{1}{\sqrt{3}}\right), \quad \boldsymbol{w}^{(3)} = \frac{1}{2}\left(0, -\frac{2}{\sqrt{3}}\right)$$

and consequently the six roots are

$$\begin{aligned} \boldsymbol{r}_{12} &= -\boldsymbol{r}_{21} = (1,0) , \\ \boldsymbol{r}_{13} &= -\boldsymbol{r}_{31} = \tfrac{1}{2}(1,\sqrt{3}) , \\ \boldsymbol{r}_{23} &= -\boldsymbol{r}_{32} = \tfrac{1}{2}(-1,\sqrt{3}) . \end{aligned} \tag{11.4.25}$$

For $\mathscr{SU}(4)$ we employ a normalization analogous to (11.4.8): $c_1 = 1/2$, $c_2 = 1/(2\sqrt{3})$, $c_3 = 1/(2\sqrt{6})$, and thus obtain the weights of the defining IR

$$\begin{aligned} \boldsymbol{w}^{(1)} &= \frac{1}{2}\left(1, \frac{1}{\sqrt{3}}, \frac{1}{\sqrt{6}}\right), \quad \boldsymbol{w}^{(2)} = \frac{1}{2}\left(-1, \frac{1}{\sqrt{3}}, \frac{1}{\sqrt{6}}\right), \\ \boldsymbol{w}^{(3)} &= \frac{1}{2}\left(0, -\frac{2}{\sqrt{3}}, \frac{1}{\sqrt{6}}\right), \quad \boldsymbol{w}^{(4)} = \frac{1}{2}\left(0, 0, -\frac{3}{\sqrt{6}}\right) . \end{aligned} \tag{11.4.26}$$

Fundamental weights and roots are illustrated in Figs. 11.2, 3. The Weyl groups are $\mathscr{C}_s$, $\mathscr{C}_{3v}$ and $\mathscr{T}_d$. Not all of the roots are linearly independent. For the definition of a root lattice, $r = n - 1$ basis roots are sufficient. These have to be chosen in an appropriate way. Possible choices would be

$$\Delta = \{\boldsymbol{r}_{i,i+1} | i = 1, \ldots, r = n-1\} \qquad \text{or} \tag{11.4.27a}$$

$$\Delta = \{\boldsymbol{r}_{i+1,i} \wedge \boldsymbol{r}_{1,n} | i = 2, \ldots, r = n - 1 \geqslant 2\} , \tag{11.4.27b}$$

i.e. the vectors $\{\boldsymbol{r}_{12}, \boldsymbol{r}_{23}\}$ and $\{\boldsymbol{r}_{12}, \boldsymbol{r}_{23}, \boldsymbol{r}_{34}\}$ or $\{\boldsymbol{r}_{32}, \boldsymbol{r}_{13}\}$ and $\{\boldsymbol{r}_{32}, \boldsymbol{r}_{43}, \boldsymbol{r}_{14}\}$ in Fig. 11.3. In Fig. 11.5, (11.4.27b) has been used, but sometimes (11.4.27a) is more convenient. In any case, the roots of the basis have to satisfy the following:

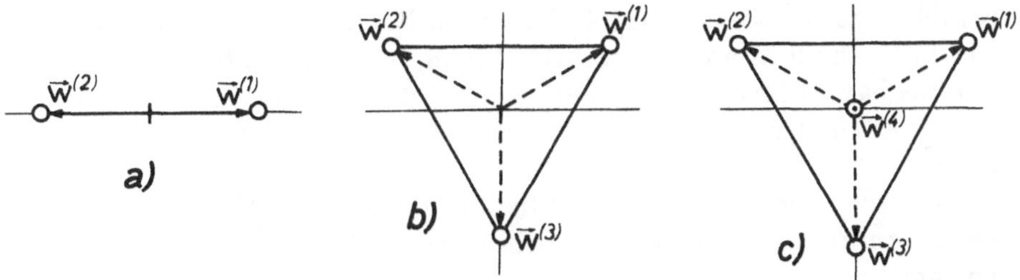

Fig. 11.2a–c. Defining IRs and multiplets of the groups $\mathscr{SU}(n)$, $n = 2, 3, 4$. The REPs are n-dimensional and the multiplets correspondingly doublets for $\mathscr{SU}(2)$, triplets for $\mathscr{SU}(3)$, quartets for $\mathscr{SU}(4)$; $\boldsymbol{w}^{(j)}$ are the fundamental weights

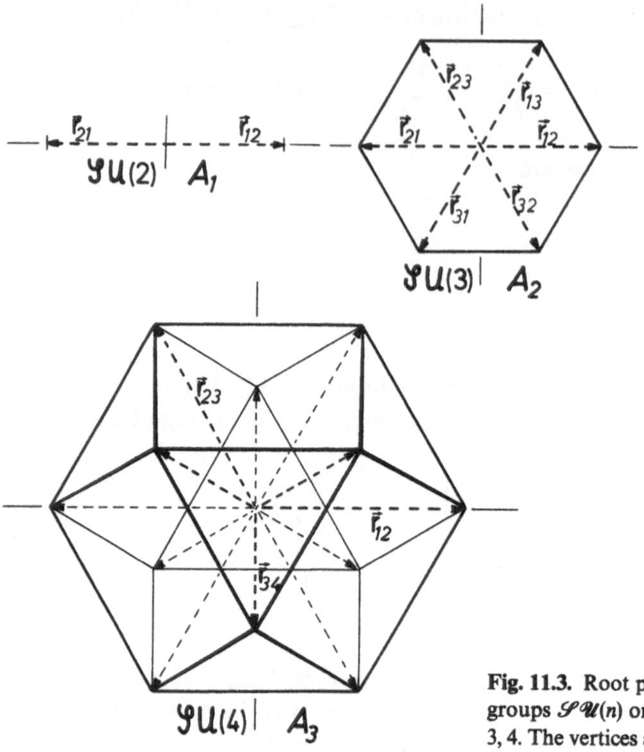

Fig. 11.3. Root polygons or polyhedra for the groups $\mathcal{SU}(n)$ or the algebras A_{n-1} for $n = 2, 3, 4$. The vertices always determine the roots r_{ij}

(i) The basis $\Delta = \{r_b | b = 1, \ldots, r = n-1\}$ spans the (Euclidean) space $\mathbb{R}_{n-1}$ of the roots r_{ij}.

(ii) Every root $r \in \{r_{ij}\}$ can be represented as

$$r = \sum_{b=1}^{r} h_b r_b \ , \qquad r_b \in \Delta \ , \qquad h_b \in \mathbb{Z} \tag{11.4.27c}$$

where *all* the h_b are either non-negative or non-positive.

In this case the roots of the basis are said to be simple and $\sum_{b=1}^{r} h_b = +1$. A root r is said to be *positive* if in (11.4.27c) all the h_b are non-negative, *negative*, if in (11.4.27c) all the h_b are non-positive. Taking (11.4.27) as a basis, all the positive roots have a first nonvanishing component that is positive. In general a root is positive if in any arbitrary basis the first nonvanishing component is positive.

If $m = n^2 - 1$ is the dimension of the Lie group $\mathcal{SU}(n)$ and $r = n - 1$ is its rank, there are $(m - r)/2 = n(n - 1)/2$ positive roots, apart from the r basis roots. There are

$$p = \frac{1}{2}(m - 3r) \rightarrow \frac{(n-1)(n-2)}{2} \tag{11.4.28}$$

further positive roots. From this it follows that $m \geqslant 3r$ always. This means, however, that for a characterization of a weight system (multiplet) apart from the r H_l operators, a further p operators may be needed.

The ladder operators defined in (11.4.6a) obviously satisfy

$$E_{ij}^{+} = E_{ji} , \tag{11.4.29}$$

i.e. the $m - r \to n(n-1)$ ladder-operators are assigned to each other pairwise just like the roots $\boldsymbol{r}_{ij} = -\boldsymbol{r}_{ji}$. The metric tensor (11.4.28) then has the structure [the proof may be given using (11.4.2)]

$$g_{kk'} = \begin{bmatrix} [\bar{g}_{ll'}] & & & & & \\ & 0 & 1 & & & \\ & 1 & 0 & & & \\ & & & 0 & 1 & \\ & & & 1 & 0 & \\ & & & & & \ddots \end{bmatrix} ; \quad \begin{aligned} &\bar{g}_{ll'} = \sum_{i,j=1}^{n} r_{l,ij} r_{l',ij} , \\ &l, l' = 1, \dots, r = n-1 , \\ &k, k' = 1, \dots, m = n^2 - 1 , \end{aligned} \tag{11.4.30a}$$

where $\bar{g}_{ll'}$ can be diagonalized. Using (11.4.5), $\bar{g}_{ll'}$ is already diagonal for the groups $\mathscr{SU}(n, \mathbb{C})$. With a normalizing constant $\bar{c}_l$ we have

$$\bar{g}_{ll'} = 2nl(l+1)\bar{c}_l^2 \delta_{ll'} \to \delta_{ll'} \qquad \text{with} \qquad \bar{c}_l^2 = \frac{1}{2nl(l+1)} , \tag{11.4.30b}$$

thus $\bar{c}_l^2 = c_l^2/n$, where the c_l's correspond to the normalization chosen in (11.4.8, 24–26). Both the normalizations are used (see also Sects. 11.4.2 and 14.5.4).

Finally we remark that all the algebras in which the generators $\{J_k\} \to \{H_l, E_{ij}\}$ are real or can be chosen to be real are said to be *real algebras*. All the relevant LAs are real in this sense. A real algebra has a *complex extension*:

$$\{J_k\} \to \{J_k + \mathrm{i}Z_k\} , \tag{11.4.31}$$

where the same structure constants for the $J_k + \mathrm{i}Z_k$ are assumed to be valid as for the J_k. Several real LAs may be assigned to such a *complex* LA or, in other words, different LAs may have the same complex extension.

11.4.2 The Cartan-Weyl Basis

In the preceding section we discussed the groups $\mathscr{SU}(n, \mathbb{C})$ using a simple basis, the Cartan-Weyl basis, for the generators. Such bases with the most simple (canonical) commutation relations according to Cartan and Weyl can also be found for other semisimple and simple LAs $\mathscr{L}$. Their construction will be shown in the following. For this we define an eigenoperator $Y \in \mathscr{L}$ of an operator $A \in \mathscr{L}$

by[5]

$$[A, Y] = \lambda Y \; ; \qquad \lambda : \text{eigenvalue} \; . \tag{11.4.32a}$$

Cartan has used these eigenoperators of an operator A, yet to be given, as a new standard basis in $\mathscr{L}$. According to (11.1.15a) $\mathscr{L}$ is spanned by the basis $\{X_i | i = 1, \ldots, m\}$ with

$$[X_i, X_j] = \sum_{l=1}^{m} c_{ij}^l X_l \; . \tag{11.4.33}$$

The operators A and Y are elements of $\mathscr{L}$ and thus can be represented as[6]

$$A = \sum_{i=1}^{m} a^i X_i \; , \qquad Y = \sum_{j=1}^{m} y^j X_j \; , \tag{11.4.34}$$

so that we have from (11.4.32)

$$\sum_{jl} \left[\left(\sum_i a^i c_{ij}^l - \lambda \delta_{jl} \right) y^j X_l \right] = 0 \tag{11.4.35}$$

and finally, because of the linear independence of the X_l,

$$\sum_j \left[\left(\sum_i a^i c_{ij}^l - \lambda \delta_{lj} \right) y^j \right] = 0 \qquad \text{or} \qquad \det\left(\sum_i a^i c_{ij}^l - \lambda \delta_{lj} \right) = 0 \; . \tag{11.4.36}$$

The eigenvalues depend on the choice of A (or the a^i) and the basis $\{X_i\}$ (or the c_{ij}^l). It has been shown by Cartan that:

(i) If A is chosen such as to maximize the number of different *roots* of (11.4.36), then in semisimple and simple LAs only one of the roots is degenerate; this belongs to the eigenvalue $\lambda = 0$ and its multiplicity (degree of degeneracy) is equal to the rank r of the LA.

(ii) The r linearly independent, mutually degenerate eigenoperators of A, belonging to $\lambda = 0$, $\{H_l | l = 1, \ldots, r\}$ according to (11.4.32a) satisfy

$$[A, H_l] = 0 \; , \qquad l = 1, \ldots, r \tag{11.4.32b}$$

and mutually commute

$$[H_i, H_l] = 0 \; , \qquad i, l = 1, \ldots, r \; . \tag{11.4.37a}$$

Altogether $\mathscr{L}$ is spanned by the r operators H_l and the $m - r$ other eigen-

---

[5] $[A, Y]$ is a mapping $F_A : \mathscr{L} \to \mathscr{L}$ with $Y \to [A, Y] \in \mathscr{L}$, i.e. $F_A Y := [A, Y]$. Thus $F_A Y = \lambda Y$ is an eigenvalue problem of the mapping F_A or the operator A with the composition $[\, ,]$.

[6] The difference between co- and contravariant components can be ignored in this section.

operators of A that belong to nonzero nondegenerate roots α_j of (11.4.36): $\{E_{\alpha_j}|j = 1,\dots,m-r\}$. According to (11.4.32a) the latter satisfy

$$[A, E_{\alpha_j}] = \alpha_j E_{\alpha_j} , \qquad j = 1,\dots,m-r . \tag{11.4.32c}$$

The operators $\{H_l, E_{\alpha_j}|l = 1,\dots,r; j = 1,\dots,m-r\}$ constitute the Cartan-Weyl standard basis (CWB) and replace the basis $\{X_i\}$.

According to Exercise 11.1, for the algebra $B_1 \cong A_1$ we have with $J_j \to X_j$ for example $[X_i, X_j] = \sum_{k=1}^{3} \varepsilon_{ijk} X_k$. Using (11.4.34) we obtain the system of equations

$$\begin{bmatrix} -\lambda & -a_3 & a_2 \\ a_3 & -\lambda & -a_1 \\ -a_2 & a_1 & -\lambda \end{bmatrix} \cdot \begin{bmatrix} y_1 \\ y_2 \\ y_3 \end{bmatrix} = 0$$

with the roots $\lambda_1 = 0$ and $\lambda_{2,3} = \pm i(a_1^2 + a_2^2 + a_3^2)^{1/2}$. Choosing $a_1 = a_2 = 0$ and $a_3 = i\alpha$, $A = i\alpha X_3$ then $\lambda_1 = 0$, $\lambda_{2,3} = \pm\alpha$; the CWB is then

$$\lambda = 0 : H_1 = y_3 X_3 ; \qquad \lambda = \pm\alpha : \qquad E_{\pm\alpha} = y_1(X_1 \pm iX_2) .$$

From $[A, A] = 0$ and (11.4.32a, b) we can conclude that

$$\mathrm{A} = \sum_{l=1}^{r} c_l H_l ; \qquad c_l \in \mathbb{C} . \tag{11.4.38}$$

Therefore we have from the Jacobi identity (11.1.15b) with $J_i = A$, $J_j = H_l$, $J_k = E_\alpha$ and (11.4.32b, c, 38)

$$[A, [H_l, E_\alpha]] = \alpha[H_l, E_\alpha] , \tag{11.4.39}$$

that means, according to (11.4.32a) $[H_l, E_\alpha]$ is an eigenoperator belonging to the eigenvalue α, and according to (11.4.32c) thus has to be proportional to E_α, since $\alpha \neq 0$ is nondegenerate. Therefore

$$[H_l, E_\alpha] := r_{l\alpha} E_\alpha ; \qquad c_{l\alpha}^{\beta} = r_{l\alpha}\delta_{\alpha\beta} ; \qquad l = 1,\dots,r . \tag{11.4.37b}$$

The vectors $\boldsymbol{r}_\alpha := \{r_{1\alpha}, r_{2\alpha}, \dots, r_{r\alpha}\}$ are the *root vectors* according to (11.4.14). With them, the eigenvalues of A can be represented by

$$\alpha = \sum_l c_l r_{l\alpha} \tag{11.4.40}$$

since according to (11.4.32c, 37b, 38)

$$\alpha E_\alpha = \sum_l c_l [H_l, E_\alpha] = \sum_l c_l r_{l\alpha} E_\alpha .$$

Using $J_i = A$, $J_j = E_\alpha$, $J_k = E_\beta$ we also obtain from the Jacobi identity with (11.4.32c)

$$[A,[E_\alpha,E_\beta]] = (\alpha+\beta)[E_\alpha,E_\beta]\ , \qquad \text{thus if} \tag{11.4.41}$$

(i) $\alpha+\beta$ is not an eigenvalue, $[E_\alpha, E_\beta] = 0$;
(ii) $\alpha+\beta \neq 0$ and is an eigenvalue, i.e. $\alpha+\beta$ is nondegenerate,

$$[E_\alpha,E_\beta] = c_{\alpha\beta}^{\alpha+\beta}E_{\alpha+\beta}\ ; \tag{11.4.37c}$$

(iii) $\alpha+\beta=0$, i.e. $\beta=-\alpha$, it follows that $[A,[E_\alpha,E_{-\alpha}]]=0$ and according to (11.4.32b)

$$[E_\alpha,E_{-\alpha}] = \sum_l c_{\alpha,-\alpha}^l H_l\ . \tag{11.4.37d}$$

The relations (11.4.37a–d) correspond to (11.4.2a–d) for $\mathscr{SU}(n,\mathbb{C})$. Several rules can be deduced from these relations restricting the possible root systems. For this purpose there exist algebraical (Cartan) and geometrical (van der Waerden, Weyl) methods. Both use a *metric of the* LA already defined in (11.1.28, 36; 11.4.30a). Here we define it in terms of the scalar product of two elements A, $B\in\mathscr{L}$, decomposing them into contravariant components according to (11.4.34). If F_A, F_B: $\mathscr{L}\to\mathscr{L}$ are two mappings according to footnote 5, we have for the mapping $F=F_AF_B$ applied to a basis $\{X_i\}$

$$\begin{aligned}F_AF_BX_i &= F_A[B,X_i] = F_A\sum_j b^j[X_j,X_i] = F_A\sum_{jk} b^jc_{ji}^kX_k = \sum_{jk} b^jc_{ji}^k[A,X_k]\\ &= \sum_{ljk} a^lb^jc_{ji}^k[X_l,X_k] = \sum_{ljks} a^lb^jc_{ji}^kc_{lk}^sX_s\ .\end{aligned}$$

Thus F has the matrix REP

$$F\to F_{si} = \sum_{jkl} a^lb^jc_{ji}^kc_{lk}^s\ .$$

Now we define the trace of $F=F_AF_B$ as the scalar product of the operators A and B

$$\langle A|B\rangle := \mathrm{Tr}\{F\} = \sum_i F_{ii} = \sum_{il} a^lb^jg_{jl} \tag{11.4.42}$$

with the metric from (11.1.28). Using (11.4.37a–d) the metric has the form (11.4.30a) with

$$\bar{g}_{ll'} = \sum_{\alpha=1}^{m-r} r_{l\alpha}r_{l'\alpha}\ ; \tag{11.4.43}$$

the E_α and thus the $c_{\alpha\beta}^{\alpha+\beta}$ in (11.4.37c) have to be normalized in an appropriate way in order to have "1" as the remaining elements g_{jl} that are different from zero (Exercise 11.9). Since for a semisimple LA $\det g\neq 0$, if α is an eigenvalue of (11.4.36) then so is $-\alpha$. The commutation relation (11.4.37d) can finally be brought into the form

$$[E_\alpha, E_{-\alpha}] = \sum_l r^l{}_\alpha H_l \quad \text{with} \tag{11.4.44}$$

$$r^l{}_\alpha = \sum_{l'} \bar{g}^{ll'} r_{l'\alpha} \quad \text{and} \quad \sum_{l'} \bar{g}^{ll'} \bar{g}_{l'k} = \delta_{kl} \; .$$

With the r-dimensional metric $\bar{g}^{ll'}$ a scalar product in the space of the root vectors may be defined by

$$(\boldsymbol{r}_\alpha \cdot \boldsymbol{r}_\beta) = \sum_{ll'} \bar{g}^{ll'} r_{l\alpha} r_{l'\beta} = \sum_l r_{l\alpha} r^l{}_\beta \; . \tag{11.4.45}$$

Here $r_{l\alpha}$ are the covariant, $r^l{}_\alpha$ the contravariant components of the root vector. For $\mathscr{SU}(n, \mathbb{C})$ according to (11.4.30b) $g_{ll'}$ may be chosen to be the unit matrix; the metric is Euclidean. The angle $\phi_{\alpha\beta}$ between two root vectors follows from

$$\cos\phi_{\alpha\beta} = \frac{(\boldsymbol{r}_\alpha \cdot \boldsymbol{r}_\beta)}{[(\boldsymbol{r}_\alpha \cdot \boldsymbol{r}_\alpha)(\boldsymbol{r}_\beta \cdot \boldsymbol{r}_\beta)]^{1/2}} \; . \tag{11.4.46}$$

In the following we shall state the most important relations for the roots and weights of semisimple LAs. The proofs (mostly of geometrical nature) can easily be found (see [11.1]).

(1) If $\boldsymbol{r}_\alpha$ is a root, then $\boldsymbol{r}_{-\alpha}$ is a root, too [see (11.4.37d, 40) and above].

(2) If $\boldsymbol{r}_\alpha$ and $\boldsymbol{r}_\beta$ are roots, then

$$\frac{2(\boldsymbol{r}_\alpha \cdot \boldsymbol{r}_\beta)}{(\boldsymbol{r}_\alpha \cdot \boldsymbol{r}_\alpha)} \in \mathbb{Z} \; . \tag{11.4.47}$$

(3) If $\boldsymbol{r}_\alpha$ and $\boldsymbol{r}_\beta$ are roots, then

$$\boldsymbol{r} = \boldsymbol{r}_\beta - 2\frac{(\boldsymbol{r}_\alpha \cdot \boldsymbol{r}_\beta)}{(\boldsymbol{r}_\alpha \cdot \boldsymbol{r}_\alpha)} \boldsymbol{r}_\alpha \tag{11.4.48}$$

is a root, too; geometrically we obtain $\boldsymbol{r}$ by reflection of $\boldsymbol{r}_\beta$ in a plane perpendicular to $\boldsymbol{r}_\alpha$ (Fig. 11.4). By means of these relations the root systems of all the semisimple LAs can be evaluated geometrically. From (11.4.46, 47) it follows that

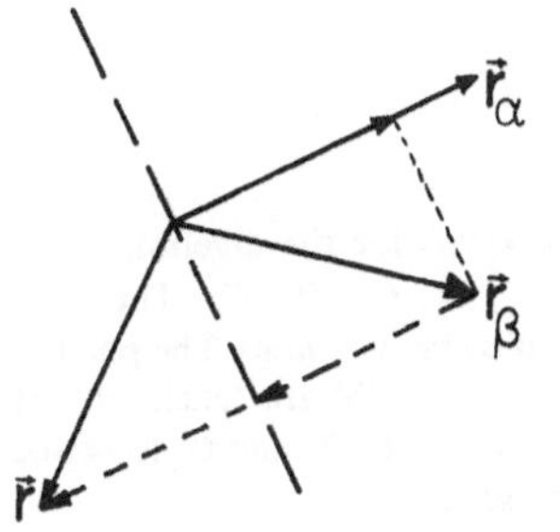

Fig. 11.4. Generation of a vector $\boldsymbol{r}$ by reflection of $\boldsymbol{r}_\beta$ in a plane perpendicular to $\boldsymbol{r}_\alpha$

$$4\cos^2\phi_{\alpha\beta} = 2\frac{(\boldsymbol{r}_\alpha\cdot\boldsymbol{r}_\beta)}{(\boldsymbol{r}_\alpha\cdot\boldsymbol{r}_\alpha)}\,2\frac{(\boldsymbol{r}_\alpha\cdot\boldsymbol{r}_\beta)}{(\boldsymbol{r}_\beta\cdot\boldsymbol{r}_\beta)} = st = 0, 1, 2, 3, 4\ ,$$

thus

$$\phi_{\alpha\beta} = 90°, 60°, 45°, 30°, 0° \tag{11.4.49a}$$

and for the relative lengths $l^2_{\alpha\beta} = (\boldsymbol{r}_\alpha\cdot\boldsymbol{r}_\alpha)/(\boldsymbol{r}_\beta\cdot\boldsymbol{r}_\beta) = t/s$ of the roots, assuming $|\boldsymbol{r}_\alpha| \geqslant |\boldsymbol{r}_\beta|$, we have

$$l_{\alpha\beta} = \text{undetermined}, 1, \sqrt{2}, \sqrt{3}, 1\ . \tag{11.4.49b}$$

Then the only possible roots for the algebra of rank $r = 1$ are $\pm\boldsymbol{r}_\alpha$, thus $\phi_{\alpha\alpha} = 0°$ or 180°. The algebras are A_1 of $\mathcal{SU}(2)$ and B_1 of $\mathcal{SO}(3)$, C_1 of $\mathcal{S}_p(2)$, the last two being isomorphic to A_1. For the algebras of rank $r = 2$ we obtain the root systems given in Fig. 11.5, where $D_2 \cong A_1 \times A_1$ (all the semisimple "90° algebras" are direct product of simple algebras of lower rank).

(4) As for the algebras A_{n-1}, the root systems can be defined by basis roots (simple roots). For the bases Δ the definitions and statements given in Sect. 11.4.1 remain valid. Additionally, we have if $\boldsymbol{r}_\alpha, \boldsymbol{r}_\beta \in \Delta$:

$$\boldsymbol{r}_\alpha - \boldsymbol{r}_\beta \notin \Delta \tag{11.4.50a}$$

$$\frac{2(\boldsymbol{r}_\alpha\cdot\boldsymbol{r}_\beta)}{(\boldsymbol{r}_\alpha\cdot\boldsymbol{r}_\alpha)} = -n\ , \qquad n \in \mathbb{N}\ . \tag{11.4.50b}$$

The possible angles $\theta_{\alpha\beta} \neq \phi_{\alpha\beta}$ between them are

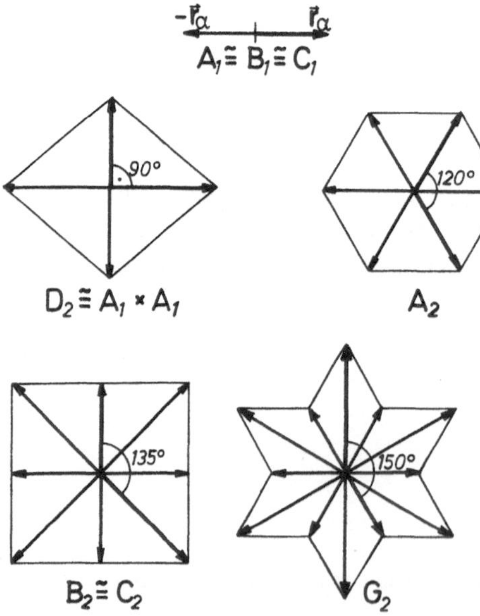

Fig. 11.5. Root systems for the algebras $A_1 \cong B_1 \cong C_1(a)$ and A_2, B_2, C_2, D_2, G_2. The basis roots are indicated by heavier lines. The positive roots according to (11.4.27b) are "enclosed" by the positive basis roots. For B_2 and G_2 long and short roots can be seen

$$\theta_{\alpha\beta} = 90°, 120°, 135°, 150° , \tag{11.4.50c}$$

and the possible relative lengths are

$$L_{\alpha\beta} = \text{undetermined}, 1, \sqrt{2}, \sqrt{3} , \tag{11.4.50d}$$

(see Fig. 11.5 and [11.1]).

As for the groups $\mathscr{SU}(n)$, using (11.4.19, 37a, b) we can show that every IR of a semisimple LA may be described by a weight system in which the weights are connected by the roots (11.4.19). The equation analogous to (11.4.11, 13) is

$$H_l E_\alpha |w, \mu\rangle = E_\alpha H_l |w, \mu\rangle + r_{l\alpha} E_\alpha |w, \mu\rangle = (w_l + r_{l\alpha})|w, \mu\rangle , \tag{11.4.51}$$

i.e. if $E_\alpha |w, \mu\rangle \neq 0$, then both $\boldsymbol{w} = \{w_l\}$ and $\boldsymbol{w} + \boldsymbol{r}_\alpha = \{w_l + r_{l\alpha}\}$ are weights. For the weights we have similar relations to those for the roots.

(5) If $\boldsymbol{w}$ is a weight and $\boldsymbol{r}_\alpha$ a root, then

$$\frac{2(\boldsymbol{w} \cdot \boldsymbol{r}_\alpha)}{(\boldsymbol{r}_\alpha \cdot \boldsymbol{r}_\alpha)} \in \mathbb{Z} \qquad \text{and}$$

$$\boldsymbol{w}' = \boldsymbol{w} - 2\frac{(\boldsymbol{w} \cdot \boldsymbol{r}_\alpha)}{(\boldsymbol{r}_\alpha \cdot \boldsymbol{r}_\alpha)} \boldsymbol{r}_\alpha \tag{11.4.52}$$

again is a weight with the same degeneracy (multiplicity) as $\boldsymbol{w}$ (see the construction in Fig. 11.4). The latter equation can be looked upon as being a mapping generated by the roots in the weight space: σ_α: $\boldsymbol{w} \to \boldsymbol{w}'$. According to Weyl, the possible σ_α's constitute a finite group (of reflections), the *Weyl group* $\mathscr{W}$, which leaves the weight system invariant. The generators of $\mathscr{W}$ are just the r simple (basis) roots. Because of (11.4.48) the roots have the same Weyl group as the weights. For A_2, for example Figs. 11.2b, 5

$$\mathscr{W} = \{e, \sigma_\alpha, \sigma_\beta, \sigma_\alpha\sigma_\beta, \sigma_\beta\sigma_\alpha, \sigma_\alpha\sigma_\beta\sigma_\alpha\} \cong \mathscr{C}_{3v} .$$

(6) Positive and negative weights are distinguished in the same way as the roots (Sect. 11.4.1). A weight is said to be larger (>) than another one if the difference between them is positive. If a weight $\boldsymbol{W} > \boldsymbol{w}$ for *all* the $\boldsymbol{w}$ of an IR, then it is said to be the *maximal weight*. In Fig. 11.2 the weights $\boldsymbol{w}^{(1)}$ are maximal.

(7) The maximal weight of an IR is nondegenerate.

(8) Two IRs are equivalent if their maximal weights coincide.

(9) If $\{\boldsymbol{r}_b\} \in \Delta$ are all of the basis roots then for $\boldsymbol{W}$ to be the maximal weight it is necessary and sufficient that it satisfies

$$2(\boldsymbol{W} \cdot \boldsymbol{r}_b)/(\boldsymbol{r}_b \cdot \boldsymbol{r}_b) = n_b , \qquad n_b \in \mathbb{N}_0 . \tag{11.4.53}$$

The term n_b is called the *root length* of $\boldsymbol{W}$ associated with the root $\boldsymbol{r}_b$. Furthermore,

$$(E_{-b})^k|W,\mu\rangle \begin{cases} \neq 0, & \text{if } k \leqslant n_b \\ =0, & \text{if } k > n_b \ . \end{cases} \tag{11.4.54}$$

From this, together with (7), (8), it follows that the finite IRs of the compact semisimple Lie groups can be characterized by their maximal weights.

(10) If $D^{(W)}$ is such a unitary IR, then its dimension is given by

$$d_W = \prod_{r_\alpha \in +} \frac{[(W+s)\cdot r_\alpha]}{(s \cdot r_\alpha)} \quad \text{with} \tag{11.4.55a}$$

$$s = \frac{1}{2} \sum_{r_\alpha \in +} r_\alpha \tag{11.4.55b}$$

(theorem of Weyl), where $r_\alpha \in +$ denotes the set of positive roots, s is half their vector sum.

(11) In an IR the $\{H_l\}$ are diagonal; the diagonal elements are the components of the weights (11.4.11, 20):

$$H_l|w,\mu\rangle = w_l|w,\mu\rangle \ . \tag{11.4.56}$$

As an example we again consider $\mathscr{SU}(n,\mathbb{C})$ with $n = 2, 3, 4$.

$\mathscr{SU}(2)$. From (11.4.30a) we choose an Euclidean metric: $\bar{g}_{ll'} = \delta_{ll'}$; then $c_1 = 1/(2\sqrt{2})$, $r_{12} = -r_{21} = 1/\sqrt{2}$. According to (11.4.53) we then have $W = n_b/(2\sqrt{2})$, $n_b \in \mathbb{N}_0$. With $n_b = 2j$ we can also write

$$W = j/\sqrt{2} \ , \qquad j = 0, 1/2, 1, 3/2, \ldots \ . \tag{11.4.57}$$

Furthermore, $s = 1/(2\sqrt{2})$, thus using (11.4.55) $d_W := d_j = 2j+1$.

$\mathscr{SU}(3)$. We choose $c_1 = 1/(2\sqrt{3})$, $c_2 = 1/6$. Then $\bar{g}_{ll'} = \delta_{ll'}$. According to (11.4.25, 27a) the basis roots are $r_{12} = (1/\sqrt{3}, 0)$; $r_{23} = \frac{1}{2}(-1/\sqrt{3}, 1)$; $r_{13} = \frac{1}{2}(1/\sqrt{3}, 1)$. Thus $s = \frac{1}{2}(1/\sqrt{3}, 1) = r_{13}$. From (11.4.53) it follows that

$$W = \left\{\frac{1}{2\sqrt{3}} n_1, \frac{1}{6}(n_1 + 2n_2)\right\} \ , \qquad n_1, n_2 \in \mathbb{N}_0 \ . \tag{11.4.58}$$

Furthermore, according to (11.4.55) we obtain

$$d_W := d_{n_1 n_2} = \tfrac{1}{2}(1+n_1)(1+n_2)(2+n_1+n_2) \ .$$

$\mathscr{SU}(4)$. $c_1 = 1/4$, $c_2 = 1/(4\sqrt{3})$, $c_3 = 1/(4\sqrt{6})$.

$$r_{12} = \tfrac{1}{4}(2,0,0) \ , \qquad r_{13} = \tfrac{1}{4}(1,\sqrt{3},0) \ ,$$

$$r_{23} = \tfrac{1}{4}(-1,\sqrt{3},0) \ , \qquad r_{14} = \tfrac{1}{4}(1, 1/\sqrt{3}, 4/\sqrt{6}) \ ,$$

$$r_{34} = \tfrac{1}{4}(0, -2/\sqrt{3}, 4/\sqrt{6}) \ , \qquad r_{24} = \tfrac{1}{4}(-1, 1/\sqrt{3}, 4/\sqrt{6}) \ ,$$

$$s = \tfrac{1}{4}(1, \sqrt{3}, \sqrt{6}) \ ,$$

$$W = \frac{1}{4}\left[n_1, \frac{1}{\sqrt{3}}(n_1 + 2n_2), \frac{1}{\sqrt{6}}(n_1 + 2n_2 + 3n_3)\right] , \qquad (11.4.59)$$

$$d_W = d_{n_1n_2n_3} = \tfrac{1}{12}(1 + n_1)(1 + n_2)(1 + n_3)(2 + n_1 + n_2)(2 + n_2 + n_3)$$
$$\times (3 + n_1 + n_2 + n_3) .$$

(12) For the determination of the degree of degeneracy (multiplicity) m_W of the single weights there exists a recursion formula. Starting with the maximal weight, which is always simple [see (7)], we are able to calculate all the multiplicities by using the Weyl symmetry. With the notation of (11.4.55) the *theorem of Freudenthal* reads:

$$m_W(w) = \frac{2}{[(W + s)^2 - (w + s)^2]} \sum_{h=1}^{\infty} \sum_{r_\alpha \in +} [(w + hr_\alpha)\cdot r_\alpha]m_W(w + hr_\alpha) . \qquad (11.4.60)$$

In Fig. 11.6 are represented, the lowest weight systems (multiplets) of $\mathcal{SU}(n)$, $n = 2, 3, 4$, together with their multiplicities. Specifically for $\mathcal{SU}(3)$, from (11.4.60) we have the following rule for the multiplicities: If the multiplets have a triangular shape then all the weights are simple. If the multiplets have a hexagonal shape then the multiplicities of the weights increase from the outside to the interior from "layer" to "layer" by *one* until a triangle of equal multiplicities is reached. In the interior of such triangles the multiplicities remain constant.

(13) Instead of characterizing the multiplets by their maximal weights W or by r numbers $n_1, \ldots, n_b, \ldots, n_r$, we can also use the (eigenvalues of the) invariant Casimir operators, see (11.1.37, 39). For the Casimir operator C_2 from (11.1.37) we have with the CWB (11.4.30a, 43)

$$C_2 = \sum_{ll'} \bar{g}^{ll'} H_l H_{l'} + \sum_\alpha E_\alpha E_{-\alpha} . \qquad (11.4.61)$$

Since C_2 is invariant, for the calculation of the eigenvalues of C_2 we may use the maximal weight. As $E_\alpha|W\rangle = 0$ (11.4.18, 54), it follows with (11.4.44) that

$$\langle W|C_2|W\rangle = \sum_{ll'} \langle W|\bar{g}^{ll'} H_l H_{l'}|W\rangle + \sum_l \sum_{r_\alpha \in +} \langle W|r^l_\alpha H_l|W\rangle .$$

Thus, using (11.4.56) and (11.4.55b)

$$\langle C_2\rangle = (W\cdot W) + 2(s\cdot W) = (v\cdot v) - (s\cdot s) = v^2 - s^2 \quad \text{with} \quad v = W + s . \qquad (11.4.62)$$

The eigenvalues of the other $r - 1$ Casimir-operators required for the complete characterization of a multiplet will not be given here.

For the groups $\mathcal{SU}(n)$, $n = 2, 3, 4$, we obtain with (11.4.57–62)

$$\mathcal{SU}(2): \langle C_2\rangle = \tfrac{1}{2}j(j + 1) ,$$

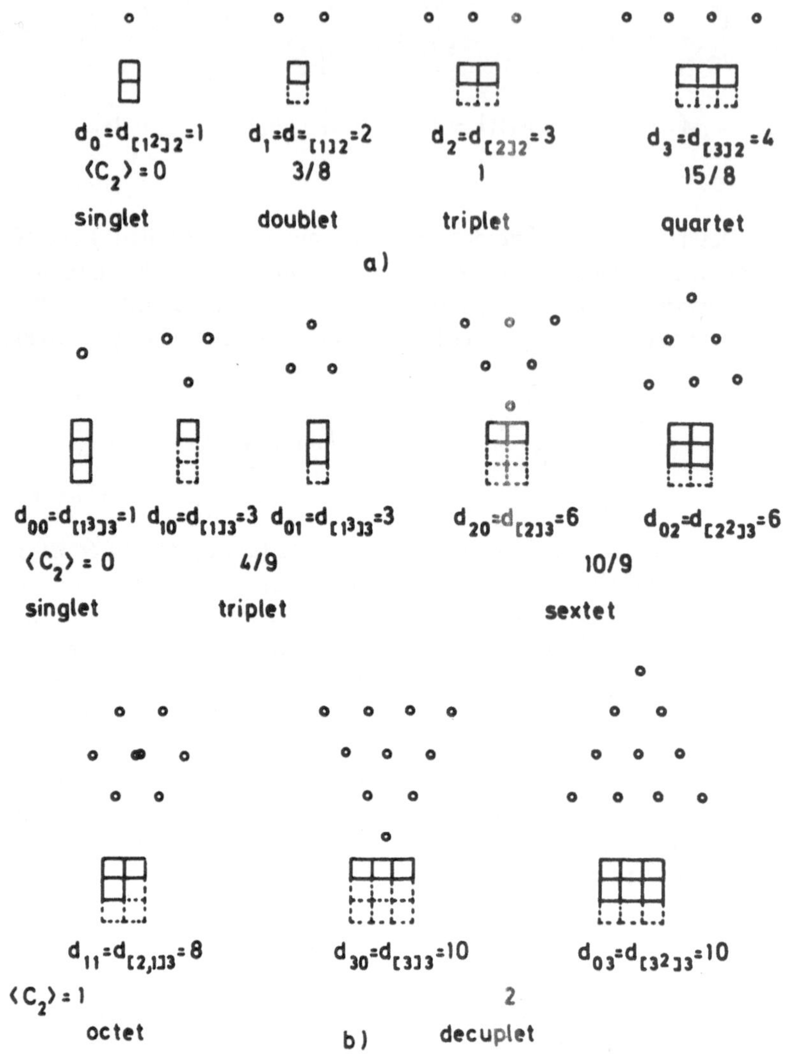

Fig. 11.6a–c. The lowest multiplets of $\mathcal{SU}(n)$, $n = 2, 3, 4$. The dimensions of the REPs and the eigenvalues of the Casimir operator C_2 are given. In elementary particle physics instead of the quantum numbers $n_1, \ldots, n_r$, different numbers are used (isospin, hypercharge, charm, etc.). For $\mathcal{SU}(4)$ the different signs (o, •, ▪) indicate different "heights". Also given are the assigned Young diagrams according to (12.1.7–9)

$$\mathcal{SU}(3)\colon \langle C_2 \rangle = \tfrac{1}{9}(n_1^2 + n_2^2 + n_1 n_2 + 3n_1 + 3n_2) \; ,$$

$$\mathcal{SU}(4)\colon \langle C_2 \rangle = \tfrac{1}{32}(3n_1^2 + 4n_2^2 + 3n_3^2 + 4n_1 n_2 + 2n_1 n_3 + 4n_2 n_3 + 12n_1 + 16n_2 + 12n_3) \; . \tag{11.4.63}$$

In some applications it is helpful to have the connection between the A_1 algebra

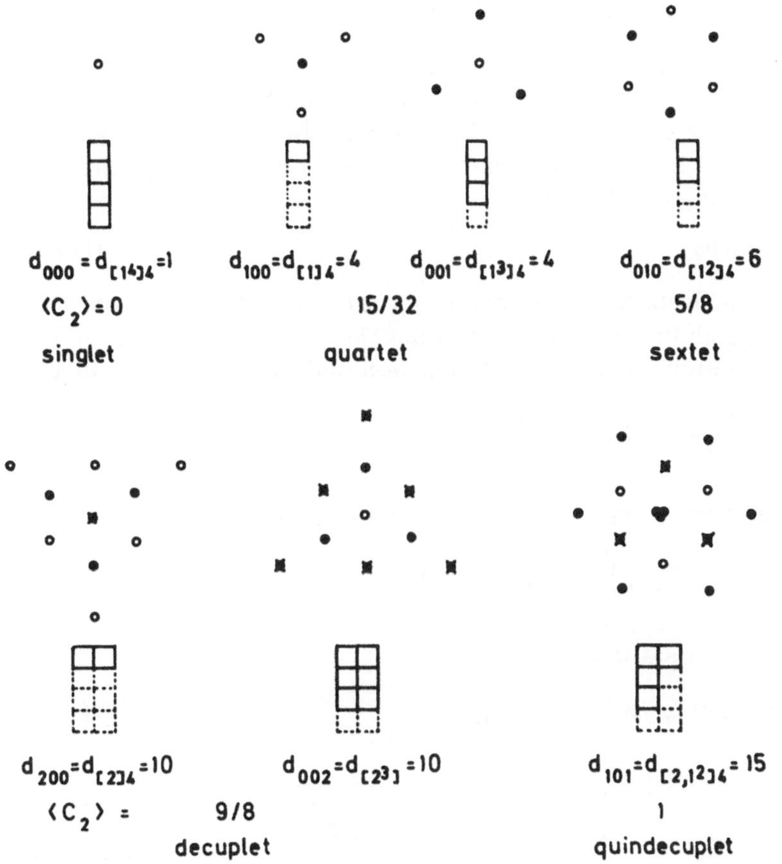

Fig. 11.6 (cont.) c)

of $\mathcal{SU}(2)$ and the well-known algebra of angular momentum. From the considerations in Sects. 11.1.2 and 11.4.1 we deduce (to within factors of $\hbar$)

$$H_1 = \frac{1}{\sqrt{2}} L_z \ , \qquad E_\alpha \to E_{12} = \frac{1}{2} L_+ = \frac{1}{2}(L_x + \mathrm{i}L_y) \ , \tag{11.4.64}$$

$$E_{-\alpha} \to E_{21} = \frac{1}{2} L_- = \frac{1}{2}(L_x - \mathrm{i}L_y) \ ,$$

$$[L_i, L_j] = \mathrm{i} \sum_k \varepsilon_{ijk} L_k \ , \qquad i, j, k = x, y, z \ .$$

Furthermore the Casimir operator satisfies

$$C_2 = \tfrac{1}{2} L^2 = \tfrac{1}{2}(L_x^2 + L_y^2 + L_z^2) \ .$$

Therefore, with (11.4.63)

$$\langle L^2\rangle = 2\langle C_2\rangle = j(j+1)\ , \qquad j = 0, 1/2, 1, \dots\ . \tag{11.4.65a}$$

By means of (11.4.57), the maximal weight j and dimension $d_j = 2j + 1$ of the IRs are completely determined and the multiplets are characterized.

(14) The characters for $\mathscr{SU}(2)$ follow from the basis functions $|W, w, \mu\rangle = |j, m\rangle$ of the IRs, where m is the weight; the index μ of the multiplicity can be dropped. According to (11.4.56)

$$L_z|jm\rangle = m|jm\rangle\ . \tag{11.4.65b}$$

By successive application of the ladder operator L_- according to (11.4.51, 54) we obtain from $|jj\rangle$ all the $2j + 1$ basis functions $|jm\rangle$, $m = j, j - 1, \dots, -j$. By an infinitesimal rotation through φ/N the function $|jm\rangle$ changes into the function

$$\left(1 - \mathrm{i}L_z\frac{\varphi}{N}\right)|jm\rangle = \left(1 - \mathrm{i}m\frac{\varphi}{N}\right)|jm\rangle\ ,$$

see (11.1.9, 22). With a finite rotation (about the z-axis)

$$P_\varphi = \lim_{N\to\infty}\left(1 - \mathrm{i}L_z\frac{\varphi}{N}\right)^N = \exp(-\mathrm{i}L_z\varphi)\ ,$$

i.e. the matrices of the IRs are

$$D^{(j)}_{mm'}(\varphi) = \langle jm|P_\varphi|jm'\rangle = e^{-\mathrm{i}m\varphi}\delta_{mm'}\ , \tag{11.4.66}$$

and thus the characters are

$$\chi^{(j)}(\varphi) = \sum_m D^{(j)}_{mm}(\varphi) = \sum_{m=-j}^{j} e^{-\mathrm{i}m\varphi} = \frac{\sin[(j+1/2)\varphi]}{\sin(\varphi/2)}\ , \qquad j = 0, 1/2, 1, \dots\ . \tag{11.4.67}$$

Generally we obtain the characters of $\mathscr{SU}(n)$ in the following way (a generalization to other groups is again obvious). Every element $u \in \mathscr{SU}(n)$ can be brought into diagonal form $S^{-1}uS$ by a unitary transformation S. Since this is a conjugation, this means that a representative element of every (conjugacy) class can be written as

$$u_\phi = \begin{pmatrix} e^{\mathrm{i}\phi_1} & & \\ & \ddots & \\ & & e^{\mathrm{i}\phi_n} \end{pmatrix}\ , \qquad \sum_{\nu=1}^{n}\phi_\nu = 0 \quad \text{because } \det u_\phi = 1\ . \tag{11.4.68a}$$

Thus the classes are characterized by the $r = n - 1$ independent angles $\phi = \{\phi_1, \dots, \phi_{n-1}\}$. Using (11.4.5, 9), Eq. (11.4.68a) can also be brought into the form

$$(u_\phi)_{ij} = \delta_{ij}\exp\left(\mathrm{i}\sum_l^r \Lambda^i_l\phi_l\right) = \delta_{ij}\prod_l^r \exp(\mathrm{i}\Lambda^i_l\phi_l) \tag{11.4.68b}$$

or

$$u_\phi = \prod_l^r \exp(\mathrm{i}H_l\phi_l) \ .$$

The characters of the IRs, specified by the maximal weight $\boldsymbol{W}$ are obtained with (11.4.56):

$$D^{(W)}_{ww'}(\boldsymbol{\phi}) = \langle \boldsymbol{W}, \boldsymbol{w}, \mu | u_\phi | \boldsymbol{W}, \boldsymbol{w}', \mu' \rangle = \prod_l^r \exp(\mathrm{i}w_l\phi_l)\delta_{ww'}\delta_{\mu\mu'}$$

thus

$$\chi^{(W)}(\boldsymbol{\phi}) = \prod_l^r \sum_{w_l\mu} \exp(\mathrm{i}w_l\phi_l) = \sum_w m_W(\boldsymbol{w}) \exp[\mathrm{i}(\boldsymbol{w}\cdot\boldsymbol{\phi})] \ , \tag{11.4.69}$$

where the sum is to be taken over all the weights including their multiplicities (11.4.60). Equation (11.4.67) again follows from this with $\phi_1 = \varphi\sqrt{2}$, $W = j/\sqrt{2}$.

(15) By means of (11.4.67) or (11.4.69), the Clebsch-Gordan expansion of the inner direct product of two IRs can be given. For $\mathcal{SU}(2)$ we have for the character of $D^{(j_1)} \otimes D^{(j_2)}$

$$\begin{aligned} \chi^{(j_1 \otimes j_2)}(\varphi) &= \chi^{(j_1)}(\varphi)\chi^{(j_2)}(\varphi) = \sum_{m_1=-j_1}^{j_1} \mathrm{e}^{\mathrm{i}m_1\varphi} \sum_{m_2=-j_2}^{j_2} \mathrm{e}^{\mathrm{i}m_2\varphi} \\ &= \sum_{m=-j_1-j_2}^{j_1+j_2} \mathrm{e}^{\mathrm{i}m\varphi} + \sum_{m=-j_1-j_2+1}^{j_1+j_2-1} \mathrm{e}^{\mathrm{i}m\varphi} + \cdots + \sum_{m=-|j_1-j_2|}^{|j_1-j_2|} \mathrm{e}^{\mathrm{i}m\varphi} \ , \end{aligned} \tag{11.4.70}$$

which can be seen immediately by counting the terms with fixed $m = m_1 + m_2$. For example, $m = j_1 + j_2 - v$ occurs $(v + 1)$ times. Thus we have the Clebsch-Gordan expansion of $\mathcal{SU}(2)$

$$D^{(j_1)} \otimes D^{(j_2)} = D^{(j_1+j_2)} \oplus D^{(j_1+j_2-1)} \oplus \cdots \oplus D^{(|j_1-j_2|)} \ . \tag{11.4.71}$$

The reduction coefficients $(j_1 j_2 | j)$ are all equal to 1; $\mathcal{SU}(2)$ is simply reducible, whereas $\mathcal{SU}(n)$ with $n \geqslant 3$ is not. Equation (11.4.71) can also be proved by using the method of Young diagrams discussed in Sect. 5.5.4. The application of the Littlewood theorem (5.5.47) is illustrated in Fig. 5.7. However, we have to take into account the relation between maximal weight (quantum number) j and dimension $d_j = 2j + 1$, while on the other hand $2j = n_1$ and $d_j = n_1 + 1$, where n_1 is the number of cells in the Young diagram. With this all the relations can be established.

The characters do not enter the Littlewood theorem explicitly. Thus the reduction with respect to IRs of $\mathcal{GL}(n)$ and its subgroups can be achieved without the use of characters, which in general is even simpler. Therefore we do not give the characters of $\mathcal{SU}(n)$, $n \geqslant 3$, explicitly.

(16) The CGC may be determined by means of the ladder operators, as will be shown using $\mathscr{SU}(2)$ as an example again. Application to $\mathscr{SU}(n)$ is similar. According to (4.4.18a) the CGC for $\mathscr{SU}(2)$ are defined by

$$|j,m\rangle = \sum_{m_1=-j_1}^{j_1} \sum_{m_2=-j_2}^{j_2} \begin{pmatrix} j_1 j_2 & j \\ m_1 m_2 & m \end{pmatrix} |j_1, m_1\rangle |j_2, m_2\rangle \ . \qquad (11.4.72)$$

Because of (11.4.70) the CGC are only different from zero for $m = m_1 + m_2$. Furthermore, with $m = j_1 + j_2$ (maximal weight) there is only *one* product function in $\mathscr{L}^{(j_1)} \otimes \mathscr{L}^{(j_2)}$, that means

$$\begin{pmatrix} j_1 j_2 & j_1 + j_2 \\ m_1 m_2 & j_1 + j_2 \end{pmatrix} = 1 \ , \qquad m_1 = j_1 \ , \qquad m_2 = j_2 \ . \qquad (11.4.73)$$

According to (11.4.15), by successive application of $E_{21} \sim L_-$ to $|j_1 + j_2, j_1 + j_2\rangle$ we can obtain all the $2j_1 + 2j_2 + 1$ basis functions belonging to $D^{(j_1+j_2)}$. Since $L_-|j_1 + j_2, -(j_1 + j_2)\rangle$ has to be zero the sequence stops with this last function. In using the ladder operators we have to take into account the normalization in

$$L_\pm |j,m\rangle = c_\pm(j,m)|j, m \pm 1\rangle$$

giving

$$c_\pm(j,m) = [j(j+1) - m(m \pm 1)]^{1/2}$$

when (11.4.64) is used.

If we now apply $L_- = L_{1-} + L_{2-}$ to the function of maximal weight, $|j_1 + j_2, j_1 + j_2\rangle = |j_1, j_1\rangle \cdot |j_2, j_2\rangle$ we obtain the linear combination for $|j_1 + j_2, j_1 + j_2 - 1\rangle$ and consequently we can determine the CGC by comparing with (11.4.72). In this case we have

$$\begin{pmatrix} j_1 & j_2 & j_1 + j_2 \\ j_1 - 1 & j_2 & j_1 + j_2 - 1 \end{pmatrix} = \left(\frac{j_1}{j_1 + j_2}\right)^{1/2} \ ,$$

$$\begin{pmatrix} j_1 & j_2 & j_1 + j_2 \\ j_1 & j_2 - 1 & j_1 + j_2 - 1 \end{pmatrix} = \left(\frac{j_2}{j_1 + j_2}\right)^{1/2} \ .$$

The function $|j_1 + j_2 - 1, j_1 + j_2 - 1\rangle$ has to be orthogonal with respect to the above linear combination and it is the starting function belonging to the multiplet $D^{(j_1+j_2-1)}$. Because of the orthogonality we have, apart from common phases,

$$\begin{pmatrix} j_1 & j_2 & j_1 + j_2 - 1 \\ j_1 - 1 & j_2 & j_1 + j_2 - 1 \end{pmatrix} = \left(\frac{j_1}{j_1 + j_2}\right)^{1/2} \ ,$$

$$\begin{pmatrix} j_1 & j_2 & j_1 + j_2 - 1 \\ j_1 & j_2 - 1 & j_1 + j_2 - 1 \end{pmatrix} = -\left(\frac{j_2}{j_1 + j_2}\right)^{1/2} \ .$$

By further application of L_- to these functions and orthogonalization of the additional functions belonging to the other multiplets in (11.4.71) we obtain successively all the CGC. However, this procedure is rather troublesome. Finally we have

$$\left(\begin{array}{c|c} j_1 j_2 & j \\ m_1 m_2 & m \end{array}\right) = \delta_{m,m_1+m_2}$$

$$\times \left(\frac{(2j+1)(j_1+j_2-j)!(j_1-m_1)!(j_2-m_2)!(j+m)!(j-m)!}{(j+j_1+j_2+1)!(j+j_1-j_2)!(j-j_1+j_2)!(j_1+m_1)!(j_2+m_2)!}\right)^{1/2}$$

$$\times \sum_k (-1)^{k+j_1-m_1} \frac{(k+j_1+m_1)!(j+j_2-m_1-k)!}{k!(j_1-m_1-k)!(j-m_1-m_2-k)!(j_2-j+m_1+k)!}\,, \tag{11.4.74}$$

where k runs over all integers which leave the arguments of the factorials non-negative. From (11.4.74) we obtain the relation

$$\left(\begin{array}{c|c} j_2 j_1 & j \\ m_2 m_1 & m \end{array}\right) = (-1)^{j_1+j_2-j} \left(\begin{array}{c|c} j_1 j_2 & j \\ m_1 m_2 & m \end{array}\right)\,. \tag{11.4.75}$$

When transferring these methods and results to the groups $\mathscr{SO}(n)$ and $\mathscr{Sp}(2n)$ their special features have to be taken into account (see also Sects. 5.5.3, 4). The Cartan-Weyl method gives for example vector *and* spinor REPs (single- and double-valued) for the $\mathscr{SO}(n)$ groups, classified by the maximal weights $\boldsymbol{W}$ or by the numbers (root lengths) n_b according to (11.4.53).

(16) When the generators in a certain REP are given by matrices, e.g. $\{H_l, E_\alpha\}$, then the substitution $H_l \rightarrow -\tilde{H}_l$, $E_\alpha \rightarrow -\tilde{E}_\alpha$ ($\sim$ denotes the transpose matrix) leaves the commutation relations (11.1.15) unchanged and thus they are again a REP, which is said to be the *contragredient* or *dual* REP. If the generators are chosen to be anti-Hermitian (11.1.23) the substitution means complex conjugation of the REP matrices. If the REPs are expressed in terms of their weights $\boldsymbol{w}$, the dual REP has the weight $-\boldsymbol{w}$ [Fig. 11.6, (11.4.11)], the weight diagram being obtained by inversion. In the case that the weight diagram is invariant under inversion, the REPs are equivalent (Sect. 5.5.3). In this case the REPs are said to be *real*. In the $(n_1, \ldots, n_r)$ notation (root lengths), real REPs have "symmetric" brackets, e.g. (1, 0, 1) or (0, 1, 0) or (1, 0, 0, 1), etc. All the $\mathscr{SU}(2)$ IRs are real in this sense.

(17) We refer to two special REPs of groups, especially for the $\mathscr{SU}(n)$ groups. These are the *defining* or *fundamental* REP already mentioned and the *adjoint* REP. In the defining REP the m generators are represented by the m linearly independent $(n \times n)$ matrices which are compatible with possible additional conditions. For $\mathscr{SU}(n)$ we can take those matrices given in (11.4.5, 6). In the adjoint REP the generators are represented by the structure constants c_{is}^r, i.e.

$$J_{i,}{}^r{}_s = c_{is}^r\,. \tag{11.4.76}$$

Inserting this into the defining relation (11.1.15), this relation becomes identical with the Jacobi identity in (11.1.17). The dimension of the REP matrices is equal to the number of generators, thus $d_w = m = n^2 - 1$ for $\mathscr{SU}(n)$. This adjoint REP is in a certain sense comparable with the regular REP of finite groups (4.3.2), but contrary to those regular REPs it is irreducible! The maximal weights of the defining REPs are $(n_1, \ldots, n_r) = (1, 0, 0, \ldots)$ and those of the adjoint REPs are (2) for $\mathscr{SU}(2)$, $(1, 0, 0, \ldots, 1)$ with $(n-3)$ zeros for $\mathscr{SU}(n)$, $n \geqslant 3$.

Exercise 11.7. Calculate the structure constants of the LA A_2 of $\mathscr{SU}(3)$ in the basis (11.4.8).

Exercise 11.8. Calculate the 12 roots for $\mathscr{SU}(4)$ from (11.4.26).

Exercise 11.9. Using (11.4.37a–d) prove the form (11.4.30a) of the metric in the CWB, where $\bar{g}_{ll'}$ is given by (11.4.43).

Exercise 11.10. Calculate the multiplicities of the weights in the octet of $\mathscr{SU}(3)$ and in the quindecuplet of $\mathscr{SU}(4)$ according to (11.4.60).

Exercise 11.11. Determine all the adjoint REP matrices for $\mathscr{SU}(2)$ and $\mathscr{SU}(3)$ starting from the structure constants (see Exercise 11.7).

12. Representations by Young Diagrams. The Method of Irreducible Tensors

In Sect. 5.5.3 we described how to reduce tensor spaces. From this we are able to obtain the basis functions belonging to the IRs $D^{[\lambda]}(n)$ of $\mathcal{GL}(n, \mathbb{C})$ and of its subgroups $\mathcal{U}(n)$, $\mathcal{SU}(n)$, $\mathcal{O}(n)$, etc. The basis functions which span the IR spaces $\mathcal{L}_{n,k}^{[\lambda]}$ of $D^{[\lambda]}(n)$ are according to (5.5.31) the symmetrized irreducible tensors of rank m $\Psi_{f,k}^{[\lambda]}$ with $f = \{f_1, \ldots, f_m\}$. This description of the IRs of continuous groups is quite different from the Lie-Cartan-Weyl method; it supplements the latter. In order to illustrate this, we recall the method for finding the basis tensors.

As in Sect. 5.5.3 we start with a product function $\psi_{f_1}(1) \ldots \psi_{f_m}(m)$ with a fixed configuration (set of numbers) $f = \{f_1, \ldots, f_m\}$. Altogether n^m configurations are possible: $f_i = 1, \ldots, n$; $i = 1, \ldots, m$. For a fixed configuration we obtain the allowed Young tableaux when we put the numbers $f_1, \ldots, f_m$ into the cells of the graph $[\lambda]$ in such a way that the numbers increase from top to bottom in a column and do not decrease from left to right in a row, see Sect. 5.5.3. Hence it is possible that two (or even more) identical numbers occur in one row, but never in one column of $[\lambda]$. Each of these Young tableaux defines a starting function $\psi_{f_1}\psi_{f_2} \ldots \psi_{f_m}$ from which we get the basis functions by a symmetry projection with the projection operators $e_{ij}^{[\lambda]} \sim P_{ij}^{[\lambda]}$ of the permutation group $\mathcal{P}_m$ according to (5.4.18a, 29):

$$\Psi_{f,k}^{[\lambda]} := e_{kk}^{[\lambda]} \psi_{f_1}(1) \ldots \psi_{f_m}(m) \ . \tag{12.1.1}$$

When this procedure is performed for all possible configurations (disregarding the order) all the basis functions belonging to $D^{[\lambda]}(n)$ are obtained in the form of irreducible tensors of rank m.

If the one-particle functions

$$\{\psi_{f_i} | f_i = 1, \ldots, n; i = 1, \ldots, m\} \tag{12.1.2}$$

are basis functions belonging to the defining IR (11.4.20) of $\mathcal{SU}(n)$, then the symmetrized tensors $\Psi_{f,k}^{[\lambda]}$ are just the eigenfunctions of the "many-particle" operator

$$H_l := \sum_{i=1}^{m} H_l(i) \ , \tag{12.1.3}$$

see (12.1.6), the defining IR is assumed to be constituted by the "one-particle" operators $\{H_l(i); E_{jl}(i) | i = 1, \ldots, m\}$ according to (11.4.5, 6); thus according to

(11.4.20)

$$H_l(i)|\psi_{f_i}(i)\rangle = \Lambda_l^{f_i}|\psi_{f_i}(i)\rangle, l = 1, \ldots, r = n-1 \ . \tag{12.1.4}$$

Due to (11.4.21) the basis function $|\psi_{f_i}(i)\rangle$ belongs to the weight $w^{f_i} = (\Lambda_1^{f_i}, \ldots, \Lambda_r^{f_i})$; the $\Psi_{f,k}^{[\lambda]}$ are basis functions of the IR $D^{[\lambda]}(n)$ and define the weights of this IR of $\mathscr{SU}(n)$. The matrix REPs of the H_l according to (12.1.3) and those of the corresponding $E_{jl} = \sum_{i=1}^m E_{jl}(i)$ in the basis $\{\Psi_{f,k}^{[\lambda]}\}$ are just generators (in matrix form) of the IR $D^{[\lambda]}(n)$ of $\mathscr{SU}(n)$. For the proof of the foregoing theorem we consider the product function formed from (12.1.2) and calculate the eigenvalues of H_l with the product function

$$H_l\psi_{f_1}(1)\ldots\psi_{f_m}(m) = \sum_{i=1}^m H_l(i)\psi_{f_1}(1)\ldots\psi_{f_m}(m) = \sum_{i=1}^m \Lambda_l^{f_i}\psi_{f_1}(1)\ldots\psi_{f_m}(m) \ ;$$

the eigenvalue thus being $w_l = \sum_{i=1}^m \Lambda_l^{f_i}$. Since for every permutation $p \in \mathscr{P}_m$

$$[p, H_l] = 0 \ , \tag{12.1.5}$$

the projector according to (5.4.20)

$$e_{kk}^{[\lambda]} \sim \sum_{p \in \mathscr{P}_m} D_{kk}^{[\lambda]}(p^{-1})P_p$$

commutes with H_l. Using (12.1.1) we have

$$\begin{aligned} H_l\,\Psi_{f,k}^{[\lambda]} &= H_l e_{kk}^{[\lambda]}\psi_{f_1}(1)\ldots\psi_{f_m}(m) = e_{kk}^{[\lambda]}H_l\psi_{f_1}(1)\ldots\psi_{f_m}(m) \\ &= w_l\,\Psi_{f,k}^{[\lambda]} \ , \end{aligned} \tag{12.1.6}$$

i.e. the basis function $\Psi_{f,k}^{[\lambda]}$ of the IR $D^{[\lambda]}(n)$ of $\mathscr{SU}(n)$ belongs to the weight (eigenvalue)

$$w_f^{[\lambda]} := \left(\sum_{i=1}^m \Lambda_1^{f_i}, \sum_{i=1}^m \Lambda_2^{f_i}, \ldots, \sum_{i=1}^m \Lambda_r^{f_i}\right) := \sum_{i=1}^m w^{(f_i)} \ . \tag{12.1.7}$$

However, according to Sect. 5.5.3 only *one* linearly independent basis tensor belongs to each allowed Young tableau. Therefore we obtain the following rule:

> From (12.1.7) all the weights of an IR of $\mathscr{SU}(n)$ belonging to the Young diagram $[\lambda]$ are obtained by inserting for $f = \{f_1, \ldots, f_m\}$ in (12.1.7) all the configurations which result from an allowed Young tableau. The fundamental weights $w^{(f_i)}$ follow from (11.4.21).

The maximal weights W, or the "quantum numbers" $\{n_1, \ldots, n_r\}$ assigned to them according to (11.4.53, 57–59), characterize the IRs of $\mathscr{SU}(n)$ as well as the partitions $[\lambda] = [\lambda_1, \lambda_2, \ldots, \lambda_n]$. With the fundamental weights $w^{(j)}$ from (11.4.21) the maximal weight is given by

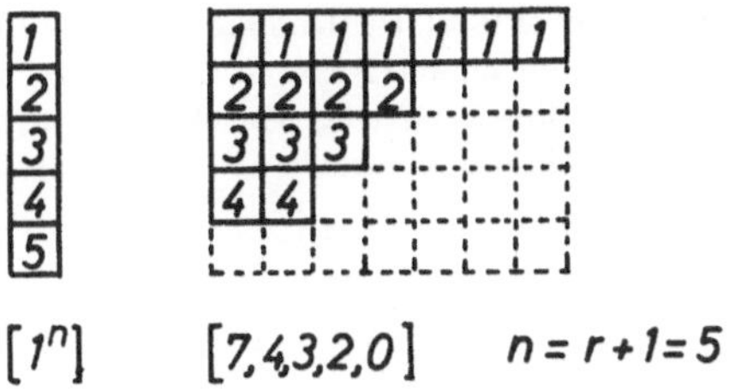

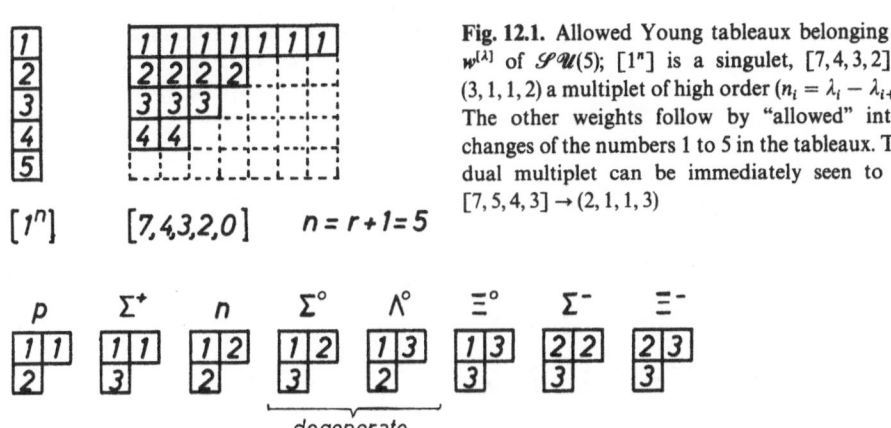

Fig. 12.1. Allowed Young tableaux belonging to $w^{[\lambda]}$ of $\mathcal{SU}(5)$; $[1^n]$ is a singulet, $[7,4,3,2] \to (3,1,1,2)$ a multiplet of high order ($n_i = \lambda_i - \lambda_{i+1}$). The other weights follow by "allowed" interchanges of the numbers 1 to 5 in the tableaux. The dual multiplet can be immediately seen to be $[7,5,4,3] \to (2,1,1,3)$

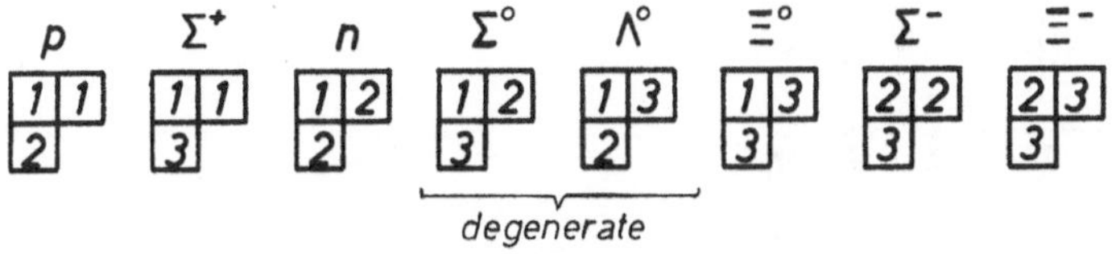

Fig. 12.2. The eight allowed Young tableaux belonging to the octet $[2,1] \to (1,1)$ of $\mathcal{SU}(3)$

$$W^{[\lambda]} = \sum_{j=1}^{n} \lambda_j w^{(j)} \ . \tag{12.1.8}$$

When the root basis or the "quantum numbers" $\{n_l\}$ are chosen appropriately, the allowed Young tableau belonging to $W^{[\lambda]}$ has in its first row λ_1 "1"s, in its second row λ_2 "2"s, ..., in its rth row λ_r cells with "$r = n - 1$" in them (Fig. 12.1). This clearly shows that this arrangement can only be realized in one way, i.e. the maximal weight is nondegenerate (Sect. 11.4.2). With the basis roots according to (11.4.27a) it follows, using (11.4.53), that

$$\begin{aligned} \lambda_1 - \lambda_n &= n_1 + n_2 + \cdots + n_r \\ \lambda_2 - \lambda_n &= \phantom{n_1 + {}} n_2 + \cdots + n_r \\ &\ \vdots \\ \lambda_r - \lambda_n &= \phantom{n_1 + n_2 + \cdots + {}} n_r \end{aligned} \qquad \text{or} \qquad \begin{aligned} n_l &= \lambda_l - \lambda_{l+1} \ ; \\ l &= 1, \ldots, r = n-1 \ . \end{aligned} \tag{12.1.9}$$

That means, the identity REP (singlet) has $\lambda_l = 1$, $l = 1, \ldots, n$ or $n_l = 0$, $l = 1, \ldots, r$. In all other cases λ_n can be chosen to be zero: $\lambda_n = 0$. Then the number of cells in the Young diagram is $\sum_{l=1}^{r} \lambda_l = \sum_{l=1}^{r} l n_l$. The defining REP has $\{n_l\} = (1,0,\ldots)$ or $[\lambda] = [1,0,\ldots] = [1]$. The adjoint REP of $\mathcal{SU}(n)$ has the partition $[\lambda] = [2, 1^{n-2}]$. Contragredient (dual) REPs have $[\lambda_1, \lambda_2, \ldots, \lambda_n]$ and $[\lambda_1 - \lambda_n, \lambda_1 - \lambda_{n-1}, \ldots, \lambda_1 - \lambda_1 = 0]$ (see Sect. 5.5.3 and Fig. 5.4) with the exception of $[1^n]$ being real. The Young diagrams for the lowest multiplets are given in Fig. 11.6.

We illustrate all these statements with the REP $D^{[2,1]}(3)$ of $\mathcal{SU}(3)$ whose allowed tableaux are given in Fig. 12.2. There are exactly 8 of these tableaux, i.e. the dimension of this REP is $d_{[2,1]3} = 8$; the assigned quantum numbers according to (12.1.9) are $(n_1, n_2) = (1,1)$. The numbers f_i in (12.1.2) run over 1, 2, 3. Instead of the generators H_l according to (11.4.8), in elementary-particle physics

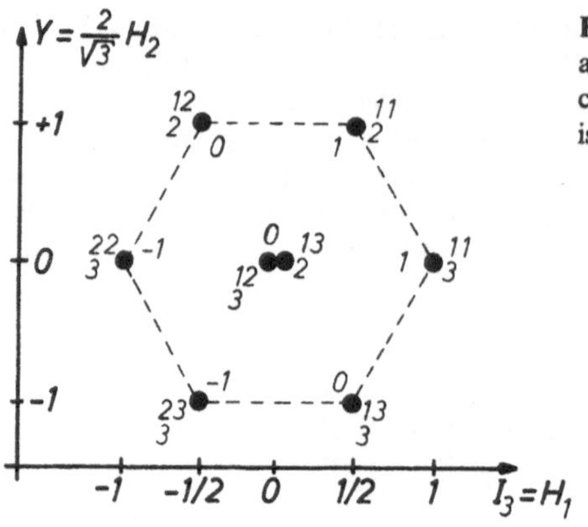

Fig. 12.3. Octet of $\mathcal{SU}(3)$ with notation of Fig. 12.2 and the quantum numbers isospin I_3 and hypercharge Y according to (12.1.12). The electric charge is $Y/2 + I_3$ and is also indicated

one uses another normalization:

$$I_3 = H_1 \ , \qquad Y = \frac{2}{\sqrt{3}} H_2 \ ; \tag{12.1.10}$$

I_3 corresponds to the 3(z)-component of the isospin I, and Y to the hypercharge. The fundamental weights according to (11.4.23) are, with $c_1 = 1/2$, $c_2 = 1/3$,

$$w^{(1)} = (1/2, 1/3) \ , \qquad w^{(2)} = (-1/2, 1/3) \ , \qquad w^{(3)} = (0, -2/3) \ . \tag{12.1.11}$$

According to (12.1.7) we have thus for the tableaux of Fig. 12.2 in the given order, with $w^{[2,1]}_{11,2} = 2w^{(1)} + w^{(2)}$, etc.,

$$\begin{aligned} &w^{[2,1]}_{11,2} = (1/2, 1) \ , && w^{[2,1]}_{13,3} = (1/2, -1) \ , \\ &w^{[2,1]}_{11,3} = (1, 0) \ , && w^{[2,1]}_{22,3} = (-1, 0) \ , \\ &w^{[2,1]}_{12,2} = (-1/2, 1) \ , && w^{[2,1]}_{23,3} = (-1/2, -1) \ , \\ &w^{[2,1]}_{12,3} = w^{[2,1]}_{13,2} = (0, 0) \ . \end{aligned} \tag{12.1.12}$$

The weight diagram (multiplet) is shown in Fig. 12.3. The weight (0, 0) is doubly degenerate, as can be seen immediately. Obviously the following rule holds: If there are $z^{[\lambda]}$ allowed tableaux belonging to a diagram $[\lambda]$ in which the number "1" occurs z_1 times, the number "2" z_2 times, etc., then these tableaux have according to (12.1.7) the same weight

$$w^{[\lambda]} = \sum_{i=1}^{n} z_i w^{(i)} \ , \tag{12.1.13}$$

i.e. this weight is $z^{[\lambda]}$-fold degenerate. The determination of the degeneracy is thus simpler than according to (11.4.60). The degenerate basis functions belonging to (0, 0) in the above example (12.1.12) are $\Psi^{[2,1]}_{13,2,k}$ and $\Psi^{[2,1]}_{12,3,k}$.

For the formation of direct products of two REPs, e.g. $D^{[\lambda]}(n) \otimes D^{[\lambda']}(n)$, the rules discussed for the $\odot$-product in Sect. 5.5.4 are valid correspondingly (Fig.

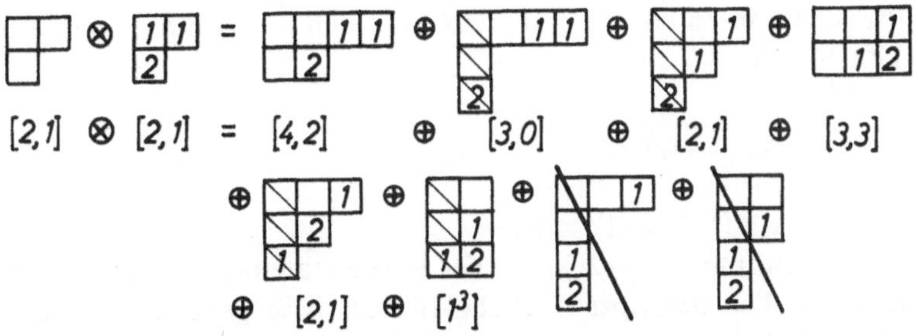

Fig. 12.4. Formation of the product $D^{[2,1]} \otimes D^{[2,1]}$ of two octets of $\mathscr{SU}(3)$ according to Fig. 5.6. For $\mathscr{SU}(3)$, the last two diagrams have to be dropped since no diagram is allowed to have more than n $(=3)$ rows. The possible reductions are indicated. In the (n_1, n_2) notation the decomposition reads $(1,1) \otimes (1,1) = (2,2) \oplus (3,0) \oplus 2 \cdot (1,1) \oplus (0,3) \oplus (0,0)$. The dimensions of the REPs are $8 \times 8 = 27 + 10 + 8 + 10^* + 8 + 1$

5.6). However, only allowed tableaux may be taken into account. For example, for $n = 3$, tableaux having more than three rows are forbidden. Figure 12.4 shows the formation of the product $(1,1) \otimes (1,1) = [2,1] \otimes [2,1]$.

We add some remarks concerning the group $\mathscr{U}(n)$, which is isomorphic to the direct product $\mathscr{SU}(n) \times \mathscr{U}(1)$:

$$\mathscr{U}(n) \cong \mathscr{SU}(n) \times \mathscr{U}(1) \; . \tag{12.1.14a}$$

Every element $u \in \mathscr{U}(n)$ can be represented by

$$u = a e^{i\phi_0} \; , \qquad u \in \mathscr{U}(n) \; , \qquad a \in \mathscr{SU}(n) \; , \tag{12.1.14b}$$

so that $\det u = \exp(in\phi_0)$. For the "class" representatives, (11.4.68a), for example, is valid without any further conditions. For the generators we may choose the H_l of (11.4.5) adding $H_n = c_n 1_n$ as a further generator. Since the condition $\mathrm{Tr}\{H_l\} = 0$ drops out, we may also use simpler generators, namely

$$(H_l)_{ij} = c_l \delta_{il} \delta_{ij} \; , \qquad i, j, l = 1, \ldots, n \; ; \tag{12.1.15}$$

these have only one "1" at the position l in the diagonal and zeros everywhere else. The remaining $n(n-1)$ infinitesimal generators may be chosen as in (11.4.6a). Analogously to (11.4.21) the fundamental weights of (12.1.15) are (with an appropriate normalization)

$$w^{(l)} = \underbrace{(0, 0, \ldots, c_l, \ldots, 0)}_{n \text{ components}} \; ; \qquad c_l = 1 \; . \tag{12.1.16}$$

Then, as in (12.1.6, 7), the weight

$$w_f^{[\lambda]} = \sum_{l=1}^{m} w^{(f_l)} \; , \qquad w^{(f_l)} \in \{w^{(l)}\} \tag{12.1.17}$$

belongs to the basis function $\Psi_{f,k}^{[\lambda]}$ of the IR $D^{[\lambda]}(n)$ of $\mathscr{U}(n)$. For example, if there are z_1 "1"s, z_2 "2"s, etc., in an allowed tableaux $[\lambda]$, then the assigned weight is according to (12.1.16, 17)

$$w = \sum_{i=1}^{n} z_i w^{(i)} = (z_1, z_2, \ldots, z_n) . \tag{12.1.18}$$

Finally, we remark that $\mathscr{U}(1) \otimes \mathscr{U}(1) \cong \mathscr{U}(1)$.

For the groups $\mathscr{SO}(n)$ and $\mathscr{Sp}(2n)$ we refer to the statements at the end of Sects. 5.5.3 and 5.5.4. The description of the IRs of these groups by Young diagrams or tableaux needs the contractions of Sect. 5.5.3. Starting from the transformations in Sect. 5.5.3, for the $\mathscr{SO}(n)$ groups we only obtain the (normal) vector REPs. The relation between root lengths (11.4.53) and partitions, given in (12.1.9) for $\mathscr{SU}(n)$, is somewhat different in this case; in particular, one has to consider the lengths of the roots assigned to the indices b of n_b [see Fig. 11.5 and (11.4.53)]:

Root lengths: $(n_1, n_2, \ldots, n_r)$
If the roots have different lengths, n_1 is associated with a long root, n_2 with a short root, and n_3 with a long root for $\mathscr{SO}(7)$ and short one for $\mathscr{Sp}(6)$.

Rank: r

Partitions: $[\lambda_1, \lambda_2, \ldots, \lambda_r]$.

Rank 1 groups:

$D_1, \mathscr{SO}(2)$: n_1 not defined

$B_1, \mathscr{SO}(3)$: $\lambda_1 = n_1/2$

$C_1, \mathscr{Sp}(2)$: $\lambda_1 = n_1$

Rank 2 groups:

$D_2, \mathscr{SO}(4)$: $\lambda_{1,2} = \frac{1}{2}(n_1 \pm n_2)$

$B_2, \mathscr{SO}(5)$: $\lambda_1 = n_1 + \frac{n_2}{2}$, $\lambda_2 = \frac{n_2}{2}$

$C_2, \mathscr{Sp}(4)$: $\lambda_1 = n_1 + n_2$, $\lambda_2 = n_1$

Rank 3 groups:

$D_3, \mathscr{SO}(6)$: $\lambda_1 = n_2 + \frac{1}{2}(n_1 + n_3)$, $\lambda_{2,3} = \frac{1}{2}(n_1 \pm n_3)$

$B_3, \mathscr{SO}(7)$: $\lambda_1 = n_1 + \frac{1}{2}n_2 + n_3$, $\lambda_2 = n_1 + \frac{1}{2}n_2$, $\lambda_3 = \frac{1}{2}n_2$

$C_3, \mathscr{Sp}(6)$: $\lambda_1 = n_1 + n_2 + n_3$, $\lambda_2 = n_1 + n_2$, $\lambda_3 = n_1$. (12.1.19)

The corresponding relations for higher ranks are similar. The occurrence of half-odd-integer λ_i's for the $\mathscr{SO}(n)$ groups always indicates spinor REPs.

Exercise 12.1. Using the method described in (12.1.8, 9) determine the assignment of Young tableaux and multiplets given in Fig. 11.6, including the degeneracies and the dimensions of the IRs.

Exercise 12.2. Calculate the following direct products of $\mathscr{SU}(3)$ by first reformulating the (n_1, n_2) notation in terms of the notation for the partitions $[\lambda]$:

$$(1,0) \otimes (0,1) \ , \quad (1,0) \otimes (1,0) \ , \quad (2,0) \otimes (1,0) \ ,$$

$$(2,0) \otimes (0,1) \ , \quad (2,0) \otimes (2,0) \ , \quad (2,0) \otimes (0,2) \ ,$$

$$(1,1) \otimes (1,0) \ , \quad (1,1) \otimes (2,0) \ .$$

13. Applications of the Theory of Continuous Groups

In the following we discuss some applications of the theory of continuous groups, but without being complete and exhaustive. It is our aim just to illustrate the possibilities of the theory with some examples. We cannot give a complete discussion of all symmetry groups used in modern particle physics. Rather, we show with some examples how to employ some groups, $\mathcal{SU}(n)$, $n \leqslant 5$, in modern physics and how to work with mechanisms of symmetry breaking. The methods given can also be transferred to an investigation of higher and supersymmetries and more fundamental (?) particles (i.e. partons, rishons, haplons and preons). We assume the physical foundations essentially to be known.

In particular, we discuss symmetry aspects of baryon and meson spectra and of the corresponding wave functions within the quark model using three or four flavour degrees of freedom. Then colour is added to the quark model. Additional topics deal with the implications of rotational symmetry in atomic and nuclear shell theory.

13.1 Elementary Particle Spectra

13.1.1 General Remarks

According to the most recent ideas, all *hadrons* (baryons and mesons) are composed of more fundamental particles, *quarks*. Baryons consist of three quarks, and mesons of a quark-anti-quark pair. The interaction is mediated by renormalizable fields (gauge-invariant Lagrangian densities, Sect. 14.3). An example of such a gauge theory is electrodynamics with gauge group $\mathcal{U}(1)$ (Sects. 14.2.1, 14.3.1).

One possible starting point for a classification of the elementary particles is the *isospin*, introduced by Heisenberg; it is necessary for the description of the charge independence of the strong interaction between nucleons. In this interaction the isospin I is a conserved quantity (internal symmetry), i.e.

$$[\hat{I}^2, H] = [\hat{I}_3, H] = 0 \ . \tag{13.1.1}$$

This allows a classification of nucleons and nuclei of the same mass, spin, parity

Table 13.1. Quantum numbers, masses and lifetimes of mesons and baryons. (ΔE values are marked by asterisks.) Table continues on following page [13.9]

a) Pseudoscalar mesons = $\bar{q}q$ *para-states*, $J^{PC} = 0^{-+}$, $B = 0$

	I^G	I_3	S	C	Mc^2[MeV]	τ[s] or ΔE^*[MeV]
$\pi^\pm$	1^-	± 1	0	0	139.6	2.6×10^{-8}
π^0	1^-	0	0	0	135.0	0.84×10^{-16}
$K^\pm$	1/2	$\pm 1/2$	± 1	0	493.7	1.24×10^{-8}
$K^0, \bar{K}^0$	1/2	$\mp 1/2$	± 1	0	497.7	0.89×10^{-10}
η^0	0^+	0	0	0	547.5	0.0012*
$D^0, \bar{D}^0$	1/2	$\mp 1/2$	0	± 1	1864.6	0.42×10^{-12}
$D^\pm$	1/2	$\pm 1/2$	0	± 1	1869.4	1.06×10^{-12}
$F^\pm = D_s^\pm$	0	0	± 1	± 1	1968.5	0.47×10^{-12}
η_c	0	0	0	0	2978.8	10.3*
η'	0^+	0	0	0	957.8	0.20*

into $\mathcal{SU}(2)$ isospin multiplets.[1] The constituents of an isospin multiplet differ in the component $I_3 = -I, \ldots, +I$; thus the proton p ($I_3 = +1/2$) and neutron n ($I_3 = -1/2$) with $J^P = 1/2^+$ form an isospin doublet; ^{14}C, ^{14}N, ^{14}O with $I_3 = -1$, 0, 1 and $J^P = 0^+$ form an isospin triplet (see Fig. 11.6a).

Considering further experimental evidence, Gell-Mann and Nishijima have introduced another conserved quantity, the hypercharge[2] (also including charm)

$$Y = 2(Q - I_3) - \tfrac{4}{3}C = B + S - \tfrac{1}{3}C \tag{13.1.2}$$

(Q: electric charge, B: baryon number, S: strangeness, C: charm).
Because of (13.1.2) the reaction

$$\left.\begin{array}{lll} \pi^- + p \to \Sigma^+ + K^- & \text{is forbidden,} & \Delta Y = 2 \\ \pi^- + p \to \Sigma^- + K^+ & \text{is allowed,} & \Delta Y = 0 \end{array}\right\} \begin{array}{l}\text{in a strong}\\ \text{interaction}\end{array}$$

(see Table 13.1). The baryon number B (+1 for baryons, -1 for antibaryons, 0 otherwise) is normally a conserved quantity, too, e.g. $B = 1$ in the decay

$$\Xi^- \to \pi^- + \Lambda^\circ \to \pi^- + \pi^\circ + n \to \pi^- + \pi^\circ + e^- + \bar{\nu}_e + p$$

(weak interaction also), whereas a further decay of the proton into lighter particles is forbidden (stability of the universe, see Sect. 14.5). According to this concept, hadrons having the same spin, parity and baryon number and approximately the same mass, may be classified in terms of I_3 and Y, i.e. in a

[1] Of course, the symmetry of this interaction is broken by the electromagnetic interaction and by small mass differences of the quarks which leads to small energy differences between the constituents of the multiplets.

[2] If necessary we distinguish between operators $\hat{Y}$ and quantum numbers (expectation values) Y. Often an explicit distinction is not necessary.

Table 13.1 (continued)

b) *Vector mesons* = $\bar{q}q$ *ortho-states*, $J^{PC} = 1^{--}$, $B = 0$

	I^G	I_3	S	C	Mc^2	τ[s] or ΔE^*[MeV]
$\rho^\pm$	1^+	± 1	0	0	769.9	151.2*
ρ^0	1^+	0	0	0	769.9	151.2*
$K^{*\pm}$	1/2	$\pm 1/2$	± 1	0	891.6	49.8*
$K^{*0}, \bar{K}^{*0}$	1/2	$\mp 1/2$	± 1	0	896.1	50.5*
ϕ	0^-	0	0	0	1019.4	4.43*
$D^{*0}, \bar{D}^{*0}$	1/2	$\mp 1/2$	0	± 1	2006.7	<2.1*
$D^{*\pm}$	1/2	$\pm 1/2$	0	± 1	2010.0	<0.13*
$F^{*\pm} = D_s^{*\pm}$	0	0	± 1	± 1	2110.0	<4.5*
J/ψ	0	0	0	0	3096.9	0.088*
ω	0^-	0	0	0	781.9	8.43*

c) *Baryons* = qqq *states*, $B = 1$

	J^P	I	I_3	Y	S	C	Mc^2	τ[s] or ΔE^*[MeV]
p	$1/2^+$	1/2	+1/2	1	0	0	938.3	∞?, $> 10^{31}$ *a*
n		1/2	−1/2	1	0		939.6	887.0
Λ^0		0	0	0	−1		1115.7	2.63×10^{-10}
Σ^+		1	+1	0	−1		1189.4	0.80×10^{-10}
Σ^0		1	0	0	−1		1192.6	7.4×10^{-20}
Σ^-		1	−1	0	−1		1197.4	1.48×10^{-10}
Ξ^0		1/2	+1/2	−1	−2		1314.9	2.90×10^{-10}
Ξ^-		1/2	−1/2	−1	−2		1321.3	1.64×10^{-10}
Δ^{++}	$3/2^+$	3/2	+3/2	1	0	0		
Δ^+		3/2	+1/2	1	0		1230	115*
Δ^0		3/2	−1/2	1	0		−1234	−125*
Δ^-		3/2	−3/2	1	0			
Σ^{*+}		1	+1	0	−1		1382.8	35.8*
Σ^{*0}		1	0	0	−1		1383.7	36*
Σ^{*-}		1	−1	0	−1		1387.2	39.4*
Ξ^{*0}		1/2	+1/2	−1	−2		1531.8	9.1*
Ξ^{*-}		1/2	−1/2	−1	−2		1535.0	9.9*
Ω^-		0	0	−2	−3		1672.5	0.82×10^{-10}
Λ^*	$1/2^-$	0	0	0	−1	0	1407	50.0*
N'^+	$1/2^+$	1/2	+1/2	1	0	0	1430	250*
N'^0		1/2	−1/2	1	0		−1470	−450*

strong interaction I_3 and Y are compatible quantum numbers[3]:

$$[\hat{I}_3, \hat{Y}] = 0 \ . \tag{13.1.3}$$

Group theoretically we thus need for the description of the states [particle (hadron) spectra] a group of at least rank 2. This leads [as an extension of the isospin group $\mathcal{SU}(2)$] to the group $\mathcal{SU}(3)$ [13.1, 2]. Since according to experimental evidence, the multiplets contain a finite number of particles, a limitation

[3] Using an $\mathcal{SU}(4)$ classification the charm C is a further quantum number.

to compact groups is a priori evident, because their unitary IRs are finite. Consequently the Hamiltonian H_S of the strong interaction has to be invariant under $\mathcal{SU}(3)$.

Generally, if a Lie group $\mathcal{G}$ is the symmetry group of a Hamiltonian H, and if $\{X_i\} = \{i\hat{X}_i\}$ are the generators of the LA of $\mathcal{G}$, then according to (11.1.7, 4.4a)

$$P_a(\alpha) = 1 + \sum_i X_i\alpha_i = 1 + i\sum_i \hat{X}_i\alpha_i \tag{13.1.4}$$

is an infinitesimal transformation which satisfies

$$[P_a, H] = 0 \quad \text{or} \quad [\hat{X}_i, H] = 0 \ . \tag{13.1.5}$$

For unitary REPs, or $P_a^+ P_a = 1$, the $\hat{X}_i$ are Hermitian, the X_i are anti-Hermitian (Sect. 11.4.1).

In quantum mechanics Hermitian operators are assigned to physical quantities and vice versa. Equation (13.1.5) expresses the fact that the physical quantity $\hat{X}_i$ is *conserved* in a system described by H. The symmetry of H with respect to $\mathcal{G}$ thus leads to the conservation of physical quantities (Exercise 13.1 and Noether's theorem, Appendix F). For the strong interaction with $\mathcal{SU}(3)$ symmetry this means $[\hat{X}_i, H_S] = 0$ with $\hat{X}_i \in A_2$, thus providing a classification of the hadrons according to IRs of $\mathcal{SU}(3)$. In the multiplets (see Fig. 11.6b, octet) certain weights may be degenerate. For a unique characterization we then need further quantities that commute with $\hat{I}_3$ and $\hat{Y}$; in general $\hat{I}^2$ is used, i.e. the states with fixed Y are classified according to isospin multiplets of $\mathcal{SU}(2)$ (Fig. 12.3). Thus a state is completely determined by four quantum numbers:

$$|\lambda, I, I_3, Y\rangle \quad \text{with } [\lambda] = (n_1, n_2) \qquad \text{IRs of } \mathcal{SU}(3) \ , \tag{13.1.6}$$

thus in Fig. 12.3 $\lambda = [2, 1] \to (1, 1)$.

The mass differences of the particles assigned to the isospin hypercharge multiplets, have to be explained by a perturbation $H_S^{(1)}$ of lower symmetry, e.g. $\mathcal{SU}(2)$, which causes a splitting of the $\mathcal{SU}(3)$ multiplets into isospin multiplets with the same Y. Finally, the electromagnetic interaction means a further splitting with respect to I_3: $Q = I_3 + Y/2$. The total Hamiltonian is then, for example, $H = H_S + H_S^{(1)} + H_{em}$.

13.1.2 Hadronic States

Gell-Mann and *Zweig* [13.3, 4] were the first to interpret the $\mathcal{SU}(3)$ multiplets of the hadrons as being many-particle states formed of one-particle states (the eightfold way). In this picture the hadrons are composed of more-fundamental particles (*quarks, partons*). But, as yet, no one has succeeded in finding these quarks as asymptotically free, i.e. free-particle states (quark confinement, see Sect. 14.3). However, in deep inelastic scattering of leptons point-like scattering centres (partons) with properties resembling those of quarks have been observed. As an

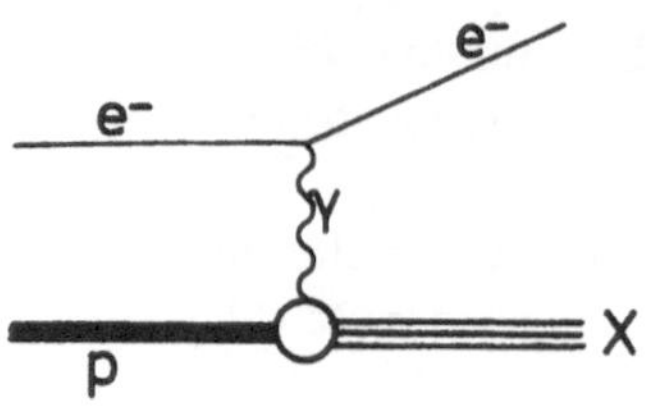

Fig. 13.1. Deep inelastic scattering processes $e^- p \to e^- X$; X represents an arbitrary hadronic final state

Table 13.2. Eigenfunctions and eigenvalues of quarks. The baryon number is 1/3. Antiquarks ($\bar{u}, \bar{d}, \ldots$) have $J^P = 1/2$ and all quantum numbers with reversed sign. M is the constituent mass, m the current quark mass, which enters the Lagrangian, e.g. [13.9, 10]

Flavour	J^P	I_3	Y	S	C	B	T	Q	M	m [GeV/c^2]
$u = \psi_1$	$1/2^+$	1/2	1/3	0	0	0	0	2/3	0.3	0.005
$d = \psi_2$	$1/2^+$	−1/2	1/3	0	0	0	0	−1/3	0.3	0.010
$s = \psi_3$	$1/2^+$	0	−2/3	−1	0	0	0	−1/3	0.45	0.200
$c = \psi_4$	$1/2^+$	0	0	0	1	0	0	2/3	1.5	1.3
$b = \psi_5$	$1/2^+$	0	0	0	0	−1	0	−1/3	5.0	4.3
$t = \psi_6$	$1/2^+$	0	0	0	0	0	1	2/3		174

example one might consider a high momentum probe like a photon (γ) emitted by an electron (e^-) which strikes a nucleon (Fig. 13.1). If the momentum of the γ is high enough, its wavelength will be smaller than the size of the nucleon and it penetrates deep inside it. It then sees that there are actually constituents inside (quarks, partons), which are freely moving point-like objects. This "confinement" of the quarks in the hadrons is a problem of quantum chromodynamics (QCD; see also the running coupling constant in QCD, which shows asymptotic freedom, Sect. 14.4).

There is some evidence that in order to explain the complete hadron spectrum further quark types (flavour states) besides the u, d, s quarks of the eightfold way are required: e.g. a c quark (charmed quark) and a b quark (bottom, sometimes beauty) . For example, the mesonic resonance J/ψ ($M = 3097$ MeV/c^2) consists of quark and antiquark, i.e. $\bar{c}c$, and the Y meson ($M = 9460$ MeV/c^2) is assumed to be a $\bar{b}b$ combination. It is conjectured that there even exists a sixth quark flavour, the t (top, truth) quark. This is derived from the quark–lepton symmetry in the grand unified theories (GUTs, see Sect. 14.4).

If six quark flavours do in fact exist, one could try to use an $\mathscr{SU}_f(6)$ flavour group for classification, instead of an $\mathscr{SU}_f(3)$ group. But there is good evidence (e.g. different masses, see Table 13.2) that this is only a very weak symmetry (if it exists at all) and that even the $\mathscr{SU}_f(3)$ group is not really an exact group (see Sect. 13.1.3, Chap 14 and above). The flavour states can be further divided into pairs u, d; s, c; b, t having almost exact $\mathscr{SU}_f(2)$ symmetry (see the GSW model in Sect. 14.5). In particular, the large difference in mass between u, d, s on one hand and c, b, t on the other justifies the use of $\mathscr{SU}_f(3)$ as an approximate symmetry in connection with the eightfold way. The latter was the first step along the lengthy way to an understanding of quarks and all these things.

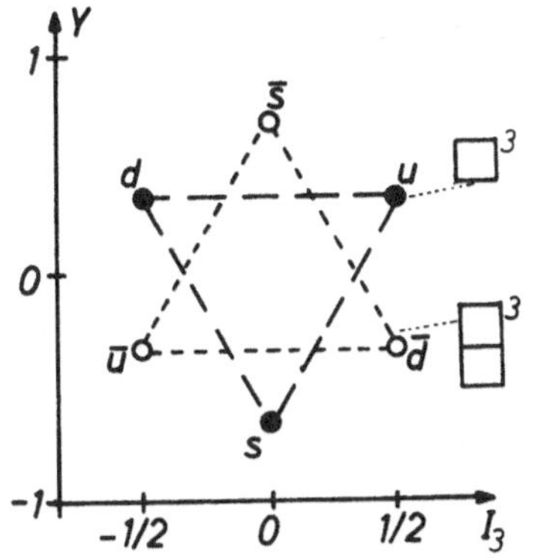

Fig. 13.2. Weight diagram of $[1,0]^3 = (1,0)$, quark states, and of $[1^2]^3 = (0,1)$, antiquark states

In the eightfold-way (*quark*) model of the baryons the baryons are built up of three quarks, each of which may occur in one of the three flavour states if the approximate $\mathcal{SU}_f(3)$ classification is used. The *quarks* are described by three basis functions $\{\psi_1, \psi_2, \psi_3\} := \{u, d, s\} := \{\text{up, down, strange}\}$ (quark-flavour; see also colour states in Sect. 13.1.3), which transform according to the defining IR $[1] := [1,0] = (1,0)$ of $\mathcal{SU}(3)$ (Fig. 11.6b). To these primarily mathematically defined objects there is assigned a real particle, which is just the quark, which may occur in three isospin-hypercharge states. This corresponds to an electron, which occurs in two spin states, transforming according to the defining IR of $\mathcal{SU}(2)$. The basis functions are simultaneously eigenfunctions of the one-particle operators $\hat{I}_3$ and $\hat{Y}$ [see (12.1.10, 11), Table 13.2 and Fig. 13.2].

The space of the baryon functions is the space of the third rank tensors with respect to $\mathcal{SU}(3)$. Group theoretically this means the Clebsch-Gordan expansion of the threefold inner product of $[1,0]^3$ with respect to the IRs of $\mathcal{SU}(3)$.[4] For this we obtain, using Exercise 12.2,

$$[1,0] \otimes [1,0] \otimes [1,0] = [3,0] \oplus [2,1] \oplus [2,1] \oplus [1^3] \ . \qquad (13.1.7)$$

The IRs $[3,0]^3 = (3,0)$ and $[2,1]^3 = (1,1)$ are of special interest because the baryon octet with $J^P = 1/2^+$ fits into the weight diagram of $[2,1]$ and the decuplet of the baryon resonances with $J^P = 3/2^+$ fits into $[3,0]$ (Fig. 13.3, Table 13.1).

In the following we will construct the baryon wave functions from the three quark functions. The method is illustrated in Sect. 5.5.3 and in Exercise 5.19. Here $\psi_f(i) := \psi_f(\xi_i)$ denotes the quark state f, ξ_i is the isospin-hypercharge variable of the ith quark ($i = 1,2,3$), which according to (12.1.11) may take the three values $w^{(1)} = (1/2, 1/3)$, $w^{(2)} = (-1/2, 1/3)$ and $w^{(3)} = (0, -2/3)$. The three flavour functions $\psi_{f_i}(\xi_j)$ are defined by

$$\psi_{f_i}(\xi_j) = \begin{cases} 1 & \text{if } \xi_j = w^{(f_i)} \ ; \quad f_i = 1,2,3 \\ 0 & \text{otherwise} \end{cases} \qquad (13.1.8)$$

[4] The superscript indicates the assignment to $\mathcal{SU}(3)$, see Sect. 5.5.3.

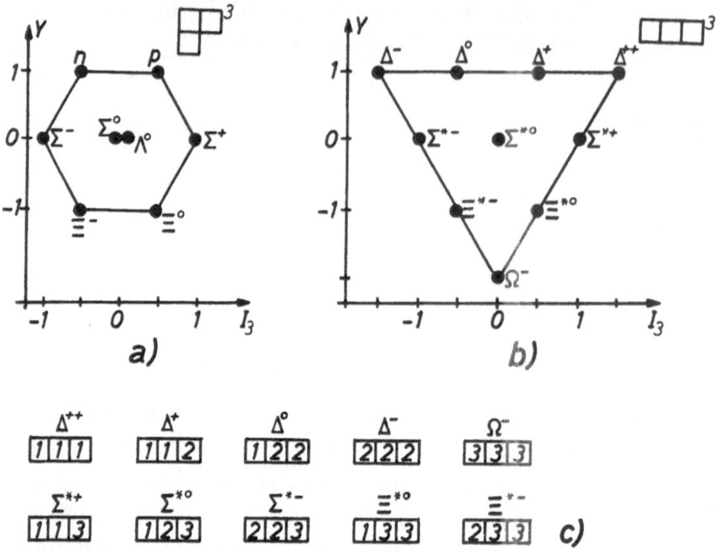

Fig. 13.3a–c. Baryon multiplets. (a) $1/2^+$ octet; (b) $3/2^+$ decuplet; (c) allowed tableaux of the decuplet according to (13.1.7). The allowed tableaux of the octet are given in Fig. 12.3

with

$$\langle \psi_f | \psi_{f'} \rangle := \sum_{\xi_i = w^{(1)}}^{w^{(3)}} \psi_f^*(\xi_i) \psi_{f'}(\xi_i) = \delta_{ff'} \ .$$

Any function of the isospin-hypercharge variables can be expanded with respect to the ψ_f. These concepts are taken from the spin formalism with $\mathcal{SU}(2)$: the spin variable s may take the values $w^{(1)} = +1$ and $w^{(2)} = -1$; the functions $\{\psi_f | f = 1, 2\} \to \{\alpha, \beta\}$ satisfy

$$\alpha(s) = \begin{cases} 1 & \text{if } s = w^{(1)} = 1 \\ 0 & \text{otherwise} \end{cases} \ ; \qquad \beta(s) = \begin{cases} 1 & \text{if } s = w^{(2)} = -1 \\ 0 & \text{otherwise} \end{cases} \ ;$$

$$\langle \psi_f | \psi_{f'} \rangle = \sum_{s = \pm 1} \psi_f^*(s) \cdot \psi_{f'}(s) = \delta_{ff'} \ . \tag{13.1.9}$$

According to Exercise 5.19 we then obtain for the two octets $[2, 1]^3$

$$\Psi_{ij,l,1}^{[2,1]} = \tfrac{1}{3}(\psi_i\psi_j\psi_l + \psi_j\psi_i\psi_l - \psi_l\psi_j\psi_i - \psi_l\psi_i\psi_j) \ , \tag{13.1.10}$$

$$\Psi_{ij,l,2}^{[2,1]} = \tfrac{1}{3}(\psi_i\psi_j\psi_l + \psi_l\psi_j\psi_i - \psi_j\psi_i\psi_l - \psi_j\psi_l\psi_i) \ ,$$

and for the decuplet $[3, 0]^3$ $(+)$ and the singlet $[1^3]^3$ $(-)$, respectively,

$$\Psi_{ijl}^{\pm} = \tfrac{1}{6}(\psi_i\psi_j\psi_l + \psi_l\psi_i\psi_j + \psi_j\psi_l\psi_i \pm \psi_j\psi_i\psi_l \pm \psi_l\psi_j\psi_i \pm \psi_i\psi_l\psi_j) \ . \tag{13.1.11}$$

In the one-particle functions ψ_i, the arguments always have the order 1, 2, 3. This also holds for the baryon functions $|d_\lambda, I, I_3, Y\rangle$ given in Table 13.3. They result from (13.1.10, 11) by using the allowed tableaux for the decuplet (Fig. 13.3) and for the octet (Fig. 12.2) as well as $\psi_1, \psi_2, \psi_3 \rightarrow$ u, d, s. We obtain the functions $|\Sigma^0\rangle$ by applying $\hat{E}_{21}$ to $|\Sigma^+\rangle$ or $\hat{E}_{12}$ to $|\Sigma^-\rangle$, respectively (see Exercise 13.2). Further, we then get the functions $|\Lambda^0\rangle$ by orthogonalizing the linear combination $|\Lambda^0\rangle = a\psi^{[2,1]}_{\{12,3\}} + b\psi^{[2,1]}_{\{13,2\}}$ belonging to the degenerate weight (0,0) with respect to $|\Sigma^0\rangle$.

In this model the *pseudoscalar and vector mesons* (Table 13.1) consist of quark-antiquark pairs. The antiquarks transform according to the REP $[1,1]^3$ contragredient to $[1,0]^3$. This REP is the complex conjugate of the defining REP $[1,0]^3$ (Sect. 11.4.2). We denote the corresponding functions by $\bar{\psi}_f(-\xi_i)$, i.e. they are defined at the positions of the negative weights, see (13.1.8). According to Exercise 12.2, the decomposition of the direct product yields

$$[1,0]^3 \otimes [1,1]^3 = [1^3]^3 \oplus [2,1]^3 \ , \tag{13.1.12}$$

i.e. a meson singlet and a meson octet. The product space spanned by the functions $\bar{\psi}_f(-\xi_i)\psi_{f'}(\xi_i)$ has to be decomposed correspondingly. Similarly to the procedure for $\mathcal{O}(n)$ in (5.5.37–41), the invariant tensor spaces may be generated by contraction. To this end we consider the behaviour of the "mixed" tensors $\bar{\psi}_f\psi_{f'}$ in transformations with $P_a \in \mathcal{SU}(3)$:

$$P_a\bar{\psi}_f\psi_{f'} := (\bar{\psi}_f\psi_{f'})' = \sum_{ll'} a^*_{lf}a_{l'f'}\bar{\psi}_l\psi_{l'} \ . \tag{13.1.13}$$

By contraction we have

$$\sum_f (\bar{\psi}_f\psi_f)' = \sum_{fll'} a^*_{lf}a_{l'f}\bar{\psi}_l\psi_{l'} = \sum_l \bar{\psi}_l\psi_l \ , \quad \text{since} \sum_f a^*_{lf}a_{l'f} = \delta_{ll'} \ .$$

Thus $\sum_f \bar{\psi}_f\psi_f$ is invariant under $\mathcal{SU}(3)$ and it has to be assigned to the singlet. Therefore, the traceless tensors (contraction zero)

$$\Psi^{[2,1]}_{ij} = \bar{\psi}_i\psi_j - \tfrac{1}{3}\delta_{ij}\sum_f \bar{\psi}_f\psi_f \tag{13.1.14}$$

have to be assigned to the octet, and the tensor

$$\Psi^{[1^3]} = \frac{1}{\sqrt{3}}\sum_f \bar{\psi}_f\psi_f \tag{13.1.15}$$

to the singlet. The single factors $\bar{\psi}_f$ and ψ_f have the arguments $-\xi_1$ and ξ_2, respectively. The states are given explicitly in Table 13.4 (again $\psi_1, \psi_2, \psi_3 \rightarrow$ u, d, s), and the multiplets in Fig. 13.4.

Using these functions, it is possible to calculate approximately the mass differences of the mesons. If the quark-antiquark interaction were completely symmetric, then all nine meson states would be degenerate in energy (mass, $E = Mc^2$). But this is not true: the singlet and octet have different masses and

Table 13.3. Baryon functions $|d_\lambda, I, I_3, Y\rangle$

$|1,0,0,0\rangle$	$= |\Lambda^*\rangle = \frac{1}{\sqrt{6}}(uds + sud + dsu - dus - sdu - usd)$		
$|8,1/2,1/2,1\rangle_1$	$= |p\rangle_1 = \frac{1}{\sqrt{2}}(uud - duu)$	$|8,1/2,1/2,1\rangle_2$	$= |p\rangle_2 = \frac{1}{\sqrt{2}}(duu - udu)$
$|8,1/2,-1/2,1\rangle_1$	$= |n\rangle_1 = \frac{1}{\sqrt{2}}(udd - ddu)$	$|8,1/2,-1/2,1\rangle_2$	$= |n\rangle_2 = \frac{1}{\sqrt{2}}(udd - dud)$
$|8,1,1,0\rangle_1$	$= |\Sigma^+\rangle_1 = \frac{1}{\sqrt{2}}(uus - suu)$	$|8,1,1,0\rangle_2$	$= |\Sigma^+\rangle_2 = \frac{1}{\sqrt{2}}(suu - usu)$
$|8,1,0,0\rangle_1$	$= |\Sigma^0\rangle_1 = \frac{1}{2}(uds - sdu + dus - sud)$	$|8,1,0,0\rangle_2$	$= |\Sigma^0\rangle_2 = \frac{1}{2}(usd - sud + dsu - sdu)$
$|8,1,-1,0\rangle_1$	$= |\Sigma^-\rangle_1 = \frac{1}{\sqrt{2}}(dds - sdd)$	$|8,1,-1,0\rangle_2$	$= |\Sigma^-\rangle_2 = \frac{1}{\sqrt{2}}(sdd - dsd)$
$|8,0,0,0\rangle_1$	$= |\Lambda^0\rangle_1 = \frac{1}{\sqrt{12}}(uds - sdu - dus + sud + 2usd - 2dsu)$	$|8,0,0,0\rangle_2$	$= |\Lambda^0\rangle_2 = \frac{1}{\sqrt{12}}(sdu - dsu + usd - sdu + 2uds - 2dus)$
$|8,1/2,1/2,-1\rangle_1$	$= |\Xi^0\rangle_1 = \frac{1}{\sqrt{2}}(uss - ssu)$	$|8,1/2,1/2,-1\rangle_2$	$= |\Xi^0\rangle_2 = \frac{1}{\sqrt{2}}(uss - sus)$
$|8,1/2,-1/2,-1\rangle_1$	$= |\Xi^-\rangle_1 = \frac{1}{\sqrt{2}}(dss - ssd)$	$|8,1/2,-1/2,-1\rangle_2$	$= |\Xi^-\rangle_2 = \frac{1}{\sqrt{2}}(dss - sds)$

Table 13.3 (continued)

$$|10, 3/2, 3/2, 1\rangle = |\Delta^{++}\rangle = uuu$$

$$|10, 3/2, 1/2, 1\rangle = |\Delta^{+}\rangle = \frac{1}{\sqrt{3}}(uud + udu + duu)$$

$$|10, 3/2, -1/2, 1\rangle = |\Delta^{0}\rangle = \frac{1}{\sqrt{3}}(udd + dud + ddu)$$

$$|10, 3/2, -3/2, 1\rangle = |\Delta^{-}\rangle = ddd$$

$$|10, 1, 1, 0\rangle = |\Sigma^{*+}\rangle = \frac{1}{\sqrt{3}}(uus + usu + suu)$$

$$|10, 1, 0, 0\rangle = |\Sigma^{*0}\rangle = \frac{1}{\sqrt{6}}(uds + sud + dsu + dus + sdu + usd)$$

$$|10, 1, -1, 0\rangle = |\Sigma^{*-}\rangle = \frac{1}{\sqrt{3}}(dds + dsd + sdd)$$

$$|10, 1/2, 1/2, -1\rangle = |\Xi^{*0}\rangle = \frac{1}{\sqrt{3}}(uss + sus + ssu)$$

$$|10, 1/2, -1/2, -1\rangle = |\Xi^{*-}\rangle = \frac{1}{\sqrt{3}}(dss + sds + ssd)$$

$$|10, 0, 0, -2\rangle = |\Omega^{-}\rangle = sss$$

Table 13.4. Meson functions

	Pseudoscalar meson $J^P = 0^-$	Vector meson $J^P = 1^-$	Quark structure function
$\|8, 1/2, 1/2, 1\rangle$	K^+	K^{*+}	$\bar{s}u$
$\|8, 1/2, -1/2, 1\rangle$	K^0	K^{*0}	$\bar{s}d$
$\|8, 1, 1, 0\rangle$	π^+	ρ^+	$\bar{d}u$
$\|8, 1, -1, 0\rangle$	π^-	ρ^-	$\bar{u}d$
$\|8, 1, 0, 0\rangle$	π^0	ρ^0	$\frac{1}{\sqrt{2}}(\bar{u}u - \bar{d}d)$
$\|8, 0, 0, 0\rangle$	η^0	ϕ	$\frac{1}{\sqrt{6}}(2\bar{s}s - \bar{u}u - \bar{d}d)$
$\|8, 1/2, 1/2, -1\rangle$	$\bar{K}^0$	$\bar{K}^{*0}$	$\bar{d}s$
$\|8, 1/2, -1/2, -1\rangle$	K^-	K^{*-}	$\bar{u}s$
$\|1, 0, 0, 0\rangle$	η	ω	$\frac{1}{\sqrt{3}}(\bar{u}u + \bar{d}d + \bar{s}s)$

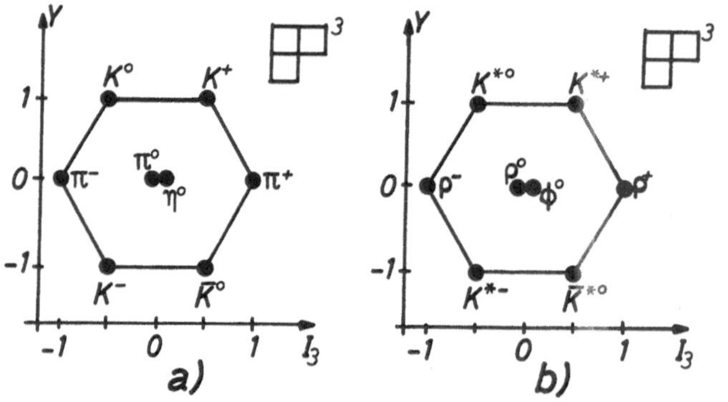

Fig. 13.4a, b. Mesonic octets. (**a**) 0^- (pseudoscalar) mesons; (**b**) 1^- (vector) mesons

even states of one multiplet with different Y differ in mass, whereas the isospin symmetry is conserved to a good approximation. This means that u and d have degenerate masses M, but s has a different mass $M + m$. The symmetry $\mathcal{SU}(3)$ of H_S is broken by a perturbation $H_S^{(1)}$ having $\mathcal{SU}(2)$ symmetry only. In the lowest approximation the masses may be calculated using Table 13.4. Let

$$\langle H_S \rangle = \langle \bar{s}u | H_s | \bar{s}u \rangle = 2M - E_B(8) \tag{13.1.16}$$

be the unperturbed mass of the octet. Then we have

$$M(K^*) = \langle \bar{s}u | H_s + H_s^{(1)} | \bar{s}u \rangle = 2M - E_B(8) + m \tag{13.1.17a}$$

with

$$m = \langle \bar{s}u | H_s^{(1)} | \bar{s}u \rangle = \langle \bar{s} | H_s^{(1)} | \bar{s} \rangle = \langle s | H_s^{(1)} | s \rangle \ ,$$

since the perturbation only acts on either $\bar{s}$ or s. Correspondingly we obtain

$$M(\rho) = 2M - E_B(8) \; ,$$
$$M(\phi) = 2M - E_B(8) + \tfrac{4}{6} \times 2m \; , \qquad (13.1.17b)$$
$$M(\omega) = 2M - E_B(1) + \tfrac{1}{3} \times 2m \; ,$$

thus

$$m = M(K^*) - M(\rho) \approx 130 \text{ MeV}/c^2 \; , \qquad (13.1.18)$$
$$4M(K^*) = M(\rho) + 3M(\phi) \; .$$

[13.5.6]. Analogously we have for the pseudoscalar mesons[5]

$$m = 3[M(\eta) - M(K)] \approx 150 \text{ MeV}/c^2 \; . \qquad (13.1.19)$$

In the baryon decuplet (Table 13.3) the states with $Y = 1, 0, -1, -2$ contain respectively $s = 0, 1, 2, 3$ s quarks. Thus there is a rule for "mass differences":

$$M(\Sigma^*) - M(\Delta) = M(\Xi^*) - M(\Sigma^*) = M(\Omega) - M(\Xi^*) = m \approx 145 \text{ MeV/c}^2 \; . \qquad (13.1.20)$$

This value is in quite good agreement with the value in (13.1.18) determined from the vector-meson data. Using Table 13.3 the baryon octet satisfies

$$M(\Sigma) - M(\text{p}) = M(\Xi) - M(\Sigma) = m \; , \qquad M(\Lambda) = M(\Sigma) \; ; \qquad (13.1.21)$$

experimentally these relations do not agree so well.

The decay of states, e.g. that of the baryon-resonance decuplet, can be well understood within this formalism, too. We ask whether a decay into a baryon (octet) and a pseudoscalar meson (octet) is possible or not and what the possible probability amplitudes for the decay are:

$$B^*_{10} \to B_8 + M_8 \qquad \text{or} \qquad |10, B^*\rangle \to |8, B\rangle |8, M\rangle \; . \qquad (13.1.22)$$

Here B^*, B, M stand for the quantum numbers I, I_3 and Y. The decay amplitudes are given by the matrix elements of the S or T operator, thus

$$T_{BM,B^*} := \langle 8, B; 8, M | T | 10, B^* \rangle \; . \qquad (13.1.23)$$

The final states may be decomposed according to the Clebsch-Gordan expansion, see (4.4.18b),

$$|8, B\rangle |8, M\rangle = \sum_{dDs} \begin{pmatrix} 88 & ds \\ BM & D \end{pmatrix}^* |d, D, s\rangle \; , \qquad (13.1.24)$$

where the sum is over the six IRs of $\mathscr{SU}(3)$ occurring in the decomposition of the product of two octets $[2, 1] \otimes [2, 1]$. This decomposition is given in Fig. 12.4.

[5] The relatively large deviations from experimental values are due to "other" interactions.

Using (13.1.24), (13.1.23) yields

$$T_{BM,B^*} = \sum_{dDs} \left(\begin{array}{c|c} 88 & ds \\ BM & D \end{array} \right) \langle d, D, s | T | 10, B^* \rangle \; . \tag{13.1.25}$$

Because the strong interaction is assumed to be $\mathscr{SU}(3)$ invariant, H, S and T are $\mathscr{SU}(3)$ invariant, i.e. T is a scalar operator in the sense of the Wigner-Eckart theorem. Equation (13.1.25) is different from zero only if $d = 10$ occurs in the decomposition. According to Fig. 12.4 this is true just once: the decay of the baryon resonances into baryons and (pseudoscalar) mesons is possible. The transition amplitude different from zero, see (6.2.5a), is

$$T_{BM,B^*} = \left(\begin{array}{c|c} 88 & 10 \\ BM & B^* \end{array} \right) \langle \| T \| \rangle_{10} \; . \tag{13.1.26}$$

This means that all the decay channels are described by a single decay constant $\langle \| T \| \rangle_{10}$ and the relative amplitudes are given by the CGC. These are (without proof)

$$(\mathrm{n}\pi|\Delta) = -(\Sigma \mathrm{K}|\Delta) = 1/\sqrt{2} \; , \qquad (\Xi\bar{\mathrm{K}}|\Omega) = 1 \; ,$$

$$(\mathrm{n}\bar{\mathrm{K}}|\Sigma^*) = (\Sigma\pi|\Sigma^*) = (\Xi\mathrm{K}|\Sigma^*) = 1/\sqrt{6} \; , \qquad (\Lambda\pi|\Sigma^*) = -(\Sigma\eta|\Sigma^*) = 1/2 \; ,$$

$$(\Sigma\bar{\mathrm{K}}|\Xi^*) = (\Xi\pi|\Xi^*) = (\Lambda\bar{\mathrm{K}}|\Xi^*) = -(\Xi\eta|\Xi^*) = 1/2 \; .$$

The sign depends on convention.

13.1.3 Colour States of Quarks

Up to now we have neglected the spin of the hadrons completely. But this, of course, leads to an incomplete discussion of the possible states. This will be illustrated with the example of the baryon decuplet. The baryon resonances have spin 3/2; according to Exercise 7.3 their spin function is totally symmetric in the spin variables and belongs to the $D^{(3/2)} = [3]^2$ REP of $\mathscr{SU}(2)$ (Sect. 7.2.1). Thus a spin of 1/2 is assigned to the quarks (they are fermions), their spin state being described, for example, by (13.1.9). For vanishing orbital angular momentum of the quark system, the total angular momentum is $J = 3/2$ for the decuplet. The baryon wave function

$$X^{[3]}(s_1, s_2, s_3) \cdot \Psi^{[3,0]}(\xi_1, \xi_2, \xi_3) \tag{13.1.27}$$

is totally symmetric in the spin (s_i) and isospin-hypercharge (ξ_i) variables. However, a system of fermions has to be totally antisymmetric in the particle variables. We therefore have to choose an antisymmetric orbital $\Phi(\boldsymbol{x}_1, \boldsymbol{x}_2, \boldsymbol{x}_3)$ for the space variables. But the common interactions (oscillator, Coulomb, etc. potentials) have a ground state described by a symmetric orbital wave function for the coordinate space $\boldsymbol{x}$. If this is valid for the quark interaction as well, and we have no reason to doubt that, we can only resolve this contradiction [13.7] by assuming a further internal property of quarks, besides flavour: "colour"[6]. According to this assumption a quark with a fixed flavour may occur in three possible colour basis states $\{r, y(\text{or } g), b\} \rightarrow$ {red, yellow(or green), blue}, thus

$$q_{\mathrm{fc}} = \left\{ \begin{matrix} u_{\mathrm{r}} & u_{\mathrm{y}} & u_{\mathrm{b}} \\ d_{\mathrm{r}} & d_{\mathrm{y}} & d_{\mathrm{b}} \\ s_{\mathrm{r}} & s_{\mathrm{y}} & s_{\mathrm{b}} \\ c_{\mathrm{r}} & c_{\mathrm{y}} & c_{\mathrm{b}} \\ \vdots & \vdots & \vdots \end{matrix} \right\} \quad \text{6 flavours}$$

3 colours

The basis states $\{r, y, b\}$ span the defining REP $[1,0]^3$ of the $\mathcal{SU}_{\mathrm{c}}(3)$ colour group, which is assumed to be an exact symmetry group. It is *different* (!) from the approximate $\mathcal{SU}_{\mathrm{f}}(3)$ flavour group discussed above, however, it has the same formal properties. Now, it is assumed that the three quarks combined in a baryon are always in a colour singlet state, the wave function thus being

$$K^{[1^3]} = \frac{1}{\sqrt{6}} \sum_{\kappa_1\kappa_2\kappa_3} \varepsilon_{\kappa_1\kappa_2\kappa_3} q_{f\kappa_1} q_{f'\kappa_2} q_{f''\kappa_3} ,$$

which is totally antisymmetric in the colour variables κ_i of the quarks (the colour of a mixture of red, blue and yellow is colourless, white). Here q_f means one of the quarks u, d, s, ... and $\varepsilon_{\kappa_1\kappa_2\kappa_3}$ is the third rank totally antisymmetric tensor.

Neglecting the space and spin degrees of freedom, the proton wave function, for example, is

$$|\mathrm{p}\rangle = \frac{1}{\sqrt{6}}(u_{\mathrm{r}}u_{\mathrm{y}}d_{\mathrm{b}} + u_{\mathrm{y}}u_{\mathrm{b}}d_{\mathrm{r}} + u_{\mathrm{b}}u_{\mathrm{r}}d_{\mathrm{y}} - u_{\mathrm{r}}u_{\mathrm{b}}d_{\mathrm{y}} - u_{\mathrm{b}}u_{\mathrm{y}}d_{\mathrm{r}} - u_{\mathrm{y}}u_{\mathrm{r}}d_{\mathrm{b}}) . \tag{13.1.28a}$$

With this new degree of freedom we are able to derive the total wave function

$$\begin{aligned} |..\rangle = {} & \Phi^{[3]}(\boldsymbol{x}_1, \boldsymbol{x}_2, \boldsymbol{x}_3) X^{[3]}(s_1, s_2, s_3) \Psi^{[3]}(\xi_1, \xi_2, \xi_3) \\ & \times K^{[1^3]}(\kappa_1, \kappa_2, \kappa_3) , \end{aligned} \tag{13.1.28b}$$

which is totally antisymmetric in the variables space $\boldsymbol{x}$, spin s, isospin-hypercharge ξ and colour κ. Since the fundamental REP of $\mathcal{SU}_{\mathrm{c}}(3)$ is assigned to

[6] This property determines the coupling of the quarks to the gluon field in the gauge theory of the strong-interaction (QCD; Chap. 14). It corresponds to the property "electric charge" which determines the coupling of particles to the photon field in the gauge theory of the electromagnetic interaction; now generally "green", i.e. g.

the colour degrees of freedom, and since the sum of the fundamental weights vanishes, the condition for colourless hadrons can be expressed as

$$J_{c,i}|\text{hadron}\rangle = 0 \; ; \qquad i = 1, \ldots, 8 \; ;$$

$J_{c,i}$ is a generator of $\mathscr{S}\mathscr{U}_c(3)$.

Of course, the property "colour" has to occur in the baryon octet, too. Assuming again a symmetric space function and an antisymmetric colour function (as above), the state spin-isospin-hypercharge has to be symmetric altogether. Since baryon octets have $J = 1/2$, the spin function X has the IR [2, 1] of the permutation group $\mathscr{P}_3$, just as Ψ has. Because, according to (5.5.43), the symmetric REP [3, 0] is contained in the Clebsch-Gordan expansion of the inner product[7] of $[2,1] \otimes [2,1]$, the construction of a function that is symmetric under an interchange of the variables s_i, ξ_i is possible. For this we only have to apply the corresponding Young operator $Y^{[3]} \sim \sum_{p \in \mathscr{P}_3} P_p$ from (5.4.5) to the product function $X_1^{[2,1]} \cdot \Psi_1^{[2,1]}$ with

$$\begin{aligned} X_1^{[2,1]}(s_1, s_2, s_3) &\sim \alpha\alpha\beta - \beta\alpha\alpha \\ \Psi_1^{[2,1]}(\xi_1, \xi_2, \xi_3) &\sim uud - duu \end{aligned} \tag{13.1.29}$$

(see Exercise 7.3, Table 13.3). For example, using (13.1.29) we get the proton function. The arguments again have to be chosen in the order 1, 2, 3 in the functions α, β as well as in u, d. Thus we obtain

$$\begin{aligned} |p, s_z = 1/2\rangle = \frac{1}{\sqrt{18}} [2\,(\alpha u \alpha u \beta d + \alpha u \beta d \alpha u + \beta d \alpha u \alpha u) - \alpha u \alpha d \beta u \\ - \alpha d \alpha u \beta u - \beta u \alpha d \alpha u - \alpha u \beta u \alpha d - \beta u \alpha u \alpha d - \alpha d \beta u \alpha u] \end{aligned} \tag{13.1.30}$$

for the proton state and a corresponding expression for the neutron state with u and d interchanged. In the same way we get the other baryon states when the appropriate function from Table 13.3 is used in (13.1.29).

These functions may be used for the calculation of the magnetic moments and other properties of baryons. The magnetic moment of a baryon is assumed to be the sum of the moments of the three quarks. We therefore set

$$\mu = \sum_{i=1}^{3} 2\mu_p Q(i) s(i) \; , \qquad Q(i) = I_3(i) + \tfrac{1}{2} Y(i) \; , \tag{13.1.31}$$

where Q_i is the charge operator and $s(i)$ the spin operator of the ith quark. The

---

[7] Here we have the decomposition of the inner product of REPs of the permutation group $\mathscr{P}_3$, which is different from the decomposition in Fig. 12.4 of inner products of REPs of $\mathscr{S}\mathscr{U}(3)$. With the characters of $\mathscr{P}_3$ we obtain

$$[2,1] \otimes [2,1] = [3] \oplus [1^3] \oplus [2,1] .$$

magnetic moment of a baryon is

$$\langle\mu\rangle = \langle\Psi, s_{z,\max}|\mu_z|\Psi, s_{z,\max}\rangle \;, \tag{13.1.32}$$

where the baryon function is that with maximal z-component of the spin. Using (13.1.30) we obtain (the proof is left to the reader)

$$\begin{aligned} &\langle\mu\rangle = \mu_p && \text{for the proton },\\ &\langle\mu\rangle = \mu_n = -\tfrac{2}{3}\mu_p && \text{for the neutron },\end{aligned} \tag{13.1.33}$$

which is in good agreement with the experimental value of $\mu_n = -0.685\,\mu_p$.

For the mesons the colour state is also a singlet (colourless, white). The complementary colour has to be assigned to the antiquarks in a way very similar to the assignment in (13.1.15). The singlet function is then (where $u_r \to r$, etc.)

$$K^{[1^3]}(f,f') = \frac{1}{\sqrt{3}}\sum_{\kappa} \bar{q}_{f\bar{\kappa}}q_{f'\kappa} = \frac{1}{\sqrt{3}}(\bar{r}r + \bar{b}b + \bar{y}y) \;,$$

where κ takes the weights from $(1,0)$, $\bar{\kappa}$ the weights from $(0,1)$, so that altogether a colourless (white) state results which has a vanishing sum of fundamental weights. The conjugate functions $\{\bar{\kappa}_1, \bar{\kappa}_2, \bar{\kappa}_3\}$ are $\{\bar{r}, \bar{y}(\text{or } \bar{g}), \bar{b}\}$. A $\pi^+(\rho^+)$ meson is then represented by

$$|\pi^+\rangle \;, \qquad |\rho^+\rangle = \frac{1}{\sqrt{3}}(u_r\bar{d}_{\bar{r}} + u_y\bar{d}_{\bar{y}} + u_b\bar{d}_{\bar{b}}),$$

where the corresponding spin function has to be included (Table 13.4).

13.1.4 A Possible $\mathscr{SU}(4)$ Classification

The discovery of the J/ψ resonance at 3097 MeV with a very small line width of 70 keV (typical line widths of hadronic resonances are 10 − 100 MeV) led to the "necessity" of supplementing the flavours of the quarks with a further degree of freedom, the *charm* C (see Table 13.2, $J/\psi \to \bar{c}c$). The flavour multiplets which include charm show a large mass splitting, so that $\mathscr{SU}_f(4)$ can be looked upon as being a symmetry group only very roughly. On the other hand, the large mass difference between c and u, d, s explains a decoupling of the dynamics and, consequently, why the hadron dynamics may be well described without charm, just using a quark model with three types of flavour [$\to\mathscr{SU}_f(3)$] or even fewer (Sect. 13.1.1). The u and d quarks transform as a doublet under almost exact $\mathscr{SU}_f(2)$; u, d and s transform as a triplet under an approximate $\mathscr{SU}_f(3)$; u, d, s and c transform as a quartet under a strongly broken $\mathscr{SU}_f(4)$, and so on.

The flavour degree of freedom "charm" is described by a further additive operator $\hat{C}$ (not to be confused with the Casimir operator, Sect. 11.1.3) commuting with $\hat{I}_3$ and $\hat{Y}$, since the quantum number C is assumed to be "good" here (conservation law). The group $\mathscr{SU}_f(3)$ has to be extended to a third-rank group, which leads to $\mathscr{SU}_f(4)$ as the simplest extension. The mesonic resonance J/ψ consists of quark and antiquark ($C = \pm 1$) and so has $C = 0$ itself (hidden charm).

Using the $\mathscr{SU}_f(4)$ classification a quark may exist in four flavour states. These isospin-hypercharge-charm states $\{\psi_1, \psi_2, \psi_3, \psi_4\} = \{u, d, s, c\}$ transform according to the defining IR $[1]^4 := [1,0,0]^4$ of $\mathscr{SU}(4)$ (Fig. 11.6c). We choose the commuting generators according to (11.4.5) to be

$$\hat{H}_1 = \hat{I}_3 \;, \qquad \hat{H}_2 = \hat{Y} \;, \qquad \hat{H}_3 = \tfrac{3}{4}\hat{B} - \hat{C} \;, \qquad \hat{Q} = \hat{I}_3 + \hat{Y}/2 + 2\hat{C}/3 \tag{13.1.34}$$

with $c_1 = 1/2$, $c_2 = 1/3$, $c_3 = 1/4$; $\hat{B}$: baryon number with $\hat{B}\psi_i = \frac{1}{3}\psi_i$, $i = 1, 2, 3, 4$. According to (11.4.23) the fundamental weights are

$$\begin{aligned} w^{(1)} &= (1/2, 1/3, 1/4) \;, \qquad & w^{(2)} &= (-1/2, 1/3, 1/4) \;, \\ w^{(3)} &= (0, -2/3, 1/4) \;, \qquad & w^{(4)} &= (0, 0, -3/4) \;. \end{aligned} \tag{13.1.35}$$

Only the c state of the quark has charm! The possible hadron multiplets follow as in Sect. 13.1.2. Equation (13.1.7) is again valid for the baryons, but the graphs have to be labelled with respect to $\mathscr{SU}(4)$. For the dimension we now have

$$4 \times 4 \times 4 = 20 + 20 + 20 + 4 \;. \tag{13.1.36}$$

For mesons, (13.1.12) has to be changed correspondingly (Fig. 11.6):

$$\begin{aligned} &[1,0,0]^4 \otimes [1,1,1]^4 = [2,1,1]^4 \oplus [1^4]^4 \\ &\quad 4 \quad \times \quad 4 \quad = \quad 15 \quad + \quad 1 \end{aligned} \tag{13.1.37}$$

From the assigned weight diagrams (Fig. 13.5) we see that $[2,1,0]^4$ comprises the baryon octet, $[3,0,0]^4$ the baryon decuplet and $[2,1,1]^4$ the meson octets.

We obtain the wave functions for the baryons from (13.1.10, 11) in the same way as for $\mathscr{SU}(3)$, but i, j, l now take the values 1, 2, 3, 4. The meson functions are essentially given by (13.1.14, 15):

$$\Psi_{ij}^{[2,1,1]} = \bar{\psi}_i \psi_j - \tfrac{1}{4}\delta_{ij} \sum_f \bar{\psi}_f \psi_f \;, \tag{13.1.38a}$$

$$\Psi_{ij}^{[1^4]} = \tfrac{1}{2} \sum_f \bar{\psi}_f \psi_f \;. \tag{13.1.38b}$$

The distinct weight diagrams can be determined using (12.1.7) and (13.1.35); they then give the multiplets represented in Fig. 13.5 (see also Exercise 13.4).

From the discussion it also follows that the flavour types (degrees of freedom) may be developed further, if necessary. This would lead to rather poor symmetry groups $\mathscr{SU}(n)$ with $n > 4$. The technique of Young diagrams and the generation of the multiplets can be easily extended (but see again the remarks in Sect. 13.1.2).

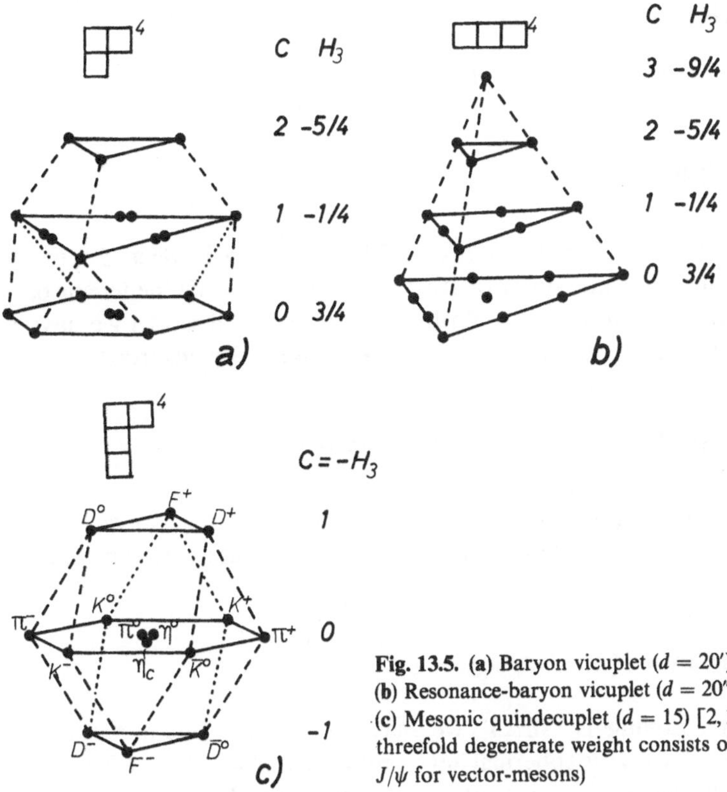

Fig. 13.5. (**a**) Baryon vicuplet ($d = 20'$) $[2, 1, 0]^4 = (1, 1, 0)$. (**b**) Resonance-baryon vicuplet ($d = 20''$) $[3, 0, 0]^4 = (3, 0, 0)$. (**c**) Mesonic quindecuplet ($d = 15$) $[2, 1, 1]^4 = (1, 0, 1)$. The threefold degenerate weight consists of π^0, η^0, η_c (or ρ^0, ϕ, J/ψ for vector-mesons)

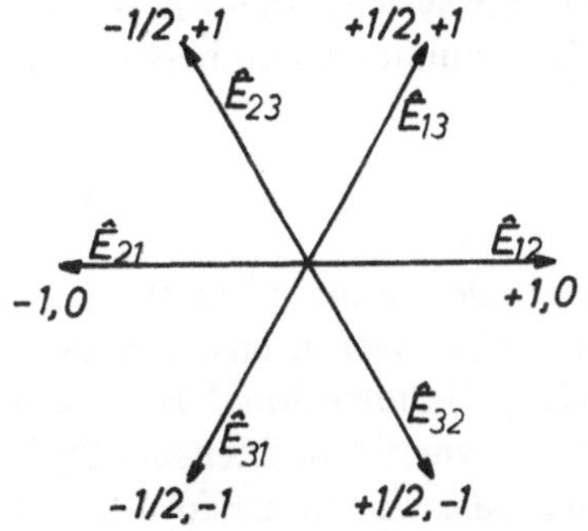

Fig. 13.6. Illustration of the effect of ladder operators $\hat{E}_{ij}$ of $\mathcal{SU}(3)$

Exercise 13.1. Show that momentum, angular momentum and energy are conserved if H is invariant with respect to: translations in space, rotations in space, translations in time.

Exercise 13.2. Show that the application of the ladder operators $\hat{E}_{ij}$ of $\mathcal{SU}(3)$ in the $I_3 Y$-plane can be illustrated by the diagram given in Fig. 13.6. Use (11.4.13–15).

Exercise 13.3. Show explicitly the validity of (13.1.28a) for the proton state and the validity of the corresponding neutron state of the octet.

Exercise 13.4. Using (13.1.35) and (12.1.7) determine the multiplets

$$[2,1,0]^4,\ [3,0,0]^4 \text{ and } [2,1,1]^4 \text{ of } \mathcal{SU}(4)\ .$$

13.2 Atomic Spectra

In order to be able to apply the tensor REPs described in Sect. 5.5 to the classification of the electronic states of atoms we have to specify the tensor basis $\{\psi_{i_1}(1),\ldots,\psi_{i_m}(m)\}$ defined in (5.5.9). Within the atomic shell model we have to distinguish whether the Coulomb interaction V between the electrons entering the electronic Hamiltonian (7.2.3, 5) or (8.2.9),

$$H_e = H_0 + V + H_{LS}\ , \qquad H_{LS} = \sum_i f(x_i)[\boldsymbol{l}(i)\cdot \boldsymbol{s}(i)]\ , \tag{13.2.1}$$

is larger than the spin-orbit coupling H_{LS} or not. This determines the approximation to be used. If necessary one has to interpolate between different approximations.

13.2.1 Russell-Saunders (*LS*) Coupling

If the spin-orbit coupling is "small" we may identify the one-particle basis functions from (5.5.9) with the spherical harmonics $Y_m^l(i)$, $m = -l,\ldots,l$. The radial function R_{nl} can be taken into account just as a factor multiplying the Y_m^l. The argument i runs over the number N of electrons in the system (an atom or only the outer-shell electrons). All the electrons with fixed quantum numbers n and l form one shell; they have the product basis

$$\Phi_{m_1\ldots m_N}(1,\ldots,N) = Y_{m_1}^l(1)\ldots Y_{m_N}^l(N)\ . \tag{13.2.2}$$

In the sense of Sect. 5.5 we may look upon the single factors Y_m^l as the basis functions of the defining IR of $\mathcal{U}(2l+1)$ or $\mathcal{SU}(2l+1)$, or even as basis functions of the IR $D^{(l)}$ of $\mathcal{SU}(2)$ or $\mathcal{SO}(3)$, which are also $(2l+1)$-dimensional. By means of the methods of Sect. 5.5 we are able to generate the symmetrized tensors $\Psi_{\boldsymbol{m},k}^{[\lambda]}$, $\boldsymbol{m} = \{m_1\ldots m_N\}$, according to (5.5.31). These are basis functions for the IR $D^{[\lambda]}(2l+1)$ of $\mathcal{SU}(2l+1)$. In general the space spanned by these basis functions is not irreducible with respect to the subgroup $\mathcal{SU}(2)$ and decomposes into irreducible subspaces $\mathcal{L}^{(l)}$, so that

$$D^{[\lambda]}(2l+1) = \sum_l a_l D^{(l)} \tag{13.2.3}$$

is valid [subduction with respect to $\mathcal{SU}(2)$ or $\mathcal{SO}(3)$]. In this way a set of $\mathcal{SU}(2)$ and $\mathcal{SO}(3)$ multiplets, characterized by the quantum number l of angular momentum is assigned to every "particle permutation symmetry" $[\lambda]$. This assignment allows one to decide which REPs $[\lambda]$ have to be taken into account when, for a given angular momentum, the wave functions are to be classified in terms of permutation symmetries (Sect. 7.3.1).

First we discuss a two-electron problem, i.e. the "shell" $Y^l_m(1)Y^l_m(2)$. These $(2l+1)^2$ functions span the basis for $D^{(l)} \otimes D^{(l)}$ with respect to $\mathscr{SU}(2)$ and the basis for $[1]^{2l+1} \otimes [1]^{2l+1}$ with respect to $\mathscr{SU}(2l+1)$. We shall compare the Clebsch-Gordan expansions for both the products; for this reason we abbreviate $D^{(l)}$ to (l). According to Fig. 5.7 and, for example, (11.4.71),

$$(l) \otimes (l) = (2l) \oplus (2l-1) \oplus \cdots \oplus (1) \oplus (0) \ , \tag{13.2.4a}$$

whereas according to (5.5.46, 47) with the dimensions according to (5.5.33)

$$\begin{aligned} &[1]^{2l+1} \otimes [1]^{2l+1} = [2]^{2l+1} \oplus [1^2]^{2l+1} \ , \\ &d\colon \quad (2l+1)^2 \quad = (l+1)(2l+1) + l(2l+1) \ . \end{aligned} \tag{13.2.4b}$$

Using the Clebsch-Gordan decomposition

$$\Phi^L_M(1,2) := \sum_{m_1 m_2} \begin{pmatrix} l & l & L \\ m_1 & m_2 & M \end{pmatrix} Y^l_{m_1}(1)\, Y^l_{m_2}(2) \tag{13.2.5a}$$

and (11.4.75)

$$\Phi^L_M(2,1) = (-1)^{2l-L} \Phi^L_M(1,2) \ . \tag{13.2.5b}$$

That means Φ^L_M is symmetric (belonging to [2]) if $2l - L$ is even, and antisymmetric (belonging to $[1^2]$) if $2l - L$ is odd. By subducing into $\mathscr{SU}(2)$ or $\mathscr{SO}(3)$ we therefore have the assignment

$$\left.\begin{aligned} &[2]^{2l+1} := \square\square \to (0) \oplus (2) \oplus \cdots \oplus (2l) \\ &[1^2]^{2l+1} := \begin{smallmatrix}\square\\\square\end{smallmatrix} \to (1) \oplus (3) \oplus \cdots (2l-1) \end{aligned}\right\} \text{if } l \text{ is an integer} \ , \tag{13.2.6a}$$

$$\left.\begin{aligned} &[2]^{2l+1} := \square\square \to (1) \oplus (3) \oplus \cdots \oplus (2l) \\ &[1^2]^{2l+1} := \begin{smallmatrix}\square\\\square\end{smallmatrix} \to (0) \oplus (2) \oplus \cdots \oplus (2l-1) \end{aligned}\right\} \text{if } l \text{ is half an odd integer} \ . \tag{13.2.6b}$$

The spin function assigned to (13.2.6a) in the LS scheme has to belong to the associated Young diagram according to (5.4.28). Consequently, if the space function (orbital) is symmetric it has to be a singlet ($S = 0$, $[\tilde{\lambda}] = [1^2]$); if the space function is antisymmetric it has to be a triplet ($S = 1$, $[\tilde{\lambda}] = [2]$).

In the case that there are more than two electrons this procedure can be generalized recurrently, where use is always made of the equivalence of associated diagrams [see (5.5.35) and Fig. 5.4b]. As an example we consider three d-electrons ($l = 2$) and use the spectroscopic notations

$$L = \begin{matrix} 0, & 1, & 2, & 3, & 4, & 5, & 6 \\ S, & P, & D, & F, & G, & H, & J \end{matrix} \ .$$

For one electron

$$[1]^5 \to D \;, \qquad \text{i.e. } l = 2 \;.$$

Using (13.2.6a) for two electrons we have

$$[2]^5 \to S \oplus D \oplus G \;, \qquad [1^2]^5 \to P \oplus F \;. \tag{13.2.7}$$

The coupling to a third *d*-electron gives according to (5.5.46, 47)

$$[2]^5 \otimes [1]^5 = [2,1]^5 \oplus [3]^5 \;, \tag{13.2.8a}$$

$$[1^2]^5 \otimes [1]^5 = [2,1]^5 \oplus [1^3]^5 \;, \tag{13.2.8b}$$

or, using (11.4.71) and (13.2.7), the reductions

$$(S \oplus D \oplus G) \otimes D = S \oplus P \oplus 3D \oplus 2F \oplus 2G \oplus H \oplus J \;, \tag{13.2.9a}$$

$$(P \oplus F) \otimes D = 2P \oplus 2D \oplus 2F \oplus G \oplus H \;. \tag{13.2.9b}$$

Because of (5.5.35), for $\mathscr{SU}(5)$ the REP $[1^3]^5$ is according to (13.2.7) equivalent to the REP $[1^2]^5 \to P \oplus F$, thus

$$[1^3] \to P \oplus F \qquad \text{with} \qquad S = 3/2 \;. \tag{13.2.10a}$$

By comparing (13.2.8b) with (13.2.9b) we obtain the assignment

$$[2,1]^5 \to P \oplus 2D \oplus F \oplus G \oplus H \qquad \text{with} \qquad S = 1/2 \;. \tag{13.2.10b}$$

By a further comparison of (13.2.8a) with (13.2.9a) we have

$$[3]^5 \to S \oplus D \oplus F \oplus G \oplus J \;, \qquad \text{forbidden} \;, \tag{13.2.10c}$$

as, according to the Pauli principle, $[3]^5$ is forbidden in atomic shells, because the associated diagram $[1^3]^5$ for the spin has more than two rows, see Sect. 7.2.1. Therefore these states do not occur.

Equation (13.2.10b) contains two ^2D terms, and thus two states with equal L and S. Hence the LS classification in systems with more than two electrons is no longer sufficient for a unique characterization. Racah [13.8] removes these ambiguities by the introduction of additional quantum numbers which are assigned to the operators (generators) of a series of groups, e.g. $\mathscr{SU}(2l+1) \supset \mathscr{SO}(2l+1) \supset \mathscr{SO}(3)$ (seniority, for details see specialist textbooks).

The total symmetry group of the Hamiltonian H_e with $H_{LS} = 0$ is the outer product $\mathscr{C}_i \times \mathscr{SO}(3) \times \mathscr{P}_N$. The IRs are the outer direct products (4.4.8)

$$D^{(\pm)} \times D^{(L)} \times D^{[\lambda]} \;. \tag{13.2.11a}$$

The eigenvalues and eigenfunctions are then classified according to

$$E_s^{(L,\pm,\lambda)} \quad \text{or} \quad \Phi_{s,d}^{(L,\pm,\lambda)} , \tag{13.2.11b}$$

where s distinguishes between different levels with equal $(L, \pm, \lambda)$ and $d = 1, \ldots, (2L+1)d_{[\lambda]}$ distinguishes between the degenerate states. The symmetry subgroup $\mathcal{SU}(2)$ or $\mathcal{SO}(3)$ or its algebra is, of course, generated by the many-particle angular momenta

$$\boldsymbol{L} = \sum_{i=1}^{N} \boldsymbol{l}(i) \qquad \text{with} \qquad [\boldsymbol{L}, H_e] = 0 . \tag{13.2.12}$$

If, starting with a one-particle approximation (Sect. 7.2.1), the Coulomb interaction of the electrons is to be treated as a perturbation, then we first have to generate functions symmetrized with respect to $\mathcal{P}_N$ from the product functions (5.5.9) and (13.2.2) using methods of Sect. 7.3.1. These functions have to be ordered with respect to the angular momenta. In this way we get the appropriate symmetry-adapted functions $\Phi_{s,d}^{(L,\pm,\lambda)}$ necessary for a perturbation calculation. Using the Wigner-Eckart theorem the matrix elements of the perturbation W then have the form

$$\langle \Phi_{s,d'}^{(L',\pm',\lambda')} | W | \Phi_{s,d}^{(L,\pm,\lambda)} \rangle = \delta_{LL'} \delta_{\pm\pm'} \delta_{\lambda\lambda'} \delta_{dd'} \langle \| W \| \rangle^{(L,\pm,\lambda)} . \tag{13.2.13}$$

When spin-orbit coupling H_{LS} is taken into account, the symmetry group is $\mathcal{C}_i \times \mathcal{SU}(2) \times \mathcal{P}_N$. However $\mathcal{SU}(2)$, or its algebra A_1, is generated by the total angular momentum

$$\boldsymbol{J} = \boldsymbol{L} + \boldsymbol{S} , \qquad \boldsymbol{S} = \sum_{i=1}^{N} \boldsymbol{s}(i) , \qquad [\boldsymbol{J}, H_e] = 0 . \tag{13.2.14}$$

In a perturbation theory for H_{LS} we have to combine the functions $\Phi_{s,d}^{(L,\pm,\lambda)}$ for $H_{LS} = 0$ with the spin functions $X_{m_s}^{[\tilde{\lambda}]}$, $m_s = -S, \ldots, +S$; $[\tilde{\lambda}] \leftrightarrow S$ according to (7.2.7). Again, from the combined functions we are able to obtain the symmetry-adapted functions for the perturbation calculation. To achieve this we have to perform the Clebsch-Gordan decomposition

$$(L \otimes S) = (L+S) \oplus (L+S-1) \oplus \cdots \oplus |L-S| . \tag{13.2.15}$$

Then the $(2L+1)(2S+1)$-fold degenerate energy levels split according to (13.2.15).

13.2.2 *jj* Coupling

This coupling occurs if the spin-orbit coupling is "large" compared to the Coulomb interaction. Then, first, the one-particle functions ψ_i have to be determined from the reduction

$$(l \otimes s) = (l + 1/2) \oplus (l - 1/2) , \tag{13.2.16}$$

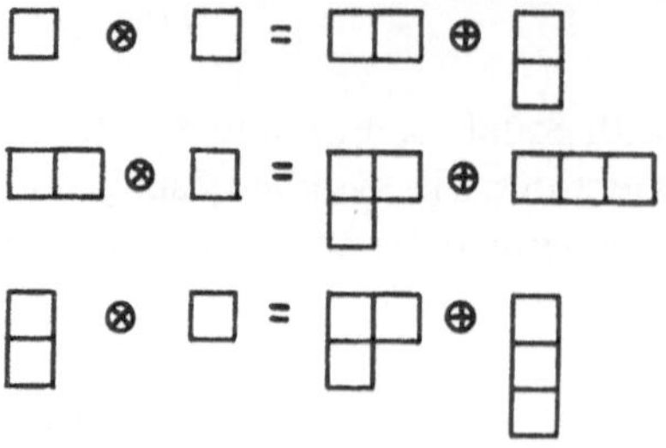

Fig. 13.7. Graphical representation of the products in (13.2.4b) for $2l+1$ and in (13.2.8a, b) for $l = 2$

yielding the spin-orbit coupled functions

$$\psi^j_{m_j} = \sum_{m_l m_s} \begin{pmatrix} l & 1/2 & j \\ m_l & m_s & m_j \end{pmatrix} Y^l_{m_l} \chi_{m_s} \; , \qquad j = l \pm 1/2 \; . \tag{13.2.17}$$

Here $\chi_{1/2} \to \alpha$, $\chi_{-1/2} \to \beta$ are the one-electron spin functions. Of course, a radial function R_{nl} needs to be added to (13.2.17). Using (13.2.17) we have to form product functions of the type (5.5.9), which must be eigenfunctions of J^2 and J_z simultaneously because $[J, H] = 0$. However, first of all we have to check which J values are allowed by the Pauli principle. The product functions built from (13.2.17) must be basis functions belonging to the IR $[1^N]$ of $\mathscr{P}_N$.

We illustrate this with the example of two equivalent electrons with angular momentum j. Then

$$(j \otimes j) = (2j) \oplus (2j-1) \oplus \cdots \oplus (1) \oplus (0) \tag{13.2.18}$$

and from this, using (13.2.6b),

$$\begin{aligned} &[2] := \square\!\square \to (1) \oplus (3) \oplus \cdots \oplus (2j) \; , \\ &[1^2] := \begin{smallmatrix}\square\\ \square\end{smallmatrix} \to (0) \oplus (2) \oplus \cdots \oplus (2j-1) \; . \end{aligned} \tag{13.2.19a}$$

Because of the Pauli principle, for $N = 2$ only $[1^2]$ is allowed; the possible total angular moments are thus

$$J = 0, 2, \ldots, 2j-1 \; , \qquad \text{i.e. they are even} \; . \tag{13.2.19b}$$

Exercise 13.5. Using the method given in (13.2.7–10), calculate the states of angular momenta for three p-electrons ($l = 1$) and determine their assignment to the Young diagrams $[1]^3$, $[2,1]^3$ and $[3]^3$ (Fig. 13.7).

13.3 Nuclear Spectra

In the shell model of nuclei, unlike the case of atomic shells, we have additionally to take into account the new degree of freedom "isospin" $\mathscr{I} = (I_1, I_2, I_3)$, which describes the charge independence of nuclear forces (Sect. 13.1.1). The formalism is identical with that of spin. The isospin basis functions corresponding to the spin functions α, β are

$$\psi_{m_I} \quad \text{with} \qquad \psi_{1/2} := p \ , \qquad \psi_{-1/2} := n \ . \tag{13.3.1}$$

They are basis functions of the defining IR $[1]^2$ of the isospin group $\mathcal{SU}_I(2)$ which is generated by I_1, I_2, I_3. Together with the basis functions of the spin group $\mathcal{SU}_s(2)$ there are four spin-isospin states

$$\psi_1 = \alpha p \ , \qquad \psi_2 = \alpha n \ , \qquad \psi_3 = \beta p \ , \qquad \psi_4 = \beta n \ , \tag{13.3.2}$$

where α, β are functions of the spin variable, and p, n are functions of the isospin variable.

13.3.1 *jj–JI* Coupling

Because of the large spin-orbit interaction of nucleons in the nucleus, nuclei have *jj* coupling, except for the very lightest nuclei. The Hamiltonian of the shell model is (A: number of nucleons)

$$H = \sum_{i=1}^{A} \Bigg(\frac{p_i^2}{2\mu_i} + \underbrace{V(\boldsymbol{x}_i) + \zeta_{ls} \frac{1}{r_i} \frac{\partial V}{\partial r_i} [\boldsymbol{l}(i) \cdot \boldsymbol{s}(i)]}_{:= V^{(1)}(i)} \Bigg) \ ; \tag{13.3.3}$$

where μ_i is the reduced mass of the ith nucleon, $V(\boldsymbol{x}_i)$ a mean effective potential (usually chosen to be an oscillator, square well etc., potential) and ζ_{ls} the spin-orbit coupling constant. Equation (13.3.3) represents a one-particle approximation, which is justified by the "success" of the shell model. For an improvement one has to expand the many-particle interaction with respect to different n-particle contributions [compare $V^{(1)}(i)$ in (13.3.3)]

$$V(1, \ldots, A) = \sum_{i=1}^{A} V^{(1)}(i) + \tfrac{1}{2} \sum_{i,j=1}^{A} V^{(2)}(i, j) + \cdots \ . \tag{13.3.4}$$

The higher terms $V^{(2)}, \ldots$, according to Sect. 7.3.1 cause an energy shift of the diagonal terms and a splitting of the one-particle levels due to exchange effects.

To solve the eigenvalue problem belonging to (13.3.3), the one-particle basis functions in (5.5.9) have to be specified. Taking into account the isospin, we have in analogy to (13.2.17)

$$\psi^j_{m_j, m_I} = \sum_{m_l m_s} \begin{pmatrix} l & 1/2 & j \\ m_l & m_s & m_j \end{pmatrix} Y^l_{m_l} \chi_{m_s} \psi_{m_I} \ , \qquad j = l \pm 1/2 \ , \qquad m_I = \pm 1/2 \ . \tag{13.3.5}$$

The product function formed from these functions has to be antisymmetric because of the Pauli principle, i.e. it has to be a basis function of the IR $[1^A]$ of $\mathscr{P}_A$. The treatment corresponds to the procedure for the space-spin functions described in Sect. 7.2.1.

The Young diagrams of the isospin functions $\psi^{[\tilde{\lambda}]}(1,\ldots,A)$ have two rows because the one-particle isospin functions transform according to $\mathscr{SU}_I(2)$. The total isospin is calculated according to (7.2.11). The Young diagrams belonging to the space-spin functions $\psi^{[\lambda]}(1,\ldots,A)$ then have at most two columns $(1,\ldots,A$ here denote the space and spin coordinate). The discussion of JI coupling in this case is very similar to that of LS coupling for the atomic shells. Here the method described in Sect. 13.2.1, which assigns the L values to diagrams $[\lambda]$, just has to be transferred to the assignment of J. We shall not give the details here.

13.3.2 *LSI* Coupling

In this case the one-particle functions have the form $Y^l_{m_l}\chi_{m_s}\psi_{m_I}$. The product functions according to (13.2.2) to be evaluated have to be eigenfunctions of L^2 and L_z with a well-defined permutation symmetry $[\lambda]$ because of the condition $[H,L]=0$. The spin-isospin functions are chosen to be eigenfunctions of S^2, S_z, I^2 and I_3 because of the invariance of H under $\mathscr{SU}_S(2)$, $\mathscr{SU}_I(2)$ and $\mathscr{P}_A$ and because of the Pauli principle, and they have to belong to the associate diagram $[\tilde{\lambda}]$ of $[\lambda]$. The Young diagrams $[\tilde{\lambda}]$ of the spin-isospin functions in this case have at most four rows, since there are four basis states (13.3.2). This implies that the graphs $[\lambda]$ assigned to the space functions possess at most four columns. Therefore the angular momentum states $[\lambda]=[3]$ (13.2.10c), forbidden for atomic shells, for example, are allowed for nucleonic functions.

The spin-isospin functions thus are basis functions of IRs

$$D^{(S)}\times D^{(I)} := (S,I)\ , \qquad \text{dimension}\ (2S+1)(2I+1)\ , \tag{13.3.6}$$

of the outer product group $\mathscr{SU}_S(2)\times\mathscr{SU}_I(2)$. These IRs are generated by the many-particle operators $\boldsymbol{S}=\sum_i \boldsymbol{s}(i)$ and $\boldsymbol{I}=\sum_i \boldsymbol{I}(i)$. To determine the assignment of the values (S,I) to the different diagrams $[\tilde{\lambda}]$ we proceed as follows:

(i) The four spin-isospin states (13.3.2) of a nucleon are looked upon as a basis of the defining IR $[1]^4$ of the $\mathscr{SU}(4)$ group, which contains the spin-isospin group $\mathscr{SU}_S(2)\times\mathscr{SU}_I(2)$ as a subgroup. The fundamental weights (S_z, I_3) assigned to this group are (11.4.24)

$$\begin{aligned} &w^{(1)}=\tfrac{1}{2}(1,1)\ , &&w^{(2)}=\tfrac{1}{2}(1,-1)\ ,\\ &w^{(3)}=\tfrac{1}{2}(-1,1)\ , &&w^{(4)}=\tfrac{1}{2}(-1,-1)\ ,\\ &\sum_j w^{(j)}=0\ . \end{aligned} \tag{13.3.7}$$

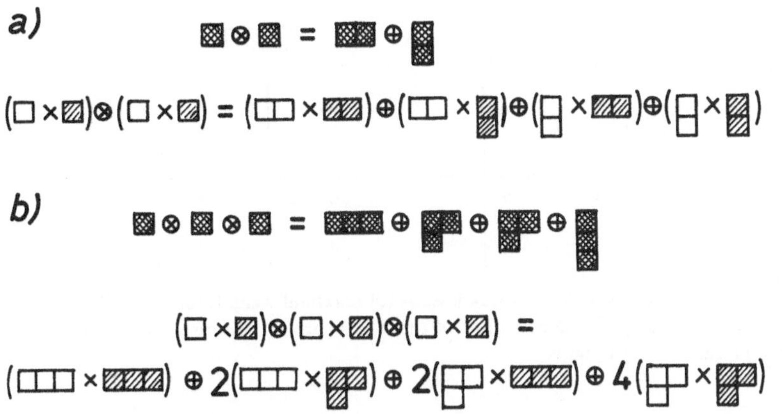

Fig. 13.8a, b. Graphical illustrations of the products in (13.3.8, 9); □ for $\mathcal{SU}_s(2)$, ▨ for $\mathcal{SU}_I(2)$, ▩ for $\mathcal{SU}(4)$. For $\mathcal{SU}(2)$ more than two rows are forbidden

(ii) We then treat the four functions (13.3.2) as the basis functions of the IR $D^{(1/2)} \times D^{(1/2)} := (1/2, 1/2) := [1]^2_S \times [1]^2_I$ of $\mathcal{SU}_S(2) \times \mathcal{SU}_I(2)$.

(iii) By comparing the Clebsch-Gordan expansions of the many-nucleon states according to (i) and (ii) we determine the multiplets subduced on $\mathcal{SU}_S(2) \times \mathcal{SU}_I(2)$ from $\mathcal{SU}(4)$.

For *two nucleons* we obtain in this way, using (7.2.11),

$$\text{(i)} \quad [1]^4 \otimes [1]^4 = [2]^4 \oplus [1^2]^4 \; ; \qquad d\colon 4 \times 4 = 10 + 6 \; . \tag{13.3.8}$$

$$\begin{aligned} \text{(ii)} \quad & ([1]^2_S \times [1]^2_I) \otimes ([1]^2_S \times [1]^2_I) \\ & = ([2]^2_S \times [2]^2_I) \oplus ([2]^2_S \times [1^2]^2_I) \oplus ([1^2]^2_S \times [2]^2_I) \oplus ([1^2]^2_S \times [1^2]) \\ & = \quad (1,1) \quad \oplus \quad (1,0) \quad \oplus \quad (0,1) \quad \oplus \quad (0,0) \\ & d\colon \quad 9 \quad + \quad 3 \quad + \quad 3 \quad + \quad 1 \quad . \end{aligned}$$

See Fig. 13.8. This also follows by direct reduction of the single components according to (11.4.71).

(iii) From a comparison of the weights and dimensions (taken from Exercise 5.19) we have the assignment using (13.3.7)

$$[2]^4 \to (1,1) \oplus (0,0) \; ; \qquad [1^2]^4 \to (1,0) \oplus (0,1) \; .$$

For *three nucleons*

$$\begin{aligned} \text{(i)} \quad & [1]^4 \otimes [1]^4 \otimes [1]^4 = [3]^4 \oplus [2,1]^4 \oplus [2,1]^4 \oplus [1^3]^4 \\ & d\colon \quad 4 \times 4 \times 4 \quad = 20 + \quad 20 \quad + \quad 20 \quad + \quad 4 \quad . \end{aligned} \tag{13.3.9}$$

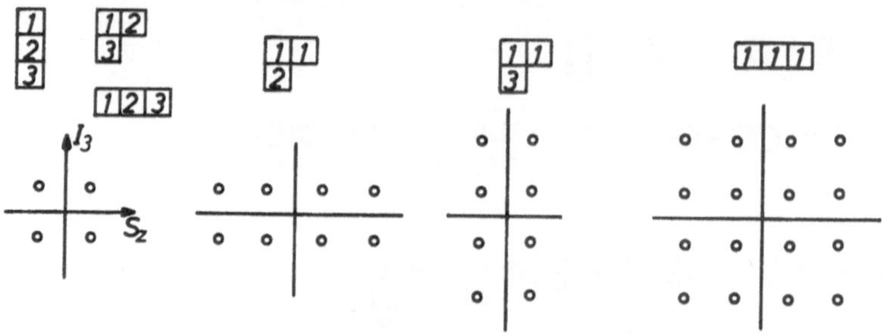

Fig. 13.9. Weight diagrams and allowed Young graphs (of maximal weight) for $\mathscr{SU}_s(2) \times \mathscr{SU}_I(2)$ and the Lie algebras $A_{1_s} \times A_{1_I}$. The dimensions of the REPs can be seen directly. (1/2, 1/2) is the defining REP with fundamental weights

(ii) $([1]_S^2 \times [1]_I^2) \otimes ([1]_S^2 \times [1]_I^2) \otimes ([1]_S^2 \times [1]_I^2)$

$$= ([3]_S^2 \times [3]_I^2) \oplus 2([3]_S^2 \times [2,1]_I^2) \oplus 2([2,1]_S^2 \times [3]_I^2) \oplus 4([2,1]_S^2 \times [2,1]_I^2)$$

$$= \quad (3/2, 3/2) \quad \oplus \quad 2(3/2, 1/2) \quad \oplus \quad 2(1/2, 3/2) \quad \oplus \quad 4(1/2, 1/2)$$

$$d: \quad 16 \quad + \quad 2 \times 8 \quad + \quad 2 \times 8 \quad + \quad 4 \times 4 \quad .$$

See Fig. 13.8.

(iii) We get the assignment by forming the allowed Young diagrams and comparing the weights; in general we can even restrict ourselves to the diagrams with maximal weight (Fig. 13.9). The corresponding summation $\sum \lambda_i w^{(i)}$, e.g. $w^{(1)} + w^{(2)} + w^{(3)} = (1/2, 1/2)$ for the diagram $[1^3]^4$ in Fig. 13.9 then directly gives the assignment. We thus obtain

$$\begin{aligned} &[1^3]^4 \to (1/2, 1/2) \;, \qquad [3]^4 \to (3/2, 3/2) \oplus (1/2, 1/2) \;, \\ &[2,1]^4 \to (3/2, 1/2) \oplus (1/2, 3/2) \oplus (1/2, 1/2) \;, \qquad \text{twice.} \end{aligned} \tag{13.3.10}$$

This case corresponds to the mirror nuclei ^{7}Li and ^{7}Be, the lowest excited states of which we shall now consider. In the regime of low excitations the magic configuration $(1s)^4$ will not be broken, so the spectrum is determined by the three further p nucleons ($l = 1$). The three space functions $\Phi^{[\lambda]}$ assigned to the spin-isospin functions are symmetrized products from the three one-particle functions with angular momentum $l = 1$. They follow from the spin-isospin associate diagram $[\tilde{\lambda}]^4$ associated with $[\lambda]$. Contrary to the case of the atomic shell, in this case all the diagrams $[\lambda]$ are allowed.

Using Exercise 13.5 and Eq. (13.3.10) we obtain Table 13.5 in analogy to (13.2.10), or with the spectroscopic notation $^{2I+1,2S+1}L_J$, we obtain Table 13.6. The lowest levels ($[3], [1^3]$) of the nuclei ^{7}Li and ^{7}Be are given in Fig. 13.10; they are in good agreement with the experimental data.

Table 13.5. Classification of spin-isospin states according to (13.3.6)

$[\lambda]$	L	$[\tilde{\lambda}]$	(S, I)
$[1^3]^3$	$\rightarrow S$	$[3]^4$	$\rightarrow (3/2, 3/2) \oplus (1/2, 1/2)$
$[2,1]^3$	$\rightarrow P \oplus D$	$[2,1]^4$	$\rightarrow (3/2, 1/2) \oplus (1/2, 3/2) \oplus (1/2, 1/2)$, twice
$[3]^3$	$\rightarrow P \oplus F$	$[1^3]^4$	$\rightarrow (1/2, 1/2)$

Table 13.6. Lowest states of ^{7}Li and ^{7}Be mirror nuclei

$[\lambda]$	$[\tilde{\lambda}]$	$^{2I+1,2S+1}L_J$
$[1^3]$	$[3]$	$^{44}S_{3/2}$, $^{22}S_{1/2}$
$[2,1]$	$[2,1]$	$^{24}P_{5/2,3/2,1/2}$, $^{42}P_{3/2,1/2}$, $^{22}P_{3/2,1/2}$ $^{24}D_{7/2,5/2,3/2,1/2}$, $^{42}D_{5/2,3/2}$, $^{22}D_{5/2,3/2}$
$[3]$	$[1^3]$	$^{22}P_{3/2,1/2}$, $^{22}F_{7/2,5/2}$

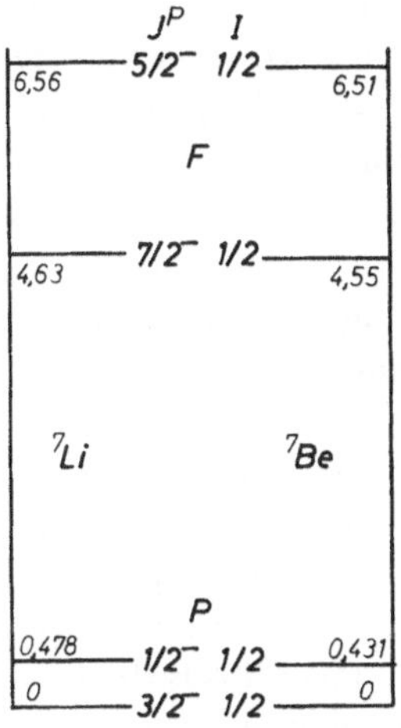

Fig. 13.10. The lowest (excited) levels of the mirror nuclei ^{7}Li and ^{7}Be according to Table 13.6. Energies are given in MeV

Exercise 13.6. Following the discussion in Sects. 13.2.1, 2, calculate the possible states for three s particles ($l = 0$); show that this is forbidden in atomic shells (electrons), but leads to a $^{22}S_{1/2}$ state in the nucleus (nucleons, isospin)

13.4 Dynamical Symmetries of Classical Systems

In Sect. 13.1 we saw the importance of internal symmetries [isospin $\mathcal{SU}_I(2)$, quark structures $\mathcal{SU}(n)$] besides that of geometrical symmetries. However, there are other internal symmetries due to the dynamics of a physical system. In general they exist only in interacting systems and are said to be dynamical symmetries (Sect. 7.1.1 and Chap. 14).

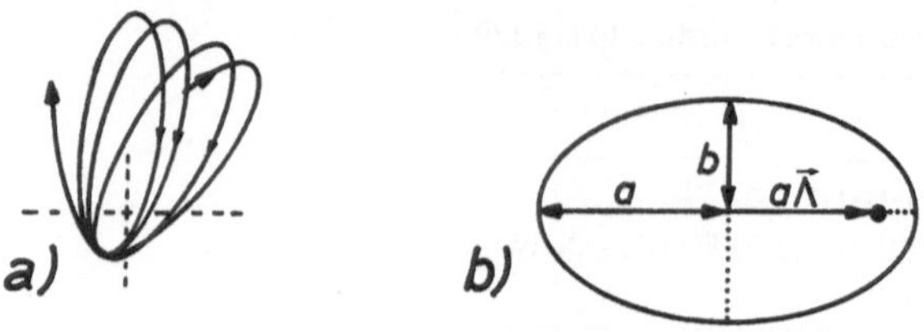

Fig. 13.11a, b. Classical orbits (qualitative) of particles in an attracting central potential. (**a**) Arbitrary potential, e.g. $V \sim -1/r + \beta/r^2$. (**b**) Elliptic orbit in a potential $V = -1/r$. The distance from the centre to the focus defines the invariant Lenz vector Λ according to (13.4.1). a: major, b: minor axis of the orbit

As we have seen from the foregoing discussions, the invariance of a system under rotations or inversion leads to a degeneracy of energy levels. This can be described by the values of J_z and of the parity $(\pm)$ if J is given. Besides geometrical transformations there may exist other, in a certain sense "hidden", symmetry transformations which, for example, simultaneously contain positions *and* momenta or represent variables of an internal space. The corresponding dynamical invariance group (DIG) of a physical system includes the geometrical transformations as a subgroup. The "free" part H_0 of the Hamiltonian $H = H_0 + W$ is *not* invariant with respect to the dynamical group, but the total H is. The symmetries thus are only present in connection with an interaction (or potential). For the classification of energy spectra including their degeneracies the DIG is appropriate. We illustrate the DIG with the example of the H atom.

The spherical symmetry $\mathcal{O}(3)$ of a central potential and of the corresponding Hamiltonian causes the angular momentum to be constant, the plane of the orbit being perpendicular to $\boldsymbol{L}$. But this is not sufficient to allow us to conclude that the orbit is a conic section (ellipse, hyperbola). This is only valid for a potential $V(r) \sim 1/r$ (see Fig. 13.11). The $(1/r)$-potential thus might have a further invariant, e.g. the vector $a\Lambda$ in Fig. 13.11, which is parallel to the main axis. This is the *Lenz vector*, which may be written in a quantum-mechanical generalization with appropriate factors as

$$\Lambda := \frac{1}{2\alpha M}(\boldsymbol{p} \times \boldsymbol{L} - \boldsymbol{L} \times \boldsymbol{p}) - \boldsymbol{x}/r \ , \tag{13.4.1}$$

$$r = |\boldsymbol{x}| \ , \qquad \alpha = GM_1M_2 \quad \text{or} \quad \frac{Q_1Q_2}{4\pi\varepsilon_0} \ .$$

It satisfies

$$\begin{aligned} &[\boldsymbol{L}, H] = 0 \ , \qquad [\Lambda, H] = 0 \ , \\ &\boldsymbol{L}\cdot\Lambda = \Lambda\cdot\boldsymbol{L} = 0 \ , \qquad \Lambda^2 = \frac{2}{\alpha^2 M}H(L^2 + \hbar^2) + 1 \ , \end{aligned} \tag{13.4.2}$$

with

$$H = \frac{p^2}{2M} - \frac{\alpha}{r} \ . \tag{13.4.3}$$

Thus there are six operators (generators) $\boldsymbol{L}$, $\boldsymbol{\Lambda}$ defining the conserved quantities. These again allow us to obtain the symmetry transformations (Sect. 13.1.1) that commute with H. In order to find the symmetry group we investigate the commutators of $\boldsymbol{L}$ and $\boldsymbol{\Lambda}$, replacing $\boldsymbol{\Lambda}$ by

$$\boldsymbol{\Lambda}' := \sqrt{-\frac{\alpha^2 M}{2E}}\,\boldsymbol{\Lambda} \ .$$

Here we restrict ourselves to bound states ($E < 0$). Then (with $\hbar = 1$)

$$[L_i, L_j] = [\Lambda'_i, \Lambda'_j] = \mathrm{i}\sum_k \varepsilon_{ijk} L_k \ ; \qquad [L_i, \Lambda'_j] = \mathrm{i}\sum_k \varepsilon_{ijk}\Lambda'_k \ . \tag{13.4.4}$$

These relations exactly correspond to those of (11.1.30a), so we may introduce new operators

$$\boldsymbol{J} = \tfrac{1}{2}(\boldsymbol{L} + \boldsymbol{\Lambda}') \ , \qquad \boldsymbol{K} = \tfrac{1}{2}(\boldsymbol{L} - \boldsymbol{\Lambda}') \ , \tag{13.4.5}$$

which have two disjoint sets of generators of the DIG. Each set represents the algebra $A_1[\cong B_1$ of $\mathscr{SO}(3,\mathbb{R})]$ of the simple Lie group $\mathscr{SU}(2,\mathbb{C})$ known from Sect. 11.1.2. The algebra is thus $A_1 \oplus A_1 \cong D_2$. The DIG of the $(1/r)$-potential is therefore locally isomorphic to $\mathscr{SU}(2,\mathbb{C}) \times \mathscr{SU}(2,\mathbb{C})$ or $\mathscr{SO}(4,\mathbb{R})$. Taking into account the invariance of H under space inversion, the DIG is

$$\mathscr{G} \cong \mathscr{O}(4,\mathbb{R}) \cong \mathscr{C}_i \times \mathscr{SU}(2,\mathbb{C}) \times \mathscr{SU}(2,\mathbb{C}) \ , \cong \text{locally.} \tag{13.4.6}$$

From (13.4.4) or (11.1.30b), it follows that there are at most two commuting generators, e.g. L_z and Λ'_z. The rank of $\mathscr{SO}(4)$ equals two and there are also exactly two Casimir operators (11.1.37–39) that commute with all six generators. According to (11.1.30b) these are J^2 and K^2, with the eigenvalues $j(j+1)$ and $k(k+1)$, respectively, with $j, k = 0, 1/2, 1, \ldots$ according to (11.4.65a). For $\mathscr{SO}(4)$ we choose rather

$$C := J^2 + K^2 = \tfrac{1}{2}(L^2 + \Lambda'^2) \qquad \text{and} \qquad C' := J^2 - K^2 = \boldsymbol{L}\cdot\boldsymbol{\Lambda}' \ , \tag{13.4.7}$$

where because of (13.4.2)

$$C' = \boldsymbol{L}\cdot\boldsymbol{\Lambda}' = 0 \qquad \text{for a } (1/r)\text{-potential} \tag{13.4.8}$$

thus $j = k$, too. With this the operator (here $\hbar = 1$)

$$C = \frac{1}{2}\left(L^2 - \frac{\alpha^2 M}{2E}\Lambda^2\right) = -\frac{\alpha^2 M}{4E} - \frac{1}{2} \tag{13.4.9}$$

has the eigenvalues

$$\langle C \rangle = 2j(j+1) \quad \text{and} \quad \langle H \rangle = E_n = -\frac{\alpha^2 M}{2(2j+1)^2} = -\frac{\alpha^2 M}{2n^2}, \tag{13.4.10}$$
$$n = 2j+1 = 1, 2, 3, \ldots .$$

The dimensions of the IRs $(j, k = j)$ of

$$\mathscr{SO}(4) \overset{\text{loc}}{\cong} \mathscr{SU}(2) \times \mathscr{SU}(2)$$

are just $(2j+1)^2 = n^2$, because $k = j$; the energy levels are n^2-fold degenerate, neglecting spin, whereas $\mathscr{SO}(3)$ symmetry only leads to $(2l+1)$-fold degeneracies. For the H atom the levels with $l = 0, \ldots, n-1$ also coincide again, resulting in a $\sum_{l=0}^{n-1}(2l+1) = n^2$-fold degeneracy. From this we may expect the subduction

$$(jj) \to (0) \oplus (1) \oplus (2) \oplus \cdots \oplus (2j) \tag{13.4.11}$$

or, more generally $(j_1 - j_2 \in \mathbb{N})$,

$$(j_1 j_2) \to (|j_1 - j_2|) \oplus \cdots \oplus (j_1 + j_2)$$

from $\mathscr{SO}(4)$ into $\mathscr{SO}(3)$. The states may be classified according to $|(jj), l, m\rangle$ or $|n, l, m\rangle$.

Here we have calculated the energy spectrum with the help of the DIG of the system. Alternatively, we might also try to find the possible DIG from experimentally known spectra, which is possible by consideration of conservation laws. In the above example this would mean determining the potential in the Hamiltonian from the energy spectrum of the H atom. Such problems occur in the physics of elementary particles. In that case, however, the fundamental dynamics and thus also the DIG of the systems is in general not known.

To conclude, we want to point out briefly that there exists a noncompact group $\mathscr{SO}(4, 1)$ which includes $\mathscr{SO}(4)$ and which, when subduced onto $\mathscr{SO}(4)$, possesses an IR that contains as a direct sum *all* the IRs of $\mathscr{SO}(4)$ describing bound states of the Hamiltonian, each IR appearing just once. This means that with the help of $\mathscr{SO}(4, 1)$ all the energy levels, including their degeneracy, can be represented within *one* IR. Of course, $\mathscr{SO}(4, 1)$ itself is not a DIG of the Hamiltonian.

Generally, physical systems can be completely described by such groups containing the energy spectrum, its degeneracies and also operators causing transitions between different states. However, such *spectra-generating groups* are not invariance groups of the systems.

Exercise 13.7. Determine the eigenvalues and the corresponding degeneracies of the rigid rotator $H = L^2/2\theta$ by means of the Casimir operator of the symmetry group $\mathscr{SO}(3)$ (θ: moment of inertia).

Exercise 13.8. The Hamiltonian of an isotropic three-dimensional harmonic oscillator can be represented as

$$H = \hbar\omega \sum_{i=1}^{3} (a_i^+ a_i + 1/2) = \frac{\hbar\omega}{2} \sum_i (a_i^+ a_i + a_i a_i^+)$$

using creation and annihilation operators a_i^+, a_j with

$$[a_i, a_j^+] = \delta_{ij} \ , \qquad [a_i, a_j] = [a_i^+, a_j^+] = 0 \ .$$

1) Show that $a_j^+ a_l$ commutes with H; further, $[H, q_{jl}] = [H, L_k] = 0$ with

$$L_k = \frac{\hbar}{\mathrm{i}} \sum_{j,l} \varepsilon_{kjl} a_j^+ a_l \ ;$$

$$q_{jl} = q_{jl}^+ = \frac{\hbar\omega}{2}(a_j^+ a_l + a_j a_l^+) \ .$$

2) Show that of the nine operators $a_j^+ a_l$, eight independent ones can be chosen commuting with H and defining the LA A_2 of $\mathcal{SU}(3)$:

$$E_{jl} := \tfrac{1}{2}(a_j^+ a_l + a_l a_j^+) \ , \qquad H_j := E_{jj} - H/3\hbar\omega \ , \qquad \sum_j H_j = 0 \ ,$$

see (11.4.2); $a_j^+ a_l$ "shifts" a vibrational quantum from the l-direction into the j-direction.
3) Which is the geometrical and which the dynamical symmetry group? What is the physical measuring of L_k and q_{jl} (in $\boldsymbol{x}$ and $\boldsymbol{p}$)?
4) Determine the energy levels and give the degeneracy of the levels, (11.4.58).

14. Internal Symmetries and Gauge Theories

In Sect. 13.1 we discussed how hadrons can be arranged in multiplets and classified by internal charge quantum numbers. The multiplets can be described by flavour symmetry groups.

In this chapter dynamical symmetries of particle fields are discussed—the groups of most interest being the gauge groups. Those of the first kind define the conserved quantities (the charges) whereas those of the second kind define the gauge fields which couple to the particle fields. Another topic is the spontaneous breaking of gauge symmetries, for which we give some examples. Finally, we consider the theory of the electro-weak interaction and the grand unified theory.

14.1 Internal Symmetries of Fields

In a field theory of particles, fields or field operators are assigned to particles:

$$\psi^{i\alpha}(x^\mu)\ , \qquad x^\mu = (x^0, x^1, x^2, x^3) = (ct, x, y, z) := x\ . \tag{14.1.1}$$

The superscript α (vector or spinor index) denotes the behaviour of the fields under transformations of the Poincaré or Lorentz group. The superscript i distinguishes between fields of different "types" of particles; it describes *internal degrees of freedom* of the fields, e.g. isospin, hypercharge, colour degrees of freedom having $\mathscr{SU}(n)$ symmetry (Sect. 13.1). We will discuss mainly these internal symmetries; the *inhomogeneous (Poincaré) and homogeneous Lorentz group*, $\mathscr{IL}z$ or $\mathscr{L}z(4)$, will be discussed only briefly (Appendix G). The REP theory of $\mathscr{IL}z(4)$ is similar to that of space groups and makes use of the *little group* (Sect. 9.2).

The vectors $\{x^\mu\}$ in (14.1.1) transform under $\{d|t\} \in \mathscr{IL}z(4)$ as

$$\{d|t\}x^\mu \rightarrow x'^\mu = \sum_\nu d^\mu{}_\nu x^\nu + t^\mu\ , \tag{14.1.2}$$

see (3.4.4). The real elements $d^\mu{}_\nu$ define the homogeneous Lorentz group $\mathscr{L}z(4)$ and leave invariant the quadratic form

$$x^2 = \sum_\mu x_\mu x^\mu = (ct)^2 - x^2 - y^2 - z^2 = \sum_\mu x'_\mu x'^\mu\ . \tag{14.1.3}$$

With (14.1.2) a field $\psi^\alpha(x)$ transforms according to

$$\psi^\alpha(x) \to \psi'^\alpha(x') := \sum_\beta T^\alpha{}_\beta \psi^\beta(x) \ ; \tag{14.1.4a}$$

$\psi^\beta(x)$ is the field measured by a first observer at the space-time point x, while a second observer (seen from his point of view) localizes this point at x' and assigns the field $\psi'^\alpha(x')$ to it. Equation (14.1.4a) is equivalent to

$$\psi'^\alpha(x) = \sum_\beta T^\alpha{}_\beta \psi^\beta(d^{-1}(x-t)) \ , \qquad \{d|t\}^{-1} := \{d^{-1}|-d^{-1}t\} \tag{14.1.4b}$$

or in the case of field operators

$$\psi'^\alpha(x) = U^\dagger(\{d|t\})\psi^\alpha(x)U(\{d|t\}) \ . \tag{14.1.4c}$$

The form of the transformation matrix $T^\alpha{}_\beta$ depends on the type of fields, e.g. scalar ($T \equiv 1$), vector ($T \equiv d$), Weyl or Dirac fields, etc.

Hamiltonians or Lagrangians have to be replaced by appropriate densities in field theory, thus

$$l = l(\psi^{i\alpha}(x), \partial_\mu \psi^{i\alpha}(x); x) \ , \qquad \partial_\mu = \frac{\partial}{\partial x^\mu} \ , \tag{14.1.5}$$

however, in general there is no explicit dependence on x (see conservation laws, Appendix F). Assumptions made about l are:

(i) It is a scalar density under transformations of $\mathscr{I}\mathscr{L}z(4)$:

$$l'(\psi'^{i\alpha}(x'), \partial'_\mu \psi'^{i\alpha}(x'))d^4x' = l(\psi^{i\alpha}(x), \partial_\mu \psi^{i\alpha}(x))d^4x \ . \tag{14.1.6a}$$

(ii) The field equations are covariant, i.e. l is "shape invariant" under $\mathscr{I}\mathscr{L}z(4)$:

$$l'(\psi'^{i\alpha}(x'), \partial'_\mu \psi'^{i\alpha}(x')) = l(\psi'^{i\alpha}(x'), \partial'_\mu \psi'^{i\alpha}(x')) \tag{14.1.6b}$$

and thus we have the symmetry condition

$$l(\psi'^{i\alpha}(x'), \partial'_\mu \psi'^{i\alpha}(x'))d^4x' = l(\psi^{i\alpha}(x), \partial_\mu \psi^{i\alpha}(x))d^4x \ . \tag{14.1.6c}$$

[Both sides of (14.1.6c) may differ by a total divergence: $\partial_\mu f^\mu(\psi^{i\alpha}(x); x)d^4x$ with f^μ arbitrary; but this is of no significance in the following.] From this condition, conservation laws related to space-time symmetry can be derived [Noether's theorem, (13.1.4, 5) with the generators of $\mathscr{I}\mathscr{L}z(4, \mathbb{R})$, Appendix G]. These laws will not be discussed in any more detail here. We will focus our attention mainly on those conservation laws which follow from the internal symmetries of the Lagrangian density l (Noether's theorem, Appendix F). Therefore we require l to be invariant under certain internal and local transformations with fixed x. Then (14.1.6c) has to be valid in the form

$$l(\psi'^i(x), \partial_\mu \psi'^i(x)) = l(\psi^i(x), \partial_\mu \psi^i(x)) \ . \tag{14.1.6d}$$

If a physical system is invariant under an m-parameter (compact) Lie group $\mathscr{G}$, then commutation relations, equations of motion, Lagrangian density, etc., remain "shape invariant" under the transformation of fields:

$$\psi^i(x) \to \psi'^i(x) = \sum_j D^i{}_j(\chi^1, \ldots, \chi^m)\psi^j(x) . \tag{14.1.7a}$$

Here $D^i{}_j(\chi^1, \ldots, \chi^m)$ is a REP of the unitary operator P_a with $a \in \mathscr{G}$ according to which the ψ^i transform:

$$\begin{aligned} &D_{ij}(\chi^1, \ldots, \chi^m) := (P_a)_{ij} , \\ &P_a(\chi^1, \ldots, \chi^m) = \exp\left(ig \sum_{l=1}^{m} J_l \chi^l\right) \approx 1 + ig \sum_{l=1}^{m} J_l \chi^l , \qquad J_l^+ = J_l . \end{aligned} \tag{14.1.8}$$

J_l are the infinitesimal generators from Sect. 11.1.1, apart from a factor i. This choice is more convenient for the applications in this section. For infinitesimal transformations (14.1.7a) reads

$$\psi'^i(x) = \psi^i(x) + ig \sum_{l=1}^{m} \sum_j \chi^l J_{l,}{}^i{}_j \psi^j(x) . \tag{14.1.7b}$$

As for (14.1.4c), the symmetry transformations of field operators can be described by unitary operators

$$\mathscr{U}(\chi^1, \ldots, \chi^m) = \exp\left(ig \sum_l Q_l \chi^l\right) \approx 1 + ig \sum_l Q_l \chi^l , \qquad Q_l^\dagger = Q_l ; \tag{14.1.9}$$

the $\mathscr{U}(\chi^1, \ldots, \chi^m)$ are also a REP of $\mathscr{G}$. Because $\mathscr{G}$ is an invariance group, the Q_l are conserved quantities. By comparing (14.1.7b) with $\psi'^i(x) = U^\dagger \psi^i(x) U$, using (14.1.9) we obtain

$$[Q_l, \psi^i(x)] = -\sum_j J_{l,}{}^i{}_j \psi^j(x) \tag{14.1.10}$$

as a condition to be satisfied by the Q_l. The quantum-mechanical description (by field operators) is in agreement with the classical one only if (14.1.10) is satisfied. The generators Q_l of the transformation correspond to the (classical conserved quantities (Sect. 13.1.1). They can be determined from the classical Lagrange formalism (see Appendix F); the condition (14.1.10) then requires at most a symmetrization (Hermitization).

In the following we only consider the gauge symmetries of l. In the examples we always choose one of the fields in the following list. Here $\kappa = mc/\hbar$; often one uses $c = \hbar = 1$. The factors in l are chosen appropriately, but are of no interest in general. We use the notation of Bjorken and Drell [14.1, 2].

1) Klein-Gordon field

$$l = \tfrac{1}{2}(\partial_\mu \psi^* \partial^\mu \psi - \kappa^2 \psi^* \psi) . \tag{14.1.11a}$$

2) Proca field or electromagnetic field

$$l = \tfrac{1}{2}(\partial_\mu \psi^{*\nu} \partial^\mu \psi_\nu - \kappa^2 \psi^{*\nu} \psi_\nu) \, , \qquad \partial_\mu \psi^\mu = 0 \tag{14.1.11b}$$

$$l = -\tfrac{1}{2}\partial_\mu A^\nu \partial^\mu A_\nu \, , \qquad \partial_\mu A^\nu = 0 \, . \tag{14.1.11c}$$

Sometimes other gauges are chosen:

$$\begin{aligned} &\partial_\mu \psi^\mu = 0 \, , \quad \mu = 0, 1, 2, 3 && : \text{Lorentz gauge} \\ &\partial_k \psi^k = 0 \, , \quad k = 1, 2, 3 && : \text{Coulomb gauge} \\ &\psi^0 = 0 && : \text{Hamilton gauge} \\ &\psi^3 = 0 && : \text{axial gauge} \, . \end{aligned} \tag{14.1.11d}$$

3) Weyl field (σ^μ: Pauli matrices)

$$l = \frac{\mathrm{i}}{2}(\psi^+ \sigma^\mu \partial_\mu \psi - \partial_\mu \psi^+ \sigma^\mu \psi) \, . \tag{14.1.11e}$$

4) Dirac field

$$\begin{aligned} &l = \frac{\mathrm{i}}{2}(\bar{\psi}\gamma_\mu \partial^\mu \psi - \partial^\mu \bar{\psi}\gamma_\mu \psi + 2\mathrm{i}\kappa \bar{\psi}\psi) \qquad \text{or} \\ &l = \bar{\psi}(\mathrm{i}\gamma_\mu \partial^\mu - \kappa)\psi \; ; \qquad \bar{\psi} = \psi^+ \gamma^0 \, . \end{aligned} \tag{14.1.11f}$$

In the following we always write ψ^* for the conjugate field. When field operators are involved, this has to be replaced by ψ^+ or $\bar{\psi}$.

We obtain the differential conserved quantities (currents) by means of local transformations using Noether's theorem (Appendix F):

$$j^\mu = -[\psi'^i(x) - \psi^i(x)]\frac{\partial l}{\partial \partial_\mu \psi^i} - [\psi^{*\prime i}(x) - \psi^{*i}(x)]\frac{\partial l}{\partial \partial_\mu \psi^{*i}} \tag{14.1.12a}$$

or, using (14.1.7b), for every l since χ^l is arbitrary

$$j_l^\mu = -\mathrm{i}\frac{\partial l}{\partial \partial_\mu \psi^i} J_l{}^i{}_j \psi^j + \mathrm{i}\frac{\partial l}{\partial \partial_\mu \psi^{*i}} J_l^*{}^i{}_j \psi^{*j} \, . \tag{14.1.12b}$$

The integral quantities (generalized *charges*) are then

$$Q_l = \int_\sigma j_l^\mu n_\mu d\sigma \, ,$$

where n_μ is the normal of a space-like surface in Minkowski space (Fig. 14.1). As the area of integration we always choose the surface $x^0 = \text{const}$; then we have

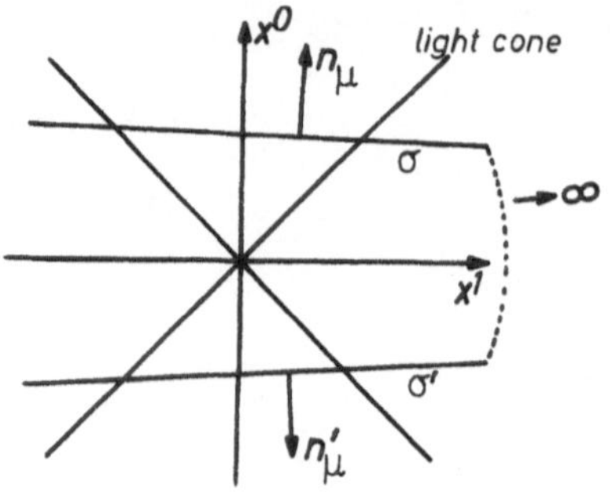

Fig. 14.1. Minkowski space with space-like surfaces σ, σ'. The normal is directed into the light cone. In particular, the surfaces x^0 = const are space-like

the generators Q_l of the symmetry transformation U (14.1.9) in the form

$$Q_l = \int \rho_l d^3x \qquad \text{with} \qquad \rho_l = j_l^0 = -\mathrm{i}\{\pi_i J_l,{}^i{}_j\psi^j - \pi_i^* J_l^*,{}^i{}_j\psi^{*j}\} \ ,$$

where $\pi_i = \partial l/(\partial \partial_0 \psi^i)$ is the density of the field canonically conjugate to ψ^i.

14.2 Gauge Transformations of the First Kind

14.2.1 $\mathscr{U}$ (1) Gauge Transformations

In the simplest case these transformations change the phase of the field

$$\psi(x) \to \psi'(x) = \mathrm{e}^{\mathrm{i}e\chi}\psi(x) \ . \tag{14.2.1a}$$

The phase change is assumed to be independent of x, thus it is a *global* transformation by a unitary one-dimensional matrix (Abelian group)

$$\mathscr{U}(1) := \{\mathrm{e}^{\mathrm{i}eJ\chi} | \chi \in \mathbb{R}; J = 1\} \ . \tag{14.2.1b}$$

Here the generator J is introduced formally for the sake of later generalizations. In electromagnetic theory, e has to be replaced by $e/\hbar c$ in order to have the usual units. The gauge transformations of the first kind thus constitute a one-parameter (one-dimensional) Abelian unitary group $\mathscr{U}(1)$, i.e.

$$\mathrm{e}^{\mathrm{i}e\chi_1}\mathrm{e}^{\mathrm{i}e\chi_2} = \mathrm{e}^{\mathrm{i}e\chi_3} \ , \qquad \chi_3 = \chi_1 + \chi_2 \ ,$$

again is a gauge transformation; identity and inverse elements obviously exist. We will sketch the properties of such transformations.

For χ infinitesimal we have

$$\psi'(x) - \psi(x) = \mathrm{i}e\chi\psi(x) \ , \qquad \psi'^*(x) - \psi^*(x) = -\mathrm{i}e\chi\psi^*(x) \ .$$

Thus the current is

$$j^\mu e\chi := \mathrm{i}\left(\psi^{*\alpha}\frac{\partial l}{\partial\partial_\mu\psi^{*\alpha}} - \psi^\alpha\frac{\partial l}{\partial\partial_\mu\psi^\alpha}\right)e\chi \ .$$

Since χ can be chosen arbitrarily, j^μ is a four-dimensional current whose component j^0 is the density of a conserved quantity (charge):

$$\rho = \mathrm{i}\left(\psi^{*\alpha}\frac{\partial l}{\partial\partial_0\psi^{*\alpha}} - \psi^\alpha\frac{\partial l}{\partial\partial_0\psi^\alpha}\right) . \tag{14.2.2}$$

Examples:

1) Klein-Gordon field

$$j^\mu = \frac{\mathrm{i}}{2}(\psi^*\partial^\mu\psi - \psi\partial^\mu\psi^*) , \qquad \rho = \frac{\mathrm{i}}{2}(\psi^*\dot{\psi} - \psi\dot{\psi}^*) .$$

2) Proca field

$$j^\mu = \frac{\mathrm{i}}{2}(\psi_\nu^*\partial^\mu\psi^\nu - \psi^\nu\partial^\mu\psi_\nu^*) , \qquad \rho = \frac{\mathrm{i}}{2}(\psi_\nu^*\dot{\psi}^\nu - \psi^\nu\dot{\psi}_\nu^*) .$$

For a (real) electromagnetic field $j^\mu \equiv 0$.

3) Weyl field

$$j^\mu = \psi^+\sigma^\mu\psi , \qquad \rho = \psi^+\psi .$$

4) Dirac field

$$j^\mu = \bar{\psi}\gamma^\mu\psi , \qquad \rho = \bar{\psi}\gamma^0\psi = \psi^+\psi .$$

On integrating over the total space, the quantities $\int \rho d^3x$ obviously give the total charges Q that are conserved:

> Invariance of the Lagrangian density under gauge transformations of the first kind thus means conservation of the (electrical) charge, and vice versa: if there are conservation laws for charges (in the generalized sense) then there exists an associated gauge transformation of the first kind under which the Lagrangian density is invariant.

Other examples of conserved charge-like quantities are baryon number B and lepton number L. The corresponding densities $\rho = \psi^+\psi$ are the particle number densities of baryons and leptons. The gauge transformations are

$$\exp(\mathrm{i}\chi_\mathrm{B}) \qquad \text{and} \qquad \exp(\mathrm{i}\chi_\mathrm{L})$$

with the gauge parameters χ_B and χ_L, respectively. The gauge group is again $\mathcal{U}(1)$, so the gauge group of baryons [or leptons] with electrical charge e is the product group

$$U_\mathrm{e}(1) \times U_\mathrm{B}(1) \qquad [\text{or} \qquad U_\mathrm{e}(1) \times U_\mathrm{L}(1)] .$$

But this is valid only if there is merely one kind of baryons, i.e. either protons or neutrons, etc.

Finally, from the transformation (14.1.9), $U(\chi) = \exp(\mathrm{i}Q\chi)$, we obtain using (14.1.10)

$$[Q, \psi^i(x)] = -e\psi^i(x) \; , \qquad [Q, \psi^{i+}(x)] = +e\psi^{i+}(x) \; . \tag{14.2.3}$$

These commutation relations express the idea that $\psi^i(x)$ annihilates a charge e, whereas $\psi^{i+}(x)$ creates a charge. This may be seen by applying (14.2.3) to an eigenstate of Q.

14.2.2 $\mathscr{SU}(n)$ Gauge Transformations

The extension of gauge symmetries to more than one degree of freeom is obvious: one has to introduce several parameters and generators in (14.2.1). This leads to non-Abelian gauge groups, since the different J_l in general do not commute.

In the strong interaction, neutrons and protons behave "identically", i.e. this interaction is independent of the electric charge, which constitutes the difference between neutron and proton. Therefore they can be looked upon as two (isospin) states of one particle. In strong and electromagnetic interactions the states are conserved (conservation of isospin) whereas in weak interactions (β-decay, $\mathrm{n} \to \mathrm{p} + \mathrm{e}^- + \bar{\nu}_\mathrm{e}$) the isospin state changes.

The mathematical properties of the isospin space are equivalent to those of the usual spin space (with $s = 1/2$, $m_s = \pm 1/2$). The neutron and the proton are two components of a state vector (see Sect. 13.1.1)

$$\psi = \begin{pmatrix} \psi_\mathrm{p} \\ \psi_\mathrm{n} \end{pmatrix} = \begin{pmatrix} \mathrm{p} \\ \mathrm{n} \end{pmatrix} . \tag{14.2.4}$$

The state $\psi = \binom{1}{0}$ is a pure proton (eigenvalue $+1/2$), $\psi = \binom{0}{1}$ a pure neutron (eigenvalue $-1/2$).* Equation (14.2.4) is a mixed state. Thus

$$\tau_3 \begin{pmatrix} \mathrm{p} \\ \mathrm{n} \end{pmatrix} = \begin{pmatrix} \mathrm{p} \\ -\mathrm{n} \end{pmatrix} \qquad \text{with} \qquad I_3 = \frac{1}{2}\tau_3 \; , \tag{14.2.5}$$

I_3 being an isospin operator. Together with two further operators τ_1 and τ_2, the τ_k are the infinitesimal generators of a $\mathscr{SU}(2)$ group (see (11.4.7)). The corresponding gauge group is defined by the operators (in isospin space)

$$P(\chi^k) := \{\exp(\mathrm{i}g\tau_k\chi^k/2);\, k = 1,2,3\} \; , \qquad P^+P = 1 \tag{14.2.6}$$

with the isospin as the corresponding charge (Exercise 14.1).

---

*Due to a superselection rule for the electrical charge, only the pure states occur in nature.

A more general example is obtained by assigning a number of internal degrees of freedom to the Dirac fields ψ. Examples of such internal degrees of freedom (Sect. 13.1) are the flavour (f) and colour (c) properties of the quark fields, the corresponding symmetry groups being $\mathcal{SU}_{\mathrm{f}}(n)$ and $\mathcal{SU}_{\mathrm{c}}(3)$, respectively. In this case we have to specify the index i in (14.1.1) by f, c, thus

$$\begin{aligned} \psi^{i\alpha} \to \psi^{fc\alpha} \;, \qquad & \alpha\text{: vector or spinor index} \;, \\ & f\text{: u, d, s, c, b, t} \;, \\ & c\text{: r, y(g), b} \;. \end{aligned} \tag{14.2.7}$$

If the quarks were free particles, the fields $\psi^{\alpha fc}(x)$ could be derived from the free Dirac equation following from the Lagrangian (14.1.11f)

$$\sum_{\alpha' f' c'} (i\gamma^{\mu}_{\alpha\alpha'}\partial_\mu - \kappa_f \delta_{\alpha\alpha'})\delta_{ff'}\delta_{cc'}\psi^{f'c'\alpha'}(x) = 0 \;; \tag{14.2.8}$$

here Dirac, flavour and colour indices are given explicitly.

The masses κ_f depend on flavour. It is obvious that (14.2.8) is invariant with respect to colour-gauge transformations

$$\psi^{fc\alpha}(x) \to \psi'^{fc\alpha}(x) = \sum_{c'} \left[\exp\left(ig \sum_{l=1}^{8} J_l \chi^l \right) \right]_{cc'} \psi^{fc'\alpha}(x) \;, \tag{14.2.9}$$

where J_l are infinitesimal generators of a group $\mathcal{SU}_{\mathrm{c}}(3)$, since $c =$ r, y, b, see (11.4.8).[1] The quark fields thus transform according to the (defining) fundamental REP of $\mathcal{SU}_{\mathrm{c}}(3)$. If the quark masses were independent of flavour (this is valid at best approximately, see Sects. 13.1.1,4), then we would also have flavour symmetry, the gauge transformation being

$$\psi^{fc\alpha}(x) \to \psi'^{fc\alpha}(x) = \sum_{f'} \left[\exp\left(ig \sum_{l=1}^{m} J_l \chi^l \right) \right]_{ff'} \psi^{f'c\alpha}(x) \;, \tag{14.2.10}$$

where the J_l are the infinitesimal generators of $\mathcal{SU}_{\mathrm{f}}(n)$. Due to Noether's theorem because of the symmetry (14.2.9) there exist colour currents and colour charges, which are conserved in interactions. Correspondingly this would be true for flavour currents and charges if flavour were an exact symmetry of the theory.

Exercise 14.1. Discuss the isospin gauge group $\mathcal{SU}(2)$ by first defining the infinitesimal generators in analogy to the Pauli matrices. Define the gauge transformation and its infinitesimal expansion and give the corresponding isospin currents.

---

[1] Gell-Mann uses $2J_l$ as the generators, see Sect. 14.5.2. See also footnote p. 331.

14.3 Gauge Transformations of the Second Kind

The invariance of equations with respect to gauge transformations of the first kind leads to the existence of (generalized) charges Q_l, J_l which are conserved in (certain) interactions. Thus it is reasonable to assume these charges to be the sources of fields causing interactions between particles (with these charges). As we shall see, these fields are a direct consequence of local gauge transformations. They form the physical symmetry groups.

If there is a local coupling between fields, e.g. a Lagrangian of the type $l \sim \bar{\psi}(x)\psi(x)\phi(x)$, then it is evident that transformations in which the fields are transformed have to be considered as locally independent. For gauge transformations this means the gauge parameters χ_l are assumed to be space-time dependent: $\chi_l(x)$.

14.3.1 $\mathscr{U}(1)$ Gauge Transformations of the Second Kind

The Abelian group

$$\mathscr{U}(1) := \{P(\chi(x)) = e^{ieJ\chi(x)} | \chi(x) \in \mathbb{R}, J \in \mathbb{R}\} \tag{14.3.1}$$

defines the simplest local gauge transformation, $\chi(x)$ being the gauge function and J the generator of the transformation, which can be taken as $J = 1$ in this simple case. The transformation of wave functions or field operators is then given by

$$\begin{aligned} &\psi(x) \to \psi'(x) = P(x)\psi(x) \qquad \text{with} \\ &\qquad P(x) = e^{ieJ\chi(x)} , \qquad J = 1 . \end{aligned} \tag{14.3.2}$$

Obviously the free Lagrangians (14.1.11) are not invariant under these transformations of the second kind. But it is known that in the case of interacting electrically charged particles the Dirac Lagrangian (14.1.11f) of free fields has to be replaced by

$$l = \bar{\psi}(i\gamma^\mu(\partial_\mu + ieA_\mu(x)) - \kappa)\psi . \tag{14.3.3}$$

1) This Lagrangian shows gauge invariance of the second kind if the wave function $\psi(x)$ and potential $A_\mu(x)$ simultaneously transform as

$$\begin{aligned} &\psi \to \psi'(x) = P(x)\psi(x) \\ &A_\mu \to A'_\mu(x) = A_\mu(x) - \partial_\mu\chi(x) , \qquad J = 1 \end{aligned} \tag{14.3.4}$$

where the latter equation corresponds to the classical gauge transformation of an electromagnetic field. The local change of phase of a particle field thus corresponds to the occurrence of an additional electromagnetic field or, in other words, all the configurations $\psi'(x)$, $A'_\mu(x)$ defined by (14.3.4) describe the same physical situation. The internal space of $\psi(x)$ (charge space) possesses a symmetry so that a local change of the basis by a phase transformation can be interpreted as a change of the gauge field $A_\mu(x)$ (principle of relativity in an internal space).

2) In quantum theory, unlike in classical electrodynamics, the potential $A_\mu(x)$ cannot be eliminated from the equations of motion [in Maxwell's equations $A_\mu(x)$ does not occur]. This has certain consequences for the motion of a charged particle, which lead to a dependence of the physical properties on $A_\mu(x)$ if motions in non-simply-connected spaces without fields are considered (flux quantization in a superconducting toroid; Bohm-Aharonov effect [14.3]).

3) Thus the substitution principle[2]

$$\partial_\mu \psi \to [\partial_\mu + \mathrm{i}eJW_\mu(x)]\psi \equiv D_\mu \psi \ , \qquad JW_\mu = A_\mu \ , \qquad J = 1 \tag{14.3.5a}$$

leads from the field equation of a noninteracting particle (14.1.11f) to that of a charged particle (14.3.4) interacting with the electromagnetic field $A_\mu(x)$. This is said to be the *principle of minimal coupling* since this field, $A_\mu(x)$, is the "minimal field" that causes particle Lagrangians to be invariant under gauge transformations of the second kind. This principle is well established in QED; since it is obvious that it can be transferred to other symmetries and interactions, we shall discuss its properties in more detail.

If the Lagrangian l is invariant under *global* gauge transformations (first kind) there is a conservation law for the charges. Because there is a derivative ∂_μ in the free Lagrangian, l according to (14.1.11) is not invariant with respect to *local* gauge transformations (second kind). Instead, according to (14.3.1, 2),

$$\partial_\mu \psi = P^{-1}(\partial_\mu - (\partial_\mu P)P^{-1})\psi' \ . \tag{14.3.6}$$

The local dependence of the phase $\chi(x)$ in $\partial_\mu P$ destroys the invariance of l, or in other words: If the invariance of l is to be maintained in gauge transformations of the second kind, then l has to contain a *vector* (*gauge*) *field* A_μ, which compensates the term $\partial_\mu P$, besides the particle field ψ. According to the principle of minimal coupling, the invariant l is then given by (14.3.3). This new Lagrangian density contains the interaction of the particle field ψ with a *gauge field*, the coupling being mediated by the charges e. Formally this is obtained by replacing the local derivative ∂_μ by the covariant derivative

$$\begin{aligned} &\partial_\mu \to D_\mu = \partial_\mu + \mathrm{i}eA_\mu = \partial_\mu + \mathrm{i}eJW_\mu \ , \\ &J = 1 \ , \qquad A_\mu = W_\mu \qquad \text{in the Abelian case} \ . \end{aligned} \tag{14.3.5b}$$

The quantities J and W_μ are introduced in view of the non-Abelian generalization. With (14.3.5b) the Lagrangian (14.3.3) is invariant provided the gauge field

[2] In differential geometry D_μ is said to be the covariant derivative or a linear connection.

transforms as

$$A_\mu \to A'_\mu = P A_\mu P^{-1} + \frac{\mathrm{i}}{e}(\partial_\mu P)P^{-1} \tag{14.3.7}$$

in $\mathscr{U}(1)$ gauge transformations. Using (14.3.1), (14.3.7) is identical with (14.3.4), where use has been made of the fact that in the Abelian group $\mathscr{U}(1)$, $P(\chi(x))$ commutes with $A_\mu(x)$. The mapping (14.3.7) satisfies the rules of group multiplication

$$P(\chi_1\chi_2)A_\mu = P(\chi_1)P(\chi_2)A_\mu \ . \tag{14.3.8}$$

Equation (14.3.3) only contains the gauge field interacting with the particle field, so we have to add a Lagrangian density of the free gauge field. In analogy to electromagnetic theory we first define the field (tensor)

$$F^{\mu\nu} = \partial^\mu A^\nu - \partial^\nu A^\mu \qquad \text{with} \qquad F'^{\mu\nu} = F^{\mu\nu} \ , \tag{14.3.9}$$

which is invariant under $\mathscr{U}(1)$ gauge transformations of the second kind. An invariant Lagrangian[3] of the free field then is

$$l_F = -\tfrac{1}{4}F_{\mu\nu}F^{\mu\nu} \ . \tag{14.3.10a}$$

This can also be expressed by the potential A_μ

$$l_F = -\tfrac{1}{2}\partial_\mu A_\nu \partial^\mu A^\nu \qquad \text{if} \tag{14.3.10b}$$

$$\partial^\mu A'_\mu = 0 \qquad \textit{and} \qquad \partial^\mu A_\mu = 0 \ , \qquad \text{thus} \qquad \partial^\mu \partial_\mu \chi = 0 \ . \tag{14.3.10c}$$

That means, we only allow for gauge transformations satisfying this condition. There is no possibility of adding a "mass term" $(-\kappa_\gamma^2 A^\mu A_\mu)$ to (14.3.10) because then l_F would no longer be invariant under local gauge transformations. In QED this means that photons are massless.

The equation of motion follows from (14.3.10a) and (14.3.3):

$$\begin{aligned} &\gamma^\mu(\partial_\mu + \mathrm{i}eA_\mu)\psi + \mathrm{i}\kappa\psi = 0 \ , \\ &\partial_\mu F^{\mu\nu} = e\bar{\psi}\gamma^\nu\psi = ej^\nu \ . \end{aligned} \tag{14.3.11}$$

Sometimes, e.g. in gravitational theory, it is useful to introduce a second field $H^{\mu\nu}$ which is dual to $F^{\mu\nu}$,

$$H^{\mu\nu} = -\frac{\partial l_F}{\partial \partial_\mu A_\nu} = -2\frac{\partial l_F}{\partial F_{\mu\nu}} \ . \tag{14.3.12a}$$

---

[3] The factor in l_F is arbitrary; it only defines the units and is chosen in analogy to electromagnetic theory (see Bjorken and Drell [14.1, 2]). We emphasize that a minus sign has to taken in (14.3.10a) unlike for other Bose fields.

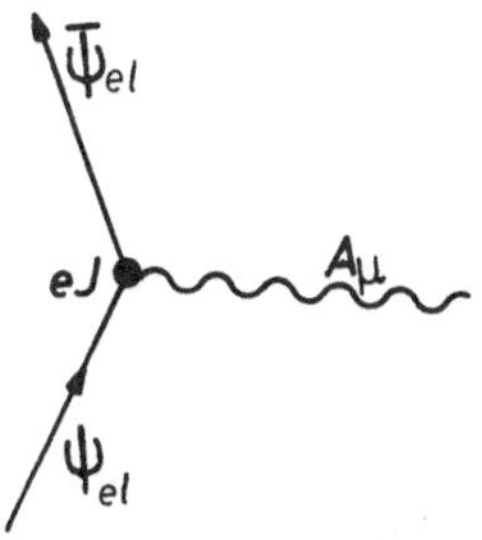

Fig. 14.2. An electron emits (absorbs) a photon with a change of energy and momentum

This is necessary, for example, when l_F is not specified or will not be specified in detail. If the particle Lagrangian does not contain derivatives $\partial_\mu A_\nu$ (which holds in general) the equation of motion is

$$\partial_\mu H^{\mu\nu} = -\frac{\partial l}{\partial A_\nu} = ej^\nu \tag{14.3.12b}$$

instead of (14.3.11). This formulation has the advantage that through the connection between $F^{\mu\nu}$ and $H^{\mu\nu}$ a further "interaction constant" or a "system of units" can be defined:

$$H^{\mu\nu} = 4\pi\varepsilon_0 F^{\mu\nu} \ .$$

In our case $\varepsilon_0 = 1/4\pi$. Equations (14.3.11, 12b) also guarantee the differential conservation law for the current since

$$\partial_\nu \partial_\mu H^{\mu\nu} \equiv 0 = \partial_\nu j^\nu \ .$$

Using (14.3.11) or (14.3.3) the interaction part of (14.3.3) can be written as

$$l_W = -ej^\mu A_\mu \ , \qquad j^\mu = \bar{\psi}\gamma^\mu\psi \ ; \tag{14.3.13}$$

it can be illustrated by the fundamental interaction graph of QED (Fig. 14.2). There is no change of the charge of the particle (electron); the gauge field (photon) itself has no charge. The latter fact is characteristic of Abelian gauge theories [see linear field equations (14.3.11)]. Thus $J(=1)$ is a 1×1-matrix describing the charge operator. The coupling strength between particle and gauge field is given by eJ.

The Lagrangians (14.3.3, 14) with interaction is invariant under (14.3.1, 7), thus the *gauge symmetries of the second kind* are *dynamical symmetries* (Sect. 13.4). One of the most important properties of these gauge theories is that they determine not only the symmetry of the interaction but *even the interaction itself*. The example with the electromagnetic field shows that the local $\mathscr{U}(1)$ gauge symmetry (second kind) requires that the principle of minimal coupling is satisfied. The Lagrangian of the free (uncharged!) electromagnetic field (14.3.10a) is also invariant with respect to these transformations of the second kind.

We summarize:

> The symmetry of a charged (particle) field under gauge transformations of the second kind requires the existence of vector fields (A_μ here, sometimes tensor fields) representing quasi-particles of zero mass (photons here) which couple to the charged particles. In other words: If the Lagrangian density shows gauge invariance of the second kind, then there must exist vector fields which universally couple to all charged (particle) fields (e.g. field to electrons, where charge means electric charge).

Just this statement is the basis of the gauge theories of the elementary particle fields and their interactions. However, it is to be expected that the gauge groups are more "complicated" than the simple Abelian gauge group $\mathscr{U}(1)$.

In connection with a possible generalization of the local gauge invariance we note that Lagrangians containing fields in the form $\psi^+ \mathcal{O}\psi$, e.g. $\bar{\psi}\gamma^0\psi$ or $\psi^+\psi$, are also invariant under local gauge transformations.

We list here the Lagrangians (14.1.11) in the $\mathscr{U}(1)$-invariant forms:

$$\begin{aligned}
&1)\ l = \tfrac{1}{2}(D_\mu^*\psi^* D^\mu\psi - \kappa^2\psi^*\psi)\ ,\\
&2)\ l = \tfrac{1}{2}(D_\mu^*\psi^{*\nu} D^\mu\psi_\nu - \kappa^2\psi^{*\nu}\psi_\nu)\ ,\\
&3)\ l = \tfrac{i}{2}(\psi^+\sigma^\mu D_\mu\psi - D_\mu^*\psi^+\sigma^\mu\psi)\ ,\\
&4)\ l = \tfrac{i}{2}(\bar{\psi}\gamma_\mu D^\mu\psi - D^{*\mu}\bar{\psi}\gamma_\mu\psi) - \kappa\bar{\psi}\psi\ ,\\
&\qquad D^\mu = \partial^\mu + ieA^\mu\ , \qquad D^{*\mu} = \partial^\mu - ieA^\mu\ .
\end{aligned} \tag{14.3.14}$$

Finally we should mention the connection between local gauge theories and charge quantization. The decisive point is that the electric charge e does not occur in the transformation of the electromagnetic potential (14.3.4). Thus no conditions for the charges e of particle fields coupled to the electromagnetic field can be derived from the invariance requirement for the theory. That means e is neither universal nor "quantized" in a $\mathscr{U}(1)$ theory. But this situation is different in $\mathscr{SU}(n)$ theories (see next section).

14.3.2 $\mathscr{SU}(n)$ Gauge Transformations of the Second Kind

Starting with the formulation of the local Abelian gauge theory in the previous section, it is possible to extend this theory to non-Abelian gauge groups. Again we start with a free Lagrangian

$$l = \bar{\psi}(i\gamma^\mu\partial_\mu - \kappa)\psi\ , \tag{14.3.15}$$

where ψ is a Dirac field forming a multiplet with respect to the gauge group. The free fields are associated with a non-Abelian charge algebra defined by the generators Q_l, J_l of the corresponding group (Sect. 14.2.2). The conservation law for the charges follows from Noether's theorem (Sect. 14.1) because of global gauge invariance.

The derivative ∂_μ in (14.3.15) is not covariant in local gauge transformations, but covariance may be achieved by introducing gauge fields with a definite transformation law. This again means a minimal coupling between particle and gauge field.

Let the non-Abelian gauge group be defined by

$$\psi(x) \rightarrow \psi'(x) = P(\chi_1(x), \ldots, \chi_m(x))\psi(x) \;, \tag{14.3.16a}$$

$$P(\chi_1(x), \ldots) := \exp\left[\mathrm{i}g \sum_{l=1}^{m} J_l \chi^l(x)\right]$$

with m gauge parameters (functions) $\chi^l(x)$ and generators J_l. Using

$$\chi(x) := \sum_{l=1}^{m} J_l \chi^l(x) \tag{14.3.16b}$$

we can also write

$$P(x) = \mathrm{e}^{\mathrm{i}g\chi(x)} \approx 1 + \mathrm{i}g\chi(x) \ldots \tag{14.3.16c}$$

in the case of infinitesimal transformations. The generalization of the covariant derivative (14.3.5a) to an m-parameter non-Abelian group is obvious:

$$D_\mu := \left(\partial_\mu + \mathrm{i}g \sum_{l=1}^{m} J_l W_\mu^l(x)\right) := \partial_\mu + \mathrm{i}gA_\mu(x) \;, \tag{14.3.17}$$

$$A_\mu(x) := \sum_{l=1}^{m} J_l W_\mu^l(x) \;.$$

Here $A_\mu(x)$ is a vector field with respect to "μ" whose "values" lie in the gauge group's Lie (charge) algebra defined by the charge operators J_l with[4]

$$[J_l, J_k] = -\mathrm{i} \sum_{j=1}^{m} c_{lk}^j J_j = \mathrm{i} \sum_{j=1}^{m} c_{kl}^j J_j \;. \tag{14.3.18}$$

The vector field $A_\mu(x)$ is the m-component gauge field; its components $W_\mu^l(x)$ are the non-Abelian generalizations of the electromagnetic potentials. The coupling constant corresponding to e is g. It determines the strength of the universal interaction of all the fields which carry the charges defined by the simple gauge group.

With (14.3.17) the Dirac Lagrangian becomes

[4] See the definition in (14.1.8) and compare with Sect. 11.1.1

$$l = \bar{\psi}[\mathrm{i}\gamma^\mu(\partial_\mu + \mathrm{i}gA_\mu(x)) - \kappa]\psi \tag{14.3.19a}$$

with the Dirac equation

$$[\gamma^\mu(\partial_\mu + \mathrm{i}gA_\mu(x)) + \mathrm{i}\kappa]\psi = 0 \ . \tag{14.3.19b}$$

The transformation properties of $A_\mu(x)$ under local gauge transformations follow in complete analogy to (14.3.7) as an invariance condition for l:

$$A_\mu(x) \to A'_\mu = PA_\mu P^{-1} + \frac{\mathrm{i}}{g}(\partial_\mu P)P^{-1} \ . \tag{14.3.20}$$

The mapping obviously shows the group properties, but because of the non-commutativity of the generators J_l, (14.3.20) is not a simple gradient transformation as are (14.3.4, 7). This can be seen in the infinitesimal limit which results in

$$A'_\mu = A_\mu - \partial_\mu \chi + \mathrm{i}g[\chi, A_\mu] \ , \tag{14.3.21a}$$

or for the components W^l_μ, using (14.3.18),

$$\delta W^l_\mu = W'^l_\mu - W^l_\mu = -\partial_\mu \chi^l - g\sum_{j,k} c^l_{kj} W^k_\mu \chi^j \ . \tag{14.3.21b}$$

Unlike (14.3.4) this transformation explicitly contains the coupling constant g. However, in Abelian groups the structure contains c^l_{jk} vanish and we again obtain (14.3.4). In non-Abelian groups the g dependence does not vanish. This automatically leads to a quantization of charges, because the gauge invariance of the theory requires the occurrence of the same coupling constant in the generalized phase transformations of all the particle fields which couple to the fields $W^l_\mu(x)$ (see Sect. 14.5).

Analogously to the electromagnetic case, the inhomogeneous gradient term means that the charges gJ_l are the sources of the W^l_μ field, the *Yang-Mills field* (*potential*, 1954) [14.4]. The second term homogeneous in W^l_μ means that the field quanta of the Yang-Mills field themselves contribute to the charges, and thus carry charges (W^l_μ is not a neutral but a charged field!). The Yang-Mills field itself is a vector in the parameter space and couples with itself, the origin being the non-Abelian character of the gauge group.

If there are additional conditions for the field components, these, in order to be reasonable, have to be covariant, too. For example, in a Lorentz gauge

$$\partial^\mu W^l_\mu = \partial^\mu W'^l_\mu = 0 \tag{14.3.22a}$$

would hold. But this also implies restrictions on the allowed gauge functions χ^l, e.g.

$$\partial^\mu \partial_\mu \chi^l + g\sum_{j,k} c^l_{kj} W^k_\mu \partial^\mu \chi^j = 0 \ . \tag{14.3.22b}$$

Only those functions satisfying (14.3.22b) are allowed. Such conditions might reduce the number of allowed gauge functions ($l = 1, \ldots, m$), which then corresponds to a reduction of gauge symmetry.

The Lagrangian (14.3.19) describes an interaction between the particle field and gauge field. Thus we have to add the Lagrangian of the free gauge field, which again is achieved in analogy to the electromagnetic case by first defining the field strength in a covariant way:

$$\begin{aligned} G_{\mu\nu} &= \partial_\mu A_\nu - \partial_\nu A_\mu + ig[A_\mu, A_\nu] \\ &= \sum_l (\partial_\mu W_\nu^l - \partial_\nu W_\mu^l) J_l + g \sum_{jkl} W_\mu^j W_\nu^k c_{jk}^l J_l \\ &= \sum_l F_{\mu\nu}^l J_l \end{aligned} \tag{14.3.23a}$$

with[5]

$$F_{\mu\nu}^l = \partial_\mu W_\nu^l - \partial_\nu W_\mu^l + g \sum_{jk} c_{jk}^l W_\mu^j W_\nu^k \ . \tag{14.3.23b}$$

Using (14.3.16a) it can be shown that the field transforms as

$$G_{\mu\nu} \to G'_{\mu\nu} = P(\chi_1(x), \ldots) G_{\mu\nu} P^{-1}(\chi_1(x) \ldots) \ . \tag{14.3.24}$$

Hence the Lagrangian of the free gauge field is formed with the Yang-Mills field strength components $F_{\mu\nu}^l$ with respect to the basis J_l, defined in (14.3.23b),

$$l_F = -\tfrac{1}{4} \sum_l F_{\mu\nu}^l F^{\mu\nu,l} \ . \tag{14.3.25}$$

The total Lagrangian[6] is the sum of (14.3.19a) and (14.3.25). From the total Lagrangian we obtain by variation with respect to $\bar{\psi}$ the equation of motion for the particle field ψ (14.3.19b) and by variation with respect to W_μ^l the Yang-Mills field equations

$$\partial^\mu F_{\mu\nu,l} = g(j_{\nu,l}^{(\psi)} + j_{\nu,l}^{(W)}) \qquad \text{with} \tag{14.3.26a}$$

$$j_{\nu,l}^{(\psi)} = \bar{\psi}\gamma_\nu J_l \psi \tag{14.3.26b}$$

$$j_{\nu,l}^{(W)} = \sum_{jk} c_{lj}^k W^{\mu,j} F_{\mu\nu,k} \ . \tag{14.3.26c}$$

In non-Abelian gauge theories there are two contributions to the current, the normal particle current (14.3.26b) and a second part originating from the gauge field (14.3.26c). The particle and gauge fields carry charges; the gauge field acts

---

[5] In differential geometry $F_{\mu\nu}^l$ is called the curvature tensor (curvature) of the linear connection.

[6] In the case of nonsimple gauge groups, e.g. semisimple gauge groups with r simple factors, the Lagrangian contains r different coupling constants.

as a source for itself. There is a *self-coupling* of the gauge fields $W_\mu^l(x)$, which originates from the nonlinear terms in the Lagrangian or in the field equations. Only the total charge is conserved, thus

$$\partial^\mu(j_{\mu,l}^{(\psi)} + j_{\mu,l}^{(W)}) = 0 \tag{14.3.27}$$

if the condition $\partial^\mu W_\mu^l = 0$ is taken into account. The charge operator is given by [see (14.1.12, 2.2)]

$$\rho_l = \psi^+ J_l \psi + \sum_{jk} c_{lj}^k W^{\mu,j} F_{\mu 0,k} \ . \tag{14.3.28}$$

Finally, we can also define a conjugated field by

$$H_k^{\mu\nu} = -\frac{\partial l}{\partial \partial_\mu W_\nu^k} \ . \tag{14.3.29a}$$

Then the equation of motion is

$$\partial_\mu H_k^{\mu\nu} = -\frac{\partial l}{\partial W_\nu^k} \tag{14.3.29b}$$

and the charge

$$\rho_k = \psi^+ J_k \psi + \sum_{j,l} c_{kj}^l W^{\mu,j} H_{\mu 0,l} \ . \tag{14.3.29c}$$

In Sect. 13.1.2 we presented an approximate $\mathcal{SU}_f(n)$ flavour classification of quarks together with a characterization according to an $\mathcal{SU}_c(3)$ group. This leads to a division of the hadrons into multiplets which are labelled by charge quantum numbers such as isospin I_3, hypercharge Y, strangeness S, charm C and also the electric charge Q, of course. In Sect. 13.1.2 the numbers were connected with the generators of the flavour group $\mathcal{SU}_f(n)$. This description is strictly valid only as long as all the quarks have the same mass, see Sect. 13.1, because then the Lagrangian is invariant under $\mathcal{SU}_f(n)$ transformations. The violation of the flavour symmetry in QCD is connected with the different quark masses.

The $\mathcal{SU}_c(3)$ symmetry, on the other hand, is thought to be an exact one. It may be used as an example of an exact Yang-Mills gauge theory and as the basis of QCD, which is constructed from the local *colour* gauge group $\mathcal{SU}_c(3)$. In this theory the fundamental (defining) REP of $\mathcal{SU}_c(3)$ is assigned to the colour degrees of freedom [colour = r, y(g), b]. The group $\mathcal{SU}_c(3)$ is assumed to be an exact (dynamical) symmetry group of the strong interaction. Bound states of the quarks (hadrons) only occur as *colourless* (*neutral with respect to colour*) colour singlets; see (13.1.28). This again shows the fundamental relation between dynamics and symmetry and indicates the importance of the colour charges in *quark confinement*. The binding of quarks in colour-neutral systems seems to be

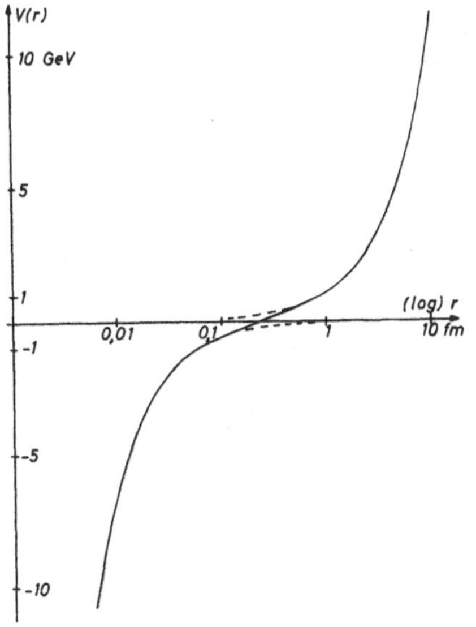

Fig. 14.3. Assumed potential describing quark confinement. It is Coulombic for small distances $r \lesssim 0.1$ fm, with $V \sim -c^2/r$, $c \sim 6.6e \sim 0.25$ (GeV fm)$^{1/2}$, and almost linear for larger distances $r \gtrsim 0.5$ fm, with $V \sim c'r$, $c' \sim 1$ GeV/fm. Note that the length scale is logarithmic but the potential scale linear

an absolute principle (Fig. 14.3). There are no experimental indications of free quarks or free gauge field particles (gluons in the $\mathcal{SU}_c(3)$ theory carry colour while photons in $\mathcal{U}(1)$ QED do not carry a charge!). There are speculations that QCD implies a confinement mechanism which can be described by a (gluon) string concept in analogy to type II superconductivity (Ginzburg-Landau theory [14.5] as an early version of the Higgs model, see Nielsen-Olesen, Nambu [14.6, 7]). The chromo (colour) flux between colour charges then corresponds to the magnetic flux tubes of type II superconductors.

The construction of QCD as an $\mathcal{SU}_c(3)$ theory starts with coloured quark fields $\psi^{\alpha,f,c}(x)$, see (14.2.7), i.e. every quark flavour type (u, d, s, c, b, t) is allowed to occur in three colour states $c = r, y, b$. The charge generators J_l of $\mathcal{SU}_3(c)$ given by the eight Gell-Mann matrices in (11.4.8) (apart from a factor of 2) act on the three components of the wave function $\psi^{\alpha,f,c}(x) \to \boldsymbol{q}(x) = \{q_r, q_y, q_b\}$ in the colour space, where now α and f are assumed to be fixed. The phase change of the quark fields in gauge transformations (second kind) is given by

$$\boldsymbol{q}(x) \to \boldsymbol{q}'(x) = P(\chi^1, \dots, \chi^8)\boldsymbol{q}(x) , \tag{14.3.30}$$

$$P(\chi^1, \dots, \chi^8) = \exp\left(ig \sum_{l=1}^{8} J_l \chi^l(x)\right) .$$

The gauge invariance means that the colours of the quarks may be mixed up in each space-time point x. The corresponding change of the gauge fields is given by (14.3.20, 21). The interaction between quarks is described by the eight massless gauge fields $W_\mu^l(x)$, $l = 1, \dots, 8$ (*gluon fields*), see Table 14.1. Their sources are the colour charges gJ_l according to (14.3.26). The interaction between quarks and

Table 14.1. Field particles [13.9, 10]

	Q	Flavour	Colour	J^{PC}	Mc^2[GeV]		
γ	0	√		1^{--}	0		Electromagnetic field
$W^{\pm}$	± 1	√		1^{-}	80.22		Weak field
Z^0	0	√		1^{-}	91.19		Weak field
Higgs particle	$0, \pm 1$	√		$0^{\pm}$	$40 < M \lesssim 200?$		Higgs field
Gluons	0		√	1^{-}	0	8 species	Strong field
Leptons	$0, -1$	√		$1/2^{+}$		6 states	
Quarks	$-1/3, +2/3$	√	√	$1/2^{+}$		6 flavour states 3 colour states	

Fig. 14.4

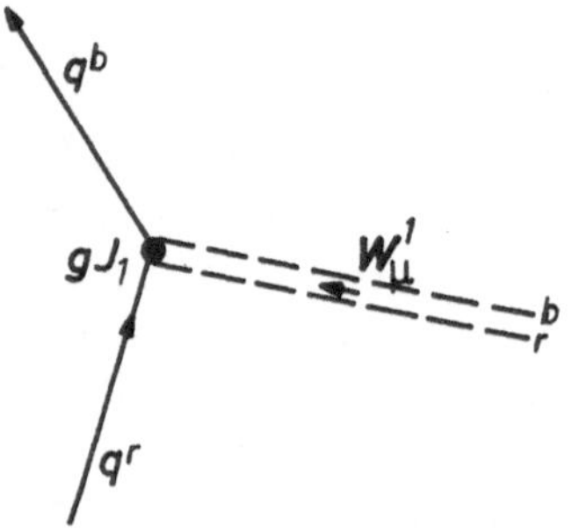

Fig. 14.5

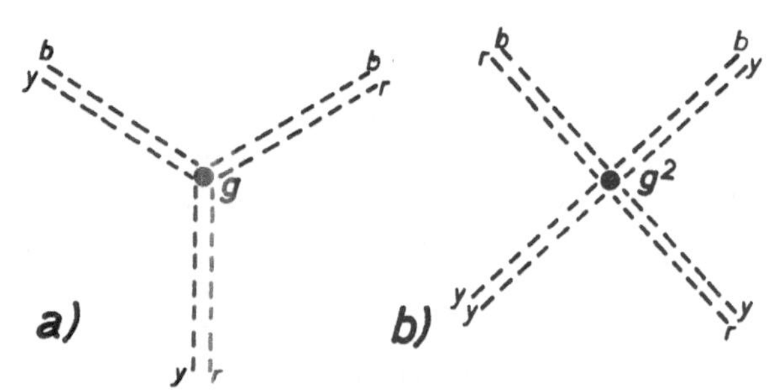

Fig. 14.4. A red quark emits (absorbs) a charged gluon W^1_μ and changes into a blue one. It also changes energy and momentum according to the usual laws

Fig. 14.5. (**a**) Vertex for a 3-gluon interaction according to (14.3.25) with (14.3.23b): $gc^l_{jk} W^j_\mu W^k_\nu \partial^\mu W^{\nu,l}$. (**b**) Vertex for a 4-gluon interaction according to (14.3.25) with (14.3.23b): $g^2 c^l_{jk} c^l_{j'k'} W^j_\mu W^k_\nu W^{\mu j'} W^{\nu k'}$

gluons is contained in the interaction part l_w of the Lagrangian (14.3.19a) together with (14.3.26b)

$$l_w = (-g)\bar{\psi}\gamma^\mu A_\mu \psi = -g \sum_{l=1} j^{(\psi)\mu}_l W^l_\mu \; , \tag{14.3.31}$$

the coupling strength between quark and gluon W^l_μ being gJ_l. The elementary interaction graph is shown in Fig. 14.4. By the interaction $\bar{\psi}\gamma^\mu J_1 W^1_\mu \psi$ with $J_1 \sim E_{13}$ according to (11.4.8) an incoming quark $\boldsymbol{q} = \{q_r, 0, 0\}$ is changed into $\{0, 0, q_r \to q_b\}$ with emission or absorption of a gluon. Contrary to the Abelian QED it may change its charge (colour), thus transferring colour to the gluon. The latter itself carries charge and thus may *interact with itself*. Such processes are shown graphically in Fig. 14.5.

The self-interaction of gauge fields (gluons) appearing in non-Abelian gauge theories also leads to an *asymptotic freedom* of the theory (Sect. 14.5.5). That means, for very large (external) momenta quark and gluon-fields may be looked upon as approximately free. Thus the colour forces decrease with decreasing distance and increase with increasing distance (running coupling constant). As a result, the colour-charged quarks and gluons cannot escape to macroscopic distances. Thus only bound states (hadrons), not free quarks, are observed. The effective interaction of these complexes, however, is only short ranged, despite the massless gluons. This fact may be understood by realizing that the colour-charged gluons which surround a colour-charged particle as a cloud of virtual particles smear out the charge of the particle, so that its effective colour charge decreases for smaller distances (from the surface region to its centre).

14.3.3 A Differential Geometric Discussion of the Yang-Mills Fields

The covariant derivatives occurring in (14.3.5a, 17) are well-known quantities in the differential geometry of curved spaces or curvilinear coordinate systems. Therefore we shall sketch the connection between local gauge invariance and differential geometry [14.8–10].

If a vector is translated along a curved line, its components change even if its absolute value remains constant. Thus, if a field vector has different absolute values at different space points, the change of the vector cannot be seen directly from the different components at different points. In order to compare vectors at different space points, one has to bring both the vectors to the same space point by a parallel translation along the curved line (Fig. 14.6a).

Let the vector $\boldsymbol{a}$ be "parallel" translated from P to P' in such a way that its "magnitude" and "direction" remain constant. Then its components change from a^j to $a^j + \delta a^j$. This vector can then be compared with the actual field vector $a^j + da^j$ at the point P' Since now both vectors are defined at the same point,

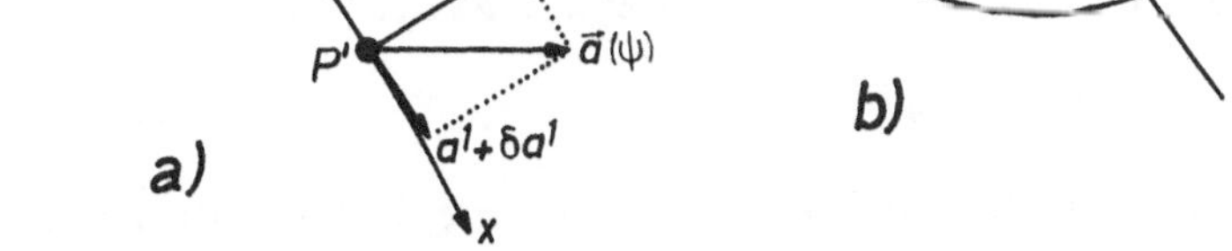

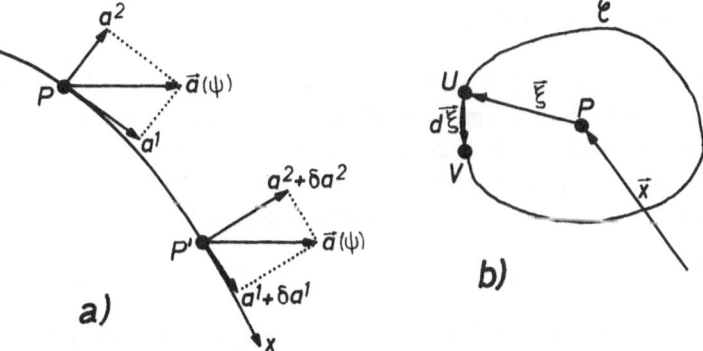

Fig. 14.6. (a) Parallel translation of a vector $\boldsymbol{a}$ along a curved line. Its components change by $\delta\boldsymbol{a}$ in this translation. The same is valid for a wave vector ψ. (b) Parallel displacement along $\mathscr{C}$ around P

their difference is also a vector and the real change of the vector between P and P' is given by

$$da^j - \delta a^j = a^j{}_{|k}\, dx^k := Da^j \ . \tag{14.3.32}$$

Here dx^k is the infinitesimal distance between P and P'. Since $da^j - \delta a^j$ and dx^k are vector components, $Da^j/Dx^k = a^j{}_{|k}$ is a tensor of second rank and is said to be the *covariant derivative* of a^j. The vector δa^j has to be determined from the infinitesimal parallel displacement. In any case, it is proportional to dx^k and to the components a^l themselves (since a sum of vectors transforms in the same way as the single terms). Thus, in an arbitrary system, we define

$$\delta a^j := -\Gamma^j{}_{lk} a^l\, dx^k \tag{14.3.33}$$

with the *Christoffel symbols* of the second kind, $\Gamma^j{}_{lk}$. They have to be determined from the actual problem depending on the coordinate system used. In Cartesian systems obviously $\Gamma^j{}_{lk} \equiv 0$. Together with the local derivative

$$da^j = \frac{\partial a^j}{\partial x^k} dx^k$$

we obtain the covariant derivative

$$Da^j = a^j{}_{|k}\, dx^k = \left(\delta^j{}_l \frac{\partial}{\partial x^k} + \Gamma^j{}_{lk}\right) a^l\, dx^k \ . \tag{14.3.34}$$

In order to see the connection between (14.3.33, 34) and (14.3.17) we have to assign to every space-time point x a local coordinate system having as many basis elements as the number of internal degrees of freedom (ψ^j, $j = 1, 2, 3$ for the colour space, for example). This system is an orthonormal basis of the charge (Hilbert) space, a unitary space in our example, $\mathfrak{U}(x)$. A gauge transformation (14.3.16a) is a mapping of $\mathfrak{U}(x)$ onto itself:

$$\psi^j(x) \to \psi'^j(x) = P(\chi_1, \dots, \chi_m)\psi^j(x) \ .$$

In other words: it is a change of the orthonormal basis of $\mathfrak{U}(x)$. Altogether we have to deal with changes in the internal space $\mathfrak{U}(x)$ and in the external space-time manifold. A parallel translation only takes place in the direction of space-time, but in this translation the components of a vector $\psi^j(x)$ change (Fig. 14.6a). When the components at x are $\psi^j(x)$, the components at $x + dx$ are [14.10]

$$\psi^j + \delta\psi^j = (\delta^j{}_k - \mathrm{i}gA^j{}_{k,\mu}\, dx^\mu)\psi^k \tag{14.3.35}$$

with conveniently chosen factors, see (14.3.33). As we are dealing with unitary gauge transformations, $A^j{}_{k,\mu}$ can be represented by the infinitesimal generators of the unitary gauge group, thus

$$A^j{}_{k,\mu} = \sum_l J_{l,}{}^j{}_k W^l_\mu(x) \tag{14.3.36a}$$

in analogy with (14.3.17). Together with the actual x-dependent change of the vector ψ^j, which is $(\partial\psi^j/\partial x^\mu)\,dx^\mu$, we obtain

$$D_\mu\psi^j\,dx^\mu = (\partial_\mu\psi^j + \mathrm{i}gA^j{}_{k,\mu}\psi^k)\,dx^\mu \ ,$$

the Christoffel symbol being

$$\Gamma^j{}_{k\mu} = \mathrm{i}gA^j{}_{k,\mu} = \mathrm{i}g\sum_l J_{l,}{}^j{}_k W^l_\mu(x) \ . \tag{14.3.36b}$$

Of course, the physical interpretation has to be adapted to this case. In this formulation we see that the fields (potentials) W^l_μ essentially determine the curvature of the space.

If the components of a vector change in a parallel translation in a curved space, it cannot be assumed that these changes vanish in the case that a complete circuit of parallel translations is made (Fig. 14.6b). Only in a Euclidean space (Cartesian system) after such a circuit the total change Δa^j does vanish. We calculate Δa^j for an infinitesimal circle around P, with ξ^j being "small" and $d\xi^j$ being even "smaller". Under a translation from U to V, a^j changes by

$$\delta a^j = -\Gamma^j{}_{lk}a^l\,d\xi^k \tag{14.3.37}$$

according to (14.3.33), where the $\Gamma^j{}_{lk}a^l$ have to be taken at U. In a translation from P to U, $\Gamma^j{}_{lk}$ changes according to

$$\Gamma^j{}_{lk} + \frac{\partial\Gamma^j{}_{lk}}{\partial x^m}\xi^m \ ,$$

where the quantities have to be taken at P. If a^l in (14.3.37) is related to P (instead of U) then

$$a^l(U) = a^l(P) - \Gamma^l{}_{rm}a^r\xi^m \ .$$

Inserting the last two equations into (14.3.37) we obtain to first order in ξ

$$\delta a^j = -\Gamma^j{}_{lk}a^l\,d\xi^k - \left(a^l\frac{\partial\Gamma^j{}_{lk}}{\partial x^m} - \Gamma^j{}_{lk}\Gamma^l{}_{rm}a^r\right)\xi^m\,d\xi^k \ . \tag{14.3.38}$$

This has to be integrated around the circle $\mathscr{C}$ in Fig. 14.6b to give the total change Δa^j. Now

$$\oint_{\mathscr{C}} d\xi^k = \Delta\xi^k = 0 \ , \qquad \oint_{\mathscr{C}} d(\xi^k\xi^m) = \oint_{\mathscr{C}} \xi^k\,d\xi^m + \oint_{\mathscr{C}} \xi^m\,d\xi^k = 0 \ ,$$

thus the first term in (14.3.38) gives no contribution, and of the second term only the part antisymmetric in k, m appears. Therefore we have

$$\Delta a^j = \tfrac{1}{2} R^j{}_{rmk} a^r \oint_{\mathscr{C}} \xi^m \, d\xi^k$$

with the Riemann-Christoffel curvature tensor

$$R^j{}_{rmk} = \Gamma^j{}_{lk} \Gamma^l{}_{rm} - \Gamma^j{}_{lm} \Gamma^l{}_{rk} + \frac{\partial \Gamma^j{}_{rm}}{\partial x^k} - \frac{\partial \Gamma^j{}_{rk}}{\partial x^m} \; . \tag{14.3.39}$$

This vanishes in Euclidean spaces and is thus a measure of the curvature of the space. In our case we obtain with (14.3.36b) (the $A^j{}_{l,\mu}$ and J_l, respectively, do not commute!)

$$\begin{aligned} R^j{}_{r\mu\kappa} &= -ig[\partial_\mu A^j{}_{r,\kappa} - \partial_\kappa A^j{}_{r,\mu} + ig(A^j{}_{l,\mu} A^l{}_{r,\kappa} - A^j{}_{l,\kappa} A^l{}_{r,\mu})] \\ &= -igG^j{}_{r,\mu\kappa} = -ig \sum_l J_{l,}{}^j{}_r F^l_{\mu\kappa} \; . \end{aligned} \tag{14.3.40}$$

Thus the field $G_{\mu\kappa}$ or $F^l_{\mu\kappa}$ determines the curvature of the charge space and vice versa.

We will conclude these remarks by briefly mentioning the general theory of relativity. In that case the curvature of the space is determined by the mass distribution, and the covariant derivative is connected with the metric tensor $g_{jk}(x)$ of space-time. The Christoffel symbols are then given by

$$\Gamma^j{}_{lk} = \tfrac{1}{2} g^{jm} (\partial_l g_{mk} + \partial_k g_{ml} - \partial_m g_{lk}) \; .$$

In this case the metric tensor $g_{jk}(x)$ describes the local symmetry, the Christoffel symbols being dynamical quantities of the theory. They correspond to the gauge potentials of the Yang-Mills theory, and the curvature tensor (14.3.39) corresponds to the field strengths, which are determined by the masses in the general theory of relativity. These correspondences also give a starting point for looking upon the theory of relativity as a gauge theory.

Exercise 14.2. Using the Schrödinger equation of a free particle, $H\psi = i\hbar\dot{\psi}$, $H = -\hbar^2 \Delta / 2m$, show that the invariance of the Hamiltonian under a local gauge transformation (second kind) requires the existence of a (electromagnetic) field.

14.4 Gauge Theories with Spontaneously Broken Symmetry

14.4.1 General Remarks

In order to explain *spontaneously broken symmetry* we consider a Lagrangian (density) l or a Hamiltonian H which is invariant under a symmetry group $\mathscr{G}$

and investigate the behaviour of the ground state (or the ground states in the case of degeneracy) under a symmetry transformation of $\mathcal{G}$.

If the ground state is single-valued (nondegenerate) it has to be invariant under $\mathcal{G}$; this means it has the same symmetry as l or H. For example, the s-state of a simple atom has the spherical symmetry of the potential. However, if the ground state is degenerate, these degenerate states transform into each other under $\mathcal{G}$. If we choose one particular state of these equivalent states, this state is not invariant under $\mathcal{G}$. It is then said that the symmetry of l or H is spontaneously broken in this state. We also refer to the Jahn-Teller effect (Sect. 8.1), which also leads to a symmetry reduction, but due to an additional term in the Hamiltonian.

Spontaneously broken symmetries are realized in connection with phase transitions in the most general sense where the state of a system changes into another one. Typically, for such a system symmetric and symmetry-broken phases occur separated by the transition. The occurrence of these transitions always depends on an external parameter like temperature, energy, pressure, etc. Furthermore, the spontaneously broken phase can be described by an *order parameter*, indicating that this phase possesses the lower symmetry. Well-known examples of symmetry-broken phases are ferromagnets, crystals and superconductors.

For a ferromagnet the Hamiltonian is rotationally symmetric, $\mathcal{O}(3)$; in the ground state, however, an arbitrary but fixed direction of direct space is distinguished and so the rotational symmetry is broken. There is a magnetization $\boldsymbol{M} \neq 0$, which is the order parameter. The symmetry-breaking field can be considered as a (small) magnetic field.

For a crystal the Hamiltonian has the continuous translational symmetry of the homogeneous space, whereas the ground state only allows for discrete translations by multiples of the lattice vectors.

For a superconductor the Hamiltonian is invariant under global gauge transformations (first kind). In the ground state the phase of the condensate wave function (order parameter) is arbitrary. It can be fixed by choice of a special gauge which then spontaneously breaks the symmetry. This also happens in the models of elementary particle theories where the symmetry is spontaneously broken by the choice of the phase of the particle fields (Sects. 14.4.2, 3; [14.11–15]). Geometrically this corresponds to a distinction of a special direction of the particle fields (or the expectation values in the vacuum state) in the (unitary) space of the internal degrees of freedom. In these cases the order parameters are the fields or their vacuum expectation values, if quantum theory is used.

Goldstone's theorem (see Appendix H) tells us that in field theories with spontaneously broken symmetries of continuous groups, *massless* Goldstone excitations (fields) may occur. In solid-state physics this means the occurrence of elementary excitations whose frequencies vanish in the long wavelength limit $[\lim_{q \to 0} \omega(\boldsymbol{q}) \to 0]$. Thus, the breaking of the translational or rotational symmetries leads to the existence of phonons or magnons.

In elementary particle physics, especially in the theory of electro-weak interactions, the inverse situation is of particular interest: by spontaneous breaking of symmetry, masses can be introduced without any violation of the gauge invariance of the Lagrangian density (see the Higgs mechanism in the model of Glashow, Salam and Weinberg [14.11, 16–23]).

14.4.2 Spontaneous Breaking of a Gauge Symmetry of the First Kind: Goldstone Model

We consider a complex scalar field whose Lagrangian is given in (14.1.11a) but we extend it by a further term describing the self-interaction of the field, thus[7]

$$l = \tfrac{1}{2}\partial_\mu\phi^*\partial^\mu\phi - V(\phi^*\phi) \qquad \text{with} \tag{14.4.1}$$

$$V(\phi^*\phi) = \frac{\lambda}{4}\left(|\phi|^2 - \varepsilon\frac{\kappa^2}{\lambda}\right)^2 = -\frac{\varepsilon}{2}\kappa^2|\phi|^2 + \frac{\lambda}{4}|\phi|^4 + \frac{\kappa^4}{4\lambda}\,,$$

$$\varepsilon = \pm 1\,, \qquad |\phi|^2 = \phi^*\phi\,.$$

The $\kappa^2|\phi|^2$ term with $\varepsilon = -1$ is the usual mass term in (14.1.11a), the $|\phi|^4$ term describes the self-interaction, while the last term is irrelevant. Obviously, l is invariant under $\mathscr{U}(1)$ gauge transformations (global, first kind) (14.2.1a):

$$\phi(x) \to \phi'(x) = \mathrm{e}^{\mathrm{i}\beta}\phi(x)\,, \qquad \beta = e\chi\,. \tag{14.4.2}$$

The parameter λ has to be positive because otherwise the energy has no "lower limit"; the "mass-term" $\kappa^2|\phi|^2$ can assume both signs.

$\varepsilon = -1$: Then the κ^2 term is a proper mass term and the field configuration with lowest energy (ground state, vacuum) is given by $\phi(x) = 0$. The symmetry of this phase is the same as that for l i.e. there is no breaking of symmetry:

$$\frac{\partial V}{\partial\phi} = \phi^*\left(-\frac{\varepsilon}{2}\kappa^2 + \frac{1}{2}\lambda|\phi|^2\right) = 0 \quad \curvearrowright \quad \phi = 0 \tag{14.4.3}$$

is the only solution for $\varepsilon = -1$.

$\varepsilon = +1$: In this case there is no true "mass term" and the potential has a double-well form (like the bottom of a bottle; Fig. 14.7). The configuration of lowest energy is given by

$$|\phi_0| = v = \sqrt{\kappa^2/\lambda}\,. \tag{14.4.4a}$$

The ground state is thus infinitely degenerate. All the fields (order parameter ϕ_0 or β) with

$$\phi_0 = v\mathrm{e}^{-\mathrm{i}\beta}\,, \qquad \beta = \text{const} \tag{14.4.4b}$$

---

[7] In a quantum-mechanical treatment, ϕ^* has to be replaced by ϕ^+.

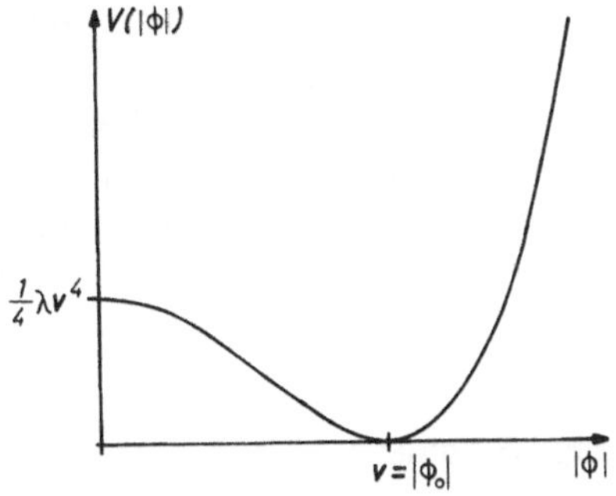

Fig. 14.7. Potential like the bottom of a bottle with a degenerate ground state at $|\phi_0| = v$. The radial curvature (mass) is $-\kappa^2$ at $|\phi| = 0$ and $2\kappa^2$ at $|\phi_0| = v$

describe configurations with lowest energy; if we fix β, e.g. choosing $\beta = 0$, we have a phase with spontaneously broken symmetry. Physically this can be achieved by introducing a small additional potential $\sim\delta(\beta)$.

The quantum mechanical expression for (14.4.4a) is

$$\langle 0|\phi|0\rangle = v = \sqrt{\kappa^2/\lambda} \neq 0 \ . \qquad (14.4.4c)$$

This means that the expectation value of the field ϕ does not vanish in the vacuum $|0\rangle$. The vacuum is filled by a Bose condensate ($\sim v$), or in other words, the field has a fixed orientation in the space of the internal degrees of freedom of ϕ.

Within this model we will discuss the general properties of spontaneously broken gauge symmetries of the first kind, showing that massless scalar field excitations occur. For this purpose we employ an expansion around the degenerate ground state (14.4.4a) ($\varepsilon = +1$), i.e. we use

$$\phi(x) = [v + \eta(x)]\exp\{-\mathrm{i}[\beta + \varphi(x)/v]\} \ ; \qquad (14.4.5a)$$

$\eta(x)$ and $\varphi(x)$ are real fields describing the deviations from the ground state (14.4.4b). Introducing (14.4.5a) into (14.4.1) we obtain the Lagrangian

$$l(\eta, \varphi) = \frac{1}{2}(\partial_\mu\eta\partial^\mu\eta + \partial_\mu\varphi\partial^\mu\varphi) + \underline{\frac{1}{2v^2}\partial_\mu\varphi\partial^\mu\varphi(2v\eta + \eta^2)}$$

$$- \kappa^2\eta^2 + \underline{\lambda v\eta^3} + \frac{1}{4}\lambda\eta^4 \qquad (14.4.6)$$

where the higher order terms (underlined) are not of interest for our discussion. They mean an interaction between the real fields η and φ.

Instead of a complex field ϕ with two components, the Lagrangian (14.4.6) now contains a massive field $\eta(x)$ with $m_\eta = \sqrt{2}\kappa$ and a massless field $\varphi(x)$; this can be seen from (14.4.6) because there is no φ^2 mass term in l. The mass m_η is determined by the nonvanishing curvature in the radial direction of the potential (Fig. 14.7). Thus $\eta(x)$ describes "radial vibrations" of the field around its vacuum value v, whereas $\varphi(x)$ represents "tangential vibrations" parallel to the minima of the potential having zero frequency because of the vanishing curvature. Thus $\varphi(x)$ represents the massless Goldstone field expected from Goldstone's theorem.

We should mention here that similar models play an important role in other theories with interacting fields. For example, laser theory where $V(\phi)$ corresponds to an effective potential for the complex laser amplitude (see Haken and Wolf [14.24]).

14.4.3 Spontaneous Breaking of an Abelian Gauge Symmetry of the Second Kind: Higgs-Kibble Model

When a *local* gauge symmetry is spontaneously broken the occurrence of massless Goldstone excitations can be avoided, which means that the Goldstone theorem is not valid in such theories. However, local gauge symmetry also means the introduction of additional gauge fields $A_\mu(x)$ which couple to the (scalar) Higgs fields ϕ according to the principle of minimal coupling. In the following we will see that the spontaneous breaking of symmetry in the "Higgs phase" ($\varepsilon = +1$) makes part of the formerly massless gauge fields massive and that there are no massless Goldstone bosons.

The Abelian Higgs-Kibble model starts with a Lagrangian, see (14.1.11a; 14.3.5a, 10a; 14.1.1),

$$l = -\tfrac{1}{4}F_{\mu\nu}F^{\mu\nu} + \tfrac{1}{2}(\partial_\mu - igA_\mu)\phi^*(\partial^\mu + igA^\mu)\phi - V(|\phi|) \tag{14.4.7}$$

with, see (14.3.9, 10c; 14.4.1),

$$F^{\mu\nu} = \partial^\mu A^\nu - \partial^\nu A^\mu \ , \qquad \partial_\mu A^\mu = 0 \ ,$$

$$V(|\phi|) = -\tfrac{1}{2}\kappa^2|\phi|^2 + \tfrac{1}{4}\lambda|\phi|^4 \ , \qquad \varepsilon = +1 \ ,$$

where the irrelevant constant term $\kappa^4/4\lambda$ has been omitted. Because $\partial_\mu A^\mu = 0$ and because the gauge field A_μ is massless, A_μ has only two independent observable components entering the Lagrangian (this can be seen by separating $A_\mu = \{\phi, A_1, A_2, A_3)$ into longitudinal and transverse parts).

Just as in the case of the Goldstone model, the degenerate ground state is again given by the field configuration

$$\phi_0 = ve^{-i\beta} \ , \tag{14.4.4d}$$

$$\beta = \text{const} \ , \qquad v = \sqrt{\kappa^2/\lambda} \ ,$$

and again we can choose a gauge so that

$$\phi_0 = v \rightarrow \langle 0|\phi|0\rangle \ . \tag{14.4.4e}$$

For a perturbation theory of the excitations we again use the expansion (14.4.5a)

$$\phi(x) = \{v + \eta(x)\}e^{-i\varphi(x)/v} \;, \qquad \beta = 0 \;; \tag{14.4.5b}$$

$\eta(x)$ and $\varphi(x)$ again being real fields. However, contrary to the Goldstone model, the field $\varphi(x)$ can now be removed by gauging and thus has no physical meaning. This is because of the local gauge invariance of the theory. So we may use the expansion

$$\phi(x) = v + \eta(x) \;. \tag{14.4.5c}$$

In this way the "would be" Goldstone boson $\varphi(x)$ is eliminated by choosing an appropriate gauge. Using (14.4.5c) we obtain from (14.4.7)[8]

$$\begin{aligned} l = {} & -\tfrac{1}{4}F_{\mu\nu}F^{\mu\nu} + \tfrac{1}{2}\partial_\mu\eta\partial^\mu\eta + \tfrac{1}{2}g^2(v + \underline{\eta})^2 A_\mu A^\mu \\ & - \lambda(v + \underline{\tfrac{1}{2}\eta})^2\eta^2 + \tfrac{1}{4}\kappa^2 v^2 \;. \end{aligned} \tag{14.4.8}$$

Now, apart from the higher-order interaction terms (underlined), l contains a mass term for the gauge field A_μ with

$$m_A = gv = g\kappa/\sqrt{\lambda} \tag{14.4.9a}$$

and another one for the Higgs field $\eta(x) \to \phi(x)$

$$m_\eta = \sqrt{2}\kappa \;. \tag{14.4.9b}$$

While the original Lagrangian contains a complex scalar field ϕ (2 real functions) and a massless gauge field A_μ (2 independent real functions, see above), after the breaking of the symmetry we have a Lagrangian which consists of an interacting real massive scalar field η (1 real function) and a massive gauge field A_μ (3 real independent functions, all having the same mass related to the coupling constant g). The degree of freedom $\varphi(x)$ which corresponds to the massless Goldstone boson has thus been converted into an additional degree of freedom of the gauge vector field (*Higgs mechanism*). Then the Lagrangian (14.4.8) is in a unitary gauge [14.10].

Using this mass-generating mechanism we are able to introduce masses for the gauge fields into locally gauge-invariant field theories (Yang-Mills theory), which themselves are not allowed to have mass terms because of their invariance. This is the basic principle for the construction of models for the electro-weak interaction. In this theory only massive gauge vector fields are expected, apart from the massless photon. This is because the weak interaction is very short ranged (about 10^{-16} cm) meaning that the field quanta exchanged in the interaction between the particles (see the analogy to the photon) must have relatively

---

[8] Remember that the gauge field term in the Lagrangian has a different sign from the Bose field term (14.1.11b, c).

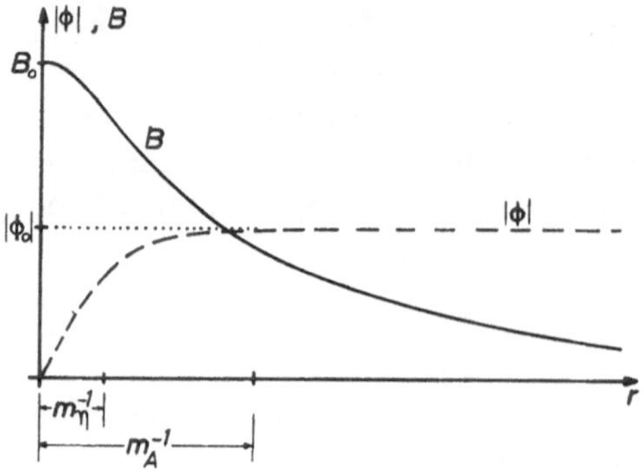

Fig. 14.8. Ginzburg-Landau solutions for a type II superconductor showing the variation of magnetic field B and phase ϕ in the string; m_η^{-1} represents the coherence length, m_A^{-1} the transverse extension of the magnetic field

large masses[9]. Recent measurements (CERN, 1994; [14.25, 26]) give masses of (80.228 ± 0.005) and (91.189 ± 0.005) GeV/c^2 for the three massive gauge fields of the weak interaction ($W^\pm$, Z^0 bosons). With the addition of the photon field all the fields of the unified electro-weak interaction are specified. Before discussing more details of this Glashow-Salam-Weinberg model, i.e. a spontaneously broken local gauge theory with the group $\mathcal{SU}_w(2) \times \mathcal{U}_e(1)$, we will mention an analogy to the Ginzburg-Landau model of type II superconductors [14.6].

Variation of the Lagrangian (14.4.7) with respect to ϕ and A_μ yields the Euler-Lagrange equations ($g \to e$)

$$(\partial_\mu - \mathrm{i}eA_\mu)(\partial^\mu + \mathrm{i}eA^\mu)\phi = \kappa^2\phi - \tfrac{1}{2}\lambda|\phi|^2\phi \ , \qquad (14.4.10)$$

$$\partial_\nu F^{\mu\nu} = \frac{\mathrm{i}}{2}\mathrm{e}\{\phi^*\partial^\mu\phi - \phi\partial^\mu\phi^*\} - e^2A^\mu|\phi|^2 \ .$$

These equations describe the gauge-invariant coupling of the electromagnetic field to the self-interacting particle field. They are identical with the relativistic form of the Ginzburg-Landau model; ϕ is the order parameter (see above) representing the coherent many-particle state of the Cooper pairs, $\rho_s = 2|\phi|^2$ is the local pair density. Because of the similarity of both theories, the static solutions of the Higgs model can be compared with the well-known solutions of the Ginzburg-Landau theory. These have a vortex-like (string-like) structure and are represented in Fig. 14.8 for $B = |\mathrm{curl}\, A|$ and $|\phi|$. In gauge theory they correspond to the assumption of gluon strings [14.7, 27–29]. The order parameter $|\phi|$ characterizes the unperturbed superconducting medium; in the centre of the flux tube the medium is a normal conductor ($|\phi| \to 0$). In the Ginzburg-Landau model, $m_\eta^{-1} = (\sqrt{2}\kappa)^{-1}$ is the coherence length describing the transition between the normal and superconducting regions for type II superconductors.

[9] The static equation $\Delta\phi - \kappa^2\phi = g\delta(r)$ gives the solution $\phi(r) = -(g/4\pi r)\exp(-\kappa r)$ for the interaction, showing that the range $1/\kappa$ of the potential is inverse to the mass κ.

In the Higgs model, m_η^{-1} is the length describing the depletion of the Higgs field out of the core region. The quantity $m_A^{-1} = \sqrt{\lambda}/e\kappa$ gives the transverse extension of the magnetic field in the superconductor. In the Higgs model this is the Compton wavelength of a massive "photon" and characterizes the transverse extension of the concentration of the vector field in the string.

14.5 Non-Abelian Gauge Theories and Symmetry Breaking

14.5.1 The Glashow-Salam-Weinberg Model of the Electro-Weak Interaction

Today, the standard model for the electro-weak interaction is that of Glashow, Salam and Weinberg (GSW, [14.16–23, 31]). With respect to fields and couplings it is a minimal model. It is based on the non-Abelian local gauge group $\mathcal{SU}(2) \times \mathcal{U}(1)$. Using the generators I_l ($l = 1, 2, 3$) of $\mathcal{SU}(2)$ and Y of $\mathcal{U}(1)$, the commutation relations are

$$[I_l, I_k]_- = \mathrm{i} \sum_{j=1}^{3} \varepsilon_{lk}{}^j I_j \; ; \qquad [I_l, Y] = 0 \; , \tag{14.5.1}$$

where $\varepsilon_{lk}{}^j$ is the totally antisymmetric tensor of third rank. Thus there are three gauge fields (bosons) $W_\alpha^l(x)$ assigned to the I_l, and the B_α field is assigned to the generator Y of the $\mathcal{U}(1)$ subgroup. (The Minkowski space index is α, β in this chapter.)

The free parameters of the model are the gauge coupling constants g and g' of $\mathcal{SU}(2)$ and $\mathcal{U}(1)$, respectively, and the parameters κ, λ of the Goldstone-Higgs potential in (14.4.1, 7). Often these independent parameters are replaced by equivalent quantities, namely e,g, θ_w, m_ϕ or e, M_Z, M_W, m_ϕ (14.5.11–13). Here e is the electric charge, θ_w the Weinberg angle and M_Z, M_W and m_ϕ are the masses of the Z, W and Higgs bosons.

In the GSW model, leptons (Table 14.2) and quarks are arranged in families of $\mathcal{SU}(2)$ doublets and singlets.

Table 14.2. Leptons [13.9]

f	Q	J^P	Mc^2[MeV]	τ[s]
e^-	-1	$1/2^+$	0.5110	∞, $>2.7 \times 10^{23}$ a
μ^-	-1	$1/2^+$	105.66	2.20×10^{-6}
τ^-	-1	$1/2^+$	1777.1	0.296×10^{-12}
ν_e	0	$1/2^+$	$<5 \times 10^{-6}$	$>1.5 \times 10^3$
ν_μ	0	$1/2^+$	<0.27	$>4.2 \times 10^6$
ν_τ	0	$1/2^+$	<31	

Doublets

$$\text{leptons:} \begin{pmatrix} \nu_e \\ e \end{pmatrix}_L , \quad \begin{pmatrix} \nu_\mu \\ \mu \end{pmatrix}_L , \quad \begin{pmatrix} \nu_\tau \\ \tau \end{pmatrix}_L \to \psi_{lL} = \begin{pmatrix} \nu_l \\ l \end{pmatrix}_L , \qquad l = e, \mu, \tau$$

$$\text{quarks:} \begin{pmatrix} u \\ d' \end{pmatrix}_L , \quad \begin{pmatrix} c \\ s' \end{pmatrix}_L , \quad \begin{pmatrix} t \\ b' \end{pmatrix}_L . \tag{14.5.2a}$$

For the quark fields, the colour index is omitted; every doublet has to be taken with $c = \{r, y, b\}$ in the most general case[10].

Singlets

$$l_R \, , \quad \nu_{lR} \, , \quad u_R \, , \quad d_R \, , \quad c_R \, , \quad s_R \, , \quad t_R \, , \quad b_R \, . \tag{14.5.2b}$$

Here L and R indicate left- and right-handed particles, which are related to spin-polarization (helicity). The massless left-handed fields projected out of an arbitrary spinor

$$\psi_L(x) = \tfrac{1}{2}(1 + \gamma_5)\psi(x) \, , \qquad \gamma^5 = \gamma_5 = -i\gamma^0\gamma^1\gamma^2\gamma^3 \, , \tag{14.5.3a}$$

transform according to the two-dimensional fundamental (defining) REP of $\mathcal{SU}(2)$, whereas the right-handed fields

$$\psi_R(x) = \tfrac{1}{2}(1 - \gamma_5)\psi(x) \tag{14.5.3b}$$

transform according to the one-dimensional identity REP. The assignment of left- and right-handed fields to different REPs allows for violation of parity. The eigenvalues of the weak hypercharge generator Y and the weak isospin generator I_3 have to be determined in such a way that the Gell-Mann–Nishijima relation (13.1.2) for the electric charge operator

$$Q = I_3 + \tfrac{1}{2}Y \tag{14.5.4}$$

is satisfied. In the following we discuss the GSW model for leptons only. Application to quarks is completely analogous [14.32].

The construction of a Lagrangian which is locally gauge invariant under $\mathcal{SU}(2) \times \mathcal{U}(1)$ starts with the Lagrangian of free, massless leptons:

$$l = \sum_{l=e,\mu,\tau} i(\bar{\psi}_{lL}\gamma^\alpha\partial_\alpha\psi_{lL} + \bar{l}_R\gamma^\alpha\partial_\alpha l_R + \bar{\nu}_{lR}\gamma^\alpha\partial_\alpha\nu_{lR}) \, , \tag{14.5.5a}$$

into which we introduce the covariant derivatives according to the principle of minimal coupling (14.3.5a, 17):

---

[10] d', s', b' indicate Cabibbo-transformed fields, but this is irrelevant for our purpose [14.33].

$$\partial_\alpha \psi_{lL} \rightarrow \left(\mathbf{1} \cdot \partial_\alpha + \frac{ig}{2} \boldsymbol{\tau} \cdot \boldsymbol{W}_\alpha + \frac{ig'}{2} Y_L B_\alpha \mathbf{1} \right) \psi_{lL} \ ,$$

$$\partial_\alpha l_R \rightarrow \left(\partial_\alpha + \frac{ig'}{2} Y_R^{-1} B_\alpha \right) \cdot l_R \ , \tag{14.5.5b}$$

$$\partial_\alpha \nu_{lR} \rightarrow \left(\partial_\alpha + \frac{ig'}{2} Y_R^0 B_\alpha \right) \cdot \nu_{lR} \ .$$

Here $\mathbf{1}$ is the unit matrix in the $\mathcal{SU}(2)$ isospin space, which will be omitted in the following, $\boldsymbol{W}_\alpha(x) = (W_\alpha^1, W_\alpha^2, W_\alpha^3)$ is the triplet of gauge fields belonging to the generators $\boldsymbol{I} = \boldsymbol{\tau}/2$, where $\boldsymbol{\tau} = (\tau_1, \tau_2, \tau_3)$ and the τ_l are the isospin (Pauli) matrices. The one-dimensional generators (just numbers, eigenvalues) Y_L, Y_R^{-1} and Y_R^0 have to satisfy (14.5.4) with $I_3 = \pm 1/2$ for the doublet and $I_3 = 0$ for the singlet, thus

$$Y_L = -1 \ , \qquad Y_R^{-1} = -2 \ , \qquad Y_R^0 = 0 \ . \tag{14.5.5c}$$

The Yang-Mills fields derived from the potentials W_α^l and B_α are according to (14.3.23)

$$F_{\alpha\beta}^l = \partial_\alpha W_\beta^l - \partial_\beta W_\alpha^l - g \sum_{jk} \varepsilon_{jk}{}^l W_\alpha^j W_\beta^k \ , \tag{14.5.5d}$$

$$F_{\alpha\beta}^B = \partial_\alpha B_\beta - \partial_\beta B_\alpha \ .$$

Thus the total invariant Lagrangian for leptons and gauge fields [vector bosons, see (14.3.19a, 25)] is

$$\begin{aligned} l = & \sum_{l=e,\mu,\tau} \bar{\psi}_{lL} i\gamma^\alpha \left(\partial_\alpha + \frac{ig}{2} \boldsymbol{\tau} \cdot \boldsymbol{W}_\alpha + \frac{ig'}{2} Y_L B_\alpha \right) \psi_{lL} \\ & + \sum_l \bar{l}_R i\gamma^\alpha \left(\partial_\alpha + \frac{ig'}{2} Y_R^{-1} B_\alpha \right) l_R + \sum_l \bar{\nu}_{lR} i\gamma^\alpha \left(\partial_\alpha + \frac{ig'}{2} Y_R^0 B_\alpha \right) \nu_{lR} \\ & - \frac{1}{4} \sum_{k=1}^{3} F_{\alpha\beta}^k F^{\alpha\beta,k} - \frac{1}{4} F_{\alpha\beta}^B F^{\alpha\beta,B} \ . \end{aligned} \tag{14.5.6}$$

The interaction between leptons and vector bosons can be expressed in terms of the lepton currents (14.3.31)

$$l_W = -g \boldsymbol{j}^{(\tau)\alpha} \boldsymbol{W}_\alpha - g' j^{(Y)\alpha} B_\alpha / 2 \qquad \text{with} \tag{14.5.7a}$$

$$\boldsymbol{j}^{(\tau)\alpha} = \sum_{l=e,\mu,\tau} \bar{\psi}_{lL} \gamma^\alpha \boldsymbol{\tau} \psi_{lL}/2 \ , \tag{14.5.7b}$$

$$j^{(Y)\alpha} = \sum_{l=e,\mu,\tau} (\bar{\psi}_{lL} \gamma^\alpha Y_L \psi_{lL} + \bar{l}_R \gamma^\alpha Y_R^{-1} l_R + \bar{\nu}_{lR} \gamma^\alpha Y_R^0 \nu_{lR}) \ . \tag{14.5.7c}$$

It is useful to introduce the "circularly polarized" components $W_\alpha^\pm$ instead of W_α^1 and W_α^2 by setting

$$W_\alpha^\pm = \frac{1}{\sqrt{2}}(W_\alpha^1 \pm \mathrm{i} W_\alpha^2) \qquad \text{and} \qquad \tau_\pm = \frac{1}{2}(\tau_1 \pm \mathrm{i}\tau_2) \,. \tag{14.5.8}$$

For the interaction part of the Lagrangian we obtain

$$l_\mathrm{W} = l_\mathrm{W}^1 + l_\mathrm{W}^0 \,, \qquad l_\mathrm{W}^1 = -\frac{g}{2\sqrt{2}}(j^{(+)\alpha} W_\alpha^- + j^{(-)\alpha} W_\alpha^+) \,, \tag{14.5.9}$$

$$j^{(\pm)\alpha} = 2 \sum_l \bar{\psi}_{l\mathrm{L}} \gamma^\alpha \tau_\pm \psi_{l\mathrm{L}} \,,$$

$$l_\mathrm{W}^0 = -g j^{(3)\alpha} W_\alpha^3 + g' j^{(3)\alpha} B_\alpha - g' j^{(e)\alpha} B_\alpha$$

with the weak isospin current

$$j^{(3)\alpha} = \tfrac{1}{2} \sum_l \bar{\psi}_{l\mathrm{L}} \gamma^\alpha \tau_3 \psi_{l\mathrm{L}}$$

and the electric current of the leptons

$$j^{(e)\alpha} = -\sum_l (\bar{l}_\mathrm{L} \gamma^\alpha l_\mathrm{L} + \bar{l}_\mathrm{R} \gamma^\alpha l_\mathrm{R}) \,.$$

Here we have used that

$$j^{(Y)\alpha} = -2 j^{(3)\alpha} + 2 j^{(e)\alpha} \,, \tag{14.5.10}$$

which can be shown using (14.5.5c) and τ_3. Note that the first term in (14.5.7c) contains a unit matrix. The term l_W^1 describes the interaction between leptons and complex vector bosons $\mathrm{W}_\alpha^\pm$ (charged bosons), whereas the term l_W^0 is an interaction with (real) neutral bosons W_α^3 and B_α. This term must contain the interaction between the (charged) leptons and the electromagnetic field, that is, $j^{(e)\alpha} A_\alpha$. To see this, a further transformation, the *Weinberg transformation*, which is a rotation in W_α^3-B_α space, is useful. We introduce the *Weinberg angle* θ_w as

$$\tan\theta_\mathrm{w} = g'/g \,, \qquad \cos\theta_\mathrm{w} = \frac{g}{\sqrt{g^2 + g'^2}} \,, \qquad \sin\theta_\mathrm{w} = \frac{g'}{\sqrt{g^2 + g'^2}} \tag{14.5.11a}$$

and

$$e = g \sin\theta_\mathrm{w} = g' \cos\theta_\mathrm{w} \,. \tag{14.5.11b}$$

The transformation then reads

$$Z_\alpha = W_\alpha^3 \cos\theta_\mathrm{w} - B_\alpha \sin\theta_\mathrm{w} \,, \tag{14.5.12}$$

$$A_\alpha = + W_\alpha^3 \sin\theta_\mathrm{w} + B_\alpha \cos\theta_\mathrm{w} \,.$$

With this we obtain

$$l_{\mathrm{W}}^{0} = -\sqrt{g^2 + g'^2}\, j^{(0)\alpha} Z_\alpha - e j^{(e)\alpha} A_\alpha \tag{14.5.13a}$$

with the *neutral current*

$$j^{(0)\alpha} = j^{(3)\alpha} - \sin^2\theta_{\mathrm{w}} j^{(e)\alpha} \ . \tag{14.5.13b}$$

The potential A_α only couples to the (charged) leptons, describing the interaction between charges and the electromagnetic field (photons). The $\mathscr{U}(1)$ invariance obviously provides no condition for the magnitudes of g' and e, which determine the coupling between leptons and the B_α and A_α fields. This freedom can be used to unify weak and electromagnetic interactions. The most recent values are

$$\sin^2\theta_{\mathrm{w}} = 0.23 \pm 0.01 \ , \qquad g = 2.09e \ , \qquad g' = 1.14e \tag{14.5.14}$$

determined by neutral-current-induced processes.

In the theory considered so far, because of the local invariance under $\mathscr{SU}(2) \times \mathscr{U}(1)$, neither the vector bosons nor the leptons (fermions) have any mass. These masses can be introduced via the Higgs mechanism discussed in Sect. 14.4.3. For this purpose we have to extend the idea of spontaneously broken symmetry to the non-Abelian case, starting from the fact that the vacuum expectation value of the Higgs field does not vanish.

14.5.2 Symmetry Breaking in the Glashow-Salam-Weinberg Model

The vacuum $|0\rangle$ is simultaneously the eigenstate of the charge generators Q_k of the gauge symmetry group defined in (14.1.9) and of the Hamiltonian H of the system, the eigenvalues being zero:

$$H|0\rangle = 0 \ , \qquad Q_k|0\rangle = 0 \ , \tag{14.5.15}$$

$$[H, Q_k] = 0 \ .$$

Furthermore, according to (14.1.7b, 10) the generators Q_k generate the symmetry transformation

$$\delta_k\phi(x) = \mathrm{i}J_k\phi(x) = -\mathrm{i}[Q_k, \phi] \ , \tag{14.5.16a}$$

where we consider scalar fields. The total change of ϕ is then according to (14.1.7b)

$$\delta\phi = \phi' - \phi = g\sum_k \chi^k \delta_k\phi = \mathrm{i}g\sum_k J_k\chi^k\phi \ . \tag{14.5.16b}$$

Spontaneous symmetry breaking now means that the vacuum is no longer an eigenstate of at least one of the generators Q_k. In general we have

$$Q_a|0\rangle \neq 0 \ , \qquad a = m_0 + 1, \ldots, m \ . \tag{14.5.17a}$$

But, of course, if there is some symmetry left there remain certain generators which have the vacuum as an eigenstate:

$$Q_j|0\rangle = 0 , \qquad j = 1, \ldots, m_0 . \tag{14.5.17b}$$

Obviously these form a subgroup or a subalgebra $\mathscr{U}$ of $\mathscr{G}$, whereas the Q_a are not necessarily a subgroup of $\mathscr{G}$. The Q_a generate a subgroup as well if $\mathscr{U}$ is an invariant subgroup (algebra) of $\mathscr{G}$, i.e. the quotient group $\mathscr{G}/\mathscr{U}$ exists.

The conditions (14.5.17a, b) can be expressed by the scalar (Higgs) fields ϕ transforming according to (14.5.16a) and their vacuum expectation value. The non-Abelian generalization of (14.4.4c) is thus

$$\langle 0|\delta_a\phi|0\rangle = \mathrm{i}J_a\langle 0|\phi|0\rangle = \mathrm{i}J_a\langle\phi\rangle_0 \neq 0 , \qquad a = m_0 + 1, \ldots, m \tag{14.5.17c}$$

because $\langle 0|\delta_a\phi|0\rangle \neq 0$ implies (14.5.17a) and thus spontaneous breaking of symmetry. According to (14.4.7) a nonvanishing vacuum expectation value is connected with an appropriate Higgs potential in the Lagrangian guaranteeing the stability of the vacuum with $\langle\phi\rangle_0 \neq 0$. On coupling the gauge fields $W_\alpha^k(x)$ to the Higgs field ϕ the Lagrangian contains the gauge invariant factor

$$D_\alpha\phi = (\partial_\alpha + \mathrm{i}gA_\alpha)\phi = \left(\partial_\alpha + \mathrm{i}g\sum_k J_k W_\alpha^k(x)\right)\phi .$$

Here J_k has to be taken in the REP defined by ϕ; at least one of these generators satisfies $J_a\langle\phi\rangle_0 \neq 0$. The gauge bosons $W_\alpha^a(x)$ corresponding to J_a then have the squared mass matrix[11]

$$\hat{M}_{ab} := g^2(J_a^\dagger\langle\phi\rangle_0, J_b\langle\phi\rangle_0) = g^2(\langle\phi^\dagger\rangle_0, J_aJ_b\langle\phi\rangle_0) \tag{14.5.18a}$$

with respect to this vacuum. There are $m - m_0$ bosons that now have mass and m_0 bosons that remain massless. Following the same procedure with (14.5.18a) as in (14.4.7, 8) leads to a term

$$l_{\text{mass}} = \tfrac{1}{2}\hat{M}_{ab}W_\alpha^a W^{\alpha,b} \tag{14.5.18b}$$

in the Lagrangian.

---

[11] Sometimes the Higgs field ϕ has to be given in a matrix REP (see Sect. 14.5.5), e.g. in the adjoint REP of $\mathscr{G}$. Then according to (14.5.16a) the expression $J_a\langle\phi\rangle_0$ in (14.5.17c) and (14.5.18a) has to be replaced by

$$\langle 0|[Q_a, \phi]|0\rangle \neq 0 , \qquad a = m_0 + 1, \ldots, m ,$$

$$\langle 0|[Q_j, \phi]|0\rangle = 0 , \qquad j = 1, \ldots, m_0 .$$

The squared gauge boson masses are then finally given by

$$\hat{M}_{ab} := g^2\,\mathrm{Tr}\{[Q_a, \langle\phi\rangle_0]^+ \cdot [Q_b, \langle\phi\rangle_0]\} .$$

The Higgs potential $V(\phi)$ has to be chosen in such a way that

$$V(\langle\phi\rangle_0) \to \text{minimum with } V'(\langle\phi\rangle_0) = 0 \ , \qquad V''(\langle\phi\rangle_0) > 0 \ .$$

We choose ϕ or $\langle\phi\rangle_0$ to have r real components. Because of (14.5.17a), the $J_a\langle\phi\rangle_0$ span an $(m - m_0)$-dimensional subspace (Goldstone space) of the r-dimensional real vector space of ϕ. The remaining r-(m $- m_0$) independent components define its complement, the Higgs space. Its dimension is equal to the number of massive Higgs bosons created by this procedure. Higgs and Goldstone spaces can always be chosen as mutually orthogonal spaces.

The mass-generating procedure using spontaneous breaking of symmetry also works for leptons (fermions), but this needs an $\mathcal{SU}(2) \times \mathcal{U}(1)$ gauge-invariant term in the Lagrangian describing the interaction between fermion fields and the Higgs field. In general this term is taken to be

$$l_Y = -\frac{1}{v}\sum_{l,l'} \bar{\psi}_{lL} M_{ll'} l'_R \phi + \text{h.c.} \tag{14.5.19}$$

where ψ_{lL} and l_R are the lepton fields (14.5.5a), ϕ is the Higgs field and $M_{ll'}$ are complex constants. Equation (14.5.19) is said to be a Yukawa interaction. The generation of masses with this term is very similar to that for the gauge fields, so we shall discuss only the latter case (see e.g. [14.32]).

In the GSW model the three gauge bosons of the weak interaction ($W^\pm$, Z) obtain masses, whereas the electromagnetic interaction is determined by the massless photon ($A_\alpha \Rightarrow \gamma$). Thus three degrees of freedom of the corresponding Higgs field have to be transformed into additional degrees of freedom of the vector fields in spontaneous symmetry breaking. The local gauge symmetry must be broken according to

$$\mathcal{SU}(2) \times \mathcal{U}(1) \to \mathcal{U}_e(1) \ . \tag{14.5.20}$$

The Higgs field has to be chosen in such a way that the vacuum expectation values

$$I_i\langle\phi\rangle_0 \neq 0 \ , \qquad Y\langle\phi\rangle_0 \neq 0 \tag{14.5.21}$$

do not vanish, whereas for the remaining $\mathcal{U}_e(1)$ symmetry generated by the electric charge Q

$$Q\langle\phi\rangle_0 = 0 \qquad \text{with} \qquad Q = I_3 + \tfrac{1}{2}Y \tag{14.5.22a}$$

has to be valid.

Thus we introduce an $\mathcal{SU}(2)$ doublet of Higgs fields

$$\phi = \begin{pmatrix} \phi_+ \\ \phi_0 \end{pmatrix} , \tag{14.5.23}$$

where each component is complex. In (14.5.23) ϕ_+ and ϕ_0 are boson fields carrying electric charge $e = +1$ and $e = 0$, respectively, with $Y_\phi = +1$ in order to satisfy (14.5.22a). This is a "minimal" choice. Since the theory including the Higgs contribution has to be $\mathscr{SU}(2) \times \mathscr{U}(1)$ invariant, the Higgs fields have to transform according to an IR of the gauge group before symmetry breaking. Furthermore, three massive gauge bosons have to be generated. Thus we have to start with at least a (complex) Higgs doublet.

The Lagrangian of the Higgs field is now, see (14.4.7),

$$l_{H,0} = +\frac{1}{2}\partial_\alpha\phi^+\partial^\alpha\phi - V(\phi^+\phi) \qquad \text{with} \tag{14.5.24}$$

$$V(\phi^+\phi) = -\frac{1}{2}\kappa^2\phi^+\phi + \frac{1}{4}\lambda(\phi^+\phi)^2 = \frac{1}{4}\lambda\left(\phi^+\phi - \frac{\kappa^2}{\lambda}\right)^2 - \frac{\kappa^4}{4\lambda}\,.$$

The vacuum expectation value (minimum of potential) then satisfies

$$\langle\phi^+\rangle_0\langle\phi\rangle_0 = \frac{\kappa^2}{\lambda} = v^2 \qquad \text{with real } v\,. \tag{14.5.25}$$

The gauge-invariant coupling between the gauge fields W_α, B_α and the Higgs field ϕ is then calculated according to the principle of minimal coupling by substituting [see (14.5.5b)]

$$\partial^\alpha\phi \to D^\alpha\phi = \left(\partial^\alpha + \frac{ig}{2}\boldsymbol{\tau}\cdot\boldsymbol{W}^\alpha + \frac{ig'}{2}Y_\phi B^\alpha\right)\phi \tag{14.5.26}$$

into the Lagrangian (14.5.24).

The vacuum expectation value $\langle\phi\rangle_0$ whose orientation is not determined is chosen by convention to be

$$\langle\phi\rangle_0 = \begin{pmatrix} 0 \\ v \end{pmatrix} \tag{14.5.27}$$

in accordance with the definition of Q:

$$Q = I_3 + \frac{1}{2}Y = \frac{1}{2}\begin{pmatrix} 1 & 0 \\ 0 & -1 \end{pmatrix} + \frac{1}{2}\begin{pmatrix} 1 & 0 \\ 0 & 1 \end{pmatrix} = \begin{pmatrix} 1 & 0 \\ 0 & 0 \end{pmatrix}\,.$$

Thus

$$Q\langle\phi\rangle_0 = \begin{pmatrix} 1 & 0 \\ 0 & 0 \end{pmatrix}\cdot\begin{pmatrix} 0 \\ v \end{pmatrix} = 0\,. \tag{14.5.22b}$$

The remaining symmetry $\mathscr{U}_e(1)$ is left unbroken.

In analogy to (14.4.5b) we represent the Higgs field as

$$\phi(x) = \begin{pmatrix} 0 \\ v + \eta(x) \end{pmatrix} \cdot \exp[\mathrm{i}\boldsymbol{\tau} \cdot \boldsymbol{\theta}(x)/2] \ , \tag{14.5.28a}$$

where $\theta_k(x)$, $k = 1, 2, 3$ are real fields which correspond to the "would be" Goldstone boson (14.4.5). Again, by choosing an appropriate gauge they can be eliminated because of the gauge invariance of l_H. In the *unitary gauge* we thus take

$$\phi(x) = \begin{pmatrix} 0 \\ v + \eta(x) \end{pmatrix} \ . \tag{14.5.28b}$$

We are left with one physical Higgs boson $\eta(x)$, while the others are used to assign masses to the gauge bosons $W^\pm$ and Z. The Higgs fields thus have to be looked upon as new fundamental fields with corresponding particles which have physical properties (see again Ginzburg-Landau theory; ϕ is the order parameter there).

Inserting (14.5.28b) into (14.5.24) after substituting for $D^\alpha\phi$ from (14.5.26) we obtain

$$\begin{aligned} l_H = {} & \frac{1}{2}\partial_\alpha\eta\partial^\alpha\eta + \left[\frac{1}{4}g^2 W_\alpha^+ W^{-,\alpha} + \frac{1}{8}(gW_\alpha^3 - g'B_\alpha)(gW^{3,\alpha} - g'B^\alpha)\right](v+\eta)^2 \\ & - \frac{\lambda}{4}\eta^2(2v+\eta)^2 + \left(\frac{1}{4}\lambda v^4\right) \end{aligned} \tag{14.5.29a}$$

or

$$\begin{aligned} l_H = {} & \frac{1}{2}\partial_\alpha\eta\partial^\alpha\eta + \frac{1}{2}(v+\eta)^2\left[\frac{1}{2}g^2 W_\alpha^+ W^{-,\alpha} + \frac{1}{4}(g^2 + g'^2)Z_\alpha Z^\alpha\right] \\ & - \frac{\lambda}{4}\eta^2(2v+\eta)^2 \ . \end{aligned} \tag{14.5.29b}$$

Comparing these functions with the usual mass terms having the form $m^2\eta^2/2$, etc. [because of (14.5.8) the factor 1/2 is missing in the $W_\alpha^- W^{+,\alpha}$ term], we obtain the masses

$$m_{\mathrm{W}}^{\pm} = \tfrac{1}{2}gv$$

$$m_{\mathrm{Z}} = \tfrac{1}{2}v\sqrt{g^2 + g'^2} \qquad \text{for the vector bosons} \tag{14.5.30a}$$

$$m_\gamma = 0$$

and

$$m_\eta = v\sqrt{2\lambda} = \kappa\sqrt{2} \qquad \text{for the scalar Higgs boson (electrically neutral, see Table 14.1).} \tag{14.5.30b}$$

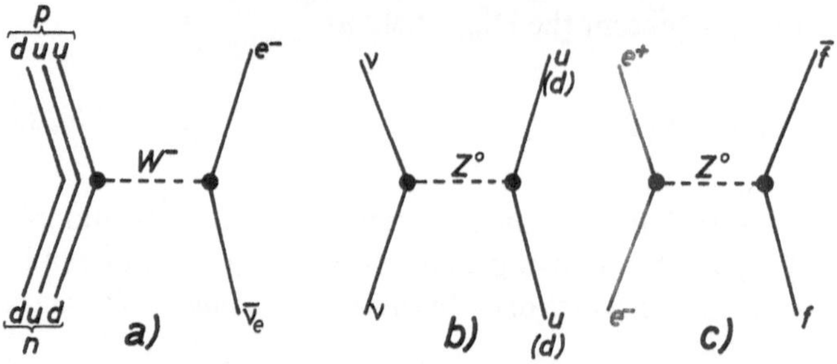

Fig. 14.9. (a) β decay. A d quark (of a neutron) emits a virtual W^- boson and changes into a u quark (of a proton); the W^- boson then decays into an electron and a neutrino. (b) Neutral current. Exchange of a virtual Z^0 boson between a neutrino and u, d, ... quarks. (c) Electron-positron annihilation. Exchange of a virtual Z^0 boson between an e^-e^+ pair and an $f\bar{f}$ pair

What remains is the determination of v and λ. Whereas so far the determination of λ has not been possible (estimates vary between $\lambda \approx 10^{-2}$ and $\lambda \approx 10^2$) the parameter v can be determined from the interaction of leptons and quarks with charged $W^\pm$ and neutral Z bosons in the low energy limit ($E \ll m_W, m_Z$). With the Fermi coupling constant G_F for the currents,

$$v^2 = (G_F\sqrt{2})^{-1} \; , \qquad G_F = 1.116 \times 10^{-5}\,(\mathrm{GeV})^{-2} \; . \tag{14.5.31}$$

Using this value and (14.5.11, 14) we obtain[12]

$$\begin{aligned} m_W^\pm &= \frac{1}{2}\frac{e}{\sin\theta_w}(\sqrt{2}G_F)^{-1/2} = 80 \pm 2\,\mathrm{GeV} \\ m_Z &= \frac{1}{2}\frac{e}{\sin\theta_w\cos\theta_w}(\sqrt{2}G_F)^{-1/2} = 92 \pm 2\,\mathrm{GeV} \end{aligned} \tag{14.5.32}$$

in good agreement with recent measurements at CERN [14.25, 26]. The mass of the Higgs particle cannot be estimated, since λ is unknown.

We will just mention that the mass-generating procedure can also be carried out for leptons, starting from (14.5.19), and, furthermore, it can be done for the quark families given in (14.5.2a). Therefore we can use the above results for quark–gauge-field interaction as well. On the basis of this idea we shall give some examples of interaction processes. Figure 14.9a shows β decay, in which a neutron $|n\rangle \sim (udd-ddu)$ emits a (virtual) W^- boson and changes into a proton $|p\rangle \sim (uud-duu)$ (see Table 13.3). Figure 14.9b illustrates a neutral current reaction which gives rise to a weak force between neutrinos and other particles. Figure 14.9c shows electron-positron annihilation into a fermion-antifermion pair.

---

[12] The numbers include radiative corrections. In the units used the elementary charge is $e = \sqrt{4\pi\alpha} = 0.3028$.

14.5.3 Grand Unified Theories: General Remarks

The unification of electromagnetic and weak interactions leads to the conjecture that the strong interactions (QCD) might be included in a unified gauge theory on the basis of a direct product group $\mathscr{G}' := \mathscr{SU}_c(3) \times \mathscr{SU}(2) \times \mathscr{U}(1)$. The particles assigned to this group, i.e. their representation within this group, can be divided into three families (see Table 14.4)

First family:

$$\begin{pmatrix} u^c \\ d^c \end{pmatrix}_L , \quad \begin{pmatrix} \nu_e \\ e^- \end{pmatrix}_L , \quad u_R^c , \quad d_R^c , \quad e_R^- ;$$

Second family:

$$\begin{pmatrix} c^c \\ s^c \end{pmatrix}_L , \quad \begin{pmatrix} \nu_\mu \\ \mu^- \end{pmatrix}_L , \quad c_R^c , \quad s_R^c , \quad \mu_R^- ; \tag{14.5.33a}$$

Third family:

$$\begin{pmatrix} t^c \\ b^c \end{pmatrix}_L , \quad \begin{pmatrix} \nu_\tau \\ \tau^- \end{pmatrix}_L , \quad t_R^c , \quad b_R^c , \quad \tau_R^- .$$

Each family comprises 15 particles (states). The $\mathscr{SU}(2)$ doublets are given in the form of column vectors, see (14.5.2), whereas the $\mathscr{SU}_c(3)$ colour triplets [IR "3" of $\mathscr{SU}_c(3)$] are denoted by their colour index $c = \{\mathrm{r}, \mathrm{y}, \mathrm{b}\}$. The right-handed fields can, if necessary, be expressed by left-handed antiparticle fields $\bar{\psi}$ when the charge conjugation operator G is used:

$$\bar{\psi}_{L,R} = \mathfrak{C}\bar{\psi}_{R,L}^{T} , \qquad \mathfrak{C} = i\gamma^2\gamma^0 , \tag{14.5.34}$$

where T indicates a transposed operator. Then, the first family reads, for example,

$$\begin{pmatrix} u^c \\ d^c \end{pmatrix}_L , \quad \begin{pmatrix} \nu_e \\ e^- \end{pmatrix}_L , \quad \bar{u}_L^c , \quad \bar{d}_L^c , \quad e_L^+ . \tag{14.5.33b}$$

Here the charge conjugate quark fields transform according to the IR "3*" of $\mathscr{SU}_c(3)$. Equation (14.5.33a) can be extended by further families or by further particles, if necessary, e.g. by ν_R, if the neutrino has nonvanishing mass.

The unification of interactions and particles in a direct product group has a disadvantage: it allows three independent coupling constants g, g' and g_s (strong interaction), whereas a true unified theory should have only one, possibly varying ("running"), coupling constant.

One possible way of true unification is the choice of a simple gauge group $\mathscr{G} \supset \mathscr{G}'$; thus $\mathscr{G}' = \mathscr{SU}_c(3) \times \mathscr{SU}(2) \times \mathscr{U}(1)$ is a subgroup of $\mathscr{G}$, where $\mathscr{G}$ comprises all the interactions with one coupling constant. Leptons and quarks then have to transform according to a REP (possibly IR, multiplets) of $\mathscr{G}$, and some of the generators of $\mathscr{G}$, or the gauge bosons assigned to them, have to mediate

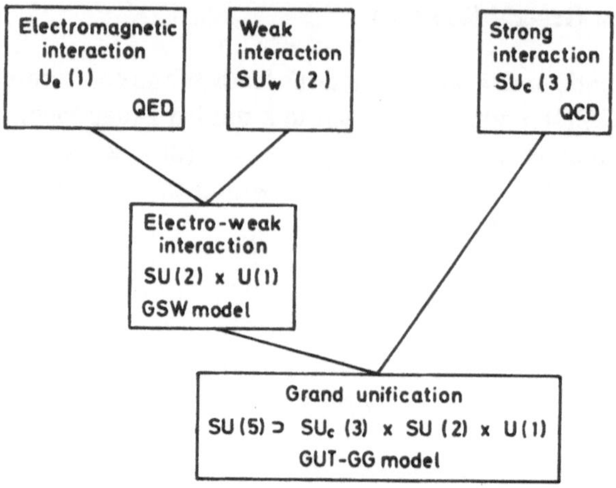

Fig. 14.10. Scheme of fundamental gauge theories

between $\mathscr{SU}_c(3)$ triplets and $\mathscr{SU}(2)$ doublets See (14.5.42b). The standard model for such a grand unified theory (GUT) is a $\mathscr{SU}(5)$ gauge group proposed by Georgi and Glashow [14.20–23]. With such an assumption we are led to the scheme of Fig. 14.10 of fundamental gauge theories or of the interactions defined by them.[13]

Obviously, at low energies strong and electro-weak interactions cannot be described by an $\mathscr{SU}(5)$-invariant coupling g_5 so that $\mathscr{SU}(5)$ symmetry is allowed to be present only as a broken symmetry in that region. However, the dependence of the interactions on energy makes it possible to start with a universal coupling constant that is energy dependent. It may be strong at low energies (below about 100 GeV) and it may become (super-) weak at high energies (probably about 10^{14} GeV; asymptotic freedom, see end of Sect. 14.5.4). Such "running" coupling constants can be extended to the region of energies just considered. The different magnitudes of coupling constants are explained by a Higgs mechanism similar to that of the GSW model.

In the Georgi-Glashow $\mathscr{SU}(5)$ model (GG model) three different regions of energy or momentum are assumed, which correspond to two scales of symmetry breaking at about m_W^2 and m_X^2 (see Fig. 14.11). The running coupling constants of $\mathscr{SU}_c(3)$, $\mathscr{SU}_w(2)$ and $\mathscr{U}(1)$, i.e. $g_3 = g_s$, $g_2 = g$ and $g_1 = \sqrt{5/3}\,g'$, are given qualitatively as a function of the energy (squared). For comparison, the corresponding lengths and symmetries are also given ($E = mc^2$, $L = \hbar/mc = \hbar c/E$).

In the region $E \gg m_X$, i.e. at very small distances $x \ll L_X$, there is an $\mathscr{SU}(5)$ gauge symmetry with only one coupling constant g_5 for all the interactions. In the region between m_W and m_X (the "great desert") the $\mathscr{SU}(5)$ symmetry is spontaneously broken into $\mathscr{SU}_c(3) \times \mathscr{SU}(2) \times \mathscr{U}(1)$, so that the coupling constants for decreasing energies (increasing distances) are allowed to become different. The $\mathscr{SU}_c(3)$ coupling constant g_3 is more strongly asymptotically free

[13] A further unification would have to include gravitational interactions, but this would lead to another category of gauge groups (supersymmetry) which we will not discuss here.

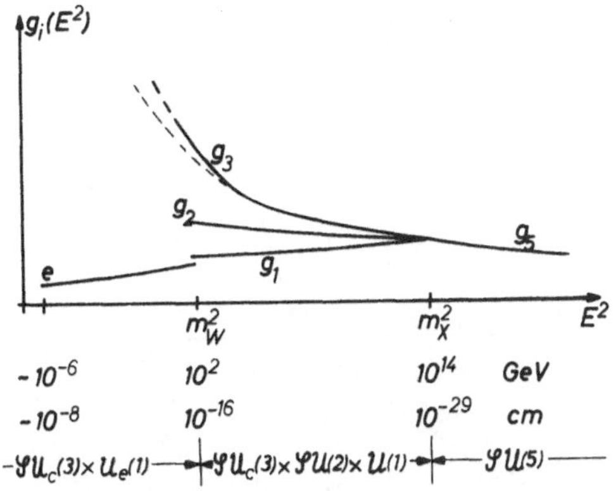

Fig. 14.11. Estimate of the coupling constants in different regions of symmetry, energy and length according to the GG model

than g_2 and thus increases most. The strong increase indicates that a QCD perturbation theory is not possible at small energies and thus the quark should be confined (Sect. 14.3.2). The $\mathcal{U}(1)$ coupling constant is not asymptotically free (the gauge bosons have no charge) and therefore it decreases with increasing distance. Finally, at m_W the $\mathcal{SU}(2)\times\mathcal{U}(1)$ symmetry is spontaneously broken into the remaining $\mathcal{U}_e(1)$ and we then have electro-magnetic coupling (e). Before a more detailed discussion of these mechanisms is given we shall first discuss the properties of the $\mathcal{SU}(5)$ group and its connection with the GG model.

14.5.4 $\mathcal{SU}(5)$ Group and Georgi-Glashow Model

The group $\mathcal{SU}(5)$ possesses 5 charge states generally denoted as *metacolour states a, b, c, d* and *e*, and 24 gauge bosons denoted by X, Y (12 bosons), G (8 bosons), W (3 bosons) and B (sometimes called V). The properties can be derived from the general relations given in Chaps. 11 and 12; they will be reviewed briefly. The fundamental weights[14], i.e. the weights of the defining REP, follow from (11.4.23) with (11.4.30b):

$$\begin{array}{ll} & \text{Metacolour} \\ w^{(1)}=\tfrac{1}{2}(1,1/\sqrt{3},1/\sqrt{6},1/\sqrt{10}) & a \\ w^{(2)}=\tfrac{1}{2}(-1,1/\sqrt{3},1/\sqrt{6},1/\sqrt{10}) & b \\ w^{(3)}=\tfrac{1}{2}(0,-2/\sqrt{3},1/\sqrt{6},1/\sqrt{10}) & c \\ w^{(4)}=\tfrac{1}{2}(0,0,-3/\sqrt{6},1/\sqrt{10}) & d \\ w^{(5)}=\tfrac{1}{2}(0,0,0,-4/\sqrt{10}) & e \end{array} \qquad (14.5.35)$$

[14] The weights and roots given here differ from those in (11.4.57–59) by a factor of $\sqrt{n}$ [for $\mathcal{SU}(n)$]; it seems that this normalization is more common in physical applications. It makes a difference of a factor n in the Casimir operators!

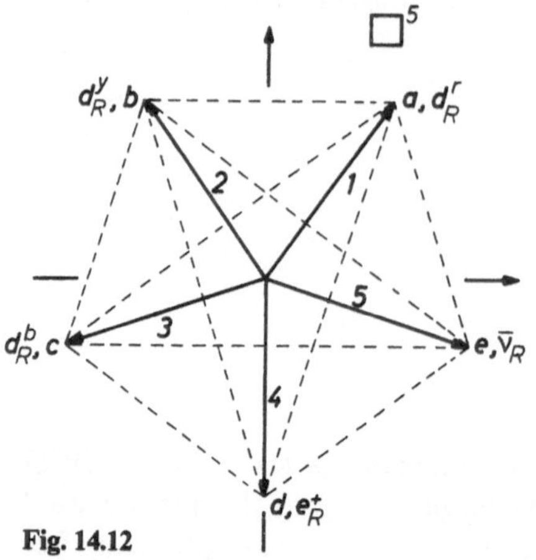

Fig. 14.12

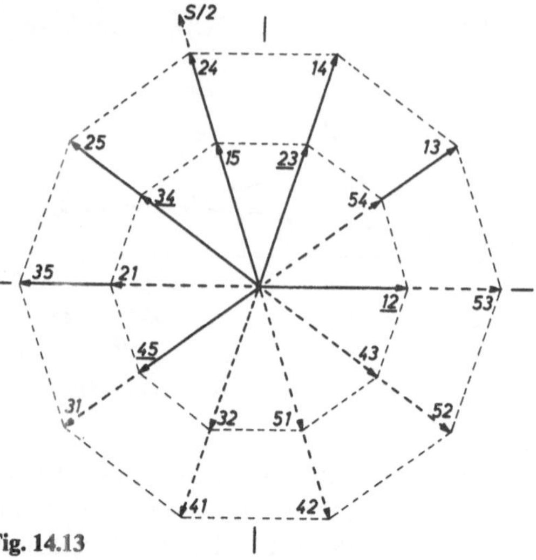

Fig. 14.13

Fig. 14.12. Defining REP (fundamental weights) of the IR 5 of $\mathscr{SU}(5)$. The figure represents a projection out of the root space $\mathbb{R}_4$ onto an $\mathbb{R}_2$ plane; this projection is different from that used by Georgi and Glashow [14.20, 21]

Fig. 14.13. Roots of $\mathscr{SU}(5)$ according to (11.4.27a). Projections of the root lattice onto an $\mathbb{R}_2$ plane. Positive roots: ———; negative roots: - - - - -; basis roots underlined

The metacolour states are assigned to them in the order given (Fig. 14.12). The basic roots[14] according to (11.4.27a) are

$$r_{12} = (1,0,0,0) \ , \qquad r_{23} = \frac{1}{2}(-1,\sqrt{3},0,0) \ ,$$
$$r_{34} = \frac{1}{\sqrt{3}}(0,-1,\sqrt{2},0) \ , \qquad r_{45} = \frac{1}{2\sqrt{2}}(0,0,-\sqrt{3},\sqrt{5}) \ , \tag{14.5.36}$$

while the other roots follow from (11.4.22). All the roots are represented in Fig. 14.13, which is a projection of the root lattice from $\mathbb{R}_4$ into $\mathbb{R}_2$. For an IR $\{n_1,n_2,n_3,n_4\}$ the maximal weight is according to (11.4.53)

$$W = \frac{1}{2}\left\{n_1, \frac{1}{\sqrt{3}}(n_1+2n_2), \frac{1}{\sqrt{6}}(n_1+2n_2+3n_3), \right.$$
$$\left. \frac{1}{\sqrt{10}}(n_1+2n_2+3n_3+4n_4)\right\} \ ,$$

$$d_w = \frac{1}{288}(1 + n_1)(1 + n_2)(1 + n_3)(1 + n_4)(2 + n_1 + n_2)(2 + n_2 + n_3)$$

$$\times (2 + n_3 + n_4)(3 + n_1 + n_2 + n_3)(3 + n_2 + n_3 + n_4)$$

$$\times (4 + n_1 + n_2 + n_3 + n_4) , \tag{14.5.37}$$

$$\langle C_2 \rangle = \tfrac{1}{5}\{2n_1^2 + 3n_2^2 + 3n_3^2 + 2n_4^2 + 3n_1 n_2 + 2n_1 n_3 + n_1 n_4$$

$$+ 4n_2 n_3 + 2n_2 n_4 + 3n_3 n_4 + 10n_1 + 15n_2 + 15n_3 + 10n_4\} .$$

According to (12.1.7–9), the IRs can also be characterized by partitions. The relations follow immediately from (12.1.9). The REPs of lowest dimension are

singlet: $n_l = 0$, $l = 1, 2, 3, 4$; $[\lambda] = [1, 1, 1, 1, 1]$,

dim: 1 , $W = 0$.

quintuplet: $\{n_l\} = (1, 0, 0, 0)$; $[\lambda] = [1] := 5$,

dim: 5 , $W = \tfrac{1}{2}(1, 1/\sqrt{3}, 1/\sqrt{6}, 1/\sqrt{10})$.

decuplet: $\{n_l\} = (0, 1, 0, 0)$; $[\lambda] = [1, 1] := 10$,

dim: 10 , $W = (0, 1/\sqrt{3}, 1/\sqrt{6}, 1/\sqrt{10})$. $\qquad$ (14.5.38)

decuplet: $\{n_l\} = (0, 0, 1, 0)$; $[\lambda] = [1, 1, 1] := \overline{10}$ or 10* ,

dim: $\overline{10}$, $W = \tfrac{3}{2}(0, 0, 1/\sqrt{6}, 1/\sqrt{10})$.

quintuplet: $\{n_l\} = (0, 0, 0, 1)$; $[\lambda] = [1, 1, 1, 1] := \overline{5}$ or 5* ,

dim: $\overline{5}$, $W = 2\,(0, 0, 0, 1/\sqrt{10})$.

The decuplet "10" can also be represented as an antisymmetrized tensor product of two REPs "5":

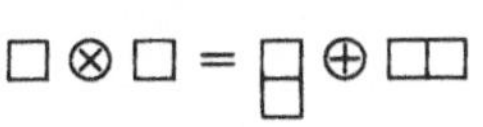

$$\square \otimes \square = \square\!\!\!\!\square \oplus \square\square = 10 \oplus 15 .$$

The generators in the defining REP "5" are given by (11.4.5 and 6a), where instead of H_l ($l = 1, 2, 3, 4$) as in the GSW model, other notations or combinations are used:

$$H_1 = T_3 , \quad H_2 = T_8 , \quad H_3 = -\frac{\sqrt{3}}{2\sqrt{2}}Q , \quad H_4 = -\frac{1}{2\sqrt{10}}(3Q - 8R_3) ;$$

$$\tag{14.5.39}$$

T_3, T_8 are the operators of the strong charge, Q and R_3 are those of the electrical and weak charges, respectively, and $Y_W = 2(Q - I_3)$ is that of the weak hypercharge. Thus we have

$$H_1 = T_3 = \frac{1}{2}\begin{bmatrix} 1 & & & & 0 \\ & -1 & & & \\ & & 0 & & \\ & & & 0 & \\ 0 & & & & 0 \end{bmatrix}, \quad H_2 = T_8 = \frac{1}{2\sqrt{3}}\begin{bmatrix} 1 & & & & 0 \\ & 1 & & & \\ & & -2 & & \\ & & & 0 & \\ 0 & & & & 0 \end{bmatrix},$$

$$H_3 = \frac{1}{2\sqrt{6}}\begin{bmatrix} 1 & & & & 0 \\ & 1 & & & \\ & & 1 & & \\ & & & -3 & \\ 0 & & & & 0 \end{bmatrix}, \quad H_4 = \frac{1}{2\sqrt{10}}\begin{bmatrix} 1 & & & & 0 \\ & 1 & & & \\ & & 1 & & \\ & & & 1 & \\ 0 & & & & -4 \end{bmatrix}, \tag{14.5.40}$$

$$Q = -\frac{1}{3}\begin{bmatrix} 1 & & & & 0 \\ & 1 & & & \\ & & 1 & & \\ & & & -3 & \\ 0 & & & & 0 \end{bmatrix}, \quad I_3 = R_3 = \frac{1}{2}\begin{bmatrix} 0 & & & & 0 \\ & 0 & & & \\ & & 0 & & \\ & & & 1 & \\ 0 & & & & -1 \end{bmatrix},$$

$$Y_W = 2(Q - I_3) = -\frac{1}{3}\begin{bmatrix} 2 & & & & 0 \\ & 2 & & & \\ & & 2 & & \\ & & & -3 & \\ 0 & & & & -3 \end{bmatrix}.$$

The other 20 generators E_{ij}, $i \neq j$, are nondiagonal, according to (11.4.6a).

In connection with $\mathcal{SU}(5)$ theories, the $n^2 - 1 = 24$ generators are often denoted by T_k, $k = 1, \ldots, 24$. [They are normalized in such a way that $\mathrm{Tr}(T_i T_k) = \delta_{ik}/2$.] In the defining (fundamental) REP they contain the 8 Gell-Mann matrices of the $\mathcal{SU}_c(3)$ group, see (11.4.8),

$$T_k = \left[\begin{array}{c|c} \lambda_k/2 & 0 \\ \hline 0 & 0 \end{array}\right], \qquad k = 1, \ldots, 8 \text{ (G bosons)} \; ; \tag{14.5.41a}$$

the 3 isospin (Pauli) matrices of the $\mathcal{SU}_W(2)$ group, see (11.4.7),

$$T_k = \left[\begin{array}{c|c} 0 & 0 \\ \hline 0 & \tau_i/2 \end{array}\right], \qquad k = 8 + i \; , \qquad i = 1, 2, 3, (W \text{ bosons}) \; , \tag{14.5.41b}$$

and the weak hypercharge matrix

$$T_{12} = \frac{1}{2}\sqrt{\frac{3}{5}}\,Y_W \,, \qquad [\gamma\text{- }(B\text{-})\text{ boson}] \,. \tag{14.5.41c}$$

The remaining 12 generators $T_{13}, \ldots, T_{24}$ correspond to the generators E^R_{ij}, E^R_{ji} (or E^I_{ij}, E^I_{ji}) with $i = 1, 2, 3$ and $j = 4, 5$ [see (11.4.6)]. They generate transformations between $\mathcal{SU}_c(3)$ and $\mathcal{SU}_w(2)$ (X, Y bosons).

Since $\mathcal{SU}(5)$ is a simple Lie group, all the charges are multiples of the defining charge generators. Thus, the charges, including electric charge, are quantized in this theory (Sect. 14.3.1). According to (14.5.40) the electric charges $Q = I_3 + \frac{1}{2}Y_W$ are integer multiples of the "elementary" charge 1/3. However, it should be mentioned here that the quantization is more or less trivial because of the commensurability of the eigenvalues of Q. In fact, the problem of charge quantization is somewhat more sophisticated and related to the existence of magnetic monopoles as possible stationary solutions of simple gauge theories [14.27, 28, 34, 35].

Local gauge invariance requires the existence of 24 gauge bosons transforming according to the adjoint REP 24, see (14.5.43). The 8 gluons of QCD couple to the generators T_k, $k = 1, \ldots, 8$, the 4 bosons $\boldsymbol{W}$, B of the electro-weak interaction couple to the T_{k+i}, $i = 1, 2, 3$ and T_{12} whereas the remaining 12 bosons X, Y, which carry flavour as well as colour, couple to $T_{13}, \ldots, T_{24}$. According to (14.3.17) the gauge fields can be decomposed into

$$A_{\alpha}{}^{i}{}_{j} = \sum_{k=1}^{24} W^k_\alpha T_k{}^i{}_j \tag{14.5.42a}$$

or schematically ($\bar{X}$, $\bar{Y}$: antiparticles)

$$A_{\alpha}{}^{i}{}_{j} = \left[\begin{array}{c|c} \text{G gluons, } B & \bar{X}\,\bar{Y} \\ \hline X & \boldsymbol{W}, B \text{ bosons} \\ Y & \end{array}\right] \,. \tag{14.5.42b}$$

In general we can say, the larger the symmetry group the larger the particle multiplets. Thus the construction of the $\mathcal{SU}(5)$ multiplets "needs" more particles, or in other words: leptons and quarks of one family are expected to be members of one REP of $\mathcal{SU}(5)$.

We start with a classification in Table 14.3 of the particle multiplets according to the subgroup $\mathcal{SU}_c(3) \times \mathcal{SU}_W(2)$. For the subduction of the IRs of $\mathcal{SU}(5)$, Eq. (14.5.38), onto the subgroup $\mathcal{SU}_c(3) \times \mathcal{SU}_W(2)$ we have the following decompositions:

$$\mathcal{SU}(5) \to \mathcal{SU}_c(3) \times \mathcal{SU}_W(2)$$

$$\begin{aligned}
&5 \to (1,2) \oplus (3,1) \,, \qquad 5^* \to (1,2^*) \oplus (3^*,1) \,, \qquad 2^* \cong 2 \\
&10 \to (3^*,1) \oplus (3,2) \oplus (1,1) \\
&24 \to \underset{G}{(8,1)} \oplus \underset{W^\pm, Z, \gamma}{(1,3) \oplus (1,1)} \oplus \underset{X,Y}{(3,2)} \oplus \underset{\bar{X},\bar{Y}}{(\bar{3},2)} \,.
\end{aligned} \tag{14.5.43}$$

Table 14.3. Classification of the particle multiplets according to the subgroup $\mathcal{SU}_c(3) \times \mathcal{SU}(2)$

Particles	Classification	Meaning	
Leptons	(1, 2)	$\mathcal{SU}_c(3)$ singlet	$\mathcal{SU}(2)$ doublet
Antileptons	(1, 1)	$\mathcal{SU}_c(3)$ singlet	$\mathcal{SU}(2)$ singlet
Quarks	(3, 2)	$\mathcal{SU}_c(3)$ triplet	$\mathcal{SU}(2)$ doublet
Antiquarks	(3*, 1)	$\mathcal{SU}_c(3)$ associated triplet	$\mathcal{SU}(2)$ singlet

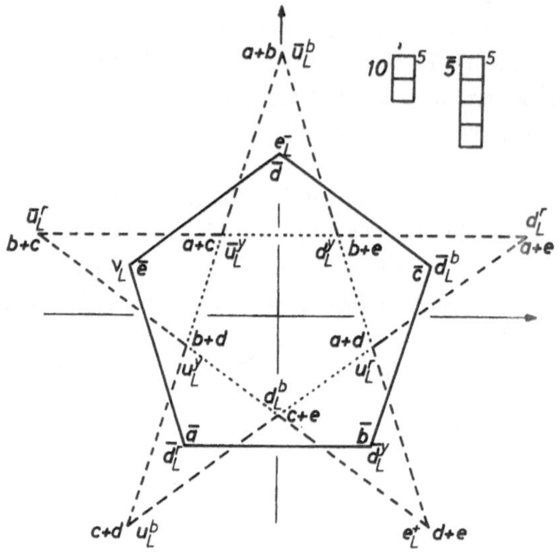

Fig. 14.14. REPs $\bar{5}$ and 10 of $\mathcal{SU}(5)$ in a projection corresponding to Fig. 14.12

Thus it is appropriate to combine each of the families (14.5.33a and b), each consisting of 15 particles (fields), into a (reducible!) (5* ⊕ 10)-REP. The family (14.5.33b) thus has the decomposition

$$
\begin{array}{cccccc}
5^* \oplus 10 \rightarrow & (3,2) \; \oplus & (3^*,1) \oplus & (3^*,1) \oplus & (1,2) \; \oplus & (1,1) \; . \\
& \downarrow & \downarrow & \downarrow & \downarrow & \downarrow \\
& \begin{pmatrix} u^c \\ d^c \end{pmatrix}_L & \bar{u}^c_L & \bar{d}^c_L & \begin{pmatrix} \nu_e \\ e^- \end{pmatrix}_L & e^+_L
\end{array}
\tag{14.5.44}
$$

The two (3*, 1) REPs can be distinguished with respect to $\mathcal{U}(1)$, i.e. to the weak hypercharge. Using (14.5.43) we obtain the correspondence (see Fig. 14.14 and Table 14.4)

$$
5^* := \bar{5} \rightarrow \psi^j{}_L \begin{pmatrix} \bar{d}^r \\ \bar{d}^y \\ \bar{d}^b \\ e^- \\ \nu_e \end{pmatrix}_L , \tag{14.5.45a}
$$

Table 14.4. Metacolor states and assignment of the elementary particles. (After [14.20, 21].)

Particle	REP.	Metacolor	T_3	$T_8 \cdot \sqrt{3}$	Q	R_3	$3Q - 8R_3$	B
d_R^r	5	a	1/2	1/2	−1/3	0	−1	1/3
d_R^y	5	b	−1/2	1/2	−1/3	0	−1	1/3
d_R^b	5	c	0	−1	−1/3	0	−1	1/3
e_R^+	5	d	0	0	1	1/2	−1	0
ν_R	5	e	0	0	0	−1/2	4	0
u_L^r	10	$a+d$	1/2	1/2	2/3	1/2	−2	1/3
u_L^y	10	$b+d$	−1/2	1/2	2/3	1/2	−2	1/3
u_L^b	10	$c+d$	0	−1	2/3	1/2	−2	1/3
d_L^r	10	$a+e$	1/2	1/2	−1/3	−1/2	3	1/3
d_L^y	10	$b+e$	−1/2	1/2	−1/3	−1/2	3	1/3
d_L^b	10	$c+e$	0	−1	−1/3	−1/2	3	1/3
e_L^+	10	$d+e$	0	0	1	0	3	0
$\bar{u}_L^r$	10	$b+c$	−1/2	−1/2	−2/3	0	−2	−1/3
$\bar{u}_L^y$	10	$a+c$	1/2	−1/2	−2/3	0	−2	−1/3
$\bar{u}_L^b$	10	$a+b$	0	1	−2/3	0	−2	−1/3
$\bar{d}_L^r$	$\bar{5}$	$\bar{a}$	−1/2	−1/2	1/3	0	1	−1/3
$\bar{d}_L^y$	$\bar{5}$	$\bar{b}$	1/2	−1/2	1/3	0	1	−1/3
$\bar{d}_L^b$	$\bar{5}$	$\bar{c}$	0	1	1/3	0	1	−1/3
e_L^-	$\bar{5}$	$\bar{d}$	0	0	−1	−1/2	1	0
ν_L	$\bar{5}$	$\bar{e}$	0	0	0	1/2	−4	0
$\bar{u}_R^r$	$\overline{10}$	$\bar{a}+\bar{d}$	−1/2	−1/2	−2/3	−1/2	2	−1/3
$\bar{u}_R^y$	$\overline{10}$	$\bar{b}+\bar{d}$	1/2	−1/2	−2/3	−1/2	2	−1/3
$\bar{u}_R^b$	$\overline{10}$	$\bar{c}+\bar{d}$	0	1	−2/3	−1/2	2	−1/3
$\bar{d}_R^r$	$\overline{10}$	$\bar{a}+\bar{e}$	−1/2	1/2	1/3	1/2	−3	−1/3
$\bar{d}_R^y$	$\overline{10}$	$\bar{b}+\bar{e}$	1/2	1/2	1/3	1/2	−3	−1/3
$\bar{d}_R^b$	$\overline{10}$	$\bar{c}+\bar{e}$	0	1	1/3	1/2	−3	−1/3
e_R^-	$\overline{10}$	$\bar{d}+\bar{e}$	0	0	−1	0	−3	0
u_R^r	$\overline{10}$	$\bar{b}+\bar{c}$	1/2	1/2	2/3	0	2	1/3
u_R^y	$\overline{10}$	$\bar{a}+\bar{c}$	−1/2	1/2	2/3	0	2	1/3
u_R^b	$\overline{10}$	$\bar{a}+\bar{b}$	0	1	2/3	0	2	1/3

while the antisymmetrized product, belonging to the 10 REP (14.5.38), $\psi^{jl}{}_{\mathrm{L}} = -\psi^{lj}{}_{\mathrm{L}}$, can be given as

$$10 \rightarrow \psi^{jl}{}_{\mathrm{L}} = \begin{bmatrix} 0 & \bar{u}^b & -\bar{u}^y & -u^r & -d^r \\ -\bar{u}^b & 0 & \bar{u}^r & -u^y & -d^y \\ \bar{u}^y & -\bar{u}^r & 0 & -u^b & -d^b \\ u^r & u^y & u^b & 0 & -e^+ \\ d^r & d^y & d^b & e^+ & 0 \end{bmatrix}_{\mathrm{L}} . \qquad (14.5.45\mathrm{b})$$

The antiparticles given in Table 14.4 then belong to the 5 or 10* REP, as can be seen from (14.5.34). Corresponding statements are valid for the leptons and quarks of the other two families.

The description of the particles within an $\mathcal{SU}(5)$ scheme is somewhat unsatisfactory because the particles are assigned to a *reducible* REP $5^* \oplus 10$. Therefore it has been proposed that $\mathcal{SU}(5)$ be considered as a subgroup of a larger (simple) gauge group in which every fermion family belongs to an IR of this new gauge group. In such a model there would exist generators which combine 5* and 10 REPs of $\mathcal{SU}(5)$. Furthermore it would be possible to consider even larger gauge groups which connect the different fermion families.

The construction of the $\mathcal{SU}(5)$ multiplets shows that quarks and leptons belong together; in gauge transformations they transform into each other (lepton-hadron symmetry). Correspondingly, the off-diagonal X and Y bosons in (14.5.42b) cause transitions between quarks and leptons. Thus transitions violating baryon-number conservation, in particular proton decay, become possible processes (see next section).

Using the particle multiplets (14.5.45) and the gauge fields (14.5.42a), covariant derivatives can be constructed for the 5* multiplet

$$(D_\alpha \psi_{\mathrm{L}})^j = \partial_\alpha \psi_{\mathrm{L}}^j + \mathrm{i} g_5 A_{\alpha,}{}^j{}_k \psi_{\mathrm{L}}^k \tag{14.5.46a}$$

and for the antisymmetrized product

$$(D_\alpha \psi_{\mathrm{L}})^{jl} = \partial_\alpha \psi_{\mathrm{L}}^{jl} + \mathrm{i} g_5 A_{\alpha,}{}^j{}_k \psi_{\mathrm{L}}^{kl} + \mathrm{i} g_5 A_{\alpha,}{}^l{}_k \psi_{\mathrm{L}}^{jk} \; . \tag{14.5.46b}$$

If we want to use particles transforming according to the fundamental REP 5 instead of 5*, we have to take an antiparticle multiplet using (14.5.34) and (14.5.45a):

$$\bar{\psi}_{\mathrm{R}}^j = \mathfrak{C}\bar{\psi}^{T,}{}_{\mathrm{L}}{}^j = \begin{pmatrix} \mathrm{d^r} \\ \mathrm{d^y} \\ \mathrm{d^b} \\ \mathrm{e^+} \\ \nu_{\mathrm{e}} \end{pmatrix}_{\mathrm{R}} \; . \tag{14.5.46c}$$

The gauge-invariant fermion Lagrangian for the kinetic energy is then

$$l_f = \bar{\bar{\psi}}_{\mathrm{R}}^j \mathrm{i}\gamma^\alpha (\delta^j{}_k \partial_\alpha + \mathrm{i} g_5 A_{\alpha,}{}^j{}_k) \bar{\psi}_{\mathrm{R}}^k + \bar{\psi}_{\mathrm{L}}^{jl} \mathrm{i}\gamma^\alpha (\delta^j{}_k \partial_\alpha + 2\mathrm{i} g_5 A_{\alpha,}{}^j{}_k) \psi_{\mathrm{L}}^{kl} \; . \tag{14.5.47}$$

Some interaction vertices of this Lagrangian are given in Fig. 14.15. Because of the transitions caused by the X and Y bosons, these are also said to be lepto-quark or di-quark bosons, respectively.

In the limit $E \gg m_{\mathrm{X}}$, i.e. in the unbroken-$\mathcal{SU}(5)$-symmetry limit, the gauge-field term of the covariant derivatives reads more explicitly

$$g_5 W_\alpha^k T_k = g_5 \Bigg(\underbrace{\sum_{k=1}^{8} W_\alpha^k T_k}_{\in \mathcal{SU}_c(3)} + \underbrace{\sum_{i=1}^{3} W_\alpha^{8+i} T_{8+i} + W_\alpha^{12} T_{12}}_{\in \mathcal{SU}(2) \times \mathcal{U}(1)} + \underbrace{\quad \cdots \quad}_{\text{additional}} \Bigg) \; .$$

Comparing this with the gauge-field terms in the Lagrangian (14.5.6) of $\mathcal{SU}(2) \times$

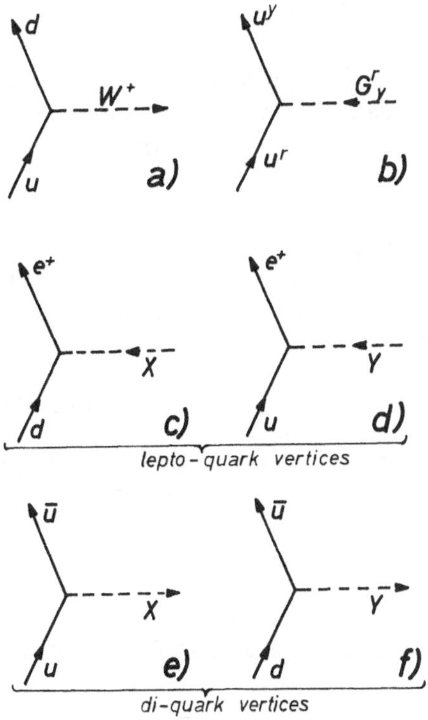

Fig. 14.15a–f. Vertices for interactions according to $\mathcal{SU}(5)$ gauge theory. (**a**) Transition of a u quark into a d quark with emission of a W^+ boson. (**b**) Transition of a red u quark into a yellow one with absorption of a gluon. (**c**) Transition of a d quark into a positron with absorption of an X boson. (**d**) Transition of a u quark into a positron with absorption of a Y boson. (**c, d**) lepto-quark processes. (**e**) Particle-antiparticle u-quark decay into an X boson. (**f**) d-ū decay into a Y boson. (**e, f**) Di-quark processes. Note that W bosons carry only flavour, gluons only colour and X and Y bosons colour and flavour (see also Table 14.1)

$\mathcal{U}(1)$ and the corresponding one for $\mathcal{SU}_c(3)$, i.e. with the terms

$$g_s W_\alpha^k \lambda_k/2 \ , \qquad g W_\alpha^k \tau_k/2 \ , \qquad g' B_\alpha Y/2 \ ,$$

using (14.5.41) we obtain

$$g_5 = g_s = g = g' \sqrt{\frac{5}{3}} \ , \tag{14.5.48}$$

since the generators at this energy are identical with some of the $\mathcal{SU}(5)$ generators. Using (14.5.11a) we get the (high-energy, running) Weinberg angle

$$\sin^2\theta_w = \frac{g'^2}{(g^2+g'^2)} \xrightarrow[E \gg m_X]{} \frac{3}{8} \tag{14.5.49a}$$

and furthermore with (14.5.11b) the ratio of the fine-structure constants

$$\alpha/\alpha_s = e^2/g_s^2 = g^2 \sin^2\theta_w/g_s^2 \xrightarrow[E \gg m_X]{} 3/8 \ . \tag{14.5.49b}$$

This prediction of the magnitude of the Weinberg angle is an interesting feature of the $\mathcal{SU}(5)$ gauge model, since the value of θ_w was not known well at that time. Later experiments (neutral currents) yield those smaller values given in (14.5.14) for $E \lesssim m_W$. From this we can estimate the E^2 dependence of the constants g_5, g and g' shown in Fig. 14.11.

Theoretically, the E^2 dependence of the coupling constants may be calculated from field theories (renormalization group equations) which we will not deal with here. Using these results and the experimental values for $\sin^2\theta_w$ and α/α_s at low E^2, the masses of the X and Y bosons may be estimated from the "unification" point of g_3, g_2 and g_1. This gives about $m_X \geqslant 10^{14}$ GeV, which also follows from the proton lifetime which is now known to be $\tau_p \gtrsim 2 \times 10^{30}$ years.

We will just mention in passing some further results of a perturbation field theory. The running coupling constant $g(E^2)$ or $\alpha(E^2) = g^2/4\pi$ of a general gauge theory of a group $\mathscr{G}$ is given by [14.32]

$$\alpha^{-1}(E^2) = \alpha^{-1}(m^2) - 4\pi b \ln(E^2/m^2) \tag{14.5.50a}$$

[m is a reference energy (momentum) defining an energy scale for the gauge theory under consideration] with

$$b = -\frac{1}{(4\pi)^2}\left(\frac{11}{3}\langle C_2\rangle - \frac{4}{3}\langle C_f\rangle\right) . \tag{14.5.50b}$$

Here $\langle C_2(\mathscr{G})\rangle$ is the quadratic Casimir operator of the adjoint REP of $\mathscr{G}$, see (11.4.76),

$$C_{2,ij} = \langle C_2\rangle\delta_{ij} = \sum_{kl} c^l_{ik} c^k_{lj} , \qquad \langle C_2(\mathscr{SU}(n))\rangle = n , \tag{14.5.51}$$

$$\langle C_2(\mathscr{U}(1))\rangle = 0 ,$$

whereas $\langle C_f\rangle$ is related to the Casimir operator of the fermion REP and can be expressed by the generators J_R or J_L of respectively left- and right-handed fields:

$$\langle C_f\rangle\delta_{ij} = \tfrac{1}{2}\mathrm{Tr}(J_{L,i}J_{L,j}) + \tfrac{1}{2}\mathrm{Tr}(J_{R,i}J_{Rj}) . \tag{14.5.52}$$

For large energies $E^2 \to \infty$ the first term in (14.5.50a) can be neglected.

If $b < 0$, g or α decreases with increasing energy E^2, and the theory becomes asymptotically free. According to (14.5.50b) this is related to $\langle C_2(\mathscr{G})\rangle$ of the corresponding gauge group. In the case of $\mathscr{SU}_c(3)$, $b < 0$ is possible, whereas in the Abelian case $\mathscr{U}(1)$, $C_2 = 0$, thus $b > 0$ and the theory is not asymptotically free; $\alpha(E^2)$ decreases with decreasing E^2 (increasing distance).

14.5.5 Some Consequences of $\mathscr{SU}(5)$ Theory

The spontaneous symmetry breaking of $\mathscr{SU}(5)$ into $\mathscr{SU}_c(3) \times \mathscr{SU}(2) \times \mathscr{U}(1)$ can be explained (Sect. 14.5.2) by a nonvanishing vacuum expectation value of a Higgs (24) multiplet $\{\varphi^k\}$ which transforms according to the adjoint REP of $\mathscr{SU}(5)$ (see for example (11.4.76), [14.32])

$$\phi = \sum_{k=1}^{24} \varphi^k T_k \tag{14.5.53}$$

with the broken vacuum expectation value

$$\langle\phi\rangle_0 = -\frac{3}{2}vY_{\mathrm{W}} = -\sqrt{15}vT_{12} = v\cdot\begin{pmatrix} 1 & & & & 0 \\ & 1 & & & \\ & & 1 & & \\ & & & -3/2 & \\ 0 & & & & -3/2 \end{pmatrix} . \tag{14.5.54}$$

An appropriate potential for the Higgs field has to be chosen, e.g.

$$V(\phi) = -\tfrac{1}{2}\kappa^2 \operatorname{Tr}\{\phi^+\phi\} + \tfrac{1}{4}\lambda(\operatorname{Tr}\{\phi^+\phi\})^2 + \tfrac{1}{2}\mu \operatorname{Tr}\{\phi^+\phi\}^2 \tag{14.5.55}$$

with $\mu > 0$ and $v^2(15\lambda + 7\mu) = 2\kappa^2$. Using the mass formula (14.5.18) and taking ϕ from the real orthogonal Higgs space with an appropriate normalization (see footnote 11) as

$$\phi = \langle\phi\rangle_0 + \sum_{k=1}^{12} \sqrt{2}\eta^k T_k \ , \qquad \text{note: } k = 1, \dots, 12 \ , \tag{14.5.56}$$

we obtain the masses $m_{\mathrm{X}} = m_{\mathrm{Y}} = g_5 v(5/\sqrt{8})$ for the X and Y gauge bosons and 12 massive Higgs bosons with masses $m_3 = v\sqrt{5\mu/2}$ (8 bosons), $m_2 = v\sqrt{10\mu}$ (3 bosons), $m_0 = \kappa\sqrt{2}$ (1 boson). Thus all the Higgs bosons have very large masses [14.32] (see bosons), (see Table 14.1).

The introduction of a second Higgs multiplet ψ transforming according to the 5 REP of $\mathcal{SU}(5)$ leads to a further symmetry breaking into $\mathcal{SU}_{\mathrm{c}}(3) \times \mathcal{U}_{\mathrm{e}}(1)$. Again an appropriate potential $V(\phi, \psi)$ has to be chosen (for details see [14.11, 14, 32]).

We will discuss some of the results qualitatively in connection with Fig. 14.11 which shows three relevant regions of energy and length.

a) $E \gg m_{\mathrm{X}}$ or $x \ll L_{\mathrm{X}}$ ($\approx 10^{-29}$ cm). All the effects of spontaneous symmetry breaking can be ignored, as $\mathcal{SU}(5)$ is an essentially unbroken symmetry group. The coupling constants $g_3 = gs, g_2 = g, g_1 = \sqrt{5/3}g'$ of the $\mathcal{SU}_{\mathrm{c}}(3)$, $\mathcal{SU}(2)$ and $\mathcal{U}(1)$-subgroups are equal to $g_5(E^2)$ of $\mathcal{SU}(5)$. All the gauge bosons, i.e. γ(photon), G(gluon), W, Z(B), X and Y, have masses which can be neglected compared to the corresponding energies.

b) $E \lesssim m_{\mathrm{X}}$ ($\approx 10^{14}$GeV) or $x \gtrsim L_{\mathrm{X}}$($\approx 10^{-29}$ cm). The symmetry $\mathcal{SU}(5)$ is spontaneously broken into $\mathcal{SU}_{\mathrm{c}}(3) \times \mathcal{SU}(2) \times \mathcal{U}(1)$ by means of the Higgs multiplet ϕ. By this mechanism X and Y bosons obtain masses of about 10^{14} GeV. They are heavier than all the other (gauge) particles, i.e. the masses of W, Z, leptons and quarks are approximately zero in the region $m_{\mathrm{W}} \ll E \lesssim m_{\mathrm{X}}$ and $\mathcal{SU}_{\mathrm{c}}(3) \times \mathcal{SU}(2) \times \mathcal{U}(1)$ is essentially unbroken. The coupling constants g_3, g_2

and g_1 develop independently of each other. In the whole of this region it is assumed that there are no further symmetry breakings, so it is a rather uninteresting plateau and is said to be a "(great) desert".

c) $E \lesssim m_W$ or $x \gtrsim L_W$ ($\approx 10^{-16}$ cm). The Higgs multiplet ψ, which has significantly smaller expectation values ($\approx 10^2$ GeV) than those of ϕ, spontaneously breaks the symmetry from $\mathscr{SU}_c(3) \times \mathscr{SU}(2) \times \mathscr{U}(1)$ into $\mathscr{SU}_c(3) \times \mathscr{U}_e(1)$. The W and Z bosons ($\approx 10^2$ GeV) as well as leptons and quarks gain masses. The X and Y bosons are too heavy to be directly produced at this energy, however, they give rise to a (super-) weak, short-range interaction. If the energies become even smaller, $E \ll m_W$ or $x \gg 10^{-16}$ cm, W and Z bosons can no longer be produced directly, but by a virtual exchange they give rise to a weak short-range interaction which shows up in β decay or in the neutral-current interaction of neutrinos (Fig. 14.9). This discussion can be extended to even smaller energies (larger distances) where geometrical aspects of the interactions become more relevant.

d) $x < 10^{-8}$ cm. This is the region inside an atom. The (Coloumb) interaction is rotationally invariant: $\mathscr{O}(3) = \mathscr{C}_i \times \mathscr{SO}(3)$. If the distances become larger the more complicated interatomic (electromagnetic) interactions are relevant causing the atoms to condense into crystals.

e) $x \gg 10^{-8}$ cm. Only the crystalline structure can be seen, the rotational (and translational) invariance of space is spontaneously broken into a discrete point group (and a discrete translational space group) symmetry. The symmetry breaking at the atomic scale is said to produce the "coarse-grained" structure of a crystal and correspondingly a coarse-grained structure of the vacuum is induced by symmetry breaking of gauge symmetries at a smaller scale (10^{-16} cm).

We should add some remarks on the super-weak interaction mediated by the virtual exchange of X and Y bosons. Such processes include a proton decay due to the lepto-quark or di-quark processes, which violate the conservation of baryon number [14.36] (Fig. 14.16). Using the diagrams of Fig. 14.15c,e we obtain a process in which within a proton an X boson is exchanged between a d and a u quark. Thus the proton is allowed to decay into a positron and a neutral meson. The baryon number changes from 2/3 to $-1/3$, thus $\Delta B = +1$. From

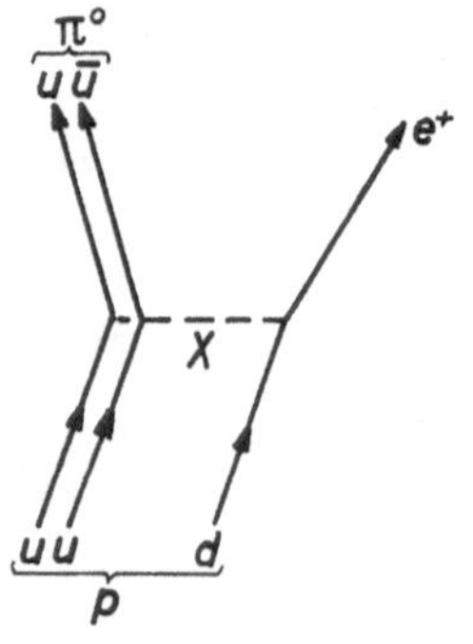

Fig. 14.16. Proton decay. Virtual exchange of an X boson between u and d quarks according to a super-weak interaction: $(d, u) \xrightarrow{X} (e^+, \bar{u})$ leads to $p \to e^+ + \pi^0$

these reactions the proton lifetime can be estimated to be [14.37, 38]

$$\tau_p \approx \frac{1}{\alpha_5^2} \cdot \frac{m_X^4}{m_p^5} \ . \tag{14.5.57}$$

Experiments give a lower limit of $\tau_p \geqslant 2 \times 10^{30}$ years. Using this value and $\alpha_5 \approx \alpha$ one obtains $m_X \geqslant 10^{14}$ GeV. The physical consequences of such a proton decay are very far reaching and even lead to cosmological or astrophysical questions. One example of which is the question why our universe consists of matter rather than of antimatter [14.32, 39]. This combines the history of the universe with the physics of the microcosm of the nucleus, mediated by the GUT.

The origin of the universe is often discussed in terms of the *big bang model.* This model also connects cosmological questions with those of unification and spontaneously broken symmetries in particle physics. The interactions of the GUT are too weak in order to be observed directly with the energies available today, but in the big bang model there was a time (about 10^{-40} s after the big bang, Fig. 14.17) in which the GUT interactions were decisive; in that cosmic laboratory there might have been processes at about 10^{19} GeV via X and Y bosons which cannot be observed today because the universe has cooled down. But there might be relics of GUT interactions which could be observed today. Such a relic for example might be the surplus of baryons (matter) compared with

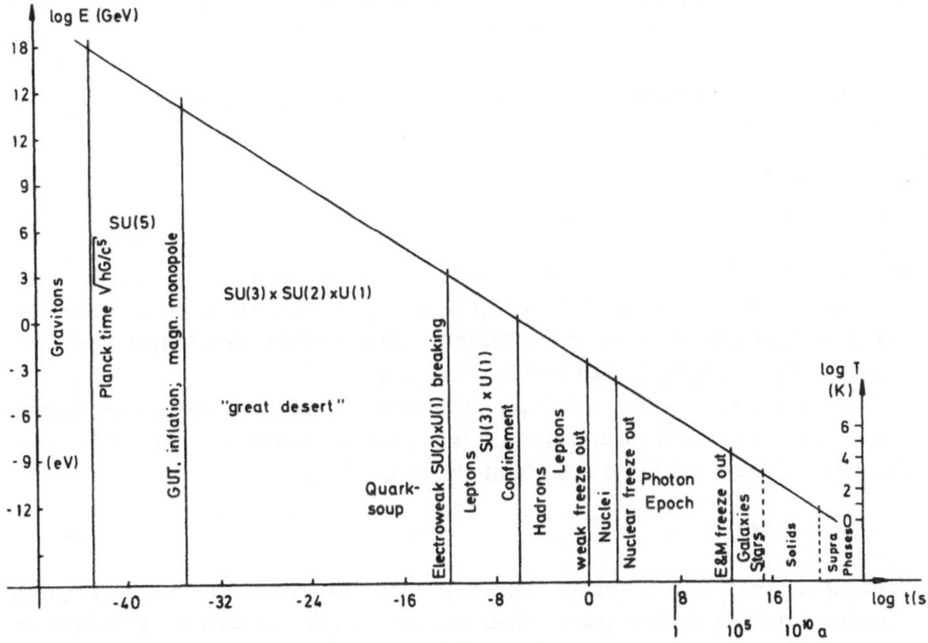

Fig. 14.17. Development of the universe according to the big bang model and gauge theories of Glashow, Salam, Weinberg and Georgi. (Adapted from [14.39])

antibaryons (antimatter). This excess might have its origin in processes mediated by X and Y bosons violating the conservation of baryon number. One could assume that after the big bang the universe started cooling down, and then one by one phase transitions (spontaneous symmetry breakings) occurred. It is possible that at 10^{-35} s after the big bang, interactions violating conservation of baryon number and CP symmetry led to an excess of quarks compared to antiquarks. Later on, at about 10^{-6} s, quarks and antiquarks combined to form baryons and anti-baryons which annihilated each other, but with an excess of baryons.

Of course, a complete theory of all the interactions, including the gravitational one, has to combine the space-time symmetries with these gauge symmetries, that means also bosons and fermions. Such an approach will lead to the theory of supersymmetry (Appendix J).

Another possible relic of the GUT interactions in the big bang might be the super-heavy magnetic monopoles. These might be stable (classical) configurations of the Higgs and gauge fields [14.27, 28] related to spontaneously broken symmetries. They should always exist in gauge theories if a semisimple gauge group $\mathscr{G}$ is spontaneously broken at an energy (mass) M into a subgroup $\mathscr{G}' \subset \mathscr{G}$ containing $\mathscr{U}(1)$ explicitly as a factor: $\mathscr{G} \to \mathscr{G}' = \mathscr{G}'' \times \mathscr{U}(1)$. If, as is the case for $\mathscr{S}\mathscr{U}(5)$, the symmetry is broken several times at different energies, then the mass M determining the monopole mass is related to that symmetry breaking in which the $\mathscr{U}(1)$ factor occurs for the first time with an unbroken subgroup. Thus monopoles are expected to exist as a consequence of all realistic GUTs, including $\mathscr{S}\mathscr{U}(5)$, since $\mathscr{U}_e(1)$ can be assumed to exist as an unbroken symmetry group in any case.

The mass m_m of the monopole and its minimal magnetic charge q_m is approximately [14.32]

$$m_m = \frac{4\pi M}{g^2} , \qquad q_m = \frac{2\pi}{g} = \frac{g}{2\alpha_g} ; \tag{14.5.58}$$

g is the minimal $\mathscr{U}(1)$ charge of a particle which transforms as a singlet under the unbroken symmetry. If $\mathscr{U}(1) = \mathscr{U}_e(1)$, then $g = e$, i.e. g is equal to the electric charge of an electron because the more fundamental quarks, which have fractional electric charges, do not exist as free particles.

That q_m is equal to $2\pi/g$ is a direct consequence of non-Abelian gauge theories without any additional assumptions. It is a special case ($n = 1$) of the well-known Dirac [14.34] condition for the quantization of charges

$$q_m q_e = 4\pi(n/2) , \qquad n \in \mathbb{Z} , \tag{14.5.59}$$

where q_e is the electric charge. Dirac derived this condition within classical electrodynamics (Abelian gauge theory) and with the additional assumptions that there exist magnetic monopoles and that the vector potential A is not unique and singular on a string ending at the monopole.

A recent experiment by Cabrera [14.40] hints at the existence of a magnetic monopole. According to this experiment a monopole with a single Dirac charge $n = 1$ ($q_m = 2\pi/e$) has been observed. As a consequence, the electric charge could only occur as multiples of e and not of $e/3$. Thus free quarks could not exist. An absolute quark confinement would then be a necessary consequence of the existence of this Dirac monopole.

Specifically, for $\mathscr{SU}(5)$ we obtain with $M \approx m_X \geqslant 10^{14}$ GeV and $g \to e$, $e^2/4\pi = \alpha_e$,

$$m_m \approx m_X/\alpha_e \approx 137\, m_X \gtrsim 10^{16}\ \mathrm{GeV}\ . \tag{14.5.60}$$

Such large masses, and consequently magnetic monopoles, could only have been generated in the first 10^{-35} s after the big bang, if at all. Finally, the monopoles got their $\mathscr{U}_e(1)$ Dirac charge after breaking $\mathscr{SU}(2) \times \mathscr{U}(1)$ into $\mathscr{U}_e(1)$, about 10^{-12} s after the big bang.

A certain problem raised by the monopole production in GUTs is connected with the question of where the monopoles have gone (possibly apart from Cabrera's one). A definite answer has not been given so far, though some possible means of elimination of magnetic monopoles have been discussed. Although the "great-desert" hypothesis (which was introduced in connection with $\mathscr{SU}(5)$ and according to which there is no further symmetry breaking, i.e. no new particles in the region between 10^2 and 10^{14} GeV) seems to be somewhat strange, the success of $\mathscr{SU}(5)$, especially in predicting the Weinberg angle, is rather remarkable. A more exact experimental limit of the proton lifetime might be helpful in this question—but things might be different, too:

"There are more things[15] in heaven and earth, ...
Than are dreamt of in your philosophy;"
Shakespeare, Hamlet I.5

Exercise 14.3. Using the relations (14.5.18, 53–56) calculate the masses of gauge and Higgs bosons given at the beginning of Sect. 14.5.5 for the symmetry breaking $\mathscr{SU}(5) \to \mathscr{SU}(3) \times \mathscr{SU}(2) \times \mathscr{U}(1)$.

[15] Symmetries, fields, particles and what else?

Appendices

A. Character Tables

The tables given in this appendix contain only those groups which do not have the inversion as an element. Groups containing the inversion can be written as the direct product $\mathscr{G} = \mathscr{N} \times \mathscr{C}_i$; their characters for point and double point groups can be determined from the scheme in Table A.1. These are the groups $\mathscr{C}_i$, $\mathscr{C}_{2h}$, $\mathscr{C}_{3i}$, $\mathscr{C}_{4h}$, $\mathscr{D}_{2h}$, $\mathscr{D}_{3d}$, $\mathscr{C}_{6h}$, $\mathscr{D}_{4h}$, $\mathscr{D}_{6h}$, $\mathscr{T}_h$, $\mathscr{O}_h$, $\mathscr{Y}_h$. For the double groups for which the elements $a \in \mathscr{G}$ multiplied by c_0 form separate classes (i.e. $\mathscr{C}_1$, $\mathscr{C}_s \cong \mathscr{C}_2$, $\mathscr{C}_3$, $\mathscr{C}_4 \cong \mathscr{S}_4$, $\mathscr{C}_{2h}$, $\mathscr{C}_{3v} \cong \mathscr{D}_3$ and $\mathscr{C}_6 \cong \mathscr{C}_{3h}$) characters are given for elements a, but not for $c_0 a$. These can also be determined from Table A.1.

For the permutation groups the classes are denoted according to their cycles and the representations according to the diagrams (partitions). The alternating groups are denoted analogously. The symbol $\mathscr{Q}$ means the group of quaternions, which are represented by Pauli matrices.

For the spinor REPs of the double groups, for $l = 1/2$,

$$\chi^{(1/2)}(\varphi) = \frac{\sin(l + 1/2)\varphi}{\sin \varphi/2} = 2 \cos \varphi/2 \ ,$$

thus for example

$$\chi^{(1/2)}(c_3) = +1 \ , \qquad \chi^{(1/2)}(c_3^2) = -1 \ ;$$

c_3 and c_3^2 can never be a member of *one* class. The different classes always contain $\{c_3, \bar{c}_3^2\}$, $\{c_3^2, \bar{c}_3\}$, and correspondingly $\{c_4, \bar{c}_4^3\}$, $\{c_4^3, \bar{c}_4\}$, $\{c_6, \bar{c}_6^5\}$, $\{c_6^5, \bar{c}_6\}$, $\{c_5, \bar{c}_5^4\}$, $\{c_5^2, \bar{c}_5^3\}$, $\{c_5^3, \bar{c}_5^2\}$, $\{c_5^4, \bar{c}_5\}$. This has to be taken into account when using the character tables of the double groups, especially for the extra REPs of the groups

$$\mathscr{C}_{3v}^{D} , \ \mathscr{D}_3^{D} ; \ \mathscr{D}_{2d}^{D} , \ \mathscr{C}_{4v}^{D} , \ \mathscr{D}_4^{D} ; \ \mathscr{D}_{3h}^{D} , \ \mathscr{C}_{6v}^{D} , \ \mathscr{D}_6^{D} ; \ \mathscr{T}^{D} ; \ \mathscr{T}_d^{D} , \ \mathscr{O}^{D} ; \ \mathscr{Y}^{D}$$

and their products with the inversion.

The IRs are labelled according to several different systems, which we illustrate in Table A.2 with the example of $\mathscr{O}_h$. For the groups $\mathscr{O}$ and $\mathscr{T}_d$ the indices, i.e. g, u, +, − and primes, have to be dropped. Some authors interchange the notations Γ_{15}, Γ_{25} for $\mathscr{T}_d$ in comparison with $\mathscr{O}$ (e.g. Koster [A.5]). Furthermore, T_1 and T_2 are often used instead of F_1 and F_2. The REPs of the space groups or the related little groups are labelled with respect to the $\boldsymbol{k}$-vector, i.e. Λ, L, Δ, X, etc.;

Table A.1. Schemes to determine the characters of groups containing the inversion element $a \in \mathscr{N}$ **(a)** and of double groups $a \in \mathscr{G}$ **(b)**

$\mathscr{G}$	a	ia	**(a)**	$\mathscr{G}^{\mathrm{D}}$	a	$c_0 a$	**(b)**
D_{g}	$\chi(a)$	$\chi(a)$		D	$\chi(a)$	$\chi(a)$	
D_{u}	$\chi(a)$	$-\chi(a)$		$\bar{D}$	$\bar{\chi}(a)$	$-\bar{\chi}(a)$	

Table A.2. Designations of the irreducible representations of the group $\mathscr{O}_{\mathrm{h}}$ according to various authors

	Notation				
IR	*M* [A.1]	*BO* [A.2, 3]	*BSW* [A.4–6]	*BL* [A.7, 8]	*Zak* [A.9]
Γ_1^+	A_{1g}	Γ_1	Γ_1	Γ_1^+	1
Γ_2^+	A_{2g}	Γ_2	Γ_2	Γ_2^+	2
Γ_3^+	E_g	Γ_3	Γ_{12}	Γ_{12}^+	3
Γ_4^+	F_{1g}	Γ_4	Γ'_{15}	Γ_{15}^+	5
Γ_5^+	F_{2g}	Γ_5	Γ'_{25}	Γ_{25}^+	4
Γ_6^+	$\bar{E}_{1g}$	Γ_6	Γ_6^+	$'E^+$	$\bar{1}$
Γ_7^+	$\bar{E}_{2g}$	Γ_7	Γ_7^+	$'E_5^+$	$\bar{2}$
Γ_8^+	$\bar{G}_g$	Γ_8	Γ_8^+	$'U_3^+$	$\bar{3}$
Γ_1^-	A_{1u}	Γ'_1	Γ'_1	Γ_1^-	6
Γ_2^-	A_{2u}	Γ'_2	Γ'_2	Γ_2^-	7
Γ_3^-	E_u	Γ'_3	Γ'_{12}	Γ_{12}^-	8
Γ_4^-	F_{1u}	Γ'_4	Γ_{15}	Γ_{15}^-	10
Γ_5^-	F_{2u}	Γ'_5	Γ_{25}	Γ_{25}^-	9
Γ_6^-	$\bar{E}_{1u}$	Γ'_6	Γ_6^-	$'E^-$	$\bar{4}$
Γ_7^-	$\bar{E}_{2u}$	Γ'_7	Γ_7^-	$'E_5^-$	$\bar{5}$
Γ_8^-	$\bar{G}_u$	Γ'_8	Γ_8^-	$'U_3^-$	$\bar{6}$

to these letters is added the index of the corresponding REP. Other groups have similar relations between the different notations, which can be guessed with the help of Table A.2.

Tables A.3–16 present the character tables as follows:

Table A.3: $\mathscr{C}_1$

Table A.4: $\mathscr{C}_2, \mathscr{C}_s, \mathscr{P}_2$

Table A.5: $\mathscr{C}_3, \mathscr{A}_3$

Table A.6: $\mathscr{D}_2, \mathscr{C}_{2v}, \mathscr{Q}$

Table A.7: $\mathscr{C}_4, \mathscr{S}_4$

Table A.8: $\mathscr{D}_3, \mathscr{C}_{3v}, \mathscr{P}_3$

Table A.9: $\mathscr{C}_6, \mathscr{C}_{3h}$

Table A.10: $\mathscr{D}_4, \mathscr{C}_{4v}, \mathscr{D}_{2d}$

Table A.11: $\mathscr{D}_6, \mathscr{C}_{6v}, \mathscr{D}_{3h}$

Table A.12: $\mathscr{T}, \mathscr{A}_4$

Table A.13: $\mathscr{O}, \mathscr{T}_d, \mathscr{P}_4$

Table A.14: $\mathscr{Y}, \mathscr{A}_5$

Table A.15: $\mathscr{P}_5$

Table A.16: $\mathscr{D}_\infty, \mathscr{C}_{\infty v}$.

Symbols used in the tables are

$$\varepsilon = e^{2\pi i/3} = \frac{-1 + i\sqrt{3}}{3}\,, \qquad \omega = e^{i\pi/4} = \sqrt{i} = \frac{1+i}{\sqrt{2}}\,,$$

$$\eta_+ = \frac{\sqrt{5}+1}{2} = 1.618034\,, \qquad \eta_- = \frac{\sqrt{5}-1}{2} = 0.618034\,.$$

Tables A.3–16: Character tables

Table A.3

$\mathscr{C}_1$		e
A	Γ_1	1
$\bar{A}$	Γ_2	1

Table A.4

$\mathscr{P}_2$				(1^2)	(2)
	$\mathscr{C}_s$			e	σ
		$\mathscr{C}_2$		e	c_2
[2]	A'	A	Γ_1	1	1
$[1^2]$	A''	B	Γ_2	1	-1
	${}^1\bar{E}$	${}^1\bar{E}$	Γ_3	1	i
	${}^2\bar{E}$	${}^2\bar{E}$	Γ_4	1	$-i$

Table A.5

$\mathscr{A}_3$			1^3	$(3)_1$	$(3)_2$
	$\mathscr{C}_3$		e	c_3	c_3^2
[3]	A	Γ_1	1	1	1
[2, 1]	${}^2E$	Γ_2	1	ε	ε^*
	${}^1E$	Γ_3	1	ε^*	ε
	${}^1\bar{E}$	Γ_4	1	$-\varepsilon$	ε^*
	${}^2\bar{E}$	Γ_5	1	$-\varepsilon^*$	ε
	$\bar{A}$	Γ_6	1	-1	1

Table A.6

$\mathscr{Q}$			e	$\pm i\sigma_1$	$\pm i\sigma_3$	$\pm i\sigma_2$	$-e$
$\mathscr{C}_{2v}$			e	$c_2, \bar{c}_2$	$\sigma_v, \bar{\sigma}_v$	$\sigma'_v, \bar{\sigma}'_v$	c_0
	$\mathscr{D}_2$		e	$c_2, \bar{c}_2$	$c'_2, \bar{c}'_2$	$c''_2, \bar{c}''_2$	c_0
A_1	A_1	Γ_1	1	1	1	1	1
B_1	B_3	Γ_2	1	-1	-1	1	1
A_2	B_1	Γ_3	1	1	-1	-1	1
B_2	B_2	Γ_4	1	-1	1	-1	1
$\bar{E}$	$\bar{E}$	Γ_5	2	0	0	0	-2

Table A.7

$\mathscr{S}_4$		e	s_4^3	s_4^2	s_4
$\mathscr{C}_4$		e	c_4	c_4^2	c_4^3
A	Γ_1	1	1	1	1
B	Γ_2	1	-1	1	-1
${}^2E$	Γ_3	1	i	-1	$-$i
${}^1E$	Γ_4	1	$-$i	-1	i
${}^2\bar{E}_1$	Γ_5	1	ω^*	i	ω
${}^1\bar{E}_1$	Γ_6	1	ω	$-$i	ω^*
${}^2\bar{E}_2$	Γ_7	1	$-\omega^*$	i	$-\omega$
${}^1\bar{E}_2$	Γ_8	1	$-\omega$	$-$i	$-\omega^*$

Table A.8

$\mathscr{P}_3$			(1^3)	(3)	$(1,2)$
	$\mathscr{C}_{3v}$		e	$2c_3$	$3\sigma_v$
	$\mathscr{D}_3$		e	$2c_3$	$3c_2'$
$[3]$	A_1	Γ_1	1	1	1
$[1^3]$	A_2	Γ_2	1	1	-1
$[2,1]$	E	Γ_3	2	-1	0
	${}^1\bar{E}$	Γ_4	1	-1	i
	${}^2\bar{E}$	Γ_5	1	-1	$-$i
	$\bar{E}_1$	Γ_6	2	1	0

Table A.9

$\mathscr{C}_{3h}$				e	$\sigma_h c_3^2$	c_3	σ_h	c_3^2	s_3
		$\mathscr{C}_6$		e	c_6	c_3	c_2	c_3^2	c_6^5
A'	Γ_1	A	Γ_1	1	1	1	1	1	1
A''	Γ_4	B	Γ_4	1	-1	1	-1	1	-1
${}^1E'$	Γ_3	${}^1E_1$	Γ_6	1	ε	ε^*	1	ε	ε^*
${}^2E'$	Γ_2	${}^2E_1$	Γ_5	1	ε^*	ε	1	ε^*	ε
${}^1E''$	Γ_6	${}^1E_2$	Γ_3	1	$-\varepsilon$	ε^*	-1	ε	$-\varepsilon^*$
${}^2E''$	Γ_5	${}^2E_2$	Γ_2	1	$-\varepsilon^*$	ε	-1	ε^*	$-\varepsilon$
${}^1\bar{E}_1$	Γ_{11}	${}^1\bar{E}_1$	Γ_{11}	1	i	-1	i	-1	$-$i
${}^2\bar{E}_1$	Γ_{12}	${}^2\bar{E}_1$	Γ_{12}	1	$-$i	-1	$-$i	-1	i
${}^1\bar{E}_2$	Γ_{10}	${}^1\bar{E}_2$	Γ_{10}	1	iε	$-\varepsilon^*$	i	$-\varepsilon$	$-$iε^*
${}^1\bar{E}_3$	Γ_7	${}^1\bar{E}_3$	Γ_7	1	iε^*	$-\varepsilon$	i	$-\varepsilon^*$	$-$iε
${}^2\bar{E}_2$	Γ_9	${}^2\bar{E}_2$	Γ_9	1	$-$iε^*	$-\varepsilon$	$-$i	$-\varepsilon^*$	iε
${}^2\bar{E}_3$	Γ_8	${}^2\bar{E}_3$	Γ_8	1	$-$iε	$-\varepsilon^*$	$-$i	$-\varepsilon$	iε^*

Table A.10

$\mathscr{D}_{2d}$		e	$2s_4$	$c_2, \bar{c}_2$	$2c_2', \bar{c}_2'$	$2\sigma_v', \bar{\sigma}_v'$	c_0	$2\bar{s}_4$
$\mathscr{C}_{4v}$		e	$2c_4$	$c_2, \bar{c}_2$	$2\sigma_v, \bar{\sigma}_v$	$2\sigma_v', \bar{\sigma}_v'$	c_0	$2\bar{c}_4$
$\mathscr{D}_4$		e	$2c_4$	$c_2, \bar{c}_2$	$2c_2', \bar{c}_2'$	$2c_2'', \bar{c}_2''$	c_0	$2\bar{c}_4$
A_1	Γ_1	1	1	1	1	1	1	1
A_2	Γ_2	1	1	1	-1	-1	1	1
B_1	Γ_3	1	-1	1	1	-1	1	-1
B_2	Γ_4	1	-1	1	-1	1	1	-1
E	Γ_5	2	0	-2	0	0	2	0
$\bar{E}_1$	Γ_6	2	$\sqrt{2}$	0	0	0	-2	$-\sqrt{2}$
$\bar{E}\bar{E}_2$	Γ_7	2	$-\sqrt{2}$	0	0	0	-2	$\sqrt{2}$

Table A.11

$\mathscr{D}_{3h}$				e	$\sigma_h, \bar{\sigma}_h$	$2c_3$	$2s_3$	$3c_2', \bar{c}_2'$	$3\sigma_v, \bar{\sigma}_v$	c_0	$2\bar{c}_3$	$2\bar{s}_3$
	$\mathscr{C}_{6v}$			e	$c_2, \bar{c}_2$	$2c_3$	$2c_6$	$3\sigma_v, \bar{\sigma}_v$	$3\sigma_v', \bar{\sigma}_v'$	c_0	$2\bar{c}_3$	$2\bar{c}_6$
		$\mathscr{D}_6$		e	$c_2, \bar{c}$	$2c_3$	$2c_6$	$3c_2', \bar{c}_2'$	$3c_2'', \bar{c}_2''$	c_0	$2\bar{c}_3$	$2\bar{c}_6$
A_1'	A_1	A_1	Γ_1	1	1	1	1	1	1	1	1	1
A_2'	A_2	A_2	Γ_2	1	1	1	1	−1	−1	1	1	1
A_1''	B_2	B_1	Γ_3	1	−1	1	−1	1	−1	1	1	−1
A_2''	B_1	B_2	Γ_4	1	−1	1	−1	−1	1	1	1	−1
E''	E_1	E_1	Γ_5	2	−2	−1	1	0	0	2	−1	1
E'	E_2	E_2	Γ_6	2	2	−1	−1	0	0	2	−1	−1
$\bar{E}_1$	$\bar{E}_1$	$\bar{E}_1$	Γ_7	2	0	1	$\sqrt{3}$	0	0	−2	−1	$-\sqrt{3}$
$\bar{E}_2$	$\bar{E}_2$	$\bar{E}_2$	Γ_8	2	0	1	$-\sqrt{3}$	0	0	−2	−1	$\sqrt{3}$
$\bar{E}_3$	$\bar{E}_3$	$\bar{E}_3$	Γ_9	2	0	−2	0	0	0	−2	2	0

Table A.12

$\mathscr{A}_4$			(1^4)	$(2,2)$	$(1,3)_1$	$(1,3)_2$			
	$\mathscr{T}$		e	$3c_2, \bar{c}_2$	$4c_3$	$4c_3'$	c_0	$4\bar{c}_3$	$4\bar{c}_3'$
[4]	A	Γ_1	1	1	1	1	1	1	1
[2, 2]	^1E	Γ_2	1	1	ε	ε^*	1	ε	ε^*
	^2E	Γ_3	1	1	ε^*	ε	1	ε^*	ε
[3, 1]	F	Γ_4	3	−1	0	0	3	0	0
	$\bar{E}$	Γ_5	2	0	1	1	−2	−1	−1
	$^1\bar{E}$	Γ_6	2	0	ε^*	ε	−2	$-\varepsilon^*$	$-\varepsilon$
	$^2\bar{E}$	Γ_7	2	0	ε	ε^*	−2	$-\varepsilon$	$-\varepsilon^*$

Table A.13

$\mathscr{P}_4$				(1^4)	$(1,3)$	$(1^2,2)$	(4)	(2^2)			
	$\mathscr{T}_d$			e	$8c_3$	$6\sigma_d, \bar{\sigma}_d$	$6s_4$	$3c_2, \bar{c}_2$	c_0	$8\bar{c}_3$	$6\bar{s}_4$
		$\mathscr{O}$		e	$8c_3$	$6c_2', \bar{c}_2'$	$6c_4$	$3c_2, \bar{c}_2$	c_0	$8\bar{c}_3$	$6\bar{c}_4$
[4]	A_1	A_1	Γ_1	1	1	1	1	1	1	1	1
$[1^4]$	A_2	A_2	Γ_2	1	1	−1	−1	1	1	1	−1
[2, 2]	E	E	Γ_3	2	−1	0	0	2	2	−1	0
$[2,1^2]$	F_1	F_1	Γ_4	3	0	−1	1	−1	3	0	1
[3, 1]	F_2	F_2	Γ_5	3	0	1	−1	−1	3	0	−1
	$\bar{E}_1$	$\bar{E}_1$	Γ_6	2	1	0	$\sqrt{2}$	0	−2	−1	$-\sqrt{2}$
	$\bar{E}_2$	$\bar{E}_2$	Γ_7	2	1	0	$-\sqrt{2}$	0	−2	−1	$\sqrt{2}$
	$\bar{G}$	$\bar{G}$	Γ_8	4	−1	0	0	0	−4	1	0

Table A.14

$\mathscr{A}_5$	$\mathscr{Y}$		(1^5) e	$(5)_1$ $12c_5$	$(5)_2$ $12c_5^2$	$(1^2,3)$ $20c_3$	$(1,2^2)$ $15c_2, \bar{c}_2$	c_0	$12\bar{c}_5$	$12\bar{c}_5^2$	$20\bar{c}_3$
[5]	A	Γ_1	1	1	1	1	1	1	1	1	1
$[3,1^2]$	F_1	Γ_2	3	η_+	$-\eta_-$	0	-1	3	η_+	$-\eta_-$	0
	F_2	Γ_3	3	$-\eta_-$	η_+	0	-1	3	$-\eta_-$	η_+	0
$[2,1^3]$	G	Γ_4	4	-1	-1	1	0	4	-1	-1	1
$[1,2^2]$	H	Γ_5	5	0	0	-1	1	5	0	0	-1
	$\bar{E}_1$	Γ_6	2	η_+	η_-	1	0	-2	$-\eta_+$	$-\eta_-$	-1
	$\bar{E}_2$	Γ_7	2	$-\eta_-$	$-\eta_+$	1	0	-2	η_-	η_+	-1
	$\bar{G}$	Γ_8	4	1	-1	-1	0	-4	-1	1	1
	$\bar{J}$	Γ_9	6	-1	1	0	0	-6	1	-1	0

Table A.15

$\mathscr{P}_5$	(1^5)	$(1^3,2)$	$(1,2^2)$	$(1^2,3)$	$(2,3)$	$(1,4)$	(5)
[5]	1	1	1	1	1	1	1
[4, 1]	4	2	0	1	-1	0	-1
[3, 2]	5	1	1	-1	1	-1	0
$[3,1^2]$	6	0	-2	0	0	0	1
$[2^2,1]$	5	-1	1	-1	-1	1	0
$[2,1^3]$	4	-2	0	1	1	0	-1
$[1^5]$	1	-1	1	1	-1	-1	1

Table A.16

	$\mathscr{C}_{\infty v}$ / $\mathscr{D}_\infty$	e / e	$2c(\phi)$ / $2c(\phi)$	$\sigma_v, \bar{\sigma}_v$ / $c_2, \bar{c}_2$	c_0 / c_0	$2c(\phi)$ / $2c(\phi)$
Σ^+	A_1	1	1	1	1	1
Σ^-	A_2	1	1	-1	1	1
Π	E_1	2	$2\cos\phi$	0	2	$2\cos\phi$
Δ	E_2	2	$2\cos 2\phi$	0	2	$2\cos 2\phi$
⋮	⋮	⋮	⋮	⋮	⋮	⋮
	E_m	2	$2\cos m\phi$	0	2	$2\cos m\phi$
	⋮	⋮	⋮	⋮	⋮	⋮
	$\bar{E}_{1/2}$	2	$2\cos\phi/2$	0	-2	$-2\cos\phi/2$
	$\bar{E}_{3/2}$	2	$2\cos 3\phi/2$	0	-2	$-2\cos 3\phi/2$
	⋮	⋮	⋮	⋮	⋮	⋮
	$\bar{E}_{m+1/2}$	2	$2\cos(m+1/2)\phi$	0	-2	$-2\cos(m+1/2)\phi$
	⋮	⋮	⋮	⋮	⋮	⋮

Note: $\mathscr{D}_\infty$ only occurs together with the inversion: $\mathscr{D}_{\infty h} = \mathscr{D}_\infty \times \mathscr{C}_i$.

Tables A.17–22 present the character tables for some of the little groups of the space groups $\mathcal{O}_h^5$ and $\mathcal{O}_h^7$. The additional elements of the double groups have the character of the corresponding simple elements for the single-valued REPs, but a minus sign in the characters of the double-valued extra REPs. For $\mathcal{O}_h^5$, $s = 0$ (symmorphic), and for $\mathcal{O}_h^7$, $s = (1, 1, 1)a/4$ (nonsymmorphic).

Tables A.17–22: Character tables for some of the little groups of the space groups $\mathcal{O}_h^5$ and $\mathcal{O}_h^7$

Table A.17. $\mathcal{G}_{0\Sigma} \cong \mathcal{C}_{2v}$, $\Sigma = \frac{\pi}{a}(\xi, \xi, 0)$

	e	$\sigma_{x\bar{y}}$	$\{c_2\|s\}$	$\{\sigma_z\|s\}$
Σ_1	1	1	1	1
Σ_2	1	−1	−1	1
Σ_3	1	−1	1	−1
Σ_4	1	1	−1	−1
Σ_5	2	0	0	0

Table A.18. $\mathcal{G}_{0\Lambda} \cong \mathcal{C}_{3v}$, $\Lambda = \frac{\pi}{a}(\xi, \xi, \xi)$

	e	$2c_3$	$3\sigma_v$
Λ_1	1	1	1
Λ_2	1	1	−1
Λ_3	2	−1	0
Λ_4	1	−1	i
Λ_5	1	−1	$-i$
Λ_6	2	1	0

Table A.19. $\mathcal{G}_{0\Delta} \cong \mathcal{C}_{4v}$, $\Delta = \frac{\pi}{a}(\xi, 0, 0)$

	e	c_{2x}	$2\sigma_d$	$2\{c_{4x}\|s\}$	$2\{\sigma_v\|s\}$
Δ_1	1	1	1	1	1
Δ_2	1	1	−1	1	−1
Δ_3	1	1	−1	−1	1
Δ_4	1	1	1	−1	−1
Δ_5	2	−2	0	0	0
Δ_6	2	0	0	$\sqrt{2}$	0
Δ_7	2	0	0	$-\sqrt{2}$	0

Table A.20. $\mathcal{G}_{0\Gamma} \cong \mathcal{O}_h$, $\Gamma = (0,0,0)$
As for the point group $\mathcal{O}_h$ (see Tables A.1a and A.13).

Table A.21. $\mathcal{G}_{0L} \cong \mathcal{D}_{3d}$, $L = \frac{\pi}{a}(1,1,1)$

	e	$2c_3$	$3\sigma_v$	$\{i\|s\}$	$2\{s_6\|s\}$	$2\{c_2'\|s\}$
L_1^+	1	1	1	1	1	1
L_2^+	1	1	−1	1	1	−1
L_3^+	2	−1	0	2	−1	0
L_1^-	1	1	1	−1	−1	−1
L_2^-	1	1	−1	−1	−1	1
L_3^-	2	−1	0	−2	1	0
L_4^+	1	−1	i	1	−1	i
L_5^+	1	−1	−i	1	−1	−i
L_6^+	2	1	0	2	1	0
L_4^-	1	−1	i	−1	1	−i
L_5^-	1	−1	−i	−1	1	i
L_6^-	2	1	0	−2	−1	0

Table A.22. $\mathcal{G}_{0X} \cong \mathcal{D}_{4h}$, $X = \frac{\pi}{a}(2,0,0)$

$\mathcal{O}_h^5$	e	c_{2x}	$2\{c_{4x}\|s\}$	$2c_{2v}$	$2\{c_{2d}'\|s\}$	$\{i\|s\}$	$\{\sigma_x\|s\}$	$2s_{4x}$	$2\{\sigma_v\|s\}$	$2\sigma_d'$
X_1^+	1	1	1	1	1	1	1	1	1	1
X_2^+	1	1	1	−1	−1	1	1	1	−1	−1
X_3^+	1	1	−1	1	−1	1	1	−1	1	−1
X_4^+	1	1	−1	−1	1	1	1	−1	−1	1
X_5^+	2	−2	0	0	0	2	−2	0	0	0
X_1^-	1	1	1	1	1	−1	−1	−1	−1	−1
X_2^-	1	1	1	−1	−1	−1	−1	−1	1	1
X_3^-	1	1	−1	1	−1	−1	−1	1	−1	1
X_4^-	1	1	−1	−1	1	−1	−1	1	1	−1
X_5^-	2	−2	0	0	0	−2	2	0	0	0
X_6^+	2	0	$\sqrt{2}$	0	0	2	0	$\sqrt{2}$	0	0
X_7^+	2	0	$-\sqrt{2}$	0	0	2	0	$-\sqrt{2}$	0	0
X_6^-	2	0	$\sqrt{2}$	0	0	−2	0	$-\sqrt{2}$	0	0
X_7^-	2	0	$-\sqrt{2}$	0	0	−2	0	$\sqrt{2}$	0	0
$\mathcal{O}_h^7$					$y\bar{z}$ yz					
X_1	2	2	0	0	0	0	0	0	0	2
X_2	2	2	0	0	0	0	0	0	0	−2
X_3	2	−2	0	0	2 −2	0	0	0	0	0
X_4	2	−2	0	0	−2 2	0	0	0	0	0
X_5	4	0	0	0	0	0	0	0	0	0

B. Representations of Generators

One-dimensional REPs are always identical with the characters and can be taken from the character tables (Appendix A). For the many-dimensional REPs, in Table B.1 we just give the REPs of the generators (but not necessarily the minimal

Table B.1. Representations of generators for the different groups

$\mathscr{D}_2, \mathscr{C}_{2v}$:	c_{2x}, σ_x	c_{2y}, σ_y	c_{2z}
$\bar{E}$:	$i\sigma_1$	$i\sigma_2$	$i\sigma_3$

$\mathscr{D}_3, \mathscr{C}_{3v}$:	c_{3z}	c_{2y}, σ_x
E:	α^2	$-\sigma_3$
$\bar{E}$:	α	$i\sigma_3$

$\mathscr{D}_4, \mathscr{C}_{4v}, \mathscr{D}_{2d}$:	c_{4z}, c_{4z}, s_{4z}^3	c_{2x}, σ_y, c_{2x}
E:	$-i\sigma_2$	σ_3
$\bar{E}_1$:	$-\beta$	$i\sigma_1$
$\bar{E}_2$:	β	$i\sigma_1$

$\mathscr{D}_6, \mathscr{C}_{6v}, \mathscr{D}_{3h}$:	$c_{6z}, c_{6z}, s_{3z}^{-1}$	c_{2x}, σ_y, c_{2x}
E_1:	α	σ_3
E_2:	α^2	σ_3
$\bar{E}_1$:	ζ	$i\sigma_1$
$\bar{E}_2$:	$-\zeta^*$	$i\sigma_1$
$\bar{E}_3$:	$i\sigma_3$	$i\sigma_1$

$\mathscr{T}$:	c_{2z}	c_{3xyz}
F:	α_2	α_3
$\bar{E}$:	$-i\sigma_3$	κ
${}^1\bar{F}$:	$-i\sigma_3$	$\varepsilon^*\kappa$
${}^2\bar{F}$:	$-i\sigma_3$	$\varepsilon\kappa$

$\mathscr{O}, \mathscr{T}_d$:	c_{4z}, s_{4z}	c_{3xyz}	c_{2xy}, σ_d
E:	$\sigma_3\alpha$	$-\alpha$	$\sigma_3\alpha$
F_1:	α_4	α_3	α_2'
F_2:	$-\alpha_4$	α_3	$-\alpha_2'$
$\bar{E}_1$:	λ	κ	μ
$\bar{E}_2$:	$-\lambda$	κ	$-\mu$
$\bar{F}$:	$\frac{1}{\sqrt{2}}\begin{pmatrix}(1-i)\sigma_3\alpha & 0\\ 0 & (1+i)\sigma_3\alpha\end{pmatrix}$	$-\frac{1}{2}\begin{pmatrix}(1-i)\alpha & -(1+i)\alpha\\ (1-i)\alpha & (1+i)\alpha\end{pmatrix}$	$\frac{1}{\sqrt{2}}\begin{pmatrix}0 & -(1+i)\sigma_3\alpha\\ (1-i)\sigma_3\alpha & 0\end{pmatrix}$

$\mathscr{D}_\infty, \mathscr{C}_{\infty v}$:	$c(\varphi)$	c_2, σ_v
E_m:	$\begin{pmatrix}\cos m\varphi & -\sin m\varphi\\ \sin m\varphi & \cos m\varphi\end{pmatrix}$	$\begin{pmatrix}1 & 0\\ 0 & -1\end{pmatrix}$
$\bar{E}_{m+1/2}$:	$\begin{pmatrix}e^{+i(m+1/2)\varphi} & 0\\ 0 & e^{-i(m+1/2)\varphi}\end{pmatrix}$	$\begin{pmatrix}0 & -i\\ -i & 0\end{pmatrix}$

$\mathscr{Q}$:				
E:	e	$i\sigma_2$	$i\sigma_1$	$i\sigma_3$

Note: $\mathscr{Q}$ is isomorphic to $\mathscr{D}_2^D$.

set, see Sect. 2.1). The REPs of the other elements can be determined with the help of geometrical considerations. The form of the REP matrices depends on the choice of the rotation (crystal) axes relative to a Cartesian coordinate system and on the assignment of x, y, z to rows and columns. For the double-valued REPs it also depends on the choice of the Pauli matrices. The REPs of the groups containing the inversion follow from the direct product $D(\mathcal{G}) = D(\mathcal{N}) \times D(\mathcal{C}_i)$. Table B.2 gives the REPs of the icosahedral group $\mathcal{Y}$. Applications in Appendix I.

Symbols used in Tables B.1 and B.2:

$$\varepsilon = e^{2\pi i/3} = \frac{-1 + i\sqrt{3}}{2} ,$$

$$\eta_+ = 1.618034 , \qquad \eta_- = 0.618034 ,$$

$$\sigma_1 = \begin{pmatrix} 0 & 1 \\ 1 & 0 \end{pmatrix} , \qquad \sigma_2 = \begin{pmatrix} 0 & -i \\ i & 0 \end{pmatrix} , \qquad \sigma_3 = \begin{pmatrix} 1 & 0 \\ 0 & -1 \end{pmatrix} ,$$

$$i\sigma_2 = \begin{pmatrix} 0 & 1 \\ -1 & 0 \end{pmatrix} ,$$

$$\alpha = \begin{pmatrix} 1/2 & -\sqrt{3}/2 \\ \sqrt{3}/2 & 1/2 \end{pmatrix} , \qquad \alpha^2 = \begin{pmatrix} -1/2 & -\sqrt{3}/2 \\ \sqrt{3}/2 & -1/2 \end{pmatrix} ,$$

$$\beta = \frac{1}{\sqrt{2}} \begin{pmatrix} 1 & 1 \\ -1 & 1 \end{pmatrix} , \qquad \zeta = \begin{pmatrix} -i\varepsilon & 0 \\ 0 & i\varepsilon^* \end{pmatrix} ,$$

$$\kappa = \frac{1}{2} \begin{pmatrix} 1 - i & -(1 + i) \\ 1 - i & 1 + i \end{pmatrix} , \qquad \lambda = \frac{1}{\sqrt{2}} \begin{pmatrix} 1 - i & 0 \\ 0 & 1 + i \end{pmatrix} ,$$

$$\mu = \frac{1}{\sqrt{2}} \begin{pmatrix} 0 & -(1 + i) \\ 1 - i & 0 \end{pmatrix} ,$$

$$\alpha_2 = \begin{pmatrix} -1 & 0 & 0 \\ 0 & -1 & 0 \\ 0 & 0 & 1 \end{pmatrix} , \qquad \alpha_2' = \begin{pmatrix} 0 & 1 & 0 \\ 1 & 0 & 0 \\ 0 & 0 & -1 \end{pmatrix} ,$$

$$\alpha_3 = \begin{pmatrix} 0 & 0 & 1 \\ 1 & 0 & 0 \\ 0 & 1 & 0 \end{pmatrix} , \qquad \alpha_4 = \begin{pmatrix} 0 & -1 & 0 \\ 1 & 0 & 0 \\ 0 & 0 & 1 \end{pmatrix} .$$

Each section of Table B.1 has the form:

group(s) : generators , REPs : matrix generators .

Table B.2. Representations of the icosahedral group $\mathscr{Y}$

	$c_2(001)$	$c_3(0\eta_+\eta_-)$	$c_5(0\sqrt{\eta_-}\sqrt{\eta_+})$
F_1	$\begin{bmatrix} -1 & 0 & 0 \\ 0 & -1 & 0 \\ 0 & 0 & 1 \end{bmatrix}$	$\frac{1}{2}\begin{bmatrix} -1 & -\eta_- & \eta_+ \\ \eta_- & \eta_+ & 1 \\ -\eta_+ & 1 & -\eta_- \end{bmatrix}$	$\frac{1}{2}\begin{bmatrix} \eta_- & -\eta_+ & 1 \\ \eta_+ & 1 & \eta_- \\ -1 & \eta_- & \eta_+ \end{bmatrix}$
F_2	$\begin{bmatrix} -1 & 0 & 0 \\ 0 & -1 & 0 \\ 0 & 0 & 1 \end{bmatrix}$	$\frac{1}{2}\begin{bmatrix} -1 & -\eta_- & \eta_+ \\ \eta_- & \eta_+ & 1 \\ -\eta_+ & 1 & -\eta_- \end{bmatrix}$	$\frac{1}{2}\begin{bmatrix} -\eta_+ & -1 & \eta_- \\ 1 & -\eta_- & \eta_+ \\ -\eta_- & \eta_+ & 1 \end{bmatrix}$
G	$\begin{bmatrix} 1 & 0 & 0 & 0 \\ 0 & -1/3 & 2\sqrt{2}/3 & 0 \\ 0 & 2\sqrt{2}/3 & 1/3 & 0 \\ 0 & 0 & 0 & -1 \end{bmatrix}$	$\begin{bmatrix} 1 & 0 & 0 & 0 \\ 0 & 1 & 0 & 0 \\ 0 & 0 & -1/2 & \sqrt{3}/2 \\ 0 & 0 & -\sqrt{3}/2 & -1/2 \end{bmatrix}$	$\frac{1}{12}\begin{bmatrix} -3 & 3\sqrt{15} & 0 & 0 \\ -\sqrt{15} & -1 & 8\sqrt{2} & 0 \\ -\sqrt{30} & -\sqrt{2} & -2 & 6\sqrt{3} \\ -3\sqrt{10} & -\sqrt{6} & -2\sqrt{3} & -6 \end{bmatrix}$
H	$\begin{bmatrix} -1/3 & 2\sqrt{2}/3 & 0 & 0 & 0 \\ 2\sqrt{2}/3 & +1/3 & 0 & 0 & 0 \\ 0 & 0 & +1 & 0 & 0 \\ 0 & 0 & 0 & -1 & 0 \\ 0 & 0 & 0 & 0 & -1 \end{bmatrix}$	$\begin{bmatrix} +1 & 0 & 0 & 0 & 0 \\ 0 & -1/2 & 0 & \sqrt{3}/2 & 0 \\ 0 & 0 & -1/2 & 0 & \sqrt{3}/2 \\ 0 & -\sqrt{3}/2 & 0 & -1/2 & 0 \\ 0 & 0 & -\sqrt{3}/2 & 0 & -1/2 \end{bmatrix}$	$\frac{1}{12}\begin{bmatrix} -4 & -3\sqrt{3} & 4\sqrt{6} & 0 & 0 \\ -4\sqrt{2} & 1 & -\sqrt{3} & -3\sqrt{3} & 9 \\ 0 & -3\sqrt{3} & -3 & -9 & -3\sqrt{3} \\ -4\sqrt{6} & +\sqrt{3} & -3 & 3 & -3\sqrt{3} \\ 0 & -9 & -3\sqrt{3} & 3\sqrt{3} & 3 \end{bmatrix}$

C. Standard Young-Yamanouchi Representations of the Permutation Groups $\mathscr{P}_3$–$\mathscr{P}_5$

We give only the many-dimensional REPs of the generating elements $p_{i-1,i}$, $i = 2, \ldots, n$ in the sequence $p_{12}, p_{23}, \ldots, p_{n-1,n}$. All other standard Young-Yamanouchi representations of $\mathscr{P}_3$ to $\mathscr{P}_5$ can be determined by appropriate multiplications. The A_s, where $s = 2, 3, 4$, are two-dimensional matrices, which intersect for $\mathscr{P}_5$.

$\boxed{2}$ denotes the submatrix A_2, etc.

$$A_2 = \begin{pmatrix} -1/2 & \sqrt{3}/2 \\ \sqrt{3}/2 & 1/2 \end{pmatrix}, \qquad A_3 = \begin{pmatrix} -1/3 & 2\sqrt{2}/3 \\ 2\sqrt{2}/3 & 1/3 \end{pmatrix},$$

$$A_4 = \begin{pmatrix} -1/4 & \sqrt{15}/4 \\ \sqrt{15}/4 & 1/4 \end{pmatrix}.$$

$\mathscr{P}_3$

[2, 1] $\begin{array}{|c|c|}\hline 1&2\\\hline 3\\\cline{1-1}\end{array}$ $\begin{array}{|c|c|}\hline 1&3\\\hline 2\\\cline{1-1}\end{array}$

$$\begin{pmatrix} 1 & 0 \\ 0 & -1 \end{pmatrix} \quad \begin{pmatrix} \boxed{2} \end{pmatrix}$$

$\mathscr{P}_4$

[3, 1] $\begin{array}{|c|c|c|}\hline 1&2&3\\\hline 4\\\cline{1-1}\end{array}$ $\begin{array}{|c|c|c|}\hline 1&2&4\\\hline 3\\\cline{1-1}\end{array}$ $\begin{array}{|c|c|c|}\hline 1&3&4\\\hline 2\\\cline{1-1}\end{array}$

$$\begin{pmatrix} 1 & 0 & 0 \\ 0 & 1 & 0 \\ 0 & 0 & -1 \end{pmatrix} \quad \begin{pmatrix} 1 & \begin{matrix} 0 & 0 \end{matrix} \\ \begin{matrix} 0 \\ 0 \end{matrix} & \boxed{2} \end{pmatrix} \quad \begin{pmatrix} \boxed{3} & \begin{matrix} 0 \\ 0 \end{matrix} \\ \begin{matrix} 0 & 0 \end{matrix} & 1 \end{pmatrix}$$

[2^2] $\begin{array}{|c|c|}\hline 1&2\\\hline 3&4\\\hline\end{array}$ $\begin{array}{|c|c|}\hline 1&3\\\hline 2&4\\\hline\end{array}$

$$\begin{pmatrix} 1 & 0 \\ 0 & -1 \end{pmatrix} \quad \begin{pmatrix} \boxed{2} \end{pmatrix} \quad \begin{pmatrix} 1 & 0 \\ 0 & -1 \end{pmatrix}$$

[2, 1^2] $\begin{array}{|c|c|}\hline 1&2\\\hline 3\\\cline{1-1} 4\\\cline{1-1}\end{array}$ $\begin{array}{|c|c|}\hline 1&3\\\hline 2\\\cline{1-1} 4\\\cline{1-1}\end{array}$ $\begin{array}{|c|c|}\hline 1&4\\\hline 2\\\cline{1-1} 3\\\cline{1-1}\end{array}$

$$\begin{pmatrix} 1 & 0 & 0 \\ 0 & -1 & 0 \\ 0 & 0 & -1 \end{pmatrix} \quad \begin{pmatrix} \boxed{2} & \begin{matrix} 0 \\ 0 \end{matrix} \\ \begin{matrix} 0 & 0 \end{matrix} & -1 \end{pmatrix} \quad \begin{pmatrix} -1 & \begin{matrix} 0 & 0 \end{matrix} \\ \begin{matrix} 0 \\ 0 \end{matrix} & \boxed{3} \end{pmatrix}$$

$\mathscr{P}_5$

[4, 1] $\quad \begin{array}{|c|c|c|c|}\hline 1&2&3&4\\\hline 5\\\cline{1-1}\end{array} \quad \begin{array}{|c|c|c|c|}\hline 1&2&3&5\\\hline 4\\\cline{1-1}\end{array} \quad \begin{array}{|c|c|c|c|}\hline 1&2&4&5\\\hline 3\\\cline{1-1}\end{array} \quad \begin{array}{|c|c|c|c|}\hline 1&3&4&5\\\hline 2\\\cline{1-1}\end{array}$

$$\begin{pmatrix}1&0&0&0\\0&1&0&0\\0&0&1&0\\0&0&0&-1\end{pmatrix} \quad \begin{pmatrix}1&0&0&0\\0&1&0&0\\0&0&\multicolumn{2}{c}{\boxed{2}}\\0&0&&\end{pmatrix} \quad \begin{pmatrix}1&0&0&0\\0&\multicolumn{2}{c}{\boxed{3}}&0\\0&&&0\\0&0&0&1\end{pmatrix} \quad \begin{pmatrix}\multicolumn{2}{c}{\boxed{4}}&0&0\\&&0&0\\0&0&1&0\\0&0&0&1\end{pmatrix}$$

[3, 2] $\quad \begin{array}{|c|c|c|}\hline 1&2&3\\\hline 4&5\\\cline{1-2}\end{array} \quad \begin{array}{|c|c|c|}\hline 1&2&4\\\hline 3&5\\\cline{1-2}\end{array} \quad \begin{array}{|c|c|c|}\hline 1&2&5\\\hline 3&4\\\cline{1-2}\end{array} \quad \begin{array}{|c|c|c|}\hline 1&3&4\\\hline 2&5\\\cline{1-2}\end{array} \quad \begin{array}{|c|c|c|}\hline 1&3&5\\\hline 2&4\\\cline{1-2}\end{array}$

$$\begin{pmatrix}1&0&0&0&0\\0&1&0&0&0\\0&0&1&0&0\\0&0&0&-1&0\\0&0&0&0&-1\end{pmatrix}$$

$$\begin{pmatrix}\multicolumn{2}{c}{\boxed{3}}&0&0&0\\&&0&0&0\\0&0&1&0&0\\0&0&0&1&0\\0&0&0&0&-1\end{pmatrix} \quad \begin{pmatrix}1&0&0&0&0\\0&\multicolumn{2}{c}{\boxed{2}}&0&0\\0&&&0&0\\0&0&0&\multicolumn{2}{c}{\boxed{2}}\\0&0&0&&\end{pmatrix}$$

[3, 1^2] $\quad \begin{array}{|c|c|c|}\hline 1&2&3\\\hline 4\\\cline{1-1}5\\\cline{1-1}\end{array} \quad \begin{array}{|c|c|c|}\hline 1&2&4\\\hline 3\\\cline{1-1}5\\\cline{1-1}\end{array} \quad \begin{array}{|c|c|c|}\hline 1&2&5\\\hline 3\\\cline{1-1}4\\\cline{1-1}\end{array} \quad \begin{array}{|c|c|c|}\hline 1&3&4\\\hline 2\\\cline{1-1}5\\\cline{1-1}\end{array} \quad \begin{array}{|c|c|c|}\hline 1&3&5\\\hline 2\\\cline{1-1}4\\\cline{1-1}\end{array} \quad \begin{array}{|c|c|c|}\hline 1&4&5\\\hline 2\\\cline{1-1}3\\\cline{1-1}\end{array}$

$$\begin{pmatrix}1&0&0&0&0&0\\0&1&0&0&0&0\\0&0&1&0&0&0\\0&0&0&-1&0&0\\0&0&0&0&-1&0\\0&0&0&0&0&-1\end{pmatrix}$$

$$\begin{pmatrix}\multicolumn{2}{c}{\boxed{3}}&0&0&0&0\\&&0&0&0&0\\0&0&-1&0&0&0\\0&0&0&1&0&0\\0&0&0&0&\multicolumn{2}{c}{\boxed{3}}\\0&0&0&0&&\end{pmatrix} \quad \begin{pmatrix}-1&0&0&0&0&0\\0&\multicolumn{2}{c}{\boxed{4}}&0&0&0\\0&&&0&0&0\\0&0&0&\multicolumn{2}{c}{\boxed{4}}&0\\0&0&0&&&0\\0&0&0&0&0&1\end{pmatrix}$$

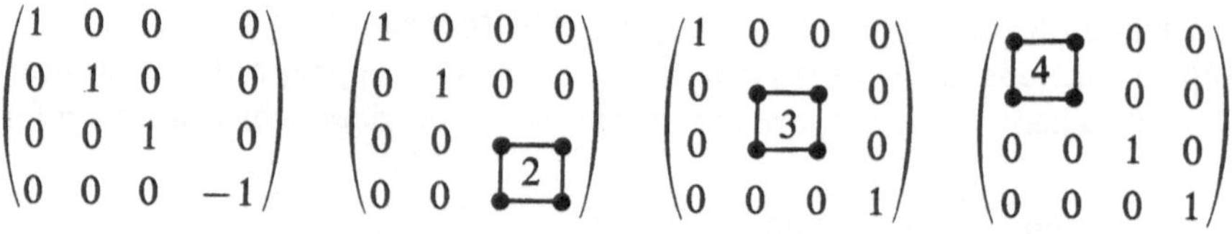

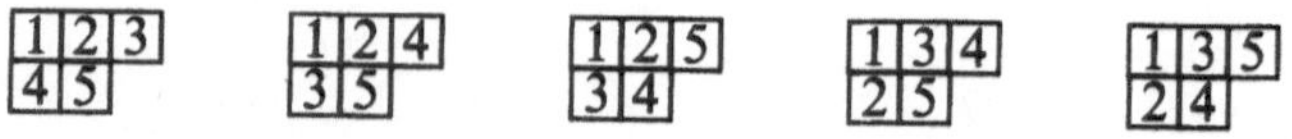

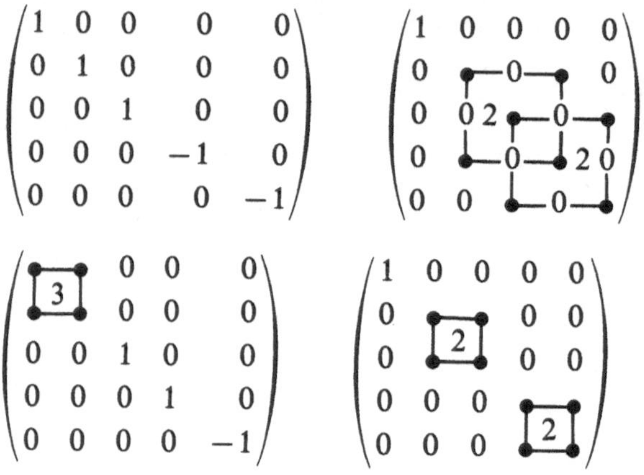

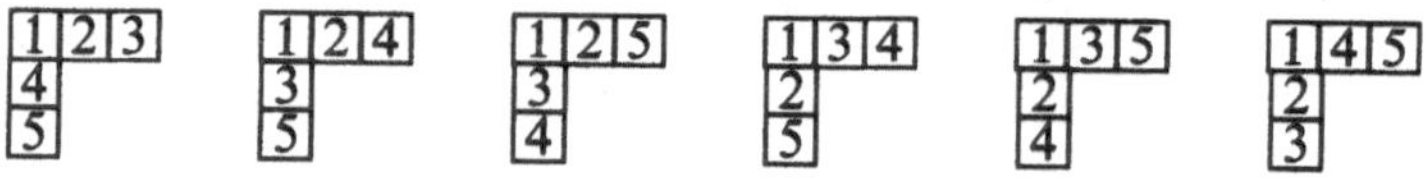

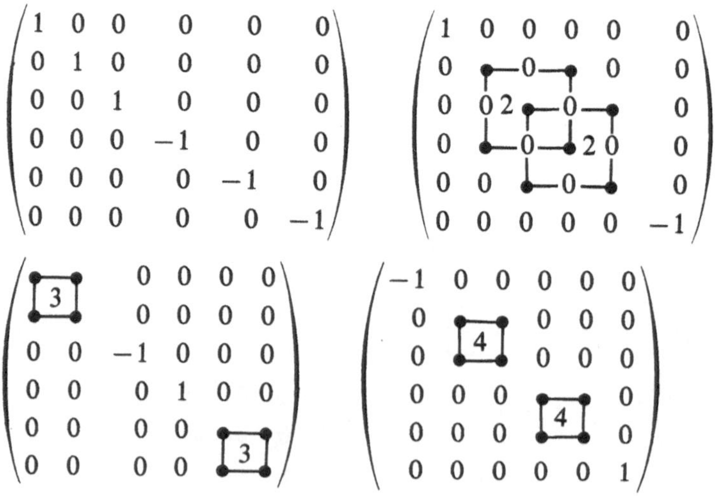

$[2^2,1]$ $\begin{array}{|c|c|}\hline 1&2\\\hline 3&4\\\hline 5\\\cline{1-1}\end{array}$ $\begin{array}{|c|c|}\hline 1&2\\\hline 3&5\\\hline 4\\\cline{1-1}\end{array}$ $\begin{array}{|c|c|}\hline 1&3\\\hline 2&4\\\hline 5\\\cline{1-1}\end{array}$ $\begin{array}{|c|c|}\hline 1&3\\\hline 2&5\\\hline 4\\\cline{1-1}\end{array}$ $\begin{array}{|c|c|}\hline 1&4\\\hline 2&5\\\hline 3\\\cline{1-1}\end{array}$

$$
\begin{pmatrix}
1 & 0 & 0 & 0 & 0\\
0 & 1 & 0 & 0 & 0\\
0 & 0 & -1 & 0 & 0\\
0 & 0 & 0 & -1 & 0\\
0 & 0 & 0 & 0 & -1
\end{pmatrix}
\qquad
\begin{pmatrix}
\bullet & 0 & \bullet & 0 & 0\\
0\;2 & \bullet & 0 & \bullet & 0\\
\bullet & 0 & \bullet & 2\;0 & 0\\
0 & \bullet & 0 & \bullet & 0\\
0 & 0 & 0 & 0 & -1
\end{pmatrix}
$$

$$
\begin{pmatrix}
1 & 0 & 0 & 0 & 0\\
0 & -1 & 0 & 0 & 0\\
0 & 0 & -1 & 0 & 0\\
0 & 0 & 0 & \boxed{3} & \\
0 & 0 & 0 & &
\end{pmatrix}
\qquad
\begin{pmatrix}
\boxed{2} & & 0 & 0 & 0\\
 & & 0 & 0 & 0\\
0 & 0 & \boxed{2} & & 0\\
0 & 0 & & & 0\\
0 & 0 & 0 & 0 & -1
\end{pmatrix}
$$

$[2,1^3]$ $\begin{array}{|c|c|}\hline 1&2\\\hline 3\\\cline{1-1}4\\\cline{1-1}5\\\cline{1-1}\end{array}$ $\begin{array}{|c|c|}\hline 1&3\\\hline 2\\\cline{1-1}4\\\cline{1-1}5\\\cline{1-1}\end{array}$ $\begin{array}{|c|c|}\hline 1&4\\\hline 2\\\cline{1-1}3\\\cline{1-1}5\\\cline{1-1}\end{array}$ $\begin{array}{|c|c|}\hline 1&5\\\hline 2\\\cline{1-1}3\\\cline{1-1}4\\\cline{1-1}\end{array}$

$$
\begin{pmatrix}
1 & 0 & 0 & 0\\
0 & -1 & 0 & 0\\
0 & 0 & -1 & 0\\
0 & 0 & 0 & -1
\end{pmatrix}
\qquad
\begin{pmatrix}
\boxed{2} & & 0 & 0\\
 & & 0 & 0\\
0 & 0 & -1 & 0\\
0 & 0 & 0 & -1
\end{pmatrix}
$$

$$
\begin{pmatrix}
-1 & 0 & 0 & 0\\
0 & & & 0\\
0 & & \boxed{3} & 0\\
0 & 0 & 0 & -1
\end{pmatrix}
\qquad
\begin{pmatrix}
-1 & 0 & 0 & 0\\
0 & -1 & 0 & 0\\
0 & 0 & & \\
0 & 0 & \boxed{4} &
\end{pmatrix}
$$

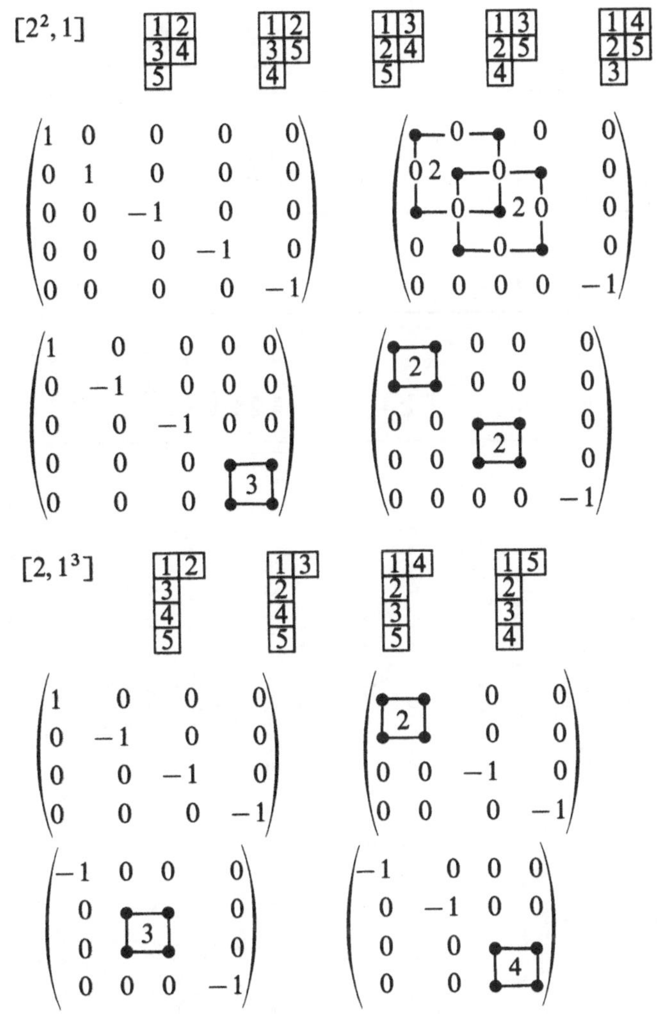

D. Continuous Groups

Table D.1 presents information about continuous groups. There are further continuous groups, which we cannot give in detail. For example, all triangular matrices ($a_{ij} = 0$ for $i > j$), and all diagonal matrices ($a_{ij} = 0$ for $i \neq j$) form continuous groups if $\det a \neq 0$. Their algebras are subalgebras of A_l, B_l, C_l or D_l.

The Lorentz groups are isomorphic to groups which are listed in Table D.1. For example, the proper orthochronous Lorentz group $\bar{\mathscr{L}}z^+(4,\mathbb{R})$ is isomorphic to $\mathscr{SO}(3,1)$ [covering group is $\mathscr{SL}(2,\mathbb{C})$], and the de Sitter group to $\mathscr{SO}(3,2)$.

For the inhomogeneous (general affine) groups there is in addition to the homogeneous part (element a) a translation part t. They describe transformations

Table D.1. Continuous groups

Group	Name	Defining relations via the finite element a	$a = 1 + \alpha$ infinitesimal, $\det a = 1 + \mathrm{Tr}\{\alpha\}$	Dimension (number of real parameters)	Rank	Associated algebra	
$\mathscr{GL}(n, \mathbb{C})$	General linear group	$a_{ij} \in \mathbb{C}$; $\det a \neq 0$	$\alpha := \{\alpha_{ij}\}$	$2n^2$			
$\mathscr{GL}(n, \mathbb{R})$	General linear group	$a_{ij} \in \mathbb{R}$; $\det a \neq 0$	$\alpha := \{\alpha_{ij}\}$	n^2			
$\mathscr{SL}(n, \mathbb{C})$	Special (unimodular) linear group	$a_{ij} \in \mathbb{C}$; $\det a = +1$	$\alpha_{ij} \in \mathbb{C}$; $\mathrm{Tr}\{\alpha\} = 0$	$2(n^2 - 1)$			
$\mathscr{SL}(n, \mathbb{R})$	Special (unimodular) linear group	$a_{ij} \in \mathbb{R}$; $\det a = +1$	$\alpha_{ij} \in \mathbb{R}$; $\mathrm{Tr}\{\alpha\} = 0$	$n^2 - 1$	$n - 1$	A_{n-1}	
$\mathscr{U}(n, \mathbb{C})$	Unitary group	$a^+ a = 1$; $\lvert\det a\rvert = 1$	$\alpha + \alpha^+ = 0$; $\alpha_{ii} \in i\mathbb{R}$	n^2	n		
$\mathscr{SU}(n, \mathbb{C})$	Special (unimodular) unitary group	$a^+ a = 1$; $\det a = +1$	As before, but $\mathrm{Tr}\{\alpha\} = 0$	$n^2 - 1$	$n - 1$	A_{n-1}	
$\mathscr{SU}(p, q, \mathbb{C})$	Pseudounitary group	$a^+ g a = g$; $\det a = +1$ $g_{ii} = -1$ for $i = 1 \dots p$, $g_{ik} = 0$ otherwise ; $p + q = n$;	$\alpha^+ g + g\alpha = 0$; $\mathrm{Tr}\{\alpha\} = 0$ $g_{kk} = +1$ for $k = p + 1 \dots p + q = n$ $p \geqslant q$	$n^2 - 1$	$n - 1$	A_{n-1}	
$\mathscr{SU}^*(2n, \mathbb{C})$	All matrices from $\mathscr{SL}(2n, \mathbb{C})$ that commute with the transformations $\{z_1 \dots z_n, z_{n+1} \dots z_{2n}\} \to \{z^*_{n+1} \dots z^*_{2n}, -z_1 \dots -1^*_n\} \in \mathbb{C}_{2n}$	$a^* g a = g$; $\det a = +1$	$a = 1 + i\alpha$; $\alpha^* g - g\alpha = 0$, $g = \begin{bmatrix} 0 & 1_n \\ -1_n & 0 \end{bmatrix}$, $\alpha = \begin{bmatrix} \alpha_1 & \alpha_2 \\ -\alpha^*_2 & \alpha^*_1 \end{bmatrix}$.	$n^2 - 1$	$n - 1$	A_{n-1}	
$\mathscr{O}(n, \mathbb{C})$	Complex orthogonal group	$\tilde{a}a = 1$; $\det a = \pm 1$	$\alpha + \tilde{\alpha} = 0$; $\alpha_{ii} = 0$	$n(n - 1)$			
$\mathscr{SO}(n, \mathbb{C})$	Special complex orthogonal group	$\tilde{a}a = 1$; $\det a = +1$	$\alpha + \tilde{\alpha} = 0$; $\alpha_{ii} = 0$; $\mathrm{Tr}\{\alpha\} = 0$	$n(n - 1)$			
$\mathscr{SO}^*(2n, \mathbb{C})$	All matrices from $\mathscr{SO}(2n, \mathbb{C})$ that leave the skew-Hermitian form $z^*_1 z_{n+1} + \dots + z^*_n z_{2n} - z_1 z^*_{n+1} - \dots - z_n z^*_{2n}$ invariant		g, α as $\mathscr{SU}^*(2n, \mathbb{C})$, but $\alpha_1 = -\alpha^+_1$, $\alpha_2 = \alpha^+_2$	$2n^2$	n	D_n	
$\mathscr{O}(n, \mathbb{R})$	Orthogonal group ($n \geqslant 2$)	$\tilde{a}a = 1$; $\det a = \pm 1$	$\alpha + \tilde{\alpha} = 0$; $\alpha_{ii} = 0$				
$\mathscr{SO}(n, \mathbb{R})$	Special orthogonal group	$\tilde{a}a = 1$; $\det a = +1$	$\alpha + \tilde{\alpha} = 0$; $\alpha_{ii} = 0$; $\mathrm{Tr}\{\alpha\} = 0$	$n(n - 1)/2$ $n(n - 1)/2$	$(n - 1)/2$ $n/2$	$B_{(n-1)/2}$ $D_n/2$	n odd n even
$\mathscr{PO}(n, \mathbb{R})$	Pseudoorthogonal group	$\tilde{a}ga = g$; $\det a = \pm 1$	$g\alpha + \tilde{\alpha}g = 0$	$n(n - 1)/2$			
$\mathscr{SPO}(n, \mathbb{R})$	Special pseudoorthogonal group	$\tilde{a}ga = g$; $\det a = +1$	$g\alpha + \tilde{\alpha}g = 0$; $\mathrm{Tr}\{\alpha\} = 0$	$n(n - 1)/2$			
$\mathscr{SO}(p, q, \mathbb{R})$	As before with $p \geqslant q$	$g_{ii} = +1$ for $i = 1 \dots p$ $g_{ii} = -1$ for $i = 1 \dots p$	$g_{kk} = -1$ for $k = p + 1 \dots p + q = n$ $g_{kk} = +1$ for $k = p + 1 \dots p + q = n$	$n(n - 1)/2$ $n(n - 1)/2$	$(n - 1)/2$ $n/2$	$B_{(n-1)/2}$ $D_{n/2}$	n odd n even
$\mathscr{Sp}(2n, \mathbb{C})$	Complex symplectic groups leave the form $z_1 u_{n+1} - z_{n+1} u_1 \dots z_n u_{2n} - z_{2n} u_n$ invariant	$\tilde{a}ga = g$; $\tilde{g} = -g$ $g = \begin{bmatrix} 0 & 1_n \\ -1_n & 0 \end{bmatrix}$	$g\alpha + \tilde{\alpha}g = 0$; $\mathrm{Tr}\{\alpha\} = 0$ $\alpha = \begin{bmatrix} \alpha_1 & \alpha_2 \\ \alpha_3 & -\tilde{\alpha}_1 \end{bmatrix}$; $\alpha_2 = \tilde{\alpha}_2$, $\alpha_3 = \tilde{\alpha}_3$	$2n(2n + 1)$			
$\mathscr{Sp}(2n)$	Unitary symplectic group	In addition: $a^+ a = 1$	$\alpha^+ + \alpha = 0$	$n(2n + 1)$	n	C_n	
$\mathscr{Sp}(2n, \mathbb{R})$	Real sympletic group	As $\mathscr{Sp}(2n, \mathbb{C})$ but $\mathbb{R}$ instead of $\mathbb{C}$		$n(2n + 1)$	n	C_n	
$\mathscr{Sp}(p, q)$	Pseudosymplectic group	$g = \begin{bmatrix} 0 & 0 & 1_p & 0 \\ 0 & 0 & 0 & 1_q \\ -1_p & 0 & 0 & 0 \\ 0 & -1_q & 0 & 0 \end{bmatrix}$	$\alpha = \begin{bmatrix} \alpha_1 & \alpha_5 & \alpha_3 & \alpha_6 \\ \tilde{\alpha}^*_5 & \alpha_2 & \tilde{\alpha}_6 & \alpha_4 \\ -\alpha^*_3 & \alpha^*_6 & \alpha^*_1 & -\alpha^*_5 \\ \tilde{\alpha}^*_6 & -\alpha^*_4 & -\tilde{\alpha}_5 & \alpha^*_2 \end{bmatrix}$ $\alpha_1, \alpha_3 : p \times p$; $\alpha_2, \alpha_4 : q \times q$; $\alpha_5, \alpha_6 : p \times q$ $\alpha_3 = \tilde{\alpha}_3$; $\alpha_4 = \tilde{\alpha}_4$; $\alpha_1 = -\tilde{\alpha}^*_1$; $\alpha_2 = -\tilde{\alpha}^*_2$	$n(2n + 1)$	n	C_n	

of the kind $x' = ax + t$; their general element $\{a|t\}$ corresponds to the element of a space group.

The Euclidean groups with $a \in \mathcal{O}(n)$ or $\mathcal{SO}(n)$ and the Poincaré group, which is equal to the inhomogeneous Lorentz group, belong to these groups.

E. Stars of $\boldsymbol{k}$ and Symmetry of Special $\boldsymbol{k}$-Vectors

Table E.1 gives the order s of the star of $\boldsymbol{k}$ and the symmetry group $\mathcal{G}_{0\boldsymbol{k}}$ for cubic lattices. Points on symmetry planes $k_x = 0, \ldots; k_x = \pm k_y, \ldots$ in the interior of the BZ have $\mathcal{C}_s$ symmetry ($s = 24$). Points on the surfaces of the BZ have $\mathcal{C}_s$ symmetry, too, but with one exception: Points at the hexagonal surfaces (planes) of the BZ of the fcc lattice only have $\mathcal{C}_1$ symmetry, like the general points in the interior (i.e. "no" symmetry, $s = 48$).

In Table E.1 the $\boldsymbol{k}$-vectors are given in units of π/a and ξ runs from 0 to the maximum value 1 or 1/2.

The order s of the star of $\boldsymbol{k}$ and the symmetry groups $\mathcal{G}_{0\boldsymbol{k}}$ for hexagonal lattices are given in Table E.2. Points on symmetry planes in the interior of the BZ have $\mathcal{C}_s$ symmetry ($s = 12$). The components k_1, k_2 of the $\boldsymbol{k}$-vectors are given in units of π/a and the components k_3 in units of π/c; ξ and ζ run from 0 to the maximal value 4/3, 1, 1/3 or 1.

Table E.1. Stars $\boldsymbol{k}$ for cubic lattices

Position	$\mathcal{G}_{0\boldsymbol{k}}$	s	Coordinates of the $\boldsymbol{k}$-vectors: $\mathcal{O}_h^1$–$\mathcal{O}_h^4$; sc	$\mathcal{O}_h^5$–$\mathcal{O}_h^8$; fcc	$\mathcal{O}_h^9$, $\mathcal{O}_h^{10}$; bcc
Γ	$\mathcal{O}_h$	1	(0, 0, 0)	(0, 0, 0)	(0, 0, 0)
H	$\mathcal{O}_h$	1	–	–	(2, 0, 0)
R	$\mathcal{O}_h$	1	(1, 1, 1)	–	–
P	$\mathcal{T}_d$	2	–	–	(1, 1, 1)
X	$\mathcal{D}_{4h}$	3	(1, 0, 0)	(2, 0, 0)	–
M	$\mathcal{D}_{4h}$	3	(1, 1, 0)	–	–
L	$\mathcal{D}_{3d}$	4	–	(1, 1, 1)	–
Δ	$\mathcal{C}_{4v}$	6	$(\xi, 0, 0)$	$(\xi, 0, 0)$	$(\xi, 0, 0)$
T	$\mathcal{C}_{4v}$	6	$(1, 1, \xi)$	–	–
N	$\mathcal{D}_{2h}$	6	–	–	(1, 1, 0)
W	$\mathcal{D}_{2d}$	6	–	(2, 1, 0)	–
Λ	$\mathcal{C}_{3v}$	8	(ξ, ξ, ξ)	(ξ, ξ, ξ)	(ξ, ξ, ξ)
F	$\mathcal{C}_{3v}$	8	–	–	$(2 - \xi, \xi, \xi)$
Σ	$\mathcal{C}_{2v}$	12	$(\xi, \xi, 0)$	$(\xi, \xi, 0)$	$(\xi, \xi, 0)$
K	$\mathcal{C}_{2v}$	12	–	(3/2, 3/2, 0)	–
S	$\mathcal{C}_{2v}$	12	$(1, \xi, \xi)$	$(2, \xi, \xi)$	–
U	$\mathcal{C}_{2v}$	12	–	(2, 1/2, 1/2)	–
Z	$\mathcal{C}_{2v}$	12	$(1, \xi, 0)$	$(2, \xi, 0)$	–
D	$\mathcal{C}_{2v}$	12	–	–	$(1, 1, \xi)$
G	$\mathcal{C}_{2v}$	12	–	–	$(2 - \xi, \xi, 0)$
Q	$\mathcal{C}_2$	24	–	$(2 - \xi, 1, \xi)$	–

Table E.2. Stars of $\boldsymbol{k}$ for hexagonal lattices

Position	$\mathscr{G}_{0k}$	s	Coordinates of the $\boldsymbol{k}$-vectors $\mathscr{D}^1_{6h} - \mathscr{D}^4_{6h}$
Γ	$\mathscr{D}_{6h}$	1	$(0,0,0)$
A	$\mathscr{D}_{6h}$	1	$(0,0,1)$
K	$\mathscr{D}_{3h}$	2	$(4/3,0,0)$
H	$\mathscr{D}_{3h}$	2	$(4/3,0,1)$
Δ	$\mathscr{C}_{6v}$	2	$(0,0,\zeta)$
M	$\mathscr{D}_{2h}$	3	$(1,1/\sqrt{3},0)$
L	$\mathscr{D}_{2h}$	3	$(1,1/\sqrt{3},1)$
P	$\mathscr{C}_{3v}$	4	$(4/3,0,\zeta)$
Σ	$\mathscr{C}_{2v}$	6	$(\xi,\xi/\sqrt{3},0)$
R	$\mathscr{C}_{2v}$	6	$(\xi,\xi/\sqrt{3},1)$
U	$\mathscr{C}_{2v}$	6	$(1,1/\sqrt{3},\zeta)$
T	$\mathscr{C}_{2v}$	6	$(\xi,0,0)$
T'	$\mathscr{C}_{2v}$	6	$(4/3-\xi,\sqrt{3}\xi,0)$
S	$\mathscr{C}_{2v}$	6	$(\xi,0,1)$
S'	$\mathscr{C}_{2v}$	6	$(4/3-\xi,\sqrt{3}\xi,1)$

F. Noether's Theorem

Since in some places we make use of Emmy Noether's theorem we shall briefly sketch its content, as far as it is used in the text. The starting point is the invariance condition (14.1.6a–c) which we may write (for every i)

$$\begin{aligned} &l(\psi'^{\alpha}(x'),\partial'_{\mu}\psi'^{\alpha}(x')) - l(\psi'^{\alpha}(x),\partial_{\mu}\psi'^{\alpha}(x)) + l(\psi'^{\alpha}(x),\partial_{\mu}\psi'^{\alpha}(x)) \\ &\quad - l(\psi^{\alpha}(x),\partial_{\mu}\psi^{\alpha}(x)) = 0 \ . \end{aligned} \tag{F.1}$$

In infinitesimal transformations

$$\begin{aligned} &\psi'^{\alpha}(x') = \psi'^{\alpha}(x+\delta x) = \psi'^{\alpha}(x) + \delta x^{\mu}\cdot\partial_{\mu}\psi'^{\alpha}(x) \ , \\ &\psi'^{\alpha}(x) = \psi^{\alpha}(x) + \delta\psi^{\alpha}(x) \ . \end{aligned} \tag{F.2}$$

Using (F.2) and $\partial_{\mu}\delta = \delta\partial_{\mu}$, the change of the Lagrangian in (F.1) is given by

$$\partial_{\mu}l(\psi'^{\alpha}(x),\dots)\delta x^{\mu} + \frac{\partial l}{\partial\psi^{\alpha}}\delta\psi^{\alpha} + \frac{\partial l}{\partial\partial_{\mu}\psi^{\alpha}}\partial_{\mu}\delta\psi^{\alpha}(x) = 0 \ . \tag{F.3}$$

Due to the equations of motion (field equations) we have

$$\frac{\partial l}{\partial\psi^{\alpha}} = \partial_{\mu}\frac{\partial l}{\partial\partial_{\mu}\psi^{\alpha}} \ . \tag{F.4}$$

Furthermore, neglecting higher-order terms we may replace $l(\psi'^{\alpha},\dots)$ by $l(\psi^{\alpha},\dots)$ in the first term of (F.3). Then we obtain

$$(\partial_\mu l)\delta x^\mu + \partial_\mu\left(\frac{\partial l}{\partial\partial_\mu\psi^\alpha}\delta\psi^\alpha\right) = 0 \ , \tag{F.5}$$

or, using [see (i) and (ii) below]

$$\delta x^\mu = x'^\mu - x^\mu \ , \qquad \partial_\mu(x'^\mu - x^\mu) = 0 \ , \qquad \delta\psi^\alpha = \psi'^\alpha(x) - \psi^\alpha(x) \ , \tag{F.6}$$

we have instead of (F.5)

$$\partial_\mu\left[l(x'^\mu - x^\mu) + \frac{\partial l}{\partial\partial_\mu\psi^\alpha}(\psi'^\alpha - \psi^\alpha)\right] = 0 \ . \tag{F.7}$$

In this equation all the fields are related to the position x. For practical purposes it is convenient to relate the fields ψ'^α to positions x' and the fields ψ^α to x. Then we have to first order

$$\psi'^\alpha(x') - \psi'^\alpha(x) = \partial_\mu\psi'^\alpha(x'^\mu - x^\mu) + \cdots = \partial_\mu\psi^\alpha(x'^\mu - x^\mu)\dots$$

and finally

$$\partial_\mu\theta^\mu = 0 \qquad \text{with} \tag{F.8a}$$

$$\theta^\mu = -l(x'^\mu - x^\mu) + \{\partial_\nu\psi^\alpha(x'^\nu - x^\nu) - [\psi'^\alpha(x') - \psi^\alpha(x)]\frac{\partial l}{\partial\partial_\mu\psi^\alpha} \ . \tag{F.8b}$$

Equation (F.8a) represents the infinitesimal conservation law. We obtain the quantities that are conserved by integrating (F.8a) over space-like surfaces (Fig. 14.1). That means, for example, that

$$\theta = \int \theta^0 \, d^3x \tag{F.9}$$

where the integration is with $x^0 = \text{const}$, is a conserved quantity.

The condition $\partial_\mu(x'^\mu - x^\mu) = 0$ in (F.6) is satisfied if the transformation is:

(i) An infinitesimal Lorentz transformation, since then

$$a^\nu{}_\mu = g^\nu{}_\mu + \omega^\nu{}_\mu \qquad \text{with} \qquad \omega^\nu{}_\mu = -\omega_\mu{}^\nu \qquad \text{and} \qquad \omega^\nu{}_\nu = 0$$

and $\bar{\mathscr{L}}z^+(4,\mathbb{R}) \cong \mathscr{SO}(3,1)$ and for this group we have $\mathrm{Tr}\{\omega\} = 0$ according to Appendix D. Therefore

$$\partial_\mu(x'^\mu - x^\mu) = \partial_\mu(\omega^\mu{}_\nu x^\nu) = \omega^\mu{}_\mu = 0 \ .$$

(ii) A gauge transformation with $x' = x$.

In these two cases we have laws of conservation. In the following we are only interested in gauge transformations of the first kind, for which we have, according to (14.1.7, 8),

$$\psi'^i(x) = P_\beta \psi^i(x) \ , \qquad P_{\beta},^i{}_j = \delta^i{}_j + i \sum_l J_{l,}{}^i{}_j \beta_l \ , \tag{F.10}$$

where the field components ψ^i are necessarily complex. In (F.10) β_l are the gauge parameters and i and j are indices for the internal degrees of freedom of the fields, see (14.2.7), which have to be completed by vector or spinor indices if the fields are not scalar fields; $J_{l,}{}^i{}_j$ are the infinitesimal generators of the gauge transformation. Using (F.10) we obtain from (F.8b)

$$\theta^\mu = -i \sum_l \sum_{ij} J_{l,}{}^i{}_j \psi^j \frac{\partial l}{\partial \partial_\mu \psi^i} \beta_l + \text{c.c.} \tag{F.11}$$

Since the β_l are independent gauge parameters, we have conservation laws for every internal component of the current

$$j_l^\mu = -i \sum_{ij} J_{l,}{}^i{}_j \psi^j(x) \frac{\partial l}{\partial \partial_\mu \psi^i(x)} + \text{c.c.} \ , \qquad \partial_\mu j_l^\mu = 0 \ . \tag{F.12}$$

This expression is Hermitian; real fields do not carry any charge (Sect. 14.2.1). The conserved quantity "charge l" then is the integral over the 0-component

$$Q_l = \int j_l^0 \, d^3x \ . \tag{F.13}$$

Using the Lagrangian density of the Schrödinger field, (F.12 and 13) lead to the well-known current and charge density of this field.

G. Space-Time Symmetry

G.1 Canonical Transformations and Algebra

In Sect. 13.1.1 as well as in Exercise 13.1, we mentioned that the generators of a Lie algebra (symmetry group $\mathcal{G}$ of a physical system) are conserved quantities. From this follows the possibility of defining physical quantities by symmetry transformations (Exercise 13.1). Here we shall show in which way the space-time-invariance group of a system determines the dynamics of this system (for example, the equations of motion).

Canonical transformations do not change the Hamilton equations of motion:

$$\dot{q}_i = \frac{\partial H}{\partial p_i} \ , \qquad \dot{p}_i = -\frac{\partial H}{\partial q_i} \ , \qquad H = H(p_i, q_i, t) \ , \qquad i = 1, \ldots, 3N \ . \tag{G.1}$$

In particular, this is valid for infinitesimal canonical transformations with

[G.1–3]

$$q_i' - q_i := \delta q_i = \sum_k \frac{\partial J_k}{\partial p_i}\alpha_k \ , \qquad k = 1, \dots, m \ , \tag{G.2a}$$

$$p_i' - p_i := \delta p_i = -\sum_k \frac{\partial J_k}{\partial q_i}\alpha_k \ , \tag{G.2b}$$

$$H' - H := \delta H = \sum_k \frac{\partial J_k}{\partial t}\alpha_k \ . \tag{G.2c}$$

The $J_k(p_i, q_i, t)$ are the m infinitesimal generators of the m-parameter transformation group. The change of a function $f(p_i, q_i)$ under a transformation (G.2) is given by

$$\delta f = \sum_{i,k}\left(\frac{\partial f}{\partial p_i}\frac{\partial p_i}{\partial \alpha_k} + \frac{\partial f}{\partial q_i}\frac{\partial q_i}{\partial \alpha_k}\right)\alpha_k = \sum_k \{J_k, f\}\alpha_k \tag{G.3}$$

using (G.2a,b). The Poisson bracket is

$$\{A, B\} := \sum_i \left(\frac{\partial A}{\partial p_i}\frac{\partial B}{\partial q_i} - \frac{\partial A}{\partial q_i}\frac{\partial B}{\partial p_i}\right) \ . \tag{G.4}$$

For $f = H$ it follows from (G.2c and 3) that

$$\{J_k, H\} = \frac{\partial J_k}{\partial t} \ . \tag{G.5}$$

The total change of a function $f(p_i, q_i, t)$ in time is

$$\frac{df}{dt} = \sum_i \left(\frac{\partial f}{\partial p_i}\frac{\partial p_i}{\partial t} + \frac{\partial f}{\partial q_i}\frac{\partial q_i}{\partial t}\right) + \frac{\partial f}{\partial t} \ . \tag{G.6}$$

or, using (G.1 and 4),

$$\frac{df}{dt} = \{H, f\} + \frac{\partial f}{\partial t} \ . \tag{G.7}$$

In the special case $f = J_k$ it follows with (G.5, 7) that

$$\frac{dJ_k}{dt} = \{H, J_k\} + \frac{\partial J_k}{\partial t} = 0 \ . \tag{G.8}$$

The m infinitesimal generators of a group of canonical transformations are constants of motion, i.e. conserved quantities.

As in (11.1.1) and Exercise 11.2, we obtain from the group multiplication the Lie algebra of the Poisson brackets:

$$\{J_i, J_j\} = \sum_k \hat{c}_{ij}^k J_k \ . \tag{G.9}$$

Here $\hat{c}_{ij}^k$ are the structure constants. Now we consider certain special infinitesimal transformations, where we divide the index i into ${}^{\nu}_{i}$: $\nu = 1, \dots, N$ (particles), $i = 1$, 2, 3 (coordinates).

(i) Translation in time:

$$J = H \ , \qquad \alpha = \tau \ , \qquad \partial H/\partial t = 0, \tag{G.10}$$

$$\delta q_i^{(\nu)} = \frac{\partial H}{\partial p_i^{(\nu)}}\tau \ , \qquad \delta p_i^{(\nu)} = -\frac{\partial H}{\partial q_i^{(\nu)}}\tau \ , \qquad \delta H = 0 \ .$$

The generator is the Hamiltonian itself.

(ii) Translation in space:

$$J_k = \sum_\nu p_k^\nu = P_k \ , \qquad \alpha_k = t_k \ , \tag{G.11}$$

$$\delta q_i^{(\nu)} = \sum_k \delta_{ik} t_k = t_i \qquad \text{for all } \nu \ ;$$

$$\delta p_i^{(\nu)} = 0 \rightarrow \delta \boldsymbol{P} = 0 \ ; \qquad \delta H = 0 \ .$$

The generator is the total momentum.

(iii) Rotation of space:

$$J_k = \sum_\nu (q^{(\nu)} \times p^{(\nu)})_k = L_k \ , \qquad \alpha_k = \omega_k \ , \tag{G.12}$$

$$\delta q_i^{(\nu)} = \sum_{jk} \varepsilon_{ijk}\omega_j q_k^{(\nu)} \qquad \text{for all } \nu \ ,$$

$$\delta p_i^{(\nu)} = \sum_{jk} \varepsilon_{ijk}\omega_j p_k^{(\nu)} \qquad \text{for all } \nu \ ,$$

$$\delta \boldsymbol{L} = \boldsymbol{\omega} \times \boldsymbol{L} = 0 \ , \qquad \text{if } \boldsymbol{\omega} \| \boldsymbol{L} \ ; \qquad \delta H = 0 \ .$$

The generator is the total angular momentum.

(iv) Motion of the center of mass:

$$J_k = \sum_\nu (t p_k^{(\nu)} - m^{(\nu)} q_k^{(\nu)}) \ , \qquad \alpha_k = v_k \ , \tag{G.13}$$

$$\left.\begin{aligned} \delta q_i^{(\nu)} &= v_i t \\ \delta p_i^{(\nu)} &= m^{(\nu)} v_i \end{aligned}\right\} \qquad \text{for all } \nu \ ,$$

$$\delta\left(\boldsymbol{P}t - \sum_\nu m^{(\nu)} \boldsymbol{q}^{(\nu)}\right) = 0 \ , \qquad \delta H = \boldsymbol{v} \cdot \boldsymbol{P}.$$

The generator is the uniform motion of the center of mass $\boldsymbol{G} = \boldsymbol{P}t - M\boldsymbol{R}$ where

$$M = \sum_{\nu} m^{(\nu)} , \qquad \boldsymbol{R} = \frac{1}{M} \sum_{\nu} m^{(\nu)} \boldsymbol{q}^{(\nu)} . \tag{G.14}$$

The transfer to quantum mechanics may be achieved by the prescription [G.4]

$$\{A, B\} \rightarrow \frac{\mathrm{i}}{\hbar} [A, B] . \tag{G.15}$$

From (G.9) we obtain the isomorphic commutator algebra (11.1.15)

$$[J_i, J_j] = \sum_k c_{ij}^k J_k \qquad \text{with} \qquad c_{ij}^k = \frac{\hbar}{\mathrm{i}} \hat{c}_{ij}^k . \tag{G.16}$$

In order to see under which conditions the (nonrelativistic) equations of motion remain invariant with respect to a transformation group $\mathscr{G}$, we start with two observers. The first one observes the system in the state $\psi(t)$, the second one in the state $\psi'(t) = P_a(t)\psi(t)$ with $a \in \mathscr{G}$, $P_a P_a^+ = 1$. If the first observer uses the Schrödinger equation

$$\mathrm{i}\hbar\dot{\psi}(t) = H(t)\psi(t) , \tag{G.17a}$$

the corresponding equation of the second one is

$$\mathrm{i}\hbar\dot{\psi}'(t) = H'(t)\psi'(t) \qquad \text{with} \tag{G.17b}$$

$$H'(t) = P_a H P_a^+ - \mathrm{i}\hbar P_a \dot{P}_a^+ . \tag{G.17c}$$

If the theory is invariant under all the $P_a \in \mathscr{G}$, i.e. if $H = H'$, then $\mathscr{G}$ is called the symmetry group of the system; from (G.17c) it then follows that

$$\frac{\mathrm{i}}{\hbar}[H, P_a] + \frac{\partial}{\partial t} P_a = 0 , \tag{G.18a}$$

or, if P_a does not depend explicitly on time,

$$[H, P_a] = 0 . \tag{G.18b}$$

If $\mathscr{G}$ is an m-parameter Lie group with the parameters α_k and the generators J_k, that is $a = a(\ldots\alpha_k\ldots)$ and

$$P_a \approx 1 + \sum_{k=1}^{m} J_k \alpha_k , \tag{G.19}$$

then, using (G.18a),

$$\frac{\mathrm{i}}{\hbar}[H, J_k] + \frac{\partial J_k}{\partial t} = \frac{dJ_k}{dt} = 0 . \tag{G.18c}$$

The infinitesimal generators of the symmetry group $\mathcal{G}$ of H are conserved quantities; (G.18c) is the quantum mechanical analogue of (G.8). Of course, (G.18c) is also valid if we choose the generators to be Hermitian (Sect. 13.1.1).

G.2 The Galilei Group and Classical Mechanics

The set of transformations (G.10–13) together with the improper reflections (inversion, time reversal) establishes the inhomogeneous Galilei group; ω has to be replaced by the matrix of finite rotations d with $d\tilde{d} = 1$ ($d \to 1 + \omega$). We have

$$\begin{aligned} t' &= d_{00}t + \tau \ , \qquad d_{00} = \pm 1 \ , \\ \boldsymbol{q}' &= d\boldsymbol{q} + \boldsymbol{t} + \boldsymbol{v}t \ . \end{aligned} \tag{G.20}$$

The proper orthochronic GG has $d_{00} = +1$, $\det d = +1$. As in Sect. 3.4.1, we denote an element from the Galilei group by $\{d|\boldsymbol{t}, \tau, \boldsymbol{v}\}$. The multiplication obeys

$$\{d_1|\boldsymbol{t}_1, \tau_1, \boldsymbol{v}_1\} \cdot \{d_2|\boldsymbol{t}_2, \tau_2, \boldsymbol{v}_2\} = \{d|\boldsymbol{t}, \tau, \boldsymbol{v}\} \tag{G.21}$$

with

$$\begin{aligned} &d = d_1 d_2 \ , \qquad \boldsymbol{t} = d_1\boldsymbol{t}_2 + \boldsymbol{v}_1\tau_2 + \boldsymbol{t}_1 \ , \qquad \tau = \tau_1 + \tau_2 \ , \\ &\boldsymbol{v} = d_1\boldsymbol{v}_2 + \boldsymbol{v}_1 \ . \end{aligned}$$

The Poisson brackets can be calculated according to the classical rules and then, using (G.15), we can also give the commutator relations. Most of them are familiar:

$$\{L_i, P_j\} = \sum_k \varepsilon_{ijk} P_k \ , \qquad \{L_i, L_j\} = \sum_k \varepsilon_{ijk} L_k \ , \tag{G.22}$$

$$\{P_i, P_j\} = \{H, H\} = \{H, P_i\} = \{H, L_i\} = 0 \ . \tag{G.23a}$$

These equations also show that H is not explicitly dependent on time:

$$\partial H/\partial t = 0 \ . \tag{G.23b}$$

From (G.13, 14) with (G.4) we have further[1]

$$\begin{aligned} &\{G_i, G_j\} = 0 \ , \qquad \{G_i, P_j\} = M\delta_{ij} \ , \qquad \{G_i, X_j\} = t\delta_{ij} \ , \\ &\{L_i, G_j\} = \sum_k \varepsilon_{ijk} G_k \ , \qquad \boldsymbol{R} = \{X_i\} \ , \end{aligned} \tag{G.24}$$

and finally, since $\boldsymbol{G}$ is a conserved quantity, with (G.8)

[1] Because $\{G_i, P_j\} = M\delta_{ij}$ in (G.24), the generators H, P_i, L_i, G_i of the Poisson algebra constitute only a projective realization of a Lie algebra and not a faithful realization. M is an invariant of the GG.

$$\{H, G_i\} = -\frac{\partial G_i}{\partial t} . \tag{G.25}$$

Using (G.13, 14, 4) we also have

$$\{H, G_i\} = -\sum_\nu m^{(\nu)}\{H, q_i^{(\nu)}\} = -P_i , \qquad \text{as } \{H, \boldsymbol{P}\} = 0 ,$$

$$= -\sum_\nu m^{(\nu)} \frac{\partial H}{\partial p_i^{(\nu)}} = -\sum_\nu p_i^{(\nu)} ,$$

thus

$$\sum_\nu \left(p_i^{(\nu)} - m^{(\nu)} \frac{\partial H}{\partial p_i^{(\nu)}} \right) = 0 . \tag{G.26}$$

This equation can be solved by

$$H = \sum_{\nu=1}^{N} \sum_{i=1}^{3} \frac{1}{2m^{(\nu)}} (p_i^{(\nu)})^2 + W \qquad \text{with} \qquad \sum_\nu m^{(\nu)} \frac{\partial W}{\partial p_i^{(\nu)}} = 0 . \tag{G.27}$$

Thus the invariance properties determine the Hamiltonian to a large extent. However, the interaction W has to satisfy all the conditions following from the invariance properties of H. As an example we assume W to be a sum of two-particle interactions

$$W = \frac{1}{2} \sum_{\mu\nu} \Phi_{\mu\nu}(\boldsymbol{p}^{(\mu)}, \boldsymbol{p}^{(\nu)}; \boldsymbol{q}^{(\mu)}, \boldsymbol{q}^{(\nu)}) . \tag{G.28}$$

From (G.23b), i.e. $\partial W/\partial t = 0$, it follows that

(i) $\Phi_{\mu\nu}$ does not depend explicitly on time.

From (G.23a), i.e. $\{H, \boldsymbol{L}\} = 0$, it follows that

(ii) $\Phi_{\mu\nu}$ is rotationally invariant, i.e. it is scalar, and only depends on scalar products of the $\boldsymbol{q}^{(\nu)}$ or/and of the $\boldsymbol{p}^{(\nu)}$ with each other.

Also from (G.23a), i.e. $\{H, \boldsymbol{P}\} = 0$, it follows that

(iii) $\Phi_{\mu\nu}$ is translationally invariant and thus only depends on the relative distances $|\boldsymbol{q}^{(\mu)} - \boldsymbol{q}^{(\nu)}|$ of the particles; Coulomb and gravitational potentials satisfy these conditions.

Finally, from the condition in (G.27) it follows that

(iv) $\Phi_{\mu\nu}$ may only depend on the absolute value of relative velocities $|\boldsymbol{v}^{(\mu)} - \boldsymbol{v}^{(\nu)}|$, if at all. Apart from the interaction, the theory has the masses $m^{(\nu)}$ as the only physical parameters, where the $m^{(\nu)}$ are defined by the special Galilei group in (G.13, 14). Because of the equivalence in (G.15) or between (G.8) and (G.18c), it is obvious that also the Hamiltonian of the Schrödinger theory is correspondingly determined by the Galilei invariance.

G.3 Lorentz and Poincaré Groups

According to the theory of special relativity, all physical laws have the same form (shape) in all systems of reference moving uniformly with respect to each other; the velocity of light is a constant with respect to all systems of reference. The invariance group (symmetry group) is the inhomogeneous Lorentz group $\mathscr{I}\mathscr{L}z(4)$, which is equal to the Poincaré group $\mathscr{P}(4)$. Space reflections and time reversal will not be considered in the following, i.e. we restrict ourselves to the proper orthochronous groups $\bar{\mathscr{L}}z^+(4,\mathbb{R})$ and $\bar{\mathscr{P}}^+(4)$ (see Appendix D).

The more general groups may be obtained from these by direct products with reflections or time reversal. First, we consider $\bar{\mathscr{L}}z^+$ and then we treat $\bar{\mathscr{P}}^+$ by the method of little groups (Chap. 9). We denote the elements of $\bar{\mathscr{L}}z^+$ by $\{d|0\}$ analogously to Sect. 14.1, and those of $\bar{\mathscr{P}}^+$ by $\{d|t\}$. The scalar product $\sum_\mu x^\mu x_\mu = c^2t^2 - x^2 - y^2 - z^2$ remains invariant under transformations of $\bar{\mathscr{L}}z^+$ (Sect. 14.1):

$$\sum_\mu x'^\mu x'_\mu = \sum_\mu x^\mu x_\mu \ , \qquad x'^\mu = \sum_\nu d^\mu{}_\nu x^\nu \ , \qquad \det d = 1 \ , \qquad d^0{}_0 \geqslant 1 \ , \tag{G.29}$$

i.e. the elements $\{d|0\}$ are defined in the neighbourhood of the identity element [G.5]. The group $\bar{\mathscr{L}}z^+$ has a subgroup which is isomorphic to $\mathscr{S}\mathscr{O}(3)$; its elements have the representation

$$d := d_{cn} := \begin{pmatrix} 1 & 0 \\ 0 & c_n \end{pmatrix} \ , \tag{G.30}$$

where c_n is a rotation matrix about the unit vector $\boldsymbol{n}$ in $\mathbb{R}_3$. A further simple transformation is the special Lorentz transformation (boost)

$$d_{bu} := \begin{pmatrix} \cosh b & -\sinh b & 0 & 0 \\ -\sinh b & \cosh b & 0 & 0 \\ 0 & 0 & 1 & 0 \\ 0 & 0 & 0 & 1 \end{pmatrix} \ , \qquad \begin{aligned} &\beta = \frac{v}{c} = \tanh b \ , \\ &-\infty < b < +\infty \ , \\ &-1 < \beta < +1 \ . \end{aligned} \tag{G.31}$$

The real parameter b (and thus the parameter space) is unbounded, i.e. $\bar{\mathscr{L}}z^+$ *is noncompact*. Analogously a special Lorentz transformation for every direction $\boldsymbol{u}$ in space can be defined. Then

$$\begin{aligned} &ct' = ct\cosh b - (\boldsymbol{x}\cdot\boldsymbol{u})\sinh b \ , \qquad \boldsymbol{x} = (x, y, z) \ , \\ &\boldsymbol{x}' = \boldsymbol{x} + [(\cosh b - 1)(\boldsymbol{x}\cdot\boldsymbol{u}) - ct\sinh b]\boldsymbol{u} \end{aligned} \tag{G.32a}$$

or with $\boldsymbol{u}\|x$

$$(ct', x', y', z') = \left(\frac{ct - \beta x}{\sqrt{1-\beta^2}}, \frac{x - \beta ct}{\sqrt{1-\beta^2}}, y, z\right) \ . \tag{G.32b}$$

Any Lorentz transformation can be represented as

$$d = d_{bu} d_{cn} \tag{G.33}$$

From (G.30–33) it results that $\bar{\mathscr{L}}z^+$ is characterized by six parameters, three of them correspond to a rotation of space and three to a special Lorentz transformation b, $\boldsymbol{u}$. We obtain the six infinitesimal generators from the corresponding infinitesimal limits, where we take into account Exercise 11.1 or Eq. (11.1.7) for the rotation:

$$d \cong 1 + \lambda = 1 + \sum_{i=1}^{3} \alpha_i J_i + \sum_{i=1}^{3} \beta_i K_i \ , \qquad i = 1, 2, 3 \text{ or } x, y, z \tag{G.34a}$$

with

$$J_x = \left(\begin{array}{c|ccc} 0 & 0 & 0 & 0 \\ \hline 0 & 0 & 0 & 0 \\ 0 & 0 & 0 & -1 \\ 0 & 0 & 1 & 0 \end{array}\right) \ , \quad K_x = \left(\begin{array}{c|ccc} 0 & -1 & 0 & 0 \\ \hline -1 & 0 & 0 & 0 \\ 0 & 0 & 0 & 0 \\ 0 & 0 & 0 & 0 \end{array}\right) \ , \quad \text{etc.} \tag{G.34b}$$

$$[J_i, J_j] = -[K_i, K_j] = \sum_k \varepsilon_{ijk} J_k \ , \qquad [J_i, K_j] = \sum_k \varepsilon_{ijk} K_k \ .$$

In order to simplify the commutators or the structure constants of the Lie algebra, we introduce, see (11.1.30) and (13.4.5),

$$\tilde{J}_j = \tfrac{1}{2}(J_j + \mathrm{i}K_j) \ , \qquad \tilde{K}_j = \tfrac{1}{2}(J_j - \mathrm{i}K_j) \tag{G.35a}$$

and thus obtain

$$[\tilde{J}_i, \tilde{J}_j] = \sum_k \varepsilon_{ijk} \tilde{J}_k \ ; \qquad [\tilde{K}_i, \tilde{K}_j] = \sum_k \varepsilon_{ijk} \tilde{K}_k \ ; \qquad [\tilde{J}_i, \tilde{K}_j] = 0 \ . \tag{G.35b}$$

These relations are very similar to those in (11.1.30) and (13.4.5); however, this does not mean that $\bar{\mathscr{L}}z^+$ is isomorphic to $\mathscr{SO}(3) \times \mathscr{SO}(3)$, since according to (G.34), in $\bar{\mathscr{L}}z^+$, J_i and K_i consist of real elements, whereas in $\mathscr{SO}(3) \times \mathscr{SO}(3)$ those quantities which correspond to $\tilde{J}_i$ and $\tilde{K}_i$ are real.

Nevertheless, we can use the labelling of the IRs of $\mathscr{SO}(3) \times \mathscr{SO}(3)$ also for $\bar{\mathscr{L}}z^+$; that is, we may denote the IRs $D^{(j,j')}$ of finite dimensions of $\bar{\mathscr{L}}z^+$ by pairs (jj'), where j and j' take the values 0, 1/2, 1, 3/2, 2, ... The dimension of these REPs is

$$d_{jj'} = (2j + 1)(2j' + 1) \ . \tag{G.36}$$

Those REPs having j and j' either both integer or both half-integer are single valued, those having one integer index and the other half-integer, are double-valued (spinor) representations. These finite-dimensional REPs are nonunitary (Sect. 11.2). This can be seen directly from (G.34b); whereas for unitary REPs [see (11.4.4a)] the generators are anti-Hermitian, the K_i from (G.34b) are Hermitian. For noncompact groups like $\bar{\mathscr{L}}z^+$, only infinite-dimension REPs are unitary (apart from the identity REP).

The REP may also be characterized by the two Casimir operators [Table 11.2, eqs. (11.1.38), (13.4.7)] instead of j, j':

$$C_2 = \sum_i \mathcal{J}_i^2 \to -j(j+1) , \qquad C_2' = \sum_i \mathcal{K}_i^2 \to -j'(j'+1) , \tag{G.37a}$$

which may also be taken as the combinations

$$\begin{aligned} \tilde{C}_2 &= \sum_i (\mathcal{J}_i^2 + \mathcal{K}_i^2) = \tfrac{1}{2}(\boldsymbol{J}^2 - \boldsymbol{K}^2) , \\ \tilde{C}_2' &= \sum_i (\mathcal{J}_i^2 - \mathcal{K}_i^2) = \mathrm{i}\boldsymbol{J}\cdot\boldsymbol{K} . \end{aligned} \tag{G.37b}$$

The reduction of inner direct products of two IRs of $\bar{\mathscr{L}}z^+$ follows the same rules as for $\mathscr{SO}(3)$. This means

$$D^{(j_1 j_1')} \otimes D^{(j_2 j_2')} = \sum_{jj'}{}_{\oplus}\, D^{(jj')} \qquad \text{with} \tag{G.38}$$

$$|j_1 - j_2| \leqslant j \leqslant (j_1 + j_2) , \qquad |j_1' - j_2'| \leqslant j' \leqslant (j_1' - j_2') .$$

Correspondingly, the subduction of the IRs $D^{(jj')}$ of $\bar{\mathscr{L}}z^+$ onto its subgroup $\mathscr{SO}(3)$ is given by

$$D^{(jj',\mathrm{sub})} = \sum_l{}_{\oplus}\, D^{(l)} \qquad \text{with} \qquad |j - j'| \leqslant l \leqslant (j + j') . \tag{G.39}$$

The Poincaré group $\bar{\mathscr{P}}^+$ consists of the elements of $\bar{\mathscr{L}}z^+$ and additionally of the translations $\{e|t\}$ with

$$\{e|t\}x = x + t , \qquad t = (c\tau, t_x, t_y, t_z) . \tag{G.40}$$

A general element of $\bar{\mathscr{P}}^+$ can be written as

$$\{d|t\} = \{e|t\}\{d|0\} . \tag{G.41}$$

Multiplication and inversion of such elements are defined by (3.4.5, 6). The set of (four-dimensional) translations is an Abelian invariant subgroup of $\bar{\mathscr{P}}^+$; the quotient group is $\bar{\mathscr{P}}^+/\mathbb{T} \cong \bar{\mathscr{L}}z^+$. The group $\bar{\mathscr{P}}^+$ can be written as a semidirect product (see Sect. 2.2):

$$\bar{\mathscr{P}}^+ = \mathbb{T} \,\boxed{\mathrm{S}}\, \bar{\mathscr{L}}z^+ . \tag{G.42}$$

This simplifies the representation theory [see (9.2.10) for the symmorphic space groups]. We obtain the infinitesimal generators $P = (P_0, P_x, P_y, P_z)$ of the translations $\{e|t\}$ through the elements $P(e|t)$ of the operator group which are isomorphically assigned. We have

$$P(e|t) \approx 1 - \sum_\mu t_\mu P^\mu = 1 - c\tau P_0 + t_x P_x + t_y P_y + t_z P_z \ . \tag{G.43}$$

As in $\mathbb{R}_3$, $\mathbb{T}$ is a group of outer products of Abelian groups, see (5.3.3) and thus Abelian itself; the IRs $D^{(k)}$ are one dimensional. In the basis

$$\left\{ |k\rangle \, | k = \left(\frac{\omega}{c}, k_x, k_y, k_z\right); \boldsymbol{k} = (k_x, k_y, k_z) \right\}$$

we find [see (5.3.13, 16)]

$$P(e|\boldsymbol{t})|k\rangle = \exp(-\mathrm{i}\sum k^\mu t_\mu)|k\rangle \approx (1 - \mathrm{i}\sum k^\mu t_\mu)|k\rangle \ ,$$

thus

$$D^{(k)}(e|\boldsymbol{t}) = \exp(-\mathrm{i}\sum k^\mu t_\mu) \ , \qquad P^\mu = \mathrm{i}k^\mu \text{ in this basis} \ . \tag{G.44}$$

A function that transforms as $D^{(k)}$ is

$$\langle x|k\rangle := \psi_k(x) = \exp(-\mathrm{i}\sum k^\mu x_\mu) = \exp(-\mathrm{i}\omega t + \mathrm{i}\boldsymbol{k}\cdot\boldsymbol{x}) \ ,$$

with

$$P(e|t)\psi_k(x) = \psi_k(x - t) \ . \tag{G.45}$$

In systems with translational invariance the energy-momentum vector P^μ is a conserved quantity, the Hamiltonian being $H = -\mathrm{i}\hbar c P_0$. Since H commutes with $P(e|\boldsymbol{t})$, $|k\rangle$ is a common eigenfunction and describes a state with given energy $E = \hbar\omega$ and momentum $\boldsymbol{p} = \hbar\boldsymbol{k}$ in such a system. The states of a system which is homogeneous only in time can be characterized by REPs with index $\omega = E/\hbar$. This leads to stationary states $\psi_\omega := \psi_E = \psi_E(\boldsymbol{x})\exp(-\mathrm{i}Et/\hbar)$. The (nonrelativistic) Schrödinger equation follows, for example, if we use the differential REP $P_0 \to -\partial/\partial ct$ and make a special assumption about H, e.g. $H = -\hbar^2\Delta/2m + V(\boldsymbol{x})$.

We obtain the only finite-dimensional IRs of $\bar{\mathscr{P}}^+$ from the finite-dimensional REPs D of the quotient group $\bar{\mathscr{L}}z^+ = \bar{\mathscr{P}}^+/\mathbb{T}$ (because of the homomorphisms $\bar{\mathscr{P}}^+ \to \bar{\mathscr{P}}^+/\mathbb{T} \to D$, [G.6]). Thus, in the following we only consider the infinite-dimensional unitary IRs of $\bar{\mathscr{P}}^+$, which we shall determine with the method of the little groups according to Sect. 9.1. The k^μ defining the IRs we again divide into *stars of* k $(*k)$ (orbits):

$$*k = \left\{ k'^\mu | k'^\mu = \sum_\nu d^\mu{}_\nu k^\nu; d^\mu{}_\nu \in \bar{\mathscr{L}}z^+ \right\} \ . \tag{G.46}$$

Since

$$\begin{aligned} P(e|\boldsymbol{t})(P(d|0)|k\rangle) &= P(d|0)P(e|d^{-1}\boldsymbol{t})|k\rangle = P(d|0)\exp[-\mathrm{i}\sum k^\mu(d^{-1}t)_\mu]|k\rangle \\ &= \exp[-\mathrm{i}\sum (dk)^\mu t_\mu](P(d|0)|k\rangle) \ , \end{aligned} \tag{G.47}$$

as in Sect. 9.1.2 it follows that an IR of $\bar{\mathscr{P}}^+$ containing the basis vector $|k\rangle$ also contains the basis vectors $|k'\rangle$; $k' = dk$ has the same "length" as k. Thus the IRs of $\bar{\mathscr{P}}^+$ can be classified according to their "length". We have to distinguish between four fundamental "regions":

$$\begin{aligned}&1)\sum_\mu k^\mu k_\mu < 0: \quad \text{space-like}\ , \qquad 2)\sum_\mu k^\mu k_\mu > 0: \quad \text{time-like}\\ &3)\sum_\mu k^\mu k_\mu = 0, \quad \omega \neq 0: \quad \text{light cone}\ , \qquad 4)\ k^\mu = (0,0,0,0)\ .\end{aligned} \tag{G.48}$$

2) and 3) may be further distinguished with respect to k_0.

The IRs of $\bar{\mathscr{P}}^+$ can be derived using the method of induction as given in Sect. 9.1.2. The little group $\mathscr{G}_{k_0}$ is a subgroup of $\bar{\mathscr{P}}^+$ and consists of all the elements $\{d_{k_0}|t\}$ for which $d_{k_0}k_0 = k_0$. Thus we choose a k_0 from one of the "regions" in (G.48) and first determine the IRs of the assigned little group $\mathscr{G}_{k_0}$. Corresponding to the four regions there are four fundamental types of REPs; as an example we discuss the time-like type 2).

According to (G.44, 47) the functions $P(d_{k_0}|0)|k_0,j,m_j\rangle$ also belong to the vector k_0 (j, m_j: index distinguishing the IRs of $\mathscr{G}_{k_0}$), i.e.

$$P(d_{k_0}|0)|k_0,j,m_j\rangle = \sum_{m_j'} D^{(j)}_{m_j'm_j}(d_{k_0})|k_0,j,m_j'\rangle\ . \tag{G.49}$$

Now, let $\{d|0\}$ be an element mapping k onto k', that means $dk = k'$, and further let $\{d_k|0\}$ and $\{d_{k'}|0\}$ map k_0 onto k and k', respectively. Finally, we assign a basis function with k via

$$|k,j,m_j\rangle = P(d_k|0)|k_0,j,m_j\rangle \tag{G.50}$$

to any basis function with $|k_0,j,m_j\rangle$. Then, as in (9.1.20), we may write for every element of $\mathscr{G}_{k_0}$

$$d_{k_0} = d_{k'}^{-1}\,d\,d_k \qquad \text{or} \qquad d = d_{k'}\,d_{k_0}\,d_k^{-1} \tag{G.51}$$

and determine the IRs of $\{d|0\} \in \bar{\mathscr{P}}^+$ from

$$\begin{aligned}P(d|0)|k,j,m_j\rangle &= P(d_{k'}|0)P(d_{k_0}|0)P(d_k^{-1}|0)|k,j,m_j\rangle\\ &\underset{(G.50)}{=} P(d_{k'}|0)P(d_{k_0}|0)|k_0,j,m_j\rangle\\ &\underset{(G.49)}{=} P(d_{k'}|0)\sum_{m_j'} D^{(j)}_{m_j'm_j}(d_{k_0})|k_0,j,m_j'\rangle\\ &\underset{(G.50)}{=} \sum_{m_j'} D^{(j)}_{m_j'm_j}(d_{k_0})|k',j,m_j'\rangle = \sum_{m_j'} D^{(j)}_{m_j'm_j}(d_{k_0})|dk,j,m_j'\rangle\ .\end{aligned} \tag{G.52}$$

taking (G.44) into consideration we finally obtain

$$P(d|t)|k,j,m_j\rangle = P(e|t)P(d|0)|k,j,m_j\rangle$$
$$= \exp\left[-\mathrm{i}\sum (dk)^\mu t_\mu\right]\sum_{m_j'} D^{(j)}_{m_j'm_j}(d_{k_0})|dk,j,m_j'\rangle \ . \qquad \text{(G.53)}$$

These equations are analogous to (9.1.21, 22) and determine the IRs of $\bar{\mathscr{P}}^+$. Since the number of k-vectors of equal "length" is infinite, these unitary representations have infinite dimensions. The different types of (G.48) differ only in the IRs of the little group (see [G.6, 7]). For further discussion we choose as a representative vector $k_0 = (\omega/c, 0, 0, 0)$. Any other vector of equal length $\sum k^\mu k_\mu$ of type 2) leads to equivalent representations. The k_0 chosen is left invariant by all elements d_{cn} according to (G.30), i.e. the little point group is isomorphic to $\mathscr{SO}(3)$; its IRs are known:

$$D^{(s)} \quad \text{with} \quad s = 0, 1/2, 1, 3/2, 2, \ldots \ ,$$
$$m_s = s, s-1, \ldots, -s \ , \qquad \text{(G.54)}$$
$$D(d_{k_0}) \rightarrow D^{(k,s)}_{m_s m_s'}(d_{cn}) \ .$$

The "constants" s and ω (from k_0) determine the IRs of $\bar{\mathscr{P}}^+$ completely. The time-like IRs are characterized by $D^{(s,\omega)}$. Thus a special $D^{(s,\omega)}$ has to be assigned to every physical state of a system that is Poincaré invariant; s and ω or k in general are the physical parameters of the states.

G.4 The Physical Quantities

The relevant quantities of invariant systems are the conserved quantities, which can be expressed by the two Casimir operators of $\bar{\mathscr{P}}^+$ characterising the IRs

$$\sum_\mu P^\mu P_\mu = P_0^2 - P_x^2 - P_y^2 - P_z^2 \ , \qquad \boldsymbol{P} = (P_x, P_y, P_z) \ ,$$
$$\sum_\mu W^\mu W_\mu = W_0^2 - W_x^2 - W_y^2 - W_z^2 \ , \qquad \boldsymbol{W} = (W_x, W_y, W_z) \ , \qquad \text{(G.55)}$$

with $W_0 = \boldsymbol{P}\cdot\boldsymbol{J}$, $\boldsymbol{W} = (\boldsymbol{P}\times\boldsymbol{K}) + P_0\boldsymbol{J}$, and $\boldsymbol{J}$ and $\boldsymbol{K}$ from (G.34). The eigenvalues of the Casimir operators are, according to (G.44) and with $k_0 = (\omega/c, 0, 0, 0)$,

$$\sum_\mu P^\mu P_\mu|k_0 s m_s\rangle = \left[-\left(\frac{\omega}{c}\right)^2 + k_x^2 + k_y^2 + k_z^2\right]|k_0 s m_s\rangle = -\frac{\omega^2}{c^2}|k_0 s m_s\rangle \qquad \text{(G.56a)}$$

$$\boldsymbol{P}|k_0 s m_s\rangle = 0 \ , \qquad P_0|k_0 s m_s\rangle = \mathrm{i}\frac{\omega}{c}|k_0 s m_s\rangle \ ,$$

$$\boldsymbol{W}|k_0 s m_s\rangle = \mathrm{i}\frac{\omega}{c}\boldsymbol{J}|k_0 s m_s\rangle \ , \qquad W_0|k_0 s m_s\rangle = 0 \ ,$$

and thus

$$\sum_\mu W^\mu W_\mu |k_0 s m_s\rangle = \frac{\omega^2}{c^2} J^2 |k_0 s m_s\rangle = -\frac{\omega^2}{c^2} s(s+1)|k_0 s m_s\rangle \ . \qquad \text{(G.56b)}$$

Here use has been made of (11.1.38) and the equations in Sect. 11.4.2.

Just like ω or k_0 and s, the eigenvalues of both the Casimir operators are suitable for a classification of the IRs and thus of the physical states. If a physical system possesses Poincaré invariance, its states (e.g. those of a free particle) can be classified according to the IRs $D^{(s,\omega)}$ of $\bar{\mathscr{P}}^+$, which themselves are described by ω and s or the eigenvalues $-\omega^2/c^2$ and $-(\omega^2/c^2)s(s+1)$, respectively, of the Casimir operators (conserved quantities).

To illustrate the physical meaning of ω, we define a mass operator by

$$M^2 = -\hbar^2 \sum_\mu P^\mu P_\mu / c^2 \ . \qquad \text{(G.57)}$$

Then it follows using (G.44, 56a) and $P_0 = \mathrm{i}E/\hbar c$ (end of previous section) that

$$M^2|k\rangle = -(k^2 + P_0^2)\hbar^2/c^2|k\rangle = \left(\frac{E^2}{c^4} - \frac{\hbar^2 k^2}{c^2}\right)|k\rangle = \hbar^2\omega^2/c^4|k\rangle \ ,$$

that is

$$\hbar\omega = Mc^2 = (E^2 - c^2\hbar^2 k^2)^{1/2} \qquad \text{(G.58)}$$

represents the rest mass of the system (particle). Equation (G.58), of course, is nothing else than the relativistic energy-momentum (mass) relation. The meaning of the second Casimir operator or of s can be illustrated best by transforming into the rest sytem of the particle. Then, using $\boldsymbol{J} = -\mathrm{i}\boldsymbol{S}/\hbar$,

$$W_0 = 0 \ , \qquad \boldsymbol{W} = P_0 \boldsymbol{J} = \mathrm{i}\frac{E}{\hbar c}\boldsymbol{J} = \mathrm{i}\frac{Mc}{\hbar}\boldsymbol{J} = Mc\boldsymbol{S}/\hbar^2$$

and with (G.56b)

$$\sum_\mu W_\mu W^\mu |ksm_s\rangle = -\left(\frac{Mc\boldsymbol{S}}{\hbar^2}\right)^2 |ksm_s\rangle = -\frac{\omega^2}{c^2} s(s+1)|ksm_s\rangle \ ,$$

thus

$$\boldsymbol{S}^2|ksm_s\rangle = \hbar^2 s(s+1)|ksm_s\rangle \ . \qquad \text{(G.59)}$$

The vector $\boldsymbol{S}$ (or $\boldsymbol{J}$) represents the internal angular momentum, which is independent of the motion of the particle, i.e. the *spin* of the system (particle).

The indices ω and s of the IRs can be interpreted as the mass and the spin of the system; the eigenvalues of the Casimir operators are related to the properties mass and spin that characterize every Poincaré-invariant state. The basis vectors

$|ksm_s\rangle$ of the IRs $D^{(k,s)}$ belong to the different states of motion of a system (particle) to which mass and spin are assigned. The absolute value and direction of $\boldsymbol{k}$ are arbitrary, the energy is given by (G.58). Similarly, it can be shown that an imaginary mass has to be assigned to the space-like REPs (region 1), whereas the REPs of the light cone (region 3) of $\bar{\mathscr{P}}^+$ have zero rest mass, thus they describe massless particles.

According to space-time symmetries, all particles can be classified with respect to the IRs of $\bar{\mathscr{P}}^+$; this leads to the invariance properties *mass* and *spin* of the particles. In (relativistic) field theories (Dirac, Weyl, electromagnetic, etc., fields) we have to introduce field operators that transform according to the corresponding IRs of $\bar{\mathscr{P}}^+$ of the (quasi) particles in question (Sect. 14.1); using these field operators a Poincaré-invariant Lagrangian density has to be constructed, from which the covariant field equations (Euler-Lagrange equations) can be determined. Finally we obtain the interactions from the gauge invariances of the second kind (Sect. 14.3). By this procedure the physical systems including their interactions are determined by symmetry. Combination of *all* the symmetries, and unification of *all* forces lead to the theory of supersymmetry and superparticles (Appendix J).

H. Goldstone's Theorem

In a number of cases, Goldstone's theorem allows statements to be made about the possible excitations when the symmetry of a system is lowered, i.e. in a spontaneous breaking of the symmetry. In general, a Lagrangian is invariant with respect to a class of symmetry transformations, but it may happen that by a definite choice of one particular state of a degenerate ground state the system which is described by the Lagrangian has lower symmetry.

Now, in the nonrelativistic limit, Goldstone's theorem states that in the case of a spontaneous breaking of symmetry there are long-wavelength ($k \sim 1/\lambda \to 0$) excitations with vanishing frequency $\omega(k \to 0) \to 0$ and the ground state is degenerate. An example is a ferromagnet with spin-wave excitations. In a certain region of size $d \ll \lambda$ all the spins are nearly parallel to a certain direction. If the interacting forces have short range, the excitation of a spin wave $k \to 0$ needs very little energy ($\omega \to 0$). Other examples are crystals (breaking of the translation symmetry $\to$ phonons) and a Bose gas (breaking of the phase symmetry $\phi \to \exp(i\alpha)\phi$ and of Galilei invariance $\to$ phonons). But, if there are long-range forces (Coulomb), the condition $\omega \to 0$ in a spontaneous breaking of the symmetry is no longer valid (plasmons)! In the relativistic case, too, there are ground state solutions of the field equations having lower symmetry than the Lagrange density. If the ground state (state of minimal energy) is unique (single-valued, singlet), it is invariant with respect to the symmetry group $\mathscr{G}$ of the Lagrangian density. However, if there is a set of degenerate states (multiplet) with minimal energy, then these states no longer transform according to the identity REP of

$\mathcal{G}$, but rather according to a multidimensional representation corresponding to the degeneracy. If one of these multiplet states is arbitrarily fixed as the ground state, then the symmetry is *spontaneously broken.*[2]

Goldstone's theorem in its relativistic form now states: If the Lagrangian density is invariant under a group $\mathcal{G}(\alpha_l)$ such that according to Noether's theorem the current densities $j_l^\mu(x)$ exist and satisfy (F.12), and if for at least one field operator $\phi^i(y)$ and one charge density j_l^0

$$\int d^3x \langle 0|[j_l^0(x), \phi^i(y)]_{x_0=y_0}|0\rangle = -\sum_j J_{l,}{}^i{}_j \langle 0|\phi^j(y)|0\rangle \neq 0 \ , \qquad \text{(H.1)}$$

then there exist massless (and spinless) particles (excited states) with the same quantum numbers as $\phi^i(y)$.

In (H.1) we have rewritten (14.1.10) for charge densities instead of charges $Q_l = \int j_l^0(x)\, d^3x$, since there are field theories having Lagrangian densities which are invariant under a group $\mathcal{G}$, for which, however, there is no unitary operator U according to (14.1.9), i.e. the integral for Q_l diverges. Thus the symmetry is spontaneously broken. Now, if $j_l^0(x)$ is a local operator with

$$[j_l^0(x), \phi^i(y)]_{x_0=y_0} = -\delta(\boldsymbol{x}) \sum_j J_{l,}{}^i{}_j \phi^j(y) \ , \qquad \boldsymbol{x} = \{x^1, x^2, x^3\} \ ,$$

then it follows from (14.1.10, H.1) that

$$\langle 0|[Q_l, \phi^i(y)]_{x_0=y_0}|0\rangle = -\sum_j J_{l,}{}^i{}_j \langle 0|\phi^j(y)|0\rangle \neq 0 \ , \qquad \text{(H.2)}$$

i.e. (H.1). Equation (H.2) states that for at least one charge $Q_l|0\rangle \neq 0$ and that at least one scalar field $\phi^j(y)$ has a nonvanishing vacuum expectation value.

On the other hand, if Q_l exists, its expectation value in the vacuum state has to vanish, i.e.

$$Q_l|0\rangle = 0 \ . \qquad \text{(H.3)}$$

Then the vacuum is invariant under the transformations (14.1.9), i.e.

$$U(\ldots \alpha_l \ldots)|0\rangle = |0\rangle \qquad \text{(H.4)}$$

is also valid. From this it can be seen that (H.1) defines a spontaneous breaking of symmetry. Equations (H.3 and 4) follow, if we assume the vacuum $|0\rangle$ to be invariant under the Lorentz group, i.e. especially under translations $\exp(\mathrm{i}P_\mu t^\mu)$ so that

$$\exp(\mathrm{i}P_\mu t^\mu)|0\rangle = |0\rangle \ . \qquad \text{(H.5)}$$

[2] We may also generate a symmetry breaking by adding a noninvariant (lower-symmetry) term to the Lagrangian density (in the previous example by introducing a magnetic field); however, the procedure of Goldstone is another way of looking at things.

Since furthermore $[Q_l, P_\mu] = 0$, $Q_l|0\rangle$ is also translationally invariant. Thus we have

$$\langle 0|j_l^0(x)Q_l|0\rangle = \langle 0|j_l^0(x-t)Q_l|0\rangle \ ,$$

independent of x. The norm of $Q_l|0\rangle$ is

$$\langle 0|Q_l Q_l|0\rangle = \int \langle 0|j_l^0(x)Q_l|0\rangle\, d^3x \to \infty \ , \qquad \text{(H.6)}$$

if (H.3) is not satisfied.

In the following we roughly sketch the proof of Goldstone's theorem in order to indicate some of the assumptions. Because of the translational invariance supposed, (H.1) is independent of $x_0 = y_0$. Furthermore, in (H.1) we replace $\phi^i(y)$ by $\phi(0)$ and then consider the four-dimensional Fourier transform

$$F_l^\mu(k) = \int d^4x \langle 0|[\phi(0), j_l^\mu(\boldsymbol{x}, 0)]|0\rangle e^{ikx} \ , \qquad \text{(H.7)}$$

in which we insert a complete set of energy-momentum eigenstates $|p, E\rangle = |n\rangle$:

$$\begin{aligned} F_l^\mu(k) &= \sum_n \int d^4x e^{ikx}\{\langle 0|\phi|n\rangle\langle n|j_l^\mu(x)|0\rangle - \langle 0|j_l^\mu(x)|n\rangle\langle n|\phi|0\rangle\} \\ &= \frac{1}{(2\pi)^4}\sum_n \{\delta^4(p_n - k)\langle 0|\phi|n\rangle\langle n|j_l^\mu(0)|0\rangle \\ &\quad - \delta^4(p_n + k)\langle 0|j_l^\mu(0)|n\rangle\langle n|\phi|0\rangle\} \ , \end{aligned} \qquad \text{(H.8)}$$

using the translational invariance. As a further additional assumption the so-called "manifested" covariance

$$\langle 0|j_l^\mu(0)|n\rangle = p_n^\mu a_l(p^2) \ , \qquad \text{(H.9)}$$

enters the problem, which is not necessarily satisfied (loophole for the Higgs mechanism). Because ϕ is a Lorentz scalar we have

$$F_l^\mu(k) = k^\mu \rho_l^{(1)}(k^2) + \Theta(k_0)k^\mu \rho_l^{(2)}(k^2) \ , \qquad \text{(H.10)}$$

$$\Theta(k_0) = \begin{cases} 1 & k_0 \geqslant 0 \ , \\ -1 & k_0 < 0 \ , \end{cases}$$

where $\rho_l^{(1,2)}$ is arbitrary here. However, the current in (H.7, 8) has to satisfy the equation of continuity (conservation law), (F.8a, 12). An analysis of (H.10) shows that this is only possible if [H.1]

$$\rho_l^{(\nu)}(k^2) = c_\nu \delta(k^2) \ , \qquad \nu = 1, 2 \ , \qquad \text{(H.11)}$$

where c_1 and c_2 are not fixed primarily. But, using (H.1) it can be shown that c_1 is different from zero; and then (H.8) implies that there exists at least one state with $p_n^2 = 0$: this corresponds to a massless particle. And this is just a statement of Goldstone's theorem.

The key point is the statement contained in (H.9). If it is not satisfied, the statement of Goldstone's theorem above is not valid either. That provides the "loophole" for deviations from this theorem in certain gauge theories (see Sect. 14.4.3 on the Higgs mechanism). The spontaneous breaking of the symmetry of a local gauge theory enters by this "loophole" and may lead to particles with spin one and nonvanishing mass.

I. Remarks on 5-fold Symmetry

Some of the most interesting substances discovered and discussed in recent years are the fullerenes. The prototype is the C_{60} molecule, which has $\mathcal{Y}_h$-point group symmetry. It consists of 60 C atoms forming a "soccer ball" structure. Its surface is built up of 12 pentagons and 20 hexagons in such a way that every pentagon is surrounded by 5 hexagons, but every hexagon by 3 pentagons as well as 3 hexagons (Fig. I.1). The vertices are occupied by the C atoms. There are double bonds on the edges between two hexagons with a length $a_{\parallel} = 1.39$ Å and single bonds on the edges between a pentagon and a hexagon with a length $a_{|} = 1.44$ Å; the maximal radius of a C_{60} molecule is about 3.55 Å. There are 6 five-fold axes through the middles of the pentagons, 10 threefold axes through the middles of the hexagons and 15 two-fold axes through the edges of neighboring hexagons. The details may be seen from Table I.1. This table also shows the multiplicities

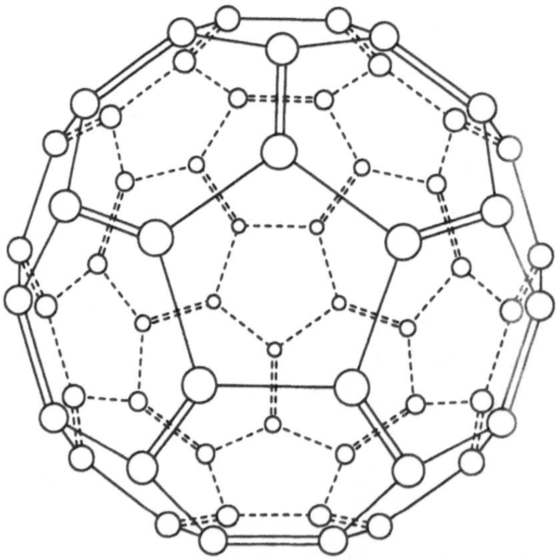

Fig. I.1. Sketch of a C_{60} fullerene with single and double bonds

Table I.1. Character table for the icosahedral group $\mathscr{Y}_h$ together with the characters of the total REPs Γ_t of molecules in the icosahedral, dodecahedral and C_{60} fullerene from. Furthermore, the multiplicities m_α of the IRs are given (*Table 5.1, last column)

$\mathscr{Y}_h$	e	$12c_5$	$12c_5^2$	$20c_3$	$15c_2$	i	$12s_{10}$	$12s_{10}^2$	$20s_6$	$15\sigma_h$	m_α Ikos	m_α Dod	m_α Full	
A_g	1	1	1	1	1	1	1	1	1	1	1	1	2	r^2
F_{1g}	3	η_+	$-\eta_-$	0	-1	3	η_+	$-\eta_-$	0	-1	1	1	4	$\omega_x, \omega_y, \omega_z$
F_{2g}	3	$-\eta_-$	η_+	0	-1	3	$-\eta_-$	η_+	0	-1	0	1	4	
G_g	4	-1	-1	1	0	4	-1	-1	1	0	1	2	6	
H_g	5	0	0	-1	1	5	0	0	-1	1	2	3	8	5 SphH 2.O.*
A_u	1	1	1	1	1	-1	-1	-1	-1	-1	0	0	1	
F_{1u}	3	η_+	$-\eta_-$	0	-1	-3	$-\eta_+$	η_-	0	1	2	2	5	x, y, z
F_{2u}	3	$-\eta_-$	η_+	0	-1	-3	η_-	$-\eta_+$	0	1	1	2	5	
G_u	4	-1	-1	1	0	-4	1	1	-1	0	1	2	6	
H_u	5	0	0	-1	1	-5	0	0	1	-1	1	2	7	
Γ_t Ikos	36	$2\eta_+$	$-2\eta_-$	0	0	0	0	0	0	4	$\eta_+ = \frac{1}{2}(\sqrt{5}+1)$			
Dod	60	0	0	0	0	0	0	0	0	4	$\eta_- = \frac{1}{2}(\sqrt{5}-1)$			
C_{60}Full	180	0	0	0	0	0	0	0	0	4				

of the IRs contained in the total REP of the C_{60} fullerene according to (4.2.31). Assuming any model for the force constants between the C atoms, we can determine the eigenvibrations of the C_{60} according to Sect. 7.1. The corresponding infrared and Raman spectra have been investigated.

The electronic (binding) states can be discussed, on the other hand, using the methods of Sect. 7.2. The wave functions of the valence electrons have to be composed of sp^2 hybrid orbitals (3 strong σ bonds in the surface and a weaker π bond perpendicular to the surface).

The C_{60} molecules are also able to form solid crystals in form of the fcc packing (lattice constant about 14 Å). "Higher" fullerenes can be obtained by cutting the C_{60} fullerene parallel to two pentagons in such a way that two equal capsules result and then inserting rings of hexagons between these capsules. With just one ring inserted, the C_{70} molecule looks more or less like an American football. It has $\mathscr{D}_{5h}$ symmetry; its characters and symmetry decomposition can be obtained from Table I.2. [I.1–3]. For further details we refer the reader to the literature, which is now available to a large extent.

Other examples of 5-fold (or even $\geqslant$ 7-fold) symmetries are the quasicrystals. These possess local symmetries of $\mathscr{C}_5$-type (planar systems) or of icosahedral or dodecahedral symmetry, respectively, but they do not have long-range periodic translation symmetry. According to (3.4.1, 2) a strong periodic translation symmetry is not compatible with a 5-fold or an n-fold symmetry ($n \geqslant 7$). The quasicrystals are a special kind of what are called incommensurate crystals or phases.

For an understanding we have to introduce some notions, mainly in connection with reciprocal space (Bragg peaks of lattices):

(i) A structure for which the scattering amplitude (intensity $I(\boldsymbol{K})$) is given by a set of discrete Bragg peaks is called a *translationally ordered structure* (TOS).

(ii) If these peaks can be described by reciprocal vectors $\{\boldsymbol{K}_i\}$ and if these vectors can be represented by integer linear combinations of a finite set of vectors $\{\boldsymbol{g}_n; n = 1, \ldots, D\}$, the minimal set then is called the *basis*; $\{\boldsymbol{g}_n\}$ spans the

Table I.2. Character table for the group $\mathscr{D}_{5h}$, the total REP of the C_{70} fullerene and multiplicities m_α

$\mathscr{D}_{5h}$	e	$2c_5$	$2c_5^{\,2}$	$5c_2$	σ_h	$2s_5$	$2s_5^{\,2}$	$5\sigma_v$	C_{70}	
A_{1g}	1	1	1	1	1	1	1	1	12	$x^2 + y^2; z^2$
A_{2g}	1	1	1	−1	1	1	1	−1	10	ω_z
E_{2g}	2	$-\eta_+$	η_-	0	2	$-\eta_+$	η_-	0	22	$x^2 - y^2, 2xy$
E_{1g}	2	η_-	$-\eta_+$	0	2	η_-	$-\eta_+$	0	22	$x, y; xz, yz$
A_{1u}	1	1	1	1	−1	−1	−1	−1	9	
A_{2u}	1	1	1	−1	−1	−1	−1	1	11	z
E_{2u}	2	$-\eta_+$	η_-	0	−2	η_+	$-\eta_-$	0	20	
E_{1u}	2	η_-	$-\eta_+$	0	−2	$-\eta_-$	η_+	0	20	ω_x, ω_y
$\Gamma_t\, C_{70}$	210	0	0	0	10	0	0	4		

reciprocal lattice and the $\{K_i\}$ form the reciprocal lattice. D is the number of elements in the basis.

(iii) The group of point operations, which leaves $I(K)$ invariant, describes the *orientational symmetry* of the TOS.

(iv) Then an (ordinary) *crystal* in d dimensions is a TOS with $D = d$ and a *quasiperiodic structure* in d dimensions is a TOS with a finite basis and $D > d$.

Incommensurate crystals have (long-range) translational order and (long-range) orientational order which corresponds to one of the crystallographic point groups compatible with translational symmetry, see (3.4.1, 2). However, the translational symmetry is quasiperiodic rather then periodic, that means there are two or even more incommensurate periodicities. Sometimes these crystals can be looked upon as composed of two interpenetrating incommensurate lattices, sometimes as crystals with an incommensurate modulation of the atom sites.

Quasicrystals have a (long-range) quasiperiodic translational order, but a (long-range) orientational order with a non-crystallographic symmetry ($n = 5, 7, \ldots$), which is the essential property (and forces the crystal to be quasiperiodic).

The properties of quasiperiodic structures can be understood best in the reciprocal space, because the investigations in this space (e.g., with X-rays) show their properties in a rather simple way. Quasiperiodic structures are characterized in the reciprocal space by densely lying, but discrete Bragg peaks which can be described by a finite set of indices. They can be looked upon (space dimension d, minimal set of basis vectors D) as the intersection of a D-dimensional periodic structure with the d-dimensional space, e.g., the electronic density in a plane different from a *lattice plane* ($d = 2$, $D = 3$) is quasiperiodic. Quasicrystals then have crystallographically not-allowed orientational symmetry (e.g., 5-fold patterns). General quasicrystals can be obtained by decoration of elementary cells of a quasilattice.

For a short discussion we start with a special diffraction pattern (Fig. I.2). A structure is called *incommensurate*, if every point (spot) of the pattern is of the form

$$K = \sum_{i=1}^{n} z_i a_i^* , \qquad z_i \in \mathbb{Z} , \qquad n > 3 \tag{I.1}$$

in such a way that

$$m_i \in \mathbb{Z} \qquad \text{with} \qquad \sum_i m_i a_i^* = 0 \tag{I.2}$$

do not exist.

For $n = 3$ the structure is *periodic*, for $n > 3$ it is *quasiperiodic*. This means: if a periodic function $f(x_1, \ldots x_n)$ is periodic in every argument x_i, then the function $g = f(\alpha_1 x_1, \ldots \alpha_n x_n)$ is quasiperiodic, if the $\alpha_1, \ldots \alpha_n$ are rationally independent.

Incommensurate phases exhibit a lattice of *main spots* which is forbidden in

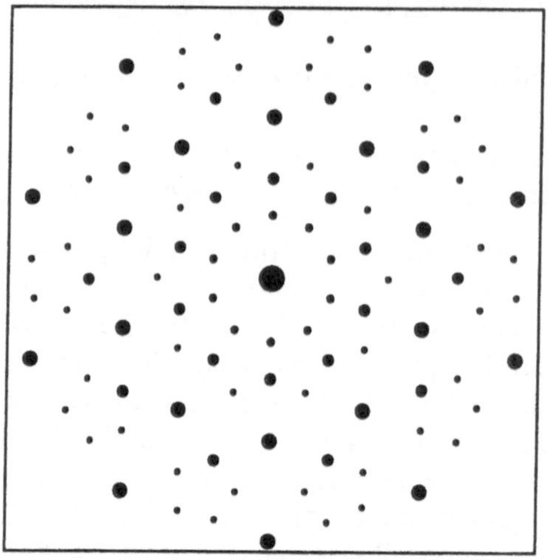

Fig. I.2. Schematic X-ray scattering pattern for an $Al_{0.86}Mn_{0.14}$ crystal

systems with crystallographically not-allowed point group symmetry (quasicrystals). In systems with main and satellite spots these are described by three basis vectors $\boldsymbol{a}_i^*$, $i = 1, 2, 3$ and additional basis vectors for the satellites:

$$\boldsymbol{K} = \sum_{i=1}^{3} z_i \boldsymbol{a}_i^* + \sum_{j=4}^{D} z_j \boldsymbol{a}_j^* \ . \tag{I.3}$$

This lattice M^* now can be described as a projection of a D-dimensional lattice onto the three-dimensional space. The basis and the metric of this space can be constructed starting with the point symmetry group $\mathscr{G}$ of the scattering pattern. An element $r \in \mathscr{G}$ transforms the pattern in such a way that

main spots → main spots

satellite spots → satellite spots ,

i.e.

$$\begin{aligned} r^{-1}\boldsymbol{a}_i^* &= \sum_{k=1}^{3} \Gamma_{E,ik}(r)\boldsymbol{a}_k^* \ , \qquad && i = 1, 2, 3 \ , \\ r^{-1}\boldsymbol{a}_j^* &= \sum_{k=1}^{3} \Gamma_{M,jk}(r)\boldsymbol{a}_k^* + \sum_{l=4}^{D} \Gamma_{I,jl}(r)\boldsymbol{a}_l^* \ , \qquad && j = 4 \dots D \ . \end{aligned} \tag{I.4}$$

The elements $r \in \mathscr{G}$ can be represented by a D-dimensional integral (reducible) REP $\Gamma(\mathscr{G})$

$$r \to \Gamma(r) \qquad \text{with} \qquad \Gamma(r) = \begin{pmatrix} \Gamma_E(r) & 0 \\ \Gamma_M(r) & \Gamma_I(r) \end{pmatrix} \ . \tag{I.5}$$

$\Gamma(r)$ is a reducible REP of $\mathscr{G}$ (4.2.16). According to Sect. 4.2.3 there exists a basis in D dimensions, so that the REP is orthogonal and the direct sum of two REPs. The basis vectors in the ordinary space V_E may be $\boldsymbol{a}_i^*$, $i = 1, 2, 3$ and those in V_I (orthogonal complement to V_E) may be $\boldsymbol{b}_4^* \dots \boldsymbol{b}_D^*$. Then the basis vectors of the D-dimensional space (spanning the lattice Σ^*) are

$$\begin{aligned} \boldsymbol{d}_i^* &= (\boldsymbol{a}_i^*, 0) \ , & i &= 1, 2, 3 \ , \\ \boldsymbol{d}_j^* &= (\boldsymbol{a}_j^*, \boldsymbol{b}_j^*) \ , & j &= 4, \dots D \ . \end{aligned} \tag{I.6}$$

Obviously the set M^* defined by (I.3) is just the projection of the lattice Σ^*. The basis of Σ^* has to be constructed in such a way that the pairs (r, r_I) of orthogonal transformations which are defined by the decomposition into the direct sum, are represented on this basis by the matrices $\Gamma(r)$.

In the space Σ, which is dual to Σ^*, the corresponding basis is

$$\begin{aligned} \boldsymbol{d}_i &= (\boldsymbol{a}_i, -\Delta \boldsymbol{a}_i) \ , & i &= 1, 2, 3 \ , \\ \boldsymbol{d}_j &= (0, \boldsymbol{b}_j) \ , & j &= 4, \dots D \ , \end{aligned} \tag{I.7}$$

Here $\{\boldsymbol{b}_j\}$ is the basis dual to $\{\boldsymbol{b}_j^*\}$ and $\Delta \boldsymbol{a}_i$ is determined by the requirement

$$\boldsymbol{d}_i \cdot \boldsymbol{d}_j^* = \delta_{ij} \ . \tag{I.8}$$

Finally, starting form a quasiperiodic function with the Fourier spectrum corresponding to the pattern M^* a periodic function in Σ (dual to Σ^*) can be constructed by

$$f_s(\boldsymbol{r}, \boldsymbol{r}_I) = \sum \hat{f}(\boldsymbol{K}) \mathrm{e}^{\mathrm{i}(\boldsymbol{K}\boldsymbol{r} + \boldsymbol{K}_I \boldsymbol{r}_I)} \tag{I.9}$$

with

$\hat{f}(\boldsymbol{K})$: Fourier component of f with vector $\boldsymbol{K}$,

$\{\boldsymbol{K}, \boldsymbol{K}_I\}$: definite vector of Σ^*, which is projected on $\boldsymbol{K}$ in M^* .

We will first explain the above procedure for quasicrystals with a two-dimensional example. We assume a diffraction pattern with 5-fold symmetry, in which all the spots can be described by linear combinations of five vectors with integer coefficients (I.1) pointing to the corners of a regular pentagon with the special symmetry group $\mathscr{D}_{5h}$. The set M^* can be generated by the five vectors (Fig. I.3).

$$\boldsymbol{a}_n^* = a(\cos 2\pi n/5, \sin 2\pi n/5) \ , \qquad \boldsymbol{a}_0^* = \boldsymbol{a}_5^* \ . \tag{I.10a}$$

One of these vectors is rationally dependent (e.g., $\boldsymbol{a}_0^*$). For a quasicrystal, see (I.1), there are no main spots; therefore every vector can be expressed by

$$\boldsymbol{K} = \sum_{i=1}^{D} z_i \boldsymbol{a}_i^* \ , \qquad D = 4 \ . \tag{I.11a}$$

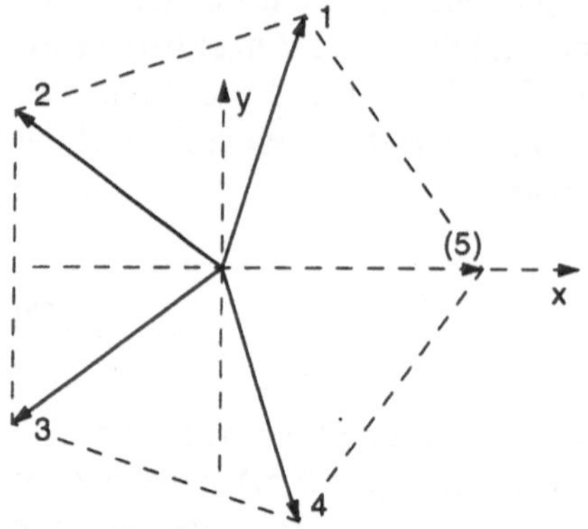

Fig. I.3. Basis vectors of a two-dimensional structure with five-fold symmetry. The basis consists of 4 independent vectors

The set M^* can be looked upon as the projection of a D-dimensional lattice Σ^*. The point group which leaves M^* invariant, is (isomorphic to) $\mathscr{D}_{5h}$ and it has to be

$$r^{-1}\boldsymbol{a}_i^* = \sum_{i=1}^{D} \Gamma_{ij}(r)\boldsymbol{a}_j^* \tag{I.12a}$$

where $\Gamma(r)$ is a D-dimensional integral ($D = 4$) reducible REP of $\mathscr{G} = \mathscr{D}_{5h}$ which is equivalent to an orthogonal one. Similarly to the case of the incommensurate structures, we have to construct a basis

$$\boldsymbol{d}_i^* = (\boldsymbol{a}_i^*, \boldsymbol{b}_i^*) \ , \qquad i = 1, \dots D \ , \tag{I.13a}$$

so that the elements (r, r_I) of orthogonal transformations can be represented by applying $\Gamma(r)$ on this basis. The generating elements of $\mathscr{D}_{5h}$ are c_5, c_2 and the inversion (Table I.2). The total REPs then are

$$\Gamma(c_5) = \begin{pmatrix} 0 & 1 & 0 & 0 \\ 0 & 0 & 1 & 0 \\ 0 & 0 & 0 & 1 \\ -1 & -1 & -1 & -1 \end{pmatrix} \ ; \qquad \chi(c_5) = -1 \ , \tag{I.14a}$$

$$\Gamma(c_2) = \begin{pmatrix} 0 & 0 & 0 & 1 \\ 0 & 0 & 1 & 0 \\ 0 & 1 & 0 & 0 \\ 1 & 0 & 0 & 0 \end{pmatrix} \ ; \qquad \chi(c_2) = 0 \ ,$$

$$\Gamma_{kl}(i) = -\delta_{kl} \ ; \qquad k, l = 1, \dots 4 \ ; \qquad \chi(i) = -4 \ .$$

Thus, according to Table I.2 the reducible REP decomposes into the two

2-dimensional inequivalent IRs: E_{1u}, E_{2u}. A basis for the representation in the 4-dimensional space Σ^* which projects onto the basis (I.10a) is

$$d_n^* = (a_n^*, c a_{2n}^*) \ , \qquad n = 1, 2, 3, 4 \ , \tag{I.15a}$$

with an arbitrary constant c. The basis $\{a_n^*\}$ of M^* is the projection of a basis $\{d_n^*\}$ of a 4-dimensional lattice Σ^*. The lattice Σ^*, which is generated by (I.15a) is a decagonal lattice. The distances are determined by the metric tensor g with

$$g_{ij} = d_i^* \cdot d_j^* = g_{ji} \ , \tag{I.16a}$$

$$g_{ii} = 1 + c^2 \qquad \text{for } i = 1, 2, 3, 4 \ ,$$

$$g_{i,i+1} (= g_{i+1,i}) = (\eta_-/2)(1 - c^2) - \tfrac{1}{2}c^2 \qquad \text{for } i = 1, 2, 3 \ ,$$

$$g_{13} = g_{14} = g_{24} = -\tfrac{1}{2}\eta_-(1 - c^2) - \tfrac{1}{2} \ .$$

The direct (dual) lattice Σ corresponding to Σ^* is then formed by the dual basis vectors d_n with respect to d_n^* from (I.15a):

$$d_n = \tfrac{2}{5}\{a_n^* - a_0^*, (a_{2n}^* - a_0^*)/c\} \ . \tag{I.17a}$$

An example of a function that is quasiperiodic in "physical" space is

$$f(r) = \sum_{n=0}^{4} \cos(a_n^* \cdot r) \ . \tag{I.18}$$

It is the restriction of a periodic 4-dimensional function in "super" space with lattice Σ. The corresponding periodic function in 4 dimensions is

$$f_s(r, r_I) = \sum_{n=0}^{4} \cos(a_n^* r + c a_{2n}^* r_I) \ . \tag{I.19}$$

The symmetry group of f_s in 4 dimensions is the symmorphic holohedral space group of Σ. The restriction to the 2-dimensional space $r_I = 0$ is quasiperiodic and may be called a quasicrystal.

Another, more realistic example of a three-dimensional quasicrystal corresponds to the best-known quasicrystal $Al_{0.86}Mn_{0.14}$ [I.5]. This can be described with the special symmetry group $\mathcal{Y}_h$ (icosahedral group with inversion) and a basis in 6-dimensional space ($D = 6$). The set M^* can be generated by the 12 vectors which point to the 12 surface midsts of a regular dodecahedron. Six of these 12 vectors are rationally independent, namely (Fig. I.4)

$$\begin{aligned} a_1^* &= (0, 0, 1) \ , \\ a_n^* &= (\sin\vartheta \cos 2\pi n/5, \sin\vartheta \sin 2\pi n/5, \cos\vartheta) \\ &\qquad n = 2, 3, 4, 5, 6 = D \ ; \qquad \cos^2\vartheta = 1/5 \ . \end{aligned} \tag{I.10b}$$

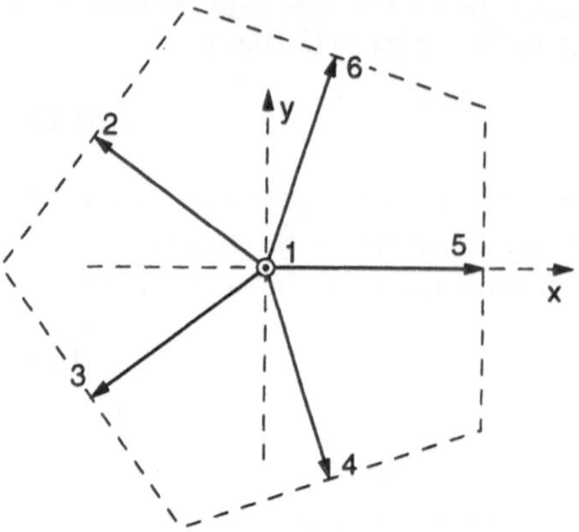

Fig. I.4. Basis vectors of a three-dimensional structure with icosahedral symmetry. The basis consists of 6 independent vectors

Since again there are no main spots, it is, see (I.1),

$$\boldsymbol{K} = \sum_{i=1}^{D} z_i \boldsymbol{a}_i^* , \qquad D = 6 > 3 , \tag{I.11b}$$

and M^* can be looked upon as the projection of Σ^*.

The point group $\mathscr{G}$ which leaves M^* invariant, is $\mathscr{Y}_h$ and it has to be

$$r^{-1}\boldsymbol{a}_i^* = \sum_{j=1}^{D} \Gamma_{ij}(r)\boldsymbol{a}_j^* , \tag{I.12b}$$

where $\Gamma(r)$ is a D-dimensional integer ($D = 6$) reducible REP of $\mathscr{G}$ being equivalent to an orthogonal one. A basis is

$$\boldsymbol{d}_i^* = (\boldsymbol{a}_i^*, \boldsymbol{b}_i^*) , \qquad i = 1, \dots D = 6 , \tag{I.13b}$$

so that the elements (r, r_I) can be represented by $\Gamma(r)$. The generating elements of $\mathscr{Y}_h$ are c_5, c_3 and the inversion (Table A.14, B.2). The total REPs then are

$$\Gamma(c_5) = \begin{bmatrix} 1 & 0 & 0 & 0 & 0 & 0 \\ 0 & 0 & 1 & 0 & 0 & 0 \\ 0 & 0 & 0 & 1 & 0 & 0 \\ 0 & 0 & 0 & 0 & 1 & 0 \\ 0 & 0 & 0 & 0 & 0 & 1 \\ 0 & 1 & 0 & 0 & 0 & 0 \end{bmatrix} , \qquad \chi(c_5) = 1 , \tag{I.14b}$$

$$\Gamma(c_3) = \begin{bmatrix} 0 & 0 & 0 & 0 & 0 & 1 \\ 1 & 0 & 0 & 0 & 0 & 0 \\ 0 & 0 & 0 & 0 & 1 & 0 \\ 0 & 0 & -1 & 0 & 0 & 0 \\ 0 & 0 & 0 & -1 & 0 & 0 \\ 0 & 1 & 0 & 0 & 0 & 0 \end{bmatrix} , \qquad \chi(c_3) = 0 ,$$

$$\Gamma_{kl}(i) = -\delta_{kl} ; \qquad kl = 1, \dots 6 ; \qquad \chi(i) = -6 .$$

Thus, according to Tables A.14, B.2 the reducible REP decomposes into the two 3-dimensional inequivalent IRs: F_1, F_2.

A basis for the representation in the 6-dimensional space Σ^* which projects onto the basis (I.10b) is

$$\boldsymbol{d}_1^* = (\boldsymbol{a}_1^*, c\boldsymbol{a}_1^*) , \qquad \boldsymbol{d}_n^* = (\boldsymbol{a}_n^*, -c\boldsymbol{a}_{2n}^*) , \tag{I.15b}$$

with an arbitrary constant c. The scalar products (metric tensor of the 6-dimensional lattice Σ^*) are given by

$$\boldsymbol{d}_i^* \cdot \boldsymbol{d}_i^* = 1 + c^2 , \qquad \boldsymbol{d}_i^* \cdot \boldsymbol{d}_j^* = (1 - c^2)/\sqrt{5} , \qquad \text{for } i \neq j . \tag{I.16b}$$

The basis $\{\boldsymbol{d}_i\}$ of the direct lattice Σ corresponding to Σ^* is then formed by the vectors, defined by (I.8),

$$\boldsymbol{d}_1 = \tfrac{1}{2}(\boldsymbol{a}_1^*, \boldsymbol{a}_1^*/c) ; \qquad \boldsymbol{d}_n = \tfrac{1}{2}(\boldsymbol{a}_n^*, -\boldsymbol{a}_{2n}^*/c) , \qquad n = 2, \dots, 6 . \tag{I.17b}$$

For $c = 1$ the basis vectors (I.15b) are of equal length and mutually perpendicular to each other because of (I.16b). Thus Σ^* is a hypercubic lattice as well as Σ and the set M^* for an iscosahedral quasicrystal is a projection of a 6-dimensional hypercubic lattice onto a 3-dimensional space. For $c \neq 1$ this is a little bit more complicated.

The construction can also be done by starting with another set of rationally independent vectors. For this and especially for other examples we refer the reader to the original literature [I.1, 4–8].

J. Supersymmetry

In Chap. 14, especially in Sect. 14.5, we have given a classification of the fundamental particles according to possible gauge symmetries. Matter consists of hadrons and leptons. The constituents of hadrons are the quarks, which occur with six flavours (Table 13.2), whereas the leptons are looked upon as being elementary; there also exist six flavours of leptons (Table 14.2). The six flavours of quarks and leptons can be combined to form three families (generations), see

(14.5.2a, 33a). Beyond these families there seem to be no others. All these particles are fermions with spin 1/2.

The interactions between these particles are described by electro-weak, strong, and gravitational forces, as discussed in Chap. 14 and are mediated by bosons. A fifth force does not exist as far as is known. The electro-weak and strong forces can be unified by certain gauge symmetries (e.g., $\mathscr{SU}(5)$) and are mediated by bosons; the electro-weak forces are mediated by photons or $W^{\pm}$, Z^0 bosons, the strong forces by gluons. Photons and gluons are massless, the heavy bosons $W^{\pm}$, Z^0 possess mass (Table 14.1). All these bosons have spin 1. The gravitational force is supposed to be mediated by gravitons with spin 2 and zero mass.

In addition, the standard model (Glashow-Salam-Weinberg) yields at least one Higgs boson η (Sect. 14.5.2) with zero spin and zero electrical charge, the mass of which is unknown. More refined models have need for five Higgs bosons (spin 0), two having electrical charge ($\eta^{\pm}$), three are neutral (η^0) (Table 14.1). These particles have not been observed until now.

The masses of the particles, especially those of the $W^{\pm}$, Z^0 bosons, are generated by symmetry breaking as explained with simple examples in Sect. 14.5. The basic concept of the standard model is *symmetry*: the Lagrangian is invariant against special symmetry (gauge) transformations as discussed in Chap. 14.

If one wants to combine all these particles (fermions) and interactions (bosons) into one model it is necessary to introduce a certain *supersymmetry* beyond that of the standard model because the standard model cannot explain why it is as it is, e.g., the origin of all masses and relative strength of the forces. Supersymmetry has to relate all the objects mentioned, i.e., the fundamental classes, fermions, and bosons, to each other. So to every fermion there has to be a bosonic superpartner, and vice versa, where the superpartners have the same properties as the particles, including the strength of the interactions between the partners, but besides the spin which differs by 1/2. The modern theories also require a certain number of Higgs bosons (see above).

Supersymmetry should also include gravitational forces, that is a unification of all four forces. If particles exist with the so-called Planck mass ($m_p = \sqrt{\hbar c/G}$; $m_p c^2 = 1.221 \times 10^{19}$ GeV) the gravitational force among them would be equal or higher then the other forces which are represented by $W^{\pm}$, Z^0 particles (masses 80–90 GeV). Supersymmetry has to explain differences in mass up to these orders of magnitude.

As mentioned above an exact supersymmetry would require that the superpartners had a mass equal to that of the particles themselves, which implies hat they should have been found in experiments already. But, since this is obviously not the case it means that the superpartners have larger masses then do the particles and, in turn, this implies that *supersymmetry* is a *spontaneously broken symmetry*.

Another point to mention is the following. If there is any supersymmetry there has to be a transformation group which leaves the corresponding Lagrangian invariant. Thus there have to be conservation laws for the superparticles (Noether's theorem). This would mean that in a process in which superparticles are produced, pairs of superparticles are always produced. On the other hand, in a decay process of *one* superparticle there must occur on odd number of superparticles. This means that there must exist one least-massive superparticle which is stable. Otherwise it would be necessary to have lighter superparticles into which it could decay. It has to be supposed that this least-massive superparticle will be the first one to be detected experimentally. But even this one has not yet been found. The preceding holds only if there is no violation of the R parity, where $R_p = +1$ for particles and $R_p = -1$ for the superparticles.

Since a more detailed discussion of supersymmetry is beyond this short Appendix J we will not go into details but only offer a simple model [J.1–3], which combines bosonic and fermionic degrees of freedom. The linear harmonic oscillator is the simplest bosonic system with the variable $x(t)$ in real space. The fermionic superpartner is described by a variable $\xi(t)$ and a complex conjugate $\xi^*(t)$, which are *anticommuting* and are the classical correspondences to the spin operators (Table B.1)

$$\begin{aligned} &\sigma_\pm = \tfrac{1}{2}(\sigma_1 \pm i\sigma_2) \ , \\ &[\xi, \xi^*]_+ = 1 \ ; \qquad [\xi, \xi] = [\xi^*, \xi^*] = 0 \ . \end{aligned} \tag{J.1}$$

This system has no classical significance, but its quantum limit must coincide with the familiar relations. The combinations of commuting and anticommuting quantities, and the corresponding numbers constitute a *Grassmann algebra* [J.4, 5], which replaces the Lie algebra of gauge transformations in the case of supersymmetry and is related to Pauli's principle.

The simplest Lagrangian of the type mentioned is

$$\mathscr{L} = \frac{1}{2}\dot{x}^2 - \frac{1}{2}\{F(x)\}^2 + \frac{i}{2}(\xi^*\dot{\xi} - \dot{\xi}^*\xi) - F'(x)\xi^*\xi \ , \tag{J.2}$$

with

$$F(x) = -\frac{d}{dx}W(x) \ , \qquad F'(x) = \frac{d}{dx}F(x) \ ,$$

where $W(x)$ is a "superpotential" in the "superspace" [J.5]. The linear oscillator has $W(x) \sim x^2$, then $F'(x)$ is a constant. For the construction of supersymmetric models the choice of the superpotential is of essential significance.

According to E. Noether, the invariance of a Lagrangian can be established by infinitesimal transformations (see Chap. 14, Appendix F). Since these transformations mix up bosons and fermions, the infinitesimal transformation parameters χ, χ^* also have to be anticommuting quantities. The infinitesimal transformations then turn out to be

$$\delta x(t) = \chi \cdot \xi(t) + \xi^*(t) \cdot \chi^* \ .$$
$$\delta \xi(t) = -[\mathrm{i}\dot{x}(t) + F(x(t))]\chi^* \ . \tag{J.3}$$

The variation of the Lagrangian is

$$\delta \mathscr{L} = \frac{1}{2}\frac{d}{dt}[(\dot{x} - \mathrm{i}F)\chi\xi + (\dot{x} + \mathrm{i}F)\chi^*\xi^*] \tag{J.4}$$

which leaves the equations of motion and the action integral covariant or invariant, respectively, if the boundary conditions are chosen appropriately.

According to Noether's theorem the associated conserved supercharges then are

$$Q = [-\mathrm{i}p + F(x)]\xi \ ,$$
$$Q^* = \xi^*[\mathrm{i}p + F(x)] \ . \tag{J.5}$$

The supercharges are the generators of a supersymmetry group (Chap. 14.1). It can clearly be seen that these generators (J.5) of the group couple a translation p (which is a special relic of a Lorentz and Galilei transformation, boson field) and a spin changing transformation ξ, ξ^* (fermion field) so that a group with elements of combined character results.

From (J.2) the Hamiltonian, the Hamiltonian equations of motion, and the quantization can be derived. In more realistic models (J.2) has to be extended to a relativistic Lagrangian together with fermionic variables. For more details we refer the reader to the large set of papers [J.6, 7].

K. List of Symbols and Abbreviations

$\sim$, $\propto$	proportional, equivalent
$\approx$	approximately equal, order of
$\simeq$	asymptotically equal
$\cong$	isomorphic
$:=$	definition
$\equiv$	identical
$\hat{=}$	corresponds to
$\times$	vector product, outer direct product of a group
$\boxed{S}$	semidirect product
$\otimes$	inner direct product of representations or matrices
$\oplus$	direct sum of spaces or representations
$\odot$	(new) outer product (Sect. 5.5.4)
$[\![\ ]\!]$	equivalence class
$[\lambda]$	partition

[,]	commutator
$[\ ]_\pm$	symmetrized, antisymmetrized sum or product
$\tilde{D}$	transpose (associate)
D^*	complex conjugate
D^+, ψ^+	adjoint
A	linear operator
$\mathscr{A}$	algebra, alternating group
a	group element
$\boldsymbol{a}$	basis vector
AO	atomic orbital
APW	augmented plane wave
$\mathscr{B}$	basis of a space
$\boldsymbol{b}$	basis vector
BZ	Brillouin zone
$\mathbb{C}$	complex numbers
$\mathscr{C}$	cyclic groups
$\mathscr{C}^{\mathrm{D}}$	cyclic double groups
C, C_n	Casimir operator
c_n	rotation elements
c_{ik}^l	structure constants
$\mathfrak{C}$	charge conjugation operator
CGC	Clebsch-Gordan coefficients
CI	configuration interaction
COR	corepresentation
CWB	Cartan-Weyl standard basis
$D, D^{(\alpha)}$	representations, irreducible representations
$\mathscr{D}$	dihedral groups
d	group element (usually rotation)
d_α	dimension of a REP α
DIG	dynamical invariance group
E_{ij}, E_α	infinitesimal generator of a Lie algebra, in particular ladder operator of $\mathscr{SU}(n)$
e	identity element, elementary charge
$e_i^{(\alpha)}, \psi_j^{(\alpha)}$	basis system for representations α
$\mathscr{F}$	Fermi group
FC	force constant
$\mathscr{G}$	general group
g	order of a group
g_{ik}	metric tensor
$\mathscr{G}_{0\boldsymbol{k}}, \mathscr{G}_{\boldsymbol{k}}$	little group (first, second kind)
GG	Georgi-Glashow
$\mathscr{GL}$	general linear group
GSW	Glashow-Salam-Weinberg
GUT	grand unified theories

H	linear operator
H_i, H_α	infinitesimal generator of a Lie algebra, in particular of $\mathscr{S}\mathscr{U}(n)$
$\mathscr{H}$	(basis of a) unitary space
i	inversion element, imaginary unit
$\mathscr{I}$	rotation-inversion group
ICOR	irreducible corepresentation
IPR	irreducible projective representation
IR	irreducible representation
J_i, $J_{i,kl}$	infinitesimal generators (or their matrices)
K, K_a	class (containing element a)
$\boldsymbol{K}$	lattice vector in reciprocal space
$\mathbb{K}$	field
$\boldsymbol{k}$	wave vector
$*\boldsymbol{k}$	star of $\boldsymbol{k}$
KKR	Kohn-Korringa-Rostoker
$\mathscr{L}$	linear space
LA	Lie algebra
LCAO	linear combination of atomic orbitals
$\mathbb{M}(\mathscr{G})$	multiplier group
$\mathscr{M}$	magnetic groups, general set
m_α	multiplicity
MO	molecular orbital
$\mathbb{N}$	natural numbers
$\mathscr{N}$	invariant subgroup
$\mathscr{O}$	octahedral or orthogonal groups
ONS	orthonormal(ized) system
OPW	orthogonalized plane wave
P_a	operator assigned to group element a
$\mathscr{P}$	permutation group
p	group element
p-equivalent	projective equivalent
PR	projective representation
PW	plane wave
$\mathscr{Q}$	quaternion group
q	group element
$\boldsymbol{q}$	wave vector
Q	charge (operator)
QCD	quantum chromodynamics
QED	quantum electrodynamics
$\boldsymbol{R}$	lattice vector in real space
$\mathbb{R}$	real numbers
$\mathbb{R}_n$	Cartesian space
$\mathscr{R}$	space group
r	colour changing-element

r_a	number of elements in class K_a
r_{ij}	roots of a Lie algebra
RC	reduction coefficient
REP	representation
S	linear transformation
$\mathscr{S}$	rotation-reflection group, subalgebra
$\mathscr{SO}$	special orthogonal groups
$\mathscr{Sp}$	symplectic groups
$\mathscr{SU}$	special unitary groups
s_n	group element rotation-reflection
SYT	standard Young tableau
T	general tensor
$\mathbb{T}$	translation group
$\mathscr{T}$	tetrahedral groups
t	translation element
U	linear transformation
$\mathscr{U}$	subgroup, unitary group
$\mathscr{V}$	special group
w	weights of a Lie algebra
Y_m^l	spherical harmonics
$\mathscr{Y}$	icosahedral groups
$\mathbb{Z}$	integer numbers
$\Delta, \Delta^{(\alpha)}$	representations, irreducible representations
ϑ, ϑ_0	time reversal
$\sigma_{\mathrm{d}}, \sigma_{\mathrm{h}}, \sigma_{\mathrm{v}}$	reflection elements
σ_i	Pauli matrices
$\chi, \chi^{(\alpha)}$	character of a REP or IRα
$\Psi_{ij}^{(\alpha)}$	basis system for representations
$(\alpha\beta\|\gamma)$	reduction coefficients
$\begin{pmatrix} \beta\alpha & \gamma s \\ j i & l \end{pmatrix}$	Clebsch-Gordan coefficients

May the God who watches over the right use of mathematical symbols in manuscript, print, and on the blackboard, forgive me this and my many other sins.

H. Weyl, The Classical Groups, p. 289

References

Chapter 4

4.1 E. Fick: *Einführung in die Grundlagen der Quantentheorie* (Akademische Verlagsgesellschaft, Wiesbaden 1972)
4.2 J.D. Bjorken, S.D. Drell: *Relativische Quantenfeldtheorie*, BI Taschenbuch 101/101a (Bibliographisches Institut, Mannheim 1967)
4.3 R.S. Mulliken: Electronic structures of polyatomic molecules and valence IV: Electronic states, quantum theory of double bond. Phys. Rev. **43**, 279 (1933)

Chapter 5

5.1 E.P. Wigner: *Group Theory and its Application to the Quantum Mechanics of Atomic Spectra* (Academic, New York 1959)
5.2 G.J. Bradley, A.P. Cracknell: *The Mathematical Theory of Symmetry in Solids* (Clarendon, Oxford 1972)
5.3 C.D.H. Chisholm: *Group Theoretical Techniques in Quantum Chemistry* (Academic, New York 1976)
5.4 I.V. Schensted: *A Course on the Application of Group Theory to Quantum Mechanics* (Neo Press, Peaks Island, ME 1976)
5.5 M. Hamermesh: *Group Theory and its Application to Physical Problems* (Addison-Wesley, Reading, MA 1964)
5.6 H. Weyl: *The Theory of Groups and Quantum Mechanics* (Dover, Mineola, NY 1950); also *The Classical Groups* (Princeton Univ. Press, Princeton, NJ 1946)
5.7 H.-W. Streitwolf: *Gruppentheorie in der Festkörperphysik* (Akademische Verlagsgesellschaft, Leipzig 1967)
5.8 D.E. Littlewood: *The Theory of Group Characters* (Oxford Univ. Press, Oxford 1950)
5.9 D.E. Littlewood: Invariant theory, tensors and group characters. Trans. Roy. Soc. (London) A **239**, 305 (1943); and On invariant theory under restricted groups. Trans. Roy. Soc. (London) A **239**, 387 (1943)

Chapter 7

7.1 W. Ludwig: *Festkörperphysik* (Akademische Verlagsgesellschaft, Wiesbaden 1978)
O. Madelung: *Introduction to Solid-State Theory*, Springer Ser. Solid-State Sci., Vol. 2 (Springer, Berlin, Heidelberg 1978)
7.2 G. Leibfried: Gittertheorie der mechanischen und thermischen Eigenschaften der Kristalle, in *Handbuch der Physik*, Vol. 7/1, *Crystal Physics I* (Springer, Berlin, Heidelberg 1955) p. 104

7.3 G. Leibfried, W. Ludwig: Theory of anharmonic effects in crystals. *Solid State Phys.* **12**, 275 (Academic, New York 1961)
7.4 S. Großmann: *Funktionalanalysis* (Akademische Verlagsgesellschaft, Frankfurt am Main 1970)
7.5 E.P. Wigner: *Group Theory and its Application to the Quantum Mechanics of Atomic Spectra* (Academic, New York 1959)
7.6 G.J. Bradley, A.P. Cracknell: *The Mathematical Theory of Symmetry in Solids* (Clarendon, Oxford 1972)

Chapter 8

8.1 H.A. Jahn, E. Teller: Stability of polyatomic molecules in degenerate electronic states: I. Orbital degeneracy. Proc. Roy. Soc. (London) A **161**, 220 (1937)
8.2 H.-W. Streitwolf: *Gruppentheorie in der Festkörperphysik* (Akademische Verlagsgesellschaft, Leipzig 1967)
8.3 H.A. Bethe: Termaufspaltung in Kristallen. Ann. Phys. (Leipzig) **3**, 133 (1929)
8.4 R.S. Mulliken: Electronic structures of polyatomic molecules and valence. Phys. Rev. **40**, 55 (1932); ibid **41**, 49, 751 (1932); ibid. **43**, 279 (1933)
8.5 L.P. Bouckaert, R. Smoluchowski, E. Wigner: Theory of Brillouin zones and symmetry properties of wave functions in crystals. Phys. Rev. **50**, 58 (1936)

Chapter 9

9.1 H.-W. Streitwolf: *Gruppentheorie in der Festkörperphysik* (Akademische Verlagsgesellschaft, Leipzig 1967)
9.2 A.P. Cracknell: *Group Theory in Solid-State Physics* (Taylor & Francis, London 1975)
9.3 W. Döring: Die Strahldarstellungen der kristallographischen Gruppen. Z. Naturforsch. **14a**, 343 (1959)
9.4 W. Döring, V. Zehler: Gruppentheoretische Untersuchung der Elektronenbänder im Diamantgitter. Ann. Phys. (Leipzig) (6) **13**, 214 (1953)
9.5 O.V. Kovalev: *Irreducible Representations of the Space Groups* (Gordon and Breach, New York 1965)
9.6 G.L. Bir, G.E. Pinkus: *Symmetry and Strain-Induced Effects in Semiconductors* (Wiley, New York 1974)
9.7 J. Zak (ed.): *The Irreducible Respresentations of Space Groups* (Benjamin, Elmsford, NY 1969)
9.8 G.J. Bradley, A.P. Cracknell: *The Mathematical Theory of Symmetry in Solids* (Clarendon, Oxford 1972)

Chapter 10

10.1 C. Falter: Phonon dispersion and symmetry of solid-J_2. Z. Physik **258**, 263 (1973)
10.2 J.L. Birman: Theory of crystal space groups and infra-red and Raman lattice processes of insulating crystals. *Handbuch der Physik*, Vol.25/2b, *Light and Matter Ib* (Springer, Berlin, Heidelberg 1974)

Chapter 11

11.1 B.G. Wybourne: *Classical Groups for Physicists* (Wiley, New York 1974)
11.2 G. Racah: Sulla caratterizzazione delle rappresentazione irriducibili dei gruppi semisimplici di Lie. Lincei Rend. Sci. Fis. Mat. Nat. **8**, 108 (1950)
11.3 E.G. Beltrametti, A. Blasi: On the number of Casimir operators associated with any Lie group. Phys. Lett. **20**, 62 (1966)

Chapter 13

13.1 Y. Ne'eman: Derivation of strong interactions from a gauge invariance. Nucl. Phys. **26**, 222 (1961)
13.2 M. Gell-Mann: Symmetries of baryons and mesons. Phys. Rev. **125**, 1067 (1962)
13.3 M. Gell-Mann: A schematic model of baryons and mesons. Phys. Lett. **8**, 214 (1964)
13.4 G. Zweig: An $\mathcal{SU}(3)$ model for strong interaction symmetry and its breaking. CERN Rep. 8419/Th.412 (1964)
13.5 M. Gell-Mann: The interpretation of the new particles as charge multiplets. Nuovo Cimento, Suppl.**4**, 848 (1956)
13.6 S. Okubo: Note on unitary symmetry in stron interactions. Prog. Theor. Phys. **27**, 949 (1962)
13.7 O.W. Greenberg: Spin and unitary-spin independence in a paraquark model of baryons and mesons. Phys. Rev. Lett. **13**, 598 (1964)
13.8 G. Racah: Group theory and spectroscopy. Ergeb. Exakten Naturwiss. **37**, 28 (1965)
13.9 Particle Data Group: Review of particle physics. Phys. Rev. D **50** (no.3, pt.1), 1173-1825 (1994)
13.10 G. Altarelli: Electroweak precision tests. A status report. CERN-TH 7464/94 (October 1994)

Chapter 14

14.1 J.D. Bjorken, S.D. Drell: *Relativistic Quantum Mechanics* (McGraw-Hill, New York 1964)
14.2 J.D. Bjorken, S.D. Drell: *Relativistic Quantum Fields* (McGraw-Hill, New York 1965)
14.3 Y. Aharonov, D. Bohm: Significance of electromagnetic potentials in the quantum theory. Phys. Rev. **115**, 485 (1959)
14.4 C.N. Yang, R.L. Mills: Conservation of isotopic spin and isotopic gauge invariance. Phys. Rev. **96**, 191 (1954)
14.5 V.L. Ginzburg, D.L. Landau: On the theory of superconductivity. Zh. Eksp. Teor. Fiz. **20**, 1064 (1950)
14.6 H.B. Nielsen, P. Olesen: Vortex-line models for dual strings. Nucl. Phys. B **61**, 45 (1973)
14.7 Y. Nambu: Strings, monopoles, and gauge fields. Phys. Rev. D **10**, 4262 (1974)

14.8 L.D. Landau, E.M. Lifshitz: *Course of Theoretical Physics*, Vol. 2, *The Classical Theory of Fields*, 4th edn. (Pergamon, Oxford 1976)
14.9 L.D. Faddeev, A.A. Slavnov: *Gauge Fields* (Addison-Wesley, Reading, MA 1981)
14.10 P. Becker, M. Böhm, H. Joos: *Eichtheorien der starken and elektroschwachen Wechselwirkung* (Teubner, Stuttgart 1981)
14.11 J. Bernstein: Spontaneous symmetry breaking, gauge theories, the Higgs mechanism and all that. Rev. Mod. Phys. **46**, 7 (1974)
14.12 Ling-Fong Li: Group theory of the spontaneously broken gauge symmetries. Phys. Rev. D **9**, 1723 (1974)
14.13 J. Goldstone: Field theories with superconductor solutions. Nuovo Cimento **19**, 154 (1961)
14.14 P.W. Higgs: Broken symmetries, massless particles and gauge fields. Phys. Lett. **12**, 132 (1964); Spontaneous symmetry breakdown without massless bosons. Phys. Rev. **145**, 1156 (1966)
14.15 T.W.B. Kibble: Symmetry breaking in non-Abelian gauge theories. Phys. Rev. **155**, 1554 (1967)
14.16 A. Salam: *Elementary Particle Physics* (Almqvist Förlag, Stockholm 1968) p. 367
14.17 S. Weinberg: Non-Abelian gauge theories of the strong interactions. Phys. Rev. Lett. **31**, 494 (1973)
14.18 S. Weinberg: Conceptual foundations of the unified theory of weak and electromagnetic interactions. Rev. Mod. Phys. **52**, 515 (1980)
14.19 A. Salam: Gauge unification of fundamental forces. Rev. Mod. Phys. **52**, 525 (1980)
14.20 S.L. Glashow: Towards a unified theory: Threads in a tapestry. Rev. Mod. Phys. **52**, 539 (1980)
14.21 S.L. Glashow, J. Illipoulos, L. Maiani: Weak interactions with lepton-hadron symmetry. Phys. Rev. D **2**, 1285 (1970)
14.22 S.L. Glashow: Unity of all elementary-particle forces. Phys. Rev. Lett. **32**, 438 (1974)
14.23 H. Georgi, S.L. Glashow: Unified theory of elementary-particle forces. Phys. Today **33**, 30 (September 1980); Unified theory of elementary-particle forces. Phys. Rev. Lett. **32**, 438 (1974)
14.24 H. Haken, H.C. Wolf: *The Physics of Atoms and Quanta*, 4th edn. (Springer, Berlin, Heidelberg 1994)
14.25 G. Arnison et al.: Experimental observation of isolated large transverse energy electrons with associated missing energy at $\sqrt{s}$ = 540 GeV. Phys. Lett. B **122**, 103 (1983); Further evidence for charged intermediate vector bosons at the SPS collider. Phys. Lett. B **129**, 273 (1983)
14.26 M. Banner et al.: Observation of single isolated electrons of high transverse momentum in events missing transverse energy at the CERN pp collider. Phys. Lett. B **122**, 476 (1983)
G. Altarelli: Electroweak precision tests. A Status Rpt., CERN-TH.7464/94 (October 1994)
14.27 G. t'Hofft: Magnetic monopoles in unified gauge theories. Nucl. Phys. B **79**, 276 (1974)
14.28 G. t'Hofft: Recent Progr. in Lagrange Field Theory and Applications. Marseille, France (1975) Proc., p. 58

14.29 J. Kogut, L. Süsskind: Hamiltonian formulation of Wilson's lattice gauge theories. Phys. Rev. D **11**, 395 (1975)
14.30 S. Weinberg: A model of leptons. Phys. Rev. Lett. **19**, 1264 (1967)
14.31 S.M. Bilenky, J. Hosek: Glashow-Weinberg-Salam theory of electroweak interactions and the neutral currents. Phys. Rep. **90**, 73 (1982)
14.32 P. Langacker: Grand unified theories and proton decay. Phys. Rep. **72**, 185 (1981)
14.33 N. Cabibbo: Unitary symmetry and leptonic decays. Phys. Rev. Lett. **10**, 531 (1963)
14.34 P.A.M. Dirac: Quantised singularities in the electromagnetic field. Proc. Roy. Soc. (London) A **133**, 60 (1931); The theory of magnetic poles. Phys. Rev. **74**, 817 (1948)
14.35 A.M. Polyakov: Particle spectrum in quantum field theory. JETP Lett. **20**, 194 (1974)
14.36 F. Reines, M.F. Crouch: Baryon-conversation limit. Phys. Rev. Lett. **32**, 493 (1974)
14.37 J. Ellis, M.K. Gaillard, A. Petermann, C.T. Sachrajda: A hierarchy of gauge hierarchies. Nucl. Phys. B **164**, 253 (1980)
14.38 A.J. Buras, J. Ellis, M.K. Gaillard, D.N. Nanopoulos: Aspects of the grand unification of strong, weak and electromagnetic interactions. Nucl. Phys. B **135**, 66 (1978)
14.39 D.N. Schramm: The early universe and high energy physics. Phys. Today **36**, 27 (April 1983)
14.40 B. Cabrera: First results from a superconductive detector for moving magnetic monopoles. Phys. Rev. Lett. **48**, 1378 (1982)

Appendices

A.1 R.S. Mulliken: Electronic structures of polyatomic molecules and valence. Phys. Rev. **40**, 55 (1932); ibid. **41**, 49, 751 (1932); ibid. **43**, 279 (1933)
A.2 H.A. Bethe: Termaufspaltung in Kristallen. Ann. Phys. (Leipzig) **3**, 133 (1929)
A.3 G.J. Ljubarski: *Anwendungen der Gruppentheorie in der Physik* (VEB Deutscher Verlag der Wissenschaften, Berlin 1962)
A.4 L.P. Bouckaert, R. Smoluchowski, E. Wigner: Theory of Brillouin zones and symmetry properties of wave functions in crystals. Phys. Rev. **50**, 58 (1936)
A.5 G.F. Koster: *Space Groups and their Representations*, Solid State Phys., Vol. 5 (Academic, New York 1957)
A.6 H.-W. Streitwolf: *Gruppentheorie in der Festkörperphysik* (Akademische Verlagsgesellschaft, Leipzig 1967)
A.7 J.L. Birman: Theory of crystal space groups and infra-red and Raman lattice processes of insulating crystals. *Handbuch der Physik*, Vol. 25/2b, *Light and Matter Ib* (Springer, Berlin, Heidelberg 1974)
A.8 M. Lax: *Symmetry Principles in Solid State and Molecular Physics* (Wiley, New York 1974)
A.9 J. Zak (ed.): *The Irreducible Representations of Space Groups* (Benjamin, Elmsford, NY 1969)

G.1 P. Mittelstaedt: *Klassische Mechanik*, BI Taschenbuch 500/500a (Bibliographisches Institut, Mannheim 1970)

G.2 H. Goldstein: *Classical Mechanics*, 2nd edn. (Addison-Wesley, Reading, MA 1980)

G.3 E. Schmutzer: *Symmetrien und Erhaltungssätze der Physik* (Akademie, Berlin 1972)

G.4 E. Fick: *Einführung in die Grundlagen der Quantentheorie* (Akademische Verlagsgesellschaft, Wiesbaden 1972)

G.5 L. Fonda, G.C. Ghirardi: *Symmetry Principles in Quantum Physics* (Dekker, New York 1970)

G.6 J.S. Lomont: *Applications of Finite Groups* (Academic, New York 1959)

G.7 M. Hamermesh: *Group Theory and its Application to the Physical Problems* (Addison-Wesley, Reading, MA 1964)

H.1 J. Bernstein: Spontaneous symmetry breaking, gauge theories, the Higgs mechanism and all that. Rev. Mod. Phys. **46**, 7 (1974)

I.1 B.K. Vainshtein: *Fundamentals of Crystals*, 2nd edn., Modern Crystallography, Vol.1 (Springer, Berlin, Heidelberg 1994)

I.2 R.C. Haddon, K. Raghavachari: In *Buckminsterfullerenes*, ed. by W.E. Billups, M.A. Ciufolini (VCH, Weinheim 1993)

I.3 W.E. Pickett: Electrons and phonons in C-60 compounds. *Solid State Physics* **48**, 225 (Academic, San Diego, CA 1994)

I.4 A. Janner, T. Janssen: Symmetry of periodically distorted crystals. Phys. Rev. B **15**, 643 (1977)

I.5 T. Janssen: Crystallography of quasi-crystals. Acta Cryst. A **42**, 261 (1986)

I.6 R. Currat, T. Janssen: Excitations in incommensurate crystal phases. *Solid State Physics* **41**, 201 (Academic, New York 1988)

I.7 P.J. Steinhardt, St. Ostlund: *The Physics of Quasicrystals* (World Scientific, Singapore 1987)

I.8 R. Lifshitz, N.D. Mermin: Space groups of trigonal and hexagonal quasiperiodic crystals of rank 4. Acta Crystallogr. A. Found. Crystallogr. (Denmak) A **50**, pt.1, 72-85 (1994); Braivais classes and space groups for trigonal and hexagonal quasiperiodic crysrals of arbitrary finite rank. ibid. A **50**, pt.1, 85-97 (1994)

J.1 N. Nicolai: Supersymmetry and spin systems. J. Phys. A **9**, 1497 (1976)

J.2 E. Witten: Dynamical breaking of supersymmetry. Nucl. Phys. B **188**, 513 (1981); Constraints on supersymmetry breaking B **202**, 253 (1982)

J.3 P. Salomonssen, J.W. van Holten: Fermionic coordinates and supersymmetry in quantum mechanics. Nucl. Phys. B **196**, 509 (1982)

J.4 F.A. Berezin: *The Method of Second Quantization* (Academic, New York 1966)

J.5 J.F. Cornwell: *Group Theory in Physics*, Vol. III (Academic, London 1989)

J.6 D. Bailin, A. Love: *Supersymmetric Gauge Field Theory and String Theory* (Inst.of Phys., Bristol 1994)

J.7 P.P. Srivastava: *Supersymmetry, Superfield and Supergravity: An Introduction* (Inst.of Phys., Bristol 1986)

Additional Reading

Texts on Symmetry in Physics

Barut A.O., R. Raczka: *Theory of Group Representations and Applications* (Polish Scientific Publishers, Warsaw 1977)

Birss R.R.: *Symmetry and Magnetism* (North-Holland, Amsterdam 1964)

Cornwell J.F.: *Group Theory in Physics* I-III (Academic, London 1989)

Elliot J.P., P.G. Dawber: *Symmetry in Physics* 1, 2 (Macmillan, London 1979)

Evarestov R.A., V.P. Smirnov: *Site Symmetry in Crystals*, Springer Ser. Solid-State Sci., Vol. 108 (Springer, Berlin, Heidelberg 1993)

Gibson W.M., B.R. Pollard: *Symmetry Principles in Elementary Particle Physics* (Cambridge Univ. Press, Cambridge 1976)

Gürsey F. (ed.): *Group Theoretical Concepts and Methods in Elementary Particle Physics* (Gordon & Breach, New York 1964)

Humphreys J.E.: *Introduction to Lie Algebras and Representation Theory*, Graduate Text Math., Vol. 9 (Springer, New York 1972)

Inui T., Y. Tanabe, Y. Onodera: *Group Theory and Its Applications in Physics*, 2nd edn., Springer Ser. Solid-State Sci., Vol. 78 (Springer, Berlin, Heidelberg 1995)

Ivchenko E.L., G. Pikus: *Superlattices and Other Heterostructures, Symmetry and Optical Phenomena*, Springer Ser. Solid-State Sci., Vol. 110 (Springer, Berlin, Heidelberg 1995)

Jagodzinski H.: In *Handbuch der Physik*, Vol. 7/71, *Crystal Physics* I (Springer, Berlin, Heidelberg 1955)

Kaplan I.G.: *Symmetry of Many-Electron Systems* (Academic, New York 1975)

Loebl E.M.: *Group Symmetries and its Applications* (Academic, New York 1968)

Parikh J.C.: *Group Symmetries in Nuclear Structure* (Plenum, New York 1978)

Petraschen M.I., E.D. Trifonov: *Anwendung der Gruppentheorie in der Quantenmechanik* (Akademie, Berlin 1969)

Sobelman I.I.: *Atomic Spectra and Radiative Transitions*, 2nd edn., Springer Ser. Atoms Plasmas, Vol. 12 (Springer, Berlin, Heidelberg 1992)

Sternberg S.: *Group Theory and Physics* (Cambridge Univ. Press, Cambridge 1994)

Taylor J.C.: *Gauge Theories of Weak Interactions* (Cambridge Univ. Press, Cambridge 1976)

Tinkham M.: *Group Theory and Quantum Mechanics* (McGraw-Hill, New York 1964)

Toda M.: *Theory of Nonlinear Lattices*, 2nd edn., Springer Ser. Solid-State Sci., Vol. 20 (Springer, Berlin, Heidelberg 1989)

Vainshtein B.K.: *Fundamentals of Crystals, Symmetry, and Methods of Structural Crystallography*, Modern Crystallography 1 (Springer, Berlin, Heidelberg 1994)

Vainshtein B.K., V.M. Fridkin, V.L. Indenbom: *Structure of Crystals*, Modern Crystallography 2 (Springer, Berlin, Heidelberg 1995)

Tables

Cracknell A.P., B.L. Davies, S.C. Miller, W.F. Love: *Kronecker Product Tables*, Vol. 1: General Introduction and Tables of Irreducible Representations of Space Groups, Vols.2, 3: Wave Vector Selection Rules and Reductions of Kronecker Products for Irreducible Representations of Space Groups, Vol. 4: Symmetrised Powers of Irreducible Representations of Space Groups (Plenum, New York 1979)

Original Papers

De Rujula A., H. Georgi, S.L. Glashow: Hadron masses in gauge theory. Phys. Rev. B **12**, 147 (1975)

Fritzsch H., M. Gell-Mann, H. Leutwyler: Advantages of the color octet gluon picture. Phys. Lett. B **47**, 365 (1973)

Gell-Mann M.: Isotopic spin and new unstable particles. Phys. Rev. **92**, 833 (1953)

Heisenberg W.: Über den Bau der Atomkerne I. Z. Physik **77**, 1 (1932)

Maradudin A.A.:, S.H. Vosko: Symmetry properties of the normal vibrations of a crystal. Rev. Mod. Phys. **40**, 1 (1968)

Nishijima K.: Some remarks on the even-odd rule. Prog. Theor. Phys. **12**, 107 (1954)

Nishijima K.: Charge independence theory of V-particles. Prog. Theor. Phys. **13**, 285 (1955)

Warren J.L.: Further considerations on the symmetry properties of the normal vibrations of a crystal. Rev. Mod. Phys. **40**, 38 (1968)

Weyl H.: Theorie der Darstellung kontinuierlicher halb-einfacher Gruppen durch lineare Transformationen. Math Z. **23**, 271 (1925); ibid. **24**, 328, 377, 789 (1925)

Wigner E.: Über die Operation der Zeitumkehr in der Quantenmechanik. Nachr. Akad. Wiss. Göttingen, Math.-Phys. Kl. 546 (1932)

Zak J.: Methods to obtain the character tables of nonsymmorphic space groups. J. Math. Phys. **1**, 165 (1960)

Subject Index

Springer Series in Solid-State Sciences

Editors: M. Cardona P. Fulde K. von Klitzing H.-J. Queisser

1 **Principles of Magnetic Resonance** 4th Edition By C. P. Slichter

2 **Introduction to Solid-State Theory** By O. Madelung

3 **Dynamical Scattering of X-Rays in Crystals** By Z. G. Pinsker

4 **Inelastic Electron Tunneling Spectroscopy** Editor: T. Wolfram

5 **Fundamentals of Crystal Growth I** Macroscopic Equilibrium and Transport Concepts By F. E. Rosenberger

6 **Magnetic Flux Structures in Superconductors** By R. P. Huebener

7 **Green's Functions in Quantum Physics** 2nd Edition By E. N. Economou

8 **Solitons and Condensed Matter Physics** Editors: A. R. Bishop and T. Schneider

9 **Photoferroelectrics** By V. M. Fridkin

10 **Phonon Dispersion Relations in Insulators** By H. Bilz and W. Kress

11 **Electron Transport in Compound Semiconductors** By B. R. Nag

12 **The Physics of Elementary Excitations** By S. Nakajima, Y. Toyozawa, and R. Abe

13 **The Physics of Selenium and Tellurium** Editors: E. Gerlach and P. Grosse

14 **Magnetic Bubble Technology** 2nd Edition By A. H. Eschenfelder

15 **Modern Crystallography I** Symmetry of Crystals. Methods of Structural Crystallography 2nd Edition By B. K. Vainshtein

16 **Organic Molecular Crystals** Their Electronic States By E. A. Silinsh

17 **The Theory of Magnetism I** Statics and Dynamics By D. C. Mattis

18 **Relaxation of Elementary Excitations** Editors: R. Kubo and E. Hanamura

19 **Solitons** Mathematical Methods for Physicists By. G. Eilenberger

20 **Theory of Nonlinear Lattices** 2nd Edition By M. Toda

21 **Modern Crystallography II** Structure of Crystals 2nd Edition By B. K. Vainshtein, V. L. Indenbom, and V. M. Fridkin

22 **Point Defects in Semiconductors I** Theoretical Aspects By M. Lannoo and J. Bourgoin

23 **Physics in One Dimension** Editors: J. Bernasconi and T. Schneider

24 **Physics in High Magnetics Fields** Editors: S. Chikazumi and N. Miura

25 **Fundamental Physics of Amorphous Semiconductors** Editor: F. Yonezawa

26 **Elastic Media with Microstructure I** One-Dimensional Models By I. A. Kunin

27 **Superconductivity of Transition Metals** Their Alloys and Compounds By S. V. Vonsovsky, Yu. A. Izyumov, and E. Z. Kurmaev

28 **The Structure and Properties of Matter** Editor: T. Matsubara

29 **Electron Correlation and Magnetism in Narrow-Band Systems** Editor: T. Moriya

30 **Statistical Physics I** Equilibrium Statistical Mechanics 2nd Edition By M. Toda, R. Kubo, N. Saito

31 **Statistical Physics II** Nonequilibrium Statistical Mechanics 2nd Edition By R. Kubo, M. Toda, N. Hashitsume

32 **Quantum Theory of Magnetism** 2nd Edition By R. M. White

33 **Mixed Crystals** By A. I. Kitaigorodsky

34 **Phonons: Theory and Experiments I** Lattice Dynamics and Models of Interatomic Forces By P. Brüesch

35 **Point Defects in Semiconductors II** Experimental Aspects By J. Bourgoin and M. Lannoo

36 **Modern Crystallography III** Crystal Growth By A. A. Chernov

37 **Modern Chrystallography IV** Physical Properties of Crystals Editor: L. A. Shuvalov

38 **Physics of Intercalation Compounds** Editors: L. Pietronero and E. Tosatti

39 **Anderson Localization** Editors: Y. Nagaoka and H. Fukuyama

40 **Semiconductor Physics** An Introduction 5th Edition By K. Seeger

41 **The LMTO Method** Muffin-Tin Orbitals and Electronic Structure By H. L. Skriver

42 **Crystal Optics with Spatial Dispersion, and Excitons** 2nd Edition By V. M. Agranovich and V. L. Ginzburg

43 **Structure Analysis of Point Defects in Solids** An Introduction to Multiple Magnetic Resonance Spectroscopy By J.-M. Spaeth, J. R. Niklas, and R. H. Bartram

44 **Elastic Media with Microstructure II** Three-Dimensional Models By I. A. Kunin

45 **Electronic Properties of Doped Semiconductors** By B. I. Shklovskii and A. L. Efros

46 **Topological Disorder in Condensed Matter** Editors: F. Yonezawa and T. Ninomiya

47 **Statics and Dynamics of Nonlinear Systems** Editors: G. Benedek, H. Bilz, and R. Zeyher

Springer Series in Solid-State Sciences

Editors: M. Cardona P. Fulde K. von Klitzing H.-J. Queisser

48 **Magnetic Phase Transitions**
Editors: M. Ausloos and R. J. Elliott

49 **Organic Molecular Aggregates**
Electronic Excitation and Interaction Processes
Editors: P. Reineker, H. Haken, and H. C. Wolf

50 **Multiple Diffraction of X-Rays in Crystals**
By Shih-Lin Chang

51 **Phonon Scattering in Condensed Matter**
Editors: W. Eisenmenger, K. Laßmann, and S. Döttinger

52 **Superconductivity in Magnetic and Exotic Materials** Editors: T. Matsubara and A. Kotani

53 **Two-Dimensional Systems, Heterostructures, and Superlattices**
Editors: G. Bauer, F. Kuchar, and H. Heinrich

54 **Magnetic Excitations and Fluctuations**
Editors: S. W. Lovesey, U. Balucani, F. Borsa, and V. Tognetti

55 **The Theory of Magnetism II** Thermodynamics and Statistical Mechanics By D. C. Mattis

56 **Spin Fluctuations in Itinerant Electron Magnetism** By T. Moriya

57 **Polycrystalline Semiconductors**
Physical Properties and Applications
Editor: G. Harbeke

58 **The Recursion Method and Its Applications**
Editors: D. G. Pettifor and D. L. Weaire

59 **Dynamical Processes and Ordering on Solid Surfaces** Editors: A. Yoshimori and M. Tsukada

60 **Excitonic Processes in Solids**
By M. Ueta, H. Kanzaki, K. Kobayashi, Y. Toyozawa, and E. Hanamura

61 **Localization, Interaction, and Transport Phenomena** Editors: B. Kramer, G. Bergmann, and Y. Bruynseraede

62 **Theory of Heavy Fermions and Valence Fluctuations** Editors: T. Kasuya and T. Saso

63 **Electronic Properties of Polymers and Related Compounds**
Editors: H. Kuzmany, M. Mehring, and S. Roth

64 **Symmetries in Physics** Group Theory Applied to Physical Problems 2nd Edition
By W. Ludwig and C. Falter

65 **Phonons: Theory and Experiments II**
Experiments and Interpretation of Experimental Results By P. Brüesch

66 **Phonons: Theory and Experiments III**
Phenomena Related to Phonons
By P. Brüesch

67 **Two-Dimensional Systems: Physics and New Devices**
Editors: G. Bauer, F. Kuchar, and H. Heinrich

68 **Phonon Scattering in Condensed Matter V**
Editors: A. C. Anderson and J. P. Wolfe

69 **Nonlinearity in Condensed Matter**
Editors: A. R. Bishop, D. K. Campbell, P. Kumar, and S. E. Trullinger

70 **From Hamiltonians to Phase Diagrams**
The Electronic and Statistical-Mechanical Theory of sp-Bonded Metals and Alloys By J. Hafner

71 **High Magnetic Fields in Semiconductor Physics**
Editor: G. Landwehr

72 **One-Dimensional Conductors**
By S. Kagoshima, H. Nagasawa, and T. Sambongi

73 **Quantum Solid-State Physics**
Editors: S. V. Vonsovsky and M. I. Katsnelson

74 **Quantum Monte Carlo Methods in Equilibrium and Nonequilibrium Systems** Editor: M. Suzuki

75 **Electronic Structure and Optical Properties of Semiconductors** 2nd Edition
By M. L. Cohen and J. R. Chelikowsky

76 **Electronic Properties of Conjugated Polymers**
Editors: H. Kuzmany, M. Mehring, and S. Roth

77 **Fermi Surface Effects**
Editors: J. Kondo and A. Yoshimori

78 **Group Theory and Its Applications in Physics**
2nd Edition
By T. Inui, Y. Tanabe, and Y. Onodera

79 **Elementary Excitations in Quantum Fluids**
Editors: K. Ohbayashi and M. Watabe

80 **Monte Carlo Simulation in Statistical Physics**
An Introduction 2nd Edition
By K. Binder and D. W. Heermann

81 **Core-Level Spectroscopy in Condensed Systems**
Editors: J. Kanamori and A. Kotani

82 **Photoelectron Spectroscopy**
Principle and Applications
By S. Hüfner

83 **Physics and Technology of Submicron Structures**
Editors: H. Heinrich, G. Bauer, and F. Kuchar

84 **Beyond the Crystalline State** An Emerging Perspective By G. Venkataraman, D. Sahoo, and V. Balakrishnan

85 **The Quantum Hall Effects**
Fractional and Integral 2nd Edition
By T. Chakraborty and P. Pietiläinen

86 **The Quantum Statistics of Dynamic Processes**
By E. Fick and G. Sauermann

87 **High Magnetic Fields in Semiconductor Physics II**
Transport and Optics Editor: G. Landwehr

88 **Organic Superconductors**
By T. Ishiguro and K. Yamaji

89 **Strong Correlation and Superconductivity**
Editors: H. Fukuyama, S. Maekawa, and A. P. Malozemoff

Springer-Verlag and the Environment

We at Springer-Verlag firmly believe that an international science publisher has a special obligation to the environment, and our corporate policies consistently reflect this conviction.

We also expect our business partners – paper mills, printers, packaging manufacturers, etc. – to commit themselves to using environmentally friendly materials and production processes.

The paper in this book is made from low- or no-chlorine pulp and is acid free, in conformance with international standards for paper permanency.

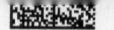